INTEGRATED NETWORK MANAGEMENT, III

IFIP Transactions C: Communication Systems

International Federation for Information Processing

Technical Committee 6

Communication Systems

IFIP Transactions Abstracted/Indexed in:
INSPEC Information Services

C-12

INTEGRATED NETWORK MANAGEMENT, III

Proceedings of the IFIP TC6/WG6.6 Third International Symposium on Integrated Network Management
with participation of the IEEE Communications Society CNOM
and with support from the Institute for Educational Services
San Francisco, California, USA, 18-23 April, 1993

Edited by

HEINZ-GERD HEGERING
Department of Computer Science
University of Munich
Munich, Germany

YECHIAM YEMINI
Department of Computer Science
Columbia University
New York, USA

1993

NORTH-HOLLAND
AMSTERDAM • LONDON • NEW YORK • TOKYO

ELSEVIER SCIENCE PUBLISHERS B.V.
Sara Burgerhartstraat 25
P.O. Box 211, 1000 AE Amsterdam, The Netherlands

Keywords are chosen from the ACM Computing Reviews Classification System, ©1991, with permission.
Details of the full classification system are available from
ACM, 11 West 42nd St., New York, NY 10036, USA.

ISBN: 0 444 89982 0
ISSN: 0926-549X

This book is printed on acid-free paper.

Printed in The Netherlands

PREFACE

The past two years, since the Second International Symposium on Integrated Network Management (ISINM), saw dramatic expansion of the field and its significance. Today, more than ever before, there is a need for a broad-based central forum for sharing knowledge among the diverse members of the network management community. Vendors and users, researchers and developers, standards developers and implementors, LAN, WAN and MAN specialists, systems and networks experts, all must find ways to share and integrate network management knowledge.

The third ISINM pursues the successful record of the first two, to build this community of knowledge. It continues the pledge to serve the diverse spectrum of interests of the network management community, bring together the leaders of the field and cover its most central developments and the state of the art. It continues the commitment to high-quality technical programs of great distinction, and to stimulate productive multilogues within the network management community.

The technical papers presented in this volume were selected from among some 100 submissions through a most rigorous review process. Each paper was reviewed by 4 referees and carefully evaluated by the program committee, to ensure the highest quality. Continuing the tradition of diverse international participation, authors represent some 11 countries, including Australia, Britain, Canada, France, Germany, Italy, Japan, Poland, Singapore, South Africa, and the USA, as well as papers involving international collaborations. Vast sections of the telecommunications, computer communications and computer industries are represented, as well as leading users, academic and industrial research labs.

The contents of the proceedings is not restricted to the 50 selected submissions. We have also included 5 papers that are related to the Technology Centers which are organized during TISINM. The Tech Centers, led by recognized experts, consist of multi-vendor cooperative demonstrations of advanced and emerging technologies implementing management architectures. Furthermore, we have added short descriptions of 8 panels involving the foremost leaders of the field.

We did not organize the table of contents using the conference chronology. We preferred to turn the proceedings into a useful reference book and a text book on current research in the field. Therefore the contributions are ordered with respect to the main topics:

- Management Architecture
- Distributed Systems Management
- Implementing Management Architectures
- Management Requirements, Experiences and Evaluation
- Telecommunications Management

- Management Applications

The contributions to each topic include selected technical papers, Tech Center papers and Panels which are grouped into sessions.

We are honored to present these proceedings of the Third ISINM'93. The work included in this volume represents the collective contributions of authors, dedicated reviewers and a committed program committee. We thank Joe Betser for soliciting the Tech Center papers and Iyengar Krishnan for coordinating the Panels. Thanks also to Branislav Meandzija and Wolfgang Zimmer for helpful comments. Last but not least, we thank Gabi Dreo, the Munich Network Management Team, and the members of the Computer Science Department of Columbia University for their help with the handling of paper submissions, conference database maintenance, and many other tasks.

We wish to extend our gratitude to these excellent members of the network management community for their contributions and commitments to the continuing excellence and success of ISINM.

Heinz-Gerd Hegering and Yechiam Yemini
Co-Chairs, Program Committee, ISINM'93

TABLE OF CONTENTS

C. THE RMON MIB - FROM IDEA TO THE MARKETPLACE

D. MANAGEMENT INFORMATION

E. APPLICATIONS OF DIRECTORY SERVICES

C. DISTRIBUTED MANAGEMENT

III. IMPLEMENTING MANAGEMENT ARCHITECTURES

A. SNMPv2 - FROM INDUSTRY STANDARD SNMP TO TOMORROW'S SNMPv2 (SMP)

B. SPECIFIC APPROACHES

C. OSI MANAGEMENT - WORLDWIDE IMPACTS

D. INTEGRATING OSI/INTERNET MANAGEMENT STANDARDS

IV. MANAGEMENT REQUIREMENTS, EXPERIENCES AND EVALUATION

A. MEETING USER NEEDS WITH A MIX OF TECHNOLOGIES

B. USER REQUIREMENTS

C. EXPERIENCES AND EVALUATION

VI. MANAGEMENT APPLICATIONS

A. PANEL

B. FAULT MANAGEMENT APPLICATIONS

C. MULTIVENDOR APPLICATIONS - THE WAVE OF THE FUTURE

D. MANAGEMENT SUPPORT TOOLS

SYMPOSIUM COMMITTEES

EXECUTIVE COMMITTEE MEMBERS

Branislav Meandzija	MetaAccess, Inc., USA, General Chair
Wolfgang Zimmer	GMD-FIRST, GERMANY, General Vice Chair
Iyengar Krishnan	The MITRE Corporation, USA, Assistant to General Chairs
Kimberly Kappel	Georgia Institute of Technology, USA, Special Events Chair
Paul Brusil	The MITRE Corporation, First ISINM Chair and Advance Planning Chair
Heinz-Gerd Hegering	University of Munich, GERMANY, Program Co-Chair
Yechiam Yemini	Columbia University, USA, Program Co-Chair
Joseph Betser	The Aerospace Corporation, USA, Vendor Program Chair
Douglas Zuckerman	AT&T Bell Laboratories, USA, IEEE/CNOM Coordinator
Keith McCloghrie	Hughes LAN Systems, USA, Internet Coordinator
Roberta Cohen	AT&T Bell Laboratories, USA, Advisory Board Chair
Melvyn Galin	The MITRE Corporation, USA, Financial Committee Chair
Michele Nessier	Action Motivation, USA, Local Arrangements

SYMPOSIUM COMMITTEES (Continued)

PROGRAM COMMITTEE MEMBERS

Lorenzo Aguilar	Taligent, Inc., USA
Robert Braden	USC Information Sciences Institute, USA
Stephen Brady	IBM Watson Research Center, USA
Will Collins	Codex Corporation, USA
James R. "Chuck" Davin	MIT Laboratory for Computer Sciences, USA
Deborah Estrin	University of Southern California, USA
Ivan Frisch	Polytech University, USA
Jörg Gonschorek	Siemens Nixdorf Informationssysteme AG, GERMANY
Rodney Goodman	California Institute of Technology, USA
Shri Goyal	GTE Laboratories, Inc., USA
Cristian Huitema	INRIA Unite de Recherche de Sophia-Antipolis, FRANCE
Dale Johnson	Merit Network, Inc., USA
Gautam Kar	IBM European Networking Center, GERMANY
Mark Klerer	AT&T Bell Laboratories, USA
Aurel Lazar	Columbia University, USA
Kenneth Lutz	Bellcore, USA
Daniel Lynch	Interop, Inc., USA
Anna Melamed	Salomon Technical Services, USA
Kazuyoshi Morino	NTT Network Systems Development Center, JAPAN
Shoichiro Nakai	NEC/C&C Systems Laboratories, JAPAN
Jan Roos	University of Pretoria, SOUTH AFRICA
Marshall Rose	Dover Beach Consulting, USA
Veli Sahin	NEC America, USA
Morris Sloman	Imperial College, UNITED KINGDOM
Chris Sluman	SEMA Group, UNITED KINGDOM
Don Stokesberry	National Institute of Standards and Technology, USA
Colin Strutt	Digital Equipment Corporation, USA
Carl Sunshine	The Aerospace Corporation, USA
Liba Svobodova	IBM Research Division, SWITZERLAND
Ole Krog Thomsen	Jutland Telephone, DENMARK
Jeremy Tucker	Logica Cambridge Ltd., UNITED KINGDOM
Jil Westcott	BBN Systems and Technologies Corporation, USA
Denis Yaro	SunConnect, USA

LIST OF REVIEWERS

L. Aguilar
S. Aidarous
H. Ambos
L. Auld
E. Aupperle
C. Aurrecoechea
B. Baer
K. Bahr
D. Bailey
M. Baker
S. Baker
J. Bannister
S. Bapat
D. Behr
G. Bonn
M. Bosch
S. Brady
T. Brunner
W. Buga
U. Bügel
H. Chung
A. Clemm
T. Coatta
R. Cohen
C. Davin
B. Delosme
J. Dinger
A. Dittrich
O. Drobnik
A. Duda
J. Duff
A. Dupuy
Y. Ebihara
S. Efrani
W. Eichler
D. Estrin
J. Ferrandiz
A. Finkel
N. Fujii
W. Fumy
K. Garbe
E. Gardner
J. George
M. Gering
J. Ghanta
G. Goldszmidt
J. Gonschorek
R. Goodman
S. Goyal
U. Gremmelmaier
P. Gunningberg
E. Haitung
J. Hall
J. Harista
H. Hegering
U. Hollberg
J. Hong
C. Horn
T. Ilse
G. Jabobson
C. Jard
T. Jeffree
C. Johnen
J. Jordaan
R. Jueneman
F. Kaplan
G. Kar
A. Karduck
U. Kohl
R. Kositpaiboon
I. Krishnan
B. Krupczak
C. Krupczak
P. Kühn
A. Law
A. Lazar
W. Lazear
B. Leon
L. Lewis
T. Lin
Y. Lin
H. Liu
K. Lor
K. Lutz
D. Matthews
S. Mazumdar
K. McCloghrie
B. Meandzija
I. Mei
K. Meyer
E. Michiels
D. Milham
H. Miyahara
J. Moffett
O. Moharram
H. Moller
R. Molva
K. Morino
G. Müller
T. Murai
S. Nakai
J. Noles
D. O'Mahony
B. Paillassa
A. Pashtan
M. Paterok
C. Petitpierre
A. Pras
O. Prohaska
P. Putter
R. Rathore
A. Rijsinghani
G. Roessler
J. Roos

LIST OF REVIEWERS (Continued)

V. Sahin
W. Schollenberger
M. Sloman
L. Svobodova
K. Tesink
L. Trajkovic
M. Tsukamoto
R. Van den Heever
R. Weihmayer
O. Wolfson
H. Yahata
S. Zatti
D. Zuckerman
S. Sakata
S. Shimojo
P. Soares
B. Tangney
M. Tobe
J. Tsay
H. Uno
C. Vissers
C. Westphall
L. Wu
S. Yamaguchi
W. Zimmer
R. Sauerwein
A. Shvartsman
C. Strutt
L. Tarouco
S. Towers
P. Tsuchiya
Y. Urano
J. Vucetic
S. Wilbur
S. Wu
Y. Yemini
G. Zörntlein

INTRODUCTION

Integrated Network Management, III (C-12)
H.-G. Hegering and Y. Yemini (Editors)
Elsevier Science Publishers B.V. (North-Holland)

From the Chairs

STRATEGIES FOR THE NINETIES

Branislav Meandzija *and* *Wolfgang Zimmer*
MetaAccess, USA *GMD FIRST, Germany*

1. Status Quo

We live in an era when computer and telecommunications networks and distributed systems are not only exploding in terms of volume and complexity, but are rapidly becoming as common-place in business as the personal computer. Such rapid changes induce growing challenges to manage an ever increasing variety of underlying network technologies and applications. These challenges are magnified by the growing demand for seamless integration of computer and communications services into overall enterprise management.

The 1993 International Symposium on Integrated Network Management (ISINM '93) offers a comprehensive response to immediate and strategic challenges of the nineties, through a week-long evaluation, review, and discussion of current research, development, technologies and products in one of the most beautiful settings imaginable, San Francisco.

For the first time, the 1993 symposium offers Technology Centers where multiple vendors will participate in live interoperability demonstrations of RMON, SNMPv2 (SMP), OSI, OMNIPoint and a variety of application technologies, products and prototypes. This highlight in the symposium program reflects our perspective on the principal element in any solution in integrated network management: interoperability across technologies, platforms and applications.

2. Reflections On History

When we held the first ISINM in Boston in 1989, standards for enabling integrated network management across multiple vendor networking resources were in the heat of development in regional and international arenas. While some thought that developing these standards was the most difficult path on the road to integrated management, many realized a few years later that standards were the beginning of a long journey.

When we held the second ISINM, in Washington D.C. in 1991, the need for enterprise-oriented management across data and telecommunications applications and distributed systems became increasingly apparent.

Principle problems related to incorporating standards into products aimed at providing coherent, integrated network management solutions across future, standards-based, multi-vendor components as well as existing proprietary components. Multi-vendor demonstrations in Europe, Japan, and North-America seemed to indicate that the time had come when users could competitively procure network management products in any of several countries and be confident that they would interoperate with comparable products in other world regions. That wasn't so.

3. Strategies For The Nineties

We have learned. We are not at the end of the road - we are not even in the middle. We are only at the beginning and will remain there probably for the greater part of the nineties. Worldwide coordinated strategies are needed to evolve network management in the best way.

The beginning of the nineties is characterized by big political, ecological and technical changes in all areas worldwide. Multimedia applications based on broadband and mobile network technology will be the future driving forces of the communications area. However, the element of uncertainty plays a dominant role in all environments. Down-sizing and up-sizing in volume and time requires flexibility to change.

There are several fundamental problem areas which will continue to characterize and challenge integrated network management throughout the nineties:

Flexibility

The rapidly changing networking environments and application areas place an ever increasing burden on today's static network management products. Flexibility to change is difficult to manage as the quantity and types of managed objects explode. The principle effect of incorporating higher flexibility into systems is increased complexity of those systems. To further complicate matters, there is a growing demand for customized management systems conforming to individual requirements.

Real-Time

As we better understand management problems today, the focus shifts towards the operations and control aspects of network management in real-time. In the future there will be a dynamic cooperation between manager systems and between manager and agent systems requiring intelligent filtering and efficient decision making in short time. Increased complexity requires the delegation of tasks to cooperating, autonomous agents. It also requires mutually agreed operations to be carried out at fast transaction processing speeds.

User-Friendliness

There is a need to speed up the introduction time for new services and to improve the quality of service for the end user. Operations and management of the dynamic changing network environments without disruption of service has to be a major goal. Therefore, strategies are needed to manage the changes of the whole networking environment of the future in a fast and flexible way.

These problems are magnified by economic and regulatory constraints, problem complexity, technology advances, standards development, product introductions, market requirements, user demands and other factors which change unpredictable over time.

The only possible strategy for dealing with these seemingly unsolvable problems is to "empower" the user and network manager with an active network management system that can act as the user's / manager's intelligent personal agent. That system would include capabilities to easily reuse, redesign, reimplement, reinstall, and reintegrate generic software modules at run-time. An obvious solution approach towards such a system is an object library of reusable generic software objects which are easily manipulated through a graphical non-programming (e.g. drag-and-drop) user interface. The overall strategy however needs more comprehensive methods and tools to answer the powerful challenges outlined above. A necessity are expert systems and automatic programming tools which will be the central element of the dynamic, integrated network management systems of the nineties which facilitate active forms of network management.

Most of the problems outlined above are addressed in many of the papers included in these proceedings. We are certain that you will find viable solution approaches to most of today's problems and future challenges.

4. Future Events

As the management world continues evolving, the ongoing series of biennial international symposia will continue to foster and to promote non-partisan fora among individuals of diverse and complementary backgrounds to encourage international information exchange about all aspects of network and distributed systems integrated management.

To broaden the scope of these symposia, the International Federation for Information Processing (IFIP) Working Group (WG) 6.6 on Network Management for Communication Networks, as the main organizer of ISINM events, has been successfully collaborating with the Institute of Electrical and Electronics Engineers (IEEE) Communications Society's (COMSOC) Committee on Network Operations and Management (CNOM). ISINM and the Network Operations and Management Symposium (NOMS) are the premier technical conferences in the area of network and systems management, operations and control. ISINM is held in odd-numbered years, and NOMS is held in even-numbered years. CNOM and IFIP WG 6.6 have been working together as a team to develop both these symposia.

NOMS '94 will take place in Orlando, Florida, February 14-18, 1994. For more information contact:
+1 415 512 1316 (tel), +1 415 512 1325 (fax), amotive@mcimail.com (e-mail).

The next International Symposium on Integrated Network Management (ISINM '95) will be held in the Spring of 1995. For more information contact:
+1 415 512 1316 (tel), +1 415 512 1325 (fax), amotive@mcimail.com (e-mail).

Starting in 1990, IFIP WG 6.6 together with IEEE CNOM has also been organizing the International Workshops on Distributed Systems: Operations and Management (DSOM) which takes place in October of every year and alternates in location internationally. DSOM '93 will take place October 5-6 in Long Branch, New Jersey and will be hosted by AT&T Bell Laboratories. For more information contact: Dorotea DeSan, Room 1L-314, AT&T Bell Laboratories, 101 Crawfords Road, Homdel, New Jersey 07733, USA, +1 908 949 5534 (tel), +1 908 949 8569 (fax), dds@arch3.att.com (e-mail).

For more information on future events (DSOM '94, ISINM '97, etc.) and other related activities of IFIP Working Group 6.6 please contact: Wolfgang Zimmer, GMD-FIRST, Rudower Chaussee 5 / G13.7, D-1199 Berlin-Adlershof, Germany, +49 30 6704-2652 (tel), +49 30 6704-5610 (fax), <zimmer@first.gmd.de> or <zimmer@first.berlin.gmd.dbp.de> (e-mail).

In order to achieve our target focus on total systems management, we hope to expand the scope of our reciprocal agreements in the future to include other major organizations involved with the many aspects of distributed systems management.

5. Acknowledgements

ISINM '93 is the result of a great coordinated effort of a number of people and organizations. First of all, we would like to thank our main sponsors, IFIP TC 6, IEEE COMSOC CNOM, CommunicationsWeek, and most of all the Institute for Educational Services (IES) for the continued financial support.

The executive committee of ISINM '93 was formed in April, 1991 and has been the main force behind the symposium. We would like to thank Joe Betser, Paul Brusil, Roberta Cohen, Mel Galin, Heinz-Gerd Hegering, Kim Kappel, Kris Krishnan, Keith McLoghrie, Lightsey Wallace, Yechiam Yemini, and Doug Zuckerman for enduring with us this two year marathon towards ISINM '93.

The program committee under the tireless leadership of Heinz-Gerd Hegering and Yechiam Yemini has once again defined the standard for conferences and journals in network management. Its creative work, represented through these proceedings, has clearly selected the main problem areas of integrated network management and the most promising solutions to those problem areas. Our deepest thanks go to Roberta Cohen, Jörg Gonschorek, Frank Kaplan, Kris

Krishnan, Aurel Lazar, Shoichiro Nakai, Jan Roos, Veli Sahin, Morris Sloman, Colin Strutt, Liba Svobodova, Doug Zuckerman, and all other members of the program committee who created the outstanding program.

The vendor program committee lead by Joe Betser did the impossible and enabled ISINM to be the first strategic conference to include live interoperability demonstrations on a live network. We would like to thank all vendor patrons, partners, and friends for making this possible. We would also especially like to thank the Technology Center leaders for going the extra mile and enabling experimental interoperability demonstrations of emerging technologies: Paul Brusil for the OSI Technology Center, Steve Waldbusser for the SNMPv2 Technology center, Mike Erlinger for the RMON Technology Center, Dave Mahler for the Applications Technology Center, and Jim Warner for the OMNIPoint Technology Center. We would also like to thank the two incredible networking gurus Philip Almquist and Peter DeVries for putting the network together.

Finally, we would like to thank our team (the ones that actually made it all happen), Michele Nessier for putting it all together, Suzanne and Michael Carey and Donna Cowan for the innovative marketing and sales efforts, and Jim Warner for the publicity and public relations work.

I.

MANAGEMENT ARCHITECTURE

A.

PANEL

Emerging Trends in Network and Systems Management

Panel Chair: Jim Warner, OSI Network Management Forum, USA

Panelists: Sharon Fisher, Communications Week, USA
Bob Brown, Network World, USA
Kella Knack, LAN Computing, USA
Mary Jander, Data Communications, USA

Journalists are constantly taking the industry's pulse. These distinguished panelists will address what they hear and see regarding the emerging trends in management. In addition, they will discuss key issues that they feel are not being dealt with by the industry. And in our version of "turnabout is fair play", here is your chance to ask questions of the questioners.

I.

MANAGEMENT ARCHITECTURE

B.

MODELING SEMANTICS AND BEHAVIOR

Integrated Network Management, III (C-12)
H.-G. Hegering and Y. Yemini (Editors)
Elsevier Science Publishers B.V. (North-Holland)

Richer Modeling Semantics for Management Information

Subodh Bapat

16441 Blatt Blvd, MS 203, Ft. Lauderdale, FL 33326

Abstract

This paper presents new modeling concepts with enhanced semantics in the area of modeling management information. The concept of virtual attributes is presented to allow externally visible attributes to be defined in terms of other attributes. A formal model of relationships, integrated with object-oriented concepts, is presented. Finally, a mechanism using virtual attributes is presented to resolve the problem of encapsulation breakdown arising from relationships between object classes.

1. Current Management Information Modeling Techniques

Current techniques for modeling management information have been specified by several Network Management Standards bodies. Object-oriented modeling techniques are used in the OSI and CCITT standards [CCITT X.720] and consortia such as the Network Management Forum [NMF 89]. These are complemented by Entity-Relationship diagrams used by national standards bodies such as ANSI T1M1 [T1M1 89]. These object-oriented models are intended for use in conjunction with OSI application-layer network management services and protocols such as the Common Management Information Service and the Common Management Information Protocol (CMIS and CMIP) [CCITT X.710, X.711]. Finally, simpler object-based schemes, not incorporating structuring principles such as inheritance or containment, are used in the Structure of Management Information specified by the Internet Society [RFC 1155]; these are typically used together with the Simple Network Management Protocol (SNMP) and have achieved a considerable measure of popularity in certain markets [RFC 1157, Rose90].

Object-oriented structuring techniques in the OSI Management Information Model allow a well-defined taxonomy of network objects to be constructed, which is especially useful for large and complex networks such as those implemented by wide-area telecommunications service providers. The simplicity, low resource demand and low deployment cost of the Internet management paradigm makes it attractive for implementation in competitive and price-sensitive markets such as subscriber premises-based data internetworking equipment.

2. Virtual Attributes

2.1. Background

While these information modeling techniques have many beneficial features, they also have some shortcomings. One disadvantage of these modeling techniques is that they do not have mechanisms to specify constraints between attributes of objects or between objects themselves. Often, many numeric attributes of objects are related by mathematical invariants which hold among them. While the concept of computed attributes and constraints among attributes has long been established in database theory, current network management modeling techniques consider constrained attribute values which can be derived from other attribute values to be redundant; only orthogonal attributes are modeled . Implicit in this approach is the assumption that the determination of the constrained attribute can be achieved using an algorithmic procedure, which, using an unspecified mechanism outside of the standard object model, is assumed to be somehow known to the management system.

As an example, consider a situation where the managed object is a subscriber-premises based multiplexer device. The purpose of this device is to accept multiple low-bandwidth data streams from various ports, which we term the *port* side, and multiplex these streams into a single moderate-bandwidth data stream for transport over a wide-area telecommunications link, which we term the *aggregate* side. Many such devices are not necessarily synchronously clocked on their aggregate port to any standard bit rate specified by the WAN service provider; for low-end devices, aggregate ports can operate at various speeds depending on the offered load from the port-side traffic.

Consider a multiplexer device with 4 ports and 1 aggregate. Because each individual port can operate at variable speeds depending on the characteristics of the devices connected to it (for example, 2400 bps, 4800 bps, 9600 bps, etc.), the aggregate can operate at various speeds as well. To represent this information for network management, we may choose to specify such a device in the multiplexer MANAGED OBJECT CLASS. For the sake of this example, assume that we have chosen a simple-minded representation for the current port operating speeds, as attributes port1speed, port2speed, port3speed and port4speed. We also choose to model the current operating speed of the aggregate port as aggregateSpeed.

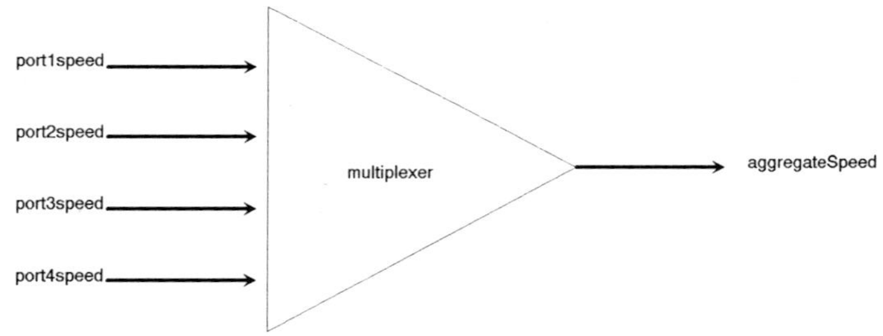

Figure 1: A Multiplexer Device

However, these attributes are not orthogonal; they are related by a mathematical constraint. This is because this version of device is not designed to transmit null codes in order to maintain synchronization on the WAN bandwidth even in the absence of adequate offered port-side load; rather, the device adjusts its aggregate speed depending on the speed characteristics of its individual ports. Therefore, the aggregateSpeed attribute is related to other attributes by the following mathematical constraint:

aggregateSpeed = port1speed + port2speed + port3speed + port4speed

In the current practice of modeling management information, the aggregateSpeed attribute would not be modeled at all. Rather, the paradigms assume that the management system would somehow understand the internal model of this particular managed object class sufficiently well that it would know how to algorithmically derive aggregateSpeed from the individual port speeds. The burden is therefore placed on the management system - rather than inside the managed object - to query the managed object for values of its port speed attributes, and externally compute the aggregateSpeed from these attributes.

This is not a satisfactory situation, because it compromises the security and privacy of the owner of the multiplexer managed object. The owner may be willing to expose the aggregateSpeed of the multiplexer object to, say, the management system of a WAN telecommunications service provider, for billing and accounting purposes. However, the owner may not be willing to expose the values of individual port speeds to external management systems, because they could reflect network usage by individual departments, which information may be considered proprietary for competitive purposes.

Therefore, a mechanism is necessary to be able to selectively present only certain attributes to external management systems, such that these attributes could be internally computed using other attributes. Furthermore, to preserve object encapsulation, such computations should be part of the managed object model, and should be executed inside the managed object itself (or in the agent representing the managed object), rather than in the external management system. This will avoid compromising the security of owner of the managed object. It is also desirable that the external management system use the same interface to query both a computed attribute value as well as a native attribute value; the external management system need not be aware which specific attributes are returned as directly stored native data values in response to a query, and which ones are algorithmically computed.

2.2. Formal Modeling of Virtual Attributes

The mechanism of *Virtual Attributes* acts as a solution to this problem. A virtual attribute is an attribute which is specified in terms of a value syntax which is the result of a computation involving values of other attributes (which may in turn themselves be virtual). Because this computation is specified as part of the attribute definition, it is assumed that the algorithm implementing this computation is implicitly encapsulated within the managed object itself, or in the agent representing the managed object. The external manifestation of a virtual attribute is the same as that of any other attribute; it is interrogated by a managing system using the same query primitives in the same network management protocols, and is returned using the same response mechanism as any other attribute. The management system need not be aware of the fact that the interrogated attribute is virtual and that it is internally computed using an algorithm involving values of other attributes.

By specifying appropriate external visibility for both native and virtual attributes, the owners of a managed object can enjoy fine granularity in their control of their security policies. In the above example, by exposing only the aggregateSpeed attribute as part of the shared management knowledge regime, and by keeping the individual port speeds hidden, the owners of the multiplexer managed object can implement a privacy policy which is mutually satisfactory both to themselves and to their WAN service providers.

The specification of virtual attributes can be accomplished in any syntax appropriate for that purpose. For example, the port speed attributes for the multiplexer MANAGED OBJECT CLASS could have been specified using the standard ATTRIBUTE syntax from [CCITT X.721] as follows:

```
port1Speed  ATTRIBUTE
        WITH ATTRIBUTE SYNTAX  BitsPerSecond;
        MATCHES FOR  Equality, Ordering;
REGISTERED AS {....}
```

and so on, for each individual port speed. The supporting ASN.1 type BitsPerSecond might be specified as

```
BitsPerSecond ::= INTEGER (0..MAX)
```

The aggregateSpeed attribute can now be specified as a virtual attribute derived using a computation involving other attributes, possibly using a template[1] similar to the following:

```
aggregateSpeed  VIRTUAL ATTRIBUTE
        WITH ATTRIBUTE SYNTAX  BitsPerSecond;
        MATCHES FOR Equality, Ordering;
        COMPUTED FROM  aggregateSpeedAlgorithm;
REGISTERED AS {....}
```

where the type of the ASN.1 value reference aggregateSpeedAlgorithm is the type reference AttributeDefinedExpression, and its value is simply specified as the sum of individual port speed attributes. The type reference AttributeDefinedExpression is expanded as a simple grammar in ASN.1 for arithmetic expressions. In this grammar, the atoms are OBJECT IDENTIFIERs representing other attributes, and the connectives are simple unary and binary arithmetic operators. Such a grammar for arithmetic expressions - allowing parenthetical expressions and operator precedence rules - is fairly straightforward to specify (and is typically a beginner's exercise in compiler design courses), and is not shown here for brevity.

Note that because virtual attributes are computationally derived from base attributes, they can never be updated. The specification of virtual attribute in any managed object class must always be qualified with a GET, and never with a GET-REPLACE. Any change to any of the native base attributes will be automatically reflected in the value of the virtual attribute, because the computation for the virtual attribute is re-executed every time the virtual attribute is queried.

[1]Because virtual attributes are not part of any standard, the template used to specify them is also non-standard.

2.3. Virtual Attributes for Protocol Objects

Virtual attributes are also useful in all situations where it is inappropriate or undesirable to expose the base values themselves to an external management station. As an example, consider the specification of various counters inside an IP (Internet Protocol) entity [RFC 791]. For management purposes, an IP protocol entity is often required to report various counter values representing the number of IP datagrams processed, the number discarded, the number forwarded, the number fragmented, and so on. These counters are formally specified in the Internet standard for the SNMP Management Information Base [RFC 1213]. A pictorial representation of these counters is provided in the Case Diagram [Rose90] of Figure 2.

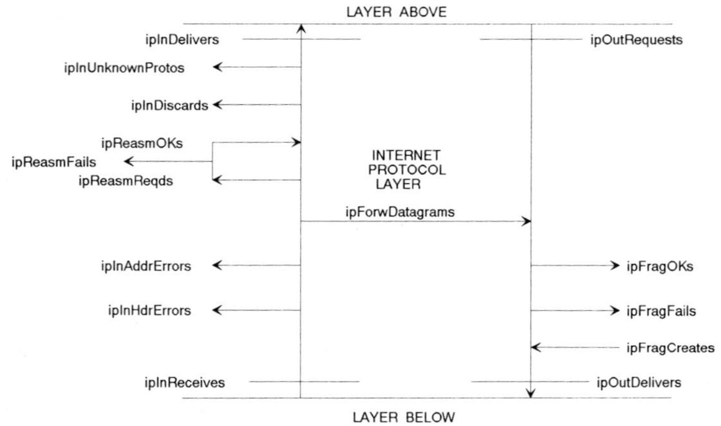

Figure 2: The Case Diagram for the Internet Protocol

In Figure 2, note that the filter counter ipOutDelivers is not formally modeled by any network management standard, because it can be inferentially derived as a linear combination of various other counters, using the expression:

ipOutDelivers =
ipOutRequests + ipForwDatagrams - ipFragOKs - ipFragFails + ipFragCreates

It may nevertheless be desirable for the owners of an IP protocol entity (that is, the owners of a host or a gateway) to make a counter similar to ipOutDelivers available to a WAN telecommunications service provider, for it may be indicative of the offered user load to an SMDS link for billing purposes. Currently, this is accomplished by having a management system query the individual counter attributes on the right hand side of the above expression, and then computing the left hand side in the management system. This arrangement assumes that the management system somehow knows the internal model of the IP entity and how the counters relate to each other. Further, it compromises the privacy of the IP protocol entity, for its owners may not be willing to let the WAN service provider query the individual counter attributes within the IP protocol entity. By specifying ipOutDelivers as a virtual attribute which internally knows how to compute itself from other attributes, the situation can be satisfactorily resolved.

3. Relationships

3.1. Background

A *relationship* is said to exist between managed object classes when the operation of one managed object class affects another managed object class. The specification of the relationship describes the nature of this influence. The knowledge of such relationships is especially important in any information model describing the network. Many types of relationships exist in a network: connectivity relationships describe the topology of network devices; interoperability relationships suggest how services could be deployed; backup relationships describe network reliability, and so on.

Current standards approach the problem of modeling relationships by using attributes to represent relationships. It is assumed that each object class possesses attributes which in some manner describe the relationships in which that class participates [CCITT X.732]. These attributes could be located in just one of the participating managed object classes (for *one-way relationships*) or in both (for *two-way relationships*). At its most simplistic level, an implementation of a relationship attribute in any managed object instance could be equivalent to an object pointer directed at another managed object instance. The Relationships Attribute Group contains all the relationship attributes of a managed object class.

There are various problems with the approach taken by the standards. When the nature of a relationship changes, we may be required to simultaneously M-SET reciprocal attributes in two different managed object instances atomically. This may not always be possible, depending on the access and authorization restrictions. If one update succeeds and the other fails for any reason, the system may be left in an inconsistent state. Protocols such as CMIP do not have the robustness of other application-layer standards such as CCR to ensure atomicity and concurrency of information modification.

Another problem is the inability to address specific managed objects to query them for their relationships. Consider a situation where an equipment managed object class participates in an is-housed-at relationship with a location managed object class. This relationship describes the semantics of identifying the physical location of any piece of network hardware. Because this is a *many-to-one* relationship (many instances of equipment may be housed at one instance of location), this would be represented as a one-way relationship with a relationship attribute in each instance of equipment, indicating at which instance of location it was housed. It is therefore possible to query an equipment instance to determine where it is housed. But it is not possible to query an instance of location to generate a report of which instances of equipment are housed there, since there is no relationship attribute to represent this information in the location managed object class.

3.2. Concepts of Relationship Modeling

In this paper, we present a fresh approach to the specification and management of relationships. In our approach, relationships are modeled as first-class constructs, and the specification of relationships is not tied to attributes of managed object classes. While attributes within managed object classes could be one implementation mechanism for relationships, the formal specification of relationships in the network information model should strive for abstraction, and leave open the possibility of other implementation mechanisms as well.

In our model, a relationship is a *binary association* between two managed object classes, and is specified as a first-class conceptual construct. The specification of a relationship does not involve instances; depending on the nature of a relationship, an instance of a managed object class may or may not choose to participate in a relationship defined on that managed object class. Each managed object class participating in a relationship is said to play a *role* in that relationship. Roles are useful both for human comprehension and for formal specification of relationships; in general, the role of a managed object in a relationship is different depending on the perspective from which a relationship is viewed. In the above example, the equipment object class could be said to play the is-housed-at role with respect to the location object class; in the inverse direction, the location object class could be said to play the houses role with respect to the equipment object class.

A *relationship instance* is a pair of managed object instances which participate in that relationship. When such a pair of instances of appropriate classes is linked using the specified relationship association, the relationship is said to be *instantiated*.

The traditional representation of relationships uses Entity-Relationship diagramming techniques. We will use these techniques as a pictorial aid only; the formal specification of relationships will be performed using notational constructs. Figure 3 provides an example of a relationship and its instance, along with roles played by the participating managed object classes.

3.3. Relationship Multiplicity

A relationship is specified with its *cardinality*, which could be *one-to-one, one-to-many, many-to-one*, or *many-to-many*. A relationship is also associated with an *optionality*, which indicates whether or not each instance of that object class is obligated to participate in that relationship with at least one instance of the related object class. In our model, we will combine the notions of optionality and cardinality to arrive at the notion of *multiplicity* [Rumb91]. In general, a different multiplicity value could be specified at either end of a relationship, specifying a *quantification* for the role played by the object class at that end of the relationship. For example, in the many-to-many **subscription** relationship between the **subscriber** object class and the **service** object class. A single **subscriber** may subscribe to one or more **services**, while each offered **service** may be subscribed to by zero or more **subscribers**.

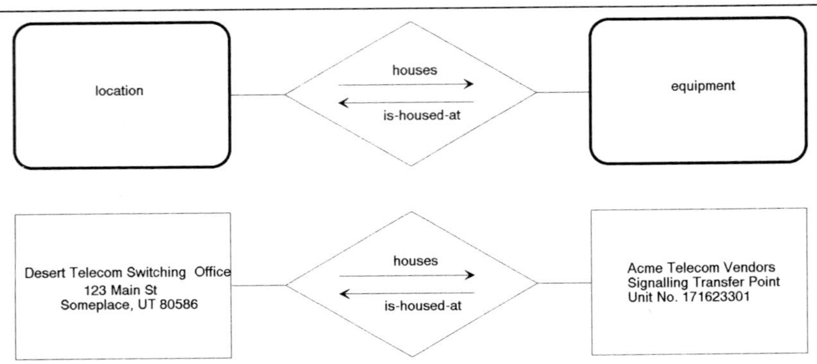

Figure 3: A Relationship Between Managed Object Classes, with a Relationship Instance

The multiplicity characteristic is specific to every relationship: the same managed object class may have one multiplicity characteristic with one related class in one relationship, and a different multiplicity characteristic with another related class in another relationship.

The most general form of a multiplicity specification indicates, at each end, both a *lower limit*, N, and an *upper limit*, M, for the number of instances which can participate in that relationship. N must be a non-negative integer and M must be a positive integer. If both the limits N and M are specified, we interpret it to mean that an instance of the managed object class must participate in the relationship role with *at least N* and *at most M* instances of the related object class. This is called the *multiplicity interval* of the role. In our diagrammatic representation, we specify the multiplicity interval of each role using the parenthetical notation **(N,M)** at the appropriate end of the relationship.

There are a number of special cases to be considered with respect to the specification of the multiplicity interval. Consideration of these special cases will provide an appreciation of the power of multiplicity as specification tool to convey the full semantics of quantification and optionality. These special cases are:

(N,M): N and M both specified. An object instance must play its relationship role with *at least* N and *at most* M instances of the related object class.

(N): Only N specified; assume N = M. An object instance must play its relationship role with *exactly* N instances of the related object class.

(N,-): N specified, M open-ended. An object instance must play its relationship role with *at least* N instances of the related object class.

(-,M): N unspecified, M specified. An object instance must play its relationship role with *at most* M instances of the related object class.

(0,1): Special case of (N,M). Participation in the relationship role is *optional.*

(0,-): Special case of (N,-). An object instance must play its relationship role with *zero or more* instances of the related object class.

(1,-): Special case of (N,-). An object instance must play its relationship role with *one or more* instances of the related object class.

none: None specified. The default is considered to be the case **(1)**. An object instance must play its role with *exactly* 1 instance of the related object class.

When we specify the syntax for relationships, we will see how all these different cases of multiplicity may be formally specified in the object model. Figure 4 demonstrates examples which use the multiplicity interval notation above. The formalized relationship knowledge encoded in this diagram may be read as follows:

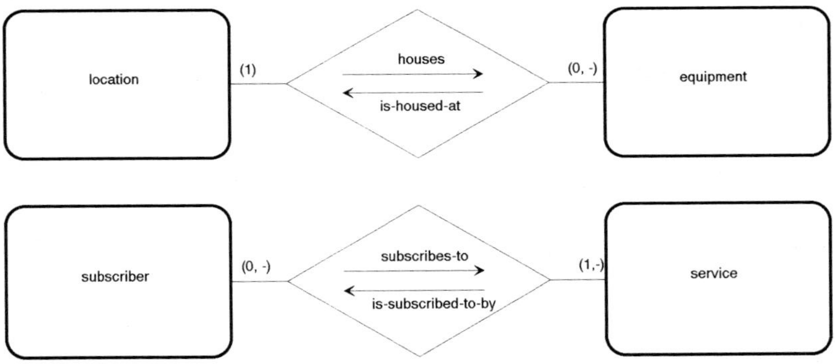

Figure 4: Some Relationships With Their Multiplicities

"A location houses zero or more (pieces of) equipment."
"A (piece of) equipment is housed at one location."
"A subscriber subscribes to one or more services."
"A service is subscribed to by zero or more subscribers."

Even though the above multiplicity specifications are adequate to cover a majority of the situations, they are still not semantically powerful enough to cover all cases. Consider a sonetCrossConnect object class, which possesses a numberOf-OC1-Ports as an attribute. For a given subclass of sonetCrossConnect - say lowEndSonetCrossConnect - this attribute may have a fixed value, say 64. This object class participates in a termination relationship with an oc1-circuit object class. Clearly, this is a one-to-many relationship: a single sonetCrossConnect object may terminate many oc1-circuit objects. To specify the multiplicity of this relationship, we may decide to use the interval (0, 64).

However, this is not sufficient, because while the general superclass sonetCrossConnect participates in a termination relationship with the oc1-circuit object class, each one of its *subclasses* may participate in the relationship with a *different* multiplicity. Depending on the number of OC-1 ports available in each class, the subclass highEndSonetCrossConnect may have a multiplicity of (0, 2048) in this relationship, the subclass midRangeSonetCrossConnect may have a multiplicity of (0, 512), while the subclass lowEndSonetCrossConnect may have (0, 64). Therefore, modeling the multiplicity of this relationship as (0, 64) for the general sonetCrossConnect superclass would be wrong.

To solve this problem, we reason that even though each subclass of sonetCrossConnect may have a different multiplicity constant in its termination relationship, we nevertheless know that this constant is available in its numberOf-OC1-Ports attribute. We therefore specify the relationship multiplicity of the oc1-circuit object class in its termination relationship with the general sonetCrossConnect superclass, not in terms of an integer *constant*, but rather in terms of an *attribute* of that superclass. We know that, no matter what value each instance of the sonetCrossConnect object class possesses for that attribute, the upper limit of the multiplicity for the number of oc1-circuit objects terminated there is defined by that attribute.

We therefore allow our relationship multiplicities to be specified not merely in terms of integer constants, but also in terms of *attribute-defined expressions* involving attributes of the related object class. In practice, this often results in relationship multiplicities being specified in terms of some *capacity-related* attribute of the related object class, such as number of ports, disk capacity, maximum forwarding rate, and so on. In an Entity-Relationship diagram, if the multiplicity for a participating object class is specified in terms of such an expression, the attributes in the expression must belong to the object class at the *other* end of the relationship.

In general, relationship multiplicities could be specified as integer constants, ranges of integer constants, and ranges in which either the lower limit, or the upper limit, or possibly both, could be specified in terms of attribute-defined expressions. In addition, because relationship multiplicities need not always conform to a continuous range but could also be a discrete set of values, we permit relationship multiplicities to be specified in terms of an enumerated set of values as well. Each value in such an enumerated set could either be an integer constant, or could itself be an attribute-defined expression.

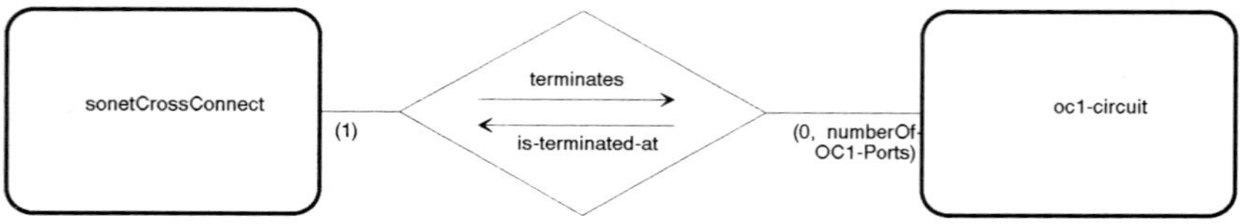

Figure 5: Multiplicities Defined By Attributes of the Related Class

3.4. Formal Modeling of Relationships

With this background, we are now ready to define the formal templates for specifying relationships. We define relationships as a first-class conceptual construct using the following template[2]:

```
<relationship-label>      RELATIONSHIP
        OBJECT CLASS                  <class-label>
        IN ROLE                       <role-label>
        [ WITH MULTIPLICITY           <multiplicity-label> ]
                RELATES TO
        OBJECT CLASS                  <class-label>
        IN ROLE                       <role-label>
        [ WITH MULTIPLICITY           <multiplicity-label> ]

        [ BEHAVIOR                    <behavior-definition-label>
                                      [,<behavior-definition-label>]* ; ]
        [ CREATE                      [create-modifier] [<parameter-label>]* ;
          DELETE                      ]

REGISTERED AS      <object-identifier, object-identifier>;

supporting productions

create-modifier -> with-reference-relationship
```

In the above template, the *class-labels* specify the two object classes which participate in the relationship. The *role-labels* are essentially role names, specified in the syntax of ASN.1 PrintableStrings, which identify for human comprehension the roles played by the object class in either direction of the relationship. The behavior of a relationship may be specified, if desired, using the *behavior-definition-label.* The WITH MULTIPLICITY clause is optional because, if omitted, it defaults to exactly 1. The CREATE and DELETE operations apply to *relationship instances*; the parameters to a normal CREATE operation will be the names of two object instances of the appropriate classes, and the operation could be optionally modified with a reference relationship which already exists. A relationship instance could also be deleted using a DELETE operation. This only deletes the knowledge of the association between the two object instances, and does not delete the instances themselves.

Each relationship is registered using *two* object identifiers. This is because an object identifier is required to identify each direction of the relationship. This is helpful, for example, during the CREATE operation. Consider a situation where **A** and **B** are two managed object classes participating in a relationship **R.** This relationship is registered with two object identifiers, $\mathbf{R\text{-}id_{AB}}$ for the forward role, and $\mathbf{R\text{-}id_{BA}}$ for the inverse role.

[2] Because relationships are not part of any standard, the template used to specify them is also non-standard.

Let *A1* and *B1* be instances of object classes **A** and **B** respectively. Then, a relationship instance can be established between objects *A1* and *B1* either by specifying the sequence {*A1* $\mathbf{R\text{-}id_{AB}}$ *B1*} as a parameter to the CREATE operation, or the sequence {*B1* $\mathbf{R\text{-}id_{BA}}$ *A1*}. These are semantically equivalent, and either could be used to establish the relationship; no particular direction is preferred. Once the relationship instance has been created using one such direction, it is also automatically considered created in the inverse direction. Similarly, the relationship instance could be deleted by specifying either directional sequence. Thus, the notion of relationship roles is formalized, because the two object identifiers correspond to the two relationship roles.

The *multiplicity-label* is a specification for the multiplicity of the object class in the relationship, specified in the following ASN.1 syntax:

```
Multiplicity ::= CHOICE
{
        set     [0]     SET OF MultiplicityValue,
        range   [1]     SEQUENCE
                        {
                                lowerlimit  [0] MultiplicityValue OPTIONAL,
                                upperlimit  [1] MultiplicityValue OPTIONAL
                        }
}

MultiplicityValue ::= CHOICE
{
        [0]     INTEGER,
        [1]     AttributeDefinedExpression
}
```

The type AttributeDefinedExpression, as specified earlier, is further expanded into a simple grammar allowing all unary and binary arithmetic expressions on attributes. As we indicated earlier, in practice the majority of such expressions will simply be the value of some capacity-related attribute of the object class at the other end of the relationship.

As an example, consider the many-to-many subscription relationship between a subscriber object class and a service object class. Each object class will be specified in its own MANAGED OBJECT CLASS template, and will have its own attributes; the subscriber object may have attributes such as name, address, and so on, while the service object class may its own attributes such as serviceType, which could possibly have enumerated values such as eightHundredService, virtualPrivateNetworkService, and so on. These templates are not shown here for brevity.

A relationship could now be defined between these two object classes as follows:

```
subscription    RELATIONSHIP
        OBJECT CLASS            subscriber
        IN ROLE                 "subscribes-to"
        WITH MULTIPLICITY       zeroOrMore
                RELATES TO
        OBJECT CLASS            service
        IN ROLE                 "is-subscribed-to-by"
        WITH MULTIPLICITY       oneOrMore

        BEHAVIOR                subscription-Behavior
        CREATE
        DELETE

REGISTERED AS       {subscribes-to, is-subscribed-to-by};
```

where the registration labels subscribes-to and is-subscribed-to-by are specified elsewhere as unique OBJECT IDENTIFIERs, and the multiplicities zeroOrMore and oneOrMore are specified in the ASN.1 syntax:

```
zeroOrMore Multiplicity ::= { range { lowerlimit 0 } }    -- upperlimit unspecified
oneOrMore  Multiplicity ::= { range { lowerlimit 1 } }    -- upperlimit unspecified
```

It is also worth explicitly mentioning the principle of *inheritance of relationships*. This principle states that any relationship (such as supervision, backup, etc) specified for a superclass, is also considered to be in effect for all subclasses of that superclass. Thus, once the MIB compiler parses a relationship template definition for a superclass, it automatically arranges for semantically equivalent definitions to be available for all its subclasses.

4. Advantages of Explicit Relationship Modeling

Our approach of modeling relationships as explicit first-class constructs makes a very clean distinction between the specification of relationships as abstract associations between object classes, and their possible implementation. It leaves open many implementation possibilities. Relationships specified as the abstract conceptual constructs above could still be implemented as pointers within object classes, with each pointer pointing to an appropriate instance of the related objects. Many other implementations are also possible. Many-to-many relationships are perhaps best implemented as tabular structures rather than mutual pointers or relationship attributes. Such a table could be resident in the memory of either object, in the memory of some third, non-participating object, or even as disk-based information in a relational database in a management station or an operations support system.

This approach to the philosophy of modeling suggests that the relationships in which an object class participates are not inherent properties of the object, and should not therefore be modeled as attributes of the object. Any change in the value of an object's attribute should reflect some fundamental change in the inherent properties of the object. An attribute value update ought not to be caused by a change in the external relationships in which an object participates with other objects. An object may be viewed as a closure, whose boundaries are crisply defined by encapsulation of its internal states. This closure nevertheless has external links with other closures; a change in its external links ought not to result in changes inside the closure. Relationships may be dynamic and transient in nature, as many relationships in communication networks frequently are. If relationships are modeled as attributes, the frequent changes in association that an object in a communications network is subjected to would cause frequent updates in its relationship attributes. Taken to the extreme, the approach of modeling relationships through relationship attributes embedded inside an object would begin to compromise the spirit of encapsulation.

Notice that relationship instances are created and deleted using CREATE and DELETE operations on the entire relationship construct, rather than by updating attributes stored inside the individual object classes. This is similar to the operations on a NAME BINDING.

A NAME BINDING, incidentally, is simply a special case of a relationship: it specifies a containment relationship between a superior object class and a subordinate object class. The following relationship specification is semantically identical to the logRecord-log name binding defined in the standards:

```
logRecordContainment RELATIONSHIP
        OBJECT CLASS                log              -- superior object class
        IN ROLE                     "contains"
                RELATES TO
        OBJECT CLASS                logRecord        -- subordinate object class
        IN ROLE                     "is-contained-in"
        WITH MULTIPLICITY           zeroOrMore

        CHARACTERIZED BY
                logRecordContainmentPackage PACKAGE
                ATTRIBUTES
                        logRecordID;    -- naming attribute for containment relationship
        DELETE;
```

REGISTERED AS {contains, is-contained-in};

The above specification preserves all the characteristics of the NAME BINDING. It uses an association attribute to indicate the naming attribute for this containment relationship. Just as name bindings specified for any superior object class are inherited by all subclasses of that object class, so is this containment relationship, due to the principle of inheritance of relationships stated earlier.

Our approach to relationship modeling is indeed far more general than name bindings, because it is able to formally model an extremely wide range of relationships. Name bindings model only containment between object classes, and can be recast as a special case of general relationship modeling.

5. Virtual Attributes to Represent Relationships

In the earlier section on virtual attributes, we demonstrated how virtual attributes of a specificand object class can be algorithmically inferred from other attributes of the same object class. In this section, we demonstrate how some virtual attributes of a class can also be inferred from attributes of a *different* object class, one which participates in a relationship with the specificand object class.

We have said before that using attributes embedded inside of managed object classes is not a desirable mechanism to represent the relationships of that managed object class. This is because many relationships are two-way and, during the execution of various network management applications, require that they be traversable in either direction, originating from either participating object. This requires embedded attributes in both participating objects which mutually point towards each other. This means that information is redundantly stored in two locations to represent the single conceptual construct of a relationship. This leads to problems during modification and deletion of relationships, because of the difficulty of implementing mechanisms to guarantee atomicity or consistency of multiple concurrent updates to management information.

In order that we may be able to easily query an object to determine all the relationships it participates in, we can use virtual attributes. Virtual attributes do not suffer from the problem of assuring update consistency with other attributes, because they can never be updated. Because virtual attributes are always qualified with only a GET and never a GET-REPLACE, any attempt to update a virtual attribute will always result in an error. Nevertheless, by changing the value of an appropriate base attribute in another object, we can in effect "update" a virtual attribute by *automatically tracking the change through its relationship.*

Consider a situation where we wish to query an equipment object to determine its current location. Conventionally, we may accomplish this by specifying that each instance of an equipment object has a currentLocation attribute, whose value reflects its current physical location. (For the time being, assume for simplicity that the syntax of currentLocation is PrintableString.) Many co-located instances of equipment could have the same value for the currentLocation attribute. If some of these pieces of equipment are moved to another location, all of those instances must be individually updated to change their currentLocation attribute. This leaves open the potential for inconsistency.

We can eliminate this problem entirely by turning the currentLocation attribute of the equipment object class into a *virtual attribute.* We know that there is an is-housed-at relationship between the equipment object class and the location object class. We further know that the location object class has identifier attributes which identify it uniquely. It may also have various other useful attributes, such as streetAddress, telephoneNumber, nearestFireStation, and so on. Suppose that we decided that the currentLocation virtual attribute of each equipment object was most usefully represented as the streetAddress attribute of the location object with which it participated in the is-housed-at relationship, as shown in Figure 6.

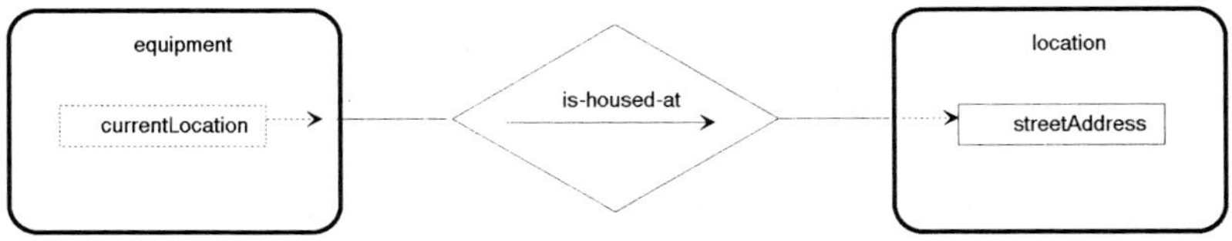

Figure 6: Tracking Virtual Attributes into Another Object Class via a Relationship

We need to inform the MIB compiler that the currentLocation attribute of any equipment instance should be dynamically computed by tracking its is-housed-at relationship link into an appropriate location object, and reading its streetAddress value. We can specify this knowledge formally as follows:

```
currentLocation  VIRTUAL ATTRIBUTE
        WITH ATTRIBUTE SYNTAX  PrintableString;
        MATCHES FOR Equality;
        COMPUTED FROM  isHousedAtStreetAddressLink;
REGISTERED AS {....}
```

The general form of a virtual attribute computation algorithm can now be specified as follows (abbreviated for succinctness to illustrate new concepts only):

```
<attribute-label>           VIRTUAL ATTRIBUTE
        WITH ATTRIBUTE SYNTAX       <syntax-label>;
        MATCHES FOR                 qualifier  [, qualifier]* ; ]
        COMPUTED FROM               <algorithm-label>;
REGISTERED AS  <object-identifier> ;
```

where the algorithm-label, as indicated earlier, is the name of an ASN.1 value reference of type AttributeDefinedExpression, which expands into a simple algorithmic grammar involving unary and binary arithmetic expressions on attributes. The atomic literals of this grammar, however, are no longer restricted to being attributes of the *same* object class; they could be attributes of a *different* object class, tracked through a relationship link. Assume that the atoms of this expression grammar are defined using the type reference AtomAttribute, defined as follows:

```
AtomAttribute ::= SEQUENCE
{
        links              SEQUENCE OF Relationship,
        baseAttribute      AttributeID
}

Relationship ::= OBJECT IDENTIFIER
```

The relationship links to compute a virtual attribute are indicated by the OBJECT IDENTIFIER which was used in the registration hierarchy to register the specific relationship between the two participating object classes in the appropriate role direction. The base attribute from the originating object class is indicated by the AttributeID, which most often is the registration OBJECT IDENTIFIER for the attribute.

We can now show that the specification for the currentLocation of the equipment object class can be computed by the syntax:

```
isHousedAtStreetAddressLink AttributeDefinedExpression::=
{
        links              is-housed-at,
        baseAttribute      streetAddress
}
```

In general, a sequence of relationship links could be traversed through many different objects to execute the computation of a virtual attribute. For example, we could formally specify an equipmentSupervisors attribute for the equipment object class, which would indicate the network operators responsible for supervising that instance of equipment. For example, we could define a supervises relationship between the networkOperator and location object classes to indicate an operator's supervision over all the equipment at a particular location. Then, for each instance of equipment, the virtual attribute equipmentSupervisors can be directly tracked into the networkOperator class, following a sequence of relationships, as shown in Figure 7.

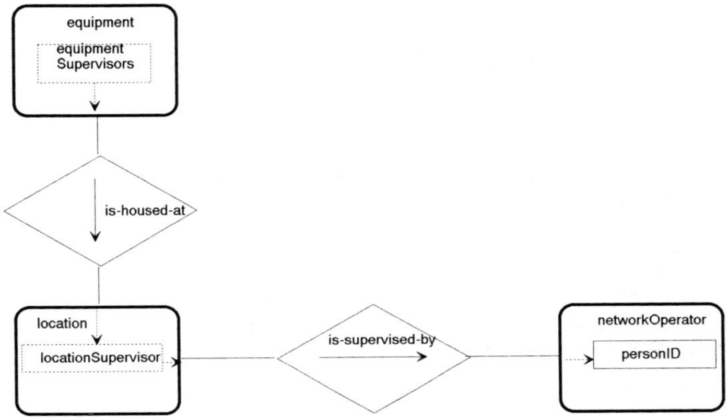

Figure 7: Tracking Virtual Attributes via Multiple Relationship Links

Virtual attributes can be set-valued even when their corresponding base attributes are single-valued. Consider the problem of querying a location object to generate a report on all the pieces of equipment located there. We could specify an equipmentHoused virtual attribute for the location object class, whose syntax would be SET OF EquipmentID, which would be dynamically computed by tracking all the instances of a houses relationship between that location object and various equipment objects:

```
equipmentHoused          VIRTUAL ATTRIBUTE
        WITH ATTRIBUTE SYNTAX                SET OF EquipmentID;
        COMPUTED FROM                        housesLinks;
REGISTERED AS  {....}

housesLinks      AttributeDefinedExpression ::=
{
        links              houses,
        baseAttribute      equipmentID
}
```

where we assume that the base attribute equipmentID has an attribute syntax of type EquipmentID. A sophisticated MIB compiler need not even be told what the attribute syntax of a virtual attribute is; it should be able to infer the type of the virtual attribute based on its knowledge of the types of its supporting base attributes, the multiplicity of the supporting relationships, and the computations involved in inferring the value of virtual attribute. A sophisticated compiler, for example, would know that because the multiplicity of the houses relationship is one-to-many, the resulting syntax of the equipmentHoused virtual attribute would be a SET OF the type of the supporting base attribute.

It should be remembered that as far as network management protocols are concerned, virtual attributes can be queried in the same manner as any other attribute. In fact, a protocol request to query an attribute value need not be concerned whether a queried attribute is native or virtual. A query to the manufacturerName attribute of an equipment object, for example, could either be reported by an agent directly reading a native attribute, or tracked to the name attribute of a manufacturer object following its is-manufactured-by relationship link. The latter mechanism would be more beneficial when certain attributes of the manufacturer object (such as address, serviceContact, etc.) need to be reflected in the equipment object, and are potentially subject to change - for example, the name and address of a manufacturer could change due to corporate merger, acquisition, or bankruptcy activity. Nevertheless, the actual mechanism whereby an attribute value is reported should be completely transparent to the protocol request.

What we have accomplished using this mechanism is the inclusion within the Management Information Base of the formal specification for our knowledge of the *value dependency* between two different attributes in two different objects. Because the Management Information Base now knows the formal specification of relationships between objects and the value dependency of virtual attributes, it can automatically recompute the value of a virtual attribute every time it is queried. Thus, we are guaranteed that the virtual attribute will always reflect a semantically correct value, irrespective of how many changes are made to its supporting base attributes in another object.

6. Conclusion

Virtual attributes are a powerful mechanism to allow the exposure of only certain attributes of managed objects which are dynamically computed from other attribute values, and so allow the enforcement of user-defined network privacy and security policies. By defining relationships as first-class modeling constructs, we raise the level of modeling abstraction by allowing the formal specification of any arbitrarily defined binary association between object classes in a single construct. Relationship multiplicities allow precise quantitative knowledge to be embedded within management information. Finally, the use of relationship links to track virtual attributes of one object class into another object class is an elegant mechanism to provide ease of information access, consistency of information values, and prevent encapsulation breakdown. All of these concepts together may act as an impetus for initiating some research directed at achieving a rich set of modeling semantics for management information.

7. References

[CCITT X.720] Information Processing - Open Systems Interconnection - Management Information Services - Structure of Management Information, Part 1: Management Information Model for CCITT Applications, CCITT, 1992.

[CCITT X.721] Information Processing - Open Systems Interconnection - Management Information Services - Structure of Management Information, Part 2: Definition of Management Information for CCITT Applications, CCITT, 1992.

[CCITT X.710] Information Processing - Open Systems Interconnection - Common Management Information Service Definition for CCITT Applications, CCITT, 1992.

[CCITT X.711] Information Processing - Open Systems Interconnection - Common Management Information Protocol Specification for CCITT Applications, CCITT, 1992.

[CCITT X.732] Information Processing - Open Systems Interconnection - Systems Management Part 3: Attributes for Representing Relationships for CCITT Applications, CCITT, 1992.

[NMF 89] Object Specification Framework, Issue 1, Network Management Forum, September 1989.

[RFC 1155] Structure and Identification of Management Information for TCP/IP-based Internets, M. Rose and K. McCloghrie, The Internet Society, May 1990.

[RFC 1157] A Simple Network Management Protocol, J. Case, M. Fedor, M. Schoffstall, and J. Davin, The Internet Society, May 1990.

[Rose90] The Simple Book: An Introduction to the Management of TCP/IP-based Internets, M. Rose, Prentice-Hall, 1990.

[T1M1 89] T1M1 Modeling Guidelines, ANSI T1M1.5/89 - 009, 1989.

[RFC 1213] Management Information Base for Network Management of TCP/IP-based Internets: MIB-II, K. McCloghrie and M. Rose, The Internet Society, March 1991.

[RFC 791] The Internet Protocol Specification, The Internet Society, September 1981.

[Rumb91] Object-Oriented Modeling and Design, Rumbaugh, James, et al, Prentice Hall, 1991.

Integrated Network Management, III (C-12)
H.-G. Hegering and Y. Yemini (Editors)
Elsevier Science Publishers B.V. (North-Holland)

A High Level Specification Technique for Modeling Networks and their Environments including Semantic Aspects

Ch. Benz and M. Leischner

Telekom c/o DETECON Deutsche Telepost Consulting GmbH, Project Digital Mobile Communications, P.O. Box 18 02 05, 5300 Bonn 1, Germany

Abstract

Integrated network management requires new sophisticated modeling concepts for management information. Using an object-oriented and relation algebraic approach we present a specification technique covering not only management information aspects but also semantic aspects. This approach allows a homogeneous and uniform treatment of the basic elements of a network description we have identified, namely objects, relations, and, in particular, consistency rules.
Examples demonstrate the applicability of our approach to major network modeling problems.

Keyword Codes: C.2.1; I.2.4
Keywords: Network Architecture and Design; Knowledge Representation Formalisms and Methods

1 Introduction

Caused by the increasing complexity of computer networks, modeling concepts in the field of integrated network management become more and more important. Standardization organizations like IAB or ISO have contributed to this tendency by establishing frameworks for modeling and structuring management information ([RFC 1155], OSI SMI (Structure of Management Information) [ISO 10165]). However, those frameworks are mainly concerned with syntactical data representation. Semantic aspects are covered only by means of informal comments or of 'behavior' definitions respectively (using ASN.1 strings). As demands for management tools are increasing and also their complexity is growing, it is obvious that it is necessary to formalize semantic aspects when modeling management information.

The following examples from the fields of standardization, management products, and research emphasize this trend:

- In [ISO 10164-3] general aspects of management relevant relations are described. First concepts for the integration of relations into OSI SMI framework by using relationship templates are proposed in [ISO N1390].
- The network management system SPECTRUM from Cabletron Systems Inc. is based on a modeling concept called IMT (Inductive Modeling Technology), a new branch of Artificial Intelligence which combines the paradigm of object-oriented modeling with the principles of Model Based Reasoning ([SheSim 90], [AbeLeiSeg 92]).
- In ([Val 91]) graph theoretical means are used to describe semantic aspects of a network configuration. The communication layering (according to the OSI Reference Model) is modeled by a Layered Attributed Graph (LAG).

All those approaches allow an adequate description of only specific semantic aspects of networks (such as the communication layering). What is needed is a general concept which is powerful enough to model *all* important semantic aspects in a uniform and homogeneous way. Such a framework has to be situated on an abstract level; thus it can serve as a starting point for a top down design of a network description.

We consider concepts for ensuring consistency a very important part in modeling semantics in the field of network management. Thus we have identified the following three elements as the main building blocks of network modeling:

- objects, by analogy with the object-oriented approach of OSI SMI,
- relations between objects to model semantic structures, and
- rules which express global properties of semantically consistent network descriptions.

Although the necessity of rules is recognized in different approaches, no concept has been found up to now which allows a homogeneous treatment of them. To achieve this goal we use a relation algebraic based approach.

This paper is organized as follows: Section 2 contains some motivations and an overview of our approach. In section 3 we introduce the notations and elements of our relational concept. Then, in section 4, we show an application of our methodology: we present a generic network model which comprises the OSI communication layering as well as the physical network environment. Possible techniques how to refine and enhance network descriptions are outlined in section 5. Section 6 briefly reviews the main ideas of the paper.

2 Overview

The current situation in network management is characterized by an increasing amount of information from different areas relevant to network management. For example: information on communication resources, distributed applications, physical environment, or organizational data (see Figure 1).

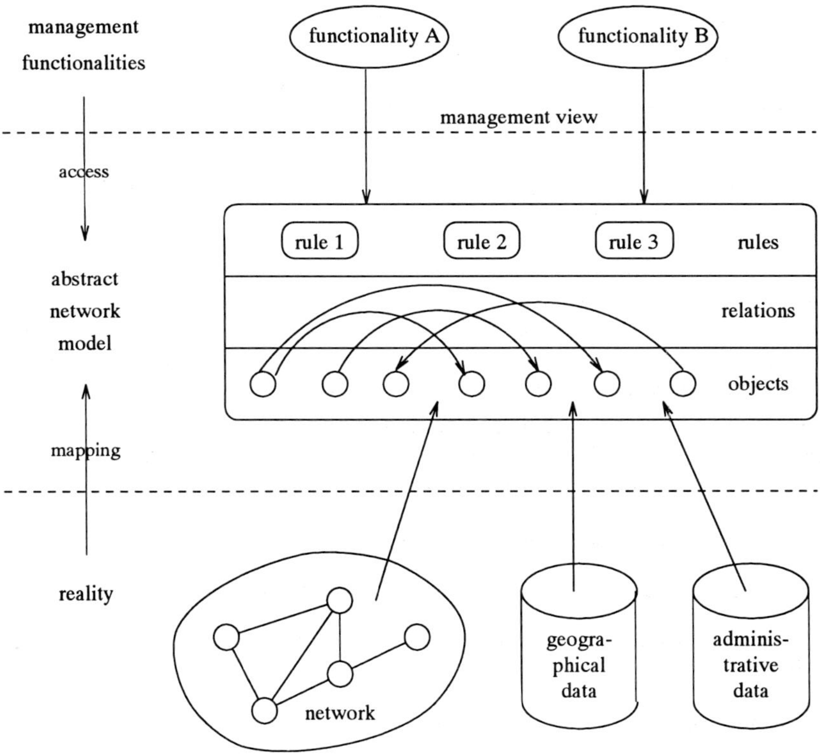

Figure 1: Relational Approach – Overview

The main task for modeling management information is to provide a management view on the real network and its environment. This view serves as a platform for management functionalities. Our approach yields a global high level view on a network. In particular, interrelations between objects together with consistency rules are regarded as an integrated part of the model.

Relations and rules are essential for sophisticated network management. Network configuration or network planning, for instance, have to take into account global interrelations. If a new node is added to a network, for example, other related nodes have to be reconfigured (e.g. routing tables). Obviously these changes have to obey certain rules such that the network continues to function correctly (i.e. the changes result in a consistent network). Consistency rules which are included in the management system can support the operator in different ways: the management system can serve as a guide for the operator identifying inconsistencies within the network configuration, or parts of the (re)configuration process can be done automatically (Network Change Control).

3 Relational Approach for Network Decriptions

The abstraction level of our approach is characterized by defining object classes without specifying the inner structure of objects, i.e. we will not consider features like attributes, notifications, actions or behavior of objects.

The object classes found in existing, established object class catalogues (e.g. [Forum 90], [ISO 10165-2]) are very specific and thus organized in rather flat inheritance trees. According to the ideas of [ISO 10165-5, Generic Management Information] we also define general classes like *protocol entity* or *device*, which are common to all networks. More specific subclasses like *transportEntity* (confer [ISO 10737]) or *transceiver* can be derived from them, but are too specific for our purposes and therefore not considered here.

According to the abstraction level of the object classes, we only define high level relations and rules. For example, we do not take into account rules which operate on object attributes or which are specific for a certain network technology (e.g. maximum cable length of an ethernet segment).

In our approach, classes and relations can be represented by diagrams, similar to Entity-Relationship (E-R) diagrams. To be precise: our information modeling approach can be termed a binary E-R model, which is interpreted in terms of object-orientation. This means that we will use the OSI-terms 'object class', 'subclass', 'object', and 'subclass relation' instead of the E-R terms 'entity type', 'entity subtype', 'entity', and 'is-a relationship', respectively. (This proximity to the field of databases supports the possibilities of an implementation of our concept.) As we use a binary E-R model, we will be able to express consistency rules by relational algebraic formulas.

We will now introduce the notations for objects, relations, and rules used for our high level network description model.

3.1 Object Classes

Object classes are denoted by boxes with the class names inside. The subclass relation between a subclass B and a superclass A is expressed by inclusion of box B in box A. According to the object-oriented principles of OSI SMI our object classes are organized in an inheritance hierarchy. Every object of subclass B can be regarded as an object of class A (allomorphism). Boxes can be nested as shown in Figure 2.

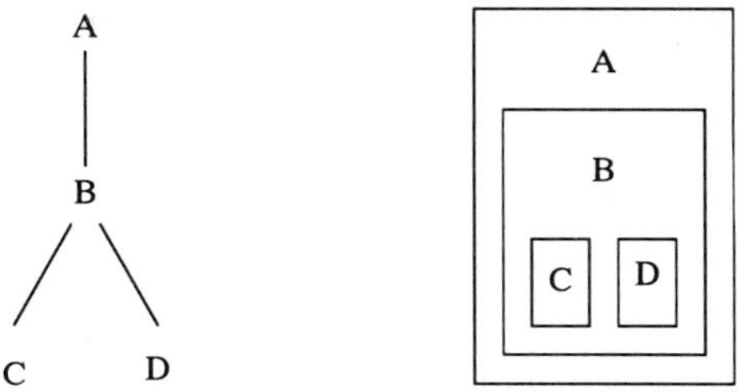

inheritance tree box notation of inheritance

Figure 2: Notations of Inheritance

3.2 Relations

Our concept of relations as well as our notation is closely related to the OSI concept ([ISO 10164-3]). A relation between two object classes is expressed as follows:

Objects of class A are related to objects of class B by the relation named *nameOfRelation*. The direction of the arrow determines how to read the relation (in this case class A is related by *nameOfRelation* to class B). The given ranges determine the number of objects of the class that can be set into relation with one object of the other class. The range $k..l$ next to class B indicates that a minimum of k and a maximum of l objects of class B are related to one object of class A; the range $m..n$ is to be interpreted accordingly.

As every subclass object can be regarded as an object of the superclass, all relations existing between two classes define a corresponding relation between the subclasses. In this sense subclasses inherit the relations of their superclasses.

The fact that in our diagrams subclass relations are expressed by box inclusions allows the representation of relation inheritance in a very elegant way which can be understood intuitively.

3.3 Consistency Rules

Our notation of relations already includes elementary consistency rules: the ranges next to the boxes determine the cardinality of objects which can be related to each other. This kind of rules are local in the sense that they refer to only one relation and thus are easy to treat.

The aim of consistency rules is to assure global consistency by describing the combination of several relations. These rules should be expressible in a formal way to guarantee precise semantics. The formalism used should fit homogeneously to the representation of networks by classes and relations. As already mentioned, we use the formalism of relation algebra for this purpose, which allows the treatment of binary relations.

Consistency rules can be expressed by relational equations. Let us first give a short introduction to the notation used in the sequel[1].

Let R and Q be relations from class A to class B, S a relation from B to C, and T a relation on A. Then we define[2]:

O	$= \emptyset$, the zero element of a relation algebra
L_{AB}	$= A \times B$, a universal element of the relation algebra[3]
$R \subseteq Q$	relation R is a subset of relation Q
$R \cup Q$	the union (supremum) of the relations R and Q
$R \circ S$	$= \{(x,z) \mid \exists\, y : (x,y) \in R \text{ and } (y,z) \in S\}$, the composition of the relations R and S
$\overline{R}$	$= L_{AB} \setminus R$, the complement of relation R
R^{-1}	$= \{(y,x) \mid (x,y) \in R\}$, the inversion (transposition) of relation R
T^*	$= \bigcup_{n=0}^{\infty} T^n$ where $T^0 = I$ (I the identical relation) and $T^n = T^{n-1} \circ T$, the transitive closure of relation T

The subclass (inheritance-)relation of a subclass B in a superclass A can be expressed by means of relation algebra by an inclusion relation, denoted by $incl_{B,A}$ or, for short, $incl$. In Figure 2, for example, the following equation is valid:

$$incl_{D,A} = incl_{D,B} \circ incl_{B,A}$$

The inversion $incl_{B,A}{}^{-1}$ of a subclass relation can be regarded as a filter which picks out the objects of subclass B in superclass A.

Using these notational elements it is possible to build up formulas expressing complex consistency rules, as will be shown in the next section.

4 Application of the Approach

In this section we want to illustrate that the formal elements introduced in the previous section are adequate to describe semantic aspects of a network in a concise and precise manner.

---

[1]A detailed mathematical foundation of (homogeneous) relation algebras may be found in [Tar 41]. However, since we have relations between different classes, we need heterogeneous relation algebras which can be treated similarly to the homogeneous algebras; a short definition of them is contained in [BerSch 89].

[2]Only to give a clear imagination of the used notation we use at the right side of our definitions set theoretic notations ($R \subseteq A \times B$) instead of the axiomatic approach usually used.

[3]Since we have a heterogeneous relation algebra there can be several universal elements for each pair of A and B. Where possible, we will denote a universal element simply by L.

For this purpose we select some basic and important aspects of a network and apply our concept of objects, relations and rules to them. In order to give an idea how our approach can serve as a universal modeling scheme – not limited to the structure of communication layering – we present an extended model which includes the physical network environment as well.

The diagram of this model can be found in Figure 3. In this diagram the upper part models the communication aspect, whereas the lower part represents the physical environment.

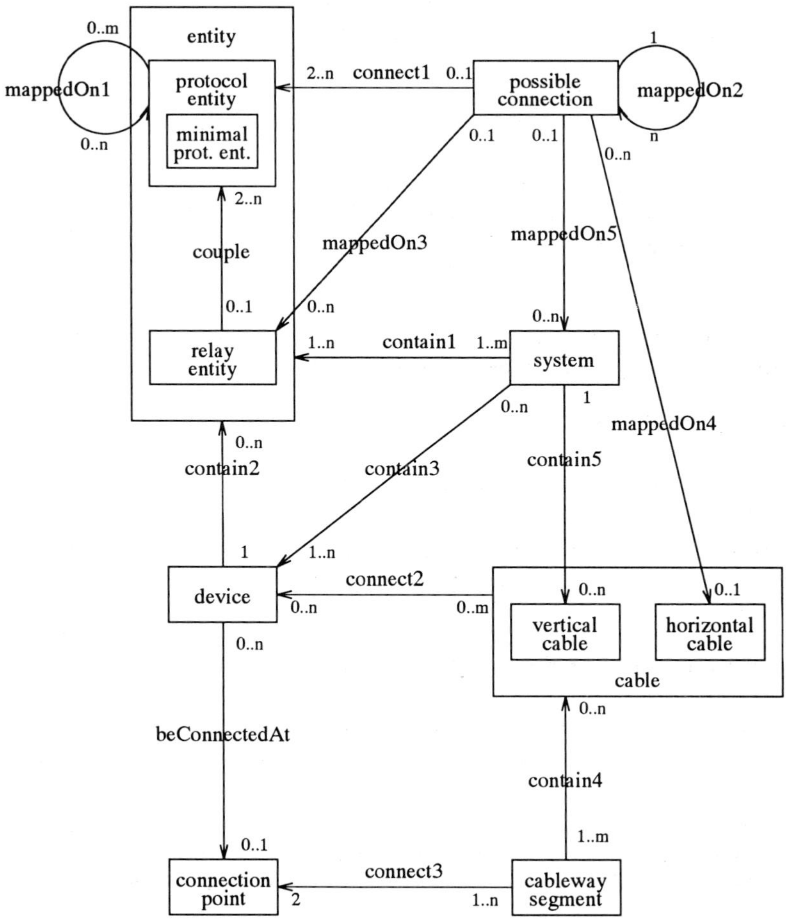

Figure 3: A Network Description Scheme

Both parts are combined by relations (and rules as will be seen in section 4.3) such that we obtain a seamless network description comprising both aspects.

4.1 Object Classes

First we describe those objects which form the basis for OSI conformant modeling of the communication layering.

- *entity*
  Entities are the active elements of a communication system. We can distinguish between:
  - Protocol entities accomplish the functionality of a layer specific protocol. In order to provide n-layer services a n-layer protocol entity can use services provided by (n-1)-layer protocol entities. Several entities can be mapped onto one entity (upward multiplexing) and one entity can be mapped onto more than one entity (downward multiplexing).
    If a protocol entity is assigned to the lowest layer (1-layer), it is called a minimal protocol entity.
  - Relay entities couple protocol entities assigned to the same layer within a transit system.
- *system*
  A (communication) system is a collection of cooperating entities which form an autonomous whole capable of communicating with other systems. A more detailed discussion of the notion *system* is given in section 4.3.
- *possible connection*
  A set of n-layer entities is connected by a possible connection if they are capable of communicating with each other. This is the case if they use the same protocol, if the corresponding (n-1)-layer entities can communicate with each other (perhaps via transit systems), and if a physical connection exists. Therefore a possible connection can be mapped onto
  - relay entities if transit systems are necessary to establish the connection,
  - possible connections which connect the underlying protocol entities, and
  - a (horizontal) cable if it connects minimal protocol entities.

The objects belonging to the physical network environment are the following:

- *device*
  Devices represent physical components of a network such as computers, but also facilities for communication purposes only like routers, bridges or transceivers.
- *cable*
  Cables represent physical connections. Motivated by the communication layering view, we distinguish between horizontal cables, which connect devices of different systems (e.g. Ethernet cable), and vertical cables, which connect devices of one system (e.g. a transceiver cable).

- *cableway segment*
  A cableway segment models the route of a cable. Cables run in a sequence of one or more than one cableway segment. At the two ends of a cableway segment there are connection points. One cableway segment can contain zero, one, or more than one cable. Only at connection points cables can leave or enter a cableway segment.

- *connection point*
  Connection points describe locations where devices can be connected to a cable (such as wall sockets or outlets). There are also connection points which are not equipped for connecting a device to a cable (potential connecting point), but can be equipped for this in future. Potential connection points allow to model transitions between cableway segments.

Obviously, object classes from many other management relevant areas such as organizational, geographical, or applicational classes could be identified and integrated if necessary.

Note that in this description of classes we have already implicitly introduced semantic relations between objects of the considered classes such as the mapping of possible connections on underlying possible connections or cables.

4.2 Relations

The meaning of all relations and their cardinalities contained in Figure 3 is obvious from the informal class description above. Therefore we do not present a list of all relations; instead we will discuss some interesting details:

- In Figure 3, there is a difference between arrows going to outer class boxes and arrows going to inner boxes. The arrow *mappedOn4*, for example, goes directly to the box of the subclass *horizontal cable*, because possible connections can only be mapped on horizontal cables. In contrast, the arrow *contain4* which expresses the containment of cables in cableway segments goes to the outer box *cable*, because cableway segments do not 'know' the difference between horizontal and vertical cables.

  Note that the class *horizontal cable*, respectively *vertical cable*, is also related to the class *cableway segment* by a *contain4* relation inherited from the superclass *cable* ('relation inheritance').

- The layering of protocol entities and possible connections is expressed by the two loops *mappedOn1* and *mappedOn2* at the corresponding classes.

- It is an interesting fact that we can not conclude from the relation arrows in the diagram whether or not a relation can be expressed by means of other relations:

  The relation *mappedOn5*, for instance, can be expressed by the composition of the relations *mappedOn3* and the inversion of *contain1* (and the inclusion relation of the class *relay entity* within the class *entity*).

In contrast, the relation *contain3* can not be derived using the composition of *contain1* and the inversion of *contain2*. The reason for this is that a system can contain (*contain3*) devices which do not contain (*contain2*) entities. Such a device could be a multiport transceiver which serves as an interface multiplier.

These facts can be expressed in terms of relation algebra:

$$mappedOn5 = mappedOn3 \circ incl^{-1} \circ contain1^{-1}$$

$$contain3 \supseteq contain1 \circ contain2^{-1}$$

- Note that the ranges are carefully chosen. For example, the range $0..m$ at the arrow *connect2* allows to include devices in the network description which are not connected to a network (for instance stand alone computers which have to be managed nevertheless): the zero in $0..m$ indicates that devices with no cables connected to them may exist. Since such devices are not capable of communicating, they are not contained within (communication) systems (refer to the zero in $0..n$ at arrow *contain3*).

We can say that the diagram in Figure 3 represents a network description scheme, because we only have considered object classes (related to each other), but not the objects themselves. A concrete network description can be generated from the description scheme by proceeding from the classes to their objects. In this sense a network description can be regarded as an instantiation of a network description scheme. Figure 4, which describes two systems connected by a transit system, represents an instantiation of the diagram in Figure 3. Some of the relations between objects are expressed implicitly in Figure 4. The 'contain' relations, for example, are expressed by inclusion of the object symbols.

4.3 Consistency Rules

In order to achieve correct network descriptions from the network description scheme in Figure 3 it is necessary to follow consistency rules. In this sense it can be said that consistency rules assure semantically correct instantiations of the description scheme.

In this section we want to give some selected consistency rules of our network description scheme to illustrate how they look like and how well the formalism of relation algebra is suited to express them in a precise way.

The first rule describes the interrelation between 1-layer connections (physical layer) and the physical media:

> *A possible connection pc can only be (directly) mapped on (mappedOn4) a horizontal cable hc if the devices which contain (contain2) the protocol entities, connected (connect1) by pc, are connected (connect2) by hc.*
>
> (In other words: in order for 1-layer entities to communicate there has to be a cable connecting the corresponding devices.)

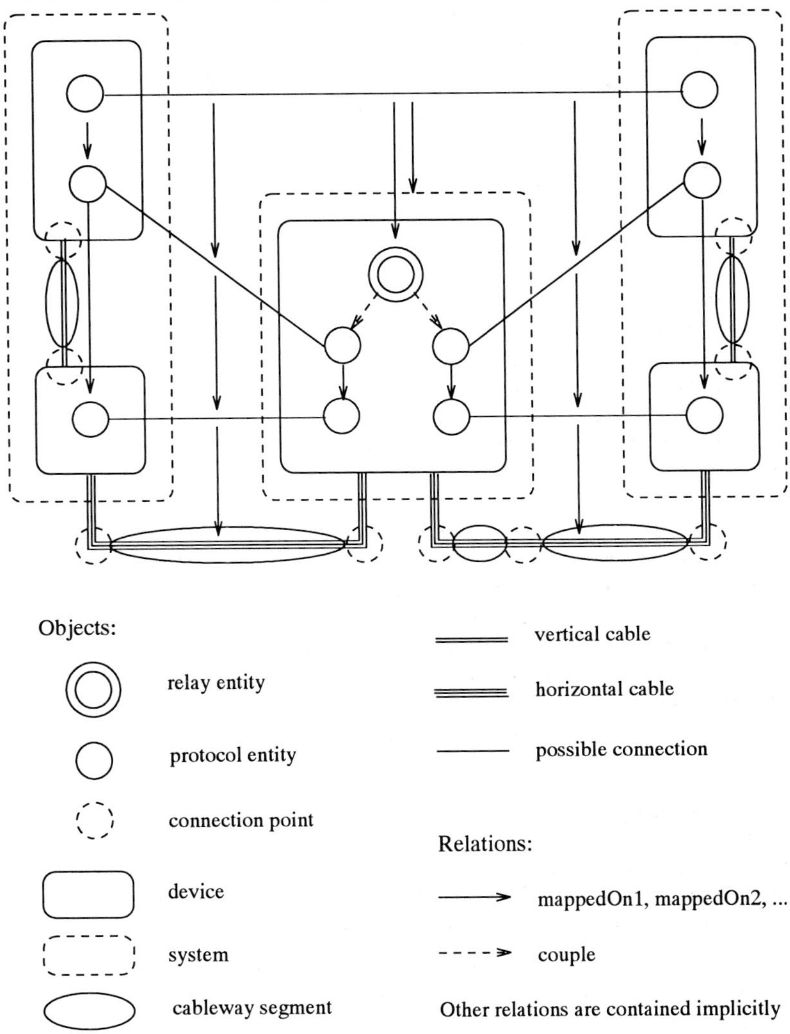

Figure 4: An Instantiation of Figure 3

Using relation calculus this rule can be expressed as follows:

$$mappedOn4 \subseteq connect1 \circ contain2^{-1} \circ connect2^{-1} \quad .$$

The following rule refers to the layering of protocol entities:

All protocol entities have to be mapped (mappedOn1) directly or indirectly onto a minimal protocol entity.

In order to be able to handle minimal protocol entities by means of relation algebra, we use the subclass relation $incl_{minimalprotocolentity,proctocolentity}$ and denote it by *min*.

Now the rule above looks like this:

($\sharp$) $mappedOn1^* \circ min \circ L = L$ (this means that $mappedOn1^* \circ min$ is total).

In the following we give some rules which refer to the notion 'system'. As mentioned in section 4.1, a system is a non-empty collection of cooperating entities. In order to form a *system* this collection has to satisfy at least the following rules:

(1) *All devices which contain (contain2) at least one entity of this collection have to be contained (contain3) in the system.*

(2) *The collection has to include at least one minimal protocol entity.*

(3) *The collection has to be closed with respect to the relations mappedOn1 and couple.* (In other word: all entities which form a protocol stack have to belong to the same system.)

These rules do not exclude the case that an entity can belong to two or more systems. An example of such an entity is an entity which is shared by several systems like a MAU (Medium Attachment Unit) entity within a transceiver which is connected to a multiport transceiver.

Using the formalism of relation algebras the three rules given above can be expressed like this:

(1) $contain1 \circ contain2^{-1} \subseteq contain3$

(2) $contain1 \circ min^{-1} \circ L = L$, (or equivalent: $contain1 \circ min^{-1}$ is total)

(3) $contain1 \circ (couple \cup mappedOn1)^* \subseteq contain1$

It is worthwhile mentioning that rule (2) is implied by rule ($\sharp$) and rule (3) (and the fact that a system has to include at least one entity). Using the calculus of relation algebras we were able to give a concise and elegant proof of this result.

5 Extensions of the Approach

In this section we want to give some ideas of how to extend our high-level modeling technique.

5.1 Refinement

Our description scheme from Figure 3 can be refined using class refinement as well as relation refinement:

Class refinement can be achieved by extending the inheritance tree by subclasses. For example, to model LANs classes like *repeater* or *MAC-level bridge* could be added as new subclasses of the class *device*.

Relation refinement means the decomposition of relations in order to get a more detailed insight into the network structure. Figure 5 shows a possible refinement of the relation

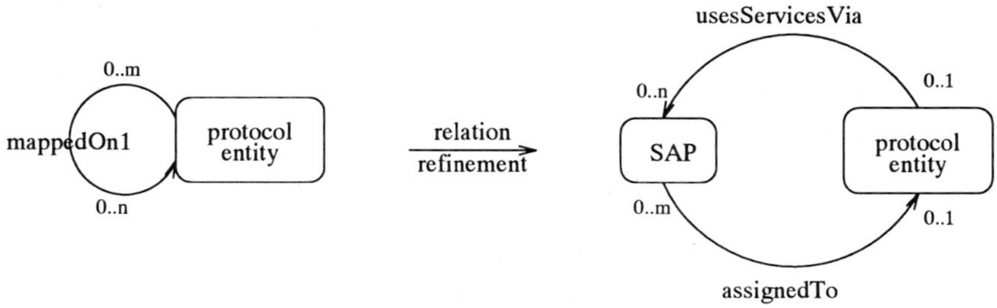

Figure 5: Refinement of a Relation

mappedOn1 by introducing the object class *SAP* (Service Access Point) and two relations. By means of relation algebra this refinement can be expressed as:

$$MappedOn1 = UseServicesVia \circ assignedTo \quad .$$

In general, to obtain more detailed and more specific network modeling schemes ('top down design') one can adapt techniques from the field of E-R models for refining E-R diagrams (for instance, the Multi-Level E-R Model ([FelMil 86]), or, the Nested E-R Model ([CarJi 90])).

5.2 Views

In practice, network descriptions will be very large and complex and thus not easy to survey. Therefore, it is necessary to have a view-mechanism to support users or applications. We can distinguish between views on class level and on object level:

A view on class level specifies a description subscheme which contains a subset of classes and appropriate relations. An example could be a subscheme comprising those classes and relations which are necessary for performance management. The class *cableway segment* would hardly be needed in that area, because there is nothing to measure.

A view on object level describes instantiations of a description (sub-)scheme by filtering certain objects (for example, all objects of a network description belonging to a certain subnet).

6 Conclusion

In this paper we have identified objects, relations, and, in particular, consistency rules as the basic elements for sophisticated modeling of network management relevant information. Using a relation algebraic approach we have given a formal semantic foundation which allows that the three elements fit together homogeneously. In this way we obtain a concept for an integrated handling of management relevant information (objects) as well as of semantic aspects (relations, rules).

Using the technique of refinement it is possible to describe networks on different levels of abstraction.

By examples we have demonstrated that our concept is well suited to deal with network modeling problems. In particular, we were able to give a set of rules describing the principles of communication layering.

Our intention was to develop a concept which is powerful enough to cover especially global aspects of such complex structures as networks. This is the reason for the high abstraction level of our approach.

Acknowledgements

The authors were members of the Munich Network Management Team. This team, a research group led by Prof. Hegering, consists of researchers of the Munich University (LMU), the Technical University of Munich (TUM), and the Leibniz Research Computing Center of the Bavarian Academy of the Sciences (LRZ). The authors wish to thank Prof. Dr. H.-G. Hegering and the Munich Network Management Team for helpful discussions and valuable comments.

References

[AbeLeiSeg 92] Sebastian Abeck, Martin Leischner and Peter Segner, "Applying the Inductive Modeling Technology to Tackle the Problem of Integrated Network Management", In *Proceedings IFIP/IEEE International Workshop on Distributed Systems; Operations & Management*, October 1992.

[BerSch 89] R. Berghammer / G. Schmidt, "Symmetric Quotients and Domain Construstions", *Information Processing Letters*, 33:163 – 168, 1989.

[CarJi 90] C. R. Carson / W. Ji, "The Nested Entity-Relationship Model", North-Holland, 1990.

[FelMil 86] P. Feldman / D. Miller, "Entity Model Clustering: Structuring a Data Model by Abstraction", *The Computer Journal*, 29(4):348 – 359, 1986.

[Forum 90] OSI/NM Forum, "Forum Library of Managed Object Classes, Name Bindings and Attributes", technical report, OSI/NM Forum, June 1990.

[ISO 10164-3] ISO/IEC, *Information Processing Systems— Open Systems Interconnection — Systems Management — Part 3: Attributes for representing relationships for CCITT Applications*, IS 10164-3, October 1991.

[ISO 10165] ISO/IEC, *Information Processing - Open Systems Interconnection - Management Information Services - Structure of Management Information, Part 1 - 6*, 10165, May 1992.

[ISO 10165-2] ISO/IEC, *Information Processing Systems— Open Systems Interconnection — Management Information Services — Structure of Management Information — Part 2: Definitions of Management Information*, IS 10165-2, September 1991.

[ISO 10165-5] ISO/IEC, *Information Processing Systems— Open Systems Interconnection — Management Information Services — Structure of Management Information — Part 5: Generic management information for CCITT Applications*, DIS 10165-5, May 1992.

[ISO 10737] ISO/IEC JTC1/SC21/WG4, *Information Technology - Telecommunications and Information Exchanges Between Systems - Elements of Management Information Related to OSI Transport Layer Standards*, DIS 10737, December 1991.

[ISO N1390] ISO/IEC JTC1/SC21/WG4 Configuration Management group, *Information Retrieval, Transfer and Management for OSI - General Relationship Model*, 2nd WD N1390, December 1991.

[RFC 1155] IAB, *Structure and Identification of Management Information for TCP/IP-based Internets*, RFC 1155, May 1990.

[SheSim 90] John W. Sheppard and William R. Simpson, "Incorporating Model-Based Reasoning in Interactive Maintenance Aids", In *Proceedings of the IEEE Integrated Diagnostic Symposium*, pages 1238–1242, IEEE, January 1990.

[Tar 41] A. Tarski, "On the Calculus of Relations", *The Journal of Symbolic Logic*, 6(3):73 – 89, 1941.

[Val 91] R. Valta, "Design Concepts for a Global Network Management Database", In *Proceedings of 2nd International Symposium on Integrated Network Management*, April 1991.

Integrated Network Management, III (C-12)
H.-G. Hegering and Y. Yemini (Editors)
Elsevier Science Publishers B.V. (North-Holland)

Formal Description of Managed Object Behavior - A Rule Based Approach

Olivier Festor(*) and Georg Zörntlein(°)

(*) CRIN, Université de NANCY I, Campus Scientifique, BP 239 F-54506 Vandœuvre-lès-Nancy France (e-mail: festor@loria.crin.fr)

(°) IBM, European Networking Center, Tiergartenstrasse 8, D-6900 Heidelberg, Germany (e-mail: gzoern@dhdibm1.bitnet)

Abstract

During the last years a lot of effort has been spent on standardizing network management in open systems. For modelling management information, an object-oriented approach has been adopted and a technique for the description of managed objects has been developed. This technique, called GDMO, provides a template notation for defining managed object classes. The definition of an object class comprises the description of object attributes together with the definition of exposed operations as well as the enumeration of all the notifications which can be issued by objects of this class. A definition is completed by specifying the object behavior.

In its current version GDMO is well-suited for some parts of an object class description. However, certain aspects are not addressed appropriately with the current standard. Most notably, no means are provided to allow a complete, precise and unambiguous description of object behavior.

To enhance GDMO so that it becomes a notation which allows a comprising description of managed objects, some extensions to the technique are proposed within this paper. Especially, the focus has been put on the behavior description part of managed objects.

Keywords: Behavior, GDMO, Managed Object, Specification.

1 Motivation

A set of standards have been defined for the management of open systems within ISO. These standards comprise a management framework [74984], a description of the management service (CMIS [9595]) and protocol (CMIP [9596]), the specification of basic management information and a technique for the definition of application specific management information.

For modelling the information concerning the network resources to be managed an object-oriented approach has been advocated. Each network resource is regarded as an object mainly characterized by four different components. These components are:

1. the exhibited attributes,
2. the exposed operations,
3. the performed behavior and
4. the notifications that can be issued.

In the current standards a set of generic object classes are defined from which specific object classes can be derived. Therefore a template notation was developed within ISO. This template notation is called Guidelines for the Description of Managed Object (GDMO [101654])

In GDMO an object class is described by denoting the object classes it is derived from (inheritance), listing all additional attributes it contains, naming the operations which can be performed and enumerating the notifications which can be issued. Currently, no formalism is provided within GDMO for a precise description of object behavior. In this paper extensions to GDMO are proposed to overcome this deficit.

Our developments have been based on three requirements. The method must provide means to

1. describe unambiguously the change of attribute values,
2. specify the pre-conditions and effects of operations, and
3. define the conditions under which notifications are issued.

Additionally, it has to be assured that the extensions adhere to the object-oriented design principles.

To motivate our proposal we will first summarize the main object-oriented design principles and, to be self-contained, introduce the GDMO template notation. Then we introduce a rule based approach and demonstrate how it fits into the object-oriented framework and how easily it can be integrated into GDMO. Throughout the paper the proposed concepts and language extensions are exemplified using the standardized log control function. Finally we draw some conclusions on the application of our method and identify some extensions for future work.

2 GDMO and the object-oriented framework

In this section the main object-oriented design principles are summarized and it is evaluated how far these principles are supported in the current management framework. Having introduced a standardized managed object, an overview of the GDMO template notation is given which provides the frame for our extensions.

2.1 Principles of object-oriented design

In the realm of software engineering, object oriented design is characterized as an approach that aims at modularity in terms of composability, decomposability, understandability, continuity and protection of software components (see [MEY88]).

An object oriented system design is a collection of classes. Each class is regarded as a module comprising both a data description and a specification of associated operations (*Encapsulation* concept). Such a module offers a set of services to the outside world through an explicit interface. Objects can interact through these interfaces.

An object (sub-)class is obtained by either extending or specializing one or more already existing (super-)classes (*Inheritance* concept). Extension or specialization is achieved by adding new features to the derived (sub-)class. These features can be either new data attributes, extensions of the range of existing data attributes, definitions of new operations or redefinitions of existing ones.

Additional facilities such as *Overloading* (ability to attach more than one meaning to a name appearing in a module, e.g. an operation name) and *Genericity* (ability to define parameterized classes, e.g. type parameters) are also objected oriented design concepts which, however, have so far not been exploited in the network management framework.

2.2 Managed Objects

Managed objects are, as described in [10040], the abstraction of data processing and data communication resources for the purpose of management. They represent the management view of network resources which can be physical (e.g. copper cable) or conceptual (e.g. association) in nature. Within an open system, all objects which represent managed resources are grouped in a Management Information Base (MIB).

In order to provide a standardized framework for the design of such management information, allowing modular development and supporting extensibility and reuse, an object oriented approach was retained for modelling the managed network resources. Following this approach, a new managed object class can be described by referring to some (super-)object class, and introducing new attributes, operations and notifications associated with the new (sub-)class.

Example of a Managed Object Class: The log control function

An example of a managed object class provided by the standards is the log control function. Its purpose is summarized in the following.

The log control function standardized within ISO provides support for creating and deleting logs which preserve information about event occurrences and allows to perform operations on logged records. The standard provides a generic specification of log classes [10164x][10165x] together with a definition of their management operations.

The whole model for the log control function depicted in Figure 1, is based on two major components: the log pre-processing function and the logs. The log pre-processing function is responsible for the acceptance of notifications issued by managed objects, their transformation into potential log records and for forwarding them to all existing logs.

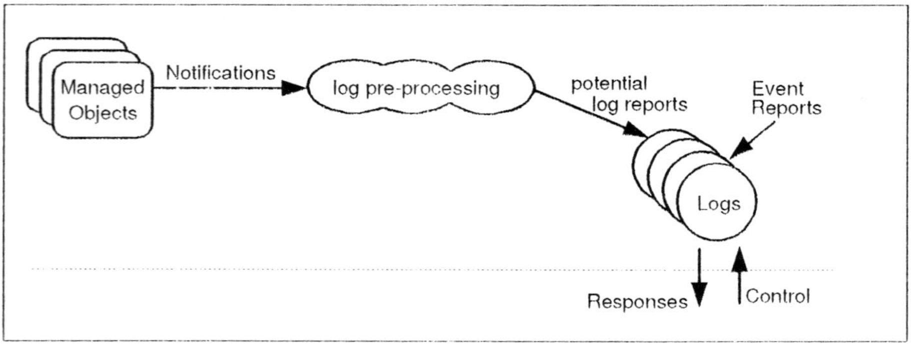

Figure 1. The conceptual view of the Log control function

The log models a repository for records with several capabilities. In the standard, a basic unsized log is defined. Conditional packages allow to extend the basic log. These packages contain sized log capabilities, a threshold alarm service and scheduling functions.

Each log object has three attributes derived from the Top object class of the inheritance hierarchy. The inherited attributes are the *operationalState*, the *availabilityStatus* and the *administrativeState*. The values of these attributes determine the behavior of a log.

If a log object has the *limitedSize* and *capacityThreshold* packages instantiated, a capacity alarm notification is emitted as soon as a threshold is exceeded. When the maximum storage capacity is reached, the log either halts or wraps.

All these behavioral aspects should be integrated in the BEHAVIOR part of a managed object definition. As mentioned earlier, currently the behavior is described via an informal model and is not reflected in the formal object description.

2.3 GDMO Notation

As basis for a coherent description of managed object classes, a notation called GDMO [101654] has been standardized within ISO. Figure 2 contains an outline of a GDMO object class description for the previously introduced example.

```
log MANAGED OBJECT CLASS
DERIVED FROM top;
CHARACTERIZED BY
 logPackage   PACKAGE
  logBehavior BEHAVIOR
   DEFINED AS
     // textual description of the behavior associated to the package
  ATTRIBUTES
   logId                GET,
   ...
  ACTIONS
   ...
  NOTIFICATIONS
   ...
  CONDITIONAL PACKAGES
   schedulingPackage PACKAGE
   ...
REGISTERED AS {smi2MObjectClass7};
```

Figure 2. An example of a GDMO managed object definition

In GDMO the keywords **MANAGED OBJECT CLASS** indicate the introduction of a new object class definition. The (super-)class(es) from which the new class is derived are listed in the **DERIVED FROM** clause. After the keywords **CHARACTERIZED BY** a set of packages is given.

Each package contains a behavior part labelled by the keyword **BEHAVIOR** a set of attributes annotated with their access status (**GET, ADD-REMOVE**, etc.), a description of operations which can be applied (**ACTIONS**), and the notifications (**NOTIFICATIONS**) which can be emitted.

A GDMO definition is completed by the introduction of a distinguished name which is provided after the keywords **REGISTERED AS**.

In the current GDMO template appropriate means for precisely describing the behavioral aspects are lacking. The use of natural language is advocated for behavior specification.

3 Rule-based approach for specifying object behavior

The rule-based approach as introduced in the course of this section is mainly based on the technique CRS (Communicating Rule Systems [MACK87],[VELT90]) which has already been successfully applied to the specification of OSI protocol behavior.

3.1 Object Interfaces

According to the presented principles, each object should be "accessible" through a well defined interface. In this context, "accessible" refers to the invocation of operations as well as emission and reception of notifications. To allow a uniform description, interfaces are modelled as a set of communication gates. Operations are invoked by events at synchronous gates, whilst asynchronous gates allow to receive and issue notifications. Before we start to introduce explicit syntactical language extensions for gate descriptions a structure for the considered example is given.

Managed Object structure of the Log Control Function

Since logs can be created and deleted, it was decided to introduce a log controller which is accessed by the management system via a special gate type (*AC* gates). The log controller is responsible for creating and deleting logs. It has also the task to forward operations and incoming notifications to the existing logs.

The *AC* gate type contains the description of all provided actions and operations. Potential event reports are accepted through another gate type, the *EV* gate. A third type of gates, *NT* gates, allow to issue notifications.

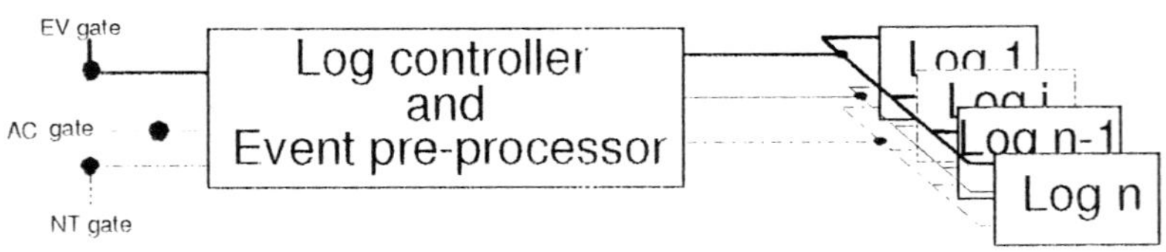

Figure 3. Object organization for the log model

Incoming events from the log controller *EV* gate are forwarded to all existing logs via gates of the same type. An operation invocation which has been received by the log controller through an *AC* gate, is sent to the corresponding log through another synchronous *AC* gate. When a log issues a notification, it is forwarded by the log controller through an asynchronous *NT* gate.

Object Interface definitions

In the proposed approach, each gate is defined by a name, a communication type (*SYNCHRONOUS* or *ASYNCHRONOUS*), the declaration of all event parameters and the enumeration of all possible event occurrences together with their parameters.

In the following example (Figure 4) the gate through which operations are invoked on the log object class is described. *PT_DELETE* is used by the log to delete stored records and *PT_SET* is the event corresponding to a *SET* operation on attributes. All event parameters are specified in ASN.1.

```
GATE SYNCHRONOUS LOG_AC_GATE;
  DECLARATIONS      //Event parameters
     deleteInfo                 : DeleteInfoType;
     getInfo                    : GetInfoType;
     setInfo                    : SetInfoType;
     ...
  EVENTS      //Parameterized events
     PT_DELETE  (deleteInfo);
     PT_SET     (setInfo    );
     PT_GET     (getInfo    );
     ...
END_GATE LOG_AC_GATE;
```

Figure 4. Example of a gate definition

The introduced interface specification of managed objects must be integrated into the GDMO template notation. Bold letters in Figure 5 indicate the proposed syntactical extension.

```
log  MANAGED OBJECT CLASS: (AC: LOG_AC_GATE; EV: LOG_EV_GATE; NT: LOG_NT_GATE );
ATTRIBUTES
  operationalState     : OPERATIONAL_STATE;      // derived from Top
  administrativeState  : ADMINISTRATIVE_STATE;
  ...
VARIABLES
  records                : SET OF RECORDS INITIALLY Ø;
  ...
ACTIONS
  addRecord;
  changeAttributeValue;
  ...
NOTIFICATIONS
  attributeStateChange : AttributeStateChangeNotification;
  ...
```

Figure 5. Extention of the GDMO notation

Each object class description is parameterized by a list of pairs of type (**gate_inst: GATE_CLASS**). This optional list is introduced immediately after the **MANAGED OBJECT CLASS** keyword and encapsulated in brackets.

3.2 Object Behavior description

For describing attributes as well as event and message types ASN.1 has been employed. To be able to denote value changes (see example Figure 6), ASN.1 has to be extended by a set of operations allowing appropriate value references (see [SCHN92]). The required operations are:

- usual logic operators for simple BOOLEAN types ($\wedge$, $\vee$, $\neg$, ...),
- arithmetic and relational operations for INTEGER and REAL types ($+$, $-$, $\times$, $>$, $\leq$, $=$, ...),
- access and composition operations for types related to strings (CONCAT, SUBSTR, LENGTH, ...),
- access functions for structured types, e.g. • (field selection in a SEQUENCE typed variable), $\in$ (membership), $\bigcup$ (union) for SET typed variables, ...

X-ASN.1 also provides some status functions like SIZE_OF or NUMBER_OF_ELEMENTS for structured and SET types. An example of their use is given in Figure 6.

```
SetInfoType ::= SEQUENCE
{ administrativeState AdministrativeState := ENUMERATED
                                              { locked   (0),
                                                unlocked (1) } OPTIONAL,
  maximalLogSize      INTEGER OPTIONAL,
  ...
}
variable declaration ch : SetInfoType;
field access ch.administrativeState
```

Figure 6. Example of ASN.1 definition of an operation parameter

The example given above shows that access and status functions for ASN.1 values are needed. So, for denoting the fields *administrativeState* and *maximalLogSize* access functions have to be available. Status functions are required to check for the presence of data values in optional fields, e.g. PRESENT(*ch.maximalLogSize*) for testing the presence of the *maximalLogSize* field.

In our method the behavior of an object, i.e. conditions and effects of operations (attribute value changes) and notifications are described by a set of rules. These rules are integrated into the BEHAVIOR template of GDMO. Each rule consists of three parts: a name, a condition and a effect formula (see example Figure 7 on page 9).

```
logBehavior BEHAVIOR DEFINED AS

delete_records;
COND: AC.PT_DELETE_evt(deleteInfo)
      AND 'operationalState = enabled;
EFFECTS: ∀ record ∈ 'records, MATCHES(record,deleteInfo)
                   => ( records = 'records \ {record}
                        AND IF maximumLogSize <> 0
                            THEN currentSize = 'currentSize - SIZE_OFrecord)
                            FI
                      );
```

Figure 7. Example of a rule

Both condition and effect descriptions are based on first order logic expressions extended by event occurrences. The condition part **(COND)** specifies the start condition associated with a rule in terms of event occurrences and attribute value constraints. In the example (see Figure 7) the rule can be started if a *PT_DELETE* event occurs at the *AC* gate and the *operationalState* attribute has the value *enabled*. The effects **(EFFECTS)** part specifies changed attribute values, invoked operations and issued notifications. The set of all rules defines precisely the behavior of the object. The description of the behavior associated with the invocation of an operation or the processing of a message is not restricted to one rule. The execution of several rules may constitute one operation.

In the following, the associated rules for additional two operations are specified. These operations are concerned with the logging of new incoming events, and the change of attribute values, respectively.

```
add_record;
COND: EV.incomming_evt(event_inst);
      AND 'administrativeState = unlocked
      AND 'operationalState = enabled
      AND 'availabilityStatus <> logFull
      AND 'availabilityStatus <> offDuty;
EFFECTS: newRecord(record,eventInst)
      AND records = 'records ∪ {record}
      AND IF PRESENT(packages.limitedSize)
          THEN numberOfRecords = 'numberOfRecords + 1
               AND currentSize = 'currentSize + SIZE_OF(record)
               AND IF currentSize = maximumLogSize
                   THEN availabilityStatus = logFull
          FI       FI;
```

Figure 8.

A log can only store records if the *operationalState* attribute has the value *enabled*, the *administrativeState* is set to *unlocked* and the *availabilityStatus* value differs from *logFull* and *offDuty* (see Figure 8). In this case, all incoming events occurrences, are stored in a record set (*records*). If the log has the *sizedLog* package instantiated then the corresponding attributes must additionally be taken into account.

```
change_attribute_status;
COND: AC.PT_SET(newLogStatus)
      AND 'operationalState = enabled
      AND PRESENT(newLogStatus.administrativeState);
EFFECTS: IF PRESENT('newLogStatus,administrativeState)
         THEN NT.attributeStateChangeNotification(attributeStateChangeInfo)
              AND attributeStateChangeInfo.changes = new
              AND new.name = "administrativeState"
              AND new.oldValue = 'administrativeState
              AND new.newValue = administrativeState
         FI;
```

Figure 9.

If attributes of a log are to be changed, a *PT_SET* action parameterized with the set of new values has to be received by the log through an instantiated *AC* gate (see Figure 9). In the basic log only the *administrativeState* attribute value can be changed. In this case an *attributeStateChangeNotification* is issued by the log at an *NT* gate.

3.3 Rules and Inheritance

In order to provide a formalism for behavior description which exploits the inheritance mechanism it has to be clarified how rules are inherited from the (super-)class(es).

The inheritance principles retained in the managed object design are the following ones:

- an object class can be extended only by adding new attributes and/or changing the range of existing ones, the addition of new actions and notifications and/or the addition of new arguments to actions and notifications;
- no characteristics from the super object class can be deleted in the sub-class. This means that the services offered by an object of a sub-class must be at least the ones offered by an object of its super-class;
- the sub-class inherits the operations, attributes, notifications and the behavior of its super-class.

As operations are described via rules, strict inheritance is guaranteed when each redefinition of an operation is done following two requirements: The condition part of each associated rule must be **weaker** and the effects part **stronger** than the ones of the corresponding rule (same rule name) in the super-class.

To obtain the complete behavior part of a derived object class the following procedures must be applied:

1. all rules defined in the sub-class which are not defined in the super-class are added without changes;
2. for each rule of the sub-class which is a redefinition of a rule already defined in the super-class, the complete condition part of the rule is obtained by $\vee$-connecting (logical OR) the sub-class condition with the super-class condition. The EFFECTS part of the complete rule results from $\wedge$-connecting (logical AND) the EFFECTS formula of the super-class rule with the EFFECTS formula of the sub-class rule;
3. the second rule must be applied recursively through the inheritance tree.

This way strict inheritance in the sense of weakening pre-conditions and strengthening post-conditions of operations is ensured.

For exemplifying the inheritance principles the derivation of an *alarm_log* from the introduced *log* object class is considered. Since the *alarm_log* object class is derived from the basic *log* object class, all attributes defined in the basic *log* are also valid in the *alarm_log*. All gates used in the *log* object class are also defined in each sub-class. Thus, in the *alarm_log* definition, only new operations and notifications (see Figure 10 *capacityAlarmNotification*) and attributes additionally used such as *capacityAlarmThreshold* have to be declared.

```
alarm_log  MANAGED OBJECT CLASS DERIVED FROM log;

ATTRIBUTES
 capacityAlarmThreshold: SET OF INTEGER (0..100);
...
ACTIONS
...
NOTIFICATIONS
 capacityAlarmNotification       : AlarmNotification;
```

Figure 10.

Like for the super-class, we will only focus on the rules related to event logging, record retrieval and attribute change.

Figure 11 on page 12 describes the new behavior of the log object resulting from the support of a capacity alarm service. All conditions for the logging of a record in the *alarm_log* are the same as the ones for the basic *log*. Thus, they are inherited from the *alarm_log* subclass. The extensions introduced in the effects part describe that a *capacityAlarmNotification* is issued as soon as a threshold value is exceeded.

```
alarmLogBehavior BEHAVIOR

add_record;
COND: EV.incomming_evt(event_type);
      AND 'administrative_state = unlocked
      AND 'operational_state = enabled
      AND 'availability_status <> logFull
      AND 'availability_status <> offDuty;
EFFECTS:
 IF ((∃ subset ⊆ capacityAlarmThreshold, ∀ threshold ∈ capacityAlarmThreshold,
      threshold ≥ 'currentSize ∧ threshold ≤ currentSize ⇒ threshold ∈ subset ))
 THEN NT.capacityAlarmNotification(alarmParameters)
      AND alarmParameters.probableCause = threshold
      AND alarmParameters.thresholdInfo.triggeredThresold = capacityThreshold
      AND alarmParameters.thresholdInfo.observedValue = currentSize
 FI;
```

Figure 11.

The *delete_record* action changes only the available logging space and the *availability_status* value. No specific features of the *alarmLog* subclass influence the operation invocation. Thus, the rule is inherited without changes from the super-class.

```
change_attribute_status;
COND: AC.PT_SET(newLogStatus)
          AND 'operationalState = enabled
          AND PRESENT(newLogStatus.capacityAlarmThreshold);
EFFECTS: IF PRESENT(newLogStatus.capacityAlarmThreshold)
         THEN capacityAlarmThreshold = newLogStatus.capacityAlarmThreshold
            AND attributeStateChangeInfo.changes = atc
            AND atc.name = "capacityAlarmThreshold"
            AND atc.oldValue = 'capacityAlarmThreshold
            AND atc.newValue = 'capacityAlarmThreshold
            AND NT.attributeStateChangeNotification(attributeStateChangeInfo)
          FI;
```

Figure 12.

Within an *Alarm_log* the *capacityAlarmThreshold* (see Figure 12) attribute as well as the *administrativeState* and *maximumLogSize* attributes inherited from the basic *log* can be changed by a management system. If one of these attributes is changed then, as in the basic *Log*, an *attributeStateChangeNotification* is issued. But in this case there is an additional parameter in the notification carrying the new *capacityAlarmThreshold*. This rule is a typical example for weakening a pre-condition and strenghening a postcondition of an operation.

4 Conclusions and future work

In this paper a rule based method has been presented which allows a complete, precise and unambiguous description of managed object classes. It has been shown that the rule concept together with means of having objects communicate can coexist with the object paradigms of inheritance and encapsulation.

We have also shown that our method can easily be integrated into the GDMO template notation. This has been demonstrated by extending GDMO with communication gates and rules integrated into the behavior part.

The proposed concepts have been convincingly demonstrated using the log control function.

In the future, various managed objects allowing an intensive and intelligent control of networks will have to be designed and integrated into MIBs. Those objects will not only represent simple passive network elements but sophisticated means of network control exposing intelligent behavior. Thus, powerful and precise behavioral notations for managed objects will be of utmost importance.

5 References

[MACK87] **L.F. Mackert, I.B. Neumeier-Mackert** *Communicating Rule Systems* Proc. 7th Int. Symp. on Protocol Specification, Testing and Validation, H. Rudin, C.H. West (editors), North-Holland 1987, pp 77-88

[MEY88] **Bertrand Meyer** *Object-Oriented Software Construction* Prentice Hall International Series In Computer Science, 1988

[SCHN92] **Jürgen M. Schneider** *Protocol Engineering: A Rule-Based Approach* Vieweg Advanced Studies in Computer Science ISBN 3-528-05243-0, 1992

[VELT90] **R.J. Velthuys, L.F. Mackert, J.M. Schneider, G. Zörntlein** *Structuring Mechanisms for the Formal Description Technique CRS* Technical Report, IBM European Networking Center Heidelberg (1991).

[74984] **ISO International Standard 7498** *Information Processing Systems - Open Systems Interconnection - Basic Reference Model Part 4: Management Framework* 1991

[8824] **ISO International Standard 8824** *Information Processing Systems - Open Systems Interconnection - Specification of ASN.1 (Abstract Syntax Notation One)* 1987

[9595] **ISO International Standard 9595** *Information Processing Systems-Open Systems Interconnection - Common Management Information Service Definition* 1990

[9596] **ISO International Standard 9596** *Information Processing Systems-Open Systems Interconnection - Common Management Information Protocol Specification* 1990

[10040] **ISO International Standard 10040** *Information Technology - Open Systems Interconnection - System Management Overview* January 1992

[10164x] **ISO Draft International Standard 10164-1** *Information Technology - Open Systems Interconnection - System Management* 1991

[101654] **ISO Draft International Standard 10165-1** *Information Technology - Open Systems Interconnection - Management Information Services - Structure of Management Information Part 4: Guidelines for the Description of Managed Objects* 1991

[10165x] **ISO Draft International Standard 10165-1** *Information Technology - Open Systems Interconnection - Management Information Services - Structure of Management Information* 1991

Integrated Network Management, III (C-12)
H.-G. Hegering and Y. Yemini (Editors)
Elsevier Science Publishers B.V. (North-Holland)

Design of Network Management Architectures for Heterogeneous Networks Using Object Oriented Approach

J. Filipiak[a], A. Lombardo[b] and S. Palazzo[b]

[a] Telecommunications Department, University of Mining and Metallurgy, 30-059 Krakow, Al. Mickiewicza 30, Poland

[b] Istituto di Informatica e Telecommunicazioni, Universita degli Studi di Catania, V.le A Doria, 6 - 95125 Catania, Italy

Abstract

The structural model of network architecture is developed and used for analysis and design of network management systems. The model is based on the object oriented approach. The considered object class is called the Decision Unit (DEU). DEUs are elements of three dimensional functional space with decision, service, and functional types as dimensions. The embedding operation is defined which allows for functional specification of network management software at various levels of abstraction. The distinctive features of the proposed method are the explicit specification of control and management interfaces and the direct method of linking objects into software packages.

Keyword Codes: C.2.1; D.1.5; D.2.9

Keywords: Computer-Communication Networks, Network Architecture and Design; Programming Techniques (E), Object-oriented Programming; Software Engineering (K.6.3), Management (K.6.3, K.6.4)

1. Introduction

Network management systems are developed to increase the revenue of network operators. In large organizations the revenue maximization is a complex task. It takes into account various factors related to a competition between service providers including a quality of service as seen by users, network reliability, tariffs, and others. Therefore, various economical and technical objectives and constraints are imposed on the designed network.

In this paper we propose the network management and control architecture which can be used to efficiently allocate network resources and monitor the network state. The proposed architecture is modular and gives a global picture of network management software.

Attempts to develop such architecture were recently presented in [1], [4], [5], and [6]. Our approach is based on the assumption that the processing (transport), control (signalling), and management function is associated with each network element. We further assume that each module implementing these functions can itself be broken-down into processor, controller, and manager. To illustrate that concept note that the management and signalling modules of the ISDN network contain themselves the transport, signalling and management elements. More precisely, the ISDN signalling nodes are connected by means of the Signalling System No. 7 network, which itself has the transport, signalling and management modules. The same comment refers to the Telecommunication Management Network (TMN), which has the transport, control, and management resources to carry management data.

Going down this embedding hierarchy allows one to explore the network management interfaces and detailed traffic control decisions. We demonstrate that the basic concept of the functional object, which is defined as a triple (Manager, Controller, Processor), applies to all levels of the developed model.

A distinctive feature of our approach is that it explicitly defines all signalling and management interfaces. Moreover, the embedding hierarchy brings an order into the whole structure allowing for automated linking of separate management modules into an integrated software product.

We stress that in this paper we concentrate on the analysis of system structure. We believe that with a present grow in system structural complexity the structural modelling is more urgent task than the analysis of management algorithms. Having a good structure with well defined system boundaries allows one to try various algorithms during network operation without additional costs.

The presented paper is organized as follows. In the next section we introduce concepts of functional space and functional object. In Section 3 the proposed method is used to present a global view of network management system. Section 4 presents a model of the B-ISDN resource management system as an example.

2. F-space concept

The layout of communication network resources is shown in Fig. 1. The basic task of communication network is to provide customer services. Usually, it is cost effective to extract a common communication function (e.g., access control function) from several applications and implement it in a separate new system. Repeating that procedure allows one to breakdown the whole communication system into functional layers. We shall assume that extracting a common function from several systems and placing it in a new software or hardware module corresponds to moving along the functional axis (F-axis). Instead, services are arranged along the service type axis (S-axis). Note that in the S-F plane each network element (software or hardware) is defined by its coordinates (s,f).

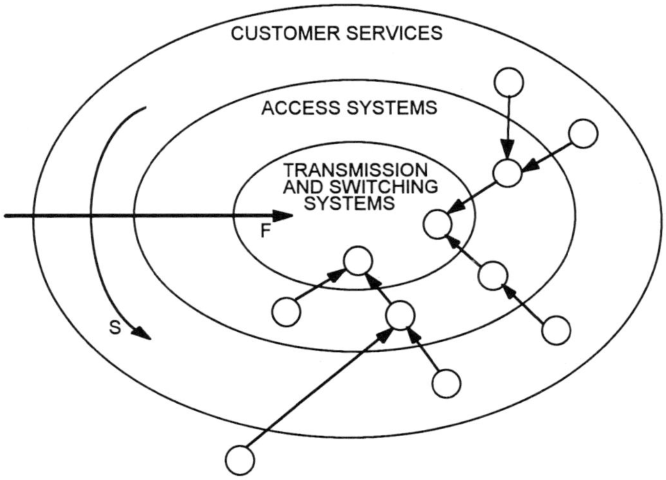

Fig. 1.

The (s,f) elements segment data blocks into smaller units, send them to other modules, reassemble received blocks, perform sequencing, detect errors, retransmit mutilated blocks, and perform other functions related to handling of user data. In our model elements performing such algorithmic actions are called the transport level Decision Units (DEUs).

We shall assume that each transport DEU is associated with one or more signalling DEUs. Signalling DEUs perform control actions. For instance, they allocate identifiers to data blocks and define Service Access Points (SAPs) through which the data blocks are exchanged, establish connections and associations, modify call control parameters and others. Signalling DEUs belong to the control or signalling plane which is situated above the transport plane.

One can envisage the third, management, plane. DEUs belonging to that plane allocate and monitor status of network resources.

From the above discussion it follows that the Decision Units can be considered as objects in a three dimensional functional space (F-space), see Fig. 2. The dimensions are:

S = service type
F = functional type
D = decisionmaking level

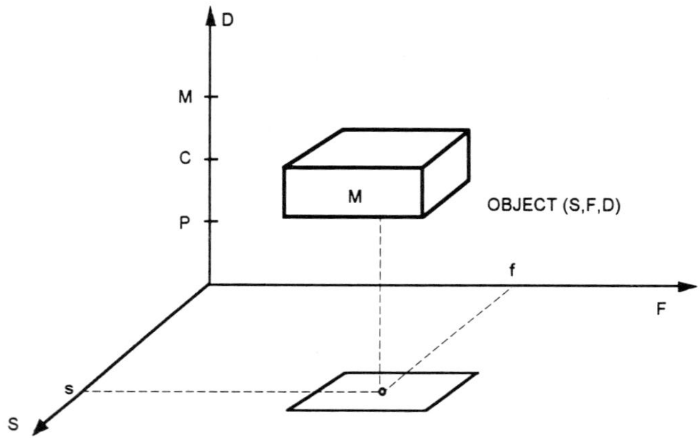

Fig. 2.

We already indicated that the decisions are taken at the transport, control, or management level. A position of each DEU in the F-space is defined by the (s,f,d) coordinates. Each DEU is an object in a sense of the object oriented design. That is, it is characterized by:

- Memory or local state that defines a set of object attributes, which are also called instance variables or data members.
- Interface which is a set of all messages exchanged by an object with the external world.
- Behavior which is a set of all algorithms, where each algorithm defines a change of the object state in response to arriving messages.

The data abstraction or encapsulation principle is often adopted which says that an object is visible to the external world only through its interface. Each object is associated with a unique object identifier, which is not changed during the object lifetime.

In what follows we impose additional structure on the F-space. For that purpose, let us first introduce the concept of the functional object (F-object).

Definition 2.1. The F-object is a triple (P_DEU,C_DEU,M_DEU), where P_DEU is a processing (transport) element, C_DEU is a controller, and M_DEU is a manager.

The F-object will be denoted by $F_{sf}(P, C, M)$. The following hypothesis underlies the modelling method proposed in this paper.

Hypothesis 2.1. Each DEU is an F-object.

In other words, each network element or subsystem can be modelled as an F-object. This is visualised in Fig. 3. According to the Hypothesis 2.1 we shall not investigate individual DEUs in isolation. We shall rather consider complexes of DEUs treating F-objects as basic elements of F-space.

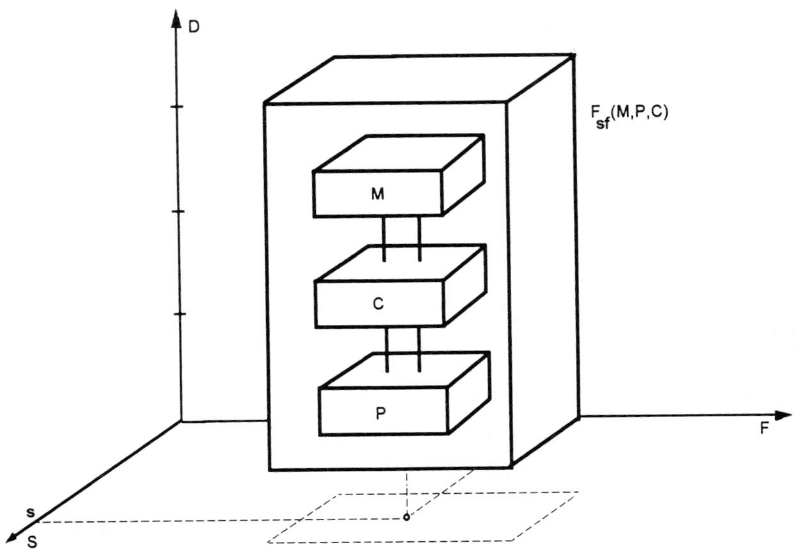

Fig. 3.

The introduced concepts provide a convenient method of network functional specification. Each DEU is composed of internal (lower level) control, manager, and processor (transport) DEUs. Treating C_DEU, M_DEU, and P_DEU as F-objects gives a meaning to symbols such as C_P_DEU, P_M_DEU, and others. For instance,

$$DEU_A = P\_C\_DEU_B \tag{1}$$

denotes a fact that DEU_A is a processor of a controller of DEU_B.

According to Hypothesis 2.1 the complex structures are modelled using the embedding concept according to which a high level DEU can be viewed as a collection of lower level DEUs. This means that the F-space can be inserted into each DEU to further investigate its internal structure.

The embedding hierarchy and general DEU structure allow one to proceed in a very systematic and straightforward way from a high level conceptual description of network architecture, through the functional network specification, to a detail design of communication protocols. High level and detail views of network software correspond to various levels of abstraction in the embedding hierarchy.

Each F-object is defined with respect to a higher level F-object to which it belongs, or alternatively, with respect to an external event (message) to which it responds. This property helps to describe the functional types of DEU elements:

- Controller monitors external events and activates procedures neeeded to serve them.
- Processor is activated by a controller to serve a particular event.
- Manager observes sequences of events and surveys the DEU operation.

In our approach we shall assume that technically the DEU elements exchange messages through common memories. M_DEU writes the management information (parameter values, control thresholds, etc.) to the common memory block (file, register). The C_DEU uses that information to update its algorithm. On the other hand, the C_DEU keeps part of its information (counter values, alarm status) in a memory block which is accessible to the M_DEU element. The same occurs at the internal interface between C and P elements. In that way, in large systems each decision level sees the lower plane as a distributed database. This exhibits some similarity to a method described in [4].

We shall distinguish between active and passive DEUs. Active DEUs respond to external events by activating passive DEUs, which are kept in the DEU memory. In that way, each algorithm can be viewed as a DEU.

In practical applications it is convenient to have a notation (syntax, semantics) for a formal specification of investigated objects. The basic element of such notation has already been introduced. Namely, we denote by $DEU_A = M\_DEU_B$ the fact that DEU_A is contained in DEU_B as its manager. Similar notation is valid for the controller and processor relationships. Sometimes, instead of using notation

$$DEU_A = M\_C\_C\_...\_DEU_B \tag{2}$$

it may be more convenient to write

$$DEU_A \in DEU_B \tag{3}$$

which means that DEU_B contains DEU_A. Relation (2) provides means of local and global addressing.

Let us denote by $OUT(DEU_A)$ and $IN(DEU_A)$ the output and input interfaces of DEU_A, respectively. We shall say that DEU_A and DEU_B interface, and denote it by $DEU_A \| DEU_B$, if

$$OUT(DEU_A) = IN(DEU_B) \tag{4}$$

or, alternatively,

$$IN(DEU_A) = OUT(DEU_B) \tag{5}$$

Finally, in some situations it may be convenient to use the following notation which directly indicates that DEU_A is a manager with respect to the same DEU_C, to which DEU_B is a controller:

$$DEU_A = M(DEU_B) \tag{6}$$

Equation (6) says that $DEU_B = C\_DEU_C$ and $DEU_A = M\_DEU_C$, and that DEU_B is a management domain of DEU_A.

Relations defined by equations (1) to (6) can be used by a compiler to link objects into a global program.

In a remainder of this paper we do not concentrate on developing formal techniques for description of DEU complexes or the analysis of abstract global models. This will be a subject of a future work. Rather, we demonstrate that the proposed method applies to variety of situations with stress on network management.

An application of this approach to functional specification of transport protocols was previously described in [2] and [3].

3. Resource management

A hierarchy of network resources in the F-S plane. Comprises: Customer Services, network products, networks, and network elements. The basic function of the highest level system consists in providing communication services. As eluded to in the previous section the lower level systems come into being by extracting a set of common functions from one or more higher level systems and placing them in one separate system.

Observe, that treating network elements as points in the F-S plane one can think about higher level systems (subnetworks, networks, network products) as $(\Delta s, \Delta f)$ sets in that plane.

We assume that each $(\Delta s, \Delta f)$ set shown in the figure is controlled or managed by F-object associated with that system. We shall distinguish between virtual (abstract) and physical DEUs. A difference between physical and virtual DEUs is explained in the examples.

Example 3.1. Consider a set of interacting subnetworks. Each subnetwork has a traffic management DEU. All DEUs exchange the management information to efficiently handle the internet traffic. From the modelling viewpoint the set of physically implemented management DEUs can be viewed as an abstract virtual management DEU.

Example 3.2. Assume that in the above example it becomes technically feasible and cost effective to implement one management DEU. That DEU exchanges information with several local controllers positioned in the subnetworks. The set of local controllers can be viewed as an abstract virtual control DEU.

Note that higher level systems may not have their specific physical resources. For instance, the management system which coordinates behavior of several network products can be the only physical system specific to the customer services system.

The management DEU (M_DEU) can be viewed as an element which interacts with higher and lower level abstract and physical DEUs. The M_DEU internal structure is shown in Fig. 4. The C_M_DEU element identifies the management task and activates the corresponding management application service P_M_DEU. The M_M_DEU element monitors the M_DEU operation and in more developed systems reports M_DEU status to a meta-management system. The management interface element constitutes a part of the P_M_DEU module. The interface is used to exchange management information with other managing systems, management agents, and management objects. The management interface structure is discussed in the next section.

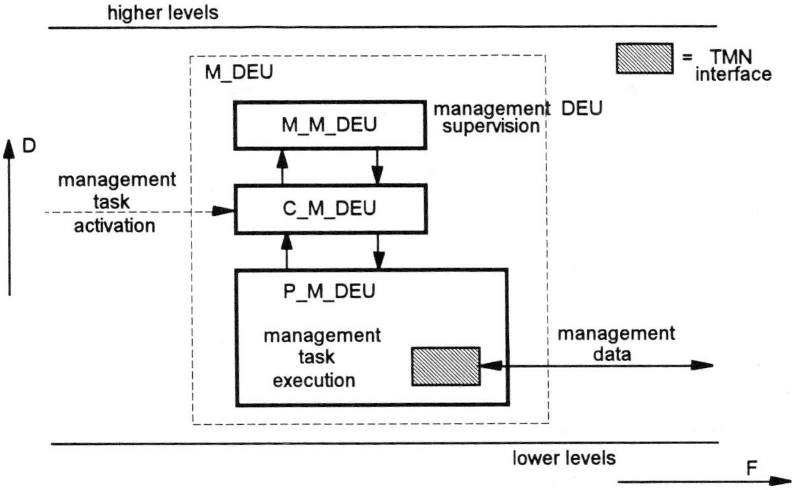

Fig. 4.

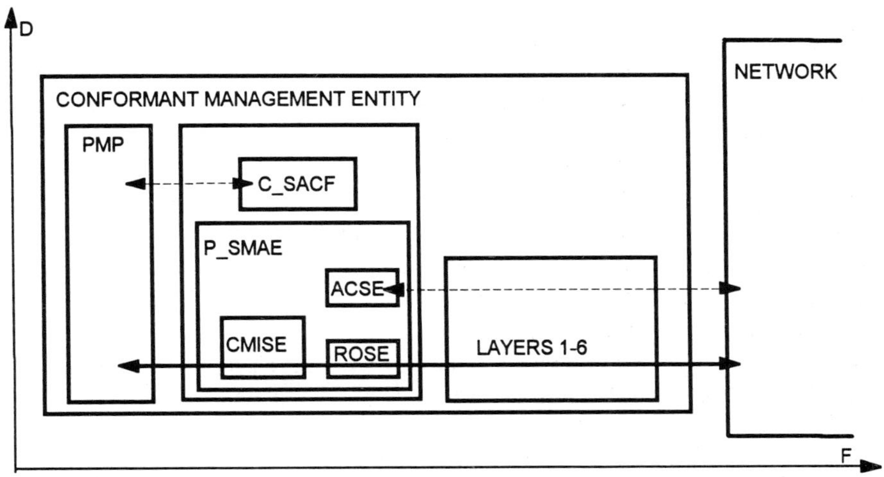

PMP = PRIOPRIETARY MANAGEMENT PROCESS

P_SMAE = SYSTEMS MANAGEMENT APPLICATION ENTITY

CMISE = COMMON MANAGEMENT INFORMATION SERVICE ELEMENT

Fig. 5.

Through the management interface the application processes exchange the management protocol data units (TMN MAPDUs) carrying information needed to coordinate the global management process.

According to ISO/OSI standards M_DEUs which interact with managed objects (MO) are called the Conformant Management Entities (CME). Examples of managed objects are local management functions associated with T1 switch, Private Branch Exchange, and others. The Managed Object is often defined as an abstract representation of a real object in software. As such, it is considered to be a part of the CME. A vertical view of the CME DEU is shown in Fig. 5.

4. B-ISDN Traffic Management

It is well understood at the moment that the traffic management in ATM networks is performed in several layers, which include flow, call, burst, and cell layer. Individual cells cannot be controlled globally at the network level because an amount of control information to be processed during a short time is too large. Therefore, the break-down of traffic management problem into levels and layers seems to be natural.

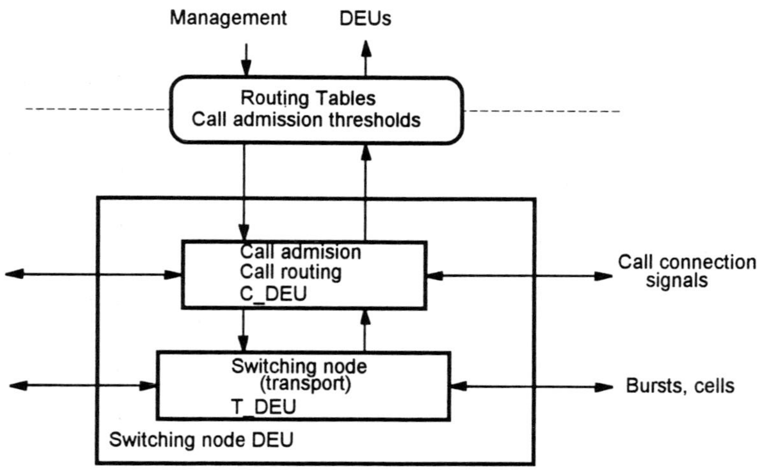

Fig. 6.

At the network level the bandwidth is allocated to various traffic types. The routing tables are also worked out at this level. At the subnetwork level, time dependent actions are performed like temporary extension/reduction of routes or skip route. Knowing the routes and bandwidth requirements that level determines a number of belonging to a particular class calls that can be supported on a given link. That information is distributed to switching nodes.

The decisionmaking unit associated with a switching node is shown in Fig. 6. The control element C_DEU reads an information from routing and flow control tables to handle individual calls. After the call connection signals are exchanged with adjacent nodes and positive acknowledgement is received the control is passed on to the transport element T_DEU. The T_DEU monitors the flow of bursts and cells as is shown in Fig. 7.

Based on information received from the C_DEU about the call type the T_DEU reads from the memory the values of credit manager (leaky bucket) parameters or other information which allows for policing of traffic parameters in the burst layer. The actual control of the cell flow is performed at the link level as is shown in Fig. 8. The C_T_T_DEU element implements the credit manager algorithm.

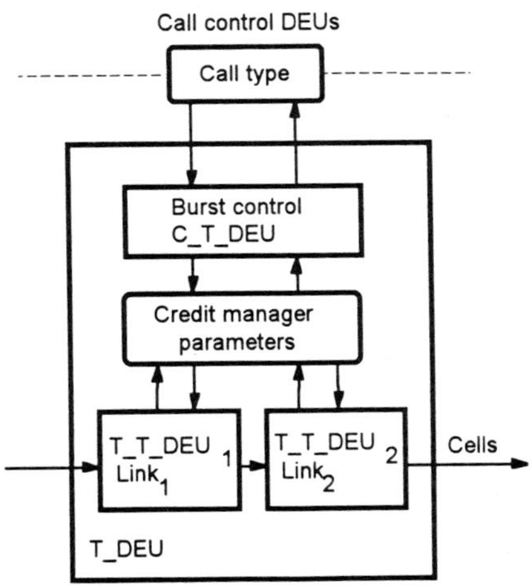

Fig. 7.

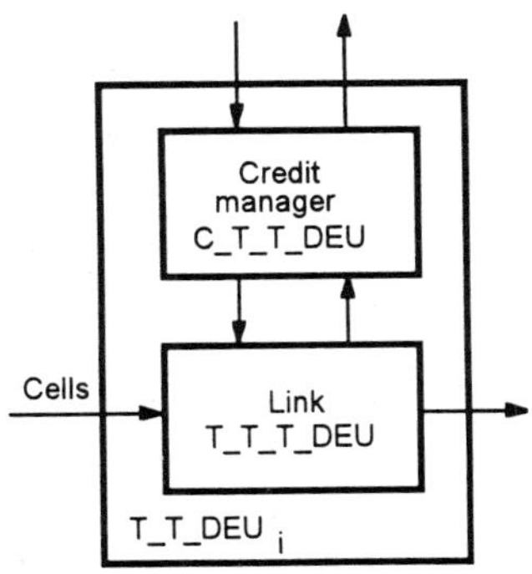

Fig. 8.

5. Concluding remarks

The approach developed in this paper helps to analyse, model, and design network management systems. A class of objects called the Functional object (F-object) has been introduced to model the network management complexity. It has been shown that F-objects can be used in a variety of situations, including modelling of OSI interfaces and functional specification of network management elements.

The concept of F-space provides a framework for structural modelling of network software. The F-space model helps to understand interdependencies between various elements of network architecture including interface protocols, resource control algorithms, as well as signalling, operations, and network management procedures. Having a good model allows one to proceed in a systematic manner. One can generate various solutions, discuss them, and choose best implementations in advance rather than arrive at them by trials and errors.

Finally, we note that the F-space approach is systematic in a sense that the DEU concept generates various accuracy models of complex software structures. The repeated use of class composition and inheritance mechanisms and DEU embedding concept developed in this paper permit to make the models so specific that they can actually be implemented in real systems.

References

1 W. Buga, Management of Telecommunications Services Provided by Multiple Carriers, in Integrated Network Management, II, I. Krishnan and W. Zimmer, Editors, pp. 343-358, North-Holland, Amsterdam 1991.

2 J. Filipiak, F-space: A method for functional network specification, ICCC '92 Proceedings, Genova, 1992.

3 J. Filipiak, Functional models of call and connection control elements in integrated broadband communication networks, Research Report, INRS-Telecommunications, Montreal, Canada 1991.

4 A. Lazar, An architecture for real-time control of broadband networks, Globecom'91, Paper 10.1, Phoenix, AZ, December 1991.

5 G. Tjaden, M. Wall, J. Goldman, and C.N. Jeromnimon, Integrated network management for real time operations, IEEE Network Magazine, 5(3):10-15, March 1991.

6 Y. Yemini, G. Goldszmidt, and S. Yemini, Network management by delegation, in Integrated Network Management II, I. Krishnan and W. Zimmer, Editors, pp. 95-108, North-Holland, Amsterdam 1991.

I.

MANAGEMENT ARCHITECTURE

C.

THE RMON MIB - FROM IDEA TO THE MARKETPLACE

Integrated Network Management, III (C-12)
H.-G. Hegering and Y. Yemini (Editors)
Elsevier Science Publishers B.V. (North-Holland)

RMON
From Concept to Specification

Michael A. Erlinger, Ph.D.

Professor, Harvey Mudd College
MTS, The Aerospace Corporation

ABSTRACT

The RMON MIB is a Proposed Standard, RFC 1271, of the Internet set of standards. Within the networking community RMON has gained a strong initial interest. This paper looks at the RMON architecture, development history, and the reasons for the community's strong interest.

1. Introduction

The Technology Centers make their initial debut during this, the Third International Symposium on Integrated Network Management. A multitude of activities are associated with the RMON Technology Center (RMON TC): this paper, a panel discussion, vendor booths, a freely available white paper, and an RMON Technology Center booth with focused demonstrations of RMON tools. This paper provides an introduction to the RMON TC by focusing on two aspects of RMON: an overview of the RMON specification and a presentation of the RMON development process.

2. What is RMON?

RMON, the Remote Network Monitoring Management Information Base - RFC 1271 [1], is currently a Proposed Standard within the Internet protocol suite. The Internet model of network management can be summarized [2,3], as a distributed model in which network management stations (NMSs) query agents for network management data. In this model the communication protocol is the SNMP (Simple Network Management Protocol), while the virtual data structure for communicating management information is defined by a MIB (Management Information Base). (Detailed discussion of this architecture can be found in [3]). Within the Internet community there are a number of MIBs at various stages in the standards process. RMON is one of these MIBs and has risen quickly to prominence.

Author's Address: Computer Science Dept, Harvey Mudd College, Claremont, CA, 91711, mike@jarthur.claremont.edu
The author is the chair of the Internet RMON working group.

2.1. RMON Architecture

The MIB designers viewed RMON RFC 1271 as accomplishing two things. First, it provides an overall structure to the concept of a MIB devoted to managing a network via remote monitoring. Previous MIBs have focused on device management (router, bridges, etc.) with minimal emphasis on management of the entire network. RMON deals solely with the networking environment itself. Second, RFC 1271 specified the particulars for management of an Ethernet environment. The MIB designers chose Ethernet as the first networking environment to have specific tables within RMON. Token Ring and FDDI were not specified in RFC 1271, but rather left to future definition with the belief that RFC 1271 would provide a sufficient structure for defining a MIB for managing these network environments.

There are nine groups defined in RFC 1271: Statistics, History, Alarm, Host, HostTopN, Matrix, Filter, Packet Capture, and Event. Compliance with RFC 1271 requires only support for every object within a selected group. The Statistics group maintains statistics measured by the agent for each monitored interface. These statistics include packets, octets, broadcasts, multicasts, and collisions on the local segment, as well as the number of occurrences of dropped packets by the agent. Each statistic is maintained in its own 32-bit cumulative counter. A real-time packet size distribution is also provided.

The History group provides periodic samples of the statistics provided in the Statistics group, with the exception of packet size distribution, which is provided only on a real-time basis. The History group features user-defined sample intervals and bucket counters for complete customization of the sampling periods.

The Alarms group provides a versatile general mechanism for setting thresholds and sampling intervals to generate events on any *integer* type variable maintained by the agent, such as segment statistics, node traffic statistics defined in the host table, or any user-defined packet match counter defined in the Filters group. Both rising and falling thresholds may be set, and both can indicate network faults. A hysteresis mechanism is provided to limit the generation of events.

The RMON Host group specifies two tables, a host table and a host-time table. Agents discover hosts by listening to all packets on the network and thus the Host group tables contain all hosts seen by the RMON agent together with host-level statistics. The host list is ordered by source and destination MAC addresses seen in good packets. For each host the traffic statistics include packets sent and received, octets sent and received, as well as broadcasts, multicasts, and error packets sent. The host time table maintains the relative order in which each host was discovered by the agent.

The HostTopN group keeps a real-time view of the top N hosts for a particular metric. The HostTopN group extends the host group by providing sorted host statistics, such as the top 20 nodes sending packets or an ordered list of all nodes according to the number of errors sent over the last 24 hours. Both the data selected and the duration of the study are configurable via MIB variables.

The RMON MIB includes a traffic matrix at the Ethernet MAC layer. A traffic matrix shows the amount of traffic and number of errors between pairs of nodes, one source and one destination address per pair. For each pair, the RMON MIB maintains counters for the number of packets, number of octets, and error packets between the nodes.

The Filter group features a generic filter engine which activates all packet capture functions and events. The filter engine fills the packet capture buffer, part of the Packet Capture group, with packets that match filters installed by the NMS via MIB variables. Any individual packet match filter can serve as a start or stop trace trigger. The contents of the trace buffer are controlled by any combination of user-selected filters, with multiple filters being combined with the Boolean *or* function. A logical data and event stream or *channel* is formed by the packets that match the filter expression.

The Packet Capture group allows NMSs to create multiple capture buffers and to control whether the trace buffers will wrap or stop capturing when full . The Packet Capture group depends upon the Filter group to define the format of the packets to be captured. Depending on the agent, the user may expand or contract the size of the buffer to fit immediate needs for packet capturing without permanently committing memory that will not be needed later. RMON includes configurable capture slice sizes to store either the first few bytes of a packet where the various protocol headers are located or, at the limit, to store the entire packet. The default slice setting specified by the RMON MIB is the first 100 octets.

Events can originate from a crossed threshold on any *integer* type variable or from any packet match count. The Event Group controls the generation and notification of events. The log includes the time of day for each event and a description of the event written by the vendor of the monitor. Traps can by delivered by the agent to a multiple number of management stations that each match the single community name destination specified for the trap.

With these nine groups the RMON architecture has proven to be adequate for Ethernet agents. It is believed that future versions of RMON — Token Ring or FDDI for example — will not require additional groups.

2.2. RMON Unique Characteristics

The RMON paradigm of remote management of networks postulated that multiple network management stations would have access to the same RMON agent. The working group felt that in a large network, various management stations would access the same RMON agent and that such access could require different levels of detail within a group. The idea of multiple management stations led to the incorporation of two features in RMON: control tables and a row-set paradigm.

The RMON developers recognized that particular groups (e.g., the statistics group) could be easily shared between NMSs. But there was also the need to have distinct instantiations for each NMS, for example, history tables with different bucket sizes. Control tables within the various RMON groups are utilized in coordinating multiple NMS activities by allowing each NMS to create its own data tables. Through the various control tables an NMS can create the related data tables with the attributes necessary for that NMS to provide management information.

The RMON row-set paradigm is also a result of trying to facilitate multiple management station access to an agent, particularly when trying to create an entry in a control table. The idea is for the management station to find an empty row and establish ownership for that row by setting the status field to 'create request'. If the row is available, the agent reserves it for the management station by setting the status field value to 'under creation'. The management station then finishes configuring the row, finally setting the status field to 'valid'. In practice this paradigm has proven workable as long as the agent and management stations are willing to cooperate. The status field can be viewed as a semaphore maintained through mutual cooperation rather than enforcement by the agent.

Another area that differentiates RMON from previous MIBs is the interdependence of various MIB groups. While control tables and data tables *within* a group are automatically interdependent, RMON also has interdependencies *between* groups (e.g. filter and packet capture). The burden placed on the agent is the necessity of ensuring that all the interdependent MIB variables are properly set prior to performing the requested actions. This issue is really a result of RFC 1271 being a specification of the MIB, not a discussion or specification of the implementation issues associated with the MIB. Thus while RFC 1271 makes reference to the interdependencies, it does not legislate a particular order to the setting of interdependent variables.

RMON also raised the issue of large data transmissions using SNMP. The packet capture capability of RMON requires that network managements stations be able to quickly and efficiently retrieve large amounts of data from RMON agents. RMON discussions led to the proposal of a bulk transmission feature for SNMP.

3. Road to Development

This paper makes no attempt to detail the Internet standards process, which has been discussed in detail in [4]. Basically, "a specification undergoes a period of development and several iterations of review by the Internet community and perhaps revision based upon experience, is adopted as a Standard by the appropriate body and is published" [4]. Usually, specification is done by a working group composed of individuals with both expertise and interest. Once consensus is reached on an initial specification, a request is made for elevation to Proposed Standard status. Internet standards then move through a three-step standards process: Proposed Standard, Draft Standard, and Standard. Progression between the official standards designations requires a minimum time period, plus implementation and interoperability experience. What is presented here is a review of the process for RMON.

3.1. RMON History

The development of RMON is an example of truly cooperative efforts among competing vendors and users. RMON began in the summer of 1990 at the Eighteenth Internet Engineering Task Force meeting, held at the University of British Columbia. A Birds-of-Feather meeting was held to discuss the interest of the community in developing a MIB focused on remotely managing a network. From this meeting a charter and working group were formed. In November of 1991, the RMON MIB specification was published as an Internet Proposed Standard, RFC 1271. RMON had progressed from concept to the first level of standardization in less than 18 months. This short development period is the result of many factors, the primary one being the interest and dedication of the working group. In general, few IETF members have job titles that include *IETF activities*. Rather, spare time is donated to Internet activities. The RMON working group had a large number of people who invested many spare hours in creating the RMON specification.

RMON developed primarily via three group activities. First, IETF meetings, which occur every three or four months, where the RMON working group met at least once. Sometimes the efforts were directed at developing a particular portion of the MIB, while at other times the group discussed global issues of remotely managing a network. Most often the IETF working group meetings were large gatherings (70+ attendees) in which consensus was attempted about major issues. These gatherings tended to focus the interested community on the issues at hand. These face-to-face activities are of primary value in the process of developing consensus and understanding within the entire Internet community.

Between each IETF meeting, the RMON working group held various meetings. These meetings, while open to all interested parties, brought out the hard-core individuals with detailed proposals — only about a dozen people attended each of these meetings (different individuals each time). Almost universally the attendees were implementors of agents or management stations from various competing vendors. One of the advantages that the working group had was the interest of organizations who had products — LAN analyzers and probes — for managing networks. Based on their real-world experiences, individuals from these organizations brought forth issues and ideas that an inexperienced group probably would have overlooked. These meetings were long and excruciatingly detailed, lasting two or three days with an occasional fine meal. The group would work line by line through proposals. Hindsight reveals RMON to be the child of these meetings, with the IETF meetings providing a venue for gathering consensus.

The third working group communication mechanism is the working group electronic mail list. The mailing list discussions have varied in both topic matter and depth. Besides specification questions, the major activity centers on implementation issues. It is important to stress that RFC 1271 is a specification, not an implementation description. It is true that implementation concerns were expressed (and continue to be expressed) by the working group throughout all phases of the MIB development, but RFC 1271 does not attempt to specify implementation concerns.

Since November 1991, RMON activities have proceeded on two fronts: progression to Draft Standard status and Token Ring RMON MIB development. The most necessary activities to continued progression of RMON within the IETF standards process, is the creation of RMON implementations and the proof of interoperability of those implementations. IETF standards move forward primarily by proof of implementation and by demonstration of interoperability among implementations [4]. Thus, RMON progression depends upon the community's ability to implement RFC 1217 and to demonstrate that a menagerie of such implementations interact. Upon publication of RFC 1271, numerous vendors put forth an immediate effort to implement RMON. (No doubt many were influenced by the marketing pressure that had grown around a standard approach to remotely managing a network.)

July 1992 was the first formal gathering of RMON vendors for an interoperability demonstration. For three days twelve vendors tried to make various agents and network management stations communicate. The group found themselves shoehorned into a small room looping on two steps: 1. try to communicate and 2. make on-the-fly changes. The individual debugging focus of the group was interrupted by a room conflict that required a two-hour interruption to the frenzied activity. During those two hours, general discussion was held. These discussions cleared up many issues. In fact, many felt that the two hours spent in conference were the most profitable of the three-day meeting. It was agreed that all future interoperability gatherings would include daily discussion periods. This first gathering was really a demonstration of partial implementations — most vendors had only implemented a few of the RMON groups. Intermittent interoperability was demonstrated, but the major accomplishment was recognition that RFC 1271 was understood, could be implemented, and that vendors were well on their way to real implementations.

A second interoperability gathering was held in September 1992. Again twelve implementors attended (not the same twelve). This time all implementations were in the 90% completed stage. Interoperability progressed rather rapidly for many of the vendors. After three days it could safely be said that full interoperability had been demonstrated between various network management station RMON tool sets and various RMON agents.

The interoperability activities, besides being required by the Internet standards process, provided the confidence that RFC 1271 was real — implementable by agents and usable by network management stations. The gatherings allowed the community to uncover problem areas and to hear about attempted solutions.

Simultaneously to RFC 1271 implementation, the working group has been developing a Token Ring specific RMON MIB, which is faithful to the RFC 1271 architecture. By January 1993 the working group plans to request Proposed Standard status for the Token Ring RMON MIB. Hopefully by the Third International Symposium on Integrated Network Management RMON Token Ring will be an Internet Proposed Standard with some available implementations.

4. Summary

4.1. Remaining Issues

Currently RFC 1271 is being reviewed prior to a request to move it to the next Internet standard level, Draft Standard. Besides minor changes being investigated by the working group, there are two major issues: SNMPv2 and implementation concerns. The SNMPv2 is currently being defined by an Internet working group. While providing general enhancements to the SNMP protocol, SNMPv2 provides specific enhancements for RMON, row-set paradigm and bulk transfer. It is clear to the RMON working group that the RMON specification will need modification to conform with the proposed SNMPv2. It is believed that these are minor modifications, but they should be made so that there are no inconsistencies between SNMPv2 and RMON.

As mentioned above, RRC 1271 is a MIB specification. Within the document a few comments refer to implementation issues, but RFC 1271 does not attempt to provide a complete discussion of implementation issues. The working group has long recognized that a companion implementation document is desired. While various private attempts have been started, there has not been a coordinated attempt to create such a document. Those implementing RMON view this as a deficiency that hopefully will be corrected in the near term.

4.2. RMON Acceptance

Basically, RMON is attaining rapid acceptance in the networking community for two reasons: it provides a forum for vendors to compete on an equal footing, the RMON MIB; and it allows network management customers to choose among competing RMON agents and applications.

For the network administrator RMON presents a unique opportunity to manage his network through remote agents with standard capabilities. As RMON agents enter the market place, they can be effectively evaluated based on their capabilities — RMON groups supported and agent performance issues. This provides the network administrator the ability to install RMON agents of varying capabilities depending on the management needs within his network. At the network management station, the network administrator can choose the best RMON applications for particular needs with these decisions being made independently of the RMON agents being employed.

Fundamentally, RMON prevents customers from being locked into a single vendor and thus it has attained wide interest and initial acceptance. The continued acceptance of RMON will depend on whether the choices made in the RMON architecture together with the application tool sets provide meaningful data for the network administrator.

5. References

[1] Steve Waldbusser. *Remote Network Monitoring Management Information Base.* Request for Comments 1271, November 1991.

[2] V. Cerf. *IAB Recommendations for the Development of Internet Network Management Standards.* Request for Comments 1052, April 1988.

[3] Marshall T. Rose, Network Management is Simple: you just need the "right" framework, *Integrated Network Mangement, II*, North-Holland, 1991.

[4] Lyman Chapin. *The Internet Standards Process.* Request for Comments 1310, March 1992.

The RMON MIB: Implementation and Network Management Issues

Panel Chair: Michael Erlinger, Harvey Mudd College and The Aerospace Corporation, USA

Panelists: Michael Watters, Hewlet-Packard, USA
Tim Lee-Thorp, Remedy Corporation, USA
Maynard Webb, Jr., Quantum, USA

The Remote Monitoring Management Information Base (RMON MIB) was the first Internet MIB to concentrate on network management. Previous MIBs were developed to manage particular application areas and devices. RMON entered the Internet standards track as a Proposed Standard in November 1991. It has subsequently become RFC 1271, an Internet standard.

The panelists will review the RMON MIB technology and its progression as a tool for network management. As drivers of the RMON technology, panelists will discuss the process of interoperability testing and the standards process. Panelists will also compare RMON with other technologies such as LAN analyzers and private MIBs.

I.

MANAGEMENT ARCHITECTURE

D.

MANAGEMENT INFORMATION

Integrated Network Management, III (C-12)
H.-G. Hegering and Y. Yemini (Editors)
Elsevier Science Publishers B.V. (North-Holland)

Design of the MANDATE MIB

Jayant Haritsa, Michael Ball, Nicholas Roussopoulos, John Baras, and Anindya Datta

Institute for Systems Research, University of Maryland, College Park, MD 20742

Abstract

A management information database is the heart of a network management system – it provides the interface between all functions of the network management system, and therefore has to provide sophisticated functionality allied with high performance. In this paper, we introduce the design of MANDATE, a database system that is designed to effectively support the management of large enterprise networks. MANDATE proposes to use special characteristics of network management data and transactions, together with recent advances in database technology, to efficiently derive its functionality.

1. INTRODUCTION

In today's global marketplace, most large-scale enterprises have widely-dispersed manufacturing and commercial operations that are connected by a communications network. The enterprise communication networks of the future are expected to be heterogeneous in several dimensions: mixed-media physical transmission facilities (e.g. copper, fiber-optic, satellite); multi-media information transmission (e.g. video, voice, data); multiple vendor sub-networks; and varied performance objectives (e.g. high throughput for data, low blocking probability for voice). As a result of this multi-faceted heterogeneity, future enterprise networks are expected to be highly complex in their transmission, performance, and communication characteristics. Due to this complexity, and due to the disparity among management systems for individual subnetworks, efficient management of enterprise networks is an extremely challenging problem.

Network researchers are generally in common agreement that a global network database, which contains all management-related data, is central to the development of an efficient network management system (e.g. [Bapa91, Valt91, Terp92]). This database, referred to in OSI parlance as Management Information Base (MIB), is required to store information on network and system configurations, current and historic performance, trouble logs, security codes, accounting information, etc. [Kler88]. While there has been intensive research on network management systems in recent years, comparatively little has been published with respect to the actual design and implementation of an MIB. In this paper, we introduce the design of MANDATE (MAnaging Networks using DAtabase TEchnology), an MIB system that is designed to effectively support the management of large enterprise networks.

The guiding principle of the MANDATE design is to have the network operator(s) interact solely with the database, that is, from the operator's perspective, the database logically *embodies* the network. Whenever the operator wishes to make changes in the network functioning, such as changing the routing scheme, for example, the operator merely updates the appropriate variables in the database. The actual implementations of these changes in the physical network are made by the database system. This design approach allows the operator to concentrate on what has to be done, rather than on the mechanics of implementing the decisions. A second important aspect of the proposed MANDATE system is that it is a bottom-up design, not a modified version of commercially available database systems. This results in a system architecture that is tailor-made specifically for network management. Finally, MANDATE plans to use special characteristics of network management data and transactions, together with recent advances in database technology, to efficiently derive its functionality.

2. ROLE OF DATABASES IN NETWORK MANAGEMENT

The ISO/ANSI standards committee [Cher87] has classified the sophisticated functionality required of network management systems into the well-known six categories of Configuration Management, Fault Management, Performance Management, Security Management, Accounting Management, and Directory Management. The functional architecture defined by these six categories clearly identifies the different facets of network management and control, and enables a modular approach to be taken towards designing network management tools. However, there is considerable overlap and interaction between the various management sub-systems. For example, fault management and performance management are closely interrelated, since poor performance is often the only visible symptom of a fault deep down in the system. Similarly, detecting a faulty resource and providing for its isolation requires both fault management and configuration management. In order for the various management modules to co-ordinate their activities, a common "public workspace" or *database* is necessary. Therefore, a logically integrated database is the *heart* of a network management system [Schw90] – it provides the interface between all functions of the network management system. This database, or MIB, is the conceptual repository of all management-related information.

2.1. Requirements on MIB

Ideally, the MIB module of an enterprise network management system should provide the following major functionalities (a more comprehensive list is described in [Hari92]):

(1) Homogeneous interface: Present a uniform interface to the operator that is independent of the individual sub-network characteristics.

(2) Fault-tolerance: Operate 24 hours on-line since the MIB is the core of the network management system.

(3) Real-time Response: Store and process in real-time the "network health" data which is continuously gathered by monitoring tools.

(4) Decision Support: Answer "what-if" questions by executing on-line simulations, thus helping the operator to evaluate the potential impacts of different control decisions.

(5) Embedded Optimization: Efficiently execute on-line optimization algorithms to adapt the network routing, configuration, etc. in response to changes in the network traffic or connectivity.

(6) High-performance: Minimize the overhead of network management on the performance of the network. In addition, the network management performance should gracefully degrade under overload conditions.

From this list, we observe that an MIB has architectural requirements, interface requirements, temporal requirements, control requirements, and performance goals. Clearly, the design of the MIB is key to providing all of these complex functionalities in an integrated fashion.

Since current database technology is fairly mature, one might think that using a popular database management package (e.g. ORACLE, INGRES) should be sufficient for implementing a network management MIB. However, this is not the case for several reasons: First, standard off-the-shelf DBMSs lack many of the required MIB functionalities such as real-time capabilities and decision support facilities. Second, conventional

DBMSs have been developed for the commercial query processing environment and are primarily geared towards applications such as banking, where the focus is on naive human users interactively performing transactions. In network management, however, software programs control the network behavior with human intervention restricted to skilled operators. Finally, the objective of conventional DBMSs is to efficiently implement a transaction model that provides the so-called **ACID** property, that is, atomicity, consistency, isolation, and durability. However, this transaction model is unsuitable for processing of network management data, since some of these properties may not be essential here, as explained in the following sections. In summary, the network management environment is a specialized application area with unique characteristics that can best be taken advantage of by a database system that is built specifically for this environment. In this paper, we introduce the design of MANDATE, an MIB that is tuned to the task of network management.

3. RELATED WORK

Issues similar to those addressed in this paper were considered in [Schw90]. The focus in that work was on evaluating how conventional DBMS packages would serve in the role of an MIB, and suggesting network-related modifications to these conventional packages. In contrast, our focus is on developing a new database system whose design is tailor-made for network management. The NETMATE project at Columbia University [Dupu91] has also investigated many of the issues discussed here. They have concentrated on the model of the network and its architectural relationship with management tools, whereas our attention is primarily on the MIB design aspect.

An overview of the issues involved in implementing the MIB interface definition laid down by the OSI standards committee was presented in [Bapa91]. The issues considered included the choice of data model, the architecture for distributing network management data, and the mechanisms for ensuring integrity of replicated data. While the paper describes several of the functionalities to be provided by a MIB, it does not, however, provide a detailed architectural design for achieving these functionalities.

4. NETWORK DATA

The first step in designing a database system is to understand the properties (semantics) of the data items that are resident in the database and to understand the properties of the tasks (or transactions) that store, process, and retrieve this data. Network management data can be broadly classified into three types: Sensor Data, Structural Data, and Control Data, as shown in Figure 4.1, which describes a high-level abstraction of the MIB data model. As explained below, the structural data describes the physical and logical construction of the network, the control data captures the operational settings of the network, and the sensor data represents the observed state of the network.

4.1. Sensor Data

The sensor (or measurement) data is the raw information that is received from the network monitoring processes, and includes variables such as node queue lengths, retransmission rates, link status, call statistics, etc. The sensor data provides the primary input for three of the six OSI network management categories: Accounting Management, Performance Management, and Fault Management. It represents the current "health" of the network in terms of the network usage and operational quality. In current large networks, the quantity of sensor data that is gathered may be as large as 20 to 30 gigabytes per day [Spri92].

Sensor data can be divided into two groups: Persistent and Perishable. The persistent data consists of sensor data whose utility is long-term and therefore needs to be maintained permanently in the database. Critical data such as customer billing

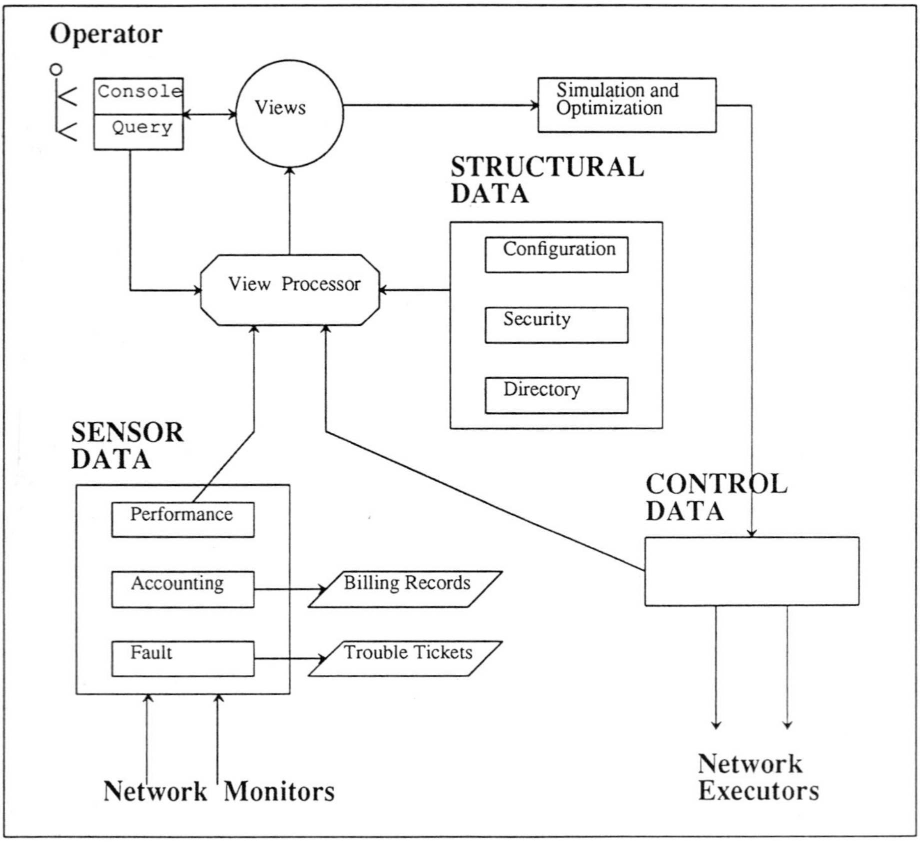

Figure 4.1: Model of MIB

information, network alarms, and security violations belong to this category. Due to the requirement of permanence, persistent sensor data requires recovery mechanisms similar to those provided by commercial DBMSs.

Perishable sensor data, on the other hand, is data that is of limited time utility in the sense that its current value is valid only until the network characteristic that is being monitored retains that value. Data such as node queue lengths, retransmission rates, and most other dynamic performance statistics fall into this category. There is no need for logging of these updates since the information will be out-of-date by the time the MIB recovers from a failure. Also, unlike the persistent sensor data, updates to perishable sensor data are not "sacred" since ignoring updates occasionally does not have serious implications. While the perishable sensor data has only limited time utility with respect to the immediate operation of the network, it is necessary to retain some *history* of the data values (using a versioning mechanism) for long-term post-mortem performance and fault analysis.

4.2. Structural Data

In contrast to sensor data, structural data is composed of static or slowly-changing network information such as the network topology, the configurations of the network switches and trunks, the data encryption keys, the customer description records, etc. This data provides the primary input for the remaining three OSI network management categories: Configuration Management, Security Management, and Directory Management. A point to note here is that unlike sensor data, structural data is valid even when the network is not in operation.

In a typical large network, the quantity of structural data depends on the detail at which the network is represented and may be of the order of several gigabytes. Most of the structural data is stored at system initiation time and is typically changed only in response to significant network events such as adding a new switch to the system or offering a new type of customer service. The structural data needs to be recoverable for monetary reasons (customer records are of vital importance), for efficiency reasons (restart quickly from a database crash), and for security reasons (accessing copies of data encryption keys remotely over the network could lead to security compromises).

4.3. Control Data

The final data category is the control data, which captures the current setting of network tuning parameters such as the maximum flows on individual trunks, the traffic split ratios on the output links of switches, the routing table, etc. In addition to the current parameter settings, the control database also stores a library of pre-defined control settings that reflect the appropriate settings for a variety of common traffic patterns and network configurations. For example, different suites of settings may be appropriate for day traffic and night traffic.

5. NETWORK TRANSACTIONS

Having discussed the characteristics of network management data, we now move on to considering the various types of transactions that operate on the sensor, structural, and control databases.

5.1. Sensor Data

Two distinct groups of transactions, "updaters" and "readers", access the performance data (perishable sensor data). The updaters are the network monitoring tools, while the readers are internal MIB processes. The updaters work in private data partitions since they update different sets of network variables and they therefore do not interfere with each other. These updates are different from typical database updates in that the updated value is independent of the current value of the data object. Such updates are referred to as "blind writes" [Bern87]. Since the performance data is versioned, readers can always read the data that they want without delay. Therefore, due to the absence of Read-Write and Write-Write conflicts, no concurrency control is necessary for the performance data.

For the accounting and fault information (the persistent sensor data), the updaters from the network monitors append records to existing tables. The MIB internal processes may both read and update these persistent records. For example, a network monitor may register a trouble ticket in the fault database. Once the fault is fixed, the trouble ticket has to be updated to reflect this fact. Due to the concurrent reading and updating, concurrency control is necessary for the accounting and fault databases.

5.2. Structural Data

Structural data can be both read and written by the network operator(s) or by MIB control processes. Since it is possible that multiple processes may access the same structural data simultaneously, concurrency control has to be implemented. However, since the structural data is updated only very rarely, concurrency control is not a major performance issue with respect to transactions having to block while accessing data objects. Yet, it is wasteful to have all transactions pay the computational overhead of invoking the concurrency control manager for each access to a data object given that regulated access is only rarely necessary. In [Hari92], a simple mechanism is described by which this problem is overcome, thereby resulting in the concurrency control protocol being used *only* when update transactions are executing.

5.3. Control Data

Control data can be both read and written by the network operator or by MIB control processes. The process for changing an existing set of control settings is usually initiated by the network operators. For example, if the operator observes from the sensor data that some links are becoming excessively utilized, he/she may decide to replace the routing scheme that is currently employed by a different scheme. Alternatively, the changes may be automatically triggered as a function of the information contained in the sensor data. For example, if there is a serious security violation at a node (e.g. introduction of a virus), the links going through the node may be automatically shut down pending investigation of the problem by the network operators. Another source of change for the control data is that produced by re-executing the optimization algorithms to reflect changes in the network configuration or activity profile, thus generating a new set of control settings.

In a properly designed network, no more than one process can update a given set of control variables at a time (it is meaningless to have concurrent updaters since the control of the network then becomes a function of the order in which the transactions are processed). Therefore, concurrency control is not required for this data category. However, the transaction construct is necessary for installing the updates in order to ensure the atomicity of the updates (half-implemented control settings can cause havoc in the network). Changes to the control data trigger off network executors (see Figure 4.1), which are processes that actually implement in the physical network the new control structure that is logically described by the updated control data.

6. Proposed MIB Design

In conventional database systems, an elaborate set of concurrency control and recovery mechanisms are utilized to provide the ACID property. However, as described in the previous sections, in the network management domain, weaker forms of the ACID property may be acceptable for certain data categories. In the MANDATE design, we propose to incorporate mechanisms that permit concurrency control and recovery to be selectively implemented on a data partition basis. By doing so, we expect to realize considerable performance improvements.

The MIB should be able at all times to provide the operator(s) with a view of the current state of the entire network. This is achieved by combining the current sensor information, the structural information, and the control settings in effect, as shown in Figure 4.1. In MANDATE, we propose to generalize this notion to allow the operators to create different views of the network by incorporating a view processor that provides the appropriate view to each operator based on the information in the database. For example, the structural database holds information about the customer sub-networks, which includes details of the physical customer access links and the logical mapping of a customer to the public shared network. An operator trying to find the cause of a customer complaint would use a view wherein the customer sub-network is superposed on the public network to determine whether the fault lies in the public network or is local to the customer subnet. The network views also serve as inputs to the embedded simulation and optimization algorithms.

In the remainder of this section, we will describe how MANDATE proposes to use recent advances in database technology to achieve some of the required MIB functionalities. Due to space constraints, we discuss the system design for only a few functionalities here; a more detailed discussion is provided in [Hari92]. In particular, we focus here on the system architectural model and the network control and decision support mechanisms.

6.1. Architectural Model

Although an MIB is logically a centralized repository of all network-management data, its *physical* implementation in large networks will have to be *distributed* for performance reasons. We assume that there is a central main database which stores all the structural data, the most critical sensor data and the major control settings, while the remaining network state data is stored in the local memories and disks of network components. As shown in Figure 6.1, the recently developed concept of a *Client-Server* architecture integrates well with this design. In this picture, the DB Server is the primary data store site where all updates are synchronized for maintenance of consistency. The network switches periodically propagate status data to the DB Server; the switches can also be queried on demand. The client modules are typically workstations that have a full DBMS functionality for cacheing views (subsets) of the MIB data. At any point in time, a particular view is maintained in each client. In many cases the client workstation will be supporting an operator who is responsible for real-time control decisions. Therefore, updates to the MIB must be propagated to the client views in real-time. The client and Server DBMSs cooperate and split the task of query processing. By placing client workstations at strategic nodes on the network, database access is performed in parallel from multiple client databases, thereby alleviating the bottleneck at the primary server.

While the above client-server architecture appears to be an attractive design choice, there are serious problems that may arise in an actual implementation: First, the activities of synchronizing and refreshing downloaded data are repeated very often in a network and therefore, as the number of sites increases, the processing throughput rapidly deteriorates due to blocking. Second, because sites collectively carry a large variety of data subsets with each subset pertinent to a particular client's function, place of deployment, response requirements, etc., the database servers waste most of their capacity in keeping track of who-needs-what-when in order to propagate changes that affect individual clients. On the other hand, if all changes are broadcast to all clients, the clients would be incapacitated due to having to check all received updates (most of them would be irrelevant to each individual client). Therefore, the intelligence, namely who-needs-what-when, must be distributed to the client workstations which would selectively request only relevant updates from the logs maintained at the servers.

6.1.1. Incremental Client-Server Architecture

The above problems associated with distributed network database processing in client-server architectures can be resolved by integrating the client-server architecture with the recently developed concept of *incremental computation models* [Rous86]. In this model, the results of subsequent accesses to the same portion of the database are realized by applying the computation to the input *differentials*, rather than re-executing the computation on the whole input. A recent study has shown that this "incremental client-server architecture" achieves two orders of magnitude performance increase over a standard client-server database architecture [Deli92]. More importantly, the incremental model seems especially appropriate for the real-time maintenance of network views which are continuously being used by the network operators and customers. This is because, with this model, keeping the views uptodate requires only incremental computations and relieves the system from the burden of having to recompute the entire view in each refresh cycle. Since, under normal network operations, the views change only slowly over time, the incremental model can realize great improvements in performance and minimize the

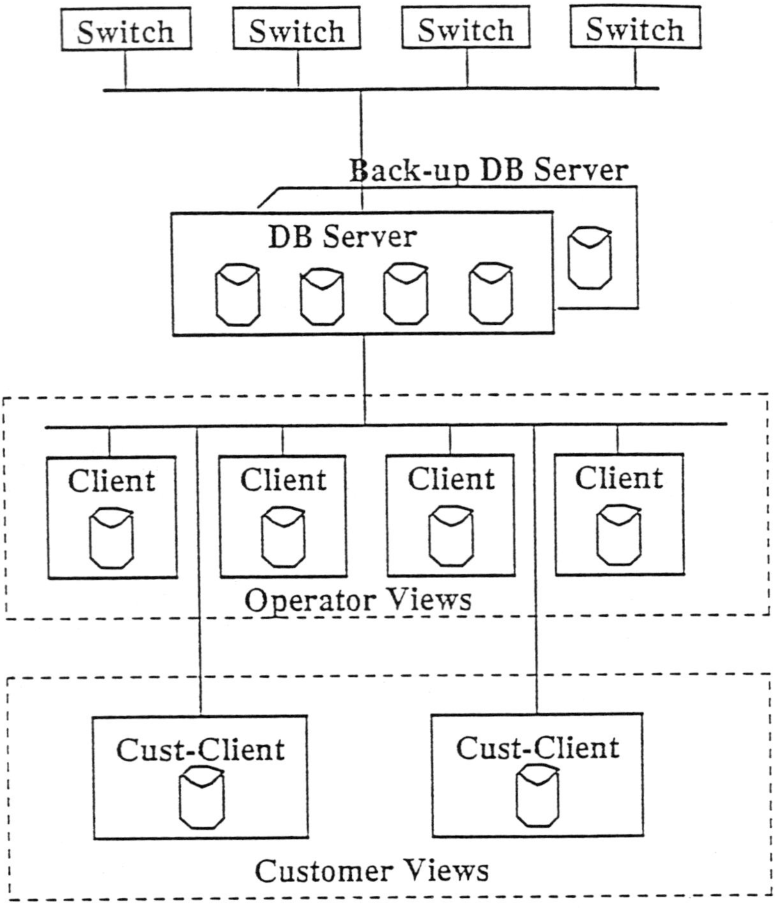

Figure 6.1: An enhanced client-server database architecture for MIB

overhead of view maintenance on the other system functions. In summary, the incremental client-server architecture provides the functional advantages of a distributed architecture while retaining the performance of a centralized system.

In MANDATE, we propose to use the incremental client-server database architecture described above to provide the following: (1) Real-time refreshing of network views; (2) Immediate update propagation from the primary server to its secondary backups – this guarantees that, in case of failure, the secondary server is uptodate for promotion to be the primary server; (3) Parallel access of dynamically distributed data on the enhanced clients and significant reduction on the servers I/O; (4) Preservation of the appropriate level of centralized control.

6.2. Network Control

The fundamental goal of network management is to be able to control the state of the network. In the context of an MIB, a network control process is any mechanism that makes the network respond to stimuli collected by various network sensors. Though most of this data is routine, events such as link failures and switch malfunctions may occur which require remedial action by the network management system. The continued operation of the network in the presence of faults is achieved through the activation of control processes which are triggered when the network is confronted with unusual circumstances. For network MIB systems, we classify network control mechanisms along two dimensions: local versus global and automatic versus manual. We present below brief descriptions of each class with illustrative examples.

6.2.1. Local Control

Local control mechanisms rely on local data collection and local decision models. By local we refer to specific components of the network as opposed to the network as a whole. An example of a local network control mechanism is the classical window-based flow control scheme for regulating traffic between a source destination pair [Ephr89]. The advantage of local controls is that they incur little or no communication overhead since decisions are made locally with local data and minimal information exchange is involved. Due to this locality of operation, local control processes are unaffected by remote network failures and network congestion. We propose to implement these local control processes in MANDATE with simple triggers or rules.

6.2.2. Global Control

Global control processes rely on network-wide data and global decision models. Examples of global control mechanisms include routing algorithms that compute routes based on network-wide traffic estimates. Clearly, global control processes are capable of optimizing network-wide performance characteristics. However, they are more vulnerable to network failures and have greater information overhead since decisions must be communicated across the network. Global controls are usually implemented as algorithms that accept input from the MIB server, perform large computations and return the results to the MIB server.

6.2.3. Automatic Control

Automatic controls monitor certain network performance characteristics. When specific conditions are met, control settings are automatically changed without operator intervention. For example, assume that a particular switch has a local control process that allocates incoming calls between a source destination pair equally among three different routes. If one of these routes becomes overloaded, an automatic control process embedded in the switch is triggered such that only twenty percent of all new traffic is sent along the saturated route and the remainder is split equally among the two other routes. Clearly, implementing automatic controls requires active database features.

6.2.4. Manual Control

Manual control processes either permit or require human intervention. Network operators alter control settings in the network using these processes. In a sense, manual controls represent one of the major justifications of the MIB. Clearly, the role of the MIB is to provide the network manager with information that supports decision making regarding the setting of control parameters. This supporting activity may be achieved passively by simply providing an interface between the network operator and network status information. Alternatively, it may be achieved actively through an alarm system that notifies the network manager of network conditions that require actions on his or her part.

6.2.5. Control Architecture

In MANDATE, we propose to provide active support for automatic control by: (1) Supporting active database elements that can monitor sensor data and automatically set local controls; (2) Providing embedded optimization algorithms that analyze global network data and automatically propagate control settings throughout the network.

We also propose to provide in MANDATE passive support for manual controls by providing the operator with: (1) Flexible, fast access to network information; (2) Embedded optimization and simulation algorithms, described in the next section, which support control decisions; (3) Facilities that implement specified control decisions.

6.3. Embedded Optimization and Simulation Algorithms

While some present-day networks employ optimization algorithms to determine global control settings, we plan to support in MANDATE the generation of alternate versions of control settings as well as their detailed analysis using simulation. These features are expected to help network operators make higher quality control decisions. Similar facilities are provided in the NETMATE system [Dupu91].

We now discuss the interaction between embedded optimization and simulation algorithms and the MIB. Figure 6.2 gives a detailed view of this interaction. Examples of optimization algorithms include algorithms that output routes, flow controls, access controls, etc. In general, the optimization algorithm draws input from the MIB and outputs control settings. The inputs to the algorithm consist of the current network state as embodied by some network view and the existing control settings. We also envision the presence of archived control settings in the MIB, which consist of old control settings and the corresponding network states for which they were used. These also may serve as input to optimizing routines which may find commonality between the current network state and some old state and may be able to use old settings.

The output from an optimization algorithm is used to update either an *active* setting or a *non-active* setting. Active settings invoke immediate execution of control action – in such a case, the algorithm would be part of an automatic control process. Implementation of active settings may be achieved through triggers or rules. Non-active settings are usually examined by users who make the final decision on implementation. That is, if the user wishes to use the control setting, he or she would transfer it from the non-active state to the active state. Whether the output of an algorithm is active or non-active is determined by the network state that invoked the execution of the algorithm. For example, a case of emergency may trigger the algorithm in an active output mode while a less critical situation may permit non active output and the subsequent examination of the proposed control settings by the operator. In the latter case, the user may wish to run the algorithm several times, possibly with different parameters, and compare the outputs. Typically, in this scenario, as a result of multiple invocations of the algorithm, several versions of control settings would be derived leading to the need for a *version control* scheme.

The role of simulation is to evaluate the performance effects of a proposed control setting. The inputs to the simulation consist of a network state, an existing control setting and the proposed control setting. The output is a detailed analysis of the effect on performance of using the proposed control setting.

The general environment that we propose to support in the MANDATE system is one in which a user can invoke a given optimization algorithm several times or can invoke alternate algorithms in order to produce different versions of the output (proposed control settings). The user would typically examine and possibly modify various versions and also may wish to invoke a simulation algorithm that produced a detailed analysis of the proposed control setting. Finally, the user would choose one of the versions as a "final version" which, in MANDATE, would correspond to making the central setting active.

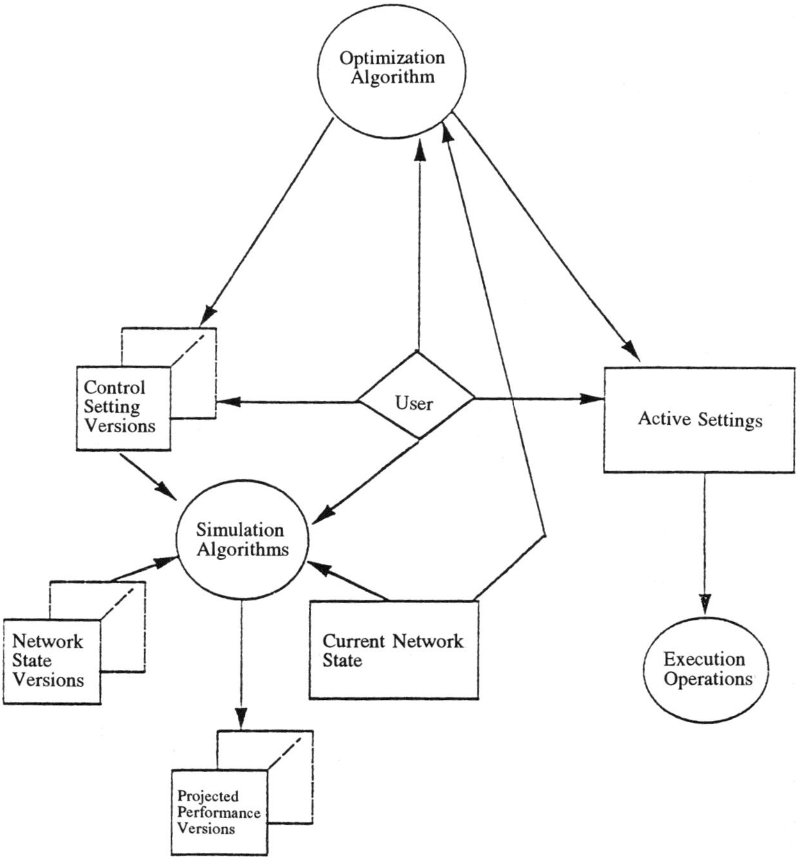

Figure 6.2: Interaction between algorithms and the MIB

7. CONCLUSIONS

In this paper, we introduced the design of MANDATE, a database system which is geared towards effectively supporting the management of large networks. In the proposed design, MANDATE provides the network operators and customers with an MIB interface that allows them to control and evaluate the network operations by interacting solely with the database. The actual implementation of the control decisions in the physical network are handled by MANDATE's internal processes.

The underlying structural framework of MANDATE is a client-server architecture which is enhanced by the use of an incremental computational model. This architecture is expected to result in a high-performance interface which provides the functional advantages of a distributed architecture while retaining the performance of a centralized system. Some of the other special features that we plan to incoporate in MANDATE are selective elimination of concurrency control and recovery overheads, variety of local and global control structures, support for embedded network simulation and optimization algorithms, specialized view-based interfaces and historical views.

In summary, the MANDATE design proposes to use special characteristics of network management data and transactions, together with recent advances in database technology, to efficiently derive its functionality. As part of our future research, we plan to test and tune MANDATE by implementing it on a "toy" network and to follow this up with a detailed performance study.

REFERENCES

[Bapa91] Bapat, S., "OSI Management Information Base Implementation," *Integrated Network Management, II*, eds. I. Krishnan and W. Zimmer, Elsevier Science Publishers B.V. (North-Holland), 1991.

[Bern87] Bernstein, P., Hadzilacos, V., and Goodman, N., *Concurrency Control and Recovery in Database Systems*, Addison-Wesley, 1987.

[Cher87] Chernick, M. et al, "A survey of OSI network management standards activities," *Tech. Report NMSIG87/16 ICST-SNA-87-01*, National Bureau of Standards, 1987.

[Deli92] Delis, A., and Roussopoulos, N., "Performance and Scalability of Client-Server Database Architectures", *Proc. of 18th Int. Conf. on Very Large Data Bases*, August 1992.

[Dupu91] Dupuy, A., et al, "NETMATE: A Network Management Environment", *IEEE Network Magazine*, March 1991.

[Ephr89] Ephremides, A., and Verdu, S., "Control and Optimization Methods in Communication Network Problems," *IEEE Trans. on Automatic Control*, 34(9), September 1989.

[Feld89] Feldkhun, L., "Integrated Network Management Systems," *Integrated Network Management, I*, eds. B. Meandzija and J. Westcott, Elsevier Science Publishers B.V. (North-Holland), 1989.

[Hari92] Haritsa, J. et al, "MANDATE: MAnaging Networks using DAtabase TEchnology," *Technical Report 92-98*, Systems Research Center, Univ. of Maryland, Sept. 1992.

[Kler88] Klerer, S., "The OSI Management Architecture: an Overview", *IEEE Network*, 2(2), March 1988.

[Rous86] Roussopoulos, N., and Kang, H., "Principles and Techniques in the Design of ADMS±," *IEEE Computer*, 19(12), December 1986.

[Rous91] Roussopoulos, N., "The Incremental Access Method of ViewCache: Concept and Cost Analysis," *ACM Trans. on Database Systems*, 16(3), September 1991.

[Schw90] Schwab, B., Wasson, L., and Sholberg, J., "Database Management for an Integrated Network Management System," *Network Management and Control*, ed. A. Kershenbaum et al, Plenum Press, New York, 1990.

[Spri92] SPRINT Network Management Center, Virginia, *Site Visit*, April 1992.

[Terp92] Terplan, K., *Communications Networks Management*, Prentice-Hall, 1992.

[Valt91] Valta, R., "Design concepts for a Global Network Management Database," *Integrated Network Management, II*, eds. I. Krishnan and W. Zimmer, Elsevier Science Publishers B.V. (North-Holland), 1991.

Integrated Network Management, III (C-12)
H.-G. Hegering and Y. Yemini (Editors)
Elsevier Science Publishers B.V. (North-Holland)

MIB Design for Network Management Transaction Processing

Shoichiro Nakai

C&C Systems Research Laboratories, NEC Corporation
1-1, Miyazaki 4-chome, Miyamae-ku, Kawasaki 216, JAPAN
e-mail: nakai@nwk.cl.nec.co.jp

Abstract

This paper proposes design alternative for Management Information Base (MIB) from the viewpoint of transaction management. In the proposed MIB, concurrency control and recovery for transaction processing are provided based on the standard OSI protocols, i.e. CMIS/P and Transaction Protocol. The concurrency control protocol uses a locking-based algorithm which eliminates transaction blocking due to data contention, and avoids cascading aborts. Hence, the protocol permits more concurrency than the well-known strict Two-Phase locking protocol, and furthermore, READ-ONLY transactions are never aborted in the proposed MIB.

Keyword Codes: K.6.4; H.2.4
Keywords: System Management; Database Management, Systems

1. Introduction

The demand for increasingly extensive functionalities in network management has resulted in the use of distributed database systems and transactions processing for Management Information Base (MIB).

A key issue for distributed transactions is concurrency control and failure recovery. This paper looks into aspects of the management information model and its manipulation from the viewpoints of concurrency control and recovery, discusses how MIB should be designed using concurrency control and recovery, and presents a locking-based concurrency control algorithm which does not restrict the execution of conflicting operations.

The capability of network systems to provide the functionalities and performance required by network customers has recently been recognized as one of the most important activities of network operations. For instance, sophisticated customers may demand the capability of dynamically alternating routes according to traffic demands or network failures, or according to communication costs.

Providing for functionalities such as dynamic routing alternation requires the execution of two or more operations, some on a locally managed object, some on remotely managed objects. Such operations may involve accessing of data from more than one database systems. Moreover, more than one process may invoke such operations concurrently. In general, a management system must be capable of distributed transaction processing.

In this paper, we classify network management transactions as follows:

Local Transactions A transaction containing a set of management operations is said to be a *local transaction* if it contains one or more operations which access data on a single database system. In current OSI management environment, any single operation involving access to one data object (such as the reading or updating of a management attribute) is a local transaction.

Global Transactions A transaction containing a set of management operations is said to be a *global transaction* if it contains operations involving more than one database system.

Figure 1 illustrates the difference of the above local and global transactions.

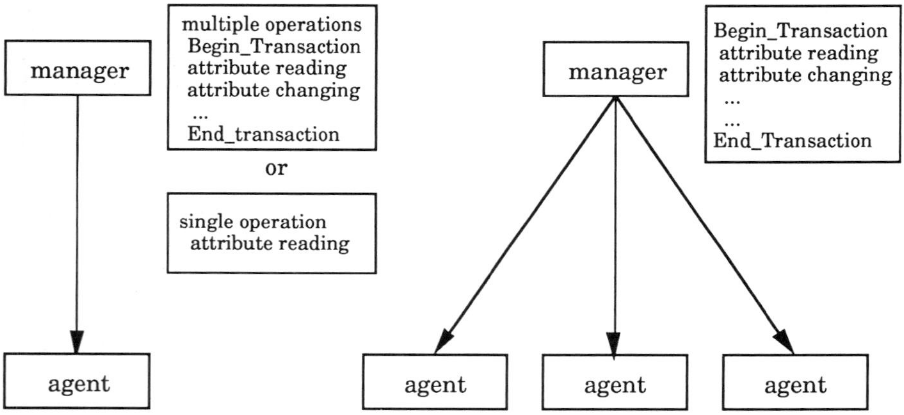

(a) Local Transactions (b) Global Transaction

Figure 1. Local and Global transactions

To support the transactions, failure atomicity and execution atomicity of transactions must be guaranteed to maintain the integrity of the database systems. In the case of system failures or processing failures which occur during the execution of transactions, the failure atomicity can be guaranteed by the well-known Two-Phase Commit (2PC) protocol [BER87]. 2PC protocol ensures that all the participating database systems either collectively commit the transaction or collectively abort it, even if failures occur during the protocol.

In executing the operations, it is also important to provide as much concurrency as possible, so as to minimize the transaction processing time. However, concurrent transactions should be executed atomically as if they are processed sequentially. Concurrent execution of transactions could lead to interference if read-write conflicts occur among the operations contained in the transactions. As a result, concurrency control is necessary to avoid or eliminate such interferences. In commercial database systems, the most prevalent concurrency control scheme is the Two-Phase Locking (2PL) protocol [BER87]. For instance, a management system which supports dynamic routing changes must provide concurrency control, otherwise the routing information may become corrupted as a result of interferences among concurrent transactions, or some process may read inconsistent information.

Although alternative correctness criteria for concurrency control have been proposed [DU89], this paper assumes that *serializability* is a sufficient consistency criterion for network management. An execution schedule for a set of concurrent transactions is said to be *serializable* if the schedule produces the same output and has the same effect as an arbitrary serial execution of the same transactions [BER87]. Modification to this criterion for network management has been suggested [YEM91] and should be the topic of future studies.

The remainder of this paper is organized as follows. First, current OSI standard protocols relevant to network management, and previous work related to MIB implementation is described. Second, we look into management information model, then discuss the MIB architecture for transaction management processing. Finally, examples of how concurrency control can be implemented using OSI standard protocols are described. The paper concludes with a discussion of future work.

2. Background

In the OSI network management model, one or more manager processes interact with one or more agent processes to access or modify objects in database systems known as *Management Information Bases* (MIBs). Transactions are issued by *manager* processes (will also be referred to as managers) to *agent* processes (will also be referred to as agents). Agents access data objects on behalf of requesting managers. Following is a very brief description of ISO standard protocols relevant to network management.

2.1. Current Standard Protocols

ISO has specified the Common Management Information Services and Protocols (CMIS/P) [BRU91]. As it stands, this protocol provides for a single manager - multiple agents environment. Although the need to do so has long been recognized, CMIS/P has not yet been extended to the distributed transactional environment [KLE88].

To realize the impending requirement of distributed transactions for management systems, ISO has recently attempted to incorporate the application service element of Transaction Protocol (TP) [UPT91] into the context of management applications [N1395][CCITT92]. TP allows transaction processing to use another ISO protocol referred to as Concurrency, Commit, and Recovery (CCR). CCR, as it currently stands, primarily addresses failure atomicity by defining 2PC as the standard commit protocol for transaction processing [CCR90-1][CCR90-2]. The need for concurrency control is recognized in the standard but has been left as an implementation issue.

2.2. Related Work

In the following some of the previous work on MIB implementation is briefly described.

Concurrency Control We address the issues of both direct and indirect conflicts of management operations in [NAK92]. *Direct conflicts* result from interferences among global transactions only, while *indirect conflicts* (also known as transitive conflicts) are due to interferences introduced by local transactions executed concurrently with global transactions [BRE90][BRE91]. The issues are simply handled if we can assume homogeneous system in which every manager and agent participating in the system conforms to the same concurrency control scheme [NAK92].

Yoda et al. [YOD92] propose concurrency control protocol which uses conservative 2PL protocol with WOUND-WAIT algorithm [BER87] for handling deadlocks. The idea is that concurrency control can be implemented using current CMIS/P without using CCR. In the protocol, the granularity of locking unit is a single managed object, so that when a transaction locks an object other transactions may not be allowed to access the object. Note that, most importantly, although not only concurrency control but also recovery shall be realized at the same time, their protocol does not take into account the need for recovery mechanism to ensure failure atomicity.

Failure Recovery Although CCR provides 2PC protocol in order to recover from transaction failures, [CCITT92] and [NAK92] state that local transactions do not require it.

Some CMIS/P operations, such as initialization and reboot operations of managed objects, assume that *un-do* operations cannot be executed on the objects [YOD92].

MIB Implementation The concept of managed object is abstraction of physical or logical managed resources. In [NAK91][HEG91][KIR91], management attributes are divided into mainly two types: one is stored and managed by repository like database system, while another is stored and managed by real resource itself, such as equipment or process. The criterion of the classification is based on the aspects of lifetime of attribute and accessing costs. The aspect of multi-process access from concurrency control viewpoint has not been taken into account, however, it should be one of the most important issues for MIB as referred to in [BAP91].

3. Analysis of Information Model and Management Transaction

In this section, we discuss the aspects of management information model and management transactions from the viewpoints of concurrency control and recovery.

3.1. Management Information Model

In each managed object, definitions of attributes, actions, notifications, etc. are included. Especially, the characteristics of attributes and actions are most important for the design of concurrency control and recovery control of MIB.

Management Attributes We classify attributes as follows:

GET-ONLY attribute: For some attributes, only the GET operation is allowed to be performed by a manager. These GET-ONLY attributes are further classified as follows: *static, pseudo-static*, and *dynamic* attributes. Static attributes, for instance serial number and vendor name of managed object, are never updated. On the other hand, for both pseudo-static and dynamic attributes, an agent or a particular process may change the values, although SET operations of CMIS/P may not be performed on the attributes by any manager. Examples of pseudo-static attribute and dynamic attribute are location information or contact information of an equipment, and any status information.

GET-SET attribute: Managers may invoke SET operations to GET-SET attributes. Multiple-access caused by either two or more managers, or a manager and an agent may result in conflicting operations.

Due to the facts mentioned above, it is obvious that GET-SET attributes, and even GET-ONLY attributes of both pseudo-static and dynamic attributes need concurrency control for possible multiple-accesses that result in conflicting operations.

ACTION operation Whereas the GET/SET operation is defined to access the specified attributes directly, ACTION operation manipulates the object, then may change the relevant attributes indirectly. As described earlier, the *un-do* operation cannot be performed for some types of action operations. We refer to these operations as *persistent* operations, since once they have been performed, all of their effects cannot be wiped out from the relevant objects.

3.2. Management Transaction

In the following, we describe the aspects of management transactions related to the design of concurrency control and recovery.

Particular Transaction Type Such transactions are used in the early stages of performance management and failure management, to retrieve relevant performance data and alarm status information etc. Also, they are used to retrieve configuration information, such as the end-to-end relationship which is essential for configuration management. Each of those activities forms READ-ONLY transaction, which means it never changes any information in the managed objects. To assume that such READ-ONLY transactions are common in management systems is quite reasonable.

Different Level of Consistency Although we assume that *serializability* is a sufficient consistency criterion, CMIS/P, in its service definition, allows us to choose *best-effort* synchronization for each operation. In the best-effort synchronization, all managed objects selected for the operation are requested to perform it, even though one or more is not able to successfully perform the operation. We now discuss the possibility of weak level of consistency in particular cases.

Degree-three Consistency: Gray et al. define four levels of consistency [GRA75]. According to the definition, *degree-three* consistency guarantees serializable history. For instance, in reconfiguration of network, the concurrent transactions must be executed with degree-three consistency. Otherwise, interferences among concurrent transactions may result in inconsistent information.

Degree-two Consistency: *Degree-two* consistency can avoid cascading abort [BER87] without necessarily ensuring serializability. In this consistency level, locking based concurrency control may release *shared lock* on reading data at any time. Note that, the transaction may read same data item twice and obtain different results.

This paper assumes *best-effort* operations should be allowed only for READ-ONLY transactions, and then, be handled with degree-two consistency. Some transactions for periodically retrieving such information as statistical data do not need consistent retrieval. In such a case, a manager may choose best-effort (i.e. degree-two consistency) for the transaction.

Different Recovery Schemes Recovery ensures the consistency of the database state, even when failures occur or when transactions abort before completion. The following are recovery schemes described in [BER87].

Recoverability: In order to ensure that aborting a transaction does not influence previously committed transactions, we must require that for every transaction T that commits, its commit operation follows the commit of every other transaction from which T read. Such executions are called *recoverable*.

Avoiding Cascading Abort: *Cascading aborts* occur when transactions read uncommitted data: if a transaction T_i read an object written by another transaction T_j, and T_j aborts, then T_i must be aborted. The cascading aborts may result in severe performance degradation. They can be prevented by requiring that every transaction reads only committed values.

Strictness: If each *read*[x] and *write*[x] operations are delayed until all transactions that had previously issued *write*[x] have either committed or aborted, such executions are *strict*. For the recovery of strict executions, we can employ a simple and commonly used technique, i.e. we save the value of the information before updating, then restore the value when the transaction aborts. This value is referred to as the *before-image*.

4. MIB Design from the Aspects of Transaction Processing

Our MIB takes advantage of (i) the assumption that the MIB systems employ *homogeneous* concurrency control [NAK92], and (ii) the OSI requirement of Two-Phase commit (2PC) in the TP standard.

4.1. MIB Basic Architecture

Considering the result of the information model analysis in the previous section, we describe the basic architecture of our MIB to which locking based concurrency control protocols are adopted.

Locking Granularity To begin with, we discuss locking granularity which refers to the size of lockable data. In [RIE77][RIE79], Ries et al. explore the desired size of a granule by simulation study. It is obvious that a small number of granules inhibits parallelism. On the other hand, a large number of granules will result in excessive overhead. In management information model, each individual attribute is too fine to handle as a locking unit, while each managed object is too coarse since it involves a lot of GET-ONLY attributes to which READ operations are never conflict with each other. Hence, it is desirable to fragment each object by means of grouping attributes. For instance, as illustrated in Figure 2 (a), we can assume each managed object consists of four fragments each of which corresponds to one of classes defined in the part of attribute analysis. Each attribute can be assigned its fragment to which it belongs when describing information model.

Logical Partitioning of MIB MIB is physically partitioned to mainly two areas: repository area and resource area [HEG91][KIR91]. We suggest that within each of these two partitions, they are further logically partioned by whether concurrency control is needed. Static fragments of managed objects, in which only static GET-ONLY attributes are included, do not require concurrency control.

Figure 2 (b) illustrates the partitioning of MIB with object fragments. Note that, all static fragments can be managed by only repository to prevent from costly access to resources, since the fragments are never changed.

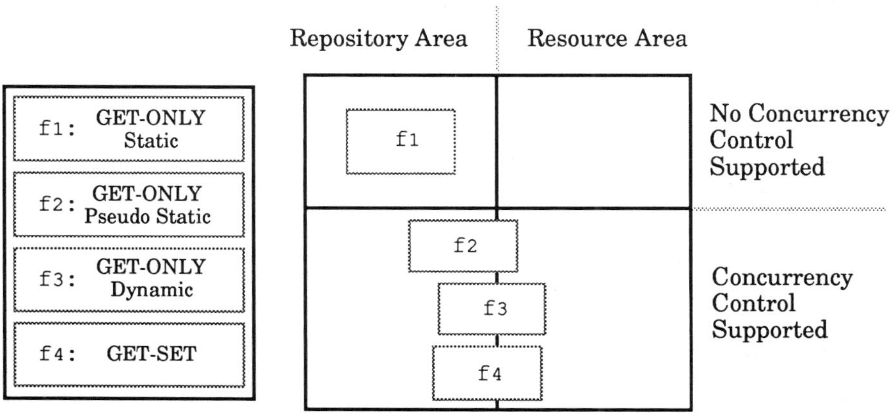

(a) Fragments, f i, of a Managed Object (b) Partitioning of MIB

Figure 2. Fragments of Managed Object and Partitioning of MIB

4.2. Transaction Management Protocol

OSI based commit protocol should be integrated with CMIS/P operation in order to recover from failures. We assume that an arbitrary number of CMIS/P operations (such as M-GET, M-SET and M-ACTION) are encapsulated by Transaction Protocol (TP) and Concurrency, Commit, and Recovery (CCR) protocols in transactions as follows [NAK92]. The transaction has the following construct:

- TP-BEGIN-DIALOGUE
- C-BEGIN
- arbitrary number of CMIS/P operations
- C-PREPARE
- C-COMMIT
- TP-END-DIALOGUE

The services prefixed with "TP-" and "C-" have been defined in TP and CCR, respectively. TP-BEGIN-DIALOGUE can establish a connection between manager and agent, then TP-END-DIALOGUE releases it. The atomicity of CMIS/P operations can be handled using C-BEGIN, C-PREPARE and C-COMMIT services. Each component in the construct is submitted from the manager to the appropriate agent. Note that, C-PREPARE operation is required only for 2PC protocol in case of global transactions.

We assume that a *persistent* operation must be placed at the end of arbitrary number of CMIS/P operations, and its execution is deferred until receiving C-COMMIT operation, since persistent operations cannot be undone when the transaction aborts.

We propose to adopt the following protocol so as to guarantee the atomicity of the transactions:

The manager submits the commit request after having received acknowledgments for the requested operations; when the agent responds with its willingness to commit, the transaction terminates correctly (it must be aborted otherwise.)

For each global transaction, a manager breaks down the transaction into groups of subtransactions, each group containing operations to be performed on a participating agent system. The manager submits a commit request only after the subtransactions have been executed by the appropriate agents. The commit request is preceded by the manager's issuance of C-PREPARE to *all* participants to confirm this condition.

4.3. The Concurrency Control Protocol

In the following we discuss the concurrency control protocol. The concurrency control protocol works on the partitions which require concurrency control in repository and resource areas. Next, we describe those protocols very briefly.

Non-Blocking Two-Phase Locking (2PL) Agrawal et al. introduce *ordered*

shared relationship between locks [AGR90], which attempts to eliminate transaction blocking due to data contention in the 2PL protocol. An ordered shared relationship between locks denoted by $pLock_i\,[x] \Rightarrow qLock_j\,[x]$, allows two transactions to share locks in a constrained manner.
The 2PL protocol with ordered sharing is summarized as follows:

1. The standard 2PL rules.

2. Lock Acquisition Rule: For any two operations $p_i[x]$ and $q_j[x]$ of transactions T_i and T_j , such that $pLock_i\,[x] \Rightarrow qLock_j\,[x]$ is permitted, if T_i acquires $pLock_i\,[x]$ before T_j acquires $qLock_j\,[x]$, then the execution of $p_i\,[x]$ must precede the execution of $q_j\,[x]$.

3. Lock Relinquish Rule: A transaction T_j is said to be *waiting for* T_i if T_j acquired an ordered shared lock held by T_i and T_i has not released *any* of its locks. In this case T_j cannot release any of its locks, and T_j is said to be *on hold.*

In the case of our MIB in a network management system, even a single CMIS/P operation may involve different objects or different object's fragments, hence, $[x]$ means a set of fragments which must be locked as the result of the operation. Combining ordered sharing with the shared and non-shared relationships between GET and SET locks leads to eight possible lock tables [AGR90]. One of the eight lock tables is illustrated in Figure 3.

Locking protocols, in general, suffer from the possibility of deadlocks due to the waiting that may occur in obtaining locks. In the above algorithm, due to the lock relinquish rule, transactions are also subject to waiting that may occur while releasing locks. Note that deadlocks arising from both types of waiting can be handled identically. Hence, any algorithm for deadlock detection can be easily adopted to handle deadlocks that may arise in the 2PL with ordered shared relationship.

Note that, in the case of degree-two consistency, i.e. processing operations with best-effort synchronization, read locks may be released at any time, even though it may violate the aforementioned lock relinquish rule.

Lock Requester T_j \ Lock Holder T_i	$GetLock_i$	$SetLock_i$
$GetLock_j$	$\Leftrightarrow$	$\Rightarrow$
$SetLock_j$	$\Rightarrow$	$\not\Rightarrow$

Figure 3. Lock Table with Ordered Sharied Relationships:
$\Leftrightarrow$ denates shared, $\Rightarrow$ denotes orderd shared, and $\not\Rightarrow$ denote non-shared relationships

Eliminating Cascading Aborts Now consider the ordered shared relationship $SetLock_i \Rightarrow GetLock_j$, which allows a GET operation $get_j\ [x]$ to read attribute after a SET operation $set_i\ [x]$ has executed and before T_i terminates, i.e. commits or aborts. Hence, if T_i aborts, T_j must be aborted. This approach requires a transaction to read the latest value of attribute even if it is written by an uncommitted transaction, potentially causing cascading aborts. The following approach is proposed in [AGR92] to avoid cascading aborts which does not require GET operations to be delayed. When a GET operation is executed, it reads the current committed version of the attribute, thus avoiding the possibility of cascading aborts. The rule of the above idea is described by modifying the lock acquisition rule as follows:

Modified Lock Acquisition Rule: For any two operations $p_i\ [x]$ and $q_j\ [x]$ of transactions T_i and T_j, such that $pLock_i \Rightarrow qLock_j$ is permitted, if T_i executes $pl_i[x]$ before T_j executes $ql_j[x]$, then

if $p_i\ [x]$ is a SET operation and $q_j\ [x]$ is a GET operation, then the GET operation is executed by returning the committed value (before-image) of x, and we say that T_i is *waiting for* T_j as long as T_j has not released *any* of its locks.
Otherwise, the execution of $p_i\ [x]$ must precede the execution of $q_j\ [x]$ and we say that T_j is *waiting for* T_i as long as T_i has not released *any* of its locks.

Note that the SET-GET conflict is being replaced by a GET-SET conflict by inverting the *waiting-for* relation between transactions. Since transactions read committed values, cascading aborts are avoided.

The Recovery Scheme For the reason of simplified implementation of recovery, we employ *unlock-on-commit* policy, i.e., all locks are released at the commit point of a transaction, with the locking protocol. Furthermore, we inhibit ordered shared relationships between SET operations to ensure strictness. Hence, the MIB maintains a maximum two values for each writable attribute, i.e. a committed value (before-image) and perhaps a non-committed value.

We believe the above 2PL protocol with ordered sharing is most desirable for MIB for the following reasons: (i) the concurrency control can be simply realized with OSI based transaction protocol, and more importantly (ii) in the case of protocols with ordered sharing and before-image, deadlocks only involve read-write transactions (i.e. UPDATE transaction), and hence READ-ONLY transactions are never aborted. This cannot be achieved in strict 2PL since READ-ONLY transactions may be involved in deadlocks.

5. Conclusion

This paper provides a basic design of Management Information Base (MIB) which supports transaction processing. In the proposed MIB the locking based protocols with ordered sharing relationship which permit more concurrency control than traditional Two-Phase locking are adopted. An efficient implementation of the proposed MIB in the framework of OSI standard protocols should be the subject of further study.

Acknowledgment The authors wish to thank Prof. D. Agrawal, Prof. A. El Abbadi and M.L. Liu of University of California, Santa Barbara, for their technical support and discussion, and would like to give our special thanks to Prof. J. Bruno of University of California, Santa Barbara, and S. Hasegawa of NEC Corp.

6. References

[AGR90] D. Agrawal, and A. El Abbadi. Locks with Constrained Sharing. 9th ACM Symposium on Principles of Database Systems, 1990.

[AGR91] D. Agrawal, and A. El Abbadi. Ordered Sharing: A New Lock Primitive for Database Systems. Technical Report TRCS 91-18, Department of Computer Science, University of California, Santa Barbara, CA 93106, 1991.

[AGR92] D. Agrawal, A. El Abbadi, and R. Jeffers. An Approach to Eliminate Transaction Blocking in Locking Protocols. ACM SIGMOD-SIGACT Symposium on Principles of Database Systems, 1992.

[BAP91] S. Bapat. OSI Management Information Base Implementation. Integrated Network Management, II, Elsevier Science Publishers B. V. (North-Holland), IFIP 1991.

[BER87] P. Bernstein, V. Hadzilacos, N. Goodman. Concurrency control and recovery in database systems. Addison-Wesley Publishing Company, 1987.

[BRE90] Y. Breitbart, D. Georgakopoulos, M. Rusinkiewicz, and A. Silberschatz. Rigorous scheduling in multidatabase systems. Workshop on Multidatabases and Semantic Interoperability, 1990.

[BRE91] Y. Breitbart, D. Georgakopoulos, M. Rusinkiewicz, and A. Silberschatz. On Rigorous Transaction Scheduling. IEEE Transaction on Software Engineering, VOL. 17, NO. 9, 1991.

[BRU91] Paul J. Brusil and Will Collins. OSI Systems Management Standards. IEEE Global Telecommunications Conference, 1991.

[CCITT92] CCITT Study Group VII, Temporary Document 5341 Rev. Atomic Transaction interaction (SGVII ref: D478/5). Geneva, April 1992.

[CCR90-1] Information technology - Open Systems Interconnection - Service definition for the Commitment, Concurrency and Recovery service element. International Standard ISO/IEC 9804, International Organization for Standardization. 1990.

[CCR90-2] Information technology - Open Systems Interconnection - Protocol specification for the Commitment, Concurrency and Recovery service element. International Standard ISO/IEC 9805, International Organization for Standardization. 1990.

[DU89] W. Du and A. Elmagarmid. QSR: A correctness criterion for global concurrency control in InterBase. 15th International Conference on Very Large Databases, 1989.

[GRA75] J.N. Gray, R.A. Lorie, G.R. Putzolo, I.L. Traiger. Granularity of Locks and Degrees of Consistency in A Shared Data Base. IBM Research, September 18, 1975.

[HEG91] H.-G. Hegering, S. Abeck, and Th. Boehnke. Converting MIB Descriptions into MIB Implementations. IFIP/IEEE International Workshop on Distributed Systems: Operations & Management. 1991.

[KLE88] S. Mark Klerer. The OSI Management architecture: an overview. IEEE Network Magazine, Vol.2, No.2, 1988.

[KIR91] Y. Kiriha, S. Nakai, K. Arima, and H. Kuriyama. An Automatic Generation of Management Information Base (MIB) for OSI based Network Management System. IEEE Global Telecommunications Conference, 1991.

[N1395] ISO/SC21 WG4, N1395. Application Context for Systems Management with TP - working document, Sydney, November 1991.

[NAK91] S. Nakai, Y. Kiriha, Y. Ihara, and S. Hasegawa. A Development Environment for OSI Systems Management. Integrated Network Management, II, Elsevier Science Publishers B. V. (North-Holland), IFIP 1991.

[NAK92] S. Nakai, and M.L. Liu. Concurrency Control for Network Management Transactions based on OSI's Systems Management and Transaction Protocols. IFIP/IEEE International Workshop on Distributed Systems: Operations & Management, 1992.

[RIE77] D.R. Ries, and M. Stonerbraker. Effects of Locking Granularity in a Database Management System. ACM Transaction on Database Systems, Vol.2, No.3, 1977.

[RIE79] D.R. Ries, and M. Stonerbraker. Locking Granularity Revised. ACM Transaction on Database Systems, Vol.4, No.2, 1979.

[UPT91] Francis Upton IV. OSI Distributed Transaction Processing, An Overview. 4th International Workshop on High Performance Transaction Systems, Asilomar, September 1991.

[YEM91] Y. Yemini, G. Goldszmidt, and S. Yemini. Network management by delegation. Integrated Network Management, II, Elsevier Science Publishers B. V. (North-Holland), IFIP 1991.

[YOD92] I. Yoda, K. Sakae, and N. Fujii. Configuration of Local Fiber Optical Network Management System based on Multiple Manager Systems Environment. IEEE Network Operations and Management Symposium, 1992.

Integrated Network Management, III (C-12)
H.-G. Hegering and Y. Yemini (Editors)
Elsevier Science Publishers B.V. (North-Holland)

Modelling Resources for Integrated Performance Management

B. Neumair
Technical University of Munich,
Postbox 20 24 20, D-8000 Munich 2, Germany
e-mail: `neumair@informatik.tu-muenchen.dbp.de`

Abstract

To meet the needs of integrated network management in a heterogeneous environment, it is essential that management information is uniformly defined and aligned with the requirements of management applications. Currently, performance related management information of this kind is not available to a sufficient extent. Therefore, a method to derive homogeneous and adequate information for the purpose of integrated performance management is presented in the following. To start with, a model for performance-related properties of resources is defined that uses a top-down-approach based on the requirements of performance monitoring. As a benefit, the model is independent of particular communication or management architectures and network components. This model serves as a basis for the derivation of unified performance related management information, which is used for the specialization of generic managed object classes. At the end, examples are given that explain how the unified information can actually be provided in an open environment by mapping it to standardized information or to information from other sources.

Keyword Codes: C.2.3; D.4.4; C.4
Keywords: Network Operations; Communications Management; Performance of Systems

1 Introduction

Network management covers methods for the planning, configuration, controlling, monitoring, fault correction and administration of computer networks. The purpose of network management is to support network providers and network users in such a way that the network and its components can be used most efficiently according to the network's quality of service objectives.

Performance management as a functional area of network management is aimed at monitoring and controlling the performance of the network and its components, gathering statistical data and acting upon resources to improve the effectiveness of communication activities. Within this task the following aspects can be distinguished ([14]):

- **Monitoring**: tracking communication activities in order to gather the appropriate data to evaluate performance;
- **Control**: setting and modifying parameters that control the measurement of performance related data (e.g. threshold values to trigger alerts) in remote systems;
- **Analysing**: evaluating the results of the measurements in order to obtain assessments of performance;

- **Tuning**: adjusting resources to improve performance (e.g. modification of resource configuration parameters or redistribution of network traffic).

Quantities that have to be monitored and analyzed in order to evaluate performance are e.g. the **workload** of network components, the **throughput** of resources, the **delay** of data elements within the network and Quality of Service parameters.

Measuring and analysing these quantities in **heterogeneous** networks, i.e. providing **integrated performance management**, is as a result of the sheer variety of devices only feasible[1] if a **common framework** for the definition of **information** basic to the quantities exists. This framework has to be applicable to all resources and has to cope with the **syntactical** and **semantical** aspect of the information. A common syntactical basis is necessary to communicate the information between open systems and uniform semantics are essential to understand and relate the quantities of the various resources in a heterogeneous environment.

Both aspects are covered by current efforts of the International Organization for Standardization (ISO) and the Internet Activity Board (IAB) in the field of network management standardization. Two different **management information models** were defined and, based on these models, concrete **management information** was standardized. Both models are centered around so-called **managed objects** which represent an abstraction of network resources for the purposes of network management, i.e. management in a heterogeneous environment operates virtually on managed objects which form the *management information base*.

The approach of the ISO is fully *object-oriented* with managed objects being instances of managed object classes ([9]). The IAB uses a less complex approach where managed objects are *simple* or *constructed* (tables, lists) *variables* ([21]).

Managed objects that are defined based on either of these two models have to provide information about resources for **all** functional areas of network management. The requirements of the functional areas are naturally different to a large extent and not yet well understood ([5]). As a consequence, a **bottom-up-approach** based on the resources to be modelled was used in most[2] cases for the definition, the result being managed objects containing primarily information that is easy to provide[3] and only secondarily information that is really demanded by management applications. Due to the work of different standardization organizations, the evolving information is mostly heterogeneous (see section 5.1 for an example) and specific with respect to communication architectures (but not to vendors or implementations!) (see figure 1).

In short, the semantic integration of management information has not been advanced to a sufficient extent. By contrast, the syntactic integration made more progress (see e.g. [17]) and allows almost unrestricted exchange of (heterogeneous!) information.

This situation yields the following problems for integrated management applications:

- Since integrated, homogeneous information which spans different communication architectures or different information models is essential for a uniform view onto the whole network, the **necessary mappings** between the information have to be made within the applications. This adds **complexity** to the implementations and **prevents a correlation** of data (e.g. for the operator) from different applications, if these mappings have not been chosen identical.

- After an introduction of new, standardized managed objects all **applications** that will support them have to be **modified** (additional mappings) even if the provided information

[1] of course, this holds for most of the other functions of network management

[2] See section 3 for a discussion of a top-down-approach of the ISO.

[3] this eases implementation and supports fast dissemination ([22]).

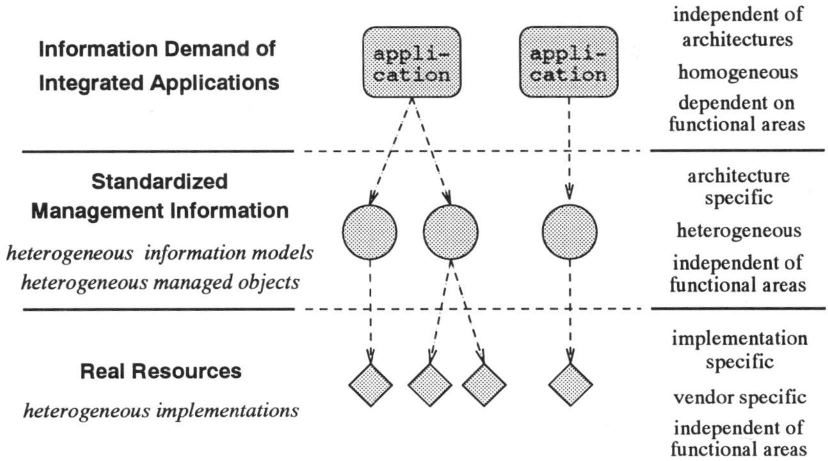

Figure 1: Management information hierarchy

is only needed at a higher degree of abstraction. (In the period from March 91 to May 92 22 Internet-Standards (*Request for Comment – RFC*) were published containing some 700 pages concerned with the definition of new managed objects.)

- Since the information contained in the existing managed objects is low-level in nature due to the bottom-up-approach and the orientation towards the entirety of management functions, **extensive preprocessing** is often necessary to obtain **significant** basic information for certain management objectives. This preprocessing has to be done in a distributed manner to keep up with performance or accuracy objectives; also the definitions of the preprocessing functions are as heterogeneous as the definitions of managed objects. Therefore additional mappings are necessary which add to the complexity of the applications (see page 11 for an example).

In short, at present the existing management information is not adequate for integrated applications. As a consequence, only few applications are available that deal with managed objects from distinct architectures.

2 General Approach

The problems mentioned above suggest the introduction of a new layer between applications and standardized managed objects being abstractions of real resources. This layer must contain **information** which is **homogeneous** and **oriented towards** the **demand**. It therefore provides a **uniform view** on heterogeneous networks and meets the needs of the applications (see figure 2). This layering has the following advantages:

- The **homogeneity** of the information yields a **separation** of the applications from **distinct information models** and **heterogeneous object definitions** for resources that have identical properties as viewed by integrated applications (see section 5.1 for an example). This results in a reduced complexity of the applications. Additionally, applications are largely independent of the introduction of new managed objects (with respect to the details contained therein) and results of different applications can easier be related.

- Information **not included** in **standardized** managed objects but obtainable from other sources (expert knowledge, specialized tools, device specifications etc. – see section 6 for an example) can thus be **provided** to applications in a **uniform manner** together with standardized management information. Hence, the extension of the standardized information and a migration towards future enriched information is eased.

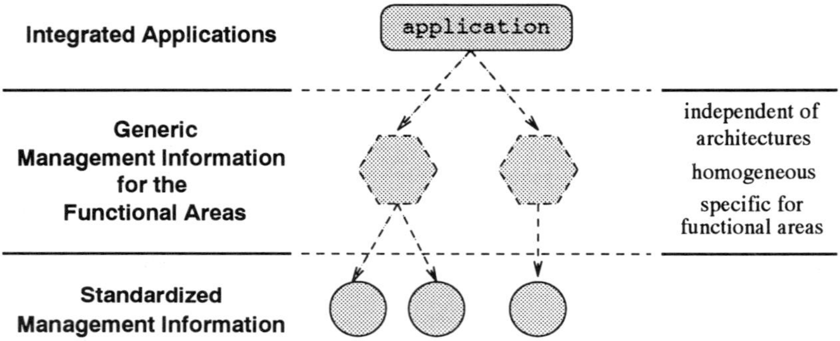

Figure 2: Architectural integration of homogeneous information

To derive information that is oriented towards the requirements of applications, a **top-down-approach** has to be used in contrast to the bottom-up-approach mentioned above. Models that describe the properties of resources from the viewpoints of the functional areas of management set the stage for complete and adequate management information. This information is used for a further refinement of generic object classes stemming from generic network models and presented to applications by extended generic classes.

In the following sections, a model for the derivation of information for performance management is presented. Steps that have to be performed are (see figure 3):

1. Selection of a generic network model which is suitable as a basis for a refinement for performance management, i.e. selection of the resources to be modelled (section 3).

2. Definition of a generic model for performance related properties of heterogeneous resources which can be uniformly applied to all performance relevant resources and is adequate for integrated performance management. The definition is based on an analysis of the information demands (section 4).

3. Derivation of concrete performance-related management information by applying the model of step 2 on the generic resources selected in step 1. This is done uniformly in different degrees of refinement (section 5). The resulting management information is presented to applications by extension of the generic classes.

Section 6 gives some hints on the mapping of the homogeneous information onto standardized information.

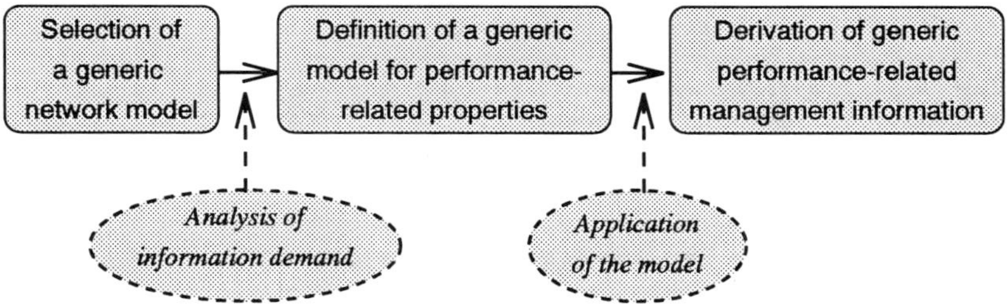

Figure 3: Derivation of homogeneous performance-related information

3 Generic model for communication resources

First of all, a generic network model has to be chosen which is suitable for further refinement for the purposes of performance management. The selection of the object-oriented paradigm is obvious since it is a widely accepted "state of the art" and simplifies the mapping onto standardized management information.

The ISO is currently defining a generic object-oriented model for communication resources ([10]). In the following, these *generic managed object classes* are used as a basis, since they are intended to serve as superclasses for further refinement[4].

These classes reflect the architecture for layered communication systems defined in the OSI reference model. Since this model can be mapped to other existing architectures, these classes are suited for management which spans different architectures ([23]).

From the viewpoint of performance management, only a subset of these classes must be further refined since certain aspects, such as pure container objects (e.g. "**`subsystem`**") or the distinction of two kinds of SAPs, are of no further interest for performance management. The class "`clAssociation`" was newly defined in analogy with "**`singlePeerConnection`**" to represent performance related characteristics of connectionless communication which can not be modelled within "`clProtocolMachine`" due to their dependencies on the communication endpoints.

Figure 4 gives an overview of the object classes used for further performance-related refinement. It should be noted that the classes as defined by the ISO do not contain any performance-related information at all.

4 Modelling performance-related properties of resources

In the following the mentioned **top-down-approach** to model the characteristics of resources from the viewpoint of performance management is shown. Central to this approach is a framework of **generic performance-related quantities** that is derived from the requirements of typical functions of performance management. These quantities are completely **independent of resources**, i.e. they are applicable e.g. to protocol machines as well as to connections. The **small number** of these quantities eases their use in the definition of concrete management information. This is accomplished by the **application** of the **quantities** to the generic managed object classes of the previous section in **several levels** of **detail**.

[4] The selection of this model from the variety of other models can not be discussed here in detail. For a more thorough description see e.g. [23].

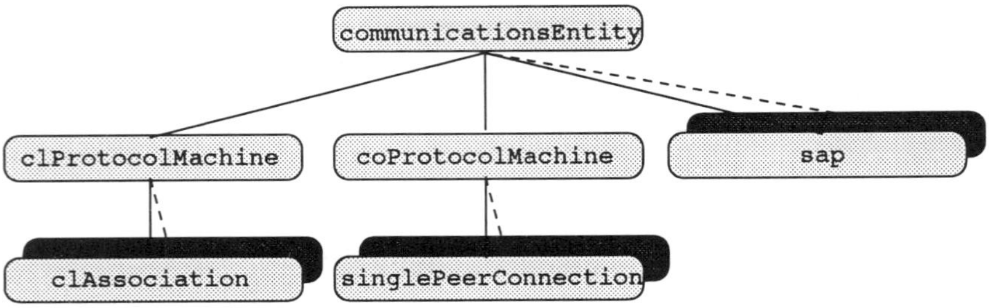

Figure 4: Generic managed object classes

4.1 Information requirements of typical performance management functions

The following list gives a brief summary of data required by functions that are mentioned in most of the publications on this topic (see e.g. [24, 15]) as the "kernel functions" of performance management.

- *Workload monitoring* needs data about the utilization of a resource and the rate of requests for services of the resource. For a (timely) detection of overload conditions, information about the capacity and the maximum tolerable request rate is necessary.
- *Throughput monitoring* is closely related to workload monitoring. As in the previous case, information about the maximum possible throughput is essential in addition to information about the actual throughput itself to recognize undesirable conditions.
- *Monitoring delay* (e.g. transit or roundtrip delay, response times) requires (from the viewpoint of **integrated** performance management) information about actual and maximum acceptable service times of services concerned with the transmission of data (transit delay) or connection management (connection establishment delay is the service time of a request for a connection). This aspect is related to the next item, since maximum tolerable values are part of the so-called "quality of service", which specifies the characteristics of a service as provided by a resource (see below).
- *Quality of service monitoring*, which is widely considered to be an important part of performance management (see e.g. [4]), is only possible if the quality of service parameters are available to management applications. This is often not true with present-day definitions of management information. In addition to the time related quantities mentioned above (e.g. transit delay or connection establishment delay) and the throughput, especially error rates related to transmission of data (residual error rate) or connection management (connection establishment failure probability) are of importance from the viewpoint of performance management.

An analysis of the standardized management information shows that especially quantities for the capacity or the maximum throughput of resources that are heavily demanded by network operators are often not included in the definition of managed object classes.

4.2 Service-related versus protocol-related properties

The OSI reference model (which can be mapped to other architectures as mentioned in section 3) is based on a hierarchical arrangement of service providers. These make use of the services of the immediately underlying layer at semantically standardized interfaces and communicate with peer entities by means of layer protocols. These architectural decisions determine the possible points of interaction at which resources (e.g. protocol machines) communicate conceptually with other resources, namely the service interface and the protocol interface.

Since properties related to these two kinds of interfaces are rather different (e.g. number of user octets sent vs. window sizes or retransmission timers), this distinction is used for structuring[5] of the generic performance related information. It results in a separated modelling of service-related and protocol-related characteristics.

Since most of the information requirements mentioned in the previous section are obviously service-related, only this part of the information is elaborated here in more detail for the sake of brevity.

4.3 Generic performance related properties of the relation between service user and service provider

In order to obtain a high degree of homogeneity for service-related management information about resources, generic quantites are introduced that are independent of a certain type of service and of the class of the resource. In a next step, by applying these quantities to classes of services and resources concrete management information is derived.

The information demand of the "kernel functions" of performance management summarized in the previous section results in the use of the quantities *utilization, capacity, request rate, throughput*, and *service time* to characterize the generic performance related properties of the relation between service user and service provider observable conceptually at the service access points. Since these quantities are used in traffic theory to describe server systems (incl. customers, see e.g. [1]), their use is additionally justified and a formal definition can be omitted here for the sake of brevity.

Both "*current values*[6]" ("cur.") and "*average values*" ("avg.") have to be provided for each of the quantities *utilization, request rate, throughput* and *service time*. The former are needed for an "online" monitoring of resources and a possible early reaction. The latter are averaged out over long periods (e.g. the period since last reset of the resource) and serve as a pragmatic approximation for the expectation to make practical use of some results of traffic theory (e.g. Littles result, see below) in network management. Additionally, "*maximum values*" ("max.") for the quantities *request rate* and *throughput* (note that *capacity* represents the maximum value of *utilization*) are necessary for setting threshold values for online monitoring.

The definition of the *error rate* ($\frac{\textit{number of executions in error}}{\textit{number of service requests}}$) and the maximum (tolerable) service time yields a convenient introduction of quality of service parameters into the model and their possible monitoring by integrated management. (The notion "quality of service" is generally used in the definitions of the layer services, but has not been deeply considered in the definition of standardized management information.)

Figure 5 gives an overview of the introduced performance related quantities.

4.4 Classification of services

For a further development and refinement of the model, the classes of services have to be determined onto which the generic quantities are to be applied. The OSI reference model

[5]Note that the large number of attributes within existing classes is often not structured at all

[6]These are average values for a "short" period. A precise establishment of this period cannot be done on the generic level since these values are dependent on e.g. network provider objectives and therefore have to be fixed by applications

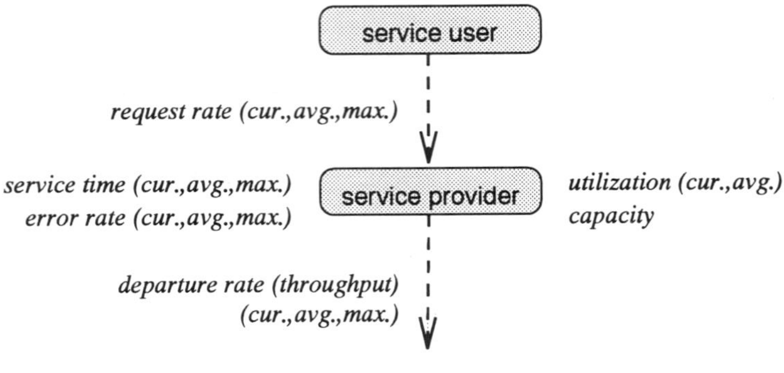

Figure 5: Performance related properties of the relation between service user and service provider

and similar communication architectures provide in general connection-oriented, connectionless and (partially) transaction-oriented modes of communication. These modes result in different groups of service elements at the service interface. Therefore this grouping is assumed for the definition of service classes. The service class related to connection-oriented communication is further elaborated in the following as an example.

With connection-oriented communication, three distinct classes of services, representing two different levels of refinement can be distinguished for performance management purposes. On the first level, a class comprises of the **whole service** "providing connection-oriented communication" with respect to its performance related properties. On the second level, two service classes, namely connection management and data transmission, represent the **phases of communication**.

In the following, the derivation of generic performance related management information for these classes will be shown.

5 Derivation of generic performance related management information

The discussion of the previous section sets the stage for a systematic derivation of concrete management information for the generic managed object classes introduced in section 3. This is done by uniformly applying the generic quantities of section 4.3 to the service classes. For the sake of brevity, the separate treatment according to the direction of communication is not shown.

In the following, two examples will be given for this method that show how:

- **additional performance related information** that is not contained in standardized managed objects can simply be derived through the use of the generic quantities,
- **quality of service parameters** can conveniently be integrated into management information, and
- the homogeneity of the generic quantities yields an **abstraction of different names** of attributes with identical meanings.

5.1 Performance related properties of protocol machines

Important properties of protocol machines are described by quantities related to the set of supported connections (e.g. the number of currently open connections). These quantities are derived by applying the generic quantities to the **whole service** "providing connection oriented communication" (see section 4.4) and relating them to **coProtocolMachine**.

The following list gives some informal, short examples for resulting quantities. Where possible, related standardized attributes as defined in [12] for a transport protocol machine are referenced.

- *Request rate (cur.,avg.)*: Current resp. average number of connection requests per unit of time as experienced by the protocol machine. These rates can be derived from the counters **localSuccessfulConnections** and **localUnsuccessfulConnections**.
- *Utilization (cur.,avg.)* and *capacity*: Number of currently or averaged open connections and maximum number of open connections as allowed by the protocol machine. These quantities relate to **openConnections** and **maxConnections**.
- *Service time (avg.)*: Average holding time of connections. Although, in general, this quantity is of interest for purposes of network planning, a related attribute can mostly not be found within standardized information. However, using the generic quantities and applying "Littles result[7]" to *utilization (avg.)* and *request rate (avg.)* yields an approximation for this quantity. A further application of this formula to the previous result and *capacity* yields a first approximation of the maximum tolerable request rate (*request rate (max.)*) which is of relevance for monitoring purposes. Note that these results are yielded using only the generic quantities — the various standardized attribute names (see table 1) have not to be taken into account.
- *Error rate*: The ratio of connections released by the service provider or otherwise incorrectly executed to the total number of connections. *Error rate (max.)* is identical to the quality of service parameter *transfer failure probability* ([13]).

Besides the convenient derivation of additional values the exclusive use of the generic quantities establishes the unification of performance parameters mentioned above. To state the present-day heterogeneity, table 1 shows the attributes related to the generic quantities *utilization*, *capacity* and *request rate* as they are defined in [17], [11] and [12] for connection-oriented protocol machines for the protocols TCP, X.25 PLP and ISO TP, respectively.

Generic quantity	Related standardized attributes: tcp	x25PLE	coProtocolMachine
utilization	tcpCurrEstab	logicalChannel–Assignments[8]	openConnections
capacity	tcpMaxConn	maxActiveCircuits	maxConnections
request rate	tcpActiveOpens tcpAttemptFails	callAttempts	localSuccessfulConnections localUnsuccessfulConnections

Table 1: Standardized attributes of protocol machines

[7] For a definition of the formula and assumptions see e.g. [1]

5.2 Performance related properties of connections

Using the model of section 4, generic performance related quantities for connections (generic class "**singlePeerConnection**") are obtained by applying the generic quantities of section 4.3 to the service class related to the data transmission phase (section 4.4).

For this, the kinds of data units have to be distinguished to which the services refer. Obviously, octets and packets have to be used. Since in general it is not known in advance whether the number of transmitted octets or the number of packets forms potential performance bottlenecks, quantities related to both classes have to be provided for monitoring. As in the previous section, some informal examples are given and reference is made (where possible) to related standardized attributes contained in [12] for a **transportConnection**.

- *Throughput (cur., avg.)*: Current and average rate of transmitted user data (octets or packets, resp.). The rate related to octets can be derived from the counter **octetsSentCounter**, whereas the rate for packets can only be approximated using **pdusSent** if no segmenting or reassembling is performed.

- *Utilization (avg.), capacity*: The application of these generic quantities yields quantities for the average resp. maximum number of octets or packets simultaneously being in transmission. These typically can not be derived directly from standardized management information. Using Littles formula for *throughput (avg.)* and *service time (avg.)*, *throughput (max.)* and *service time (max.)* respectively approximations for these quantities can be derived which could provide for example hints on configuration errors concerning window sizes[9].

- *Throughput (max.)*, *service time (max.)* and *error rate (max.)* correspond to the quality of service parameters *throughput*, *transit delay* and *residual error rate* (see [13]).

  (Additional quality of service parameters (e.g. *connection establishment delay*) are obtained by applying the generic quantities to the service class concerned with connection management in relation to protocol machines. Note that these quantities, together with the ones previously mentioned, cover all performance related quality of service parameters given in [13]. With this, the quality of service parameters become an integral part of management information.)

6 Mappings and implementational aspects

To make practical use of the generic management information it has to be shown how it can be provided in a heterogeneous environment using standardized management information or additional sources (e.g. specialized tools or expert knowledge).

In the previous section for example, a rate repeatedly had to be derived from standardized attributes of type counter. In a heterogeneous environment, the conversion of counters to rates in the **agent system** is supported by the standards [8] and [25]. Note that due to the publication of different standards that cover identical or similar functionalities, an additional variant of heterogeneity is introduced. This adds to the complexity of applications as knowledge about details of these standards is necessary to make use of them. If the applications make use of the generic performance related management information and appropriate mappings are defined, the implications of this heterogeneity are limited to the layer of mapping between generic management information and standardized information and are therefore transparent to applications.

[9]These are in general important for the performance of a network.

In general, the mapping of a generic quantity can e.g. consist of:

- a simple reference to an equivalent standardized attribute,
- a method to compute the value on the basis of standardized attributes (possibly using additional expert knowledge, see example below) or
- an invocation of a specialized function that provides the value. (This eases **migration** towards **integrated** network management, since e.g. the exchange of the invocation of a specialized function for the invocation of a standardized *systems management function* is not visible to an application.

Instead of giving more details for these mappings (refer to [19]), in the following an example is provided which shows some benefits of the approach.

Example

Monitoring bridges or brouters is an essential part of performance management due to the central location of these devices within the topology of a network. Specialized management information for this class of devices has been defined in [2]. The example shows how these attributes can be used to evaluate the *throughput (cur., avg.)* of the "forwarding entity", which forms the central part of a bridge, and how it can be related to the maximum possible throughput (*throughput (max.)*).

The entity can be modelled as a generic resource of class **communicationsEntity**. The current situation is not appropriate from the viewpoint of performance monitoring in the sense that in [2] no attribute is defined that directly models the throughput of the forwarding entity. Only counters are provided that refer to single, isolated ports (**dot1dTpPortInFrames**, **dot1dTpPortInDiscards**, **dot1dTpPortOutFrames**).

The number of frames forwarded by the bridge therefore can be obtained by

$$\sum_{port=1}^{dot1dBaseNumPorts} (dot1dTpPortInFrames(port) - dot1dTpPortInDiscards(port))$$

or by

$$\sum_{port=1}^{dot1dBaseNumPorts} dot1dTpPortOutFrames(port)$$

By periodic sampling ([8] and [25], see above) and the mentioned calculations, the generic quantity *throughput* for the bridge can be derived.

Quantities related to *throughput (max.)* of a bridge have not been considered in [2]. Nevertheless, this quantity is essential for a significant interpretation of *throughput (cur.)* or *throughput (avg.)*. Using the specification of a bridge (e.g. [20], "transfer rate") a mapping function can be defined that provides the proper values for the generic quantity *throughput (max.)* for the installed bridges depending on the attribute **sysDescr** (see [18]).

The benefit of the approach is that applications have to use **only** the **generic quantities** to perform **integrated** performance monitoring of bridges (and other resources). **Details** like the standardized attribute **dot1dTpPortInDiscards** whose name and semantics are specific to the filter process within the bridging between LAN segments are **not visible** for **integrated** applications.

The present lack of an attribute for the maximum throughput in the standardized information makes it necessary to derive this quantity within integrated applications from other sources, i.e.

details of **specific resources** (the types of bridges) have to be considered. By the introduction of the generic quantities and the associated mappings, these **specifics** are **restricted** to the **layer** where the **mapping** is performed. Therefore, integrated applications are **not affected** e.g. by the installation of new bridge types or an upgrade of existing ones.

Implementational aspects

Due to the application of the object-oriented paradigm, existing infrastructures within **Management Platforms** like the *Communications Infrastructure* within the *HP OpenView Network Management Server* ([7]) can be used to support the access of applications to the generic quantities and the realization of the mapping onto standardized management information or information from other sources. This is done by the implementation of *Object Managers* for the generic classes and their performance related quantities. If an application requests a generic quantity, the *Object Manager* executes the appropriate mapping function depending on the actual (standardized) class of the resource.

Object Managers have been implemented and tested in cooperation with the *Leibniz Rechenzentrum München* that uses existing specialized tools to provide generic quantities such as *throughput* or *service time* e.g. for TCP connections (see [3]).

7 Summary and conclusion

A model was presented that supports the definition of **homogeneous** management information which is **oriented** towards the **needs** of **integrated** performance management applications. The **top-down-approach** is guided by the requirements of performance monitoring and yields management information that is independent of concrete communication or management architectures and resources. This information is particularly suitable to provide a **uniform view** onto the heterogeneous present day information that is defined by different standardization organizations in a vast number of managed object libraries.

The approach **reduces** the **complexity** of applications, since the mappings onto the heterogeneous standardized information no longer have to be integrated into the applications. The practicability of a **correlation** of **results** of **different** applications is improved, since the basic information is homogeneous and the **dependency** of applications on the **progress** of network management **standardization** is reduced. The latter **eases** the **migration** towards open, integrated management. The homogeneity of the information frees network operators from the necessity to **interpret** highly **specialized** information. The approach is however not limited to the field of performance management but can be applied to most functional areas, due to its generic technique.

The usage of results of traffic theory on the level of the generic information gives a means for an **easy extension** of the basic information by additional performance related quantities. The model also provides a convenient **integration** of the *quality of service parameters* into the management information, which is not available up to now.

Acknowledgements

The Munich Network Management Team (MNMT) is a group of researchers of the Munich Universities and the Bavarian Academy of Sciences. Some projects of the team are supported by Siemens AG Munich. The author wishes to thank Prof. Hegering and all team members for helpful discussions and Rene Wies for carefully reading the manuscript.

References

[1] D. Bertsekas and R. Gallager. *Data Networks.* Prentice-Hall, 2nd edition, 1992.

[2] E. Decker, P. Langille, A. Rijsinghani, and K. McCloghrie. Definitions of Managed Objects for Bridges. RFC 1286, 1991.

[3] M. Ebert. Analyse vorhandener Werkzeuge des Performance Management auf ihre Eignung für eine Einbindung in HP OpenView. Master's thesis, TU München, 1992.

[4] K. Geihs and A. Mann. Service Management for Integrated Broadband Communications. In Hegering [6].

[5] H.-G. Hegering. How to find significant management information in a heterogeneous network environment. In *Proc. of the Joint SAIEE/CSSA Intern. Symp. on Network Management,* May 1991.

[6] H.-G. Hegering, editor. *Proc. of the IFIP/IEEE Intern. Workshop on Distributed Systems: Operations & Management.* IFIP, Oct. 1992.

[7] HP Open View Network Management Server Programmers Guide. Manual E1090, 2nd ed., Hewlett Packard, 1990.

[8] Systems Management – Part 11: Workload Monitoring Function. DIS 10164-11, ISO/IEC, 1992.

[9] Structure of Management Information – Part 1: Management Information Model. IS 10165-1, ISO/IEC, Aug. 1991.

[10] Structure of Management Information – Part 5: Generic Management Information. DIS 10165-5, ISO/IEC, May 1992.

[11] Elements of MI related to OSI Network Layer Standards. DIS 10733, ISO/IEC, Nov. 1991.

[12] Elements of MI relating to OSI Transport Layer Standards. DIS 10737, ISO/IEC, Dec. 1991.

[13] Transport Service Definition. IS 8072, ISO, 1986.

[14] Systems Management – Performance Management Working Document – 7th Draft. WD N6306, ISO/IEC, Dec. 1991.

[15] W. Kiesel and K.H. Deiretsbacher. Communications Network for Manufacturing Applications (CNMA). In Krishnan and Zimmer [16].

[16] I. Krishnan and W. Zimmer, editors. *Proc. of the 2nd Intern. Symp. on Integrated Network Management.* IFIP, North-Holland, 1991.

[17] L. LaBarre. OSI Internet Management: Management Information Base. RFC 1214, 1991.

[18] K. McCloghrie and M. Rose. Management Information Base for Network Management of TCP/IP-based internets: MIB-II. RFC 1213, IAB, 1991.

[19] B. Neumair. Towards integrated performance management: an object-oriented approach. In Hegering [6].

[20] Retix. *Local Bridge Installation Guide – RetixGate Bridge Models 2244 and 2255,* 1989.

[21] M. Rose and K. McCloghrie. Structure and Identification of Management Information for TCP/IP-based Internets. RFC 1155, 1990.

[22] M. T. Rose. Network Management is Simple: you just need the right framework! In Krishnan and Zimmer [16].

[23] B. Schott, A. Clemm, and U. Hollberg. An ISO/OSI based Approach for Modelling Heterogeneous Networks. In *Proc. of the 4th Int. Conf. on Information Networks and Data Communication,* 1992.

[24] M. Siegel and K. Sauer. Efficient Network Management for a Highly Distributed System. In *Local Communication Systems: LAN and PBX II, Proc.,* 1991.

[25] S. Waldbusser. Remote Network Monitoring MI Base. RFC 1271, 1991.

Integrated Network Management, III (C-12)
H.-G. Hegering and Y. Yemini (Editors)
Elsevier Science Publishers B.V. (North-Holland)

An Historical Object Base in an Enterprise Management Director

Alex Allister Shvartsman

Digital Equipment Corporation, 550 King Street, Littleton, MA 01460, USA

Abstract

''Who controls the past controls the future'', George Orwell.

A family of integrated systems has been developed within Digital's Enterprise Management Architecture (EMA). DECmcc™ is the part of the implementation of EMA that is both an enterprise management director, i.e., a distributed resource manager, and a framework for developing new incremental integrated management solutions within EMA. An important component of DECmcc is its time-oriented object base that models both the observable state of managed objects as they evolve through time and the historical state of the objects and their attributes. Generic applications developed on the basis of the framework provide polymorphic methods for recording, retrieving and processing historical object data.

Keyword Codes: K.6.4, H.1.m, C.2.4.
Keywords: Management of Computing and Information Systems, System Management; Models and Principles — Time-Oriented Object Data; Computer-Communication Networks, Distributed Systems — Object-Based Resource Management.

1 Introduction

As the size and complexity of distributed enterprise environments continue to grow, the importance of being able to manage such complex environments increases correspondingly. Significant research and development is being directed to the area of enterprise management. The importance of the area manifests itself through standardization, e.g., ISO [10], and open systems, e.g., X/Open [23], activities dedicated specifically to enterprise management.

Distributed resource management in an enterprise of significant size must deal with numerous managed object classes, a large number of managed objects, a very large number of attributes and the correspondingly large number of events and changes in the values of managed objects' attributes. Enterprise must not only operate within the prescribed parameters at any given instant of time, but also evolve in time while dealing with both the planned and unplanned changes in its networks, computing resources and other network-based and sometimes-connected resources.

Enterprise managers need to be equipped with the tools that help them predict the behavior of the resources in the future based on both the snapshot observations and on the cumulative behavior of the available resources in the past. The managers need to be able to understand and explain why certain undesirable states were reached by analysing the conditions that led to the undesirable state. The ability to accumulate historical data and the ability to retrieve and process well-integrated historical data are indispensable to managers who have the goal of controlling the evolution of an enterprise — ''who controls the past controls the future'' (the quote applies whether one intends to control the future for either the malevolent and sinister or benevolent and constructive purposes).

A family of integrated systems has been developed within Digital's Enterprise Management Architecture (EMA) [6]. DECmcc™, the product, is a part of EMA implementation that is both an enterprise management director, i.e., a distributed resource manager, and a framework on which consistent and integrated management solutions can be developed within EMA [4]. New management functions are being implemented on the basis of the DECmcc framework within EMA architecture.

An important aspect of this enterprise management system is its object-based framework that provides for time-oriented dispatching of operations. This serves as a foundation for modules that support past, present and future time views of managed objects. Such fully integrated modules implement a conceptual *diachronic* object base that allows time-oriented object-based modeling of the resources being managed. To maintain the historical part of this diachronic object base, DECmcc provides comprehensive facilities for collection, processing, purging, archival, and export of historical attribute data [3]. Users and applications can maintain and share access to on-line information about the past.

The managed object instances for the existing objects that need to be globally and directly accessible are stored using DECdns (essentially DCE/CDS). The historical data is stored on the domain basis locally (DECmcc domains are user defined sets of global entities [16]). The recorded historical data as well as the "live" enterprise data can be exported to a relational database, e.g., Ultrix/SQL or Rdb/VMS. This is performed in a uniform fashion for any combination of past time, current and future data. The exported database could be co-resident with the director or it could be remote. Polymorphic recording, exporting and analysis methods can be used with (1) the existing managed object classes, (2) any future classes that will be dynamically added to the enterprise environment, and (3) classes that might be dynamically extended with new attribute definitions and the corresponding methods.

In this paper we overview the management model used by DECmcc director and we present the implementation of an historic object base and services. The structure of the paper is as follows. Managed object and management methods definitions and the time-oriented framework are discussed in Section 2. In Section 3 we present the historical object base and we overview the storage and retrieval of historical data. Historical services and object-based modeling of historical services are given in Section 4. In Section 5 we conclude by discussing certain additional usages of historical data and some future plans.

2 Managed object and time definitions

DECmcc models resources as *objects* and it models management operations on the resources as managed object *methods* [6,19]. The objects are "viewed" through a time-oriented *conceptual* object base that models the observable state of the managed objects and as they evolve through time (Fig. A). Any method used to examine or act on an object can be invoked using time parameters to specify method scheduling and to supply scope of interest time to the methods [18]. The object-based framework allows for the historical states of objects to be recorded for subsequent retrieval and processing. It also allows one to implement modules dealing with future time, e.g., future time can be used to indicate interpolation, forecasting or planning. Therefore, this framework provides facilities for implementing a *diachronic* (i.e., past, present and future time-oriented) object base.

2.1 Modeling managed resources and management operations

Objects and methods are known respectively as *entities* and *directives* in the EMA parlance. DECmcc defines an object-based framework for defining classes of managed objects, their attributes and methods. Managed object classes are specified using the Management Specification Language (MSL) [4]. This language is sufficiently expressive to, for example, enable algorithmic translations of the SNMP MIBs ([13]) into MSL [3].

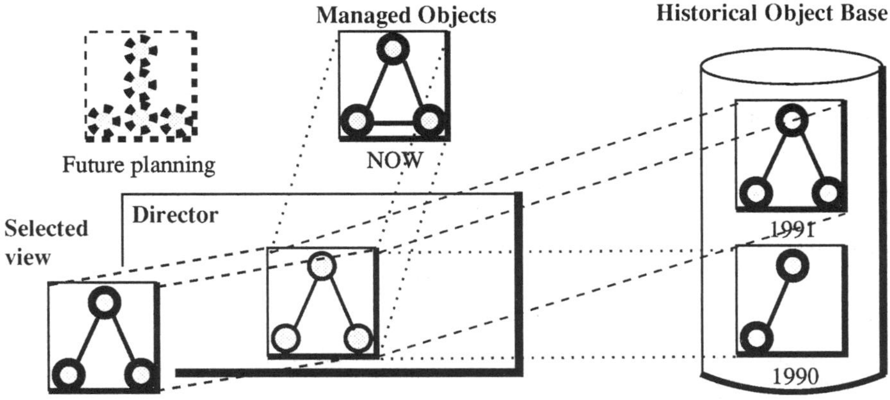

Figure A: User-selected view of objects via Director facilities

Example A: Users can select views of an enterprise based time as illustrated in Figure A. The enterprise management director provides transparent access to historical views stored in an historical object base, to the current state of the enterprise and to possible future views using planning tools or interpolation as well as scheduling actual future views (in the future). All these views are made accessible by qualifying user requests with appropriate scheduling and scope of interest time specifications. Therefore, the director provides facilities for implementing a conceptual diachronic object base. **[end]**

Management Specifications (MS) contain definitions of classes and class interfaces specified in MSL. Class specifications define attributes that are the properties of all managed objects of the class. The attributes are grouped in *attribute partitions*. There are standard partitions, e.g., identifiers, and class-specific attribute partitions defined to model unique features of managed objects. *Event partitions* dealing with events can also be defined. Named methods are defined in terms of *requests* that define the arguments supplied with the invocation message, *responses* that define the arguments that are returned to the caller upon normal completion and *exceptions* that specify reasons for a method failing to perform its operation. This is very similar to the ISO Structure of Management Information [11,21].

2.2 Polymorphic methods: generic applications

In DECmcc, the managed object specifications (which includes attributes and methods) are stored in a managed class dictionary. The class dictionary is used for management module development purposes and it also can be used by methods that need to learn class definitions at run-time. It is also required that certain standard methods, e.g., SET and SHOW, are to be supported by all managed classes. By taking advantage of the class dictionary, it is possible to develop methods that are polymorphic when such method implementations send standard method messages to the managed objects. A polymorphic method can be dynamically *projected* onto the definition of an existing object class by augmenting the MS, i.e., the *specification*, of the existing class with the MS that defines the polymorphic method. Note that a class *implementation* need not be concerned with any future augmentations to the class specification since there is no impact on the original implementation of the class.

The management modules that implement historical functions in DECmcc are specified in terms of *dynamic incremental modifications* to existing object specifications. The methods provided by the historical modules are parametrically polymorphic. That is the directives are implemented as generic functions that can be used in conjunction with *any* objects classes defined using MSL. We return to this subject in Section 4 on Historical Services.

2.3 Specifying time and types of time

The framework defined by DECmcc provides integrated facilities for specifying *time* [4]. Time parameters are commonly specified as method invocation *qualifiers* that refine the methods with respect to a variety of timing requirements. There are four standard and integrated ways in which time can be specified:

- *Timestamps* that indicate the instant of time at which a data value is a valid indication of the corresponding "real" value. It can also indicate the instant of time at which the event described by a data item happened.
- *Scheduling Parameters* that specifies when a method should be invoked. This provides for immediate dispatching of the method, or future dispatching according to the one or more future time points specified. When invoking methods (either by programs or at a user interface) scheduling parameters are specified using the *AT qualifier* on method invocation or at user interface.
- *Scope of Interest* that indicates the time instant(s) or interval for the method being invoked. When used to retrieve information, the scope of interest is used as a lookup key. Scope of interest in the past implies historical information, while future scope of interest may be interpreted by methods as extrapolation. Scope of interest parameters are specified using the *FOR qualifier* on method invocation or at user interface.
- *Algorithmic Parameters* that optionally may be used by methods to qualify their operation in a manner other than those defined above. In general, any of the time specifications given above can be used as the value of an algorithmic time parameter. Algorithmic parameters are specified as method-specific arguments on method invocation and user interfaces.

Comprehensive facilities and data structures are made available to specify time in a variety of ways [4]. Time may be specified in terms of a sequence of (one or more) time instants, or in terms of an interval of time. Scheduling and scope of interest time specifications can be used whenever methods are invoked for any of the managed objects.

2.4 Managed objects, methods and ubiquitous time

Within DECmcc, the methods are invoked on the basis of the <*Verb, Entity, Partition*> tuple, where *Verb* specifies the *method* to be invoked, *Entity* specifies the *object* (or a set of objects using scoping) and *Partition* specifies an *attribute partition* (partition is only specified for some methods). The invoker of any method (a user or another method) can supply *AT* and *FOR* qualifiers to specify the scheduling and/or scope of interest time specifications that was overviewed in the previous section.

Scheduling time and scope of interest time specifications are uniformly processed within the director. The scheduling time is interpreted prior to the actual method invocation. The scheduling of future actions or the retrieval of data at a time in the future is handled by issuing each request to the target method when the appropriate time arrives.

When retrieving historical data, scope of interest is used to select the relevant timestamped data. Requests for scope of interest time in the past result in retrieval of stored historical information. If there is no appropriate historical information to be retrieved, the request will be passed on to the appropriate management module for processing. This allows for seamless integration of modules that implement and provide local historical database services. In all other cases, time specifications are passed on to the method being invoked enabling further enriching of the integrated historical facilities.

As implemented, *time* is ubiquitous. The scheduling and scope of interest time components are *always* present, explicitly or implicitly, when a method is invoked. For example, when a user wishes to examine the COUNTERS attribute partition of the entity HOST XYZ might issue the following command at the user interface (using a simplified command line syntax):

```
MCC> SHOW HOST XYZ ALL COUNTERS
```

This is exactly equivalent to the following:

```
MCC> SHOW HOST XYZ ALL COUNTERS, AT START NOW, FOR START NOW
```

Where scheduling qualifier AT START NOW means that the SHOW method is to be invoked immediately, and the scope of interest qualifier FOR START NOW means that current values of the counters attribute partition are to be obtained. To show past time data, a FOR scope of interest is specified in the past. For example,

```
MCC> SHOW HOST XYZ ALL COUNTERS, FOR START 11-MAY-1992 15:00:00
```

invokes the SHOW method to obtain counters data for 3 PM on May 11th, 1992 (in the user's current timezone). Next example sets up a future method invocation:

```
MCC> SHOW HOST XYZ ALL COUNTERS, AT START 1-JAN-1994 00:00:00
```

These examples show only the simplest uses of scheduling and scope of interest qualifiers when examining managed objects. The important point is that the ''live'' data access, future data access and access to historical data are implemented consistently and are performed transparently to the users and other applications needing this access. The available time-oriented facilities simplify the development of applications that need to deal with time. In the current implementation, the time used to specify the scope of interest and scheduling of requests is specified in terms of the time at the system issuing the request. This assures that the recorded timestamps reflect consistent chronology of events as observed by a given system.

3 An historical object base

Time-oriented databases attracted a fair amount of research and development in the past decade as it has been recognized that time is an important dimension of modeling reality. A taxonomy of temporal databases was surveyed by Snodgrass and Ahn [14]. Such databases provide temporal extensions to the conventional ''snapshot'' databases that do not provide integrated facilities for recording and processing past configurations of the environment. A comprehensive bibliography of temporal databases is being catalogued by Snodgrass and Stam [15]. Historical relational databases have been surveyed by McKenzie and Snodgrass [12]. The authors present some of the reasons why it is not sufficient to merely represent time as an attribute of a relation in order to implement an historical relational database.

Clifford and Croker present a database that explicitly captures the *lifespans* of objects and they present an algebra based on lifespans [2]. Our historical object base uses similarly defined lifespans to record object creation and retirement times and we also use timestamps to tag historical snapshots of objects' attribute partitions.

For the purposes of enterprise management, a distributed environment consisting of objects is best modeled using the object-oriented paradigm [24]. An in-depth treatment of a comprehensive object-oriented (but not time-oriented) database is given by Bancilhon et al. [1]. Our historical object base combines some aspects of time-oriented databases and object-oriented databases in an integrated package specifically geared for enterprise management.

3.1 The Management Information Repository

The database utilized by DECmcc for all its persistent storage is called the Management Information Repository (MIR). The main categories of information stored in the MIR are the managed object class data, object instance data, historical attribute, event, etc. data and miscellaneous data. In this paper we are primarily concerned with the object instance data and historical data that together constitute an historical object base within the DECmcc director.

This historical object base implements a time-oriented information store that incorporates time as a property of objects and their attributes. The MIR stores the *lifespan* data for each stored object instance and a timestamp for each observation of the object's attributes. It also provides the ability to retrieve object instances and attribute values, handling scope of interest time specifications of instants or intervals entirely in the past.

When accessing the MIR, time is specified in the **time**, **lookup_time**, and **actual_time** parameters. The MIR uses the **time** parameter as an optional input timestamp when creating entity instances, writing historical data, and retiring entity instances. The MIR uses the **time** parameter as an optional input parameter when dealing with and acting on historical data.

The MIR uses the required input **lookup_time** parameter when reading historical data. The MIR uses the **actual_time** parameter as an output timestamp when reading historical data. The precise definitions of the object base primitive operation can be found in [4].

For existing objects, the identification attributes and some static attributes (that extend the set of original attributes provided by the objects) are stored in DECdns portion of the MIR. We now address the structure of the past time object base. Historical object and attribute data is stored within the MIR in sub-repositories of two types:

1. Historical Instance Repository (HIR) containing instances and associated lifespans
2. Historical Attribute Repository (HAR) containing timestamped attribute partitions

Each pair of HIR/HAR repositories is associated with a domain and is stored on a per-domain basis in domain-defined locations [16]. Among other advantages of domain-based historical repositories is that this allows for a natural way of securing data, assigning ownership of the data and accounting for storage used (see Fig. B).

3.2 Historical Instance Repository (HIR)

The historical object instance data in HIR contains information about the configuration of the environment, i.e., which objects exist and when. The instance data allows to maintain an internal representation of the configuration, which may or may not match reality at any given time. However, this presents a view of the configuration to the user without needing to access the network. This is particularly useful where some of the network components are inaccessible and the manager is attempting to diagnose the problem.

Information about object instances needs to be related to the times during which objects are known to exist. By providing a lifespan for the instance data, we maintain information about past configurations, the current configuration, and allow future configurations to be planned.

Object instances stored in the HIR are keyed on a symbolic object specification. Each instance entry contains an internal repository key into the Historical Attribute Repository (HAR) where the attribute data for the particular instance is stored. When, in the context of a given domain, an object becomes known, entries in the corresponding HIR are created that contain the identifiers of the object. The lifespan data about the object will contain the time of object creation, while the object retirement time is initialised to indicate that the object exists (a symbolic timestamp of infinity is used). When recording of historical data for a given entity is started, timestamped attribute records are stored in HAR for that domain. When it is detected that an object identifier is no longer present in the name space, then the corresponding identifier is retired in HIR by timestamping its retirement time with the current time.

3.3 Historical Attribute Repository (HAR)

The historical object attribute data in HAR contains information about selected objects in the enterprise. This attribute data may be obtained by polling the network periodically for the required information, by receiving events from the network, or by observing the results of operations requested by the director or user. Historical data is timestamped, to facilitate subsequent retrieval and processing.

The historical data is collected by a specific management modules. This allows the manager to define the information to be stored, the objects for which the information is to be stored, and the frequency of data collection where polling is employed. The data thus stored is made available to other modules via the standard mechanisms that include time specification. This means that stored historical data may be processed multiple times and by multiple management modules without the need to refer to the network for the raw data more than once.

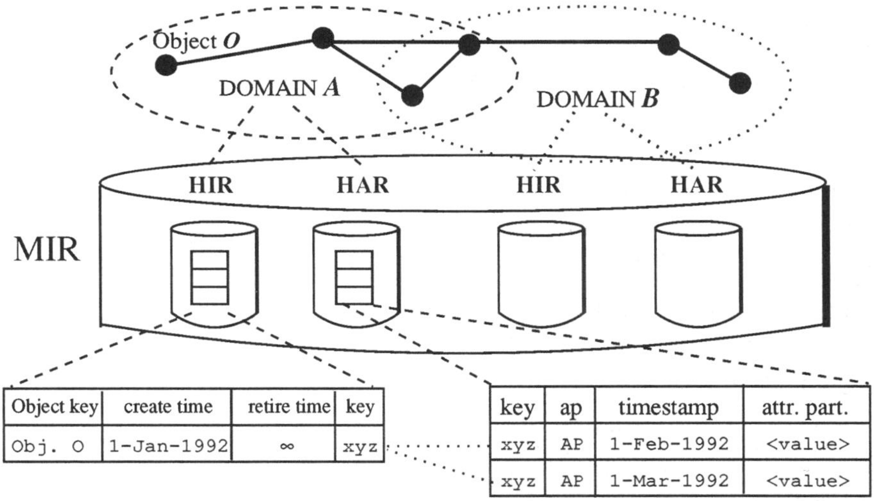

Figure B: MIR and Historical Repositories

Example B: Figure B shows a simplified view of an enterprise consisting of several interconnected network based resources. The enterprise is managed as two overlapping domains, *A* and *B*. Object *O* is in domain *A*. The historical object base is represented by HIR and HAR repositories for each domain. For domain *A*, HIR indicates that object *O* was created on 1-Jan-1992 and that it currently exists (retire time is infinity). HAR contains two historical sets of attributes. The key *xyz* is generated internally to map historical entries in HIR to historical attribute partitions stored in HAR. Entries in HIR are accessed using a composite key consisting of the internal key (*xyz*) and the registered code for the particular attribute partition (in this case the one named *AP*). **[end]**

3.4 Retrieving historical data

Historical attribute data is retrieved using the standard SHOW method. Retrieval returns a (possibly empty) set of timestamped attribute data. Consider the following archetypal query for the attribute partition `<Ap>` of `<Object O>` between the times `t1` and `t2`:

```
MCC> SHOW <Object O> ALL <Ap> FOR START t1, UNTIL t2, IN DOMAIN D
```

First the HIR for domain D is searched for the object(s) whose identifier specification matches the object in the query and such that the stored *lifespan* has a non-empty intersection with the range $[t1, t2]$. For any retrieved object instances O in the domain D, let D^O denote the set of historical attribute partitions stored in HAR. The set H of historical attribute partitions Ap_t that is returned as the result of the query is defined as follows:

$$H = \{\ Ap_t \in D^O\!: \ t1 < t < t2 \ \vee \ t = \max_t\{t \leq t1 \wedge Ap_t \in D^O\} \ \vee \ t = \min_t\{t \geq t2 \wedge Ap_t \in D^O\}\ \}$$

That is, H is a set timestamped attribute partitions that include all partitions with timestamps within the open range $(t1, t2)$ as well as one partition just outside the range (if any).

Example C: To illustrate the scope of interest time specifications and historical attribute data retrieval, we consider the following. Let the attribute partition Ap contain the data of interest for object O in domain D, given the data in HAR as shown in Figure C:

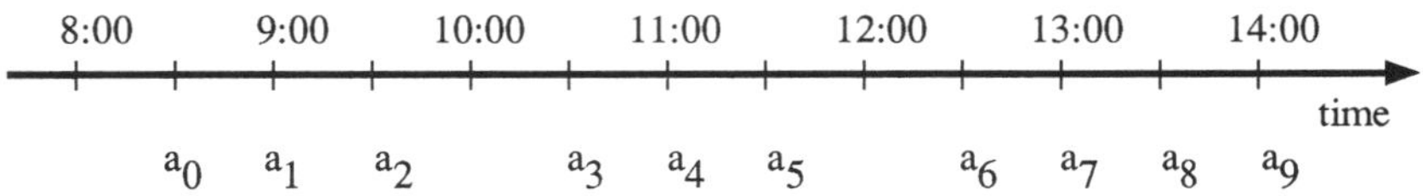

Figure C: Historical attribute partition entries in HAR for object O in domain D

Query: `SHOW O ALL Ap FOR START 12:00, IN DOMAIN D`
Result: $\{a_5, a_6\}$. Here, no attribute partition exactly matches the timestamp 12:00.

Query: `SHOW O ALL Ap FOR START 8:00, UNTIL 13:00, IN DOMAIN D`
Result: $\{a_0, a_1, a_2, a_3, a_4, a_5, a_6, a_7\}$.

Query: `SHOW O ALL Ap FOR START 8:30, UNTIL 14:00, EVERY 1:00, IN DOMAIN D`
Result: $\{a_0, a_2, a_3, a_5, a_6, a_8\}$.

In the next section we present an overview of the historical services and give examples of historical data retrieval defined above.

4 Historical services

The available DECmcc historical services include: (a) periodic recording of object instances, lifespans and historical data, (b) retrieval of historical data (defined above), and (c) exporting data and information, i.e., processed data, to external databases.

Here we overview the historical methods provided for all managed objects. All services are accessible as managed object methods via DECmcc API and standard user interfaces via either the command line or iconic network visualisation maps. The historical services can be characterized as either *periodic* and *one-time* services:

- Periodic services are those that require persistent background processing at defined intervals to record or export of data. This includes any associated services needed to examine and affect this background processing; these services either create and manage background processes or modify data structures associated with these processes.

  The basic periodic service is either a *recording* or an *exporting* request. Periodic services allow the users to initiate recording, and to manage recordings using stop, suspend, resume, show and modify services. Similar features are provided for exporting current and future data. The size of recorded data is controlled using auto-purge requests.

- One-time services are those that are performed once in response to a method invocation and involve no persistent processing. This includes archive and historical extract requests. The services in this group allow the users to purge, archive, restore, and export (past time).

4.1 Modeling historical services

The Management Specification (MS) that defines historical services consists of the definitions of polymorphic methods. The MS does not define any objects. These methods are then used in conjunction with any class MS that defines objects with standard examine methods for any attribute partition, including any non-standard attribute partitions.

In order to make historical services available for all objects, the historical methods MS is included in a *common* class that is inherited by all other classes. Any future object class will also automatically inherit these methods. The dynamic nature of such incremental modification is facilitated via the use of run-time class dictionary provided by the director and by the late binding of *self* in the parametrically polymorphic methods. In the future, we also foresee a special MS that optionally modifies any interested object classes via multiple inheritance.

4.2 Recording historical data

The methods that implement historical recording services provide periodic storage of timestamped object attributes data for subsequent examination or processing. It is not possible to know *a priori* what attribute data for what objects may need to be collected and stored, and how often, on behalf of what user (though it is quite reasonable for us to establish sensible defaults). Only the users understand the policies that they need to apply in their environment; for example, which router is connecting the president's office, versus the developers' offices, to the rest of the network and is therefore deserving of attention.

The users define the specific attribute partition for the given object(s) to be recorded and the frequency of collecting that data. The data is stored in an historical object base associated with the domain and this data is subsequently available for retrieval as already described.

Once set up, recording requests can be displayed, modified, for example changing the polling period, suspended, resumed or terminated. The recording operations take place in a background process and hence continues after the user exits DECmcc. In addition, recording requests are maintained across system reboots, so once the user enters a recording request, it will be in effect until either the request expires or a user terminates the request.

4.3 Exporting services

The exporting services permit users to export historical and live attribute data that is available from or can be produced by Management Modules (MMs) supporting standard examine methods. The exported data is stored externally in Ultrix/SQL or Rdb/VMS using an automatically defined schema to enable the use of a query language to analyze/reduce the data according to user needs, for example for regular reports. The external export databases can be local or remote using standard database technology.

The user can issue a request for a periodic export of data according to the policies set as the result of the decisions made by the users. The user may also request a one time extract of historical data for ad-hoc studies. Export requests can combine the export of past time data and data scheduled for future export.

Once the external database is set up, it is used by exporting services in a WRITE-ONLY mode, i.e., these services only write to that database. This enables the user to have full flexibility in developing custom means for dealing with the data in this external database, e.g., the user can reduce data without affecting the exporting services..

4.4 Purging and archiving

The purging and archiving services permit the user to manage the volume of historical data maintained on-line. For many configurations, it is not reasonable to store all attribute data about all components in the network at any useful interval (such as every hour). This would consume too much disk space. For example, on a DECstation 3100, the system may performs up to 120 network polls per minute. Each poll returns an attribute partition that typically contains several hundred octets of data. At this rate, over 2,500Mbytes of data will be accumulated in one month. Thus the users need to be able to specify to purge and archive data.

The volume of historical attribute data is managed in one of two ways: (1) Data can be automatically purged. Users can optionally set a parameter for a recording request. For example this can be used to maintain a 30 days' worth of data. Defaults are also provided for the users to not unwittingly consume large quantities of disk space. (2) Data can be selectively purged. This allows the user to remove data that is older than some specified time or according to a specified time interval for selected objects and attribute partitions. In the future, an integrated set of archiving services may also be provided.

Should the users need to maintain very long selected histories on-line, custom MMs can be developed to process and reduce historical data that can then be exported to an external database in a more compact form. Alternatively, the data can be first exported and then reduced in the external database.

5 Discussion and futures

We have described the historical object base and the diachronic services defined within the framework of the DECmcc director. Most of the features have been implemented in a product [3,4], and additional development is underway. Here we give additional examples of how the fully integrated historical object base can be used to advantage by enterprise manager. We close with some future development opportunities.

5.1 Historical data, alarms and performance analysis.

A generic management module (MM), called Alarms [3], provides a set of rule-based services that can be used to further advantage with historical services. Suppose a system manager detects an unusual state in a system. To monitor any future conditions such as the one observed, the manager might use the Alarms module and create one or more rules that will alert the manager, e.g., by changing a color of an object depicted on a map or by sending a mail message, when a similar condition would have occurred in the future. If the manager also would like to check whether this condition occurred in the past, the manager can use the same set of rules, but to apply it to historical data using appropriate scope of interest time. The manager can also test any rules with past time data to check whether the resulting notifications are adequate based on the past performance of the system.

In the historical repositories, it is preferable to store ''raw'' data, rather than data that has been processed by a MM, for example, to store counters as obtained from the objects rather than computed statistics. The stored historical data can be re-processing by the same MMs that process "current" network data. This is of course guaranteed by providing uniform access to current and historical data rather than requiring special code in each MM. In addition, by storing raw data, other MMs can be used to reprocess the data subsequently. For example, users can compute performance statistics using Performance Analyzer (PA) MM [3] using duration periods of varying granularity to converge on a period where performance deviated from the norm. Any raw or processed data can be exported to local or remote external databases for specialised processing or reduction.

5.2 Historical services and evolution of enterprise

The definition and implementation of the historical object base and historical services take into account that enterprise need to evolve and grow and that in the future new managed object classes will need to be managed. All historical recording and exporting services were designed so that any future implementations of managed object could *inherit* the historical service methods in their entirety. Thus, the data and computed information produced by future management modules and managed objects can be recorded and exported to external databases. For example, such is the case with PA MM that uses historical data to compute statistics of varying granularity, e.g., hourly, daily. The information produced by PA can be exported. In general, any future MM that augments any class definitions with new (or computed) attribute partitions can be taken advantage of by the exporting services to export the new type of information to external databases.

5.3 Future possibilities and ideas

Future versions of historical services might address the following areas of opportunity:

- Historical recording of events or operation results, e.g., test, can be developed as a natural extension to the existing recording methods. In fact the historical repositories have been designed for such incremental addition of new types of historical data. We may also address intelligent retrieval and correlation of test results or results of other past operations.
- The stored historical data allows the distributed resource manager to create a natural way of interpreting and visualising historical data. This might be implemented along the lines of *time lapse photography* or using a historical visualisation of processes approach such as that developed by Halstead et al. [9].
- A more extensive set of queries might be provided with future versions. Also, more flexible time specifications than simple intervals, e.g., to permit recording data (via a single request) every hour 8am to 5pm, Mondays to Fridays, and every four hours otherwise.
- Dynamic definition of historical recording: for example, a dynamic collection of entities can be defined such that historical recording is done for all entities in the collection that itself evolves over time. Alarm rules might be used to generate events that affect the frequency and extent of recording using user-defined policies related to overall performance.
- The object based framework might be evolved to extend the available inheritance features to permit greater flexibility in defining polymorphic methods and in specializing types.
- Finally, DECmcc is poised to offer rich distribution facilities as described by Strutt [17]. Historical object bases based on domains have been designed with future distribution in mind and along with director distribution they will provide an extensible and comprehensive collection of historical services for the enterprise managers.

Acknowledgments

The list of people who contributed to the definition and implementation of the systems that are the subject of this report is too long to provide here. Yet any list would be incomplete without the names of Sam Shmuylovich, Matt Guertin, Ted Hupper, Roy Lane and Peter Savage. I also thank Steve Wong for valuable discussions and Colin Strutt and Jim Swist for detailed feedback on the previous draft. Finally, I thank the referees for the thoughtful comments that lead to an improved paper and for suggesting the adjective *diachronic* that captures the *bitemporal* character of our object base.

References

[1] F. Bancilhon, C. Delobel, and P. Kanellakis, *Building an Object-Oriented Database System, The Story of O_2*, Morgan Kaufmann, San Mateo, CA, 1992.

[2] J. Clifford and A. Croker, "The Historical Relationship Data Model (HRDM) and Algebra Based on Lifespans", in *Proc. of the Internat. Conf. on Data Eng*, IEEE Computer Society, pp. 528-

537, 1987.

[3] Digital Equipment Corporation, *DECmcc -- Developers Toolkit Reference*, Order No. AA-PD5MC-TE, Maynard, 1992.

[4] Digital Equipment Corporation, *DECmcc -- Historical Data Services Use, Alarms and Notification Services*, Orders No. AA-PJWPA-TE, AA-PD1DC-TE, 1992.

[5] Digital Equipment Corporation, *DECmcc System Reference Manual*, Vols. 1 and 2, Order No. AA-PD5LC-TE, AA-PE55C-TE, Maynard, 1992.

[6] Digital Equipment Corporation, *Enterprise Management Architecture -- General Description*, Order No. EK-DEMAR-GD-001, Maynard, 1989.

[7] A. Dittrich, ''Composite Managed Objects,'' in *Proc. of 2nd IFIP Symp. on Integrated Network Management*, pp. 789-800, 1991.

[8] L. Fehskens, ''An Architectural Strategy for Enterprise Management,'' in *Proc. of 1st IFIP Symp. on Integrated Network Management*, pp. 41-60, 1989.

[9] R. H. Halstead, D. A. Krantz, P. G. Sobalvarro, ''MulTVision: A Tool for Visualizing Parallel Program Executions'', extended abstract, CRL, Digital Equip. Corp., 1992.

[10] ISO/IEC 7498, *Information Processsing Systems, Open Systems Interconnection, Basic Reference*, Part 4: *Management Framework*, International Standard, 1989.

[11] ISO/IEC 10165-1,4, *OSI Management Info. Services — Structure of Manag. Info. — Part 1: Management Info. Model* and *Part 4: Guidelines for the Definition of Managed Objects*, International Standards, 1990.

[12] L. E. McKenzie, Jr. and R. T. Snodgrass, ''Evaluation of Relational Algebràs Incorporating the Time Dimension in Databases'', *ACM Comput. Surveys*, Vol. 23, No. 4, pp. 501-543, 1991.

[13] RFC 1156, *Management Information Base for Network Management of TCP/IP-based Internets.* K. McCloghrie and M. T. Rose (Eds.), 1990.

[14] R. Snodgrass and I. Ahn, ''Temporal Databases'', *IEEE Computer*, pp. 35-46, vol. 19, No. 9, 1986.

[15] R. Snodgrass and R. B. Stam, ''A Bibliography on Temporal Databases'', *Data Engineering*, IEEE Computer Society, vol. 7, no. 4, pp. 231-239, 1988.

[16] C. Strutt, "Dealing with Scale in an Enterprise Management Director," in *Proc. of 2nd IFIP Symp. on Integrated Network Management*, pp. 577-593, 1991.

[17] C. Strutt, "Distribution in an Enterprise Management Director," in this volume, 1992.

[18] C. Strutt and D. Shurtleff, ''Arhcitecture for an Integrated, Extensible Enterprise Management, '' in *Proc. of 1st IFIP Symp. on Integrated Network Management*, pp. 61-73, 1989.

[19] C. Strutt and J. Swist, "Design of the DECmcc Enterprise Management Director," to appear in *Digital Technical Journal*, 1992.

[20] M. Sylor, ''Guidelines for Structuring Manageable Entities,'' in *Proc. of 1st IFIP Symp. on Integrated Network Management*, pp. 169-184, 1989.

[21] M. Sylor, ''DECnet/OSI Phase V Network Management,'' in *Proc. of 2nd IFIP Symp. on Integrated Network Management*, pp. 57-69, 1991.

[22] R. F. Valta, ''Design Concepts for a Global Network Management Database,'' in *Proc. of 2nd IFIP Symp. on Integrated Network Management*, pp. 777-788, 1991.

[23] X/Open, *Systems Management: Problem Statement*, X/Open Snapshot, Doc. No. XO/SNAP/91/010, 1991.

[24] X/Open, *Systems Management: Reference Model*, X/Open Guide (draft), Doc. No. G207, 1992.

I.

MANAGEMENT ARCHITECTURE

E.

APPLICATIONS OF DIRECTORY SERVICES

Integrated Network Management, III (C-12)
H.-G. Hegering and Y. Yemini (Editors)
Elsevier Science Publishers B.V. (North-Holland)

A Service for Administering Management Information

- VPN Inter-Domain Management Support using X.500 and X.700 -

Ursula Meyer zu Natrup, Michael Tschichholz and Susanne Waßerroth

GMD FOKUS, Hardenbergplatz 2, D - 1000 Berlin 12, Germany
E-Mail: tschichholz@fokus.berlin.gmd.dbp.de

Abstract

This paper describes the concept of the Management Information Service (MIS) that supports the management of globally distributed applications and telecommunication services. It is based on an integrated global information model which combines the features of the X.500 and X.700 series of CCITT recommendations. It provides information handling capabilities for human administrators and management applications required for end-to-end service management of, for example, Virtual Private Networks (VPN) in an Integrated Broadband Communications (IBC) environment.

Keywords

IBC, VPN, X.500, X.700, TMN, End-to-End Service Management, Inter-Domain Management

1. Introduction

In the nineties the management of globally distributed networks and services will become a major factor of the telecommunications industry. Several standardization bodies, international or national research programs or projects (ISO, CCITT, ETSI, EURESCOM, RACE, ESPRIT, BERKOM, etc.) as well as industry are going to great effort to find solutions for the management of distributed systems. During the last few years three important sets of international standards have been developed which can be used to solve parts of the given management problems:

- Telecommunication Management Network (TMN) [1]:

  The TMN recommendation describes the general architectural principles for a Telecommunications Management Network (TMN) to support the management requirements of administrations to plan, provision, install, maintain, operate and administer telecommunication networks and services.

- Systems Management [2]:

  OSI management (X.700) is designed to support users and operators of communication services within an open system environment in the planning, organization, monitoring and control of communication. It defines an information model, management information services and several systems management functions.

- Directory [3]:
  The X.500 Directory, which may basically be regarded as a globally distributed database, represents an important management component because it provides a globally unique namespace and location-independent access to globally relevant management information.

A set of useful architectural concepts, information and service models are provided by each of the above mentioned ISO standards or CCITT recommendations. Because of the independence of these groups, the concepts, models, terminology, etc. differ and no full integration has yet been carried out. Fortunately the basic concepts or models can be combined, which has already been shown by the integration of X.700 into the TMN architecture. An integration of management and directory is also required but has not been worked out in detail by standardization bodies.

The Management Information Service (MIS) is an attempt to integrate X.500 and X.700 in order to provide the advantages of both systems. For cooperative management of end-to-end services [4] a first implementation of the MIS [5] will be done in the framework of the PREPARE project[1]. The goal of PREPARE is the development of an end-to-end service management system for an Integrated Broadband Communication (IBC) environment based on the TMN architecture. The MIS will be used to allow access to relevant information for end-to-end service management via a unique interface. Within PREPARE, Virtual Private Networks (VPN) are used as an example to demonstrate end-to-end service management.

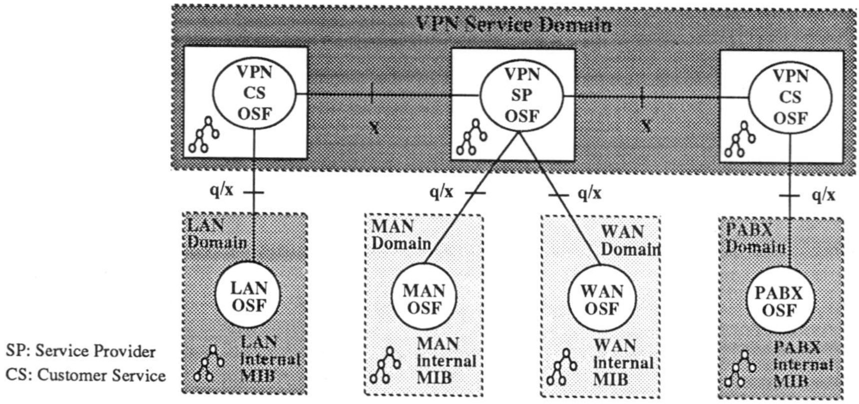

Figure 1: Management architecture for end-to-end service management

The VPN management architecture shown in Figure 1 is based on TMN [1]. It covers the different public and private network domains as well as the VPN service domain [6]. It shows how VPN management components can be decomposed into different functional building blocks (*Operation System Function - OSF*) according to TMN´s functional architecture. Regarding

1. The project "PREPilot in Advanced REsource management" is funded by the European Community as part of the RACE II program.

management-relevant information, a VPN is most suitably separated along the lines of existing administrative structures using the domain concept. The architecture distinguishes four different types of public or private network domains and one or more end-to-end service domains, which will all be important for future telecommunication systems.

Since VPNs are defined as telecommunication services, the management of VPNs is mainly related to service management. In a global IBC environment, VPN management also has an impact on the business level (global naming, bilateral contracts, service migration, etc.).

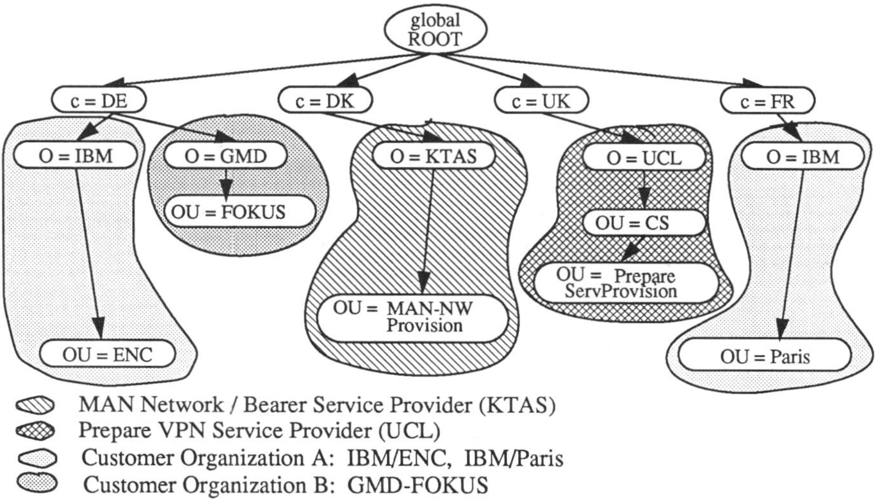

Figure 2: Representation of VPN Domains in the X.500 Directory

As shown in Figure 2 the directory is used to link the globally distributed domains of the network and VPN service provider and the customer premises [7] and stores information about the service provider and customer organizations, domains, persons, service descriptions and addresses of operation systems or application processes. Service management is of course related to network and network element management. Operation on network MOs will only be executed autonomously in the specific network domains (*intra-domain management*). Operations executed via the X-interface on service MOs (*inter-domain management*) can be mapped internally to network MOs.

This paper presents an extended TMN architecture which covers the problems of inter-domain end-to-end service management. An integrated information model and the management information service (MIS) [4] which combines the features of the Directory (X.500) and of OSI Systems Management (X.700) for the cooperative administration of globally distributed management information has been developed. Section 2 provides a motivation for an integrated information model and the MIS. A description of the integrated information model and the definition of a unique service interface is given in section 3. The summary contains an example of an information tree comprising directory objects and managed objects of the service provider domain which will be implemented in one PREPARE VPN scenario.

2. Motivation

Network administrators as well as service users require an information service which provides details about services, subscribers, operators, networks, network elements, etc.

The various kinds of information required in the global management environment can be classified as:

- long lasting data (for example, information about services, networks, network elements, communication addresses) which needs to be published globally or made available to a large sector of the operator/user community and is therefore suitable to be administered by the X.500 Directory.
- dynamic management data that can change very rapidly (e.g. a packet counter in a transport entity) and which is maintained by the object, or process, in which the information is created. This data can be accessed or modified by OSI Systems Management services and protocols.

The information administered by such an information service needs to be available in a location-transparent manner to all those requiring it. In addition, cooperation has to take place if end-to-end services are to be managed in a distributed environment. This cooperation can, in turn, require the provision of information by an end system which has to obey certain conventions in order to be understood by other end systems. In this case, the method of information handling and modelling is no longer under the authority of one end system but has to carried out in mutual agreement.

A supporting service is therefore required which is able to provide human managers and managing applications with the information they need to perform their management tasks and which can act as a handler for the various types of information required. This requires both the modelling of a common interface (Integrated User Interface) to the Directory System (via a DUA - directory user agent) and to the MIB and the development of combined operations that query both Directory and MIB.

This service is called the Management Information Service (MIS) and is intended to supply managers (both human managers as well as managing applications) with management information. The Management Information Service enables data covering all aspects of service and network management to be stored and retrieved in a location-independent manner and acts as an information handler in that it provides combined operations enabling both MIB and Directory to be queried together. The main areas where the MIS is applied are *identification*, and *location* and *addressing*. Unambiguous identification is achieved by assigning unique names to the individual information objects. Location problems can be solved by the comprehensive facilities of the MIS for searching the global information base for the required information. Addressing can be simplified through the MIS if the global information base also contains the addresses needed for communicating the management information, such as the *presentation addresses* of end systems.

Since the Management Information Service has to deal with information to be communicated throughout the environment, a common information model for this global management information has to be built. The information model is constructed in a pragmatic manner, bearing in mind the benefits and characteristics of the principal components of the MIS: the X.500 Directory and X.700 Management. The X.500 Directory is primarily designed to store and administer objects for longer term static information whereas according to X.700 Management the information maintained in MIBs is more dynamic in nature.

If end-to-end services are to be managed, the information required to manage them may well be distributed over several domains and located in various end systems. Management operations will need to be carried out in all the relevant domains with information from several MIBs being accessed. End-to-end service management therefore requires a global information architecture which considers the following:

- accessibility to end-to-end service information throughout the distributed environment,
- identification and location of objects relevant to provide this end-to-end service information, i.e. unique and global naming of these objects,
- realization of a common name space, and
- cooperative administration of this global name space[1].

The X.500 Directory is adequate to satisfy these requirements. Additionally, it is well suited for storing globally relevant information which does not change too often, i.e. objects representing organizational or addressing knowledge.

On the other hand, the dynamic end-to-end service information exhibiting some specific manageable behaviour (e.g. resources) has to be represented in terms of Managed Objects (MOs). These objects are part of the different OSI MIBs dispersed throughout the distributed environment.

This leads to the need for two different kinds of systems (X.500 Directory and OSI Systems Management) for storing the information, each with its own characteristics, interfaces, and operations. Fortunately, the information models are similar in both systems so that transparent integration can be achieved by means of the MIS.

3. The Management Information Service (MIS)

This section describes the Management Information Service. Firstly it shows the concept of integrating X.500 and X.700 Management. Furthermore, it describes the functionality offered by the MIS.

3.1 The Integrated Information Model

The global name space, being the ultimate precondition for the MIS, is set up by defining some naming conventions for X.700 managed objects. These establish a virtual link between, on the one hand, the unique and unambiguous name space of the X.500 Directory and, on the other, the local MIBs held by the administrative domains. Furthermore, the differences between Directory objects (DOs) and managed objects (MOs) will be adjusted. The result is a logical view of all of the objects abstracting from their realization as a DO or MO and, thus, from their location.

3.1.1 Building a Common Name Space

Since the mechanisms for naming objects are comparable in both the DIB and the MIB, it should be feasible to establish a common name space. Both X.500 and X.700 use a hierarchical structure in which *Distinguished Names* (DNs) are composed by concatenating *Relative Distinguished Names* (RDNs). The main distinctions are the extent of the area in which the name is

1. Restricted access to information specific to closed user groups can be guaranteed by the use of standardized authentication and access control mechanisms.

valid and the uniqueness of the names necessary to identify the objects unambiguously throughout the distributed environment. Whereas the global uniqueness of names for Directory objects is guaranteed by X.500 (i.e. it provides one global root for naming objects), the names of the objects kept within the dispersed parts of the end-to-end service MIB[1] only unambiguously identify the objects within the local extension of each OSF. The starting point for the construction of an MO's *Distinguished Name* is always the root of the containment tree of a single OSF's MIB (so-called "local MIB"), i.e. an MO of the class (or subclass of) *system* in order to be in line with the standards. Consequently, such a DN only refers unambiguously to one object within the system itself.

In order to obtain global names for the latter, all local MIBs have to be combined into one name space. Since the DIB is also to be integrated, the best solution is to select the root of the DIB as the global root for all objects (i.e. DOs and the MOs needing to be externally accessible) as it is already the root for a unique name space. That method implies that a virtual association has to be established between the DIB and every locally held MIB which is to be integrated by this concept.[2]

Since we are concerned with the naming hierarchy, this conceptual linking can be done by defining a specific, singular RDN with the following special characteristics:

- The RDN may only occur between an DO and an MO (see example below). It must not appear in any other namebinding, i. e. not between two Directory objects.
- It has to be exclusive in the sense that the attribute type of the RDN must not be applied in any other RDN within the naming tree. That is, it may not be used in a structure rule or namebinding other than the virtual namebinding between the DO and subordinate (managed) system MO.
- This mechanism (together with a rule specifying the namebinding) constitutes a common naming convention that has to be obeyed by all managed systems storing a part of the end-to-end service MIB.

By defining this convention, a globally unique and homogeneous structured, Directory-like name is implicitly assigned to each of the integrated OSFs. This conforms to X.700, since there it is possible to identify an OSF by means of a Directory name.

1. The end-to-end service MIB comprises the inter-domain related information.
2. Of course, it cannot be the aim to include all MOs "physically" in the DIB by "copying" them into Directory objects. That would cause a lot of problems due to redundancy, performance etc. So, a purely conceptual integration is the goal in order to provide a unified structuring of and transparent access to the information.
   Even the duplication of the MIB's root MO, which is the system MO, leads to problems, because in that case the two objects, MO and its equivalent DO (the "copy" of the system MO), have the same name and cannot be differentiated at the MIS interface.

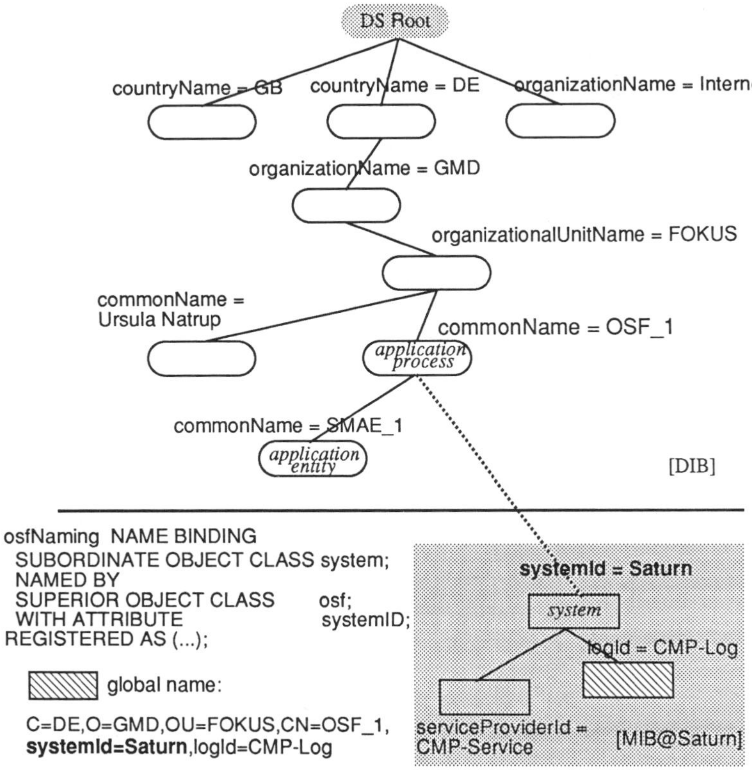

Figure 3: Linking a MIB to the DIT using a unique RDN

Focusing on the point of where within the Directory information tree (DIT) the local MIBs should be virtually linked, two possibilities exist in principal. Regarding the Directory standards, two object classes (or their subclasses) are specified for representing information of applications:

- An entry of object class *application process* contains information about communication-independent aspects of the application (the *common name* attribute needed for constructing the DN, optionally a description, the locality, etc.)
- An entry of object class *application entity* stores the information related to the communication aspects (e.g. the *presentation address*).

These DOs are paired with each other by a structure rule defining that an *application entity* entry is always subordinate to its corresponding *application process* entry.

That is, these two standardized Directory objects contain the general and addressing information of the managed system's OSF. They are embedded within the DIT according to their administrative context, which is made up by countries, organizations, organizational units, etc.

Thus, the "natural" environment is reflected by the name. When considering these two Directory entries - which have to exist if a managed system, i.e. its OSF, is to be represented in the DIB, and if one wants to conform to the standards - as a given precondition, it is logical to associate the MIB to the superior application process DO that represents the OSF holding the MIB. Note that it is a namebinding "between two kinds of information bases", i.e. the DIB and the MIB. Thus, a convention is defined through the *osfNaming* that all VPN domains must observe in order to ensure interoperability.

3.1.2 Information Objects

The concept discussed above introduces a common name space that provides each object with a unique name and guarantees that it can be identified throughout the whole distributed environment. However, there are still two kinds of objects: Directory objects, which have only attributes, and managed objects which are described by attributes but which also exhibit some management behaviour, i.e. they can offer actions and emit notifications. These differences should be hidden by the MIS so that the user does not need to be aware of two separate systems, i.e. X.500 and OSI Systems Management (DOs and MOs).

In order to abstract away from these differences the concept of an Information Object (IO) is introduced. IOs represent the information aspects of objects made accessible by the MIS and so constitute the totality of the global, integrated information base.

The entire set of IOs, no matter whether MO or DO, constitutes the Integrated Information Base (IIB) as illustrated in Figure 4. The IIB is hierarchically structured by X.500 structure rules in the upper layers, OSI Systems Management namebindings in the lower layers and through the naming convention for integrating both systems explained above.

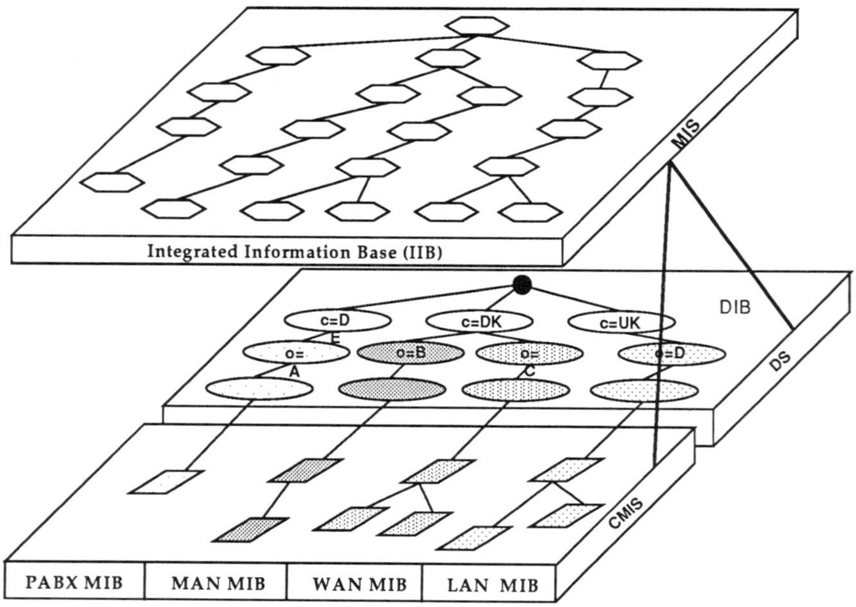

Figure 4: The Integrated Information Base

3.2 MIS Services

The operations covered by the MIS can be classified according to the following three categories:

- *Information Retrieval,* i.e. listing subordinates of a specific objects within the information tree, searching etc.
- *Information Control,* i.e. creating new objects, modifying existing objects etc.
- *Event Notification,* i.e. functions that enable reception of event reports which arrive asynchronously.

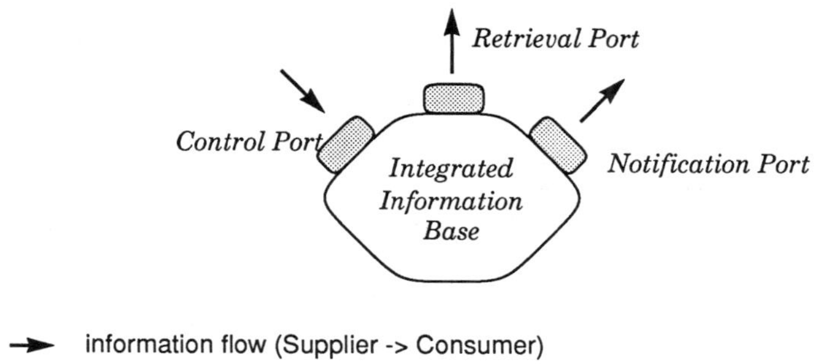

Figure 5: Logical Ports to the Integrated Information Base

Because most of the operations can comprise more than only a single object by scoping and filtering[1], the MIS operations are to be seen as being performed on the IIB as a whole. Thus, the information base forms a logical component offering the MIS through the three ports: the *Retrieval Port, Control Port* and *Notification Port* (see Figure 5).

Nevertheless, a component is needed which possesses the knowledge of the hierarchical structure of the information base, i.e. the virtual linking between DIB and MIBs. This component, termed the *MIS Agent,* is a logical part of the integrated information base. The MIS Agent is responsible for mapping the comprehensive operations onto (possibly a whole sequence of) operations upon the information objects. The MIS Agent also has to reassemble the results.

<u>*The Retrieval Port*</u>

The retrieval port offers operations for obtaining information about objects, attributes and attribute groups, e.g. reading attributes of certain objects, listing of subordinates, searching objects with special characteristics, comparing attribute values etc. At the *Retrieval Port* the following operations are offered:

- *MIS-Read,*
- *MIS-Compare,*
- *MIS-Search,*
- *MIS-List,*
- *MIS-Cancel.*

1. That is, for instance, a significant dissimilarity to the Directory Services.

All of these operations can affect more than only one object at a time, i.e. it is possible to refer to several objects with a single operation call. In this case, a base object must be specified in the operation which is the root in the information tree for all of the objects to be handled by the request. This base object has to be identified by its *Distinguished Name*. As an option, its object class can also be given. But that serves only to verify the validity of the operation. If name and class do not match (i.e. the object belongs to an object class other than the class presented within the operation call) the operation is refused. Furthermore, a *scope* and a *filter* parameter have to be determined to handle several objects at once:

- the scope defines the fragment of the information base hierarchy to which the objects pertain, e.g. the first level beneath the base object or the whole subtree.
- the filter qualifies the objects in terms of preconditions the objects have to satisfy in order to receive the operation. For instance, in terms of attributes the objects have to possess or even by some special attribute values they have to show.

Coincidently, a condition to *synchronize* such a multiple operation is required. There are two possibilities:

- *Atomic* means that the operation has to be successful on all objects which are included through the description by specifying base object, scope and filter. That means *atomic* defines the operation to be regarded as a transaction on the entire set of objects.
- *Best effort* calls for the trial on every object, but does not demand a rollback if the operation fails on any of them.

The Control Port

Operations offered at the *Control Port* serve to manipulate the integrated information base. This includes functions to create, delete and modify objects and to carry out actions. In order to manipulate the integrated information base, the *Control Port* comprises:

- *MIS-Create*,
- *MIS-Delete*,
- *MIS-Change*,
- *MIS-Action*.

The Notification Port

The notification port provides reports to a MIS user about asynchronously occurring information. Two operations are available which the user can use to control event reports. The third operation available is invoked by the MIS Agent when an IO reports an event about which a user has requested a notification. At the *Notification Port* the following operations are offered:

- *MIS-InitReport*
- *MIS-TerminateReport*
- *MIS-Report*

4. Summary

The Management Information Service (MIS) presented in this paper provides information handling capabilities to support end-to-end service management of distributed applications or services. As one example, Virtual Private Networks (VPN) in an Integrated Broadband Communication (IBC) Environment are used to demonstrate the benefits of the integrated use of X.500 and X.700 standards.

The global information architecture establishes an integrated information base (IIB). It includes a common naming scheme, rules and conventions allowing the structuring and access of VPN related information within the IBC environment.

The following example of (some) information provided by the VPN service provider domain shows how VPN relevant information (on the service provider organization, customer contact person, service provider operation system, etc.) is stored as a directory or managed object depending on the nature of the information.

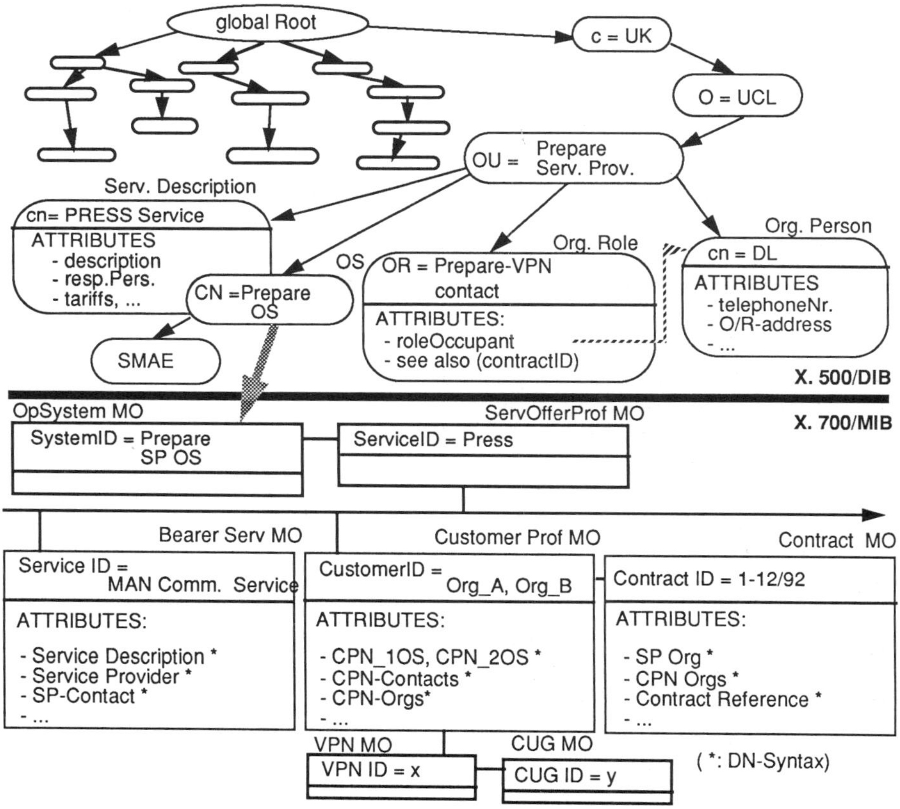

Figure 6: Information provided by VPN Service Provider Domain (Directory and MIB)

To handle the objects within the IIB easily, the advantages of the Management Information Service are:

- A conceptual integration between the X.500 DIB and the X.700 MIBs is realized by virtually linking the name spaces.
- The physical distribution and location can be hidden from the user.

- It provides access transparency throughout the inter-domain testbed, i.e. the user only has to know the name of an object and not where it is stored.
- The objects are identified by unique names within a universal, common structure.
- Information retrieval and updating can be performed via a unified interface. Also, the user (human manager or management application) is relieved from dealing with particular details of the underlying systems, e.g. how is a connection to be established, special control parameters for Directory queries (*prefer chaining, don't use copy* etc.).
- Easy-to-use retrieval functions (e.g. a search operation involving several domains) can be offered.
- Some control functions to modify the information objects, i.e. to insert new information into the information base, to remove information from it, or to modify existing objects are supported.

Summarizing the functionality offered by the Management Information Service, leads to the conclusion that the MIS is a very comprehensive and convenient service, somewhat similar to an enhanced Directory service, but also integrating different systems and, moreover, covering several information bases.

Consequently, the MIS enables the transparent administering and retrieval of end-to-end management information, wherever it is stored, while hiding whether the Directory Access Protocol (DAP) or the Common Management Information Protocol (CMIP) has to be addressed.

5. Acknowledgments

The authors would like to thank their colleagues at GMD-FOKUS and the partners of the RACE/PREPARE consortium for their ideas and comments which improved the concepts presented in this paper, especially to Jane Hall (FOKUS) and Thomas Magedanz (TU-Berlin).

6. References

[1] CCITT Draft Recommendations M.3010,
"Principles for a Telecommunications Management Network" (1991)

[2] CCITT Recommendations series X.700,
"Information processing systems - Open Systems Interconnection (OSI)", Management Framework, CMIS/CMIP, Systems Management Overview, System Management Functions, Management Information (1991)

[3] CCITT Recommendations series X.500,
"Information processing systems - Open Systems Interconnection (OSI) - The Directory" (1992)

[4] Waßerroth, Susanne,
"DIMIS - The Directory and Management Information Integrating Service", Studienarbeit, TU-Berlin, August 1991

[5] PREPARE project Deliverable 2.2B -
"Management Information Service" (1992)

[6] PREPARE project Deliverable 2.2A -
"Architecture & Interfaces" (1992)

[7] PREPARE project Deliverable 2.3A -
"TMN Information Model" (1992)

Integrated Network Management, III (C-12)
H.-G. Hegering and Y. Yemini (Editors)
Elsevier Science Publishers B.V. (North-Holland)

Integration of the Directory Service in the Network Management Framework

James W. Hong, Michael A. Bauer and J. Michael Bennett
Department of Computer Science
University of Western Ontario

Abstract

This paper examines the potential role and feasibility of using the X.500 Directory Service in the network management framework. We examine the network management requirements and investigate how the Directory Service can satisfy them. A simple network load monitoring tool was adapted to use a prototype X.500 Directory Service (Quipu) for the storage and retrieval of (1) network domain information, (2) host information, (3) network monitoring tool application information, and (4) network management information. We report on our experiences using Quipu and assess the appropriateness of using the Directory Service to provide support for network management information.

Keyword Codes: C.2.4; C.2.3; H.3.4
Keywords: Distributed Systems; Network Operations; Information Systems and Software

1 Introduction

Managing networks involves generating and collecting a vast amount of information about a variety of managed objects throughout the network. Information gathered may relate to current device configurations, recent faults, accounting information, security and performance. Information about the managed objects must be stored somewhere in the network so as to be accessible to the management tools. Some of the collected information about these managed objects must be stored and updated periodically in order to provide information for other network applications or services (*e.g.*, network load monitoring).

The Directory Service, proposed by the X.500 Standards [CCIT88, CCIT92], has been designed as a distributed repository of information. Recently, many commercial and non-commercial implementations of the Directory Service have been made available to users [Lang92]. However, their primary uses have been limited to providing directory services of human information (*e.g.*, names, phone numbers, addresses). In this paper, we investigate the possibility of using the Directory Service as a repository of information other than human information, network management information for example.

Some attempts have been made to design and implement databases specifically for network management information [Bapa91, Valt91]. Unlike these attempts, we examine the feasibility of using the existing Directory Service for providing the storage and retrieval of managed information in the network management framework. Since we do not envisage that all types of management information can possibly be stored in the Directory, we investigate what types are suitable. We also investigate whether the Directory Service can be used in a reasonably efficient way such that it does not become the bottleneck in network management operations.

We have implemented an integrated network management tool that uses the Directory Service as a managed information server for network management information. The network management tool monitors network traffic load in an interconnected network environment, and uses Quipu [Kill91] as a repository of management information. We report on our experiences using Quipu and assess the appropriateness of using the Directory Service to provide support for network management information.

The rest of this paper is organized as follows. Section 2 presents a set of requirements for managing networks. Section 3 presents a brief overview of the X.500 Directory Service and how it satisfies the network management information repository requirements. Section 4 presents an integrated network management architecture using Directory Services. The components of our prototype implementation are given in Section 5. The performance of each component is examined and compared in Section 6. Section 7 presents the summary of our work along with some possible future work.

2 Requirements for Network Management

In this section we examine the components of and requirements for network management. We particularly focus on the types of network management information and their repository service requirements.

2.1 Components of Network Management

Traditionally, managing networks involves one or more network management stations (NMSs) trying to monitor and control one or more (typically hundreds or thousands) of network elements (NEs) (*e.g.*, hosts, gateways and bridges). An NMS is typically a management application program (or a set of programs) running on a host, acting as a monitoring and controlling point for a set of managed NEs.

How these NMSs and NEs communicate is defined by network management protocols (*e.g.*, CMIP [ISO91a] and SNMP [Case90]). Management information exchanged between NMSs and NEs is defined in the management information base (*e.g.*, OSI-SMI [ISO91c], MIB-I [McCl90] and MIB-II [McCl91]). The standards also specify what kinds of services should be provided. For example, in the OSI management framework, the create and delete, get and set, event-report and action information access operations are provided [ISO91b]. In the Internet management framework, four operations are provided: get, get_next, set and trap [Case90]. However, they do not specify how these protocols, services or MIBs should be implemented. Below, we take a close examination of the characteristics of information in the MIB.

2.2 Management Information

The MIB is a conceptual repository of information on objects being managed. It specifies a set of objects which should be implemented by NEs. The structure of management information (SMI) [ISO91c, Rose90] defines the rules for defining the structure of objects and the relationships between them. It is interesting to note that most (if not all) MIBs defined thus far (whether experimental or private) form a hierarchical name space.

The contents of MIB can be largely divided into three groups: *static*, *dynamic* and *semi-dynamic*. The static information in the MIB consists of those objects whose contents hardly change (if at all) during the lifetime of the operation of a NE. Examples of such information are hardware and software configurations of a NE. Hardware configuration information may include the CPU type, type of network interface or the amount of physical memory installed. Software configuration information may include the name of operating system and its version. Such information needs to be updated only when there is a hardware upgrade (*e.g.*, replace CPU to more powerful one or add more memory) or a software update (*e.g.*, update OS to the latest version). But occurrences of these events are very infrequent – it may take place (if at all) once in a few months or years.

Other examples of the static information in the MIB include information pertaining to the NEs themselves. The name of a technical contact person responsible for a NE and where he/she can be found, where the managed NE is physically located, the type of NE (*e.g.*, host or gateway), the NE's logical name, the NE's addresses (*e.g.*, Ethernet and IP) and address translation mappings (*e.g.*, IP addresses to media specific addresses) are some of such examples.

On the other hand, the dynamic information in the MIB consists of those objects whose contents change quite frequently (*i.e.*, every few microseconds or milliseconds). Incoming and outgoing network traffic information on a host are examples of such information. For instance, the number of packets that have been sent and the number of packets that have been read by the interface to and from the network are constantly being updated. These statistics are also updated in the counters maintained in the upper layers of the communication protocol stack. For example, in the network layer various statistics on the IP packets are constantly updated and in the transport layer those on the UDP and TCP packets are maintained.

The dynamic information in the MIB mentioned above are raw data that are maintained by the managed devices themselves. These are frequently polled/reported from/to an NMS, which may collect and perform some analysis on them. For example, in order to determine the average network load for a time interval, the NMS may simply collect the total output packets sent to the network from each device on the network, sum the numbers and then divide by the number of devices (a more elaborate example of the network load monitoring application will be given in Section 5). The historic or trend data such as this is dynamic but the frequency of it being updated is much smaller than the kinds of dynamic raw data described above. We will call such kind of MIB information *semi-dynamic*.

2.3 Management Information Repository

Management information that is collected as part of network monitoring needs to reside somewhere on the network. It is desirable that a repository for such management information have the following set of characteristics. The repository should (1) provide a single logical view but be physically distributed, (2) provide a general naming structure so that the managed objects can be easily stored and retrieved in a uniform way across heterogeneous systems, (3) provide a simple and efficient access interface through which application programs or processes can easily access management information stored in the repository, (4) have an international recognition so that it has the potential for growth and acceptance worldwide, (5) should provide a reasonable performance so that the use of it by management tools does not become the bottleneck, (6) provide some form of security in order to regulate the access of important information, and (7) support the replication of management information to increase availability, reliability and performance. In the following sections, we examine how the X.500 Directory Service can satisfy these repository requirements.

3 Overview of X.500 Directory Service

The X.500 Standards [CCIT88, CCIT92] specifies a Directory Service which provides and manages information about entities. The X.500 Directory Service (or simply the Directory Service) and its Directory information is distributed physically, yet provides a unified, single logical functional view.

The information held by the Directory is called the Directory Information Base (DIB). The DIB consists of entries (or objects) which contain information about entities. Each entry consists of a set of *attributes* each with a type and one or more values. The types of attributes which are present at each entry are dependent on the *class* of entities which the entry describes. Classes can be defined by user, and thus various types of information can be stored in the Directory. Entities in the DIB are represented by entries in a global, hierarchical name space called the Directory Information Tree (DIT). Entries are placed in the DIT according to the organizational relationships between the entities which they represent.

The X.500 Abstract Service Definition [CCIT88, CCIT92] defines abstract ports and operations which provide the user functionality for retrieving, searching and modifying directory information. These operations form a simple interface protocol called Directory Access Protocol (DAP). The current X.500 Standards support the read, compare, list, search and abandon interrogation functions and the basic add, remove and modify entry manipulation functions.

The Directory Service (including the DIT) is distributed over physically separated entities (such as computer hosts) called Directory Service Agents (DSAs). This distribution is transparent to the user through the use of Directory Service Protocol (DSP) operations between DSAs. Each user or user-process is represented by a Directory User Agent (DUA) which is responsible for interacting with the Directory. Thus, users can easily and efficiently access the Directory information through local or nearby DSAs. The functional model of the Directory Service is illustrated in Figure 1.

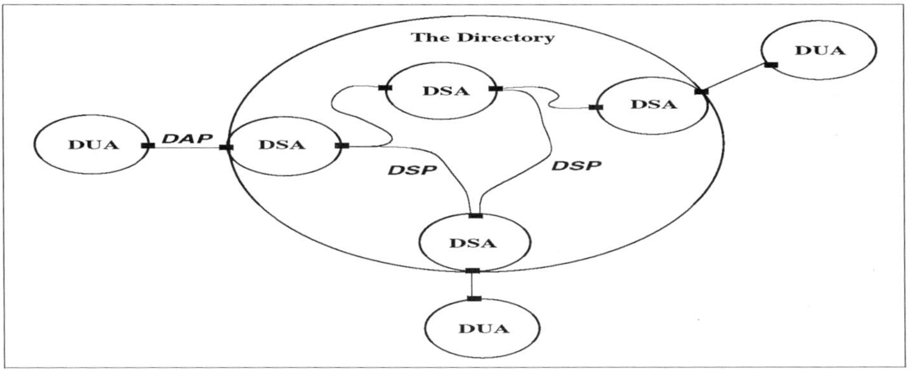

Figure 1: Functional Model of the Directory Service

It is assumed that some service will be provided regardless of network partitioning resulting from events such as non-local site failures. This is achieved by the distributed nature of the DSAs and replication of Directory information [CCIT92]. X.500 also provides a security mechanism so that directory information access by unauthorized users can be prevented. Caching is also recommended to increase the performance of the Directory Service but it is a local issue. That is, the X.500 Standards do not provide caching algorithms and it is up to the implementor of the local DSA to do so.

Thus, the Directory Service satisfies the requirements for the storage and retrieval of network management information. In Section 6, we examine the performance aspect of the Directory Service.

4 Integrated Network Management Architecture

4.1 The Architecture

In this section, we introduce an architecture that integrates the Directory Service into the network management framework. The architecture, shown in Figure 2 uses the Directory Service as a repository for network management information.

Figure 2: An Integrated Network Management Architecture

As described in Section 2, the Directory Service provides a repository of information. The Directory is distributed and replicated for increased availability, performance and reliability. The fact that there may exist multiple DSAs distributed over the network or the fact that some (or all) information is replicated is transparent to the user. What the Directory provides to the user is the ability to insert, remove, update and search information entries.

The NM-DUA (Network Management - Directory User Agent) is required as an interface between the NM tools and the Directory. Management information that is to be stored in the Directory is passed from a management entity (either NMS or NE) to the NM-DUA, which then communicates with an appropriate DSA in the Directory Service to complete the task. The Directory Access Protocol (DAP) [CCIT92] is used between NM-DUA and DSA. The NM-DUA must at least support the operations to add, update and retrieve information entries.

4.2 Management Information for the Directory

Network management involves a great deal of management information. We have classified management information into three groups in Section 3: *static*, *dynamic* and *semi-dynamic* MIB. We envisage that the static management information should be stored in the Directory for various management applications. Since the static MIB information is updated infrequently but will mainly be retrieved for the lookup purpose, this type of data suits the intended usage characteristic of the Directory Service quite well.

The dynamic MIB information which is constantly being updated is not quite suitable for being stored in the Directory because of its update frequencies. As mentioned earlier, the dynamic MIB information is usually updated every few microseconds or milliseconds. That rate of update frequency to the Directory is, in general, expected to be more than what the Directory Service can handle. This is mainly because of the communication cost involved in accessing the Directory Service. Even if the Directory Service is not distributed (*i.e.*, the Directory Service consisting of a single DSA), the same problem would still exist.

The semi-dynamic MIB information which is in general updated a lot less frequently than the dynamic MIB, however, should also be suitable for being stored in the Directory. Typically the semi-dynamic MIB information contains historic data or data that is updated every few minutes or hours. This rate of update frequency should be adequately handled by the Directory Service.

In order to store MIB information in the Directory, MIB objects and attributes must be defined in the Directory Service. The Directory Service allows the user to define the objects and attributes of information which is to be stored in the Directory. In the next section, we describe our experience of defining objects and their attributes for storing static and dynamic MIB information in the Directory.

5 Implementation

In this section, we describe a prototype implementation of the integrated network management architecture. We first describe the network management tool and the Directory Service implementation used in our prototype implementation. We then describe the object and attribute definitions for management information that is to be stored in the Directory. A description of the NM-DUA interface as well as the overall operation is also given.

5.1 Network Management Tool

The network management tool used is a network traffic load monitoring system, which uses the *aggregate agent* concept [Grat92]. The tool has been designed and implemented to overcome two major problems that arise when managing large heterogeneous interconnected networks. The first problem deals with communication between two or more NMSs involved in network management. When managing a group of subnets, some of the data collected by an individual NMS will need to be made accessible to other NMS managing nearby subnets and to NMS responsible for groups of subnets. The NMS-to-NMS communication is required to identify the scope of a current subnet anomaly as local to that subnet or a symptom of a more global factor, or simply to provide statistical information on the health of a subnet.

The second problem arises from the desire to impose a hierarchical structure on the flow of management information. This hierarchical structure allows a central authority that is responsible for overall network management to monitor changes occurring in various subnets and take appropriate action. In this hierarchical scenario, any intermediate-level NMS could be monitoring and controlling groups of subnets. In this way, departmental or regional network administrators not only have direct access to their networks for management but also participate in the overall management hierarchy by passing their subnet information to higher levels.

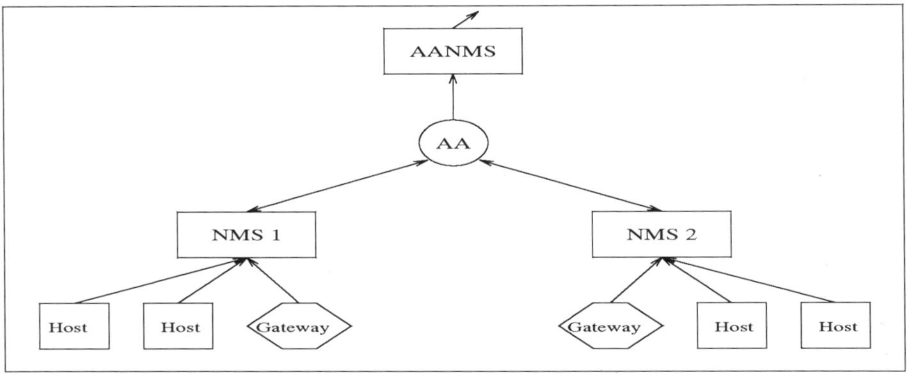

Figure 3: An Example of Hierarchical Network Traffic Load Monitoring System

An example of a network load monitoring system using the aggregate agent concept is shown in Figure 3. The system basically works as follows. An NMS manages a number of NEs, each of which monitors network traffic information from its interface(s). The number of packets that have been sent to the network from the device and the number of packets that have been read from the network from the device are examples of such traffic information. An NMS periodically polls each NE and collects the traffic information. The collected traffic information is then averaged out (per second, for example) and then passed to the aggregate agent (AA) which collects network load information from one or more NMSs. Aggregate agents are then polled by higher-level managers called *aggregate agent managers* (AANMS). In this way, the network load information can be passed up hierarchically all the way up to the root of the hierarchy.

5.2 The Directory Service

The implementation of the X.500 Directory Service we are using is Quipu version 7.0 [Kill91]. The Directory Service testbed has been installed in our systems laboratory which consists of a heterogeneous network of five hosts (2 Sun 3's, 2 Sun 4's and a MIPS), and a DSA runs on each host. For the purpose of our uses, these DSAs are connected among themselves but are isolated from DSAs running in the rest of the world.

The NM-DUA (shown in Figure 2) interface was implemented to provide access to the Directory. All operations (*e.g.*, store, update and retrieve) from the management tool to the Directory is through this interface. It is also responsible for setting up and tearing down the connection to and from a DSA in the Directory.

5.3 MIB Definitions

Since generic definitions of objects and attributes in most X.500 implementations (including Quipu) are for information about people, they do not meet the requirements for network management. Thus, we defined a new set of objects and attributes and added to the existing Quipu definitions. The following two objects and their attributes are two most relevant ones that need to be mentioned here.

5.3.1 Domain Object

A *domain* is a logical entity within an interconnected network, and consists of a set of NEs. The Quipu definition for the domain object is as follows.

```
csdDomain:  csdLocalObjectClass.4 : csdLocalTop : \
            domainName : \
            description, subnetAddress, subnetMask \
            netload, hostList, adminContact, techContact
```

`CsdDomain` is the name of the *domain* object class whose object identifier is `csdLocalObjectClass.4`. The domain object inherits attributes of the `csdLocalTop` object class and has a single mandatory attribute called `domainName`. The `description`, `subnetAddress`, `subnetMask`, `netload`, `hostList`, `adminContact` and `techContact` attributes are optional.

The `hostList` attribute contains the names of NEs in the domain. In our prototype implementation, a domain is the basic unit that an NMS manages. The network load information for this particular domain for some time interval is stored in the `netload` attribute, which can be retrieved by other applications (such as aggregate agent managers) for various uses. For example, overall network load of an entire departmental (possibly consisting of one or more domains) or campus entire network can be analyzed. In the domain object definition, only the `netload` attribute is a semi-dynamic MIB; the others are static.

5.3.2 Network Element Object

A network element object is a physical or logical computing device on a network. It appears to the user as a distinct entity that can be used for running user applications (*e.g.*, management agent programs). The Quipu definition for the network element object is as follows.

```
csdMachine:  csdLocalObjectClass.6 : csdAdmin : \
             machineName : \
             machineType, description, networkAddress, operatingSystem, \
             applications, adminContact, techContact, location
```

`CsdMachine` is the name of the NE object class whose object identifier is `csdLocalObjectClass.6`. The NE object inherits attributes of the `csdAdmin` object class and has a single mandatory attribute called `machineName`. The `machineType`, `description`, `networkAddress`, `operatingSystem`, `applications`, `adminContact`, `techContact` and `location` attributes are optional.

The `applications` attribute contains the names of applications that a particular network element can run. It also tells where the application programs can be found. For example, it contains the file path of the agent program that collects network traffic information. This is where the NMS fetches the file paths of the agents when activating them. The network element object definition consists of *static* MIB information only. As discussed earlier in this paper, *dynamic* MIB information is best kept locally within each device.

In this section, we described our prototype implementation of integrating the Directory Service in a network load monitoring system. In the next section, we examine the performance of the Directory Service in the prototype.

6 Performance

In order to determine the performance of the Directory Service in the network load monitoring tool example given in the previous section, we have compared the times taken for accessing the Directory with the rest of the integrated management system. To simplify the performance evaluation, we only included one manager (NMS) and five agents in our prototype implementation. The NMS interacted with the Directory for accessing and updating management information. The prototype management system is divided into four phases.

1. retrieve domain, host and application information from the Directory,

2. start up the agents on managed devices,

3. poll the agents to collect network load information, and

4. store network load in the Directory.

Figure 4 shows the actual times taken for each phase and the percentage of each compared to the overall time taken. The performance measurements were taken from running our prototype implementation on our systems laboratory described earlier. The first phase took only three seconds to retrieve the domain, host and application information from the Directory. This phase involved initializing the NM-DUA (including the connection set-up to a local DSA), getting the list of hosts by searching the `hostList` attribute in the Directory for a given domain (our systems laboratory subnet consisting of five hosts), and then subsequently querying the agent application information (*e.g.*, where the executable is located) for each host. The second phase involved starting up the agent on each remote host in the domain using the UNIX `rsh` command. The third phase started as soon as all the agents got started and were able to reply to polling by a management station. The final phase involved updating the network load information for the domain in the Directory by modifying the `netload` attribute.

Phases	Information Retrieval	Agent Startup	Collection	Netload Update
Time (seconds)	3	22	60	0.011
%	3.53	25.88	70.58	0.01

Figure 4: Comparison of Four Phases in the Network Load Monitoring Application

Surprisingly, the phase for updating the netload information to the Directory took only 0.011 second (or 0.01% of the overall time). In our detailed analysis, we found that the NM-DUA set-up took approximately 1 second or one-third of the first phase. The other two-thirds involved six search operations (1 domain and 5 hosts) to the Directory as well as capturing the replies and processing them. We found that the actual search operation itself is as fast or faster than the update operation. However, capturing and processing the replies from the Directory and making them meaningful to the management tool was taking most of the first phase. Nevertheless, the total time for accessing the Directory took merely less than 4% of the overall time.

Sixty seconds or 70% taken by the collection phase is the time taken to poll five agents and receive the replies from them. The collection phase included 55 seconds of idle period (*i.e.*, polling interval) and 5 seconds of actual polling time (approximately a second per agent). In practice, however, the polling interval may be a lot larger than what we used in order to avoid flooding the network with the polling traffic (which will contribute to the degradation of network performance). If we were to poll agents every few minutes or tens of minutes (which is more common and practical), the times for the first, second and fourth phases would remain constant but the collection phase will dominate the total time taken for the application. Thus, the percentages for accessing the Directory would be very small. In another words, the cost of using the Directory for the information retrieval and storage is quite minimal in the overall management operations.

We also ran some tests to determine the maximum rate of update frequency that can be handled by the Directory Service. In order to generate as many updates to the Directory as we can in our implementation, we eliminated the idle period between the polls and polled only one agent. Since it takes approximately one second to poll a single agent, the maximum rate was an update at every 2 seconds or less. The result was that the Directory Service handled the update operation at that rate without any problem. We envisage that this performance by the Directory Service is more than sufficient to provide the storage and service to most (if not all) types of semi-dynamic MIB information.

In summary, the preliminary performance results obtained from our prototype implementation suggest that the X.500 Directory Service can be used for the storage and retrieval of network management information. The results also show suggest that the Directory Service can be used without it being the bottleneck in certain management operations.

7 Concluding Remarks

We have examined the feasibility of using the Directory Service for the purpose of storing and retrieving management information in the network management framework. We have examined the requirements for network management, particularly the management information repository aspects. We showed how the X.500 Directory Service can satisfy these requirements. Preliminary results from our prototype implementation show that the Directory Service can be used for the storage and retrieval of management information in the network management framework.

Our prototype implementation retrieved from the Directory not only the information about hosts on a given domain but also information about agent applications on each host. Although it was quite minimal, the Directory was also being used for what we would call *distributed applications management.* Managing distributed systems involves just more than managing networks. It needs to manage not only the networks and devices on them but also the applications running on the system. As an extension of our work, we are also looking at the feasibility of using the Directory Services for the purpose of distributed applications management.

Acknowledgements

The authors wish to thank the reviewers of this paper for their helpful comments and suggestions.

References

[Bapa91] S. Bapat, "OSI Management Information Base Implementation", *Integrated Network Management, II*, Elsevier Science Publishers B. V., I. Krishnan & W. Zimmer (Editors), pp. 817–832, North-Holland, 1991.

[Case90] J. D. Case, M. S. Fedor, M. L. Schoffstall, and J. R. Davin, "A Simple Network Management Protocol", *Internet Request for Comments 1157*, Network Information Center, SRI International, Menlo Park, CA, 1990.

[CCIT88] CCITT, "X.500 Directory Services (Parts 1-8)", CCITT, 1988.

[CCIT92] CCITT, "X.500 Directory Services, 1992", CCITT, 1992.

[Grat92] B. E. Gratto, J. W. Hong, and J. P. Black, "The Aggregate Agent Concept: A Paradigm for Hierarchical Network Managment", *The First International Conference on Computers and Communication*, pp. 148–152, San Diego CA, June 1992.

[ISO91a] ISO, "Information Processing Systems - Open Systems Interconnection - Management Information Protocol Specification - Part 2: Common Management Information Protocol", International Organization for Standardization, International Standard 9596, 1991.

[ISO91b] ISO, "Information Processing Systems - Open Systems Interconnection - Management Information Protocol Specification - Part 2: Common Management Information Service", International Organization for Standardization, International Standard 9595, 1991.

[ISO91c] ISO, "Structure of Management Information Part 1: Management Information Model", International Organization for Standardization, International Standard 10165-1, 1991.

[Kill91] Stephen E. Kille, "Implementing X.400 and X.500: The PP and QUIPU Systems", Artech House, Boston MA, 1991.

[Lang92] R. Lang and R. Wright, "A Catalog of Available X.500 Implementations", *Internet Request for Comments 1292*, Network Information Center, SRI International, Menlo Park, CA, January 1992.

[McCl90] K. McCloghrie and M. T. Rose, "Management Information Base for Network Management of TCP/IP based internets", *Internet Request for Comments 1156*, Network Information Center, SRI International, Menlo Park, CA, May 1990.

[McCl91] K. McCloghrie and M. T. Rose, "Management Information Base for Network Management of TCP/IP based internets: MIB-II", *Internet Request for Comments 1213*, Network Information Center, SRI International, Menlo Park, CA, March 1991.

[Rose90] M. T. Rose and K. McCloghrie, "Structure and Identification of Management Information for TCP/IP based internets", *Internet Request for Comments 1155*, Network Information Center, SRI International, Menlo Park, CA, May 1990.

[Valt91] R. Valta, "Design Concepts for a Global Network Management Database", *Integrated Network Management, II*, Elsevier Science Publishers B. V., I. Krishnan & W. Zimmer (Editors), pp. 777–788, North-Holland, 1991.

Integrated Network Management, III (C-12)
H.-G. Hegering and Y. Yemini (Editors)
Elsevier Science Publishers B.V. (North-Holland)

Junction Objects

A solution to a problem in naming and locating OSI managed objects

Mark Sylor

Consulting Engineer, Enterprise Management Architecture,
Digital Equipment Corporation, 110 Spit Brook Road, ZKO1-3/J35, Nashua, NH 03062
e-mail: sylor@blumon.enet.dec.com phone: 603 881-2692

A key to any management architecture, including OSI management, is how managed objects are named and located. Managed object names must unambiguously refer to a single managed object that could be located anywhere in the network. Given an object's name, a managing system must be able to locate and communicate with its agent. In OSI, the X.500 Directory service provides just such a distributed naming service. OSI management uses names with are syntactically compatible with it. But, the OSI management standards do not describe how to use the X.500 Directory to name and locate managed objects. This omission is a barrier to interoperation of managed and managing systems conforming to OSI. This paper describes how the X.500 Directory can be used to name and locate managed objects based on the concept of junction objects. It describes junction objects in detail, and describes their advantages.

Keyword Codes: C.2.1; C.2.4; K.6.4
Network Architecture and Design; Distributed Systems; System Management

1 Introduction

All management architectures use some sort of name to identify managed objects. In the OSI System Management standards, managed objects have names based on *Distinguished Names* (DN). Distinguished Names were developed as part of the OSI Directory service which is also the CCITT X.500 series of recommendations. However, while both standards use the same type of name, the standards are silent on how (or even if) the two naming schemes are intended to interact. This is unfortunate, since OSI Management can clearly make good use of the X.500 Directory service.

A key problem in any management architecture is that of locating a managed object. Essentially, when a human manager (or an application acting on his behalf) presents a managing system with the name of a managed object, and says for example, "Get this attribute", the managing system must locate the agent for that managed object, and establish a CMIP association to that agent as shown in figure 1. The managed object which is the target is shaded in the figure. To establish that association, the managing system needs to determine the address of that agent. Soon, virtually all networked computer systems will have an agent supporting many managed objects. This is already the case with DECnet/OSI [1] where every system has an agent with a rich collection of managed objects, and it is becoming common in the Internet community for each router, bridge, host, workstation, and PC to have an SNMP agent supporting a rich collection of MIBs. The number of systems involved will make finding the right agent a non-trivial task.

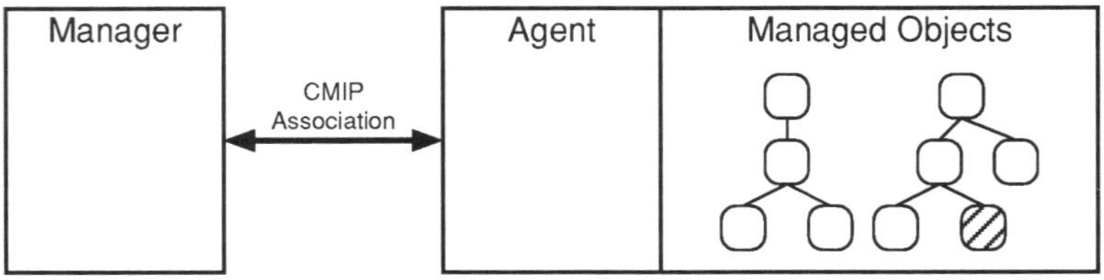

Figure 1 CMIP Manager/Agent Association

A *naming service* such as the X.500 Directory service can be used to translate a name into an address. Clearly, for managing and managed systems to interoperate, they must agree on whether or not the X.500 Directory Service is used as a naming service, and if so, exactly how is it used, and what information is stored in the Directory. Yet the OSI standards in general, and the OSI system management standards in particular do not describe how this is done. This paper argues that OSI management should use the X.500 Directory service as a naming service, and describes how it is used, based on the concept of *junction objects.*

2 Distinguished Names

Distinguished names used in OSI Management and X.500 Directory form a typed, hierarchical naming scheme. Hierarchical naming schemes are familiar to computer users in the form of file names and in ordinary life with postal addresses. The terminology surrounding distinguished names is a little daunting, and is described and discussed in this section. A distinguished name is defined as an ASN.1 type as:

```
DistinguishedName ::=             SEQUENCE OF RelativeDistinguishedName
RelativeDistinguishedName ::=     SET OF AttributeValueAssertion
AttributeValueAssertion ::=       SEQUENCE { AttributeType, AttributeValue }
AttributeType ::=                 OBJECT IDENTIFIER
AttributeValue ::=                ANY DEFINED BY AttributeType
```

The DN sequence describes the position of the named thing in the hierarchy. Hierarchical naming provides a convenient, easy to understand, and easy to navigate structure. For example, in an office, a file folder might be named by the sequence of the file cabinet name, the drawer number, and the name of the folder, e.g.,

{cabinetName="B", drawerNumber=3, folderName="Junction Objects Paper"}

This is shown in figure 2. The DN of each object, and the kind of thing it represents is shown inside the object. The arcs showing the hierarchy are labeled with the relative distinguished names (RDNs) they represent. Dashed objects show other objects in the hierarchy.

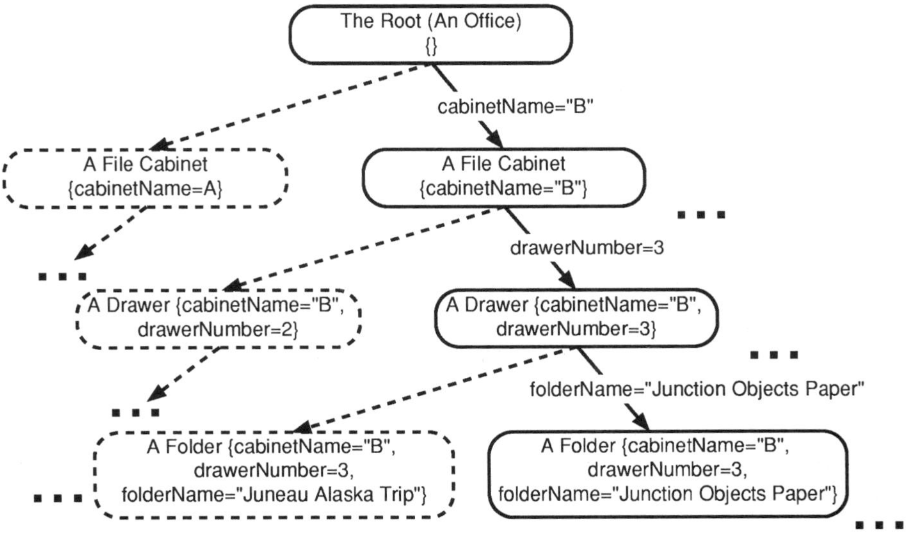

Figure 2 Example Hierarchical Name

Any initial subsequence of a DN is also a DN. Thus {cabinetName="B", drawerNumber=3} names a particular drawer, and {cabinetName="B"} names a particular file cabinet. When discussing these named objects, we often describe one object as being the *superior* of another, for example, the superior of the folder is the drawer. Conversely, we describe the folder as being *subordinate* to the drawer. This superior/subordinate relationship defines a tree. The empty DN sequence names an fictional object called the *root*. It exists only to turn a forest of trees into a single tree.

Hierarchical names make it easier to unambiguously name an object by limiting the scope of potential ambiguity. An object has an unambiguous DN if its superior has an unambiguous DN, and no other object subordinate to that (superior) object has the same final RDN as the object being named. Thus, one file folder can have the folder name "Junction Objects Paper" within drawer 3 although other folders in other drawers could have the same name.

2.1 Relative Distinguished Names and Attribute Value Assertions

In OSI Management and the X.500 Directory, objects have attributes and each attribute has a value. Each attribute of an object is identified by a distinct *attribute type* which is an Object Identifier. The definition of an attribute includes its data type, the legal set of values the attribute can have. This allows objects to be named using data types that are natural to the object, rather than forcing all objects to use some artificial name type. In the example, a file cabinet name is a string while a drawer number is a small positive integer.

Most attributes do not help in providing unambiguous names. The name of a file cabinet's manufacturer will not distinguish file cabinets. An attribute which does provide an unambiguous value is called a *distinguished attribute*. An attribute value assertion (AVA) is an assertion, or predicate, that a particular attribute has a particular value. Thus the AVA drawerNumber=3 states the value of the attribute drawerNumber is 3. For each drawer, that assertion may or may not be true. An RDN is defined as a set of AVAs because some objects require more than one attribute to distinguish them. For example a person might be distinguished by two attributes, a first name and a last name. Current practice limits an RDN to a single AVA. This rarely causes a problem, as a new distinguished attribute, can be defined as a sequence, or record, with fields for the component attribute values.

Including the attribute type in an RDN allows different kinds of objects to have the same superior object in the hierarchy. For example, one could store floppy disks in file cabinets, and I could have a disk whose DN was

{cabinetName="B", drawerNumber=3, diskLabel="Papers-92"}

For disks and folders to have different DNs, it suffices that they have different attribute types for their distinguished attributes, even though they have the same attribute value (e.g. "Papers-92"). The ability to name different kinds of objects with distinguished names and to place them appropriately within the hierarchy is an important feature for management, since management must deal with hundreds or thousands of different kinds of objects.

2.2 Logical Naming Hierarchies

The hierarchy in figure 2 is based on physical properties or location. A hierarchy can also be based on a purely logical structure. For example, that same folder could be named

{year=1992, project="Symposia", subProj="3rd IFIP ISINM", folderNo=2}

The logical structure poses the question of how one locates the folder given its name. Somewhere, a table is kept that lists for each folder, where it can be found. This location information can be thought of as a form of address.

3 Distinguished Names as Managed Object Names

OSI managed objects are named using distinguished names. The CMIP protocol supports other name forms (based on raw octet strings), however, their use is discouraged and SMI does not provide a means of defining those other name forms.

The hierarchy of managed objects is called the *management information tree* (MIT) [2]. Within that tree, we say that a subordinate object is *contained* in its superior. The structure of the MIT is defined by *name bindings*. GDMO [3] defines the form of the name of any instance of a managed object class with a NAME BINDING template. While individual objects have names, the name bindings are defined in terms of classes of managed objects. A name binding defines one possible form that the name of an instance of that class could take. Name bindings do not define the names of managed object classes, rather they define names of instances of a class.

A name binding relates three things:

1. The subordinate managed object class whose name form is being defined.

2. The superior object class which can be either a managed object class or a directory object class.

3. The attribute type of the distinguished attribute of the subordinate managed object.

Figure 3 shows part the naming hierarchy defined for the OSI Transport layer [4]. That hierarchy of name binding templates could be instantiated in the management information tree also shown in figure 3. In that figure, boxes representing classes have square corners, while boxes representing instances have rounded corners.

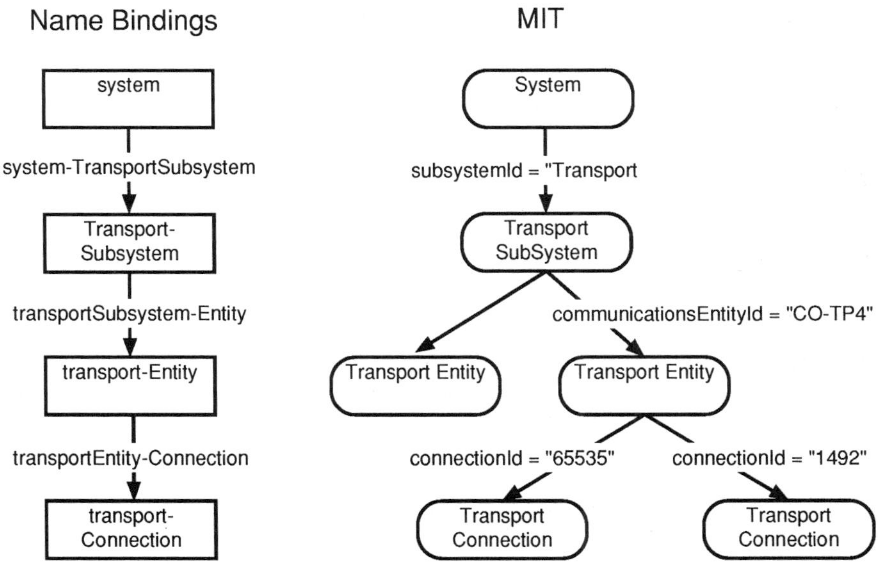

Figure 3 Name Bindings between MO Classes and the Management Information Tree (MIT)

4 Distinguished Names as Directory Entry Names

The X.500 Directory service provides access to information about real world objects. The information in the Directory is organized into *directory entries*, each of which represents an object. Directory entries have attributes that hold values. Each entry (and hence each real object represented) is named by a distinguished name. That hierarchy organizes the entries into a *directory information tree* (DIT).

The X.500 Directory service is a distributed service with client *Directory User Agents* (DUAs) and server *Directory Service Agents* (DSAs) distributed and replicated through the network. The DUAs and DSAs cooperate via application protocols in looking up information stored in the DIT and in modifying the information stored in the DIT [5].

While the X.500 Directory can be used to provide a number of services, the simplest use of the Directory is to look up a directory entry, and to return the values of its attributes. By storing an address in an attribute in a directory entry, the X.500 Directory service can be used as a simple naming service. A naming service translates a name, in this case a distinguished name, into an address.

5 Junction Objects

A *junction object* is an object that:

a) is represented as a managed object (the *junction managed object*), and
b) is represented as a directory entry (the *junction directory entry*),

such that:

1. the Distinguished Name of the junction managed object and the junction directory entry are identical,
2. the junction managed object may be the root of a local containment tree of other managed objects,
3. the junction managed object has an agent that makes it and its local containment tree visible (these agents are said to *serve* the junction managed object), and
4. the address of the agent serving the junction managed object and all its subordinates is stored in the junction directory entry in the agent address attribute.

An example of a junction object is a System. The System (colloquially called Capn) might have the distinguished name

{c=US, o=DEC, ou=ZKO, systemId="Capn"}

A transport connection managed object contained within that system might have the name

{c=US, o=DEC, ou=ZKO, systemId="Capn", subsystemId="Transport",
CommunicationsEntityId="CO-TP4", ConnectionId="1492"}

A managing system wishing to invoke an operation on that transport connection object, for example to Get its State, would first pass the entire name to the Directory which would resolve as much of the name as possible. In this case, it would resolve the system name

{c=US, o=DEC, ou=ZKO, systemId="Capn"}

The managing system would look up (in the junction directory entry with that distinguished name) the address of the agent serving the System Capn. That address would be used to establish an association to the agent. Note that in the example the agent would run on Capn, but that is not required. The managing system could then invoke a Get operation over that CMIP association using the managed object (instance) name of

{c=US, o=DEC, ou=ZKO, systemId="Capn", subsystemId="Transport",
CommunicationsEntityId="CO-TP4", ConnectionId="1492"}

A junction managed object must know its Distinguished Name (among other reasons so that it can generate a notification). Strictly speaking, the managed object need not expose that distinguished name as an attribute; however, it is recommended that it do so.

5.1 GDMO Templates and Directory Macros

The GDMO template language used to define managed objects provides no easy way to defining a junction object. For each junction object, the definer must define:

- the junction directory entry,
- the junction managed object,
- a name binding from the junction directory entry to the superior directory entry, and
- a name binding from the junction managed object to the superior directory entry, so that both name bindings reference the same distinguished attribute.

The following GDMO templates and directory macros show how the system could be defined as a junction object. The GDMO template for system is from DMI [6]. The name binding template from the system to the organizational unit [7,8] is an example.

```
system MANAGED OBJECT CLASS
    DERIVED FROM top;
    CHARACTERIZED BY
        systemPackage PACKAGE
            ATTRIBUTES
                systemId          GET,
                systemTitle       GET,
                operationalState  GET,
                usageState        GET;;;
    REGISTERED AS { smi2MObjectClass 13 };
```

```
m-system-OU NAME BINDING
    SUBORDINATE OBJECT CLASS System AND SUBCLASSES;
    NAMED BY
        SUPERIOR OBJECT CLASS
            "X.521 | ISO/IEC 9594-7:1988":organizationalUnit AND SUBCLASSES;
        WITH ATTRIBUTE "X.721 | ISO/IEC 10165-2":systemId;
REGISTERED AS {junctionGDMOdefns 1};
```

The following X.500 Directory macros define the system directory entry, and associated name bindings. The definition of the System ID attribute for the directory assigns the same object identifier as that assigned in DMI. The syntax (data type) of the Agent Address attribute is a TowerSet which is defined in figure 4.

```
systemDirEntry OBJECT CLASS
    SUBCLASS OF top
    MUST CONTAIN {systemId, agentAddress}
::= {junctionDirDefns 1}

systemId ATTRIBUTE
    WITH ATTRIBUTE SYNTAX systemId
::= {smi2AttributeID 4}

agentAddress ATTRIBUTE
    WITH ATTRIBUTE-SYNTAX TowerSet
::= {junctionDirDefns 2}
```

The 1988 X.500 Directory standards do not provide an ASN.1 macro for defining the structure rules, which are the X.500 equivalent of a name binding. Structure rules are left as implementation specific. The following STRUCTURE-RULE macro is patterned on the GDMO name binding is used as an example to demonstrate one possible implementation.

```
d-system-OU STRUCTURE-RULE
    SUBORDINATE OBJECT systemDirEntry
    SUPERIOR OBJECT organizationalUnit
    WITH ATTRIBUTES systemId
::= {junctionDirDefns 3}
```

Clearly it would help to have a GDMO extension to make it easier to define a junction object. What is needed is a a way to "equate" a GDMO managed object class definition and a X.500 Directory object class. The following example syntax would serve, although, a single syntax to define both the managed and directory object classes would be even better.

```
jo-System NAME BINDING
    JUNCTION OBJECT CLASS system AND SUBCLASSES;
    EQUATES TO DIRECTORY CLASS systemDirEntry AND SUBCLASSES;
    WITH ATTRIBUTE systemId;
REGISTERED AS {junctionDefns 2}
```

5.2 Junction Objects and Agents

An agent can serve more than one junction object as is shown in figure 1. The example above showed how the system the agent is running in could be represented as a junction object. Other examples include distributed applications and proxy or element managers as described below.

5.3 Distributed Applications

Many distributed applications are built on a simple client/server model where the server provides a location independent service. I.e. the service is not tied permanently to the system it runs on. Users of the service locate it with a naming service such as the X.500 directory. These distributed applications must also be managed, and like the service, the managed objects that represent the application should be location independent. They can be made location independent by defining a managed object representing the server as a junction object. In fact, the junction object can have the same name as the service. The directory entry would hold both the address of the service and the address of the agent, although the addresses would be different.

An example of such a service might be a computer notes conference or bulletin board run over the network. The notes conference might have as its name

{c=US, o=DEC, Conference=Sailing}

and may actually be running on System Capn. By using the X.500 Directory as a name service, users communicating through the notes conference need not be aware of the system the conference is running on. Similarly, by defining the conference managed object as a junction object, the manager need not be aware of the conference's location.

A particular kind of distributed application is a management proxy, or element manager or mediator. These terms all describe *management gateways* of varying levels of sophistication. A managed object representing some network element that does not have an OSI agent, for example a piece of equipment in a telephony network could be modeled as a junction object served by the agent residing in the element manager.

5.4 Junction Objects contained in other Junction Objects

The definition of a junction object allows one junction object to be contained in another. In that case both are represented in the management information tree and the directory information tree as managed objects and directory entries respectively. For example if

{c=US, o=DEC, ou=ZKO}

represents a site (a facility or building) in a corporation, then that site might be a managed object that can be monitored and controlled. For example the power and lighting could be managed. The agent for that site might likely reside on one of the computer systems located at that site (say on System Capn), but the site would not be contained in the system (indeed, it is the other way around). The agent address stored in the directory entry would then point to the agent on System Capn.

6 Enhancements to Junction Objects

This basic concept of a junction object can be elaborated in a number of ways, including:

- Tower Sets
- Aliases
- Automatic Directory Updates

6.1 Tower Sets

In the simplest case, the agent address stored in the junction directory entry is a single P-Address. But, an agent could have multiple P-Addresses and an agent could support more than a pure CMIP/OSI stack. The CMIP Protocol (or a variant) has been run over a 1006/TCP/IP stack, a DECnet stack, and others. Further, agents may support other management protocols based on SNMP, RPC, or other protocols [9].

To deal with these other stacks, the notion of an agent address can be generalized to a set of protocol towers. Each protocol tower identifies a protocol stack and provides any addressing information needed for that protocol to establish an association.

```
TowerSet ::=            SET OF Tower
Tower ::=               SEQUENCE OF Floor
Floor ::=               SEQUENCE {
                            protocol  ProtocolId,
                            address   AddressingInfo }
ProtocolId ::=          OBJECT IDENTIFIER
AddressingInfo ::=      OCTET STRING
```

Figure 4 ASN.1 Definition of the TowerSet data type

6.2 Aliases

The X.500 Directory allows alias names (sometimes called soft links) to be established. A junction object is unlikely to know its alias names. Thus, when a managing system detects that a name is an alias, the managing system should resolve the alias and use the resolved name in the CMIP operation, not the alias name.

6.3 Automatic Directory Entry Updates

The junction directory entry and junction managed object represent two bodies of information about the junction object that must be synchronized. In particular, the agent address stored in the directory entry must equal the actual address of the agent serving the junction managed object. The agent can automate this by automatically updating the agent address in the junction directory entry for junction objects it serves should the agent's address change.

7 Alternatives

A number of alternative ways of dealing with the problem of how to find an agent for a managed object have been proposed including:

1. naming is local to manager
2. every managed object is named in the directory
3. encapsulate directory names as values in an AVA
4. the superior of a managed object is directory entry

Each of these is described in a section below.

7.1 Naming is local to manager

Instead of using a distributed naming service, each managing system could have a local, private name to address translation service, for example through a local database. A local service will perform better, however, because the data is private keeping them synchronized and updating them when changes occur are difficult. In particular, the agent cannot automatically update them as described in §6.3.

7.2 Every managed object is named in the directory

It has been proposed that every managed object has a directory entry. One problem with this approach is the sheer number of managed objects. Moreover, many managed objects have a short lifetime, and it is impossible to create and destroy directory entries in that time, for example a Transport Connection.

7.3 Encapsulate directory names as values in an AVA

Junction objects *graft* the management information tree onto the directory information tree. An alternative approach treats the directory and the management names as distinct, and encapsulates the directory distinguished name as the value in a management RDN. The approach can best be considered by comparing the grafted name for a transport connection used in the examples above

{c=US, o=DEC, ou=ZKO, systemId="Capn", subsystemId="Transport",
communicationsEntityId="CO-TP4", connectionId="1492"}

with a name constructed by encapsulating a directory name for the system

{systemDirectoryName={c=US, o=DEC, ou=ZKO, systemCommonName="Capn"},
subsystemId="Transport", communicationsEntityId="CO-TP4",
connectionId="1492"}

Encapsulation adds an extra level of nesting which makes interpreting the name slightly harder. Worse, it makes the separation of the directory name from the managed object name explicit. If the split is in the wrong place changing it will invalidate the old names. Changing how much of the name is stored in the directory can be done without affecting the names of existing managed objects. Encapsulation also prevents implementing one junction object as being contained in another as described in §5.4.

7.4 The superior of a managed object is directory entry

As noted earlier, GDMO does allow a name binding to define a directory entry as a superior of a managed object. Indeed, that feature was needed to define a junction object and allows managed object DNs to be grafted onto the DIT. Grafting and encapsulation concepts are described in [10]. By itself, establishing a name binding from a managed object to a superior directory entry does not define how that managed object is located. A directory lookup of the managed object name will produce the directory entry for the superior, not the managed object. That directory entry would either:

- point to something (a database) which would translate names to addresses, which introduces the problems of a local naming database described in §7.1, or
- provide the name of a "gateway" agent (an element manager or a proxy agent) that would forward the operation to the right system.

In either case, an extra level of indirection and an extra point of failure have been added.

8 Summary

Junction objects are based on experience with closely related architectures. In particular:

1. The concept of a junction object is primarily based on lessons learned with DECnet/OSI [11,1] and the EMA Entity Model [12,13]. EMA includes the concept of a *global entity* which is directly analogous to a junction object. There are two minor differences. The naming service used in EMA is DECdns[14], which has hierarchical, but untyped names. DECdns names were encapsulated within management names much as described in §7.3, with the attendant problems described there. However, in all other respects, using a naming service to locate an agent for a managed object, the support of TowerSets, the support of a automatic directory update service, the basic concepts behind a junction object have all been successfully implemented in DECnet/OSI.

2. OSF DCE uses a concept of *junctions* in DCE/RPC. There as here, the name of an object (a file object or a application object) is grafted onto a CDS name.

The concept of a junction object, as described here, adds a missing piece to the OSI management standards, the procedure a managing system uses to locate the agent for a managed object based on its name. Standards must provide enough information to allow conforming implementations to interoperate. Without an agreed upon means of establishing an association to the correct agent, OSI management implementations cannot interoperate. The concept of a junction object provides the best solution to this problem.

9 References

1. M. Sylor, "DECnet/OSI Phase V Network Management", *Integrated Network Management II*, I. Krishnan & W. Zimmer (ed.), Elsevier Science Publishers, 1991, pp. 57-69.
2. ISO/IEC DIS 10165-1 | CCITT Rec. X.720, *OSI Management Information Services - Structure of Management Information - Part 1: Management Information Model.* (MIM)
3. ISO/IEC DIS 10165-4 | CCITT Rec. X.722, *OSI Management Information Services - Structure of Management Information - Part 4: Guidelines for the Definition of Managed Objects.* (GDMO)
4. ISO/IEC DIS 10737, *Transport Layer Management.*
5. CCITT Rec. X.500 | ISO/IEC 9594-1, *The Directory - Overview of Concepts, Models and Services : 1988.*
6. ISO/IEC DIS 10165-2 | CCITT Rec. X.721, *OSI Management Information Services - Structure of Management Information - Part 2: Definition of Management Information.* (DMI)
7. CCITT Rec. X.520 | ISO/IEC 9594-6, *The Directory - Selected Attribute Types : 1988.*
8. CCITT Rec. X.521 | ISO/IEC 9594-7, *The Directory - Selected Object Classes : 1988.*
9. O. Newkerk, M. Nihart, S. Wong, "The Common Agent", to be published in *IEEE JSAC*, forthcoming special issue on network management.
10. S. Zatti, J. Ashfield, J. Baker, E. Miller, "Naming and registration for IBM distributed systems", *IBM Systems Journal,* vol. 31, no. 2, 1992, pp. 353-380.
11. Digital Equipment Corporation, *Digital Network Architecture (Phase V) Network Management Functional Specification,* 1991, No. EK-DNA02-FS-001.
12. Digital Equipment Corporation, *Enterprise Management Architecture (EMA) Entity Model*, 1991, No. EK-EMAEM-SP-001.
13. M. Sylor, "Managing DECnet Phase V: The Entity Model, *IEEE Networks*, Mar 88, p 30ff.
14. S. Martin, J. McCann, D. Oran, "Development of the VAX Distributed Name Service", *Digital Technical Journal*, vol. 1, no 9, Jun 89, pp. 9-15.

I.

MANAGEMENT ARCHITECTURE

F.

DISTINGUISHED EXPERTS PANEL

International Integrated Network Management – Are We There Yet?

Panel Chair: Roberta Cohen, AT&T Bell Laboratories, USA

Many in the industry would claim that integrated Network Management has proven to be a worthy but an elusive goal for those operating networks throughout the world. Much attention has been directed at protocol issues, perhaps at the expense of application capabilities and system usability. Five distinguished representatives of the Network Management community will address themselves to the issues and challenges of making integrated network management a reality worldwide.

II.

DISTRIBUTED SYSTEMS MANAGEMENT

A.

DOMAINS AND POLICIES

Integrated Network Management, III (C-12)
H.-G. Hegering and Y. Yemini (Editors)
Elsevier Science Publishers B.V. (North-Holland)

The DOMAINS Management Language

Axel Fischer[a], Martine Herpers[b], David Holden[c] and Stephan Sievert[d]

a ITK Informationstechnologie Kiel GmbH, Holzkoppelweg 5, 2300 Kiel 1, Germany, <axel@itk.de>

b Philips Kommunikations Industrie AG, P.O. Box 35 38, 8500 Nürnberg 1, Germany, <martine@pki-nbg.philips.de>

c AEA Industrial Technology, Harwell Laboratory, Building 521.2, Didcot, Oxon OX11 ORA, United Kingdom, <holden@harinst.aeatech.uk>

d ITK Informationstechnologie Kiel GmbH, Holzkoppelweg 5, 2300 Kiel 1, Germany, <stephan@itk.de>

Abstract
The DOMAINS Management Language has been developed during the ESPRIT II DOMAINS project and was required to be: conform to GDMO; compatible with the DOMAINS architecture (based around management domains and separation of managed from managing parts through shields); object-oriented; easy and safe to use. DML achieves these aims by taking the GDMO template language and extending and enhancing it: by adding Managing Objects; by providing a formal behaviour language for specifying the complete behaviour of managed and managing objects; adding a Notification Handler template for Managing Objects; replacing the GDMO Name Binding scheme with a domain-based Hierarchical Binding scheme for defining hierarchical management structures; providing a means of specifying dynamic behaviour through the use of Policy.

The International Standard on Guidelines for the Definition of Managed Objects (GDMO) is a standard which defines a method of defining objects involved in network management activities and is used by many other International Standards. Because of the importance of standards, and because of its applicability to non-standard management approaches (due to GDMO's necessarily general nature), the GDMO template language is being adopted by commercial systems producers to support their systems management. However, GDMO is limited to the scope of Open Systems Interconnection (OSI) standards and consequently has a number of drawbacks when it is applied to real systems management: GDMO allows the specification only of the managed system, not any system doing the managing; GDMO allows the specification only of the externally visible aspects of the managed system's behaviour; the specification of behaviour is informal (natural language).

This paper describes the key issues involved in designing DML and gives examples of the syntax and semantics of this GDMO-conform language.

Keyword Codes: D.1.5; D.3.3; K.6.4
Keywords: Object-oriented Programming; Language Constructs and Features; System Management

1. Introduction

This paper looks at the results of a study into architectures and methods suitable for building managed distributed systems, and in particular at work carried out to implement the findings of this study in the form of a language (and its associated compiler) for managing the system. This work was carried out during the ESPRIT II project DOMAINS. The DOMAINS project (Distributed Open Management Architecture In Networked Systems) is strongly architecture based, but also involves implementation to prove that the architecture can be realised. Furthermore, the implementation is oriented towards international standards in accordance with current trends. The DOMAINS Management Language (DML) [DOM_D2e] is therefore influenced mainly by the DOMAINS architectural principles, described in section 2 of this paper, and by the International Standard on Guidelines for the Definition of Managed Objects (GDMO) [GDMO].

GDMO is the ISO OSI standard which underpins many other standards concerned with management. This is because it defines a "language" for use by other standards for defining objects involved in the management activity. Since these objects can be quite varied GDMO has to be a very general standard. Thus the language it defines is applicable not only to standard objects, but can also be used to define private, system-specific objects. This has led to the GDMO language being adopted by commercial systems producers to support GDMO for use in systems management, whether based on the standards which use GDMO or not.

The GDMO language is defined by a set of templates, each one of which is used for defining a different aspect of an object. The main template is the Managed Object Template, which defines an object in a management system which is used as the interface point between the manager and some object in the system being managed. All other templates are used for defining a part of the Managed Object Template, for instance the Attributes of the object (measurable and/or settable values in the object), the Behaviour of the object, what Actions can be performed on it, and what Notifications it can generate. However, because GDMO is an OSI standard, and because OSI standards are necessarily limited to defining only the interactions between entities reachable through OSI protocols and not any activity within them, GDMO is similarly limited to defining how the objects appear from outside the open system in terms of interfaces to objects and the syntax and semantics of information passed to and from them. GDMO provides a standard way of specifying and communicating with Managed Objects, but only a natural language way of specifying their behaviour. Thus although it is possible to write a program which interacts with such objects and programs which emulate them, the manager programs, the actual behaviour of the managed objects and the management structure are all decided by the programmer and cannot be described as standard. The DOMAINS implementation language, DML, aims to overcome these problems by providing:

1. a *Managing Object Template*, used to define the behaviour of an object which manages other parts of the management system;
2. a *formal Behaviour language* for programming the complete behaviour of managed and managing objects;
3. an *Hierarchical Binding Template* for defining the structure of the management system in accordance with the DOMAINS architecture: this replaces the GDMO Name Binding Template.

DOMAINS also provides a means to specify the dynamic behaviour of objects through the use of Policy [DOM_D2c], though this is not currently supported by DML.

This paper starts with the principles of the DOMAINS Architecture since the language design has been heavily influenced by that architecture. Requirements coming from that architecture as well as other requirements we considered for the language design are presented in section 3. In section 4 we compare DML with the GDMO template language and point out the enhancements DML provides. A quick overview over the DOMAINS

Management Language is given in section 5, split between the DML Template Language and the DML Behaviour Language.

2. Principles of DOMAINS

The DOMAINS Management Architecture (DMA) [DOM_D2c] maintains a distinction between the management part and the resources that are subject to management. This architecture supports hardware resources as well as software resources and simple components as well as complex systems.

To realize such an approach, the DMA leans on two basic structuring principles. The first one is to master complexity and scale by construction of logical units aiming at the separation of various management concerns. These units are called *domains* which comprise the allied managing and managed components. Corresponding to specific management needs there may exist totally disjoint, partially overlapping, or hierarchically nested domains. The second structuring principle is related to dynamic configuration of management systems. Its aims are the clear separation of the managing and managed parts on the one hand, and their structured interconnection on the other hand. This is realized by mutually shielding the managed and managing part from each other by a concept called a *Shield*. This leads to the internal structure of a domain having a triple partioning:

domains = (Kernel, Shield, Target)

The Kernel and Target (a set of Target Resources or again domains in case of hierarchical domains) are the real managing and managed parts, respectively, whilst the Shield provides precisely the relevant and/or expected view to and from each of them.

Structuring principles for domains
Management domains are formed by grouping managing and managed components according to one of several possible criteria. A domain may be built restricting the Kernel management activity to one applied policy. Alternatively, a so-called feature criterion can be used to build a domain. In such a domain several policies may be applied simultaneously to (overlapping) subsets of the contained resources. Examples of criteria for feature domains are geographical distribution or the ownership of the Targets.

Kernel and Manager activities
There is exactly one authorized kernel per domain [DOM_D2c], and this contains all of the management activity of the domain. A kernel may (conceptually) consist of several managers.

Management control is directed in only one way from the higher-level Kernel to the lower level ones. A lower-level Kernel may only use notifications to trigger management activities in kernels of the higher level.

Higher -level Kernels send management goals to a lower-level Kernel who derives Policies from the goals. Plans are derived according to the used Policy. Thus, Kernels execute dynamically interchangeable Plans to fulfil the management goals given by higher-level Kernels.

Shield functionality
The Shield provides the Kernel with an abstract view of the Target resources that it is managing. These resources may be real resources outside the DOMAINS system (e.g. modems, switches or printers) or other domains lower down the DOMAINS management hierarchy. The view may be:

- A uniform view of a number of different resources.
- A selective view which restricts visibility to the managing Kernel to that part of the resources interface relevant to the management goals of that Kernel.

The Shield also provides the Target resource with an abstract view of the Kernels which are managing it.

The Shield comprises a set of Shield elements, each one interrelating precisely one Target resource to the same Kernel. However one and the same Target resource may be related to several Kernels and thus be represented by several Shield elements, each one providing another view tailored to the needs of the related Kernel.

The Shield has a wide range of functionality covering:

- Visibility Control
- Message Forwarding
- Protocol Transformation
- Polling and Notification
- Computation and Aggregation

3. Requirements on DML

The design of DML is influenced by the requirements from the DOMAINS management architecture, as well as requirements from the aim of the DOMAINS project to be conform to and to influence management standards, the DOMAINS object architecture and the management application programmer.

DOMAINS Management Architecture
The requirement coming from DOMAINS management architecture are:

- DML has to guide management application programmers in using the DMA concepts correctly. Thus it should mirror the concepts of domains, kernels configurable by plans, shields and targets. Using the DMA concepts will conflict with the following requirement to be conform to standards.
- The language must support both reliability/safety and dynamic (re)configuration wrt. DOMAINS domains.
- DML has to be a strongly typed language, with static type checking.

Standards
An important requirement for the design of DML is GDMO conformity: DML has to be conform as far as possible to the GDMO. This implies that existing GDMO Templates should be usable as a framework for management applications. At the same time, DML Descriptions of managed objects should be usable by other systems conform to GDMO.

DOMAINS Object Architecture
DOMAINS Management Applications are running on a virtual object machine with its own object-oriented architecture. The requirements of the architecture itself upon DML are:

- Object Management: only the specifications of object classes appear. Creation and deletion is extra-linguistic.
- Communications management: DML will have explicit statements to indicate transaction groupings from non-transactional groupings. It will also distinguish between *ordinary* invocations inside DOMAINS and communication with real resources.

Management Application Programmer
To ease the work of the management application programmer, the following requirements are established:

- Inheritance: DML will include statements for inheritance of object classes with the possibility of re-defining and renaming of operations. DML permits multiple and repeated inheritance but not selective inheritance.

- Polymorphism: DML will include polymorphism constrained with inheritance (at least). If possible allomorphism should be included.
- Attributes: The scope and the visibility of attributes is at the object level.
- Object set-up: DML will have statements for the specification of object set-up operations.
- Assertions: DML will include statements for the specification of pre- and post-conditions to improve the correctness of an operation. The conditions have to stay valid also if the operation is redefined in an inheriting class. They can be extended in an inheriting class by adding boolean expressions with the logical and operator.
- Complex data types: In DML it is possible to define complex data types with ASN.1 [ISO8824].
- Invocations: DML supports synchronous as well as asynchronous invocations.
- Objects: DML includes the possibility to define private (nested) objects in addition to public (shared) objects.

4. Key Features

This section gives a short summary of the main key features of DML [DOM_D2e] in comparison with GDMO [GDMO].

GDMO provides a standard way of specifying and communicating with Managed Objects (MOs), but only a natural language way of specifying their behaviour. DML enhances GDMO in 3 ways:

1. allows the specification of Managing Objects as well as Managed Objects in a standard way;
2. provides a procedural language (Behaviour Language) for specifying the behaviour of Managed and Managing Objects;
3. provides means to specify the dynamic behaviour of Managing Objects through the use of policy.

With respect to the DOMAINS concepts, the Managing objects are described in KERNEL OBJECT CLASS templates, while Managed Objects are given by the SHIELD ELEMENT CLASSES which provide a certain view to the MO (see subsection 5.1). Thus existing GDMO templates for MOs have to be adapted to the Shield Element syntax which causes some slight syntactical changes.

In providing a formal *behaviour language* (see subsection 5.2), DML gives the means for programmers to program the management objects in their entirety, in contrast to GDMO definitions which only specify those aspects of behaviour relevant to the external view of the objects. For instance, in DML the Notification Handler template may be used to specify the *reaction to a certain incoming notification.* And because DML also provides templates for Managing Objects and *Hierarchical Binding*, it is possible to define a complete management system, including the behaviour of managers and the structure of the management system in terms of a *hierarchy of managers* in a formal and standardized way.

Policy is a general term for the method by which a manager will achieve a particular goal or set of goals. Goals which are always applied to an object lead to static policies, which can be encoded in a static behaviour of the Managing or Managed Object using DML [DOM_D2c]. Such policies, however, are a kind of static plans running the whole lifetime of an object. To make plans exchangeable during the lifetime of an instance, DML provides *Plan Object Classes* to which the Managing or Managed Object can refer. A formal goal and policy language allowing the automatic derivation of policies from goals is still under study.

5. Realization

This section deals with the template and the behaviour language of DML. While parts of the syntax of the templates will be presented in subsection 5.1, the Behaviour language is discussed in a more theoretical way, because the procedural syntax of Eiffel [Meyer 88] is adopted for this part of DML.

5.1 The Template Language

The DML provides four different Object Class templates: the *KERNEL, SHIELD ELEMENT, PLAN* and *SUPPORT OBJECT CLASS* templates. According to their different meanings with respect to the DOMAINS architecture, all these templates are composed in slightly different ways.

It should be noted that all of the non-Object Class templates and the ASN.1 data descriptions may be included inline or be included in a remote module and referred by the name of the template.

Enhancements comparing DML with GDMO on the object level are:

- The DERIVED FROM clause may only refer to the respective kind of object classes. Thus a Kernel object may only inherit from another Kernel object.
- RENAMING and REDEFINITION Clauses are included in each kind of object class template. (eg. see KERNEL OBJECT CLASS template, subsection 5.1.1)
- Each kind of object class template provides an optional OBJECT CONTROLLER CONFIGURATION clause.

Subsection 5.1.1 describes the KERNEL and SHIELD ELEMENT Object Class template, while subsection 5.1.2 discusses the differences of the templates for Attributes, Group Attributes, Actions, Notifications and Notification Handlers. The last subsection (5.1.3) deals with the means to describe to hierarchical linking of the DML objects, the Hierarchical Binding template.

5.1.1 Object Classes

The Kernel Object Class contains three description parts:

- The optional *managed part* which is compatible as far as possible to the GDMO. Here all Packages which have to be provided for the higher management are included.
- An optional *management part*, where all packages which have to be provided for the management of the targets are included.
- An optional *peer-to-peer part*, where all packages which have to be provided for the peer-to-peer communication between Kernel Objects are included.

A description of a Kernel Object Class Template is shown in the following box.

The Shield Element, Plan and Support Object Class are similar to the MO template of GDMO enhanced by Renaming, Redefinition and Object Controller Configuration Clauses. The difference between these different kinds of objects is of a purely semantical nature.

A Shield Element object represents one Shield Element of the DMA.

A Plan object should be used for the dynamic algorithmic part of the management activities.

A Support object template shall be used for non management objects only. Thus it is not allowed to define any management behaviour in a support object. The purpose of these objects is to allow the reuse of existing objects for data bases or statistical computations.

```
<kernelclasslabel>    KERNEL OBJECT CLASS
[ DERIVED FROM       <kernelclasslabel> [RENAMED <name> AS <name> [ <name> AS <name> ]*]
                     [, <kernelclasslabel> [RENAMED<name> AS <name> [ <name> AS <name> ]*] ]*
;]
[ REDEFINED                <name> REDEFINITION beh-decl [, <name> REDEFINITION beh-decl]* ;]

[ ALLOMORPHIC SET          <kernelclasslabel> [, <kernelclasslabel>]* ;]

[ OBJECT CONTROLLER CONFIGURATION  obj-controller-decl; ]

[ MANAGED PART
[ CHARACTERIZED BY         package-decl [, package-decl]* ;]
[ CONDITIONAL PACKAGES          package-decl PRESENT IF <cond-def>
                           [, package-decl PRESENT IF <cond-def>]* ;]
]
[ MANAGING PART
[ CHARACTERIZED BY         package-decl [, package-decl]* ;]
[ CONDITIONAL PACKAGES          package-decl PRESENT IF <cond-def>
                           [, package-decl PRESENT IF <cond-def>]* ;]
]
[ PEER TO PEER PART
[ CHARACTERIZED BY         package-decl [, package-decl]* ;]
[ CONDITIONAL PACKAGES          package-decl PRESENT IF <cond-def>
                           [, package-decl PRESENT IF <cond-def>]* ;]
]
[ REGISTERED AS            <obj-identifier>]
;
```

5.1.2 Non-Objects Templates

The DML supports the use of *Packages* in the same way as GDMO. In addition to the GDMO templates for Behaviour, Attributes, Group Attributes, Actions and Notifications, the DML provides a Notification Handler template. In order to include Notification Handler templates in a Package the keyword Notification Handler in the Package template is provided.

The most important enhancement over GDMO is the possibility to include procedural code in the Behaviour template making DML to a implementation language (whereas GDMO may only be used as means for MO specifications).

Beside the textual and procedural Behaviour description, the Behaviour can be externally defined in other programming languages (keyword: EXTERNAL) or may be denoted as deferred (keyword: DEFERRED) or may be left empty (keyword: EMPTY).

There are several restrictions for the use of the Behaviour clause in pure DML parts of the language. Thus the Behaviour description of pure DML parts are not allowed to contain plain text. The Behaviour of Notification templates in DML is used to define the condition to transmit this Notification.

The template shown below may only be used for the functional description (which has to be enclosed by commercial at (@) signs). In case of textual description the DEFINED AS keywords are followed by a string enclosed by any of the delimiters defined by GDMO for this purpose except @.

```
<beh-label>  BEHAVIOUR
DEFINED AS
@
[ VARIABLES              <var-def>          [,<var-def>]* ;]
[ EXCEPTION HANDLER      <stmt>   [ <stmt> ]* ;]
[ DO                     <stmt> [ <stmt> ]*                 END ]
@
;
```

Three parts are included in the DML procedural Behaviour definition: the declaration part of local VARIABLES, the EXCEPTION HANDLER part providing (like Eiffel) the possibility to define procedures as reactions to arising exceptions. The statement RETRY may be used in this Behaviour part. The third part contains all necessary Statements to control the management activities. For further details see subsection 5.2.

For the other templates, the following differences between DML and GDMO should be noted:

- Actions are enhanced with pre- and post-condition Clauses (keywords REQUIRE, ENSURE). These conditions may not be altered by the redefinition of the Action. The same holds for the Notification Handler template.
- Notifications may not be included in packages of Plan or Support Objects.
- Notification Handler may not be included in packages of Support Objects.

The DML concept of the *Notification Handler* is a special kind of Action. Each time a Notification is received by an object the respective Notification Handler has to be invoked. Notifications are always asynchronous calls.

```
<notif-handler-label>  NOTIFICATION HANDLER
BEHAVIOUR                              beh-decl;
[ REQUIRE                        <invariant> ]
[ ENSURE                         <invariant> ]

[WITH INFORMATION SYNTAX         [<attr-label> :]<syntax-decl> ;]
[ REGISTERED AS                  <obj-identifier> ]
;
```

5.1.3 Hierarchical Binding template

The Hierarchical Binding template is the means to describe the hierarchical relationships between Kernel and Shield Element Objects. It is also used for the definition of Plan / Kernel links and the linking of Support Objects to any other DML Object. The Name Binding template of GDMO will not be supported by DML because of the different naming schemes of OSI and DOMAINS.

```
<hierarchical-binding-label>    HIERARCHICAL BINDING

SUPERIOR OBJECT CLASS                  <class-label>;
WITH ATTRIBUTE                         <attr-label>;
CONNECTED TO
SUBORDINATE OBJECT CLASS               <class-label>;

[ REGISTERED AS                        <obj-identifier> ]
;
```

5.2 The DML Behaviour-Language

Behaviour descriptions according to the GDMO may be documented either by use of natural language text or by the use of formal description techniques. The DOMAINS Management Language has been defined to allow specification of DOMAINS management systems which can be executed on top of a DOMAINS object machine. This implies usage of a *formalized* behaviour description. The formalism used in DML behaviour descriptions is partly adopted from the Eiffel [Meyer 88] language. In order to be conform to the GDMO natural language definition are accepted as well but obviously cannot be executed.

5.2.1 Implications of formalized Behaviour

Behaviour is described in DML in a procedural manner. Therefore the description results in most cases in threads executing statements. This implies that information on the required behaviour can only be distributed over several templates when concurrent execution is possible. Consequently, DML allows in most cases only one formalized behaviour where GDMO allows to specify a list of behaviour definitions.

5.2.2 Key Features of the Behaviour Language

Beside standard features like conditional Clauses, assignments and loops which have been adopted from EIFFEL, and operators for all supported ASN.1 data types, DML features the following properties:

Formalized behaviour
DML not only supports formalized behaviour but also natural language behaviour which is treated as a comment.

Local variables
DML provides *local variables* which can only be accessed from the relevant Behaviour. For local Variables, properties similar to the attribute properties can be specified, e.g. the permitted value-ranges can be restricted.

Assertions and exception handling
In DML, constraints on operations can be specified in order to satisfy consistency rules. A violation of these rules as well as other runtime-errors are handled by a disciplined, EIFFEL-like exception mechanism. Raised exceptions may be handled by behaviour-specific *Exception Handlers*.
A *Retry* statement is available as well as operators determining the originator and the reason of an exception.

Deferred Classes
DML adopts the concept of *deferred classes* from EIFFEL. This means that the behaviour of an action or notification handler needs not be specified for each class but can declared as *deferred.* A class offering such deferred actions cannot be instantiated but only used as super-class. The relevant behaviour is defined in a sub-class using redefinition.

External Operations
In order to allow reuse of existing software or specification of properties which cannot expressed in DML, behaviour may be declared as *external.* External routines can be specified in any other programming language as long as they can be linked to the compiled DML-code.

Empty Behaviour
For actions and notification handlers occurring in a Shield Element Object it is possible to declare a behaviour as *empty.* This states that the behaviour belongs to a virtual routine which simply passes information without processing to the target resource or the manager respectively.

Invocations
DML supports four kinds of invocations: synchronous invocations *(Call)*, asynchronous invocations with or without results *(Fork* or *Cast)* and explicit generation of notifications *(Notify)*.

Transactions
To increase the reliability of the management-system, actions can be performed as transactions, i.e. the system keeps track of the ACID properties (Atomicity, Consistency, Isolation, Durability).

Interface to the DOMAINS Object Machine
DML programs are to be executed on an environment we call the *DOMAINS Object Machine*. This machine for instance is responsible for creating and deleting instances and for communication between different objects. Communication between DML programs and the Object Machine take place in two different ways:

- By explicit invocation of operations offered by the Object Machine,
- By Clauses integrated in the DML

5.2.3 Behaviour Descriptions in GDMO and DML

In this section we investigate where behaviour descriptions occur in GDMO, what information is given in those definitions and how this information is expressed in DML. Differences result mostly from the fact that we try to avoid distributing information over several templates, where possible.

Package-Behaviour
A GDMO package definition provides a specification of all behaviour aspects concerning that package. This includes

1. The circumstances under which Notifications are generated. DML offers two possibilities to encode this information:
   - The Notification Behaviour or
   - The *Notify*-statement which explicitly generates a notification.

   Both options can be used simultaneously in the same Package definition.
2. Identification of attributes corresponding with information in notifications, if any. In DML this information is either taken from the *AND ATTRIBUTE IDS* clause of the NOTIFICATION Template or explicit specified as parameter of a *Notify* statement depending whether the Notification is generated implicitly or explicitly.
3. The effect of the Operations upon a managed object. In DML that information is found only in the behaviour description of the Action- or Notification-Handler-Template respectively.
4. Any constraints that are placed on operations in order to satisfy consistency rules. In DML this information is not encoded in a behaviour description but in Pre- and Post-conditions defined in the Action- or Notification-Handler-Template respectively. This avoids unnecessary distribution of Action-related information over several templates.
5. Rules under which Creation and Deletion of managed objects may be performed and the consequences of these operations. Creation and Deletion of objects are not controlled within DML, since this task is performed by the DOMAINS Object Machine. Package Behaviour however may contain an Object Set-up clause. In this clause those (initialization) operations are defined to be executed before the object will be accessible by other objects.
6. A Specification of how instances of (managed) object classes interact with other objects. This information is encoded in the relevant Behaviour descriptions referred directly or indirectly by the object class.

Parameter-Behaviour and Name-Binding Behaviour
DML does not support the *Parameter*-Templates of GDMO. They will be accepted by the DML compiler but treated as a comment. Therefore the Behaviour may still be defined in plain English.

Name Binding Templates are not supported in DML due to a different naming scheme required by the DOMAINS object model. Therefore Name-Binding Behaviour is not of concern.

Attribute-Behaviour
The GDMO provides an Attribute Behaviour to describe how the chosen set of matching rules are applied to the attribute definition. DML does not provide a formalized Attribute Behaviour. The matching rules are implemented by certain Actions of the object holding the attribute. This behaviour, although predefined in DML, can be rewritten if complex matching rules are required. However, Attribute Behaviour definitions containing text-behaviour are allowed in DML. These definitions are treated as comments.

Notification-Behaviour
GDMO uses Notification Behaviour to specify the circumstances under which a notification of that type is generated. To express that information, Notification Behaviour in DML contains a boolean expression, relying only on attribute values of the relevant object. A

notification of that kind is generated when the expression evaluates to *true*. The information which is sent out with the information is determined by the *WITH ATTRIBUTE IDS* clause of the Notification Template.

Since this concept might not be powerful enough to cover all situations Notifications might also be generated explicitly by *Notify* requests.

Action-Behaviour and Notification Handling
Behaviour of Actions and Notification-Handlers (the latter are not specified in GDMO) are described in the same manner.

In contrast to the GDMO, DML demands that all aspects of the action behaviour are specified in the same template. GDMO for example allows the programmer to specify class-specific aspects in the package behaviour. To specify class specific action behaviour in DML, the behaviour for that class must be redefined.

The following BEHAVIOUR Template defines an Action Behaviour where another object is polled continuously. If polling fails, a notification *pollingFailed* is generated and polling is retried at most 5 times. After 5 unsuccessful retries no further exception handling is performed which means that execution of the *pollBehaviour* has failed and that this situation has to be handled by the invoker.

```
pollBehaviour          BEHAVIOUR
DEFINED AS
@
VARIABLES              retries INTEGER;
EXCEPTION HANDLER
    NOTIFY pollingFailed;
    IF (retries<5) THEN
            retries:=retries+1;
            RETRY;
    ENDIF;
DO
    FROM retries:=0; UNTIL FALSE LOOP
            CALL otherObject.doIt(4711->result);
    ENDLOOP;
END
@
;
```

6. Summary

In this paper we presented the DOMAINS Management Language (DML) which is an enhancement of the GDMO language for definition of managed objects. Since DML has been developed mainly for definition of DOMAINS Management Systems, we presented the key features of the DOMAINS Management Architecture and other requirements we followed during language definition.

The DOMAINS Architecture aims to reduce the complexity of management by introducing domains which may be recursive (each manager may be the subject of management as well) and overlapping. Each domains consists of three parts: A Kernel which is the managing part of the domain, a Target Resource which is subject to management and a Shield which provides precisely the relevant and/or expected view between the Kernel and higher domains.

Since DML is an object-oriented language designed to describe complete management systems instead only managed objects, we support different kinds of object classes. Instances of the same or different object classes communicate with each other either by invocations or by notifications. DML provides an EIFFEL-like formalism to describe the internal behaviour of an object. This allows DML-programs to be compiled and executed on the DOMAINS Object Machine which is an implementation of the DOMAINS Object Model on top of ANSAware.

We achieved compatibility to the GDMO in so far as we adopted most GDMO templates and extended them where GDMO is insufficient for description of DOMAINS Management Systems. Except for the naming mechanism which is totally different from GDMO, all GDMO templates are accepted by DML compilers. Conversely, DML is close enough to the GDMO, that most templates would be accepted by GDMO Template Parsers as well.

Key features of the DOMAINS Management Language are:

- Object Orientation,
- Inheritance which may be multiple and repeated,
- Conformance to the GDMO which includes usage of ASN.1 data types,
- Static type checking and semantics rules which support programmers in writing consistent DOMAINS Management systems,
- Access to external routines which can be linked to DML programs.

An example compiler for DML has been implemented during the DOMAINS project generating DPL/IPL code for the ANSAware platform. The compiler has been used to implement a demonstrator management application proving that the DOMAINS Architecture is applicable to non-trivial real-life problems.

7. References

[DOM_D2c] ESPRIT Project 5165 DOMAINS
DOMAINS Management Architecture
Deliverable D2c,
May 1992

[DOM_D2e] ESPRIT Project 5165 DOMAINS
DOMAINS Management Language
Deliverable D2e,
May 1992

[Fink] Barbara Fink et al
DOMAINS - Basic Concepts for Management of Distributed Systems
Proceedings of the 5th RACE TMN Conference
London, November 1991

[GDMO] Open Systems Interconnection - Management Information Services
Structure of Management Information Part 4
Guideline for the Definition of Managed Objects
ISO/IEC 10165 - 4
July 1991

[ISO8824] ISO IS 8824
Specification of Abstract Syntax Notation One (ASN.1)
May 1987

[Meyer 88] Bertrand Meyer
Object-oriented Software Construction
Prentice Hall International, Series in Computer Science
1988

Integrated Network Management, III (C-12)
H.-G. Hegering and Y. Yemini (Editors)
Elsevier Science Publishers B.V. (North-Holland)

Domain Management and Accounting in an International Cellular Network

Morris S. Sloman[a], Barry J. Varley[b], Jonathan D. Moffett[c] and Kevin P. Twidle[a]

[a] Department of Computing, Imperial College, 180 Queen's Gate, London SW7 2BZ, UK
Email: mss@doc.ic.ac.uk

[b] Sema Group, 233 High Holborn, London, WC1V 7 DJ, UK.

[c] Department of Computer Science, University of York, Y01 5DD, UK

Abstract

This paper discusses the application of management domains to management of an international cellular network. The Pan European Cellular Network is made up of multiple network vendors in different countries offering a variety of services. Management Domains provide the means of grouping network resources or subscribers in order to specify service and control the services that a subscriber is entitled to use. We explain how such a system can be used in an environment which permits roaming of mobile subscribers between networks in different countries, while a call is in progress, even though the current Pan European Network proposals do not permit this. The paper also discusses accounting issues in such an environment and describes the mechanisms needed to cater for roaming during a call.

Keyword codes: C.2.4; K.6.4; K.6.5

Keywords: Distributed Systems; System Management; Security and Protection;

1. INTRODUCTION

The concept of a Domain as a means of grouping objects for management purposes has been identified in a number of papers [Sloman 89, DOMAINS 91, Strutt 91]. The DOMINO Collaborative Project has been investigating the application of domains to the management of Open Distributed Systems. The objectives of the project were to refine the domain concept and to test out the concepts and techniques developed within the project by applying them to a case study. The International Cellular Communications Network was selected as a suitable application as it had many natural domains corresponding to countries and to supplier organisations within countries. In addition the issues involved in accounting for calls which crossed international boundaries, with different tariff policies, were very complicated. It appeared that domains would provide a means of structuring the accounting management as well as providing a means for specifying what services are accessible to subscribers.

It was not practical within the resources available to the project to experiment with a real cellular network so a simplified distributed simulation was used.

This paper describes how domains can be used as a means of grouping objects to which a common policy applies within a large inter-organisational environment such as the International Cellular Network. We first give a brief introduction to the Pan-European Cellular Network in order to understand the complexity of the management issues to be solved. In section 3 we give an overview of the domain concepts developed in the DOMINO project. In section 4 we discuss how the domain concepts benefit the management of cellular networks and then describe how domains can be used to specify and control what services a subscriber is entitled to use. In section 6 we explain how accounting could cope with calls which roam between international networks.

2. INTRODUCTION TO THE PAN-EUROPEAN CELLULAR NETWORK

2.1 Cellular Communication Concepts

In the cellular network, a subscriber's **mobile station** (referred to as a mobile) communicates by radio with a fixed **base station** which covers an area called a **cell.** Cells may range in size from 500 metres diameter in an urban environment to about 30 Km in rural areas. A group of base stations is connected to a **Mobile Switching Centre** (MSC), usually by fixed-wire links. We refer to a MSC as a switch in this paper. A switch provides inter-base station communication, control and accounting. They are connected together in the same manner as a conventional telephone network, with one or more of the switches also acting as gateways into other telecommunication networks.

When a mobile is switched on, it locates and connects to the base station that is providing it with the best radio signal. If the subscriber is moving while a call is in progress it may move out of one cell and into an adjoining cell. When this happens the radio connection is switched to a base station in the adjoining cell. This is known as a **handover** and is hardly detectable by the user.

2.2 Services Provided

The services provided by a vendor network are based on the Integrated Services Digital Network (ISDN) concepts suitably adapted for cellular radio. There are three types of service provided:

i) **Teleservices** include speech, short message (similar to alphanumeric paging), message handling service (based on X.400), facsimile and emergency calls.

ii) **Bearer services** allow a user to access applications resident on other networks. They include synchronous switched packet circuits, both synchronous and asynchronous switched data circuit and asynchronous terminal access to packet networks (PAD service).

iii) **Supplementary services** are used in conjunction with a bearer service or teleservice. They include services such as calling number identification and presentation, call forwarding when a mobile station is busy, notification of a busy called station's completion, advice of charge for a call, restriction of particular types of calls e.g. outgoing international or multi-party conference calls.

Subscribers may have different directory numbers for each teleservice or bearer service for which they are registered. However each subscriber has a single **Mobile Identity Number.** This information could be stored in an electronic module or smart card and plugged into a mobile phone, for example in a hired car in a different country, so that calls can be re-routed to the subscriber. Unlike the telephone system, a subscriber's identity is not linked to a particular mobile station.

2.3 Using Multiple Networks

The home network will always be used in preference to a competitor's, but users may request service from another network vendor under the following circumstances:

- Coverage of a country may be shared between different vendor networks so when users are in parts of a country not covered by their home network, they have to use another vendor's base stations to gain access.
- The home network may be temporarily inaccessible in particular areas due to faults or congestion.
- The user may be visiting another country.

The pan-European cellular system was developed to allow network operators to run compatible networks across Europe, creating a single cellular equipment market and enabling a subscriber to use the same mobile station on all the networks. This was achieved by all the participating cellular networks conforming to GSM (Groupe Speciale Mobile) technical and administrative recommendations and by all the participating network operators agreeing commercial practices through a Memorandum of Understanding [ETSI/GSM 2.13].

Within the pan-European cellular system, a subscriber will contract for service with one network operator, and this will be deemed his **home network**. A subscriber may contract, via his home network, for services on other networks, for example if he regularly visits another country.

Using services on a visited network is known as **roaming**. (Moving around on a single network is not called roaming). Roaming raises two important issues: one concerning mobility management and the other accounting.

If a subscriber may roam to any network within the pan-European system then locating a subscriber and determining the services that the subscriber is entitled to use becomes difficult. Controlling this information is known as mobility management. To support mobility management every cellular network has a **Home Location Register (HLR)** which contains information on all subscribers, such as their service entitlement and the location in which they last registered. Each network also has a number of **Visitor Location Registers (VLRs)** which contain information about visiting subscribers, such as the subscribers' service entitlement. The HLRs and VLRs allow subscribers to be located quickly and their service entitlement to be verified.

The other issue is that if a subscriber uses services on a visited network then that network operator will want recompense for the service usage but will not want to have to keep billing information for, or issue bills to, potentially every user in the pan-European system. This dilemma is solved by the visited network sending usage information of the visiting subscriber's call to the subscriber's home network. The home network pays the visited network and charges the home subscriber, with the effect that a subscriber will only ever be billed by his home network.

3. DOMAIN CONCEPTS

Domains provide a flexible and pragmatic means of specifying boundaries of management responsibility and authority [Sloman 89]. A **domain** is an object which represents a collection of **member** objects which have been explicitly grouped together to apply a common management policy. From the user's view, objects exist in a domain, but domains in fact hold references to member objects as shown in figure 3.1. In addition, domains provide a flexible structuring of the name space for identifying objects, as explained below. Domains are a generalisation of the concepts of a tree structured directory system.

Domains provide the means of naming and applying policy to managed objects so all managed objects must exist within a domain. Objects should always be created within a domain. Any policy specified in terms of the domain will apply to all its members, which are the objects referenced by its **policy set**. The domain is a **parent** of its members.

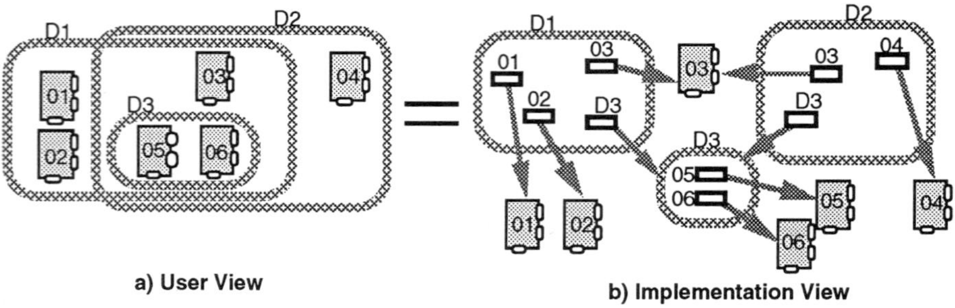

Figure 3.1 User and Implementation Views of Domains

Most management applications require persistent domains, and it must be possible to create an empty domain and later include objects in it. This permits the definition of a management structure for a system and then populating it with managed objects. Since domains are themselves objects, they may be members of other domains, and are then called **subdomains** of their parent domains. D3 is a subdomain of both D1 & D2 in figure 3.1. Through membership of subdomains, objects may be **indirect members** of an **ancestor** domain, e.g. the members of D3 are indirect members of D1 which is an ancestor of them. Subdomains form the mechanism for partitioning a large group of objects and applying different policies to subsets of the original group or assigning responsibility to different managers. Note that a subdomain's direct members are not a subset of the parent's direct members, as a change of membership of the subdomain affects the indirect, but not the direct membership of its parent.

3.1 Access Rules

An **access rule** is a generalised means of specifying access control policy which can be used for management [Moffett 90]. Access rules (figure 3.2) provide a flexible means of specifying management authorisation policy as a relationship between a manager domain and managed domain in terms of the operations managers are authorised to perform on the managed objects. Access rules can also specify constraints such as the permissible time of access or location of managers.

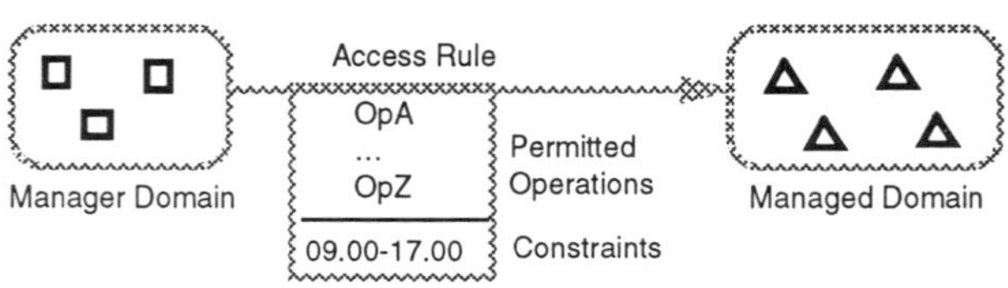

Figure 3.2 Access Rules

Objects in a subdomain, by default, 'inherit' the access rules applying to direct and indirect parents. In figure 3.3 manager objects in domain D2 inherit the access rule applying to the parent domain D1, and so have authority to perform operations (OpA & OpB) on objects in domains D3 and D4 as the latter is a subdomain of D3. There may be more than one access rule which satisfies a particular request, for example as a result of an object's membership of overlapping domains. This indicates that authority has been granted via more than one route. The implication of access rule inheritance is that the domain service must provide a means of determining the parent and ancestor domains of an object in order to determine what policies apply to that object.

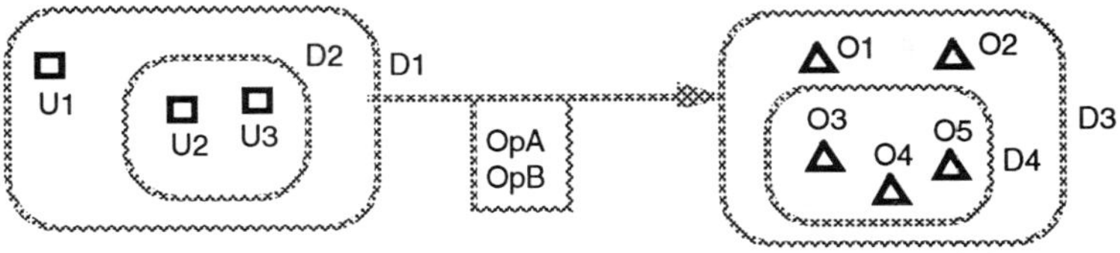

Figure 3.3 Inheritance of Access Rules

Some operations affect more than one object which implies the need for multiple access rules to authorise the operations. For example the operation to include an object in a domain requires an access rule for the source object to be included as well as one for the domain into which it is to be included. This can be seen in figure 5.2 below where, in addition to Include/Remove authority for their service domains, the Registrars have Include_Source authority for the mobiles to which they may give service entitlement.

4. BENEFITS OF DOMAIN MANAGEMENT FOR CELLULAR NETWORKS

The pan-European cellular system has all the characteristics of a large complex distributed system:

- Each network may have millions of objects to manage, such as mobile switching centres, base station controllers and transceivers, fixed communication links, authentication servers, visitor or home location registers, mobile stations and subscribers.
- Inter-network communication requires management interaction between different organisations for accounting information. The home network is billed for services used in a visited network and this includes both calls made and calls received while within the area of a visited network. Each network forms an accounting domain with a particular accounting policy and tariff structure.
- Each vendor network in the pan-European cellular system is an autonomous commercial organisation. There is no overall, central authority.
- There are multiple different services including third party services and so it is necessary to restrict subscribers' access to only those services for which they are registered. The domain framework can be used as a flexible means of specifying and controlling service entitlement as will be explained in section 5.

The remainder of this paper will concentrate on two aspects of management: specification and control of service entitlement; and accounting. However there are many other aspects of management which could benefit from the domain framework. For example domains could be used as a means of partitioning responsibility for maintenance of base stations and switches, between different engineers within a network.

5. SPECIFICATION AND CONTROL OF SERVICE ENTITLEMENT

5.1 Service Entitlement Domains

Figure 5.1 shows how access rules can be used to define the service entitlement for a particular network. Any subscribers who are members of *Network A Speech Service* domain can use the speech service offered by any of the base stations in *Network A Base Stations* domain. Similarly subscribers to *Network A Email Service* domain can use the email service as defined by the Email service access rule. Subscribers can be members of one or more **service entitlement domains** as they may be registered for more than one service. Subscriber *M1* can use both speech and email, but *M2* & *M3* can use only speech and *M4* & *M5* can use only email. (Note that in figure 5.1 subscribers are represented by mobiles). A service entitlement subdomain can be set up to represent another network or a company. Trusted external managers can be given access to this subdomain so that they can register subscribers on behalf of the external organisation as described below. In figure 5.1 Network B subscribers can use both speech and email services offered by Network A. The service entitlements for *Companies XYZ* and *LMN* have been specified in terms of domains (i.e. groups of unidentified subscribers). The specification of which employees (subscribers) from the particular companies are permitted to use the services can then be delegated to the individual companies to manage as explained in 5.2.

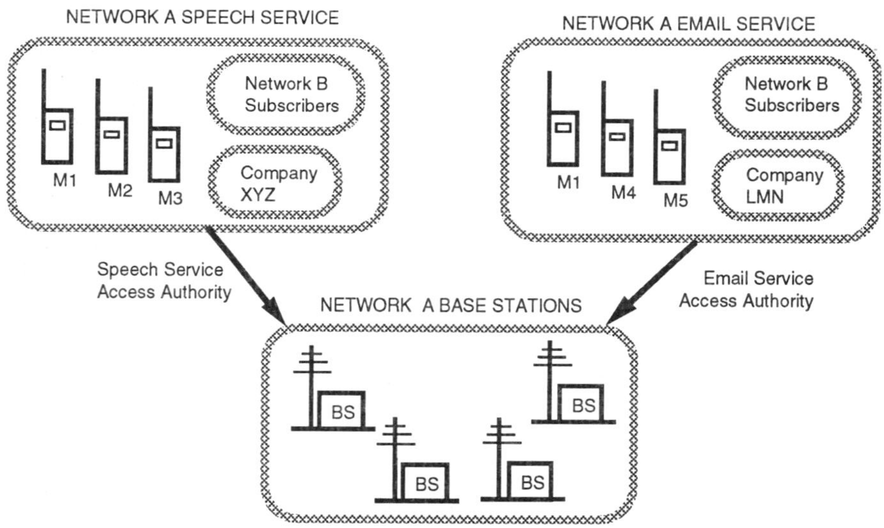

Figure 5.1 Service Entitlement Domains

In addition to the service entitlement domains, each network has a **subscriber registration domain** (see figure 5.2) which holds information about each subscriber such as account number, billing address, unit charge rate, etc. In figure 5.1, mobiles and subscribers are represented as one and the same object, to simplify the simulation. However subscribers may use different mobile stations by plugging a "smart card" into the mobile. The mobile then inherits the service entitlement of the subscriber. One way of modelling this would be to represent the subscriber as a domain and the mobile is included in the subscriber's domain when the card is plugged in and removed when the card is removed.

5.2 Registering Subscribers

A **registrar** is a manager who is authorised to register subscribers for a particular service. This requires the ability to include a subscriber, who is already registered in the network's subscriber registration domain, into the relevant service entitlement domains or remove a subscriber from that domain. Figure 5.2 shows the two access rules to permit Network A's registrars to include subscribers into Network A Speech Service domain. Note that the ability to include or remove Network B subscribers into Network A Speech Service has been delegated to Network B's registrars. However Network B Registrars have no access to other

Network A subscribers as the include/remove access rule specifies only Network B Subscriber Registration domain as the source from which they can include service users.

Company XYZ can also have authority over which of its mobile users may use Network A in a similar way to that for Network B's registrar. However, in this example Company XYZ would not have the authority to add subscribers to its subscriber sub-domain; this would be done in the normal manner by Network A's subscription service with the request for the new subscribers to be placed in the company's sub-domain.

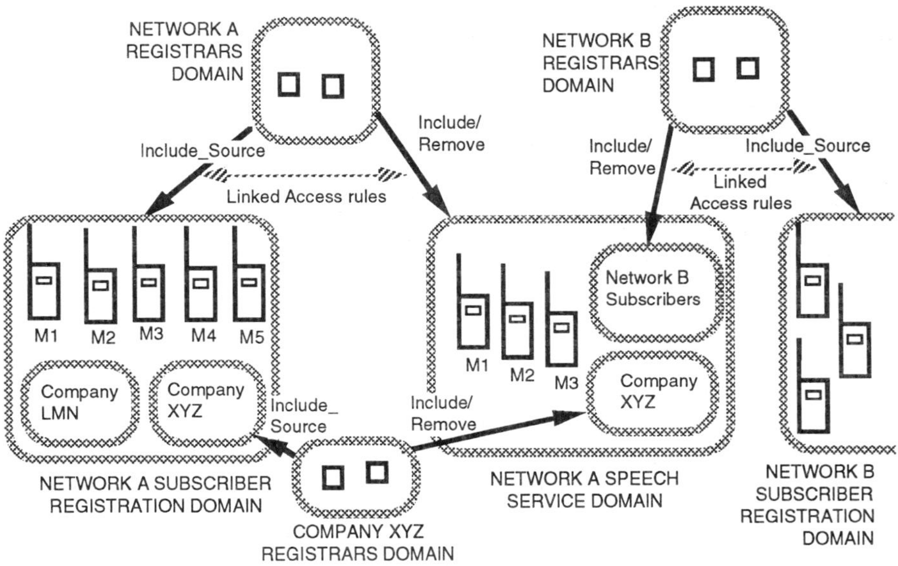

Figure 5.2 Registrars Managing Service Entitlement Domains

5.3 Inter-Network Roaming

In order to make use of the services of another network, a subscriber must be registered as a member of one or more of the service entitlement domains of that network. For example to permit Network B subscribers to use the speech service of Network A, a subdomain of *Network B Subscribers* must be set up as shown in figure 5.1. Any network A subscriber which is included in this domain, can then use the Network B service.

This structure is sufficient to control dynamic inter-network roaming during a single call even though the current GSM specification does not permit it because this would require accounting information for a single call to be accumulated in different accounting domains. Insistence on calls being dropped when roaming to a new networks means a single vendor accounts for each call. The issues of how accounting can deal with inter network roaming are discussed in section 6 below.

5.4 Implementation

A distributed simulation of this system has been implemented using Conic [Magee 89, Bannister 90]. Base stations, switches and mobiles are implemented as Conic nodes running on a number of workstations connected by a local area network. Conic supports both bidirectional (request-reply) and unidirectional interactions between an exitport on one node and an entryport on another. An exitport to entryport binding corresponds to a simulated communication channel which can be fixed for base station to Switching Centre links and dynamic for mobile to base station links.

In the simulation, the handover from one base station to another is instigated by a user typing a command to the mobile telling it to select a new base station. The mobile then executes a protocol to inform the base stations and to relink the relevant ports. The same mechanism works for roaming between base stations belonging to different networks. The Conic configuration language is used to define the required instances of component types, the allocation of instances to workstations and the binding between ports.

The concept of an access rules maps onto an entry in an access control list (ACL) associated with a managed (target) domain. There will be an entry for each applicable access rule, including those inherited from ancestor domains. When a mobile attempts to bind its ports to the base station, the domain service is used to check whether there is an ACL entry which permits the binding, in the base station's parent domain for any ancestor domain of the mobile. If not the binding is refused and the call would be dropped.

A graphical monitoring tool can be used to monitor the system to show to which base stations mobiles are connected and which domains they are members of. Work is in progress on extending this tool to give a graphical interface for registering subscribers for the use of particular services and setting up service entitlement domains. This could be accomplished by means of "drag and drop" of a subscriber icon onto a service domain icon, and a suitable menu for creating new service domains, subdomains and access rules.

6. ACCOUNTING ISSUES

6.1 Accounting Concepts

The accounting management work concentrated on usage accounting, specifically the management process that collects, collates and logs usage information that is pertinent to the cost of using or providing a service. To perform these management tasks requires information on the activities of the managed system, and this is obtained by monitoring specific parts of the system, shown as monitored objects in Figure 6.1 [Varley 1991]. Monitoring information is collected by **accounting meters**, which may process it before storing it in an **accounting log** which holds information on many instances of service or resource usage. Applications such as billing can use this information to generate bills for subscribers.

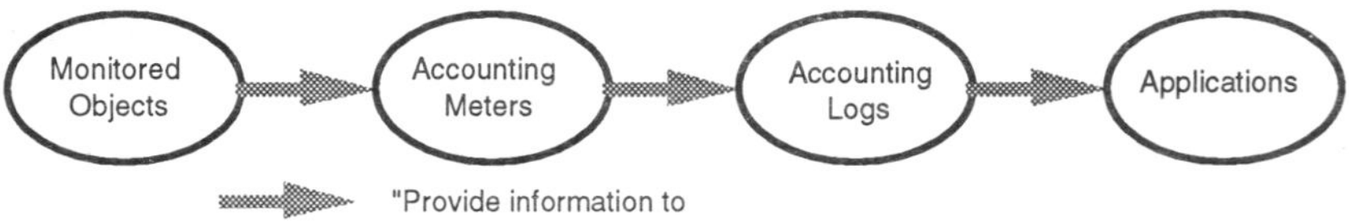

Figure 6.1 Information flow for Accounting

Before looking at the usage accounting implementation it is necessary to describe two new terms: call segment and call section. A **call segment** is a portion of a call where the services used remain constant. When either a calling mobile or a called mobile roam to another network, while the call is in progress, there will be a change in the services used and a new call segment will be started. Because the services used in different segments will be different there may be a need to cost segments separately, hence the reason for breaking the call down into call segments. Obviously, all calls will have at least one call segment.

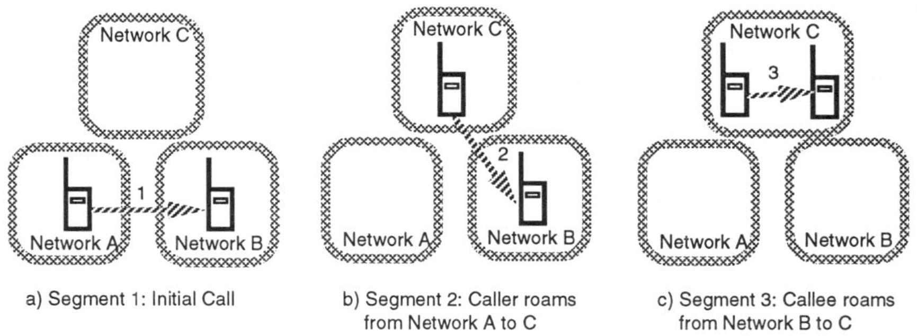

Figure 6.2 Call Segments

Figure 6.2 shows a call with three call segments. Call segment 1, shown in Figure 6.2a, is where a subscriber on network A calls a subscriber on network B. The calling subscriber then roams to network C, Figure 6.2b, setting up call segment 2 of the call. Finally the called subscriber also roams to network C, Figure 6.2c, setting up call segment 3 of the call. The service usage for this call will be the sum of the service usage of each of the three segments.

In the existing Pan-European cellular system, only one call segment is permitted because roaming while a call is in progress is expressly prohibited. The simulation permits calls with multiple segments, but this adds significant complications to accounting management.

A call segment can be broken down into a number of **call sections** representing the service used on one network in the call segment. There are three types of call sections: an originating section, a transit section, and a terminating section. All call segments will have at least two call sections: an originating and a terminating section. This holds true for calls that originate and terminate on the same network (in which case both sections relate to the same network). Figure 6.3 shows a call which has an originating section, three transit sections and a terminating section.

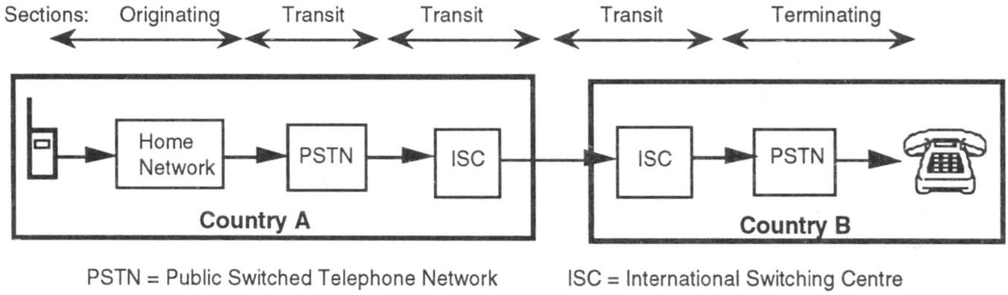

Figure 6.3 Call Sections

In the simulation each network operator meters and logs information on all calls that use the network's services and passes this information to the operator's administration centre. Calls could be originating, terminating or in-transit. The basic information recorded about an instance of usage (an individual call) is: a user identifier (the caller); a called number identifier (the called party); the service requested and the service provided (including qualities of service if they were appropriate); the usage start time and stop time; the number of units used; the call type indicator (originating, terminating or transit); a call termination identifier (e.g. normal termination or call drop-out); an instance of unit identifier; and a call segment identifier.

6.2 Fixed-wire Telephony Reconciliation

Before examining how different network operators can reconcile usage accounting in the pan-European cellular system, it is helpful to review how the problems are dealt with in international fixed-wire telephony. In an international call, each service provider involved in the call will receive some payment for the use of their services. The current methods of international inter-administration settlement originated when the telephony services of a country were provided by a single organisation. These are set out in the CCITT recommendation D.150 on accounting procedures in international traffic [CCITT D150]. In the simplest case, two countries are directly connected, with no transit carriers. Call traffic would flow both ways and at regular intervals the two organisations would settle their accounts according to their bilateral agreement. The CCITT recommends that they use one of three different procedures for settlement:

(1) *Accounting Revenue Division* is the simplest method and is widely practised. The two countries pool the revenue generated from international calls between them, and then divide equally or according to a negotiated ratio.

(2) *Flat-Rate* is an access charge method which is based on the principle that operator of the call originating network pays for the facilities used on the call terminating network. The terminating network charges a flat rate per incoming circuit that takes into account the international circuits and exchange, and the national network extensions. The settlement will be the difference in charges between outgoing and incoming circuit cost.

(3) *Traffic-Unit Price* is a usage charge method which works on the principle that the terminating network operator is paid per traffic unit received. The traffic-unit price will be related to the facilities made available to the originating network operator.

If two countries are not directly connected then calls between them have to be routed via intermediary networks, owned by transit carriers. In international calls the simple rule is that a carrier receives payment from the network that passed it the call, and pays the network it passed the call to. This means that a network operator does not have to know where a call originated, only the network that passed it the call. It also means that network operators only need bilateral agreements with those network operators with which they directly interconnect.

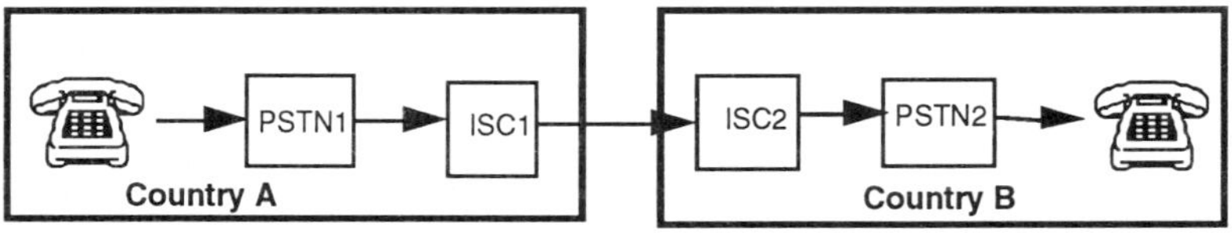

Figure 6.4 International Telephone Call

Figure 6.4 shows a call that is passing through a number of Public Switched Telephone Networks (PSTN) and International Switching Centres (ISC). Using the traffic-unit price procedure, the settlement would work as follows:

PSTN2 charges ISC2 x for terminating the call.

ISC2 charges y for using its facilities, so the charge to ISC1 is x + y.

ISC1 charges z for using its facilities, so the charge to PSTN1 is x + y + z.

PSTN1 operator charges the subscriber for the call, based upon PSTN1's tariffs.

A different way of settling this call would be for each network operator to identify where the call originated. The charge for using their services would then be sent to the originating network operator. In the above example, the cost to the PSTN1 operator in country A would still be $x + y + z$, but it would know what using each network cost and the accounting procedure for all but the originating operator is simplified. Instead of having to pay one operator and get paid by another, each network operator is paid by the originating operator alone. However, there would be a great increase in the number of bilateral agreements for each network operator and these would need to be far simpler than those currently in use.

6.3 Pan-European Reconciliation

In the pan-European cellular system all charges a subscriber incurs through using GSM services will be billed by the subscriber's home network, including any charges for roamed calls. This raises the requirement for inter-administration service reconciliation.

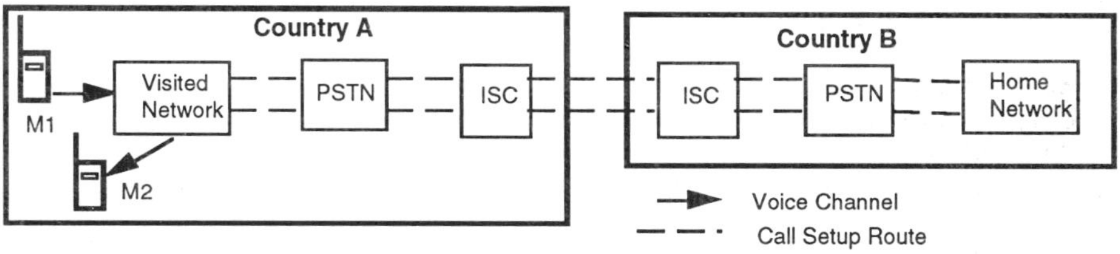

Figure 6.5 : Mobile-to-Mobile Call

Figure 6.5 shows a call between two roamed mobiles. The calling subscriber (M1) will first request service from the visited network, which identifies M1's home network (from its Mobile Identity Number) and checks its service entitlement. If the request is cleared, the next step is to locate the called mobile (M2) by interrogating its home network. In our example, M2 has roamed to the same network as M1. The cellular operator in country A will generate two call records for this call. One will be for the originating call section and will be used to charge the calling subscriber's home network for the call; the calling subscriber will then be charged by the home network. The terminating section call record will be passed to the called subscriber's home network. The home network operator will not be charged for the terminating call section but may choose to charge the called subscriber for receiving a call while on another network.

For each roamed mobile receiving a call or making a call, a usage record will have to be sent from the visited network to the home network. There are many cellular networks, so if roaming service becomes popular the amount of usage information they exchange could be rather high.

Another problem is that rogue users could move between cellular networks and run up large bills in the process. The user's home network operator is contracted to pay the visited network operators for this usage, but may have difficulties in recovering the debt from the user. So the network operators would ideally like usage charges to be exchanged frequently: at least daily. Each operator would potentially need to maintain an account with every other operator and run this process daily for each network.

A simpler method would be for service providers to identify roamed calls, cost them, then send accounting records to a central **clearing house**. The clearing house would handle the exchange of usage information between the network operators. It would also handle the exchange rates between the European currencies, using Standard Drawing Rights, and apply the appropriate credits and debits to the network operators' accounts. This concept is partly met by the CEPT clearing house, but it does not yet handle the exchange of usage data.

6.4 Reconciliation for a Roamed Call

As stated earlier, the current GSM networks [ESTI/GSM 2.11] drop calls when a subscriber roams from one network to another, forcing the user to redial, which is inconvenient. We will now show how accounting could cater for inter-network roaming during a single call, which may be permitted in the future.

Roaming during a call raises problems if the PSTN is involved. Figure 6.6a shows the route of a call from a mobile to a PSTN subscriber. Figure 6.6b shows how the call could now be routed if the call could be handed-over between cellular networks, in this example from Base Station 3 (BS3) to BS4. The call now has an additional leg between the two gateway MSCs. The cellular network to PSTN interconnection point that was set up when the call was initiated must remain throughout the call's duration as it is very unlikely that the circuits used in the fixed line portion of the call can be changed.

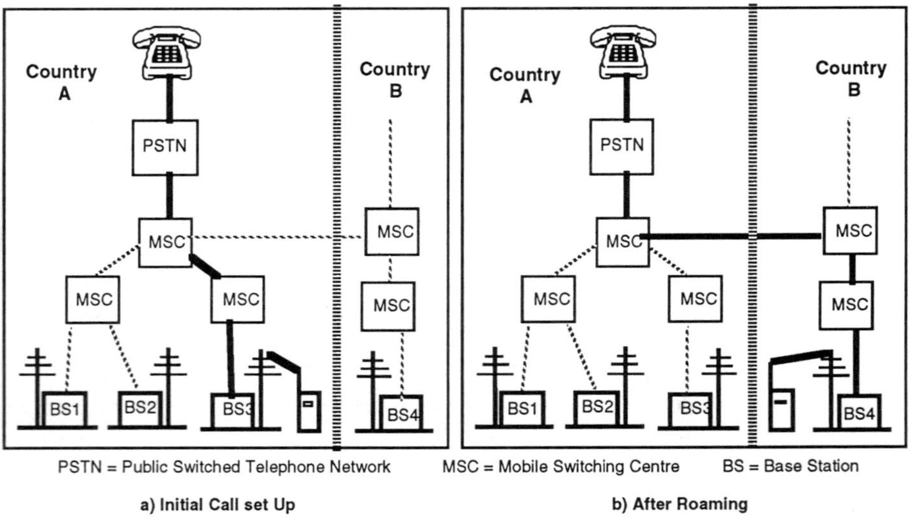

Figure 6.6 : Roaming While a Call is in Progress

Roaming during a call may involve many network operators in a single instance of service provision, and so requires call section records to be generated by different operators. Let us suppose that the call lasted x units before hand-over (figure 6.6a) and y units after hand-over (figure 6.6b). The three call section charges would be:

1) PSTN: PSTN unit charge * (x + y)

2) Network A: network A unit charge * (x + y)

3) Network B: network B unit charge * (y).

Each vendor would send a bill to the subscriber's home network which would then bill the subscriber, probably with a consolidated call charge i.e. the subscriber does not see the call charge being made up of 3 components. This can be used in conjunction with the clearing house. The intermediate service providers send call section accounting records to the clearing house, which would handle the redistribution of this information to the home network and the settlement of inter-administration accounts.

The current CCITT accounting model is based on a traffic-unit price procedure which assumes that a single service provider is involved for the entire duration of a call and so cannot cope with call section accounting.

6.5 Domains and Accounting Management

At the outset of the project, we had intended to use domains for accounting, but eventually realised they were not appropriate for the implementation of these accounting concepts. This is due to the transient nature of usage accounting information, with domains proving more useful when used to manage durable information. Accounting policy applies to complete networks i.e. an accounting domain correspond to the whole network operated by a single network vendor. In addition, the simulation assumes that the same charging policy applies to all subscribers.

Future work will investigate different accounting policies applying to groups of subscribers and how to relate the accounting policy specification to accounting implementation.

7. RELATED WORK

The OSI security framework [ISO 10181-1] defines a subdomain as a subset. If domain B is a subset of domain A then any member of Domain B is a direct member of domain A. If an object X is inserted in domain B, then X is not a direct or indirect member of A and the relationship of A and B is now overlapping domains, unless X is also inserted in Domain B. With a subdomain as a subset any policy applying to the superset (A) must, by definition apply to the subset (B), whereas our definition permits inheritance of policy as an option. Another problem with the OSI definition is that domains include both managers and managed objects. This reduces flexibility of the domain concepts as it is impractical for inter-organisational management where managers have limited management rights over objects in another organisation (different domain). It does not permit different policies to apply to managers and the objects they manage. The OSI Management working group have some initial proposals on domains that have indirectly been based on our work, but these are still rather confused and inconsistent [ISO N7118]. In our opinion the OSI concept of domains has not been fully thought through, and there is no implementation experience to validate it.

The DEC Enterprise Management Architecture [Strutt 91] uses a domain to define a 'sphere of interest of a set of managed objects for a manager' - managers are not part of the domain. Only the managers are aware of the domains and objects do not know which domains they are members of. This model is compatible with the Domino approach as it implements a subset of our concepts. There are no domains of managers and relationships between a manager and domain is implicit in the knowledge of the domain name rather than the use of explicit access rules. They do not use the domains for access control. Domain membership information is held by the name service.

There is an Esprit funded project [DOMAINS 91]) which is defining an architecture for managing distributed systems. Their domains encapsulate manager and managed objects such that the objects in a domain are only accessed via their manager. This makes it difficult to dynamically rearrange the domain structure by including or removing objects to reflect changes in service entitlement as described in this paper.

8. DISCUSSION AND CONCLUSIONS

This paper has shown the flexibility of the use of domains. Registering a user for a particular service could be done very easily by means of a “drag and drop” graphical user interface. Whether a “Drag and Drop” or more conventional interface is used, the registrar must be a member of the appropriate Registrars Domain before being allowed to control registration. Domains permits very flexible structuring of service entitlement and delegation of responsibility for managing subscriber groups such as company employees to registrars of their own company.

However, when it comes to accounting the domain structuring gives less benefit in that there are fewer domains. The domains really correspond to complete networks managed by a single vendor, and a common accounting policy applies to the whole domain. Although there may be different tariff rates applying to different subscribers, this information is only required when it comes to generating bills and is more easily maintained and used as part of the information held in a database of subscribers.

ACKNOWLEDGEMENTS

The initial implementation of the Cellular Network simulation was done by David Bannister. We acknowledge the contribution of colleagues in the DOMINO project in stimulating and criticising the ideas of this paper. This work was carried out with the support of the UK DTI/SERC (Grant Nos IED 3/1/1409, GR/F 35197). We also acknowledge support from DTI ESF Project and ESPRIT Project 7023 SysMan.

REFERENCES

Bannister 91 Bannister D.K. Simulation of a Cellular Radio Network, Imperial College of Science, Technology & Medicine, MSc Thesis, Sep. 1991

CCITT D.150 CCITT Blue Book, Volume II, Fascicle II.1 - General Tariff Principles - Charging and Accounting, in Internatinal Telecommunication Services, Series D.150 Recommedations, 1989

DNS 91 DECnet Digitial Network Architecture Name Service Functional Specification V2.2.0, Order Number EK-DNANS-FS-002, May 1991

DOMAINS 91 Esprit Project 5165 - DOMAINS Basic Concepts, version 2.0, Nov 1991, Philips Gmbh, PO Box 1980, W 5100 Aachen, Germany.

ESTI/ GSM 2.11 Service Accessibility, Versions 3.4.0, ETSI PT12, March 1990.

ESTI/ GSM 2.13 Subscription to the Service of a GPLMN-Version 3.1.0, ETSI PT12, Jan. 1990.

ISO 10040 Systems Management Overview, ISO/IEC JTC1/SC21 WG4, 1991.

ISO 10181-1 ISO Security Framework I: Overview, ISO/IEC JTC1/SC21 (May 1991).

ISO N7118 Third Working Draft For Management Domain Architecture ISO SC21/WG4 (May 1992).

Magee 89 Magee J., Kramer J. & Sloman M.S. (1989). Constructing Distributed Systems in Conic, IEEE Transactions on Software Engineering, 15(6),1989, pp. 663-675.

Moffett 90 Moffett J.D., Sloman M.S. & Twidle K.P. Specifying Discretionary Access Control Policy for Distributed Systems, Computer Communications, 13(9),1990, pp. 571-580.

Sloman 89 Sloman M.S. & Moffett J.D. Domain Management for Distributed Systems, in Integrated Network Management I, (Meandzija B and Westcott J. eds), North Holland, 1989, pp. 505-516.

Strutt 91 Strutt C. Dealing with Scale in an Enterprise Management Director, in Integrated Network Management II, (Krishnan I. and Zimmer W. eds.), North Holland, 1991, pp. 577-593.

Varley 91 Usage Accounting in Distributed Systems, IEE Colloquium on Network Management 1991.

Integrated Network Management, III (C-12)
H.-G. Hegering and Y. Yemini (Editors)
Elsevier Science Publishers B.V. (North-Holland)

Modelling Management Policy using enriched Managed Objects

Jan D Roos, Phillip Putter & Carel Bekker

Unit for Software Engineering, Department of Computer Science, University of Pretoria, Hillcrest Pretoria, 0083. Email: cbekker, pputter & jroos@cs.up.ac.za

Abstract

Managed Objects (MOs), contributing to management by performing management tasks, are regarded as critical components of distributed management systems. This paper proposes a model for enriching MOs with components supporting management policy and describes a prototype implementation of the model. The main use of the idea is to add management functionality dynamically to MOs as required by management policy. It could also be used to create that part of MOs that interact with real resources. The prototype implementation was done in Smalltalk-80 and was used to test the ideas discussed.

The paper stresses the roles of management domains and management relationships. The model uses these concepts to create components representing management functionality.

Although the model is described for a Network Management (NM) environment the same concepts apply to the general area of Distributed Systems.

Keyword Codes: K.6.4
Keywords: Network Management, Managed Objects, Management Domains, Management Policy, Management Relationships, Distributed Systems.

1. INTRODUCTION

It is well accepted that object-oriented modelling should be used for the specification and development of Network Management (NM) systems. The concept of a managed object (MO) is a familiar one in all NM standards/recommendations. A MO represents an abstraction of a physical item of equipment, logical components, or abstract collections of information in a NM system. MOs representing the management characteristics of real resources are mapped to the real resource during the management process. The term MO is used in the rest of the paper to refer to either a MO class or instance. If the difference is important, explicit reference will be made to *class* or *instance*.

A MO is defined by the attributes visible at its boundary, management operations which may be applied to it, notifications emitted by it and its behaviour. A MO's behaviour is defined in terms of the management operations which can be used to change its state as well as its reaction to other environmental stimuli. More information on the characteristics of MOs can be found in [OSI91a].

As the size of NM systems increases, MOs proliferate. Large numbers of MOs require structuring mechanisms allowing MOs to be grouped together and structured hierarchically to simplify the information model of large systems. The structuring of MOs is generally determined by management policy. Management policy also determines the management tasks to be performed. This paper proposes that it should be possible for MOs to be enriched with management structure and functionality. Enriched MOs may be able to perform a local

---

Part of this work was supported by the Foundation for Research Development, RSA management task more optimally and flexibly compared to a higher layer manager. The enriched MOs can then provide their management results to the higher layer manager. It is envisaged that MOs enriched with management functionality will be used extensively in future distributed management systems. This paper proposes a general model supporting such MO enrichment.

In a management environment a manager will delegate refined subsets of the management policy defining his responsibility. This could happen recursively and at each level the applicable management policy will result in management tasks to be performed. In network management the management tasks result in the managing MO, of that management level, performing some management functionality over some other MOs (In all, but the lowest management layer, the managed MOs will be managers themselves.) As management policy can change dynamically, the need exists to dynamically add new management functionality to a MO or modify its existing management functionality. The latter problem is the topic of this paper. The methodologies for refining and delegating management policy and the generally transformation of management policy into management operations, are left for further study.

The remainder of the paper is structured in the following fashion:

- section 2 describes the management domain, relationship and policy concepts,
- section 3 discusses the design of management policy and domains,
- section 4 describes the general support environment for management policies and domains,
- section 5 discusses the implementation of a prototype of the support environment, and
- section 6 concludes the paper and indicates remaining issues under investigation.

2. MANAGEMENT DOMAINS, RELATIONSHIPS AND POLICIES

This section discusses objects which support the creation and functioning of MO groups. Such objects can, themselves, be considered to be MOs. For the duration of this paper, support objects will be referred to as *objects* to clearly distinguish them from the real MOs.

An important concept addressed in this section is the enrichment of MOs through the addition of management structure and functionality. This is needed to allow MOs to support management policy. It enables the dynamic structuring of management tasks (applications) into hierarchical and distributed components which can be located where appropriate.

2.1. Domains

Domains are normally used to group MOs to which a common policy applies. This paper argues that the reason for the formation of a domain is not known by the domain itself. The phrase managed domain refers to a domain of MOs being managed. A manager domain refers to a domain of MOs with management responsibilities towards MOs. When referring to either a managed or a manager domain the term management domain or simply domain is used.

A domain does not encapsulate MOs, but only groups MOs. It is believed that encapsulation will rarely occur in network management or distributed systems. Encapsulation hides component MOs which makes them invisible to other domains which may want to include them or managers that may want to access them.

Overlapping management domains can exist where MOs may be members of more than one management domain [MOF91b]. Guaranteeing consistency of MOs in such environments is important but is not discussed in this paper.

If an object representing a domain includes an object representing another domain, the resulting domain consists of the union of the original domains. A domain can also include manager MOs of lower layer domains, resulting in hierarchies of MOs grouped in domains. A domain object offers operations [OSI91b] to:

- create and destroy domain instances, and
- add MOs to the domain and remove MOs from it.

2.2. Management relationships and policies

It is envisaged that management relationships and policies will be used extensively in the management of distributed and heterogenous systems. A management policy is an abstract description of the management functions applied by a manager to a certain grouping of MOs (managed domain). A management policy may result in management tasks which collectively effect the MOs in respective managed domains. These management tasks depend on the management functionality required by the management policy. Figure 1 shows a simple management example where a user interacts with a Manager which manages some elements. The two management policies implemented by the Manager are specified by the view towards the user and the view towards the elements. The policy statements of these two views are given in Table 1. Note that the view towards the user corresponds to the normal agent-type interface of a MO. In this case the interface represents the attributes and operations of the management task representing the management policy as viewed towards the elements.

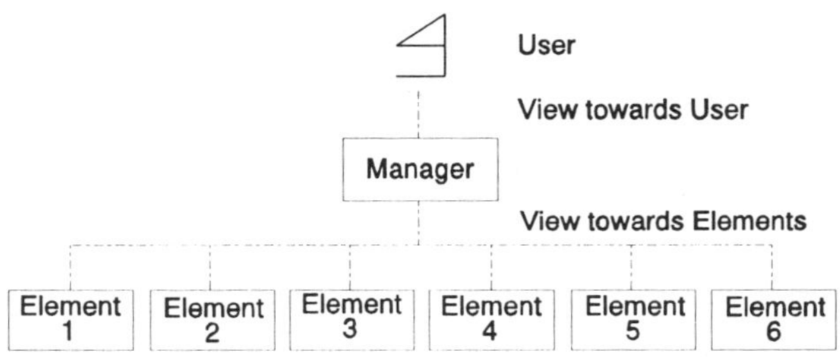

Figure 1. Management Environment

The manager's policy as viewed towards the user can be stated as:

Represent agent interface to MO and manager functionality.

The manager's policy as viewed towards the elements can be stated as:

Monitor the elements and determine number active.

Table 1. Policy Statements

A policy relationship exists between the management policy's subject and the management policy's target. A management policy's subject represents the body responsible for applying the policy, while the management policy's target represents the MOs to which the management policy applies. If a management policy applies to a domain of MOs this domain is referred to as the target domain. The policy relationship links the target domain with its manager. The management function of a management policy is referred to as the policy goal [MOF91a].

A management policy consists of a passive and active part. The passive part serves the purpose of representing the management policy relationship while the active part performs the policy goal. The passive and active management policy parts add the manager functionality to a MO.

Management policy leads to the establishment of a policy relationship between manager and managed MOs. A policy relationship between MOs should not be considered part of either MO, but is treated as a separate object. Failure to treat relationships as such leads to hidden assumptions and dependencies. Relationships form the glue of the object model and provides access paths between related objects [RUM91].

3. DESIGNING A MANAGEMENT POLICY

It is convenient to split the passive and active parts of a management policy into separate objects for design purposes. An important consideration for the separation of these parts is to keep the passive part of a management policy as generic as possible to enable the general re-use of the class for other MOs. The passive part of the management policy is referred to as a policy relationship (PR) object for the duration of this paper. The policy relationship object also represents the structure of the relationship [KLE91]. The active part of a management policy is application specific and is therefor removed from the policy relationship object. The active part of a management policy is referred to as the active policy (AP) object for the duration of this paper. Active policy objects can be developed and changed independently from the policy relationship objects. The separation between the policy relationship and active policy objects simplifies the design of a management policy. An active policy may be designed as various independent active policy objects as long as they all interface adequately with the policy relationship object. The subject, target and goal of the management policies, of the manager shown in Table 1, are given in Table 2.

Manager policy 1: Towards the User

Represent agent interface to MO and manager functionality.

```
Subject: manager
Target:  user
Goal:    if requested
           perform agent role to original MO
           and new objects representing the
           management functionality
         endif
```

Manager policy 2: Towards the elements

Monitor elements and determine number active.

```
Subject: manager
Target:  domain containing elements 1 to 6
Goal:    repeat forever
           no_elements_active = 0

           for all elements in domain
             poll(element)
             if receive(element,response)
               no_elements_active ++
             endif
          endfor

          MO_agent_set(no_elements_active)          /* request to agent */
         endrepeat
```

Table 2. Policy Statements Refined

Figure 2 shows how the management environment of Figure 1 can be represented by adding policy relationships. The first part of the figure shows a traditional Manager that implements the functionality of both the policies stated in Table 1. Policy relationships are shown between the Manager and the User, and the Manager and the elements to be managed. The element

MOs to be managed are grouped into a domain. The Manager, shown in the first part of Figure 2, fulfil the management goals of the two management policies given in Table 2.

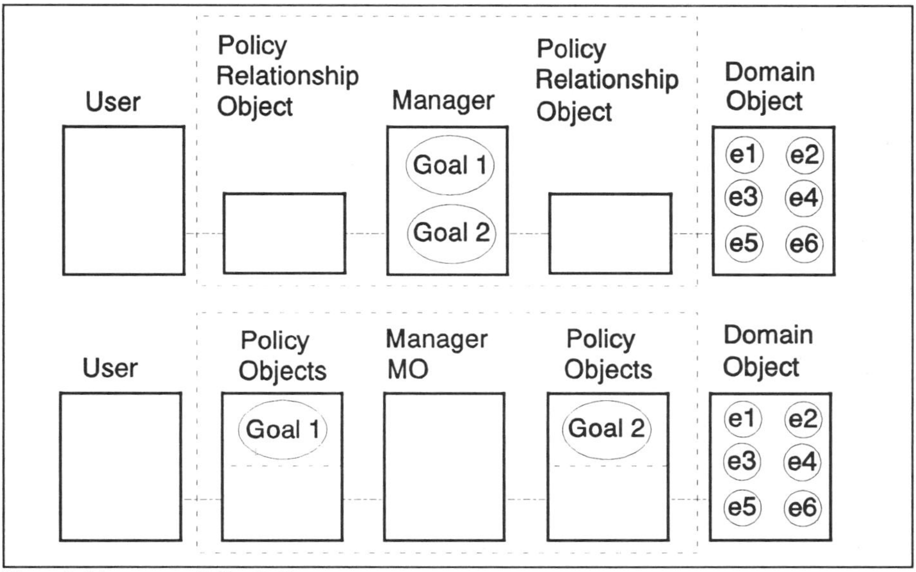

Figure 2. Management Policy Refinement

The purpose of this model is to support MO enrichment in a generic and simple way. This is done by partitioning the management policy in separate objects as shown in the second part of Figure 2. Splitting the management policy into separate objects has two advantages:

- the MO, to be enriched, is almost not modified (only its agent part needs to be informed), and
- the management policy objects can be located at any convenient location, with available processing power or where communication overhead can be minimized.

The manager functionality in Figure 2 is represented by the objects grouped together in the dashed-line boxes. The manager implementing the two goals of Table 2 along with its policy relationship objects as shown in the first part of Figure 2 can be changed to that shown in the second part of Figure 2 by moving the goals to the respective policy relationship objects. Moving the management functionality from the manager to the policy objects simplifies the manager. This paper refers to the remaining manager component, without management policy functionality, as a Manager MO. Note that the policy objects performing goal 1 are equivalent to the normal agent functionality.

For the remainder of this paper we shall focus on the policy between the manager and the element MOs (Management Policy 2 of Table 2). The policy between the manager and the user will not be discussed further.

The detail design of management policy 2 is shown in Figure 3. The Manager MO is linked with the policy relationship object which bind it to the active policy object and relates it to the other objects/MOs. Many-to-many policy relationships between objects can also be supported by a policy relationship object although Figure 3 shows a one-to-one relationship between the Manager MO and domain object.

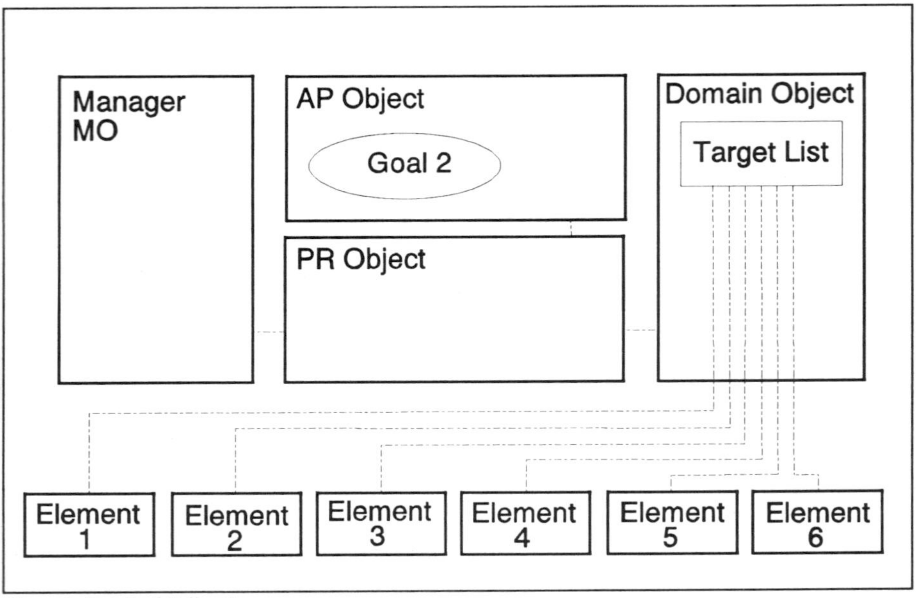

Figure 3. Detail of Policy 2

The important contribution of Figure 3 is that it shows the structure of a management policy. After the policy relationship object has been established active policy objects can be added.

In our example the active policy object polls all MOs in the management domain to determine whether or not they are active. The active policy object contains the management functionality required to achieve the goals of the management policy and makes use of the policy relationship object to access the domain object. The active policy and policy relationship objects may be changed independently to accommodate changes in the management policy without effecting each other.

The target list in the domain object of Figure 3 may contain references to other domains or MOs to which the grouping applies. The management policy's active policy object performs its regular management functions and reports the results when requested by the agent.

The active policy object in Figure 3 manages the element MOs in the related domain object. The domain object is not linked to the active policy object and is fully re-useable in another context.

The policy relationship object needs not be aware of the activities of the active policy object and it is used to represent relationships between related objects. The user also interacts with the policy objects and domain object through the agent.

All traditional MOs may be enriched by adding management policy. This implementation of management policy allows for elegant addition or modification to management functionality. Once the policy objects and domain object are in place, as shown in Figure 3, they can be modified dynamically through their defined operations. Although the example used was simple it serves to illustrate the concepts used to represent management policies.

4. THE GENERAL SUPPORT ENVIRONMENT

An important objective of the support environment is to allow the dynamic creation and modification of MO structures and MOs with management functionality. This is done by supporting the addition of management policy to MOs. To achieve this the support environment has to allow structures, as explained in section 3, and management functionality to be created and added with ease. This is supported through service requests from the support environment and takes place through MO operations.

In the support environment each object class has a unique id and each object instance has a unique handle which can be used to refer to the object instance. If the object instance handle is known the class id of the object can be obtained. From the support environment's point of view there is no difference between MOs and objects.

The functional requirements of some of the generic services offered by the support environment and their mapping to the concepts introduced in the paper so far, are discussed in more detail in the following sub-sections.

4.1. Object Creation / Deletion Services

This service creates object instances of given object classes and returns a handle to the newly created instance. A similar service deletes an object instance and releases all resources occupied by it. These object instances could be ordinary MOs, domains, policy relationship, active policy or user defined objects.

4.2. Registration Services

Registration services are important to keep track of object instances in a system. When an object instance is created, it is automatically registered, and de-registered when it is deleted. A list of object instances can be obtained from the register service according to selection criteria, e.g. creation period, class id, and / or object id. The register created by registration services is used by the agent to access all objects created in a system.

4.3. Domain Services

Domain services provide a mechanism to group objects spread over one or more systems. The reason for the grouping is of no importance to the basic domain service. Domains could be created to group objects into managed or manager domains or for any other reason required by the service user.

The domain services enable a user to:

- include and remove objects from domains,
- obtain a list of all objects from a domain.

Objects in a domain could represent other domains.

4.4. Management Policy Services

An organization's management policy is a statement of plans, rules, constraints and goals [MOF90]. Network management policy can be implemented by MOs and applies to MOs. NM policy plays an important role in determining the structure and behaviour of NM systems, and influences the formation of domains and relationships between MOs/domains [PUT91]. It is therefore vital that the general support environment offers services for the creation, addition and modification of management policy. Currently all management policy classes have to be defined beforehand but they can be instantiated when needed.

Policy services may consist partly of methods giving guidelines on how management policy objects can be implemented and services to manipulate implemented objects. To achieve this the management policy service provides operations to:

- create policy objects,
- to associate implemented active policy goals with policy relationships,
- provide information on the structure of a implemented management policy (bindings between policy relationship, active policy and domain objects), and

- support the relocation of policy objects.

4.5. Export Services

The new MO interfaces of MOs enriched with management functionality are advertised through the export services. It is generally accepted that interactions required between objects to exchange and negotiate interfaces require the involvement of a third party. This third party is referred to as a Trader in the ODP arena [GEI92] and is accessed through a service like the Management Request Broker in the DME proposal by the OSF [OSF91].

In the general support environment a MO/object that wants to advertise its service must export its interface by providing a specification of its interface to the trader service.

5. IMPLEMENTATION

Smalltalk-80 [GOL89] was used for prototyping the general support environment while experimenting with the concepts described in this paper. The general support environment will be moved to a C++ development environment as soon as some remaining prototyping issues have been resolved. Some further general remarks are:

- Notification filtering could also be handled by active policy objects. Generally all requirements that deviate from traditional MO characteristics should be handled by active policy objects.
- Normally every leaf MO (a MO that interfaces with real resources) has an active part which interacts with the real resources and could be implemented following these concepts.
- The enrichment of a traditional MO with management functionality will result in policy objects being added which can be accessed through the agent using the normal MO interface concepts.
- The management functionality added to a MO can be used to perform any application function, like user interfaces, expert system support, etc.
- These concepts can be used to create a network of related objects representing more complex management policy.
- At implementation time a pragmatic decision must be made regarding the separate or combined implementation of separately designed objects.

6. CONCLUSION

This paper describes a model which supports the structuring of MOs and the adding of management functionality to MOs. This is done by proposing objects to represent domains and management policy. This model is supported by a prototype environment which is object based and provides for a maximum amount of generality and re-useability.

Current remaining prototyping issues are:

- Circumstances when interactions between the objects of a constructed MO, become a problem.
- More general use of the export services.

It is clear that a critical component required to make this approach viabie is the automation of the transformation of policy statements to management tasks. This area requires further research.

Many of the problems experienced in the development of NM systems also exist in other application areas where object models are used. The support environment offers a powerful mechanism that can be used in an area much wider than NM, in particular in the areas of distributed systems and their management.

7. REFERENCES

[GEI92] Geihs K, Trader Interaction Models and Infrastructure Implications, Fourth International Conference on Information Networks and Data Communication (INDC-92), Espoo, Finland, 1992.

[GOL89] Goldberg Adele & Robson David, Smalltalk-80, The Language, Addison-Wesley Publishing Company, Reading, Massachusetts, 1989.

[KLE91] Klemba K. & Kosarchyn M., A Model for Object Relationship Management, Proceedings of the Second IFIP TC6 / WG6.6 International Symposium on Integrated Network Management, Washington DC, 1991.

[MOF90] Moffet J.D. & Sloman M.S. & Twidle K., Specifying a Discretionary Access Policy for Distributed Systems, Computer Communications, 13(19), 1990.

[MOF91a] Moffet J.D. & Sloman M.S., The Representation of Policies as System Objects, Proceedings of the Conference on Organizational Computer Systems, Atlanta, Georgia, 5-8 Nov 1991.

[MOF91b] Moffet J.D. & Sloman M.S., Delegation of Authority, Proceedings of the Second IFIP TC6 / WG6.6 International Symposium on Integrated Network Management, Washington DC, 1991.

[OSI91a] Open Systems Interconnection - Systems Management Overview, IS 10040, 1991.

[OSI91b] Open Systems Interconnection - Working Document on Management Domain Management Function, N1387, Joint CCITT - ISO/IEC JTC1/SC21/WG4, Dec 1991.

[OSF91] Open Software Foundation: OSF Distributed Management Environment Rationale, 1991.

[PUT91] Putter P. & Roos J.D., A Methodology for Deriving a Network Management Specification from a Network Management Policy, Proceedings of the IFIP / IEEE International Workshop on Distributed Systems: Operations and Management, Santa Barbara California, 1991.

[RUM91] Rumbaugh James & Blaha Michael & et al: Object-Oriented Modeling and Design, Prentice-Hall Inc., Englewood Cliffs, New Jersey, 1991.

II.

DISTRIBUTED SYSTEMS MANAGEMENT

B.

PANEL

Implementing and Managing the Distributed Computing Environment

Panel Chair: Kathryn DeNitto, Open Software Foundation, USA

The task of managing stand-alone systems from multiple vendors as well as a growing number of distributed systems has led to unpredictable and complicated system management procedures that result in costly training and other expenditures. Systems administrators and users can no longer count on their familiarity with a few administrative approaches to maintain a diverse computing environment. Instead, they must contend with inconsistent management schemes resulting from the lack of a common approach based on standards. The panel will address the following topics:

- Management Framework
- Requirements for products that manage distributed systems
- Industry specifications for distributed systems management, and
- Managed object definitions

The panel will provide end-user, service provider, and system vendor perspectives on the importance of these initiatives for managing distributed computing environments.

II.

DISTRIBUTED SYSTEMS MANAGEMENT

C.

DISTRIBUTED MANAGEMENT

Integrated Network Management, III (C-12)
H.-G. Hegering and Y. Yemini (Editors)
Elsevier Science Publishers B.V. (North-Holland)

Distribution in an enterprise management director

Colin Strutt

Digital Equipment Corporation, 550 King Street (LKG2-2/T2), Littleton, Massachusetts 01460, USA

Abstract

As the environments that customers need to manage are getting ever larger and more diverse, there is a requirement to treat management of these environments as a distributed application, using distributed systems technology. Some situations may permit the use of a single management system, or multiple isolated management systems. However, the trend is towards multiple inter-operating management systems. Using DECmcc™ as an example, we investigate some key aspects of distribution. The techniques allow us to migrate easily from a scenario employing multiple independent DECmccs to one where DECmcc becomes a distributed application supporting multiple cooperating managers simultaneously.

Keyword Codes: C.2.4; K.6.3; K.6.4
Keywords: Computer Systems Organization, Distributed Systems; Management of Computing and Information Systems, Software Management; System Management

1. THE PROBLEM

Early network and system management solutions involved independent programs, acting in isolation to provide some limited management application applied to a limited set of managed resources, with a user interface and database support appropriate to the program. Recently we have seen the emergence of a number of management products that provide a wider set of management applications, applying to a broader set of managed resources, with a user interface and underlying database intended to provide consistent, integrated management of the environment. However, even these newer approaches have tended to operate as independent programs. In this paper we will see how one approach to enterprise management (DECmcc[1][2], Digital's director implementation of the Enterprise Management Architecture[3]), can be evolved to provide a distributed management solution.

Consider the following situation — the user has a small self-contained network environment, consisting of a few dozen components. A single management system may suffice to allow one user (the manager) to deal with this environment. There is no need for a distributed solution here.

Consider a second case — the user has a larger network, interconnecting a small number of sites. Each site has one person, perhaps part time, that looks after the equipment (hardware and software) at that site. The site managers coordinate with each other by means of regular face-to-face or teleconference meetings to deal with issues that span management of the inter-site links, and to set name/address assignment policies for the company. In this case, multiple independent management systems may suffice, since each manager runs a separate management system, and there is no need, or desire, to have these management systems interoperate. Any information sharing or coordination is done via personal contact, rather than by having the management systems directly involved. There is no need for a distributed solution here, either.

Each of the above configurations would be likely to be changing to an environment where there is a diverse set of components from different vendors to be managed, overlapping management responsibility among a broader set of people, and a desire to integrate the management by sharing of information and data. It is this problem that calls out for a solution based on distributed systems technology.

One might define an ideal distributed management environment as being one that allows multiple managers to work cooperatively towards providing complete management of their enterprise in a mutually supportive manner. This environment is further characterised as one which allows the management applications to operate in a mutually symbiotic relationship to support the cooperative work of the managers.

2. OVERVIEW OF SOLUTIONS

DECmcc (hereinafter referred to as MCC) is an extensible, integrated management platform supporting heterogeneous, multivendor management environments. It comprises a management kernel supporting management modules which provide the form, function and access features of the director (or 'management system' in OSI terminology). The modules are known as Presentation Modules (PMs), Function Modules (FMs) and Access Modules (AMs) respectively. Intermodule communication is via an object-oriented style of procedure call known as *mcc_call*. [1] and [2] provide a more complete description of MCC.

The initial versions of MCC provided a solution addressing the two scenarios described above. Indeed, it already had much of the infrastructure in place to solve many aspects of the ideal distributed management environment. In this section we overview the various options. In later sections we cover each topic in more detail.

The interface between a user and an MCC PM can be distributed, using the X-windows client-server model, rlogin or other remote terminal connections. The interface between an MCC AM and a managed object can be distributed based on the underlying connectivity afforded by the management protocol employed.

Before we consider evolving MCC to be a distributed application there are some important topics to address. The name space used by MCC provides a network accessible, replicated, universal name space for all managed objects. This provides consistent network-wide naming for all managed objects. Information derived from stored or real-time data from the managed objects may be exported from MCC into a network accessible relational database, allowing roll up of information into single database from a number of different management systems around the network.

Where multiple different management systems are used, there is a need to arrange the systems so that one acts as a central focal point and another represents some set of managed objects. Where the relationships are fixed, one management system might act as an "integrator" and another as an "element management system". MCC can act in either or both roles via the use of the appropriate management modules (i.e., AM or Agent PMs). Where the relationships between management systems are not fixed, then the management systems take on both roles simultaneously, acting as peer management systems. We refer to this communication between management systems as Management System to Management System.

Where the MCC modules involved in an operation are distributed across systems, there is a need to provide for module inter-communication. *mcc_call* is defined to operate across machine boundaries, thus permitting a module on one machine to use the services of a module on another machine. We refer to this communication, and the related protocols, as Kernel to Kernel, supporting the director as a true distributed application.

In cases where there are multiple management systems, representing multiple managers, and each attempts to manage the same managed object (or objects), there is a desire to coordinate the management operations to ensure there is no possibility of the managed objects attaining an undesirable state due to the interference between conflicting operations from each of the management systems. In order to coordinate between directors, we refer to an inter-director communication which we call Director to Director. However, details of this work is beyond the scope of this paper.

3. USER TO PM DISTRIBUTION

The communication between the user and an MCC PM may be direct, when the user is directly connected to the system or workstation that is running the PM. However, it is possible for the user to be remote from the PM, using one of a number of remote access capabilities as shown in Figure 1 — the specific capabilities to access particular PMs remotely depend on the user interface devices supported by the PMs.

Among the possible remote access capabilities for PMs are X-windows and remote terminal access. With X-windows client-server, the user accesses a workstation (the X-windows server), and the PM (e.g., the Iconic Map PM) runs in a separate machine (the X-windows client). With remote terminal access to a command line-based PM — the user accesses a terminal (or emulator) on one machine and the PM (e.g., the Forms and Command Line PM), runs on MCC in a separate machine. Communication may be via TELNET, Kermit or other remote terminal protocols.

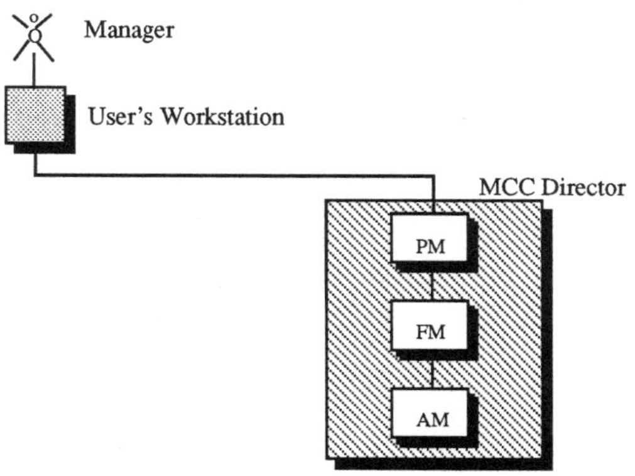

Figure 1. User accessing MCC remotely

4. AM TO MANAGED OBJECT DISTRIBUTION

Each MCC AM implements, or uses the services of, the appropriate management protocol for the managed objects it represents. Each management protocol (or more precisely, the management protocol's transport) defines the extent to which the AM and the managed object may be distributed. Figure 2 illustrates this for a LAN-based protocol.

Among the remote access capabilities for AMs are:

- OSI CMIP, using OSI Transport, so the AM and the managed object must be in the same OSI network.
- SNMP, using UDP over IP, so the AM and the managed object must be in the same IP network.
- NICE, the DECnet Phase IV management protocol, using NSP (the DECnet Phase IV transport) so the AM and the managed object must be in the same DECnet network.

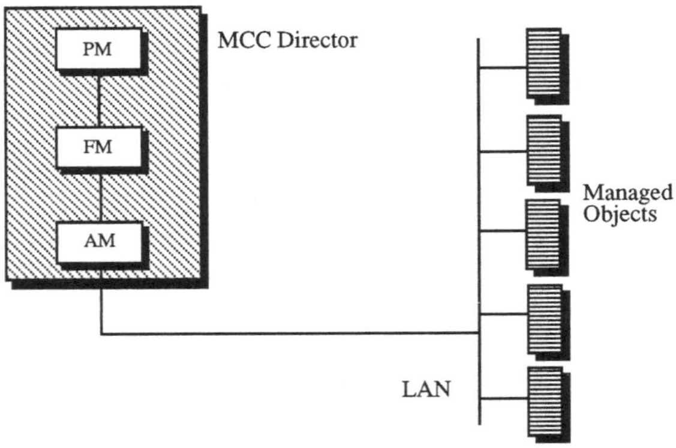

Figure 2. MCC managing remote objects

5. COMMON NETWORK-WIDE NAME SPACE

An important consideration when using multiple management systems is ensuring that all managers are able to refer to a given managed object by the same name. (This does not preclude each manager using a personal "nickname" as a shorthand.) However, unless there is a common network-wide namespace[4], there is a risk that managers may get confused when communicating with each other about devices common to their management environment.

MCC employs a network-wide directory service as the implementation of its Instance portion of the Management Information Repository. This has a number of benefits:

- Every global entity (i.e., uniquely addressable managed object) has a globally-unique name.
- The global name may be known to the managed object, or may be registered by a manager with the management system on behalf of the managed object.
- Once a managed object has been registered by one manager via an MCC, the name and other associated information is available to all other managers on all other MCCs in the same network.

- To deal with large numbers of names, name space names are defined hierarchically, so that the name space can be structured according to the needs of the manager.
- To allow global accessibility of the names of global entities, the name space administrator can distribute name space directories to appropriate places in the network, and replicate directories to provide redundancy and reliability of access to the name space.

6. COMMON NETWORK-WIDE REPOSITORY

In networks where there are multiple managers and multiple management systems in use, there is typically a need to consolidate information from each management system into a single database in order to provide different kinds of summary reports.

Each MCC currently keeps data, such as historical data, local to itself[5]. It also provides a mechanism to allow exporting of management information to a relational database. Such a database may be located on the same machine as the management system, or it may be remote from the management system. It is possible to have a relational database set up on one system, and to export information from multiple management systems into that database, so that the information can be processed to provide various management reports.

Consider a situation where a user has a central site; a few dozen remote sites, each comprising a single LAN; and a few dozen devices on each LAN. Let us further suppose that there is a person at each site, part of whose responsibilities include management of the LAN. In addition, there is a management staff at the central site, backing up the remote site's personnel, and running MCC to manage the central site in addition to overseeing the operation of the whole network. Each site manager runs a separate MCC to manage the LAN devices — of course they all use the common name space so there is no confusion between the names for distinct devices. Each MCC site is collecting data for historical purposes about the LAN devices, and is running alarm rules to check for problems on the site.

Periodically, say once per week, the information from each site, computed from the raw data stored for historical purposes, is exported from each site into a single database at the central site. This information might contain information such as traffic or error statistics, or information about alarms and the resolution of problems detected. Once the information has been collected from each of the sites, the central site personnel can produce a set of management reports based on the information from all sites, including the central site.

7. MANAGEMENT SYSTEM-TO-MANAGEMENT SYSTEM

Not every management system is based on open technology such as MCC. Thus, in many environments, multiple management systems exist. In this section we see the different ways in which these systems may interact. We examine the cases where the MCC director is either the management system closer to the user, or where it is the management system closer to the managed object — we ignore the case where neither management system is based on MCC.

Where the relation between the (non-MCC) management system and the MCC director is static, there are two cases to consider, viz:

- As shown in Figure 3, MCC is acting as the enterprise management system; is interacting with the user via a normal PM; and an appropriate AM is used to communicate with the element management system, which in turn communicates with the managed objects — presumably using a proprietary management protocol. The management system might, for example, be a modem management system or a T1 management system, specialised towards management of a particular set of managed objects.

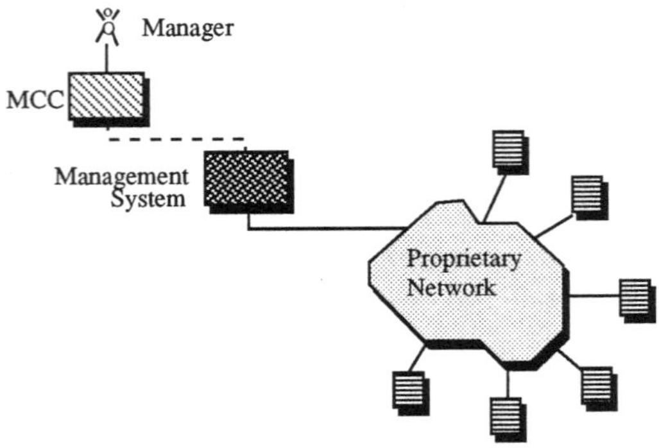

Figure 3. MCC, as an enterprise manager, interacting through another manager

- As depicted in Figure 4, MCC is acting as the element management system, using an appropriate AM for the managed objects for which it is responsible, communicating with the integrating management system through a specialised PM. To the management system, MCC appears to be an agent for the managed objects it represents, so this PM is referred to as an Agent PM. As MCC is a complete management system itself another manager, or the same manager, can be using MCC directly as a management system.

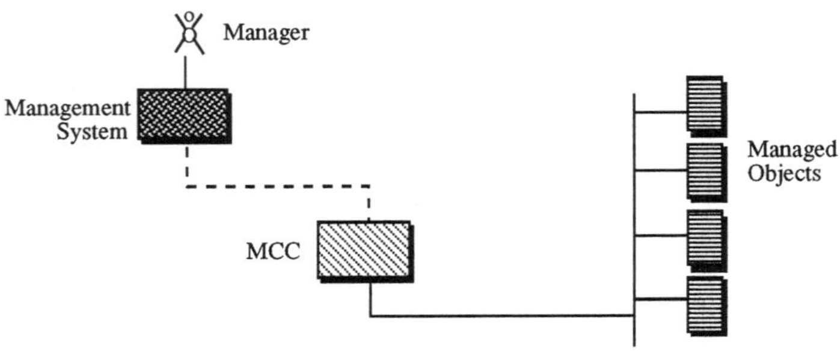

Figure 4. MCC interacting with another enterprise manager

In these two cases, the naming and addressing has to be different between the two management systems and from the element management system to the managed objects. The user refers to the managed object by a name which is translated to an address by their management system; this address refers to the management system closer to the managed object. The second management system, when receiving a request to perform an operation on the target managed object translates to an address referring to the managed object itself.

Both scenarios presume a static relationship between the management system and MCC

— with MCC either being the top level management system, or acting as the element management system to some other top level management system.

In the dynamic case, it is possible for the roles of top level management system and element management system to be determined on an operation by operation basis. Thus for one operation, MCC may use the services of an element management system, and in another case, MCC may be the element management system to another top level management system. Where a single management system is capable of operating in either role (as can MCC), then the relationship between management systems is peer-to-peer. [6] describes one such use of the peer-to-peer approach of interworking between open systems.

8. KERNEL-TO-KERNEL

In almost all cases, the manager and the managed object are not co-located. In some situations the user-to-PM or AM-to-managed object distribution allows the manager to deal with the managed object using a single MCC management system. However, some situations exist in which a single MCC management system working in isolation is insufficient. Consider the situation where a particular managed object's native management access protocol dictates the location of an AM that can access the object. Yet we may wish to manage it from different locations. This is the case, for example, with datalink layer devices, such as MAC Bridges that implement an 802.1-based management protocol — here the AM has to reside on the same extended LAN as the MAC Bridge. Consider also a PM running on the manager's workstation, using the high-bandwidth graphics capability of the workstation; executing the PM on a different system may degrade the user interface performance unacceptably. Finally, we may wish to have one or more FMs executing on a server system that has sufficient computing resources to service a number of managers simultaneously.

Thus, we would like to be able to run PM, FM and AMs in different machines — the PM located near the user, the AM located near the managed object, and the FMs on yet another machine. The simple call arrangement of MCC with all modules executing on the same system, is not sufficient; instead, we would like to have each module, for a given operation, able to execute on a separate system, as shown in the Figure 5.

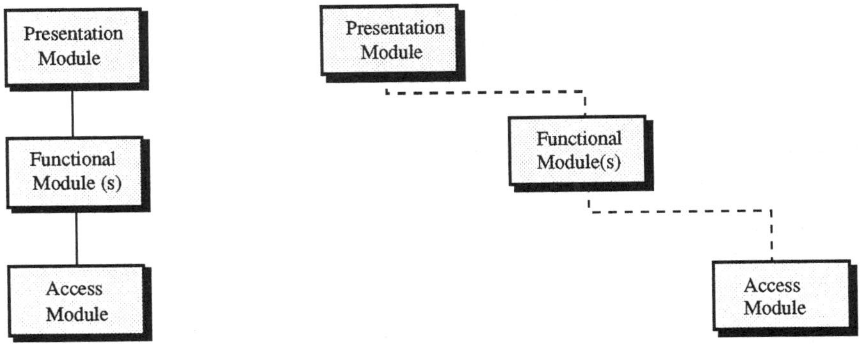

Figure 5. Applying remote procedure call to MCC's inter-module calling scheme

We arrange that the inter-module communication (i.e., *mcc_call*) be set up to operate in a distributed manner, rather than just in the context of a single machine. Indeed, such an arrangement was foreseen with the original design of the *mcc_call* inter-module communication mechanism, and has been documented in the DECmcc System Reference Manual[7] since its inception.

What we discuss here are further details of how the remoting of *mcc_call* might be accomplished, to provide for distributed access to managed objects, distributed and shared processing, and distributed access to management information. Other approaches might be used to achieve the same goals.

8.1 Remoting *mcc_call*

We first observe that, while it is initially appealing, a standard RPC mechanism in its entirety, such as that defined in the OSF DCE, does not provide the complete solution for our needs. This is because there are only two calls defined which are used for all services accessed between MCC management modules (*mcc_call_function* and *mcc_call_access*). RPC uses the name of the procedure to determine where the remote procedure is located.

In MCC, services are not defined in terms of their procedure names. Instead, an alternate method was defined that allowed definition and implementation of management module services to proceed independently. Management Module services are accessed by means of service definitions, i.e., identifying the operation and the target managed object. Thus, MCC services are defined in terms of the VEP tuple, namely the verb (operation), entity (managed object) and (for non-action directives) the partition (either an attribute or an event partition) [7].

Hence, the key to locating the service to be executed is not in the name of the procedure called (*mcc_call*), but in the parameters of the procedure call, i.e., the Verb, Entity and Partition parameters. Thus, to be able to remote an *mcc_call* using an RPC-like mechanism, we could perform the binding not on the procedure name itself, but on the arguments of the procedure. Elements of the DCE RPC can still be used to handle much of the work; however, the binding needs to be handled differently.

In the single system case, the MCC dispatcher is used to bind the requester (issuing the *mcc_call*) with the service provider (the procedure that implements the requested service) — this is done by means of two dispatch tables (one for *mcc_call_function* and the other for *mcc_call_access*). To support the different needs of remoting *mcc_call_function* and *mcc_call_access*, two slightly different methods of binding seem to be appropriate.

8.2 Remote *mcc_call_access*

It is the job of *mcc_call_access* to get to the AM that supports the real managed object (or the representation of that object through another management system). The AM that must be used to get to a particular managed object instance has to be located within the "extent" of the transport supporting the appropriate management protocol implemented by the AM and the managed object.

Remoting *mcc_call_access* provides distributed access to managed objects. To implement remoting of *mcc_call_access* the global entity portion of the *in_entity* parameter of the *mcc_call_access* is important. The access dispatcher examines the global portion to decide to which system to dispatch the request. Based on the different networking access technology, there can be three separate ways in which an AM might be chosen:

- Network-wide accessibility — some managed objects are accessible through a network protocol that is widely, if not universally, available, such as OSI or IP; *mcc_call_access* can invoke a service in the AM on any machine in the network. Specifically, the request can stay within the current machine, as this is the most efficient.

- Limited-area accessibility — some managed objects' management protocol is limited in extent, such as to a LAN; in this case the set of MCC systems that can access a given managed object must be derived. That information can be stored as part of the name space entry for the global entity when the global entity is registered. Access dispatch chooses an entry from the list of systems, and invokes the AM on that system. More than one remote system may be used; if one system is inaccessible, another may be chosen to invoke the desired AM.
- Single system accessibility — some special purpose managed objects may only be accessible from a single system. Consider a modem with an asynchronous management port connected to an asynchronous port of a particular machine. Here, the AM must be accessed on that system. This is a special case of limited-area accessibility, where there is only a single system. The access dispatch works in a similar manner, though there can be no backup if that system is inaccessible.

8.3 Remote *mcc_call_function*

One of the additional capabilities of an *mcc_call* is to access historical data that has been stored previously. The MCC Historian Services[5] arranges to store data associated with a domain — this is for reasons of securing the data as well as identifying ownership of the stored data, and accounting for the storage used. Also, any user, access rights permitting, may access any domain, showing the domain's managed objects on their map, and requesting data about them, either real time or historical data. In addition to data, executing background code, such as evaluating conditions as the basis for alarms, is set up in the context of a domain. It is important to be able to access stored data, or code that is executing, in the context of the domain that owns the data, or executes the code.

Remote *mcc_call_function* provides for distributed and shared processing, as well as distributed access to management information. The remoting of the *mcc_call_function* should not be on the global entity, but on the domain name, passed in the *in_q* (input qualifier) parameter of *mcc_call_function*. Thus, when an operation is initiated at the PM within the context of a domain, the operation can be executed by one or more FMs at the system where the domain is defined. This includes the retrieval of historical data, and access to any background executing code.

Note that, even where multiple FMs are not used for a given management operation, any historical management data may be accessed either on the *mcc_call_function* from PM to FM or on the *mcc_call_access* from FM to AM. Assuming all three modules are on different systems, the PM-FM call will attempt to look up historical data at the PM system (and will fail, as the domain will not be located there), and then on being dispatched to the FM system (where the domain is located) will attempt to look up historical data on the appropriate system. However, the processed data may not have been stored for historical use, so when the FM calls the AM, the historical data lookup will still be attempted on the system where the FM is located, and hence where the domain's historical data is located.

By simply dispatching based on the domain, in *mcc_call_function*, a number of capabilities accrue as a consequence:

- Multiple users, each with a separate PM, can be set up to receive notifications of alarms within one domain, where the evaluation of the alarms takes place once; the notification is distributed to the interested parties. Where nested domains are used, each domain's alarm rules are evaluated once, and notifications are fed "up" to the containing domain, where there are interested users to receive the results.

- Historical management data is effectively distributed around the network, on a per-domain basis, and access is transparent, through the appropriate domain context.
- Domain definitions can be shared by multiple managers across the network, and the resultant stored historical data, and executing background processes are available network-wide.

Note that domains were defined to allow exactly this type of network wide operation, to exploit the power of a distributed, cooperative solution to the management problem. An appropriate configuration of MCC systems, and the definition of, and location for, domains is important to the efficient operation of the total management solution[8].

9. DIRECTOR-TO-DIRECTOR

As multiple management systems are employed to provide management of the entire environment, often the set of managed objects with which one manager needs to be concerned overlaps with the set of managed objects used by another manager.

While one of the managers is only involved in monitoring the overlapping managed objects, and presuming that the managed objects' agents are implemented to ensure that each managed operation is atomic (so a single operation from one management system cannot interfere with a management operation from another management system), there is no conflict. One manager can monitor and change a managed object, and any number of other managers can just monitor that managed object.

When the situation arises where more than one manager can change a given managed object, conflicts will occur. Individual management operations, implemented atomically in the managed object's agent, will not conflict but the desired operation of the managed object for one manager, may not be the same as that desired by another manager. While the managers communicate out-of-band (e.g., via telephone or other inter-personal communication) the chances for conflicting operational requirements are reduced, but not avoided completely.

Certainly, access control implemented at the managed object's agent, and the appropriately granular set of access rights will ameliorate the situation, so that each manager is restricted in the set of change operations they may initiate. Where two or more managers share the same access right and yet have different ideas of how to change a managed object, conflicts can still arise. While we are unable to solve all these problems (some come down to human processes) we would like to be able to minimise the chances for inadvertent misoperation.

Appropriately implemented agents will ensure that multiple operations addressed to the same managed object, (whether from the same or different management systems), will not interfere with each other. The most obvious example is that an operation to read datalink layer counters must always return consistent information — thus a message count of 0 and a byte count of 10 is inconsistent, as the agent should not return counters that have been partially updated.

We would like to achieve an additional level of guarantee so that a management operation from the manager's viewpoint (as opposed to an individual management protocol request) is atomic — so that it occurs completely or not at all. This requirement is further complicated in that the manager's single operation may require not only multiple protocol exchanges for a single managed object (the simple case) but perhaps dealing with multiple managed objects on different systems.

Conceptually, we think of the directors as being able to coordinate their operations, so that where two or more directors are dealing with the same managed object, they ensure that they achieve the desired effect, as shown in Figure 6.

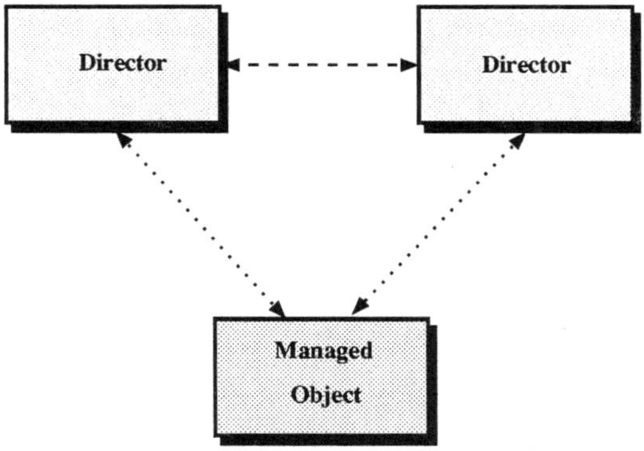

Figure 6. Directors cooperating while managing the same object

The communications between directors (here shown as two, but could be any number) is referred to as Director-to-Director communication. The communication between the directors and the managed object (again shown as one, but could in the general case be any number of managed objects) could be the normal management protocol.

A complete solution to this problem is beyond the scope of this paper, but we identify some of the pertinent aspects. Work proposed to be done in the standards organisations will address some of these aspects[9]. What follows are some initial thoughts.

We observe that transactions could be applied to management operations. This would make available across, or alongside, the management protocol between director and managed object, at least start transaction, commit transaction and rollback transaction operations. In addition, a managed object's agent, on receiving a management protocol directive, or start transaction request, for a managed object that is currently involved in a transaction, can choose to either reject the request, or delay until any extant transaction is completed. None of today's management access protocols have, to the author's knowledge, transaction semantics available — certainly not the two standard protocols, SNMP and CMIP.

Also, as an integral part of the ability to undo a partially completed transaction (the rollback) there is a need to define a set of compensating operations — i.e., for each operation define how that operation is to be undone. While some systems that support managed objects are capable of supporting a complete transaction manager, there will always be some systems that are unable to, or simply don't, provide the requisite support. Any solution to management transactions needs to take into account systems that do not provide transaction support without requiring the managed objects' agents to implement that support.

10. SUMMARY

In its initial implementations DECmcc provided much of the underlying mechanisms to enable multiple management systems to support multiple managers in the management of a user's environment. In addition, care was taken during the design of MCC to ensure that operation of the director as a distributed application could take place with the minimum of disruption — either to the managers involved with management, or to the developers of the

MCC management modules themselves.

MCC allows the user to be remote from a Presentation Module, and allows an Access Module to be remote from the managed objects. By using a global name space, each manager has the same name for the managed objects, thus reducing confusion in coordinating multiple managers. By allowing information to be exported into a remote relational database, network-wide management reports can be generated. With the appropriate management modules (Agent PMs and AMs), MCC can interoperate with other management systems, either as a subordinate, as a superior, or in a peer-to-peer relationship, thus allowing for coordinated management via multiple management systems.

With the implementation of Kernel-to-Kernel support, allowing modules from different MCCs to be involved in providing a single management operation, MCC's ability to become a true distributed application will be realised. Managed objects that have limited accessibility due to their management protocol's transport requirements will become visible network wide. Management data gathered by one MCC, such as historical data, and results of background operations, such as alarm processing, can also be made available network-wide, obviating the need to duplicate data or operations for accessibility reasons.

As management becomes widespread, the need to provide coordinated management operations will become increasingly important. Further work on Director-to-Director communication will set the stage for coordinated management among multiple managers across the distributed system.

11. REFERENCES

[1] Strutt, C., and Shurtleff, D., Architecture for an Integrated, Extensible Enterprise Management Director, in Integrated Network Management I (1989) 61-72.

[2] Strutt, C., and Swist, J., Design of the DECmcc Management Director, Digital Technical Journal (1993) Volume 5, No. 1.

[3] Fehskens, L., An Architectural Strategy for Enterprise Management, in Integrated Network Management I (1989) 41-60.

[4] Sylor, M., Junction Objects — A Solution to a Problem in OSI Managed Object Naming, this volume.

[5] Shvartsman, A., An Historical Object Base in an Enterprise Management Director, this volume.

[6] Borden, J., Digital's Telecommunications Network Management Program, in Network Operations and Management, IEEE (1992) 102-111.

[7] Digital Equipment Corporation, DECmcc System Reference Manual, two volumes, Order No. AA-PD5LC-TE, AA-PE55C-TE (1992).

[8] Strutt, C., Dealing with Scale in an Enterprise Management Director, in Integrated Network Management II (1991) 577-593.

[9] ISO/IEC JTC1/SC21 N7122, Application Context for Systems Managementt with TP, working draft, October 1992.

Integrated Network Management, III (C-12)
H.-G. Hegering and Y. Yemini (Editors)
Elsevier Science Publishers B.V. (North-Holland)

The management of monitoring in object-based distributed systems

Y. Hoffner
ANSA, Architecture Projects Management Ltd., Poseidon House, Castle Park, Cambridge CB3 0RD, UK

Abstract
A general model of management is introduced and used in the development of a model of the management of monitoring. This model is subsequently used to show how monitoring and its management can be implemented in an object-based distributed system.
The structures necessary for conducting a monitoring session with multiple objects are presented. The problem of managing a monitoring session, where the set of objects under observation changes dynamically, is addressed. Finally, the problem of management across federation boundaries is discussed.

Keyword Codes: C.2.4; D.1.5; D.2.5
Keywords: Distributed Systems, Object-oriented Programming, Testing and Debugging

1. INTRODUCTION

Distributed systems need monitoring facilities to support varied activities such as management, debugging, testing and accounting [JOYCE 87], [MCDOWELL 89].

Distribution and more specifically, dealing with issues such as heterogeneity, autonomy, physical separation and concurrency, complicates the process of monitoring. The design and development of monitoring facilities needs to overcome these problems.

In an object-oriented distributed system based on principles of encapsulation, each object will have its own monitoring and management facilities with an appropriate management interface.

Monitoring often involves correlating concurrent events at multiple objects. Management structures are therefore necessary to maintain information on the objects which are participating in a monitoring session, and manage the monitoring facilities in each of them. In addition, appropriate structures for collecting monitoring information are required.

In a distributed system, a monitoring session can evolve dynamically, as activities related to an application spread throughout the system. The management of a monitoring session must therefore be able to extend and contract the set of objects under observation. The distributed system must supply facilities for obtaining access to management functions within objects given a reference to one of the object's service interfaces.

When monitoring across federation boundaries, differences in monitoring and management facilities must be accommodated either by prior agreement on common facilities, or by supplying the appropriate translators which allow interworking. There is also a need to provide channels and mechanisms for resolving policy conflicts across federation boundaries.

2. THE APPROACH

This document proposes a model of management of monitoring processes for federated distributed systems. It is addressed to a technical audience concerned with the design of management structures for distributed systems. A federated system arises when two or more autonomous systems are interconnected whilst retaining their individual control. In such a situation, gaining access to management and monitoring interfaces must be by negotiation and discovery, rather than by previous agreement.

The approach described requires that the underlying distributed systems architecture has the means to represent monitored entities as encapsulated objects and the means to transmit references to object interfaces as parameters of management operations. Such a capability is inherent in the ISO Basic Reference Model for Open Distributed Processing [X.900 92], the OMG Combined Object Request Broker Specification [OMG 91] and ANSA.

The focus of the document is on the means to discover and negotiate monitoring interfaces. It is not concerned with the management functions provided at those interfaces: this is a separate topic well-addressed in other places (e.g. OSI management standards).

3. THE GENERIC MODEL OF MANAGEMENT

Implied in most if not all discussions of management, are two fundamental notions:

- that management is a continuous activity
- that management is a closed loop activity [WIENER 75].

These notions necessitate three generic management functions (Figure 1);

- *monitoring* - the process of observing the managed system
- *control* - the process by which changes in the managed system are effected
- *decision making* - the process of determining what changes should be effected in the managed system.

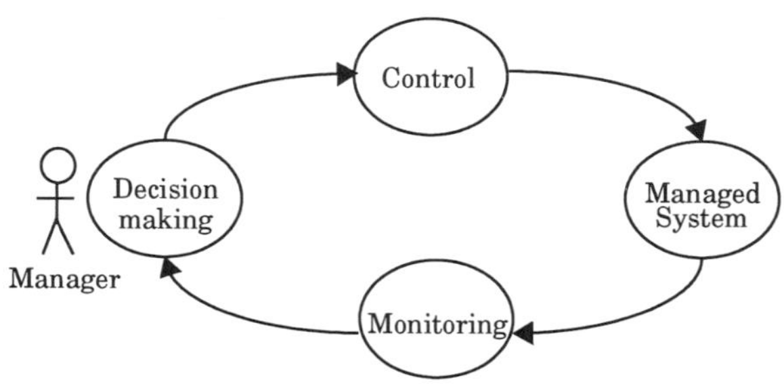

Figure 1: The generic model of management - monitoring, decision making and control

4. APPLYING THE MODEL OF MANAGEMENT TO THE MONITORING PROCESS

When the managed system is large, complex and dynamic, the processes of monitoring and control will themselves be complex activities. They may themselves, therefore, require appropriate management. The model of management presented in

Figure 1 can be applied to the monitoring process in order to show how its management is done [HOFFNER 92c], as shown in Figure 2.

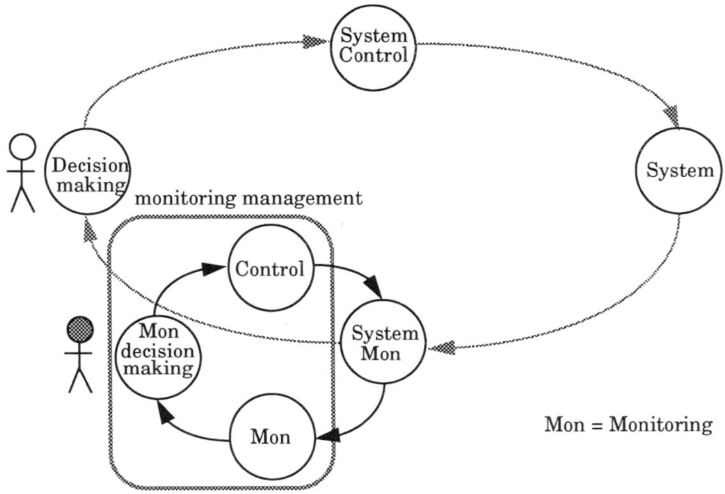

Figure 2: The management of the monitoring process

In Figure 2, the roles of the manager of the system who also observes the system through the monitoring facility, and the manager of the monitoring process, are shown separately. In practice, the system manager may also be the monitoring manager.

5. DEVELOPING THE MODEL OF MANAGEMENT OF MONITORING

Figure 2 can be simplified by combining the control and monitoring of the monitoring process into a single management component (Figure 3).

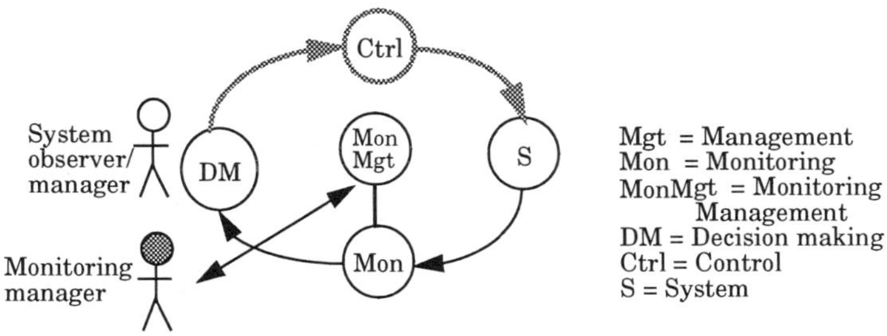

Figure 3: Combining the control and monitoring components

The simplified model shown in Figure 4 results from concentrating solely on monitoring whilst ignoring the system control process.

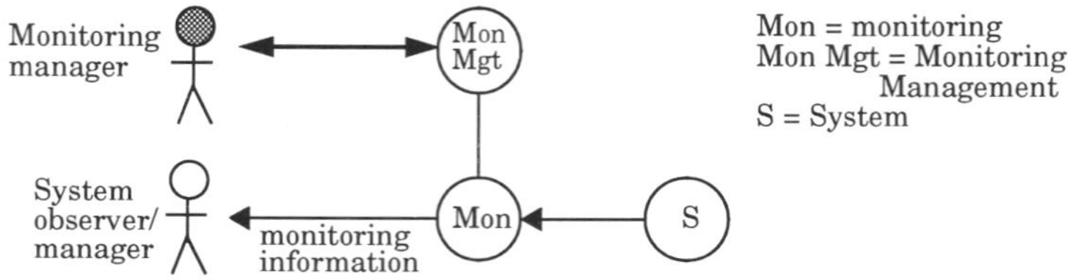

Figure 4: Concentrating on monitoring and its management

5.1 The generic model of the management of monitoring

The model of monitoring and its management shown in Figure 3 assumes that all management is directed from outside the system being monitored. However, the monitored system may itself use the monitoring facilities for obtaining information about itself. Moreover, some of the management of monitoring may be automated and embedded in the observed system [HOFFNER 92b]. This is shown in Figure 5.

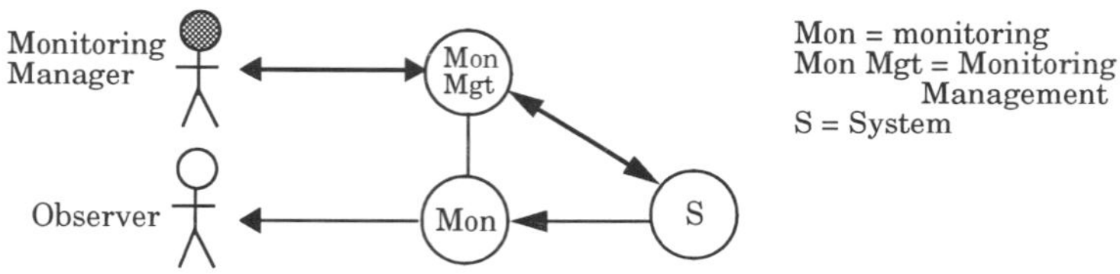

Figure 5: The generic model of the management of monitoring

The monitoring management component provides the manager with:

- a view of what monitoring facilities are available, e.g. what is the set of monitoring events defined in the system
- a view of the state of the monitoring facilities, e.g. are they on/off, where is the monitoring information sent to, whether the system is participating in more than one monitoring session
- control facilities for changing the state of monitoring.

6. MONITORING IN OBJECT-BASED DISTRIBUTED SYSTEMS

The computational model [AR.001 91] is an application writer's view of a distributed system without the concerns for the engineering required to implement it. The engineering model [HERBERT 91] is concerned with the entities, and the relations between the entities, which animate computational model objects, and allows resource issues to be discussed. Capsules are engineering entities which provide processing, storage and communication facilities for one or more objects [AIM 92].

Developing the facilities necessary for management of monitoring entails two areas of work:

- monitoring resources inside objects and capsules
- monitoring resources outside capsules.

6.1 Computational model of monitoring facilities in an object

In a distributed object-based system the principles of encapsulation and autonomy means that each object will have its monitoring and management facilities and an individual management interface [HOFFNER 92d].

The model of monitoring in objects follows from the general model of monitoring and its management, developed in section 1.4.1, and is shown in Figure 6.

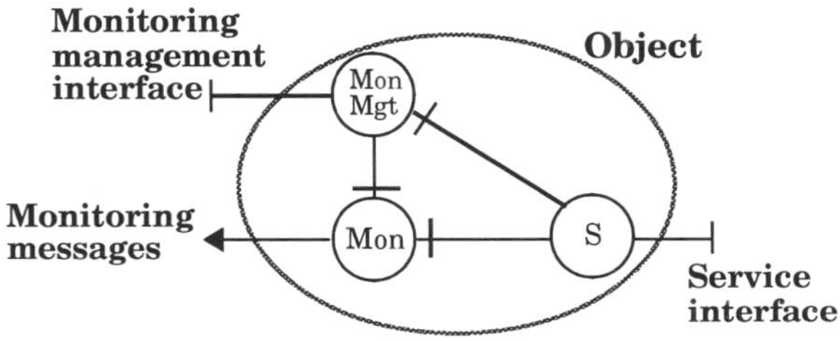

Figure 6: Objects have their own monitoring management service and interface

6.2 Engineering model of monitoring facilities in a capsule

When implementing objects it is often necessary, for resource utilization reasons, to aggregate several objects into a single capsule. The model of the management of monitoring can be applied recursively to provide the monitoring of each object as well as the integrated monitoring of the capsule in which they reside (Figure 7).

7. MONITORING AND MANAGEMENT FACILITIES IN CAPSULES

The monitoring facilities in capsules are responsible for:

- generating monitoring messages when certain events take place within the capsule
- forwarding the monitoring messages to the appropriate destination either within or without the capsule.

To achieve this, the monitoring management service has to provide the information and operations, as discussed in the following section.

7.1 Query operations

Three general types of queries can be made on a management interface. These are illustrated below:

1) *'which'* objects exist in the capsule. The response to this query is a list of management interfaces of the objects contained within the capsule being queried:

- `listManagementInterface[MgtIfRef:InterfaceRef]RETURN [SEQUENCE of InterfaceRef]`

2) *'what'* monitoring facilities does an entity have?

Having obtained the management interface of the required entity it is now necessary to find out what monitoring facilities are provided:

```
showManagementfacility[Mgtifref:InterfaceRef]RETURN [SEQUENCE of mEventClassRecord]
```

3) to enable the MMgr to find out the list of collation destinations to which monitoring messages are currently sent to:

```
listDestinIfrRef[ ] RETURN [SEQUENCE of IfRef].
```

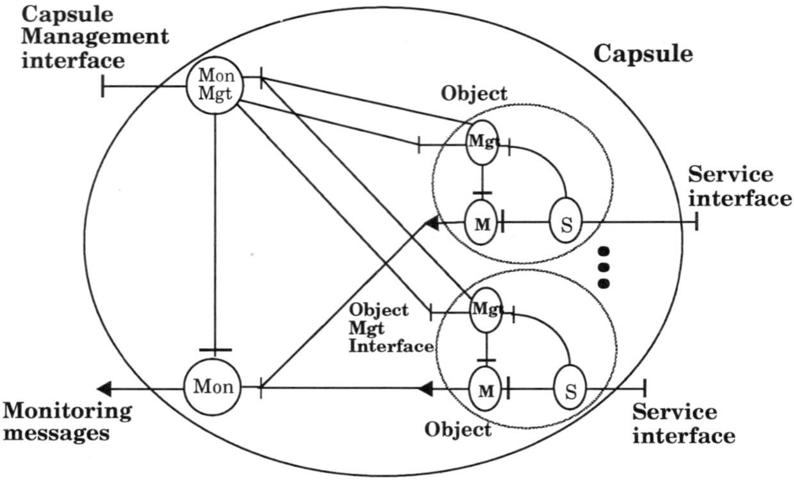

Figure 7: Monitoring in a capsule consisting of several objects

4) to enable the monitoring session to be extended to include service objects and client objects bound to the capsule under observation, it is necessary to have the following query operations:

(i) `findServiceInterface[mmgtifref:InterfaceRef]RETURN [SEQUENCE of InterfaceRef]`

(ii) `findClientInterface[mmgtifref:InterfaceRef]RETURN [SEQUENCE of InterfaceRef].`

7.2 Activation operations

Having obtained access to the monitoring management interface of an entity, the activation of specific monitoring event generators can be requested:

```
setManagementFacility[mmgtifref:InterfaceRef;
      class-typemEventClassType;event-type mEventType;on-off:Boolean]
```

8. MONITORING CONFIGURATION OF OBJECTS

8.1 Managing an on-going monitoring session

Most applications in a distributed system can be characterised as having multiple concurrent activities where:

- the activities may be distributed over different physical locations; and
- the activities dynamically:
  - interact with other activities
  - initiate or terminate other activities
  - suspend or terminate themselves

The dynamic initiation and termination of activities leads to situations where the activities stemming from an application may spread throughout the system. The extent of the initiated activities may not be known in advance. Monitoring such activities will require management facilities to extend and contract the scope of monitoring.

8.2 A monitoring session in a distributed system

In order to manage a monitoring session dynamically it is necessary to provide the observer/controller of a monitoring session with the following (Figure 8):

(i) a management view

(ii) a monitoring view

(iii) a control facility.

The monitoring view is based on the monitoring messages collected by the collation setup from the objects under observation [HOFFNER 92a]. The management view consists of the information which is necessary in order to manage the monitoring session, for example, information on the scope of monitoring.

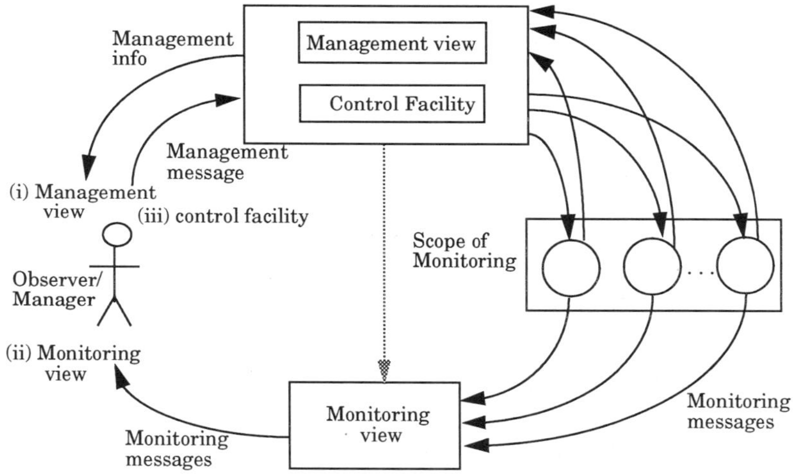

Figure 8: Management and monitoring views in a monitoring session

It is assumed throughout this document that each monitoring session has its own monitoring management and collation configuration. This assumption does not contradict the option of a single capsule participating in more than one session, but ensures that outside a capsule the monitoring management and collation management is on a "per monitoring session" basis.

8.3 Scope of monitoring

The Scope of monitoring consists of the set of objects which are known to the Monitoring manager, whose monitoring management service can be accessed, and whose monitoring information can be directed to the appropriate point of collation. In a monitoring session, it is the management view as shown in Figure 9, which provides this information.

8.4 Monitoring session management facilities

The proposed structure of the monitoring session components which provides the monitoring view, management view and the control facility mentioned has the following components (Figure 9):

A Monitoring manager (MMgr) provides a monitoring session with the management view and with the necessary functions for managing the objects which are within the scope of monitoring. Thus, the MMgr has to keep track of the objects within the scope of monitoring, and to take the necessary actions when the scope of monitoring requires modification.

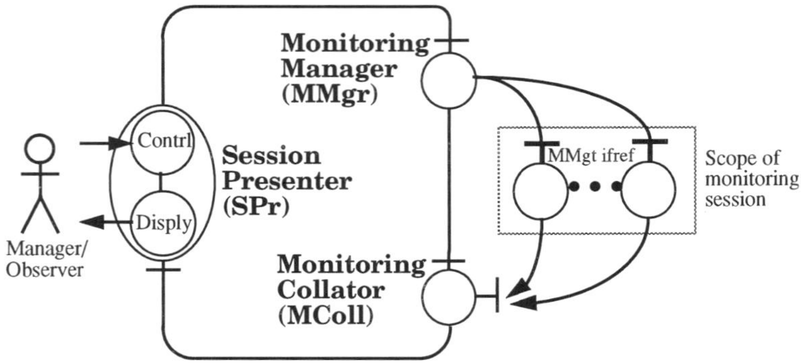

Figure 9: The configuration of a monitoring session

A Monitoring Collator (MColl) is responsible for collecting the information from the monitored objects or from other monitor-collators, necessary to provide the monitoring view. When monitoring more than a single object, more than one monitor-collator may be necessary. The MColl may store the information locally or forward it to another MColl, a file, or a display agent such as a terminal or Session Presenter.

The Session Presenter (SPr) provides the front end to the MMgr and the MColl of a monitoring session. It presents the monitoring information to the observer and allows the observer to control the progress of the monitoring session. It allows the integration of management and collation display and control functions. In its simplest form this could be a terminal with textual output, although graphical tools are more likely to be appropriate in a distributed environment.

Although not shown in Figure 10, the components themselves (Spr, MMgr and the MColl), as well as the objects under observation can be distributed throughout the system. In such cases, a hierarchy of local management and collation structures will be linked together to cater for objects distributed over long distances [HOFFNER 92a]. This is outside the scope of this paper.

9. MANAGING A MONITORING SESSION

9.1 The phases of a monitoring session

There are four phases in the life of a monitoring session:

(i) the setup of the monitoring session components: namely, the SPr, MMgr and MColl

(ii) the setup of the initial scope of monitoring: creating the link between the MMgr, MColl and the objects which the observer wishes to monitor and has knowledge of

(iii) managing an ongoing monitoring session: changing the scope of monitoring as the need arises includes expanding and contracting the scope of monitoring. This may also include extending the scope of monitoring across management domains (federation boundaries). Also, the activation and de-activation of monitoring facilities in objects within the scope of monitoring

(iv) closing down a monitoring session.

The rest of this paper deals with the management of an on-going session and in particular with the problem of extending the scope of monitoring dynamically.

9.2 Extending the scope of Monitoring during a monitoring session

The problem of dynamically extending the scope of monitoring requires two things:

- the MMgr to find out about any objects outside the scope of monitoring with which objects within the scope of monitoring are interacting with (Figure 10).
- the MMgr to gain access to the management facilities of those objects

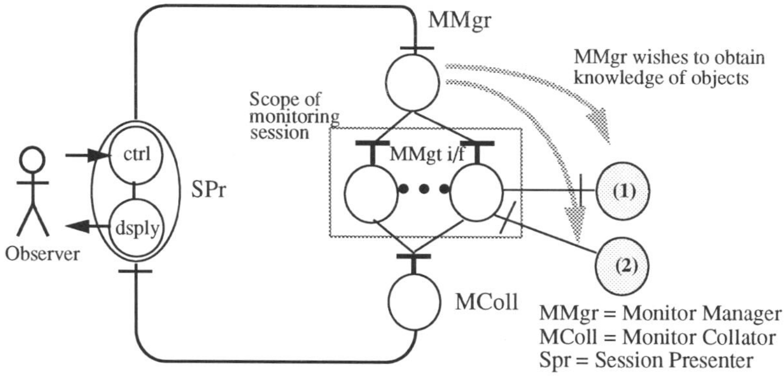

Figure 10: Finding out about object(s) outside the scope of monitoring

Objects within the scope of monitoring may interact with objects outside it. In such cases it may be necessary to activate the monitoring of those objects. This can be done in a number of ways (Figure 11):

(i) from the objects outside the scope of monitoring to the MMgr

(ii) from the objects which are within the scope of monitoring to the MMgr

(iii) from the MColl to the MMgr through the monitoring information channel

(iv) through the SPr by user intervention.

9.3 Obtaining access to management

Given that each object provides a management interface containing management operations, it is necessary to provide facilities for obtaining a reference to this interface. Using trading [LINDEN 91], [DESCHREVEL 92a], [OMG 91] for searching for these interfaces does not solve the problem entirely since not all interfaces are traded. Also, there is a need to be able to associate different interfaces of the same object to one another, so that given a service interface to an object, it is possible to obtain the management interface of the same object. Including information which associates different interfaces to each other is unlikely to scale, particularly in cases where objects may provide large numbers of interfaces.

A possible solution is to have every interface contain an operation that, when invoked, returns the references to its management interfaces (Figure 12). This can be implemented by forcing all interfaces to conform to a special management interface. In practice this means that any service interface of an object, will in addition to its

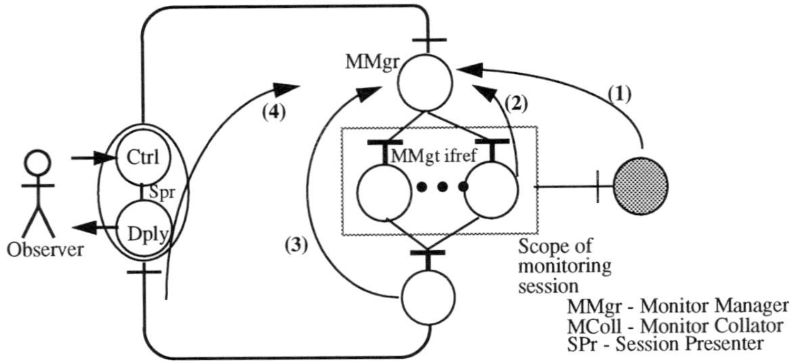

Figure 11: Possible solutions to finding out about objects outside the scope of monitoring

service operations provide an agreed operation. This operation can be used to obtain a reference to the object's management interface, and will be of the form:

```
getManagementInterface [] RETURN [MgtRef]
```

Figure 12: Gaining access to management interfaces using the getManagementInterface[] operation

Having obtained the management interface, the client can now invoke the management operations of the server.

10. USING INDIRECTION TO CONSTRUCT MANAGEMENT DOMAINS

The getManagementInterface[] scheme can be used so that the result from its invocation may be an interface to some other object which can act as an agent. The agent can be made responsible for the management of the object (Figure 13).

An agent can be provided with the management interfaces of several objects, so that by restricting the access to the management interfaces of the individual objects, except through the agent, the agent can be made responsible to ensure that a common policy will be applied to all the objects. This is a way of creating management domains [SLOMAN 89a, b].

11. MANAGEMENT ACROSS DOMAIN BOUNDARIES

Extending the scope of monitoring may take place across federation boundaries [DESCHREVEL 92b], which may have conflicting policies concerning monitoring. In such cases, resolving policy conflicts between domains requires a dialogue between the client and its management, and the management of the server. If the management facility in the server cannot resolve the conflict, the client may request

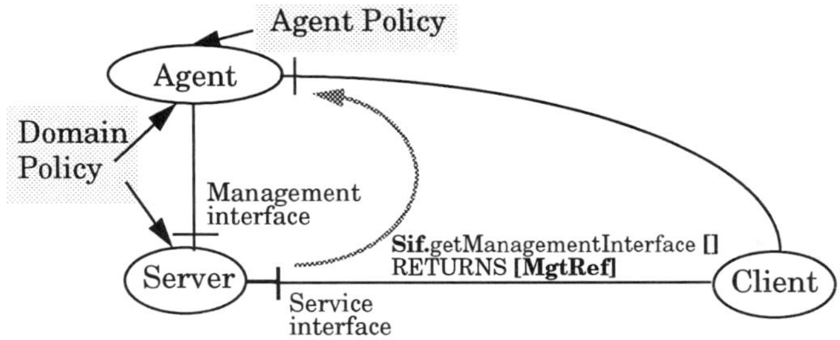

Figure 13: Indirect access to management interfaces

access to a management agent of the server. The getManagementInterface[] operation can be used to request access to successive levels of an objects agents, until the appropriate level for policy clash resolution can be achieved. In order to facilitate such a dialogue agreement on common management interfaces and operations is required [HOFFNER 92a].

12. CONCLUSIONS

A general model of management has been introduced and used to develop a model of management of monitoring. It was shown that object-based distributed systems need management facilities both inside and outside objects in order for monitoring to be implemented. Special facilities are required to facilitate dynamic monitoring sessions and to allow monitoring across federation boundaries.

The ad-hoc inclusion of monitoring facilities in distributed systems is not likely to solve the problems of dynamic monitoring and monitoring across federation boundaries. These issues have to be taken into account at the architectural level, as well as in the design and development stages of distributed systems.

There is also a need for standard management interfaces in order to implement distributed monitoring service.

13. ACKNOWLEDGEMENTS

This work was carried out as part of the ISA-DEMON Esprit Project No. 5199.

The authors would like to thank the reviewers of the ISA-DEMON project for useful discussions: Marc Shapiro (INRIA), Sean Baker (Trinity College Dublin), Ulrich Boes (Commission of the European Communities), Helmut Meitner (FHG/IAO) and Gilbert Cockton (University of Glasgow).

Thanks are also due to the other members of the ISA-DEMON project: Susan Thain (MARI), Daniel Trigo (University of Aviero), Guillaume Coutrot (Telesystem), Guo Zhi-Yi (Telesystem), Nora Bokahri (Telesystem) and Emanuel Duarte (University of Aviero).

The authors would also like to acknowledge the contribution of Andrew Herbert, David Iggulden and Rob van der Linden from the ISA team.

Many thanks are also due to Erling Lindholm and Leif Bentell from Ellemtel, Sweden, for their contributions to the early part of this work.

14. REFERENCES

[AIM 92] "ANSAware Implementation Manual", Architecture Projects Management Ltd., Poseidon House, Castle Park, Cambridge CB3 0RD, UK (1992).

[AR.001 91] Architecture Report, The Computational Model, APM., Poseidon House, Castle Park, Cambridge CB3 0RD, UK (1991).

[DESCHREVEL 92a] Deschrevel, J. P. “RC.324: An Overview of Trading”, Architecture Projects Management Ltd., Poseidon House, Castle Park, Cambridge CB3 0RD, UK (1992).

[DESCHREVEL 92b] Deschrevel, J. P. “RC.367: Trading Across Boundaries”, Architecture Projects Management Ltd., Poseidon House, Castle Park, Cambridge CB3 0RD, UK (1992).

[HERBERT 91] Herbert, A. J. “The Engineering Model”, RC.282.00, APM, Poseidon House, Castle Park, Cambridge CB3 0RD, UK (1991).

[HOFFNER 92a] Hoffner, Y. “RC.315: Monitoring in Distributed Systems”, Architecture Projects Management Ltd., Poseidon House, Castle Park, Cambridge CB3 0RD, UK (1992).

[HOFFNER 92b] Hoffner, Y. “RC.352: Introduction to Management”, Architecture Projects Management Ltd., Poseidon House, Castle Park, Cambridge CB3 0RD, UK (1992).

[HOFFNER 92c] Hoffner, Y. “RC.377: Applying the Model of Management to the Management of the Monitoring Process”, Architecture Projects Management Ltd., Poseidon House, Castle Park, Cambridge CB3 0RD, UK (1992).

[HOFFNER 92d] Hoffner, Y. “RC.389: Management in Object-Based Federated Distributed Systems”, Architecture Projects Management Ltd., Poseidon House, Castle Park, Cambridge CB3 0RD, UK (1992).

[JOYCE 87] Joyce, J. Lomow, G., Slind, K. & Unger, B. "Monitoring Distributed Systems", ACM Transactions on Computer Systems, 5(2), 121-150 (May 1987).

[LINDEN 91] Linden, R. V. D. “RC.101: Trading in the Five Projections”, Architecture Projects Management Ltd., Poseidon House, Castle Park, Cambridge CB3 0RD, UK (1991).

[MCDOWELL 89] McDowell C.E. & Helmbold, D.P., "Debugging Concurrent Programs", ACM Computing Surveys, 21(4), 593-622 (December 1989).

[OMG 91] Object Management Group, “The Common Object Request Broker: Architecture and Specification”, OMG Document 91.12.1 (1991).

[SLOMAN 89a] Sloman, M. S. “A Framework for Distributed System Management”, Distributed Processing, ed. by Barton, M. H. et al. IFIP (1989).

[SLOMAN 89b] Sloman, M. S. & Moffett, J. D. “Domain Management for Distributed Systems”, In Integrated Network Management I, Eds. Meandzija, B. & Westcott, J., North-Holland.

[WIENER 75] Wiener, N. “Cybernetics: or Control and Communication in the Animal and the Machine”, The Massachusetts Institute of Technology Press, Cambridge, Massachusetts, USA (1975).

[X.900 92] ISO 10734/CCITT X.900 Basic Reference Model for Open Distributed Processing Parts 1-5, ISO/IEC JTC1/SC21/WG7 (June 1992).

Integrated Network Management, III (C-12)
H.-G. Hegering and Y. Yemini (Editors)
Elsevier Science Publishers B.V. (North-Holland)

Evaluating Management Decisions via Delegation *

Germán Goldszmidt and Yechiam Yemini

Distributed Computing and Communications Lab., Computer Science Department, Columbia University, New York, NY 10027, USA

Abstract

A central problem of network management is compressing vast amounts of real-time operational data to accomplish management decisions. A *health function* provides such efficient compression by combining managed data linearly into a single index of network state. This paper describes a computational theory of network behaviors, their observations, and threshold-decisions via health functions. The model is used to examine feasibility of various methods for computing health functions. Health functions cannot be included as part of a static MIB design, as they may vary from site to site and over time. Nor can they be usefully computed at centralized management platforms, since this can result in excessive polling rates, lead to errors due to perturbation introduced by polling, and miss the very goal of compressing data maximally at its source. Instead, we propose the use of the management by delegation paradigm to support flexible and effective evaluation of health functions and linear threshold decisions at devices.

Keyword Codes: C.2.2; C.2.3
Keywords: Network Protocols; Network Operations

1 INTRODUCTION

A network management system need to support effective management decisions based on vast amounts of real-time operational data. One may view such decisions processes as methods to compress this vast management information into a simpler decision information. Often, much of the compression needed to evaluate management decision is accomplished through manual processes. Most network management systems are passive and offer little more than interfaces to raw or partly aggregated and/or correlated data in MIBs [1]. Developing effective technologies to support compression of management information is a central problem of network management. Given the increasing plethora of manageable resources, the explosion of standard and vendor-specific private MIBs and managed information, and the increasing scale of networks, this task is extremely difficult.

One method to compress operational data is to compute functions of managed information, reducing a large number of observed operational variables to a single indicator of

*Research supported by NSF contract # NCR-91-06127.

the network state. This is not unlike the use of different indexes to reflect the state of the securities market and the economy. Such indexing typically utilizes linear aggregation of a large number of variables, each providing a different microscopic measure of state. We call such linear weighted function of managed variables a *health function.*

In the context of SNMP [2], observations of operational variables are typically accomplished via counters and gauges. A counter represents a cumulative (integral) of an operational variable. Typically, however, only the change in the counter versus its value provides useful indication of the network state. For example, the MIB-II [3] counter representing ifInOctets accounts for the total number of bytes received by an interface since device initialization. Only the rate at which this counter changes contributes an indication of the network state. Health functions will, typically, utilize a linear combination of such rates at which status indicators vary. MIB variables, too, typically provide primitive independent measures. It is often useful to combine this variables to more useful status indicators. For example, the utilization of an interface at time t=sysUpTime can be defined as

$$\mathrm{U(t)} = [(\mathrm{ifInOctets} + \mathrm{ifOutOctects}) * 8] \; / \; (\mathrm{ifSpeed} * \mathrm{SysUptime} * 100)^1$$

This measure provides an average sense of utilization over a time window since boottime. A useful indication of the instantaneous network state is provided by the derivative $u(t) = U'(t)$, as per the discussion above. This derivative may be approximated by frequent sampling of the respective managed variables and computations of the changes in U. Similarly to utilization, one can establish measures of instantaneous error rates to capture additional network micro-state indication. For example, the percentage of input errors to packets delivered, can be evaluated as

$$\mathrm{E(t)} = \mathrm{ifInErrors} \; / \; (\mathrm{ifInUcastPackets} + \mathrm{ifInNUcastPkts}).$$

Again, only the derivative $e(t) = E'(t)$ is of interest.

A health function can be used to linearly aggregate these micro-measures of local network state. For example, the state of the network seen by a hub, including multiple interfaces, can be indexed by a health function $\mathcal{H}(\vec{e}, \vec{u}) = \vec{A}\vec{e} + \vec{B}\vec{u}$ where $\vec{e}$ and $\vec{u}$ represent the evaluation of e and u at all managed entities and $\vec{A}$ and $\vec{B}$ are weight vectors. This health function can provide a useful aggregate measure of the network state as viewed by the hub. By compressing information at the hub, the tasks of management centers and operators could be significantly simplified and management could be scalable over a wide spectrum of network sizes and complexity. Contrast this with management which requires continuous polling of vast MIB information into centers, and its manual interpretation.

Unfortunately, current network management paradigms do not support the flexibility and decentralization required to compute health functions effectively. A health function could not be usefully incorporated as part of a static MIB design. The specific function used and its parameters may vary among installations, device configurations and even time of the day. Similarly, a health function could not be usefully incorporated as part of an OSI managed object. While the OSI permits encapsulation of functions within managed objects and their remote invocation by managing entities, these functions must

[1] IfInOctets (IfOutOctets) gives the total number of bytes received (sent).

be statically bound to a managed object at its design time. Again, designers of managed objects may not usefully provide configuration/installation/site/time-independent health functions.

Nor can a health function be usefully computed by managers at centralized management platforms. First, the rates of polling required to aggregate the variables used may far exceed platform processing capability. In the example above, suppose a device is polled every 0.1 seconds. Suppose too, that the total number of interfaces aggregated by health functions is n (say n=200). The aggregated polling rate is then 10n (e.g., 2000) SNMP PDUs per-second. Second, the very goal of compression is to reduce information at the source. Third, polling through the network introduces random perturbations in approximating temporal derivatives of managed variables, leading to errors and potential hazards in decisions.

To resolve the difficulty, this paper proposes the use of the *management by delegation* paradigm [4] to compute health functions. Delegating health functions to distributed agents enables compression of data at the source. It permits flexible changes in health functions to reflect specific network behaviors at different sites and times. It allows the computations of health functions to accomplish great precision through direct access of minimal delay to observations of operational values. It improves scalability of management by reducing the rates of polling needed and restricting it mostly to times when problems are identified via aggregated health measures. It permits local evaluation of health functions and decisions by device agents, when management platforms have difficulties in accessing devices (during critical stress times). A prototype implementation of a health function application [5] over a delegation platform, MAD [6], is described.

The rest of this paper is organized as follows. In Section 2 we introduce a notation to describe the behavior of managed objects, their observations by manager applications, and management decisions. Section 3 describes problems related to derivation of MIBs, Section 4 provides an example of using observation operators for health functions, Section 5 describes a prototype health application, and we conclude the paper in Section 6.

2 BEHAVIORS AND OBSERVATIONS

Let x denote a managed entity (variable), $t_x(n)$ denote the time at which the n-th change in the state of x occurs through a sample computation/communication involving x, and $x(n)$ denote the value of x after the n-th change. A *sample behavior* of x is a sequence $X = \{[t_x(n), x(n)]\}_n$. To associate behavior values with a given time, define $n_x(t) = max\{n|t_x(n) \leq t\}$ as the most recent event occurrence prior to time t.

For example, x may denote an attribute of a PDU (Protocol Data Unit), received at an interface, and n counts PDU arrival events. In the case of IP frames, for instance, $X_{IP}(n)$ can be the number of data octets in the n-th frame, while $n_x(\tau)$ is the number of IP frames arriving by time τ.

Observations of sample behaviors compute some function over the history of the behavior. For example, the total number of data octets arriving via IP integrates the sample behavior over X_{IP} up to a given point in time. Let us denote the initial history of a sample behavior X up to time t as $X^t = \{[t_x(n), x(n)]\}_{t_x(n) \leq t}$. An *observation* of a managed variable is a computable functional $\mathcal{F}[X^t, t] = y$, where y is the value observed. An *observation process* is given by a mapping $\mathcal{F} : \{[t_x(n), x(n)]\} \mapsto \{[t_y(n), y(n)]\}$ where $y(n) = \mathcal{F}[X^{t_y(n)}, t_y(n)]$.

Sample behaviors are often observed via counters. A counter of X may be defined as $\mathcal{C}[X^t, t] = \sum_{j \leq n_x(t)} x(j)$. The MIB-II variable ipInUnknownProtos, for instance, counts (modulo 2^{32}) *"the number of locally-addressed datagrams received successfully but discarded because of an unknown or unsupported protocol"* [3]. Thus, if u(x(n)) is defined as 1 when the protocol is unsupported or unknown, and 0 otherwise, then

$$\text{ipInUnknownProtos} \approx \sum u(x(n))$$

Alternatively, define $Y = \mathcal{C}X$ recursively via

$$t_y(n) = t_x(n), \quad y(n+1) = y(n) + x(n+1).$$

The first part means that the counter is updated (observed) at times when x changes, and the second part defines the counter values recursively. In a similar manner, derivatives of behaviors may be defined recursively. As an example, let us define a moving average observation over a window (number of events) of length k:

$$Y = \mathcal{E}^k X \text{ is defined by } \quad t_y(n) = t_x(n), \quad y(n) = \sum_{n-k<j\leq n} x(j)/k.$$

Observations of temporal behaviors typically depend only on the temporal aspects of a behavior X, and are, therefore, generic. Let us consider a few examples (in all cases $t_y(n) = t_x(n)$):

- $Y = \mathcal{N}X$ is defined by $y(n) = n$ and measures the total number of events. This observation is approximated in SNMP MIBs by counters.

- $Y = \mathcal{A}X$ is defined by $y(n) = t_x(n) - t_x(n-1)$ measures the interarrival time for $x(n)$. For example, $\mathcal{A}$(ifInErrors) measures the time between errors in an interface input packets.

- $Y = \mathcal{R}X$ is defined by $y(n) = 1/[t_x(n) - t_x(n-1)]$ measures the rate of events arrivals. For example, $\mathcal{R}$ (IfInErrors) measures the input error rate.

The last two observations are not managed objects computed in MIB-II, since they are derivable from other MIB objects by arithmetic means [7]. In general, SNMP MIBs do not include such temporal observations. Only by polling the managed object sysUptime[2] a manager may obtain an approximate clue on event arrival times. Thus, temporal observations may only be approximated. Furthermore, the quality of such approximations can depend on MIB and agent implementation decisions. Proxy agents, in particular, may need to poll remote devices over slow communication links to observe managed entities. The polling delay may be significant and random, thus introducing a substantial error in the sysUptime approximation.

[2]sysUptime measures the time in hundredths of a second since the network management portion of the system was last re-initialized

Observation operators may also be composed to produce new observations. For example, $\mathcal{E}^k\mathcal{A}X$ provides the average interarrival time of X over a window of length k.

Let $Y = \mathcal{P}X$ denote the observations produced by the polling operation over X. The polling operator $\mathcal{P}$ should approximate the identity operator $X = \mathcal{I}X$. Unfortunately, polling may result in three kinds of perturbations of the behavior being observed: delays, loss and replication of values. Observations are delayed by $\delta(n)$ which may randomly depend on many independent factors, such as network traffic, the load on the server's host, and so on. Some of the original values may be lost when polling is not performed sufficiently frequent to track changes in x. When polling times are too frequent, the same values of x may be polled, resulting in duplication. As a result, a polling operator $\mathcal{P}$ may substantially differ from $\mathcal{I}$. The next Section describes some of the problems of applying polling to MIBs.

3 DERIVING MIBS AS OBSERVATIONS OF BEHAVIORS

MIBs define the data accessible to applications via a management protocol. Given a real managed entity α, with behavior X_α, a management agent applies an observation operator $\mathcal{T}_\alpha$ to produce a MIB value $\mathcal{T}_\alpha X_\alpha$. A MIB is, therefore, a directory of managed entities observations $\{\mathcal{T}_\alpha X_\alpha\}$.

When a MIB is maintained locally where the sample behaviors are observed, the delays associated with $\mathcal{T}$ will often be relatively insignificant. However, this may not always be the case. For example, when a MIB is implemented by a proxy agent which must poll remote devices to observe managed entities. Also, if $\mathcal{T}$ is applied too infrequently, some of the changes in the sample behavior X may be lost.

Management protocols (e.g. SNMP [2]) support a polling operator $\mathcal{P}$ to bring observed data from the MIB to management applications at remote platforms. Therefore, MIB data access by management applications can be described as the composition of the respective operators, $\{\mathcal{P}\mathcal{T}_\alpha X_\alpha\}$. Computations by management applications often apply additional operators to these polled values, say $\{\mathcal{U}\mathcal{P}\mathcal{T}_\alpha X_\alpha\}$. The management paradigms pursued by current standards typically assume that the observed behaviors $\{\mathcal{T}_\alpha X_\alpha\}$ are separated by the network from the locus of the applications computations $\mathcal{U}$. Since polling can introduce observation perturbations, the fidelity of $\mathcal{U}$ computations may suffer.

A static approach to collect operational data defines the collection of observations $\mathcal{T}$ a priori and independently of their use. Since the need for operational data cannot always be predicted, many observations are collected and stored for potential access by management applications. This results in the collection of large amounts of data that may never be used. Furthermore, in most MIBs, applications can not control the execution of the observations $\mathcal{T}$. Thus, the exact meaning of $\{\mathcal{T}_\alpha X_\alpha\}$, depends on the MIB implementor's design decisions.

How good is the static observation paradigm in supporting the computations of network decisions? There are two major problems of concern: the rigid definition of MIB objects and the use of polling to access management data.

The first problem is the rigidity of pre-defined MIB objects. Observational resources are statically allocated to collect information that may never be used, and relevant new objects can not be dynamically added. Thus, applications must compute approximate observations, which may have significant errors due to MIB implementation decisions. For example, MIB-II variables capture error observations via counters, e.g., ifInErrors. Let X denote the behavior captured by an error counter. The error arrival rate is given by the observation

$$Y = \mathcal{R}X \text{ where } t_y(n) = t_x(n), \quad y(n) = [x(n) - x(n-1)]/[t_x(n) - t_x(n-1)].$$

Since, however, the MIB only records the values of $x(n)$, the values of $t_x(n)$ are not directly observable. Note that it is possible to approximate $t_x(n)$ by polling sysUptime. For example, the interface input error rate can be approximated as:

$$\text{E} = \Delta\text{ifInErrors} \;/\; \Delta\text{sysUptime}.$$

However, sysUptime may differ from the time at which the errors occurred, for instance, if the agent is acting as a proxy. Thus, error rates and similar observations cannot be computed directly, but must be approximated.

A second problem is that of computing linear decisions based on complex observations obtained via polling. Even if all the behavior variables accessed by a health function were available in a MIB, it is still necessary to apply polling over the network to retrieve them to a manager which can compute $\mathcal{H}$. Since, in general, $\mathcal{PH} \neq \mathcal{H}$, the perturbation introduced by polling corrupts the values available to compute $\mathcal{H}$. Limitations on the rate of polling and randomness in polling delays bound the computations of management decisions. Moreover, they may result in management applications taking wrong decisions, or not taking action when required.

For example, security management applications use MIB variables to monitor and report to which foreign systems a device has allowed connections. To track which remote systems access resources via TCP, for instance, tcpConnTable can be used [8]. An intruder, however, may need only a brief connection to gather information. If polling does not occur within this period, the record of the intrusion may be lost, and management actions will not be taken.

The limitations of static observations were partially addressed by dynamic control of observations in the design of RMON [9]. Observations may be invoked and terminated, in RMON, through side-effects of setting appropriate MIB entries. This enables applications to control the allocation of probe resources to observation tasks. However, the observation operators are statically defined. An application cannot compose observation operators and establish new observations. It must use observations pre-defined by the standard and coded as part of the device agent. A fully dynamic approach to observations, as in the delegation paradigm, permits applications to configure observation processes to flexibly monitor managed information of their interests.

4 THRESHOLD DECISIONS

Network management decisions may be described as boolean functions defined over the universe of management observations. Therefore, they may need to compress large amounts of real-time observations. Applications use *decision processes* to trigger actions in order to handle problems, for example in fault management.

Network faults, for instance, are often determined through threshold decisions which are functions of managed variables. For instance, when the error rate in input PDUs exceeds a certain threshold, this may indicate significant faults in the underlying medium. Often, a number of indicators must be simultaneously considered in deciding a symptom. For instance, excessive error rates may also result from an increase in traffic. Thus, both the error and utilization rates must be input to the decision process.

A linear weighted measure is an efficient method to combine managed variables in a threshold decision. For example, consider $\vec{x} = (x_1, x_2, ..., x_k)$ as a collection of managed variables involved in a decision. A simple linear health function h is defined by the scalar product $h(\vec{x}) = \vec{w} \cdot \vec{x} = \sum_{1 \leq i \leq k} w_i x_i$, where $\vec{w} = (w_1, w_2, ..., w_k)$ defines a set of weights. A health function is, therefore, an aggregate measure of network behavior. A *linear threshold decision* is defined by $h^+ = \{\vec{x} | h(\vec{x}) = \vec{w} \cdot \vec{x} \geq 0\}$. Values of $\vec{x}$ in h^+ define exceptional events possibly indicating symptoms of problems[3]. Linear decisions are simple and fast to compute, thus they are particularly attractive for providing fault indications which can be used as input to health functions.

Fault indications may vary from network to network and from time to time. Health functions need, therefore, be properly established to reflect these differences. To define a health function requires computing the linear weights $\vec{w}$ used in threshold decisions. This is essentially the problem of perceptron training [10]. A number of algorithms may be used to train such a linear health function. The Least Mean Square (LMS) algorithm, for example, adapts the weights after every trial based on the difference between the actual and desired output.

The single layer perceptron model is appropriate when computing linearly separable regions. However, a more complex structure is required when decision regions can not be separated by a hyperplane [11]. In a more general scenario, multiple health functions may be simultaneously employed, each providing indications of different possible problems. Threshold decisions for these functions can be combined using multi-layer perceptrons.

A health function h (possibly vectorial-valued) defines an observation over behaviors X as follows:

$$Y = \mathcal{H}X \quad \text{where} \quad t_y(n) = t_x(n), \quad y(n) = 1 \quad \text{if} \quad h(x(n)) \geq 0 \quad \text{else } 0.$$

The operator $\mathcal{H}$ results in a value 1, whenever the values of x fall within the unhealthy domain h^+. When the observed behavior X wanders into the unhealthy region, defined by $\mathcal{H}$, it may be necessary to invoke appropriate management procedures to handle potential problems.

For example, let x_1 denote the input error rate, and x_2 the utilization rate associated with a given interface. A health function that correlates utilization and error rates may

[3]Decisions of the type $h(x) = \vec{w} \cdot \vec{x} \geq h'$ may be easily reduced to the case above by considering $h' = \vec{w}' \cdot \vec{x}'$ where $\vec{w}' = ((w_1, w_2, ..., w_k, h')$ and $\vec{x}' = (x_1, x_2, ..., x_k, -1)$

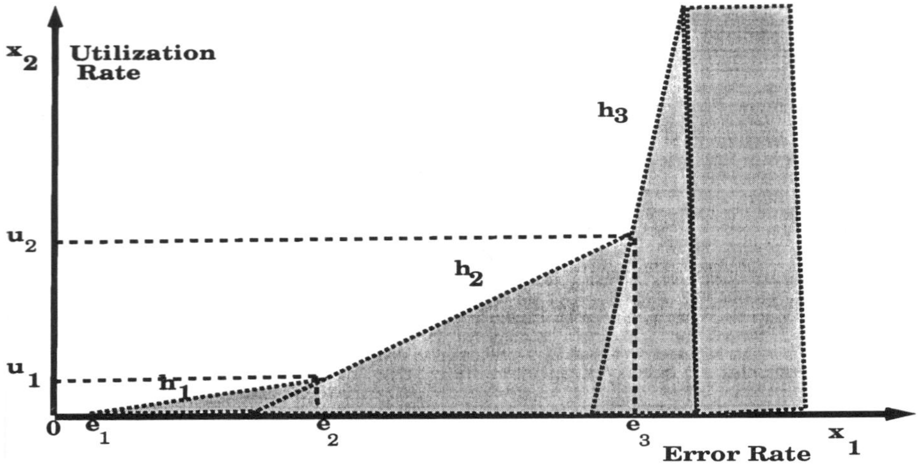

Figure 1: Utilization and Error Rates Domains

be defined as follows.

$$H = \begin{cases} 0 & x_1 < e_1 \\ h_1(x_1, x_2) & e_1 \le x_1 < e_2 \\ h_2(x_1, x_2) & e_2 \le x_1 < e_3 \\ h_3(x_1, x_2) & e_3 \le x_1 \end{cases}$$

Error rates lower than e_1 are considered insignificant. An error rate in the ranges defined by e_1 and e_2, for example, is considered significant if the utilization rate is lower than $h_1(x_1, x_2)$. The domain where faults are indicated is bounded by 3 linear functions as depicted in figure 1. The fault indication region is defined by the intersection of the half-planes decided by each of these health functions. In other words, if $h \equiv (h_1, h_2, h_3)$, the fault indication domain is $h^+ = \{\vec{x} | h(\vec{x}) \ge \vec{0}\}$.

Sometimes, management actions need to be invoked only when unhealthy behavior is sustained for a while, or when it is repeated intermittently. For example, when operational mechanisms provide temporary relief from sustained problems, intermittent problem indications may arise. To avoid spurious alerts, threshold excess must be sustained over a sufficiently long time window. For example, a hysteresis mechanism should be implemented to limit the generation of alarms [9]. If the observed behavior fluctuates, an alert should not be generated.

These problem indicators may be captured by appropriate observation operators applied to the output of the $\mathcal{H}$ observation. Sustained problems, such as unhealthy behavior for a period of duration δ, may be detected by:

$$Z = \mathcal{P}Y \text{ where } t_z(n) = t_y(n), \quad z(n) = 1 \text{ if } y(j) = 1 \text{ for } n_y(t_y(n) - \delta) \le j \le n \text{ else } 0.$$

Applying then $Z = \mathcal{PH}X$ will provide observations of sustained unhealthy behaviors. Intermittent health problems may similarly be detected.

Standard network management approaches require a priori knowledge of what algorithms are mapped into statically defined objects. For example, the OSI Workload Monitoring Function [12] specifies metric objects for determining resource performance and utilization. The parameters that define what constitutes a *healthy* network depend on the particular installation configuration, usage, and administrative policies. These parameters vary among different networks, and during different times within the same network. Therefore, health functions can not be statically defined, but should be dynamically bound to agents when needed.

Centralized, platform-based management, is unsuitable to compute health functions due to its performance limitations and the inaccuracies introduced by polling. Thus, the evaluation of health functions must be distributed. Consequently, we propose the use of the *management by delegation* paradigm [4] to compute health functions. We describe The application of this paradigm for the definition and evaluation of health functions is described in the following section.

5 HEALTH OF A NETWORK

Delegating health functions to distributed agents enables direct observation of network behaviors at sufficient precision. Health functions may be dynamically changed to reflect varying behavior patterns at different times. By maintaining health indicators locally, vast amounts of real-time data can be significantly compressed.

To study the application of management by delegation to evaluate management decisions, a simple health application for an Ethernet has been prototyped [5]. Figure 2 depicts the relationships between the components of the application. It consists of manager processes, a health process and a collection of observers. Manager processes can dynamically reconfigure the distributed application. They receive evaluation reports and present them on a graphical user interface to the operator. The observers compute observations such as operational state, interface utilization, rate of collisions and interface errors. They also perform corrective actions. The health process receives reports from the observers and evaluates a higher level abstraction ($\mathcal{H}$) of the state of the network.

A manager process can specify (and later modify) a list of observers with which health will communicate, and the relative weights of their evaluations in the overall score or index. The generic health object performs a scalar product using a vector of weights $\vec{w}$ and a vector of observation functions $f_i(\vec{x})$. Fine tuning $\vec{w}$ and the f_i is an interactive process that takes advantage of the dynamic reconfiguration capability of MAD [6]. Health may report its evaluated index to managers, by answering an explicit request, via setting a private MIB variable, or as event reports. Delegation enables managers to replace the code of any object, to instantiate and kill processes instances, and to configure their communication.

Diagnostic procedures and corrective actions can be dynamically delegated to be executed when necessary, without manager intervention. For example, if the level of an Ethernet utilization becomes too high, an observer process may automatically disconnect (temporarily) the device which is the source of the largest amount of packets from the network.

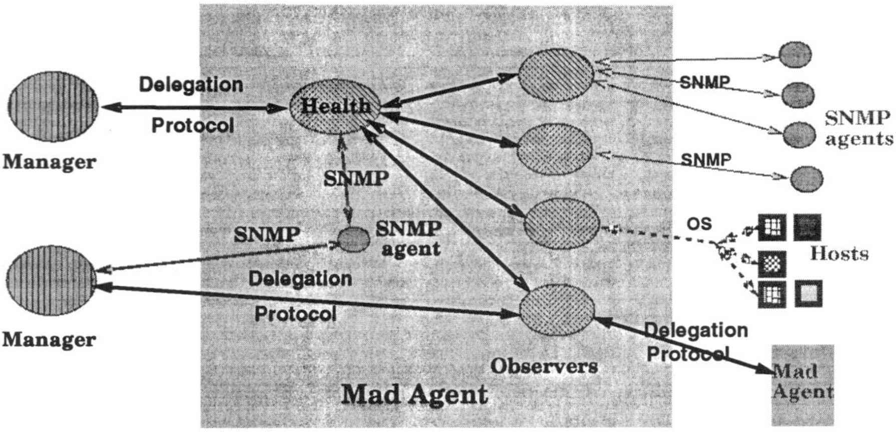

Figure 2: The Components of the Health application

Applying dynamic delegation of health functions resulted in several gains. Management decisions, such as to temporarily disconnect a device, are executed efficiently, without the need for manager platform intervention. Real time operational data is effectively compressed at the delegation agent probe. Managers can define observation operators tailored to their changing needs.

6 CONCLUSIONS

Network management applications must assist managers in making effective decisions. Health functions provide an efficient method to overcome the volume and complexity of data which characterize large heterogeneous distributed systems. Management decisions can be obtained by defining families of linear functions via observation operators.

The definition of what constitutes a *healthy* network can not be standardized or fixed for all networks, since it is installation and time dependent. Thus, it is not sufficient to provide fixed definitions as part of specific programs or MIB variables.

Current standards define observation and collection of MIB data independently of its use. Since the need for MIB data cannot be predicted, many observations are collected and stored but never used. This approach wastes management resources. In contrast, management by delegation supports a dynamic approach to defining observations. This dynamicity permits applications to configure observation processes to flexibly monitor information of their interests.

Acknowledgements. We benefited from discussions of health functions with Amatzia Ben Artzi and Cristina Aurrecoechea who implemented the health application.

References

[1] B.N. Meandzija, K.W. Kappel, and P.J. Brusil. Integrated Network Management and The International Symposia. In *The Second International Symposium on Integrated Network Management*, Washington, DC, April 1991.

[2] Jeffrey D. Case, Mark S. Fedor, Martin L. Schoffstall, and James R. Davin. A Simple Network Management Protocol (SNMP). RFC 1157, May 1990. DDN Network Information Center, SRI International.

[3] K. McCloghrie and M. Rose. Management Information Base for Network Management of TCP/IP-based internets: MIB-II. RFC 1213, March 1991.

[4] Yechiam Yemini, Germán Goldszmidt, and Shaula Yemini. Network Management by Delegation. In *The Second International Symposium on Integrated Network Management*, Washington, DC, April 1991.

[5] Cristina Aurrecoechea and Germán Goldszmidt. Evaluating the Health of a Distributed Environment. Unpublished Report, Columbia University, 1992.

[6] Germán Goldszmidt and Yechiam Yemini. The Design of a Management Delegation Engine. In *Proceedings of the IFIP/IEEE International Workshop on Distributed Systems: Operations and Management*, Santa Barbara, CA, October 1991.

[7] Marshall T. Rose. *The Simple Book, An introduction to Management of TCP/IP-based Internets.* Prentice Hall, 1991.

[8] Allan Leinwand and Karen Fang. *Network Management A Practical Perspective.* Addison-Wesley, 1993.

[9] S. Waldbusser. Remote Network Monitoring Management Information Base. RFC 1271, November 1991.

[10] R. O. Duda and P.E. Hart. *Pattern Classification And Scene Analysis.* John-Wiley & Sons, New York, 1973.

[11] DARPA. *Neural Network Study.* AFCEA International Press, Fairfax, Virginia, November 1988.

[12] International Standards Organization - ISO. Information Processing - Open System Interconnection - Systems Management - Part 11: Workload Monitoring Function. Sydney, Australia, December 1991.

III.

IMPLEMENTING MANAGEMENT ARCHITECTURES

A.

SNMPv2 - FROM INDUSTRY STANDARD SNMP TO TOMORROW'S SNMPv2 (SMP)

Integrated Network Management, III (C-12)
H.-G. Hegering and Y. Yemini (Editors)
Elsevier Science Publishers B.V. (North-Holland)

An Introduction to the Simple Management Protocol

Jeffrey D. Case
SNMP Research, Inc., USA
University of Tennessee, Knoxville, USA

Keith McCloghrie
Hughes LAN Systems, USA

Marshall T. Rose
Dover Beach Consulting, USA

Steven Waldbusser
Carnegie Mellon University, USA

The Simple Network Management Protocol (SNMP) has established itself as the basis for interoperable network management of heterogeneous devices in local and wide area networks. The Simple Management Protocol (SMP) is the next generation of this network management framework. SMP provides extensions to SNMP, but remains an evolutionary -- not revolutionary -- advance in the state of the art. This paper discusses how SMP came about, what it provides, and what impact it will have.

1 Introduction

The Simple Network Management Protocol (SNMP) and Framework was introduced in 1988, and has since established itself as the basis for integrated network management of heterogeneous devices in local and wide area networks. As experience has been gained with SNMP, the SNMP community has wished to evolve the SNMP Framework without disrupting the installed base.

The Simple Management Protocol (SMP) is the next generation in the Internet-standard network management framework. SMP provides extensions to SNMP but provides an evolutionary -- not revolutionary -- advance in the state of the art.

This paper discusses the environment in which SMP was created, the architectural principles that guided SMP, the goals that were achieved with SMP, and finally, the results of the process.

2 History of the Internet-standard Network Management Framework

SNMP [1],[2],[3] was introduced as the Internet-standard Network Management Framework in early 1988, when it was created to solve the pressing network management needs of the rapidly growing Internet. Since then, SNMP has enjoyed wide commercial success and provided stable and effective network management of the Internet.

SNMP has been successful for many reasons. The most recognized and important of these reasons is SNMP's simplicity. The simplicity inherent in SNMP lowers the cost of entry into SNMP network management. This allows SNMP to be ubiquitously implemented on a wide variety of platforms, even those with little resources to devote to network management functions. The effects of this have been seen quite clearly as SNMP has been implemented on products ranging from small embedded controllers to supercomputers.

Another reason for SNMP's success has been its stability. Many enhancements have been added to the SNMP framework through expansion of the Management Information Base (MIB) with new MIB objects. However, until recently, the SNMP community, embodied in the Internet Engineering Task Force (IETF), chose to preserve the stability of the SNMP framework by resisting any changes or improvements to the protocol. It was felt that changes to the framework would slow the continued acceptance of SNMP and would put the installed base at risk.

In the midst of this stable environment, there was a lot of activity. The SNMP community was gaining experience with the framework and was beginning to perceive a need for improvements to SNMP. As SNMP was applied in more new situations and more experience was gained, the pressure for changes to SNMP increased. In addition, the SNMP Security working group was working on much-needed security enhancements to the Internet-standard Network Management Framework.

The Spring of 1992 brought the conclusion of the Secure SNMP standardization effort [4],[5],[6], and with it came the first major change to the SNMP framework beyond the addition of new MIB objects. This long-awaited effort specified a new administrative model for SNMP that provided authentic-ation, access control and privacy.

In addition, the SNMP community succumbed to the pressures to change SNMP by issuing a call for proposals in March of 1992 to address perceived deficiencies of the SNMP framework.

These two imminent changes to the SNMP framework in the spring of 1992 threatened the stability of SNMP by introducing two transitions into the future of the Internet-standard Network Management Framework. The first transition, to SNMP Security, was to begin very soon, in the summer of that year, while the second transition would be expected to begin after a year of standardization activities. The second transition might well begin immediately after the dust settled from the first transition. The network management industry was likely to be distracted from making progress in applications and other important areas of network management for up to three years.

This environment motivated four individuals who had made key contributions to the SNMP framework to form a team and to conceive, document, and implement a successor to SNMP, which included SNMP Security. They called this proposal the Simple Management Protocol (SMP) and Framework. The release of SMP was timed so that implementors and vendors could produce one set of highly desirable changes, i.e., to carry out other changes at the same time as SNMP Security is added to products. In addition, the effort was intended to be evolutionary in nature to minimize the impact of the transition. Users and vendors would benefit from this single smooth transition.

3 Design Principles

The SMP design process was guided by the same architectural principles that were key to the design of the SNMP framework. The most important goals were thus:

- To minimize the cost and complexity of network management agents
- To foster operational robustness by centralizing control of network management operations in management stations
- To protect the existing investment and achieve an orderly transition by keeping the changes evolutionary

Because there are many more agents than management stations on a network, it is important to lower the memory and CPU requirements of the agents. The function of a bridge is to forward packets, and as more resources are devoted to network management, fewer resources are available to forward packets. In addition, when the burden is shifted from the many agents to the few management stations, the total cost of network management in an enterprise is lowered without overburdening the management stations. Finally, a simple, consistent design ensures that both management stations and agents may be as cost-effective as possible.

The second design principle was to increase the robustness of the network management system by retaining control in the management stations. These management stations are typically under the direct control of network management personnel and enjoy enough resources to act intelligently. These factors reduce the likelihood of failures of the network management system, or worse, failures of the network caused by the network management system. Putting too much control in agents could result in a loss of control over management resources which in turn could cause these failures. For example, if many agents on an Ethernet were to notify a management station of a high collision rate, the notifications would collide with each other, potentially causing the collision rate to spiral until a network operator intervenes.

The third design principle is to protect the significant investment in SNMP management stations and agents. In addition, a smooth transition would guarantee the continued progress of state-of-the-art SNMP network management. Careful engineering of the SMP framework was necessary to preserve the highest degree of interoperability possible with products already deployed in existing networks. Fundamental to the design of SMP was a plan for the periods of coexistence and transition undeniably necessary for SMP's success.

It is worth noting that the second principle of retaining the control and intelligence in the management station goes hand in hand with the first principle of pushing the burden of network management into the management station. Similarly, the continued simplicity and consistency of the architecture helps to guarantee that a viable transition plan from SNMP to SMP is available. Finally, a ruthless commitment to solid engineering principles helps to achieve these goals and to ensure that the specification is implementable.

4 Design goals

Within the constraints of the agreed upon architectural principles, the four authors had certain goals in mind for new capabilities to be added to the SMP framework. These goals included documenting the "oral tradition", fixing problems with SNMP, adding manager to manager capabilities, and enabling systems and application management.

As experience was gained with SNMP over the years, a list of perceived problems was compiled. This list included the need for more speed and efficiency, refined SET capabilities, new data types, coexistence with multiple protocols, and security.

Another goal was to find the sources of interoperability and implementation problems in the SNMP specification and to resolve them either through clarification or development of better mechanisms. The authors hope that they have not replaced these problems with new (but different) ones.

The authors agreed that SMP must address the important area of manager to manager communications. While there have always existed mechanisms by which manager to manager communications have been possible, a more straightforward mechanism was needed. This would open the door for hierarchical and peer to peer management architectures to be built with the SNMP architecture. An example of a hierarchical management strategy built with SNMP is the Remote Network Monitoring MIB (RMON MIB) [7]. While this MIB has been successful, a more open-ended manager to manager solution was needed.

Another goal of SMP was to move beyond network management into systems and application management. A fundamental requirement for these areas was a more refined strategy for performing SET operations on remote agents. Experience with the SNMP SET operator had uncovered the deficiencies that existed, and had shown how they must be solved. In addition, the security capabilities of SMP would enable these new SET capabilities to be deployed safely.

5 Results

For several months the SMP authors met and eventually fulfilled the design goals of the project within the constraints of the design principles. The SMP specification was published as eight documents:

- Introduction to the SMP Framework [8];
- Structure of Management Information for the SMP Framework [9];
- Textual Conventions for the SMP Framework [10];
- Protocol Operations for the SMP Framework [11];
- Transport Mappings for the SMP Framework [12];
- Management Information Base for the SMP Framework [13];
- Manager to Manager MIB for the SMP Framework [14]; and,
- Coexistence between SNMP and the SMP Framework [15]

This specification included many new features and enhancements. These included:

- Security enhancements
- Improvements to the protocol in the areas of bulk data retrieval, manager to manager communication, SETs, error processing, and new data types
- Richer documentation
- Clear strategies for transition and coexistence with SNMP and non-TCP/IP protocols.

5.1 Security

Secure SNMP provides the basis of the security of SMP. However, some characteristics of Secure SNMP have proven in trials to be costly to implement or have proven unworkable. In light of this, changes were made to the Secure SNMP to improve these characteristics without sacrificing the security of the original specifications.

Because of the worldwide export restrictions on the Data Encryption Standard (DES), the realities of the global marketplace dictated that DES would not be widely available, even within the country of origin. Many network management vendors felt that they would be wasting time developing a feature that effectively could not be sold to the majority of their customers. In recognition of this, DES encoding was made optional instead of mandatory.

SMP relaxed the Secure SNMP restriction that all packets must arrive in order and applied it only to SET requests. Although reordered packets can be a sign of malicious tampering, they are also encountered innocently when the network reorders packets. Given that this is most likely to happen under times of network stress, it was burdensome for Secure SNMP to throw away all of these reordered packets precisely when they are needed the most. Recognizing that protection from reordering is only useful for SET requests, this restriction was removed from all other types of commands.

With the changes that SMP makes to Secure SNMP, secure implementations of SNMP will become available and will allow managers of networks to use network control operations with confidence, using SMP's improvements to the Set command that allows SNMP systems to make changes to managed devices.

5.2 Protocol improvements

Many changes have been made to the protocol that add new features or refine existing ones. This is the area where changes have the most potential for noticeable improvements. This is also the area in which backwards compatibility and other coexistence issues are the most important.

A new protocol operation that will result in significant user-visible improvements is the GetBulk operator. The GetBulk operator allows ranges of variables to be retrieved in a single PDU. Formerly, SNMP management stations often had to send a GetNext PDU for each column retrieved. This speeds up data retrieval more than 10 times when compared with SNMP transfers, and results in more efficient use of the network because the data is packed into a fewer number of larger packets.

The speed enhancements due to the GetBulk operator are most noticeable when downloading large amounts of data such as large routing tables or an RMON host table. Besides the obvious user-visible improvements, this will also make MIBs simpler because they don't need to specify special representational techniques that facilitate fast download of large tables. For example, the RMON MIB provided an additional view of a large table that was optimized for fast retrieval.

The other significant new protocol operation is the Inform PDU. The Inform PDU introduces support for confirmed information transfer between two management stations, including confirmed event notifications. With this command, a management station can send information, such as a routing table, to another management station and receive a confirmation that the information was delivered.

Besides providing a general information transfer mechanism for management stations, the Inform PDU allows the construction of hierarchical management systems. A hierarchical system might have a mid-level management station polling devices local to a LAN. When it detects a fault, the mid-level management station can transmit an event notification across the backbone to a high level management station. This reduces the amount of network management traffic across the backbone, without compromising the reliability inherent in the polling strategy. The new SMP Manager to Manager MIB defines objects that can be used to configure such a mid-level manager.

The SNMP Set PDU has been clarified and improved, especially in the area of row creation in MIB tables. A clear mechanism has been specified that allows a management station to create new instances of variables in a table, such as a new route in a routing table. This mechanism arbitrates between multiple managers that might be performing the same operation simultaneously and only allows one to succeed. It also allows the management station to discover the existence of unimplemented variables in an agent and to continue to interoperate with that agent. Quite a few new error codes have been added as well so that a management station may perform better error recovery when it issues bad SET requests.

SNMP has error codes for different error conditions in Get, GetNext, and Set operations. SMP adds a new concept, the exception, to the Get, GetNext, and GetBulk operations. An exception is associated with a particular variable binding in a PDU when there is a problem accessing that variable binding. The notable change from SNMP is that the other variable bindings in the PDU can return values. The most significant effect of this is that it is easier, and thus more likely, for a management station to achieve interoperability even when an agent does not implement a certain variable because another PDU exchange is not required. This will lessen the pressure on agent developers to implement "dummy" variables when they cannot conform to the standard. Thus, these developers will no longer have to make a decision between interoperability and correctness.

The last significant change to the protocol is the addition of new base data types such as 64 bit counters. The former is useful for counting events that occur so frequently that a 32 bit counter might "wrap" and start over from zero between reasonable management station polling periods. This is most often a problem when counting bits on a high speed network, where a 32 bit counter might roll over in minutes or even seconds. As networks get faster and faster this problem threatens to become troublesome.

5.3 Richer Documentation

Over the last few years, there has been an evolution in the way MIBs and other network management documentation have been written. SMP continues that evolution by making further changes to these formats.

Many of these new features improve how standard MIBs are written. Data modeling issues continue to be handled by the object type macro in the MIB. However, there is now a mechanism to describe, in machine-readable format, the grouping and conformance characteristics of objects in the MIB. Subtle standardization and conformance issues are now adequately expressed in these new formats.

Another new documentation format allows an agent developer to document the agent. This machine-readable format describes the capabilities of the agent and how it differs from the MIB. This documentation is useful for two reasons: for a customer to learn the capabilities of a manageable product before purchasing it; and for a management station to learn of an agent's unique characteristics so that it can interact with it accordingly.

5.4 Coexistence

A major design goal was backward compatibility whenever reasonably possible. Great care has been taken to ensure compatibility with the installed base and to facilitate the period of coexistence that must occur while the industry transitions to SMP. It was felt that this subject was so important that it deserved its own document [14].

Of course, SMP can run on top of TCP/IP, AppleTalk, IPX, and OSI networks. Further, as SMP devices are deployed in networks with existing SNMP platforms, the SMP devices will be able to interoperate (albeit using a subset of their capabilities).

6 Implementation experience

Days after the specifications were completed and released, the four SMP authors released four independently developed, interoperable implementations of the new specifications. Two of these were made freely available so that implementors and users could experiment with the new features and gain experience with them. In addition, this code was available for vendors to begin porting the new SMP framework to products.

These implementations were developed side by side with the specifications. This helped to ensure that the specification was indeed implementable and clear enough that interoperable implementation could be written from the documents. Much was learned in the process of developing these prototypes. Some problems were discovered with the draft specifications and were fixed before these specifications were published. In addition, the prototypes helped to confirm that the design goals had been met.

6.1 Speed and Efficiency

The set of design goals that were particularly important was the size and speed of the resulting code. Due to some optimizations in the protocol and some in the prototype implementation, the resulting agent was only marginally larger than the equivalent SNMP implementation, despite all of the additional capability.

Because the speed improvements would be the feature most immediately apparent to users, it was particularly interesting to see the real-life results. For one implementation, the addition of the GetBulk PDU provided more than ten times faster performance when downloading a large table. In fact, this shifted the performance bottleneck from the network delay to the protocol processing in the management station and agent. After optimizing the protocol processing on both management station and agent, the overall data download speed was further increased to more than 9,300 variables per data download speed was further increased to more than 9,300 variables per second. This is fast enough to download 6 columns of a routing table with 1500 routes in 1 second.

6.2 Benefits for Applications

Further experience in writing applications with the SMP framework showed that they were easier to write with SMP. The addition of exceptions to data retrieval operations allowed applications to remain simple and still recover gracefully when an agent didn't implement a requested variable. The fact that the PDU would return the information for other requested variables in the packet meant that the application didn't have to have special case code to recover from the error and retransmit a modified packet.

Other changes also benefit application writers. Because the specifications are more clearly written and loopholes are closed, there are fewer exceptional conditions that the applications would have to recover from in order to be robust. The GetBulk command allows fast, efficient data transfer without specialized MIBs or complex, special case code in the application.

7 Future - SNMP version 2

In July, 1992, the four SMP authors presented their results at a session of the Internet Engineering Task Force. The audience of 220 listened to the presentation and asked questions during the 3 hour session. At the end of the session, the concensus of the room was that work on a second version of SNMP should begin immediately and should use the SMP proposal as a starting point. It was felt that this work should progress quickly, and a goal of a March, 1993, publication date was later agreed upon. Further, the concensus was that due to the improvements that SMP made to the SNMP Security, that the community should not undergo an interim transition to the (just-released) SNMP Security specifications, but should implement the security features of this second version of SNMP.

This work was later named SNMP version 2, or SNMPv2 for short. The SNMP Security working group was reconvened to discuss and standardize the security changes. In addition, the SNMPv2 working group was chartered to standardize the bulk of the new framework.

The groups performed their work during many days of meetings and via electronic mail. Many clarifications and improvements were made as a result of this effort. As of this writing, the group is nearing concensus on the final changes before submitting the specifications for standardization.

8 Conclusions

The Simple Management Protocol was designed in response to a need for an evolutionary step in the SNMP framework. This step was undertaken based upon several years of experience with SNMP and provided new features while staying true to the architectural principles that had guided SNMP.

The SMP Framework provided better documentation, better security, and improved protocol operations over its predecessor, SNMP. This resulted in a faster and more efficient protocol, as well as a protocol that can take on manager to manager and systems management tasks.

The SMP Framework has become the foundation for the Internet Engineering Task Force's SNMP version 2. Upon the conclusion of this standardization effort, SNMPv2 promises to become as stable and effective as its forebear and to move the Internet-standard Network Management Framework into still more effective network management.

References

[1] Marshall T. Rose and Keith McCloghrie. *Structure and Identification of Management Information for TCP/IP based internets.* Request for Comments 1155, Network Information Center, SRI International, May, 1990.

[2] Keith McCloghrie and Marshall T. Rose. *Management Information Base for Network Management of TCP/IP-based Internets*, Request for Comments 1213, Network Information Center, SRI International, March, 1991.

[3] Jeffrey D. Case, Mark S. Fedor, Martin L. Schoffstall, and James R. Davin. *A Simple Network Management Protocol*, Request for Comments 1157, Network Information Center, SRI International, May, 1990.

[4] James R. Davin, James M. Galvin, and Keith McCloghrie. *SNMP Administrative Model,* Request for Comments 1351, Network Information Center, SRI International, July, 1992.

[5] James M. Galvin, Keith McCloghrie, and James R. Davin, . *SNMP Security Protocols,* Request for Comments 1352, Network Information Center, SRI International, July, 1992.

[6] Keith McCloghrie, James R. Davin, and James M. Galvin. *Definitions of Managed Objects for Administration of SNMP Parties,* Request for Comments 1353, Network Information Center, SRI International, July, 1992.

[7] Steven Waldbusser. *Remote Network Monitoring Management Information Base,* Request for Comments 1271, Network Information Center, SRI International, November, 1991.

[8] Jeffrey D. Case, Keith McCloghrie, Marshall T. Rose and Steven Waldbusser. *Introduction to SMP,* Internet Draft, July, 1992.

[9] Jeffrey D. Case, Keith McCloghrie, Marshall T. Rose and Steven Waldbusser. *Structure of Management Information for SMP,* Internet Draft, July, 1992.

[10] Jeffrey D. Case, Keith McCloghrie, Marshall T. Rose and Steven Waldbusser. *Textual Conventions for SMP,* Internet Draft, July, 1992.

[11] Jeffrey D. Case, Keith McCloghrie, Marshall T. Rose and Steven Waldbusser. *Protocol Operations for SMP,* Internet Draft, July, 1992.

[12] Jeffrey D. Case, Keith McCloghrie, Marshall T. Rose and Steven Waldbusser. *Transport Mappings for SMP,* Internet Draft, July, 1992.

[13] Jeffrey D. Case, Keith McCloghrie, Marshall T. Rose and Steven Waldbusser. *Management Information base for SMP,* Internet Draft, July, 1992.

[14] Jeffrey D. Case, Keith McCloghrie, Marshall T. Rose and Steven Waldbusser. *Manager to Manager MIB for SMP,* Internet Draft, July, 1992.

[15] Jeffrey D. Case, Keith McCloghrie, Marshall T. Rose and Steven Waldbusser. *SNMP / SMP Coexistence,* Internet Draft, July, 1992.

SNMPv2

Panel Chair: Steven Waldbusser, Carnegie Mellon University, USA

Panelists: Jeff Case, SNMP Research, Inc., USA
Keith McCloghrie, Hughes LAN Systems, USA
Marshall Rose, Dover Beach Consulting, USA

Simple Management Protocol (SMP) and its Framework represent the next generation of SNMP that provides extensions to the initial version of SNMP and the entire Internet-standard Network Management Framework. Although the new Framework is substantively richer, it remains an evolutionary advance in the state-of-the-art technology. The SNMP and related standards have become de facto standards, as evidenced by the widespread support by vendors and popularity among managers of networks.

Panelists will discuss why SMP Framework was designed, identify how it differs from SNMP, and report on its current progress in standardizing and implementing the new Framework.

III.

IMPLEMENTING MANAGEMENT ARCHITECTURES

B.

SPECIFIC APPROACHES

Integrated Network Management, III (C-12)
H.-G. Hegering and Y. Yemini (Editors)
Elsevier Science Publishers B.V. (North-Holland)

A Computational Model for OSS Multivendor Software Interoperability

A. Dionisi Vici[a], S. Gugliermetti[b] and G. Sofi[b]

[a]SIP Headquarters - Information Technology Area
Via Vignaccia, 45 - 00163 Roma, Italy

[b]CSELT - Centro Studi E Laboratori Telecomunicazioni, Via Borgaro, 21
10149 Torino, Italy

Abstract

Telecommunication Operating Companies (TELCOs) are committed to provide new services to customers with increasing quality levels; to fulfil this goal a key role is played by Operations Support Systems (OSSs), i.e. application programs which automate an increasing range of Operations, Administrations and Maintenance functions. Automation alone is not effective, unless OSSs do not exploit a high degree of integration: sharing of corporate information and interoperability of distributed OSS components. SIP has addressed both integration issues in the context of a strategic "OSS Plan", whose offspring are the projects for Corporate Data Integration and for OSS Components interoperability.

This paper illustrates the foundation of both projects and the current results achieved on OSS Components interoperability; the approach is based on TMN CCITT M.30 Recommendation and on the Bellcore OSCA Architecture. The innovative contribution presented in the paper addresses issues left open by TMN and OSCA with a perspective based on the framework adopted by ISO for Open Distributed Processing and derived from the ANSA project, in particular the "Computational Viewpoint".

Keyword Codes: H.1.0
Keywords: Models and Principles, General

1. INTRODUCTION

Realising integrated OSSs is a large and complex undertaking, with many software challenges: multivendor procurement, software reuse, continuous availability, corporate data base.

As far as multivendor procurement is concerned, future integrated OSSs need to address two types of it. First, the OSSs have to support network elements produced by multiple network equipment vendors; this is subject to CCITT TMN standardisation. Second, the OSSs themselves have to be developed according to an integration framework that allows OSSs to be provided by multiple suppliers. An example of such an architecture is Bellcore OSCA.

Integration of Operations Support Systems (OSSs) in distributed processing environments can be addressed in the perspective defined by ISO Open Distributed Processing (ODP). This paper uses the ISO ODP framework to tackle open issues of TMN CCITT M.30 Recommendation and Bellcore OSCA architecture.

SIP project on OSS Components Interoperability has defined a Computational Model for OSS components interaction as the basis for a declarative language, the Contract Definition Language, used by application programmers to specify OSS interfaces in a simplified high level style. The language allows to define OSS components, arbitrarily structured by application architects and designers, as black boxes with homogeneous interfaces. The language supports a simple repertoire of communication primitives, including transaction support and TMN manager/agent dialogues.

Open Distributed Processing (ODP) [1] is a major ISO research and standardisation initiative aimed at developing distributed processing systems based upon open standards. The scope of ODP encompasses all potential applications of distributed processing systems, including (tele)communication services, networks and their support environments. The foundation for this work was derived from the Advanced Networked Systems Architecture (ANSA) [2].

In developing distributed processing applications, different aspects can be captured by different engineers with different vocabularies.

ISO ODP attempts to recognise these different aspects and vocabularies by defining five Viewpoints. Each viewpoint represents an abstraction of a complete system that emphasises a particular set of concerns. The viewpoints are:

- *Enterprise* viewpoint
- *Information* viewpoint
- *Computational* viewpoint
- *Engineering* viewpoint
- *Technology* viewpoint

and will be discussed in the next section. The use of the ODP Viewpoints to structure the analysis of distributed systems problems allows a clear separation of concerns and consequently a better understanding of the issues involved and their interrelationships.

The purpose of the TMN is the management (operation, control and maintenance) of all kinds of telecommunication networks and services, where TMN provides management functions to the telecommunications network and offers communication between a TMN and a telecommunication network [3].

The TMN defines a functional hierarchy: the complexity of management is partitioned into four layers. Each layer restricts the rank of management activities within the boundary of the layer itself, in this case namely business, service, network and element management.

Each layer is represented by a corresponding Operations System Function (OSF) which can use the functionality of the OSFs of lower layers in the same TMN via a Q3 interface.

The issues left open by TMN are relevant to the service and business management Functional Areas: their interworking needs are not well covered by the TMN Q3 interface specifications, since Q3 does not support co-ordination among distributed transactions. Furthermore, the object oriented model is too restrictive for the type of exchanged information.

The Bellcore Operations Systems Computing Architecture (OSCA™) [4] is defined to promote the interoperability of large-scale software products throughout life-cycle; it is a set of principles on the structure and interoperability of OSSs defined to enable building software systems from products developed and/or modified by different suppliers at different times. At the highest level of abstraction, a software product that conforms to the OSCA architecture provides software for one or more of three logical layers: a *corporate data* layer, a *processing* layer, and a *user* layer. The corporate data layer provides functionality to support the semantic integrity of corporate data. The user layer provides functionality to support human users. The processing layer provides functionality for business processes. Each layer is made up of one or more well-defined deliverable, interoperable, functional units called *Building Blocks*. Well documented interfaces called *Contracts* are used between the Building Blocks. For all types of Building Blocks and for Contracts OSCA defines principles, not discussed in this paper.

The issue intentionally left open by Bellcore in defining OSCA is that OSCA does not prescribe a rigourous model of interoperability, as needed by an architecture for distributed computing, since OSCA principles are not interface specifications.

2. SIP APPROACH TO OSS SOFTWARE

SIP has addressed OSS integration using the framework posted by ISO ODP viewpoints; below we briefly summarise SIP assumptions and use of each viewpoint:

- Enterprise Viewpoint. This viewpoint is assumed by SIP as the critical factor in the company for aligning the business strategy with the information technology strategy. The enterprise viewpoint, used in the strategic planning of Operations Support Systems, captures the business activities, goals and information needs in the Enterprise Model; the model represents business processes and entities of the whole company and allow to detect the information shared by different processes which create, read, update, delete the same entities.
- Information Viewpoint. In SIP this is the Corporate Data Administrator's viewpoint: it is used in representing the Conceptual and Logical Models of all Corporate Data, whose integration needs are detected by the enterprise viewpoint; the information viewpoint is used to develop integration-critical information, since the Corporate Data Administrator analyses data-flows to identify the "legal states" of corporate data and captures all semantic integrity rules, which will be triggered by information updates.

- Computational Viewpoint. Defines the structure and behaviour of OSS Building Blocks in a distributed processing environment. SIP has developed the OSS Computational model as a declarative language, the Contract Definition Language (CDL), used to specify OSS interfaces in a simplified high level style, plus a high level API (Application Programming Interface) to invoke the run-time services.
- Engineering Viewpoint. It defines the services supporting interoperability, manageability, security in a distributed environment. A major feature of the engineering model is the kind of transparencies it supports with respect to complexities arising from distribution, i.e. location and access transparency, failure transparency, concurrency and replication transparency. The engineering viewpoint has been used in SIP to map the Contract Definition Language and the high level API onto services available in run-time environments.
- Technology Viewpoint. It is the machine-oriented viewpoint, and is based on basic technologies belonging to the following areas :
  - Operating System Technology
  - Networking (Data Communication) Technology
  - Data Management Technology
  - User Interface Technology
  - Transaction Processing Technology

The following figure depicts the different ODP views onto the overall SIP OSS software architecture.

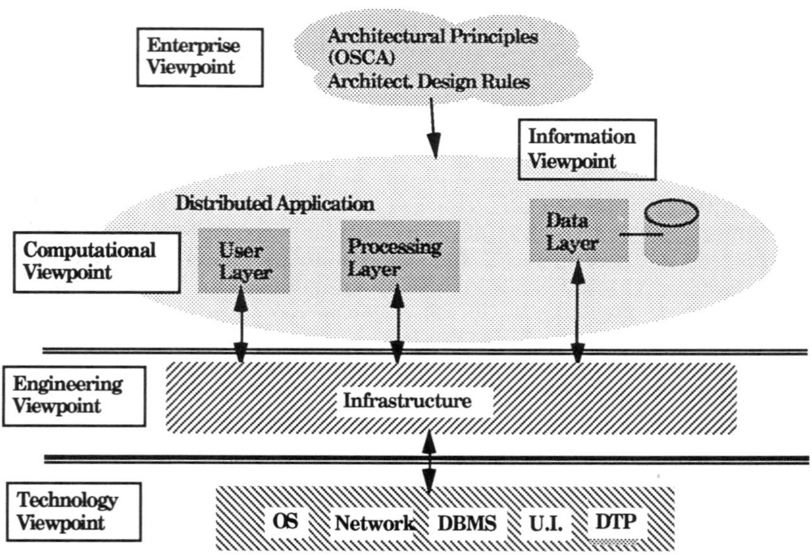

Figure 1. ODP Viewpoints on the overall OSS software architecture.

3. THE COMPUTATIONAL MODEL

The Computational Model supports the OSS interaction in a distributed environment. The objectives SIP had in defining it were :

- Meeting the application requirements. In particular :
  - ° Supporting Distributed Transactions
  - ° Supporting TMN Applications and protocols
- Compatibility with the existing software
- Minimising training costs. The model should be easily understood both by :
  - ° Designer's
  - ° Programmer's
- Documentation quality

The computational model directly supports the structuring concepts defined in OSCA: applications consist of a set of logical components, *Building Blocks*, (physically distributed on the various network nodes) and each *Building Block* can provide or invoke services to/from other components via well defined interfaces called *Contracts*. A *Transactional Construct* allows the Invoker to coordinate the execution of multiple Contracts guaranteeing the ACID (Atomicity, Consistency, Isolation and Duration) properties.

The computational model provides the application programmer with a declarative language CDL (Contract Definition Language) for both Building Blocks and Contracts definition and, in association, a high level API for Contract invocation and Transaction demarcation and co-ordination.

3.1. Contracts

Contracts are the only way allowed of communicating among Building Blocks. They are classified into three types, according to the underlying interaction mechanism:

- Send Contract
- Call Contract
- Dialogue Contract

Figure 2 shows six Building Blocks interoperating by invoking/offering Contracts in a heterogeneous distributed environment; the Contract invocation API abstracts from the complex communication details of the underlying hardware and software, since these details belong to the Engineering and Technology viewpoint.

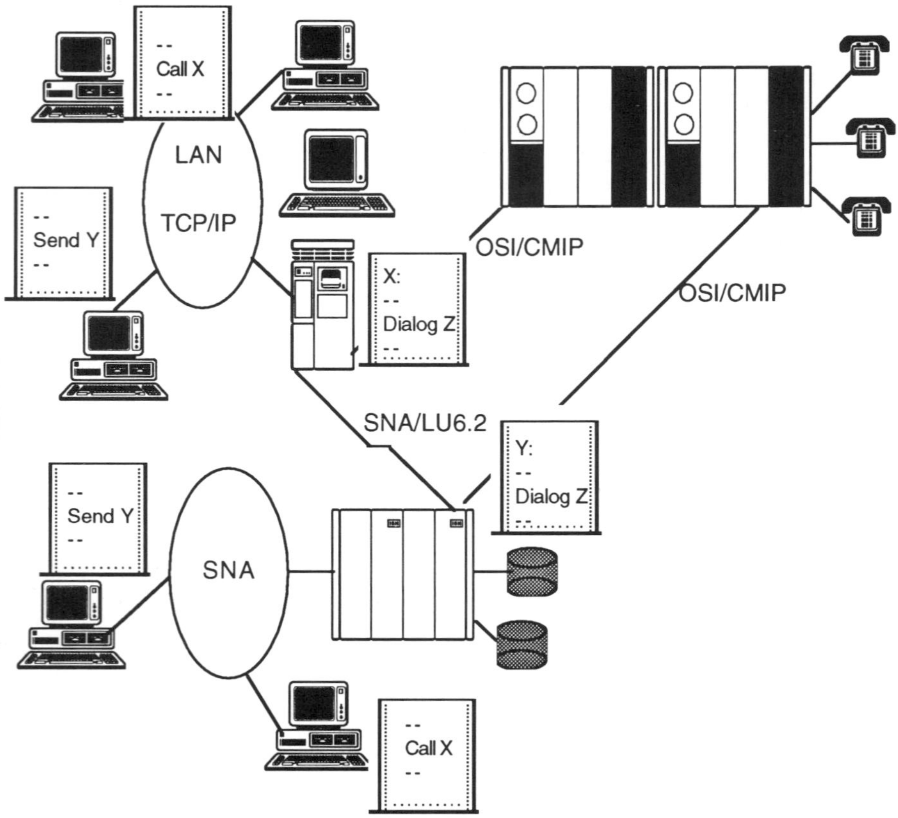

Figure 2. Abstraction of the underlying hardware/software.

3.1.1. Send Contract

The Send Contract semantically corresponds to a send of a message. The invoker of the Contract goes on in processing without waiting for a reply (non blocking asynchronous invocation). The Contract Provider successively executes the invoked Contract. This corresponds to a loose coupling, because Invoker and Provider are not required to be contemporary active. This Contract type is typically used for queuing batch processing requests.

3.1.2. Call Contract

The Call Contract semantically corresponds to a remote procedure call. The invoker of the Contract waits for a reply (blocking synchronous invocation). The Contract Provider executes the invoked Contract and responds to the Invoker. This corresponds to a tight coupling, because Invoker and Provider are required to be contemporary active. This Contract type can be used for simple interactions (queries on data, ...) among BBs.

3.1.3. Dialogue Contract

A dialogue Contract enables two BBs to communicate in a sequence of interactions (operations and notifications) during a connection. The invocation of a dialogue Contract starts an interaction mechanism with both the BBs active.

Operations are sub-functions offered by the Contract Provider to the Invoker and defined in the Contract.

Notifications are one-way asynchronous messages sent by the Contract Provider to the Invoker when some internal or external events occur. These messages are not solicited by the Invoker. Notifications are specified in the Contract.

Operations and Notifications are expressed as procedures (written in a programming language like C or Cobol).

A Dialogue Contract consists of three phases :

- opening of the dialogue (set-up of the association)
- dialogue (calling of the operations, emission of notifications)
- closing of the dialogue (release of the association)

A particular case of dialogue Contract is an OSI management Contract. In this case, the Invoker of the dialogue takes the Manager role, and the Provider takes the Agent role [5]. The CDL syntax supports the GDMO (Guidelines for the Definition of Managed Objects) from OSI Management and the Contract invocation API can map on CMIS/CMIP [6] [7].

3.2. Transactions in the Computational Model

A transaction is a state transition, from an initial to a final one, which satisfies the properties of Atomicity, Consistency, Isolation and Duration (ACID).

A transaction is called *local* if the set of actions satisfying the ACID properties is performed by a single application on a single resource.

A transaction is called *distributed* if the set of actions satisfying the ACID properties is performed by one or more applications on one or more resources. This implies that a transaction is certainly distributed if it operates on more than one physical node, but it can be distributed even if it operates on a single physical node.

As said before, in the areas of service and business management the different applications have to cope with transaction definition and for them interworking means also distributed transactions co-ordination.

Our computational model supports this through a transactional construct.

3.2.1. Transactional construct

A transactional construct is provided by the computational model to co-ordinate distributed transactions.

It groups sets of local function calls (in the same BB) with remote ones (in other BBs), and it guarantees the ACID properties of the transaction.

Access to remote resources is possible only through Contracts.

A transaction is opened in a Contract body with a *transactional block* :

```
BEGIN TRANSACTION
   .....
   .....
END TRANSACTION
ON ROLLBACK
   .....
END
```

From a functional point of view, the following actions are allowed in a transactional block :

- any kind of actions that have not to satisfy the ACID properties (e.g. printing, ...);
- local actions on protected (by the transaction) resources;
- invocations of Contracts in other BBs which access non protected resources;
- invocations of Contracts in other BBs which access protected resources;
- explicit rollback of the global transaction.

The *transaction handler* (the section beginning with the keyword ON ROLLBACK) handles the actions to be done in case of failure of the transaction.

3.2.2. Composable and non composable Contracts

Contracts can be classified in *Composable* and *non Composable*, on the grounds of their behaviour when invoked during transactions.

° *Composable Contract:* is subject to the transaction active at the invocation.

° *Non Composable Contract* : it is not subject to the transaction active at the invocation (if any); in its body it can start a new transaction with the transactional block construct.

Composable Contracts are executed in the scope of the transaction active at the invocation, and shall be capable to rollback the internal protected resources in conformity with the global transaction result (a two-phase-commit is required among the parties).

Non Composable Contracts are not subject to the transaction active at the invocation, hence the local resources accessed are not protected by that transaction. A two-phase-commit protocol is not required.

Transactional blocks cannot be nested and can appear only in a non-Composable Contract.

The Contracts of type "call" and "dialogue" can be either Composable or not. The Contracts of type "send" can only be non Composable.

For Contracts of type "dialogue", the attribute Composable applies to the Contract as a whole, not to the single operations. So, all the phases of the dialogue (opening, calling of the operations, closing) must belong to the same transaction.

3.3. Building Block example

A Building Block is structured in a declarative part, written in CDL, and a body, containing the code associated to the Contracts the Building Block provides.

In the following we give an example of what a Building Block looks like (the syntax is simplified)

```
BB Manage Assignable Inventory
     IS Multiinstance
     Version 1.0
     Constant C1                    ;Global declarations
     . . .
     Datatype Netw_serv_t      RECORD  (
                               Bandwith INTEGER
                               . . .)
     . . .
Export  Contract  Ass_inventory
     Type Call  Composable
     INPUT (
          Netw_serv
          . . .)
     OUTPUT (
          Invent_list
          . . .)

Import  Contract  Set_service_state
     Type Dialogue  Non  Composable
     Operation Set_state
          INPUT (
               New_state)
     Operation Set_Event_Forw_Discr
          INPUT (
          . . .)
          OUTPUT (
          . . .)

BB  BODY
Ass_inventory :
     . . .
     . . .
     Open_Dialogue Set_service_state ( ... )
        Status = Call_Op Set_state ( ... )
        if Status = OK
           . . .
           Call_Op Set_Event_Forw_Discr
        else
           . . .
     Close_Dialogue

End  BB
```

4. CONCLUSIONS

This paper has presented a Computational Model providing concepts to organize distributed and multivendor OSS software applications in a way which enables maximum interoperability among them. The model supports concepts from CCITT's TMN and ISO's OSI System Management, together with distributed computing and transaction processing ones.

The Computational model provides the application programmer with a declarative language CDL (Contract Definition Language) for both Building Blocks and Contracts definition and, in association, a high level API (Application Programming Interface) to run-time services for Contract invocation and Transaction demarcation and co-ordination. Current work has concluded design of an Engineering model and has initiated experimental development of the Technology model on IBM/MVS-CICS and UNIX platforms.

Mapping of CDL and Contract invocation API onto the Engineering model services is supported by a compiler.

Future work aims to :

- design of services for managing the architecture
- extension to the Computational Model to handle asynchronous interrupts within a transaction.

5. ACKNOWLEDGEMENTS

The authors wish to thank all the people of CSELT taking part to the development of this work. Particular thanks are due to Lucia Agostini, Paola Benetti, Piergiorgio Bosco, Gianfranco Botto, Maria Lorenza Demarie and Nicoletta Lumello, who participated in the definition of the Computational Model, and to Laura Cerchio and Carlo Castellano for comments and revision of the paper.

6. REFERENCES

1 "An ODP Analysis of OSI Systems Management" S. Proctor - BT,TINA 92 Conference Record.

2 Advanced Networked Systems Architecture Reference Manual ARM 1.00, A. Herbert (editor) - Architecture Project Management, Cambridge U.K., March 1989.

3 CCITT Recommendation M.30: "Principles for a Telecommunications Management Network", Blue Book, Vol. IV. Fascicle IV.1, 1989.

4 Bellcore TA-STS-000915 Issue 2, July 1990 "The Bellcore OSCA™ Architecture".

5 ISO/IEC IS 7498-4, Information Processing Systems - Open System Interconnection - Basic Reference Model - Part 4: Management Framework.

6 ISO/IEC IS 9595-2, Information Processing Systems - Open System Interconnection - Management Information Service Definition - Part 2: Common Management Information Service.

7 ISO/IEC IS 9596-2, Information Processing Systems - Open System Interconnection - Management Information Protocol Specification - Part 2: Common Management Information Protocol.

Integrated Network Management, III (C-12)
H.-G. Hegering and Y. Yemini (Editors)
Elsevier Science Publishers B.V. (North-Holland)

UNIX Systems Management via SNMP

B. Krupczak
Network Management Laboratory
College of Computing
Georgia Institute of Technology
Atlanta, GA. 30332

Abstract

The Simple Network Management Protocol (SNMP) has become an Internet standard protocol for *network* management. This paper discusses the strengths and weaknesses of using the SNMP to perform UNIX *systems* management in addition to *network* management. The discussion is motivated by the experiences gained from implementing a UNIX systems management SNMP agent. Some knowledge of UNIX, systems management, and SNMP is assumed.

Keyword Codes: C.2.0; C.2.2; D.4.9
Keywords: Computer-Communication Networks, General; Network Protocols; Operating Systems, Systems Programs and Utilities

1 Introduction

The SNMP has become an Internet standard protocol for *network* management. It has been implemented by a multitude of vendors on a wide variety of hardware platforms, ranging from network bridges and routers to workstations and mainframes. SNMP is used to manage everything from network interface cards to protocol stacks.

This paper discusses using the SNMP to perform UNIX *systems* management in addition to network management. The paper discusses an SNMP agent and *systems* management MIB implementation on SUN SparcStations running SunOS. The discussion centers on the applicability of SNMP to *systems* management.

The next section defines *network* and *systems* management and outlines differences between the two. Section 3 gives a brief overview of the SNMP while Section 4 discusses the systems management MIB implementation. Section 5 discusses limitations of SNMP with respect to UNIX systems management. Section 6 concludes the paper.

2 Network and Systems Management

OSI network management literature [3, 5, 4] defines *systems* management as the management of the OSI environment or components providing OSI services. It does not necessarily include the hardware platform and operating system that supports the OSI services. Management of the OSI services includes the network interface cards, network protocol stacks, and software used to provide OSI services. Management is accomplished using an object based approach. A managed object is the OSI management view of a resource that is subject to management. Managed objects are defined in terms of their attributes, allowable operations, notifications they may emit, and relationships they share with other managed objects.

While much of the literature uses the terms *network* and *systems* management interchangeably, systems management has taken on a different meaning than that used by the OSI literature. Network management is the management and operation of a communications subnetwork and the services it provides. Network interface cards, network protocol stacks, routers, and bridges all qualify as network devices managed by a network management system.

However, end-user systems in which the network interface cards and protocol stacks reside are not considered network components and are not usually managed using a network management system. These end-user systems often have complex operating systems and software packages requiring constant attention by system administrators. The act of managing these non-network components is referred to as *systems* management. OSI management is applied to non-OSI systems (for example, operating systems and software) by modeling those systems as OSI managed objects. The act of creating, modifying, and deleting OSI managed objects which model non-OSI systems causes systems management to be performed.

3 Overview of SNMP

The SNMP is a client-server model: the client, a network management station, sends a request to the SNMP agent, the server. The SNMP agent services the request and sends the result back to the client management station. This technology consists of three parts, each defined in an RFC.

The **first part**, the Internet Structure of Management Information (SMI), defines the common structures and identification scheme for the definition of management information. It provides the syntax for defining managed objects. The SMI includes a description of an object information model for network management as well as a set of generic types used to describe management information. It does not dictate the protocol used to manage objects, nor those objects themselves. The goals of the Internet SMI are simplicity and extensibility [11].

The **second part** is the management information base (MIB). Managed objects are accessed via a virtual information store known as the MIB. An initial MIB containing object types used to manage TCP/IP is defined in [8]; A second version, MIB-II is defined in [6]. The MIB can be visualized as an inverted tree with managed objects residing at the leaves. Each object type, defined using ASN.1, has an object identifier or name specifying its position in the MIB tree.

The object identifier can be thought of as the names of the branches that would be traversed in moving from the root of the MIB tree to the managed object, located at a leaf. The object identifier together with an object instance uniquely identifies a specific instantiation of an object.

A base MIB tree is defined by the Internet SMI and contains several branches. The *experimental* branch or subtree, is used to identify objects used in Internet experiments. Another branch, the *private-enterprise* subtree, is used to permit parties providing networking subsystems to register management objects for their products.

The **third** and final part of the technology is the actual protocol used to exchange management information – the SNMP [2]. The SNMP models all management functions as alterations or inspections of managed object values. Interaction between a management application and a managed agent consist of alterations or retrievals of existing managed objects. SNMP defines administrative relationships between managed agents and management applications as well as how managed object instances are specified. Further, SNMP defines the structure, content, and nature of message exchange. The SNMP communicates using a connectionless datagram service: each message is independent of previous messages and is contained entirely in one datagram. The SNMP defines the following message types: *GET*, *GET-NEXT*, *SET*, *TRAP*, and *GET-RESPONSE*. They are used for managed object value retrieval, object value modification, event notification, and message response.

4 System Management MIB

SNMP UNIX systems management was implemented by adding additional object types or classes to an existing SNMP agent for SunOS. These additional objects constitute the systems management MIB. By creating new object instances, deleting existing object instances, and modifying existing object instance values, the SNMP agent performs UNIX systems management. Because SunOS represents one of the major strands of UNIX, it was felt that all UNIX variants could be managed similarly.

The base SNMP agent implemented MIB-I objects for SunOS systems. Systems management object types added to the agent were placed under the *experimental* branch. The managed objects can be grouped into the following categories or MIB "branches": System Information, Mounted Devices, User Information, Group Information, Process Information, *Who* Information, Remote Command Execution, Performance Monitoring, and Interprocess Communication Information.

Object types in the systems management MIB were chosen based on two criteria. First, the object type had to represent a generic UNIX resource. The resource could not be present on one "flavor" of UNIX while not existing on others. Second, the resource had to be accessible by C-language system calls. The reason for this criteria is to avoid a problem that plagues some systems management agents. Some SNMP agents fork/exec UNIX system commands to gather much of their information. Those SNMP agents must then parse through text output for the desired information. Fork/exec is considered an expensive way (in time and system resources) to gather information. This problem was avoided by only managing resources that could be manipulated programmatically through the C-language system calls. In the end, most, if not all, of the same resources could be managed without using the *fork()* and *exec()* system calls.

4.1 System Information

The *System Information* MIB branch contains objects representing the node name, OS release, OS version, CPU type, hardware serial number, and the amount of physical memory. System Information was obtained through a variety of means including system calls and operations on the device */dev/kmem.*

System information fits nicely into the SNMP management model because it supports only SNMP *GET* primitives. SNMP *SET* primitives are not supported because the system information objects do not represent user configurable information. CPU type, hardware serial number, and the amount of physical memory should not change during the operation of a UNIX kernel. While the node name of a machine is configurable, changing its value dynamically during a UNIX machine's operation would cause considerable problems and therefore is not allowed.

4.2 Mounted Devices

The *Mounted Device* MIB branch contains a table representing the machine's mounted devices. Each row in the table represents a currently mounted device and contains columns representing a device's mount point, block size, total number of blocks, total number of free blocks, total number of files, and total number of free files. The mounted device MIB table does not fit well into the SNMP management model because devices are frequently mounted and unmounted requiring the creation and deletion of new rows into the table; SNMP does not contain primitives for creation and deletion.

Remote unmounting of a device could be accomplished by deleting its associated device MIB table row. Since SNMP lacks an explicit *DELETE* primitive, another method had to be devised. To do so, a special column, "devUnmount", was added to the table. Writing a certain value to this column in a given row, causes the SNMP agent to unmount the device represented by that row.

Remote mounting of a file system could be accomplished by the creation of an additional row in the device MIB table. Since SNMP lacks an explicit *CREATE* primitive, some implicit method had to be developed. In order for an SNMP agent to mount a device, it would have to know two things: the particular device, and its associated mount point. One method of instructing the SNMP agent would be to send it a *SET* primitive containing the desired information for the file system to be mounted. The *SET* primitive must specify an unused instance so that the agent does not mistake the request for an existing file system. The manager must therefore know how the instance for an object in the mounted device MIB table is formed, and know an unused one. Due to these complications, the systems management agent did not support remotely mounting devices.

4.3 User Information

The *User Information* MIB table contains rows, each representing a particular system user. The table's columns represent login name, password, user ID, group ID, user name, user home directory, and a login shell. Most systems store their user information in the file */etc/passwd*.

The *User Information* MIB table also does not fit well into the SNMP management model because users can be added and deleted. Since SNMP lacks explicit *CREATE* and *DELETE* primitives, an unused instance must be specified when creating a user. To delete users, an extra column could be added as described earlier.

4.4 Group Information

The *Group Information* MIB table contains rows, each representing a particular system group. The table contains columns representing a group name, group password, and group ID. Most systems store their group information in the file */etc/group*.

The *Group Information* MIB table does not fit well into the SNMP management model for exactly the same reasons as the *User Information* table. In addition, group membership cannot be modeled as a column within the *Group Information* MIB table since groups may contain multiple members. There are several ways to organize group and group member information. One such way is to add an additional column in the group MIB table. This additional column would itself be a table. This table would contain one row per group member. Adding and removing members from a given group would be accomplished by adding and removing rows from the group membership table. The Internet SMI does not allow tables within tables – that is, one MIB table cannot be contained in another MIB table.

4.5 Process Information

The *Process* MIB table contains rows, each representing a process currently running on the system. The table contains columns representing a process ID, name, state, flags, owner uid, owner gid, and nice value. Further, the table contains a column representing signal values for each process. Writing a value to this column for a given row, causes the SNMP agent to send the applicable signal to the process represented by the table row. Writing the value corresponding to the UNIX signal SIGKILL, via the SNMP *SET* primitive, causes the SNMP agent to send the *SIGKILL* signal to the associated process – effectively killing the process. The SNMP agent ignores invalid signal values written to this column.

Since SNMP models all management functions as the alteration or retrieval of existing managed object values, the process MIB table fits nicely into the SNMP management model. One restriction is placed on the MIB table though. Row creation is not allowed because SNMP lacks an explicit *CREATE* primitive. The process MIB table does not warrant an explicit *DELETE* primitive because the process signal column supports killing processes via the *SET* primitive.

4.6 Who Information

The *Who Information* MIB table contains rows, each representing a user who is currently logged on to the system. The table contains columns representing user name, user login device, login PID, login time, and from where the user is logged in. Table information is retrieved from the file */etc/utmp*. The *Who Information* table fits well into the SNMP management model. It highlights one of the strengths of SNMP – its efficiency in retrieving information – not creating or deleting it.

4.7 Remote Command Execution

The *Remote Command* MIB branch contains objects which can be used to initiate binary executables and shell scripts on the machine on which the SNMP agent is executing. This branch contains three objects – *shellOutput*, *shellCmd*, and *shellExitStatus*. The instance of the *shellOutput* object represents the name of a file that contains command execution output. The *shellCmd* object instance represents the actual command that is executing or has executed. Finally, the *shellExitStatus* object instance represents the exit code that the command has returned.

The manager first writes the name of an output file to the *shellOutput* object using the SNMP *SET primitive*. Then, upon writing the desired UNIX command in the *shellCmd* object, the UNIX command is executed. When the SNMP agent receives a SET request for the *shellCmd* object, it first validates the value of the *shellOutput* object. If valid, the SNMP agent then constructs a command line to be passed to the *system()* function.

The *Remote Command Execution* MIB branch does not fit well into the SNMP management model for several reasons. First, the objects are single-value objects – multiple instances do not exist. Therefore, only one remote command execution can be invoked at a time. Secondly, representing remote command execution as an SNMP table brings about problems discussed earlier – namely, the problems associated with creating and deleting rows in a table. Lastly, command output cannot be retrieved or viewed using SNMP since every SNMP PDU must be entirely contained in one datagram. Command output can be, and often is, quite lengthy.

4.8 Performance Monitoring

The *Performance Monitoring* MIB branch contains objects representing the outcome of kernel performance monitoring. The MIB branch contains objects representing the percentage of time over some interval that the kernel is executing in various modes. The SNMP agent implements CPU performance monitoring through several objects. First, there is a *sampleInterval* object. Writing a value to the instance of this object causes the SNMP agent to begin monitoring system performance. The value of this object represents the duration in seconds of the sampling period. When completed, the SNMP agent calculates the appropriate statistics and places the values in corresponding objects.

The *Performance Monitoring* MIB branch fits well into the SNMP management model because object values are being retrieved and set, not created and deleted. More sophisticated performance measurements could be added using the same approach as mentioned above. Disk I/O, buffer cache efficiency, and network performance objects could all be modeled similarly. Performance monitoring could be implemented more intuitively if the SNMP supported an *M_ACTION*-like primitive. This primitive could be used to explicitly start performance monitoring. The outcome of performance monitoring could then be sent back to the initiator via an event report or *TRAP* mechanism.

4.9 Interprocess Communication: Queues, Shared Memory, and Semaphores

The *IPC* MIB branch contains tables representing the three forms of UNIX System V interprocess communication mechanisms – message queues, shared memory, and semaphores. Each table represents one IPC mechanism. The tables contain one row per instance of IPC usage. For example, for each instance of shared memory, there exists a row in the *shmem* MIB table. All *IPC* MIB tables contain a column that allows an instance of an IPC mechanism to be deleted. These columns make use of the *SET* primitive since SNMP does not have an explicit *DELETE* primitive. The semantics of this column are identical to those mentioned earlier.

The message queue table contains columns representing a message queue ID, key, mode, owner, group, and size. The shared memory table contains columns representing shared memory segment ID, key, mode, owner, and size. The semaphore table contains columns representing semaphore ID, key, mode, and owner. The *IPC* MIB table, aside from the clumsiness of row deletion, also fits well into the SNMP management model because existing object instance values are retrieved or modified.

5 Limitations of SNMP

Since SNMP was specifically designed for network management, its applicability to UNIX systems management is questionable. Although SNMP has successfully met the goals for which it was designed, it appears to fall short in several areas when it comes specifically to UNIX systems mangement. These areas include: creation and deletion of managed object instances, event notification, file transfer, security, and managed object relationships.

5.1 Creation

The creating of object instances in the SNMP is ill-defined in that management operations are modeled as modifications or retrievals of **existing** object instance values. Because of this model, no explicit *CREATE* primitive exists. SNMP agent implementations have worked around the lack of an explicit *CREATE* mechanism by creating object instances when new instances are encountered in an SNMP *SET* PDU. That is, when an agent receives a *SET* PDU for an object instance that does not exist, it creates that object instance. This *CREATE* implementation is ambiguous though – what if a manager did not really intend to create an object instance when it tried to set the value of a non-existent object instance?

The addition of new users is an example of a systems management function that would benefit from an explicit *CREATE* mechanism. In current implementations, a manager would have to know a user ID not already in use, then issue a SNMP *SET* request containing that unused ID. Issues of data consistency are not addressed, in this case, if two managers try to use the same ID simultaneously.

An explicit create mechanism could solve the above problems because the manager would not be required to specify an object instance. The agent, instead, would choose an appropriate unused object instance specifier. The agent could then guarantee consistency since it alone would specify the instance of newly created objects. In addition, an explicit *CREATE* mechanism that hides the object instance details from the manager could provide greater interoperability than current object creation implementations. Currently, a manager wishing to create objects must know how each different agent implements object creation.

5.2 Deletion

In addition to lacking an explicit *CREATE* mechanism, SNMP also lacks an explicit *DELETE* primitive due to its network management model. SNMP agent implementations, including this one, have also worked around this shortcoming. In the case of SNMP tables, an extra column is usually added. Writing a predefined value to this column for a particular row causes the agent to delete the row. SNMP agent implementations have also used predefined values to allow managers to delete non-tabular managed object instances. Non-tabular objects that are deletable have some predefined predefined value that is "hard-wired" into the agent. A *SET* of an object to a predefined value causes the agent to delete the object instance. This approach for deleting object instances has its shortcomings. How do management stations discover all these predefined values? Data consistency issues are also not addressed. *CREATE* and *DELETE* primitives as side-effects of the *SET* primitive seems intuitively a bad idea.

5.3 Event Notification

SNMP also lacks a method or primitive for reporting events in a confirmed fashion. Its *TRAP* mechanism provides primitive, unconfirmed event reporting. Much, if not all, of systems management functions require confirmed events. For instance, if newly installed software requires the kernel to be rebuilt, the systems manager definitely wants to know if the kernel build failed or succeeded.

Confirmed traps have been implemented by some SNMP agents. One such implementation requires the recipient of a *TRAP* to inform the SNMP agent of the trap's receipt by writing the *TRAP*'s unique number into a managed object in that SNMP agent's MIB. The SNMP agent times-out and re-transmits the *TRAP* after some period of time if it has not received a "confirmation". The SNMP agent implementation allows the network manager to configure how many times the SNMP agent re-transmits each *TRAP* and how long the interval is between transmits. Clearly, the requirement of a priori knowledge of *TRAP* confirmation method for every SNMP agent is not a good, scalable solution.

Since SNMP places limits on the amount of information that can be placed in a protocol data unit (PDU), the amount of information that can be reported using a *TRAP* primitive is also limited. What happens if the occurrence of an event necessitates more information transfer than one SNMP PDU can handle? Should an agent send multiple *TRAP* PDUs? What happens if one of a sequence of related *TRAP* PDUs is lost?

5.4 File Transfer

SNMP was never intended to be used as a file transfer mechanism. File transfer is needed to distribute software and operating system configuration files. Although there are more efficient mechanisms, it is technically possible to use SNMP to transfer files. One such method of transferring files using SNMP is to use a table with each row containing a block number and data. The block number would define offsets inside a file. Successive *SETs* could then write the individual blocks to the file.

While a connectionless oriented protocol may be suitable for network management, it is not clear whether it is also suitable for systems management. A connectionless protocol was chosen for SNMP for reasons of efficiency and low overhead. Further it was felt that a connectionless protocol would have a higher level of success in an overloaded network. Systems management will not take place during periods when the network is not functioning properly because network managers will be busy trying to "fix" the network. Therefore, the overhead of a connection-oriented protocol may be acceptable. Systems management information exchange is best served by a connection oriented paradigm (for example, event reporting). Current connection oriented protocols, utilizing congestion control and sliding window algorithms, are much more efficient than datagram protocols for reliably transferring the large amounts of data required by systems management.

5.5 Security

Commands and operations performed by the agent are executed with root privileges because the SNMP agent runs as a root process. Anyone with some knowledge of SNMP can issue commands to the systems management agent that will then be executed as root. The only security in SNMP, at the moment, is the use of the community name which acts somewhat like a password. The community name is a trivial security method at best since it is sent across networks unencrypted.

Security is always an important issue in managing internetwork environments. Systems management requires strong security because of its very nature. Adding and deleting users, killing processes, and mounting and unmounting file systems all require some level of security to prevent anyone from performing these functions. SNMP security concerns are being addressed by the Internet Engineering Task Force (IETF) which is in the process of standardizing a security and authentication mechanism within the SNMP.

5.6 Managed Object Relationships

SNMP also lacks sophisticated object/attribute relationships. For example, SNMP does not allow tables within tables – that is, one column in an SNMP object table cannot contain another table. For instance, a system may contain several printer queues, each possibly containing multiple print jobs. One way to represent this information would be to have a printer queue table. One column of this table could then be a table containing printer jobs for that particular queue. Since SNMP does not support tables within tables, this method of organizing information is not possible. SNMP agents have worked around this restriction by creating two related tables – in this example, one table containing the printer queues, and another table containing **all** printer queue jobs. However, a network manager would have to examine and relate two tables in order to manipulate jobs in a certain queue. The ability to define tables within tables would simplify management in this example.

6 Conclusion

The implementation of an SNMP agent with a systems management MIB successfully demonstrated that SNMP can be used to perform UNIX systems management. The current MIB implementation contains over 80 managed objects. More managed object types could be added to the current MIB, thereby increasing its systems management functionality. For instance, system managers also want to manage printer queues, printer jobs, and software. More sophisticated performance monitoring could also be added to track disk I/O, cache, and virtual memory performance.

SNMP does have its strengths when used to perform systems management. Its simple, efficient design is easy to implement while requiring little processing overhead. Its object management model is easily extensible through the private-enterprise and experimental MIB branches. The Internet SMI provides powerful yet simple management relationships that have been widely implemented and tested. Further widespread agent and management station implementation and interoperability testing has been accomplished demonstrating that the SNMP can be an integrated solution for both systems and network management.

During the course of the SNMP systems management agent development, testing, and use, it became clear that SNMP is not perfect in the role of a systems management protocol. The lack of explicit *CREATE* and *DELETE* primitives can be overcome, but presents a barrier to interoperability and efficiency. Its *TRAP* mechanism appears ill-suited to the needs of systems management because of its unconfirmed nature and its restrictions placed on the quantity of information that can be sent. The addition of a confirmed event reporting mechanism would greatly simplify an SNMP agent's ability to act independently. An independently acting agent would then inform management only when a problem exists. As networks grow larger and larger, the desire for "self" management will also grow. Further, more sophisticated object/attribute relationships would allow greater flexibility. Finally, SNMP lacks the ability to efficiently transfer files, hampering its ability to perform system configuration tasks. SNMP has proven sufficient for network management but does not provide a satisfactory solution for UNIX systems management.

References

[1] Maurice J. Bach. *The Design of the UNIX Operating System.* Prentice Hall, 2nd? edition, 1986.

[2] J.D. Case, M. Fedor, M.L. Schoffstall, and C. Davin. Simple network management protocol. Anonymous FTP, May 1990. RFC 1157, obsoletes RFC 1098.

[3] Information processing systems: Open systems interconnection basic reference model part 4: Management framework, October 1989. ISO/IEC 7498-4.

[4] Information technology: Open systems interconnection: Systems management - part 1: Object management function, August 1991. ISO/IEC 10164-1, CCITT X.730.

[5] Information technology: Open systems interconnection: Systems management overview, August 1991. ISO/IEC 10040, CCITT X.701.

[6] Management information base for network management of tcp/ip-based intenets: Mib-ii. Anonymous FTP, March 1991. RFC 1213, obsoletes RFC 1158.

[7] Samuel J. Leffler, Marshall Kirk McKusick, Michael J. Karels, and John S. Quaterman. *The Design and Implementation of the 4.3 BSD UNIX Operating System.* Addison-Wesley, 2nd? edition, 1989.

[8] K. McCloghrie and M.T. Rose. Management information base for network management of tcp/ip-based intenets. Anonymous FTP, May 1990. RFC 1156, obsoletes RFC 1056.

[9] Marshall T. Rose. *The Open Book: A Practical Perspective on OSI.* Prentice Hall, 1st edition, 1990.

[10] Marshall T. Rose. *The Simple Book: An Introduction to Management of TCP/IP-based internets.* Prentice Hall, 1st edition, 1991.

[11] M.T. Rose and K. McCloghrie. Structure and identification of management information for tcp/ip-based internets. Anonymous FTP, May 1990. RFC 1155, obsoletes RFC 1065.

Integrated Network Management, III (C-12)
H.-G. Hegering and Y. Yemini (Editors)
Elsevier Science Publishers B.V. (North-Holland)
1993 IFIP.

The NAS Hierarchical Network Management System

Jude A. George, Leslie E. Schlecht
jude@nas.nasa.gov, schlecht@nas.nasa.gov

Computer Sciences Corporation
M/S 258-6, NASA Ames Research Center, Moffett Field, CA 94086, USA

Abstract

This is a case study of the development of a network management system for the Numerical Aerodynamics Simulation facility at NASA Ames Research Center. Issues relevant to IP network management are identified and examined. Experience has shown that traditional monolithic network management systems cannot address all of these issues in a large network; in many cases, a distributed, hierarchical system is more suited to this task. With these issues in mind, a scalable architecture for monitoring large networks is presented: the Hierarchical Network Management System. Methods for gathering, storing, and transporting network information within this system are described. A user interface incorporating design principles for displaying and managing large networks is also presented. We expect that the presentation of this architecture will be of value to other members of the Internet community who would participate in the development of an open standard for distributed network management.

Keyword Codes: K.6.0; C.2.4; H.5.2

Keywords: Management of Computing and Information Systems, General; Distributed Systems; User Interfaces

1.0 INTRODUCTION

The Hierarchical Network Management System (HNMS) project is a result of the continuing installation of large, high speed local and wide-area networks for the Numerical Aerodynamic Simulation (NAS) facility at the NASA Ames Research Center. The complexity of these new networks has forced us to look into a single complete network management solution in lieu of separate management tools specific to each type of network hardware.

1.1 Requirements for NAS

There are a number of issues involved in the selection of a network management system. For NAS, we developed the following criteria:

- *use open published standards where available (SNMP and X Window System):* To facilitate the integration of new network hardware and reduce reliance on any one vendor, data collection should be carried out through a standard protocol. For availability on a wide range of platforms, the user interface should be based on a standard window system. SNMP was chosen for data collection and X for display because of their wide vendor support.
- *present a clear, precise representation of network elements:* The state of a network element must be defined meaningfully, and with no assumptions. Additionally, there must be no misleading or confusing information introduced through editing or group representation.

- *scale to thousands of managed network elements:* At the start of our project, NAS provided network access to over 500 hosts. It was our goal to provide a means of monitoring all of these systems at once, presenting us with a challenge in the areas of data collection and intuitive displays.
- *minimize required human input by using automation:* The dynamic nature of a network in a research and development facility creates a network management editing nightmare. If editing were required to create a management display at NAS, the display would always be inaccurate. This makes automatic discovery and display of network elements desirable.
- *do not succumb to the problems being monitored or have a single point of failure:* The management system should be available and responsive at the most critical times and not be subject to network problems.
- *do not cause additional network problems:* The system should not overload the network with management traffic.
- *provide logging of network events:* In order to generate reports for troubleshooting and trend analysis, the system must provide a method of logging network data into a database.

1.2 Existing Management Software

At the outset of this project (late 1990) we conducted a survey of a dozen existing network management software packages and examined them according to our criteria (details can be found in reference [2]). Most of the commercial products used SNMP for data collection and X for display. However, most did not provide automatic discovery or display of elements, instead providing a graphical editor for the operator to create and update diagrams. With this editor, the operator could introduce meaningless and confusing objects into the diagram reducing the tool's usefulness and authority. In addition, none of the products addressed the issue of scalability in data collection or display; that is, none of the products could have displayed the entire NAS network in a readable form. For these reasons we decided to develop HNMS.

1.3 Development Approach

The size and complexity of the NAS local and wide-area networks are the most important issues in management, creating problems in the areas of data collection and display, and network load. We have approached these problems through a distributed and hierarchical strategy. Since no standard currently exists for a true distributed network management architecture, our design builds on existing standards currently used in network management. With the aforementioned issues in mind, this paper presents an implementation of a distributed NMS and examines how architectural features of the system allow for better fault-tolerance, creation of less management traffic, and a more structured scheme for cataloging network elements, as opposed to that provided by a conventional NMS.

2.0 ARCHITECTURE

2.1 General

The system consists of four types of modules. The modules are software entities, typically residing on separate hosts throughout the network. The *server module* is the hub of the management system. It provides a center for dissemination of topology and status information. *User Interface (UI) modules* reside on graphical workstations and provide users with access to real-time or logged data. *Input/Output (IO) modules* reside on hosts located at strategic points within the local-area and wide-area networks and handle the actual data collection. IO modules use SNMP and layer 2 monitoring for data collection. The IO modules pass filtered management data up to the server. The server, in turn, sends relevant information to the UI modules participating in the management system. Data are sent only when their values have changed. Thus the hierarchical approach allows us to avoid flooding a network with management traffic and creating bottlenecks where management information is processed. The fourth type of module, the *database module*, is a place to log long-term traffic studies, as well as keep current network state data. If the server machine should crash or become unreachable, a new server is started on a different machine; the server will then load the network state from the database. If the database module should become unreachable, data are queued until the database is again available. A fully functional HNMS consists of a server, a database, at least one IO module, and at least one UI module.

At this point, it will be useful to define a few terms. We refer to *network elements* as the actual entities that comprise a network -- routers, hosts, serial links, LAN segments, and individual interfaces. *HNMS objects* are the representations of these elements, and other abstractions, within an HNMS system. Each object is identified by a unique integer, its *HNMS id*, which is assigned by the server. There are, in total, eleven classes of HNMS objects:

Internet	the Internet
Network	an IP network
Subnet	an IP subnet
Lan	a layer 2 medium, part of a local-area network
Link	a layer 2 medium, part of a wide-area network
Interface	a device with a layer 2 address, layer 3 address optional
Processor	a host, router, or bridge
Site	a location, usually associated with a particular organization
Equipment	miscellaneous communications equipment
Administrator	a human, typically a network operator or site contact
PostalAddress	a postal mailing address or demarcation point

In addition to its class, each object has a subclass, and certain objects have a status. The subclass serves to further identify the type of object. Possible subclasses for the Interface class include Ethernet, FDDI, and serial link. The status identifies the operational and administrative state of the element, and is used with all HNMS object classes except Site, Equipment, Administrator, and PostalAddress. Values for the operational state include Responsive, Non-Responsive, and Unreachable. The operational state is ultimately set by the server, depending on whether the element has been heard from within certain preset time periods. Values for the administrative state include Alarm, Unresolved Problem, and Normal. The use of administrative state is covered in more detail in the section explaining the user interface.

Object data are the collection of information which describe a particular HNMS object; they consist of variables represented as name:value pairs which provide information apart from the class, subclass, and status. Representation of object data requires, in addition to the standard SNMP MIB variables [4], the use of a new HNMS MIB. This MIB contains separate groups of variables dedicated to describing the eleven classes of network objects. It also contains a "log entry" group whose variables are used for reporting alarms as well as for storing free-form textual data about any object.

For example, the object data for an Interface may include its Ethernet address (if it's subclass is Ethernet), its IP address, and its netmask. Object data also include variables defining the relations of an object. A relation exists if a variable contains a reference to the HNMS id of another object. For example, the object data for an Interface contains variables referring to the Interface's parent Processor, parent Subnet, and parent Lan or Link. Although we do not have space to reproduce the MIB here, we can provide a concise relational mapping between object classes and the other object classes which they are allowed to reference:

An object of class:	*has relations of class:*
Internet	*
PostalAddress	*
Administrator	PostalAddress
Site	Internet, Administrator, PostalAddress

Network	Internet, Administrator
Subnet	Network
Lan	Administrator
Link	Administrator
Processor	Site, Administrator
Equipment	Site, Administrator
Interface	Processor, Subnet, Lan, Link

Objects of class Internet and PostalAddress do not have references to any other objects.

All inter-module communication is done via a new Hierarchical Network Management Protocol (HNMP). HNMP, like SNMP, is built on top of UDP/IP, and ostensibly falls at layer 7 in the OSI network model. The protocol is specified in the Abstract Syntax Notation 1 (ASN.1) using its Basic Encoding Rules (BER); data types exchanged between the modules conform to the same subset of ASN.1 data types used by SNMP, as defined in the Structure of Management of Information (SMI) [3]. The protocol requires the use of a new HNMS MIB, which defines a set of variables (in addition to the standard SNMP variables [4]) in which object data are represented. In addition to transporting object data, HNMP facilitates a higher layer of network management with paradigms such as management domains and subscriptions, explained below. These allow a given module to receive a select set of management information about an arbitrary group of objects without repeatedly asking for it. More importantly, the protocol allows the system to uniquely identify data about any given HNMS object at any point in time, for the lifetime of the system, as each variable binding has a timestamp associated with it. We deemed that this level of flexibility was not possible with SNMP, even with solutions such as RMON (remote monitoring) devices [5] and other proxy agents [6], because of some fundamental differences in high-level versus low-level management. SNMP, by itself, provides only two levels of hierarchy: server (device agent) and client (management station). The HNMS, in conjunction with SNMP, uses four levels: device agent, IO module, server module, and UI or database module. Managing without such a scheme may not be difficult for a small network, but it is not practical for a network of the scope of the combined set of NAS networks.

2.2 Establishing Connections

The server and IO modules run as daemons on separate machines. In reality, the server is a special case of an IO module; at start-up, one IO module is given a flag telling it to be the default server.

Each server is distinguished by an HNMS community name. This is not to be confused with an SNMP community name, although the motivations behind both are similar; they are used to define an administrative classification of the managed devices and to provide a trivial form of authentication. The HNMS community name identifies the server and the object data which it disseminates. IO and database modules also identify themselves with an HNMS community name; either module type may only connect to a server whose community name matches its own. The community name is the only configuration that is given directly to any of these modules.

The UI modules initiate their connections to the server by sending an HNMP *Hello* message. The server responds with a *Welcome* message, assigning the module a module-id; this is an integer which is unique to that module, for any instantiation of an HNMS server. At start-up, a UI module presents the user with a pre-configured list of HNMS IO module hosts, one of which is the current location of the server. If the user selects a host on which an IO module is running, the IO module will respond with a *Redirect* message informing the UI module of the true location of the server. We are planning for the server module to be relocatable. If the host on which the server runs should crash, a new server is selected from the IO modules. We are also considering extensions to the protocol to allow peer to peer exchange of information between IO modules, such that there would be no definite server. A UI module would then be able to connect to the nearest IO module instead of a particular one identified as being the server.

The next message that deals with connection management is the *Ack* message. HNMP requires that its messages be delivered reliably. Since the protocol is built upon UDP, it has provisions for lightweight connections between modules, including a sliding window retransmission scheme. The Ack message is used to provide an acknowledgement of each message that is received by a destination module in an HNMP connection, as well as to verify that a connection is still up when a module is deemed to have been silent for too long.

A UI module sends a *Goodbye* message to the server when it wishes to disconnect. Once the Goodbye has been acknowledged, the module may consider itself unconnected.

2.3 System Management

The HNMS has associated with it a number of operational parameters that can be set by users through a UI module. These parameters concern polling intervals, timeout intervals for determining state changes, intervals and buffer sizes associated with HNMP, and other aspects of module control. Some parameters are module-specific, while others are system-wide; the latter are maintained by the server although copies are kept within each module. If a user chooses to set a parameter on another module, or parameters concerning the entire system, a *SetParameters* message is employed. Each parameter is a triple of a parameter name, its value, and an integer defining the scope. The scope can identify a certain module or the entire system. If a module wishes to get the current value of a set of parameters, it uses a *GetParameters* message, which is similarly defined.

Problems may result if multiple UIs request that a certain parameter be set to conflicting values. For this reason, the notion of a *key* is employed, to ensure that only one UI at a time may be modifying a parameter. The key is a fictitious object which is maintained by the server. When a UI module wishes to send a SetParameters, it must first obtain the key by use of a *GetKey* message. If the key is not currently in use, the server responds with a *SendKey* message. At this point, the UI module has free reign to change as many system parameters as it desires. When it is finished, it uses a SendKey to return the key to the server. All key requests and parameter changes are logged by the server. Additional authentication measures may be employed; however the design of such measures are outside the scope of this paper. For the initial release of the HNMS, we have settled on a scheme in which only certain UI modules have the ability to use the key; these modules are used by select members of the network administration staff.

When the server refuses a request or sends informational messages to another module, it uses a *SendExplanation* message. This message is a combination of reference fields for internal use and a descriptive text block. The text block is a printable message which the UI may display on screen, at its discretion.

2.4 Management Domains

Before a server can begin its duties as a dissemination point for network data, it must be told which portions of the Internet it will be responsible for managing, and where it may find IO and database modules to enable the system to gather and store data. To facilitate this, the HNMS employs the concept of a *management domain.* A management domain is a quadruple comprising an IP address, an address mask, an SNMP community name, and a class, where class is one of the following: Internet, Network, Subnet, Interface, IO Module, Database Module. The first four class values are used for representing some entity that will be managed; in this case, the SNMP community name is used by the IO module which is eventually assigned the task of monitoring that section of the network via SNMP. The last two are used to inform the server of IP addresses where it can expect to find a module of the appropriate type.

A UI module sends the server a sequence of management domains by means of an *AddDomains* message. For example, the server may be given a list requiring it to monitor the network 138.178.0.0, the subnets 129.200.23.0 and 129.200.48.0, as well as the individual interfaces 135.102.10.1 and 135.102.10.2. In addition, the list may tell it to find IO modules at the addresses 129.200.23.1 and

138.178.92.10, and a database module at 129.200.33.4. This is typically done by the network administrator who connects the first UI module after starting the server. The UI module gets the key, sends an AddDomains message, and returns the key.

Each IO module, in turn, is assigned a set of management domains by the server, based on the proximity of that IO module to the managed elements. In this case, the domain corresponds to a set of IP networks, subnets, and individual interfaces for which that IO module is directly responsible for monitoring. It is up to the server to determine which modules will monitor which domains. Currently, the algorithm is simply to assign all of the domains to the first IO module that connects. Subsequent modules receive subsets of the original set of management domains based on their location in the network, as determined by the IP addresses of their hosts. In most cases there is a one-to-one correspondence between IP subnets and IO modules.

A UI may request the server to stop managing a set of domains with a *DropDomains* message. This message is also used between the server and IO modules.

2.5 Data Exchange

Processor and Interface objects are first generated by the IO modules when they are discovered. These objects are given temporary HNMS id's and are sent to the server via an *Announce* message. An Announce message always includes the HNMS object and a set of object data. The variables used in an announcement are known as definitive variables, since they define the object; they represent the minimal set of variables required to distinguish an object from other objects of its class. Examples of definitive variables for an Interface object are its IP address and its layer 2 (e.g. Ethernet) address. Other variables are known as descriptive variables, which provide incidental information. Examples of these are the SNMP variables ifInOctets and ifOutOctets.

Once the server receives an announcement about an object, it checks the object's definitive variables against those of all of the objects it currently knows about. If the object appears to be unique, the server assigns it a permanent HNMS id and allows the object to be released to other modules. It immediately sends the object back to the announcing IO module, which replaces its temporary object with the one created by the server. From that point onward, the server receives data about the object by means of a *subscription* through the use of a *SubscribeData* message. A subscription consists of an HNMS id and a list of variables which the server would like the IO module to fill in. Each variable binding consists of a name:value pair and a timestamp. The HNMS objects and their object data are sent back to the server in *SendData* messages. A subscription is filled only when the value of a subscribed variable changes. Furthermore, only the changed variables are reported back to the subscribing module. If the server subscribes to 30 variables concerning an object and three of them change at any given time, those three will be sent back.

IO modules send announcements about Interface, Processor, Lan, and Link objects. Site, Administrator, PostalAddress, and Equipment objects are created by network administrators at UI modules and are announced to the server in a similar fashion. The server, being the final authority on object creation, does not allow duplicate objects to be created, from either the same module or separate modules. The rules for determining duplicates vary with each type of object, but all rely on the definitive variables in the object announcements. Finally, the server creates certain objects on its own, based on information it has gathered about IP addresses. The Internet, Network, and Subnet objects are created in this fashion.

Once an object has been created, it is announced to all UI modules. UI modules also retrieve information by means of subscriptions. The UI will choose some subset of the set of announced objects based on what the user wishes to monitor, and subscribe to certain variables regarding those objects. The UI may cancel variables in a subscription, or the entire subscription, with an UnsubscribeData message.

In any subscription request, the timestamp field in each variable binding specifies the requested "currentness" of the data (e.g. 20 seconds). For each variable on every HNMS object, the server keeps track of the intervals requested by all UI modules. The lowest interval value is used as the actual polling interval by the IO module for that variable. Note that some variables are not gathered via SNMP but are created within the HNMS itself; for these variables, the interval field is left unused. In the replies, the timestamp field refers to the actual time that the variable held that value, as determined by the polling time. This is expressed in a standard time format.

In addition to the subscription method, a UI may ask for a set of variables to be sent to it just once. This is accomplished by means of the *GetData*, *GetNextData*, and *GetWalkData* messages. The GetData and GetNextData messages are similar to the SNMP Get and GetNext messages. The GetWalkData message asks for a complete walk of a branch of the MIB starting at each variable specified in the message, and is very effective for saving bandwidth. In typical SNMP management systems, a walk of a MIB branch requires the SNMP client to send a separate message for each requested variable, using the powerful GetNext operator [6]. In return, the client will receive a separate message containing the value of each variable. The GetWalkData message allows the request and the response to each be encoded in one message. When an IO module receives a walk request from the server, it performs the sequence of GetNext operations required to retrieve the entire branch. Since the IO module is typically situated on the same subnet as the entities within its management domain, the heavy SNMP polling of each set of elements is restricted to their immediate area and does not travel over the entire network.

The last message type dealing with object data is the *SetData* message. This message is a request by the UI to set the values of variables associated with an object, and requires the use of the key in a fashion similar to the SetParameters message. The server sets the appropriate values and forwards a SetData containing the new values to all modules subscribed to those variables. If the server denies the request, it sends an explanation to the UI module. In the case of SNMP variables associated with a processor, the server forwards the SetData message to the appropriate IO module, which in turn sends an SNMP Set message to the processor. The IO module then sends the server a SendData with the new value to confirm that it has been changed. Once the server updates its own object data, all subscribing modules are sent the new values.

There are three other messages which deal with entire HNMS objects, not with specific object data: *DeleteObject*, *DeactivateObject*, and *ActivateObject*. When an object is announced by the server, it is automatically assumed to be an active object. However, situations arise where we want to make the object inactive, meaning that it no longer represents an element currently on the network. In other instances, we may want to remove an object altogether from the server's memory. The DeactivateObject and DeleteObject messages are provided for this purpose. The final message, ActivateObject, is only provided for the sake of orthogonality. It is very rare that an object, once made inactive, will be reactivated.

2.6 Device Discovery and Data Collection

The primary discovery method used by the IO module is a three-stage process. Each IO module listens to traffic on its local Ethernet to learn Ethernet and IP addresses of other hosts. If it finds a new IP/Ethernet address pair which falls within one of its management domains, it enters the addresses into a cache and attempts to contact the source device via SNMP. The definitive SNMP variables for the processor (e.g. sysName, ifNumber) are then requested. If the device responds, the IO module then asks it for detailed information about all of its interfaces by simultaneously retrieving selected variables from the ifTable and walking the ipAddrTable. The variables are used as the input to a set of rules for determining whether the processor and its interfaces are unique objects not currently known by the IO module. If the processor has not been previously detected, the IO module announces it to the server. If the server already knows of the object, it sends its existing object data to the IO module; otherwise it assigns the new object an HNMS id and incorporates it into the HNMS system.

A second discovery method is a more directed version of the above. If an IO module is assigned a management domain which specifies a single interface IP address and SNMP community name, the IO module takes the initiative and immediately polls the address for a processor's definitive variables. This method is used for all of the wide-area routers in the NAS network, as they have separate community names.

A third discovery method utilizes the "next hop" entries in the routing table of each processor to walk through the network. By default, this is used only for routers, as determined by the ipForwarding variable.

The last discovery method simply consists of the reception of a trap from a previously unknown entity.

Regular polling for determining operational state is done using the sysUpTime and ifOperStatus variables. The polling used by the IO module is asynchronous; the module does not wait a given time, e.g. 4 seconds, for a reply. Rather, it sends out the request PDU and simply updates a timestamp for the object when a response is received. Once per second, the module checks the timestamps of all objects in its memory; if a timestamp is older than the time period set for a particular state change, it updates the operational state from Responsive to Non-Responsive, or from Non-Responsive to Unreachable, and informs the server of the fact. A network object may also change state if the IO module is directly informed that there is a state change; this may happen if a processor returns an ifOperStatus of down for one of its interfaces. In this way, an interface may change from Responsive to Unreachable directly, without going through the Non-Responsive state.

Special polling schemes may be used for certain managed entities. In particular, plans are being made to support the RMON MIB. While RMON agents are still limited by the current version of the SNMP protocol, their usefulness increases considerably when incorporated into a complete network management system which allows data gathering abstractions such as subscriptions.

2.7 Management Traffic

Most traffic between IO modules and the server, as well as between the server and UI modules, consists of replies to SubscribeData messages. Since a subscription is filled only when the value of a variable changes or when the status of an object changes, the inter-module HNMP traffic in a steady state is typically very low compared to the SNMP polling traffic between IO modules and their managed agents. If we compare this system to a conventional SNMP NMS, which continuously polls all managed agents from one point, it can be seen that the sum of SNMP traffic generated by each scheme is the same. However, the benefits of the HNMS architecture are apparent when the system is used at multiple locations. The HNMS makes it feasible to monitor a network that has multiple managerial points of control. The NAS wide-area network, AEROnet, connects six NASA centers (including NAS) and 34 other sites. The HNMS architecture provides the capability for the network to be monitored from any remote location. Under a typical network management architecture, each site would run a complete NMS. Each of these systems would contact every single network device through SNMP in order to build a representation of the network. Not only will this result in slightly different network representations at each station, it will generate an inordinate amount of network management traffic. The total number of polling sessions is the number of management stations times the number of managed objects. If there are 300 devices on the network, there would be 12000 SNMP polls going on during the time interval of one polling period. Under the HNMS scheme, there would be only 300 SNMP sessions and a few dozen HNMP connections, each of which is very lightweight compared to a continuous SNMP polling session since only changes are sent. These are obviously extreme examples, since not every site will have a desire or a need to monitor the entire network; they may subscribe to a subset of the information, or none at all. However, it is apparent that the distributed architecture will provide a savings in bandwidth in an extended network. Even under situations where the network is in a state of fluctuation and the status values of objects are changing rapidly, each HNMP connection is no more of a burden than an

SNMP session, since the frequency of subscription replies would match the frequency of polling by the IO modules.

When setting up the HNMS architecture, care must be taken to choose intelligent locations for the IO modules, especially on a wide-area network. Optimally, SNMP polling is restricted to the local network at each site, with HNMP traffic traveling between modules over the WAN itself. The way to accomplish this is to situate the IO module on the boundary between the local network and the wide-area link. Many sites have a special LAN segment at this junction called the *isolation LAN*. We currently have an IO module running on a workstation situated on the isolation LAN at each of five AEROnet hub sites. Thus, no SNMP polling traffic is required to travel over the backbone.

A quantitative analysis of the performance benefits of a distributed NMS as opposed to a monolithic NMS is outside of the scope of this paper, but a document detailing such an analysis is available from the authors.

3.0 USER INTERFACE

3.1 Requirements

The most important requirements for the HNMS user interface include the following:

- *Provide automatically generated, accurate displays*: This is fundamental in a large, constantly changing network environment. With minimal instruction from the user, the UI should be able to automatically convert information from the server into a readable display. For accuracy, all of the elements in a diagram should be represented individually, not iconified as a group, which makes assumptions about an element's status and can be misleading. In addition, within the limits of the graphics hardware, displayed elements should not overlap or obscure each other.
- *Provide the observer with an immediate perception of the state of a large network without sacrificing readability:* The observer should be able to determine the general health of the network in a glance.
- *Provide notification of important events to the user.* The UI should offer an effective means of informing a network administrator that a problem has occurred.
- *Provide detailed information about a specific network element.* Provide access to the SNMP MIB variables.

3.2 Diagrams

HNMS provides four types of status diagrams, each representing the state of a network element by color. These diagrams are updated from the server as element status values change. When a new element is discovered, it is added to the diagram. Aside from object state, other useful information such as routing paths and response times may be displayed.

The *WAN* diagram depicts the state of a network and its routers over a geographical reference (e.g. a map of the US). Routers in the same site are clustered together and special consideration is taken to avoid obscuring one router with another. The routers are represented by dots and the links between them by lines.

The *Site* diagram can be created by selecting a site on the WAN diagram. This diagram represents all LANs that are connected to the routers at a given site. We determined that the interface is the primary informational element in this case, and all interfaces are displayed. The LANs are shown as concentric rings with their interfaces represented as dots, and important hosts and routers are placed around the periphery. This polar form makes a compact, centralized, representation for the observer.

The *Custom* diagram allows the user to construct a diagram with any set of network elements they wish to observe. When a Network is requested, individual Subnets within that network will appear when they are announced by the server; alternatively, the user could ask for the individual Subnets by name, and only those Subnets will be shown. This diagram is more detailed than the Site diagram and space is not a constraint. However, organization is important. The design of the Custom diagram must facilitate auto-generation while maintaining readability and avoiding obscurement. Our approach is a basic cartesian system with Subnets along the Y-axis, Processors along the X-axis, and Interfaces at intersection points. This method avoids the problem of crossed lines and provides a simple reference framework for the observer.

Any network element may be selected to create an *Object* diagram. This diagram is a textual display of the element's variables which is updated when the element changes. The Object diagram is the lowest level display in a hierarchical system of diagrams which allows the user to obtain more and more detail about a network and/or its elements.

In addition, HNMS offers several performance diagrams for visualizing network variables in troubleshooting and analysis. These real-time displays (updated at a regular interval), take the form of histograms for instantaneous single-element observation, line graphs for historical single-element observation, and an adaptation of the Site diagram for multi-element observation.

Since the UI may subscribe to any SNMP variable for any object, adding functionality to any of the diagrams is simply a matter of subscribing to different sets of MIB variables and providing graphical representations of them.

3.3 Color, Event Notification, and Customization

Color is an important visual cue in a network management system. In order to simplify diagrams and provide for a wider range of graphics workstations, we use only a few colors easily distinguishable from each other. An element's operational state is represented by four colors: *magenta* indicates Unknown, *green* indicates Responsive, *yellow* indicates Non-Responsive, and Unreachable is displayed as *red.* These states are determined by the server and are updated when they change. *White* is a neutral color used for displaying text and site boundaries and *orange* is used for routing paths.

In addition to operational state, HNMS displays the network administrative state: Normal (with no problems), Unresolved-Problem state, and Alarm state. Trouble ticket administration is carried out in conjunction with the administrative state designation, and operates in a distributed fashion. The system incorporates cross-references between the UI modules (through the server) such that network administrators at separate sites are able to participate in the tracking of problems spanning the entire network. The Alarm state, initiated by an operational state change to Unreachable, signals an operator of a problem. All UIs monitoring the questionable object will flash the object. An operator selects the object and requests the server to change the state to Unresolved Problem, flagging the object. The server validates the request and forwards the state change to the other UIs, thus informing other interested parties. When the problem is resolved, the operator again selects the element to request a change to Normal and removal of the flag, and the server informs the other UIs. At any time the operator or other users may open the current trouble log on the element to check progress or add comments. The log entries, like all other object data, consist of timestamped HNMS MIB variables and are maintained by the server and disseminated via subscriptions.

4.0 DATABASE

The database module is currently being developed to allow use of an SQL (Structured Query Language) database engine within the HNMS. Because of the large amounts of data generated by the HNMS and by the real-time nature of network management, it is essential that the database be able to complete transactions quickly. The database module itself is little more than a translator to convert HNMP messages to SQL commands. Such a module can be quickly written for any database engine. When a user needs information to be retrieved from the database, it will be done in one of two ways: 1) directly from a database client application, or 2) from an HNMS user interface module asking the server module for the data, which in turn will retrieve it from the database module and send it to the UI. To the HNMS server, the database module is just another client requesting or supplying information about the network. In practice, the server and database modules should never be located on the same host, as that would invalidate the database's use as a backup if the host were to go down.

Active Data. The database is used for two purposes by the HNMS. For the task of maintaining an up-to-date working set of information, the database maintains a set of tables containing records about HNMS objects. Each record is identified and indexed through its HNMS ID. These data must be maintained in a fail-safe location in case the HNMS server moves, or terminates unexpectedly, and needs to quickly recover its state. Active data can also be used for producing reports about the network, as it exists at that moment in time.

Inactive Data. There are also certain types of data which become aged; that is, they do not represent anything in the current state of the network. They are kept because they represent information that is useful in doing long-term traffic analysis, or simply for maintaining a record of events that have occurred in the network. There are two ways that data can become inactive. One way is to have an entire record become inactive. For example, a Processor may be removed from the network, or an Administrator may be retired. We cannot always remove these entries from the database because we may want to keep them for historical reasons. Inactive data are also created when we log specific fields from an active record. For example, an Interface object may have its ifInOctets recorded, along with a timestamp, every time it is reported to have changed (with a frequency up to its requested polling interval). In addition, every object will have a log of events associated with it. An event will consist of a timestamp, a status value, and a free-form text field. Events logged to an object can be generated by the HNMS (example: 5/5/92 05:01:00 4 "router ARC1 interface 3 down"), or by a human operator (example: 5/5/92 06:10:00 1 "There is a circuit problem on RFS62826, and I've called PSCN. --Bob."). Each log entry will be related to an object that is currently active.

All object data, whether active or inactive, consist of sets of timestamped MIB variables. A user of the HNMS may call up data for an active Interface record and some subset of its past (inactive) event log entries. Although it seems more likely that a network administrator will make such a query directly to the SQL database, situations may arise where it will be necessary to go through a UI module (and thus the server) for the information.

5.0 ADDITIONAL MODULE TYPES

One of the advantages of a modular distributed system is that additional module types may be incorporated with relative ease. We are considering the development of a Rules-Based Intelligent Processor (RBIP) module. Once the HNMS system has been put to use at a sufficient number of sites, research into the automation of network management will be conducted. The RBIP is an expert system application which will adjust various HNMS parameters based on overall network traffic and usage of the system. The RBIP may also make short-term and long-term predictions about the network load and even make configuration changes in certain types of devices. We do not anticipate that the RBIP will require significant changes to the protocol. Rather, it will be similar to a UI module with a network administrator "built in."

6.0 IMPLEMENTATION OF HNMS

Our prototype system was completed in April of 1991. We chose to implement this system in C and LISP to facilitate rapid prototyping of the protocol and internal data structures. The server and IO modules were implemented on Sun Sparcstation and Silicon Graphics Iris hardware; the UI was implemented on the Iris using the GL graphics library. System 1.0 is scheduled for release within NAS by the end of 1992. This version of HNMS is written entirely in C and is portable to any UNIX system with a Berkeley sockets library and ISODE 7.0. The protocol definition and MIB are available from the authors. The user interface module for System 1.0 uses the X Window System and the Motif tool kit for all displays. All modules in this release are developed on the Sparcstation and Iris platforms.

7.0 CONCLUSION

Our intent is to have shown that there are definite performance and administrative benefits inherent in a distributed network management architecture. As we near the completion of development for this release, we can state that this is a successful implementation of a first generation distributed NMS. This case study should provide information useful to others who are interested in developing such systems. The development of a successor to SNMP is in progress. To date, what information has been released still does not describe an approach to distributed management of the type we are proposing [7][8]. We propose an object-based structure (for devices, not just variables), multi-administrator management, and efficient ways of cataloging and displaying data for a hierarchy of network elements. We hope that the development of this system will spur interest in the Internet community toward specifying a powerful, yet simple, standard for distributed network management and practical graphical representations for large networks.

8.0 REFERENCES

[1] Case, J., Fedor, M., Schoffstall, M., and Davin, J., "Simple Network Management Protocol", RFC 1157, SNMP Research, Performance Systems International, MIT Laboratory for Computer Science, May 1990.

[2] McCabe, J., Schlecht, L., George, J., "NGMS Project Plan", Technical Report RND-91-004, NAS Systems Division, NASA Ames Research Center, Moffett Field, CA, April 1991.

[3] Rose, M., and McCloghrie, K., "Structure and Identification of Management Information for TCP/IP-based Internets", RFC 1155, Performance Systems International, Hughes LAN Systems, May 1990.

[4] McCloghrie, K., and Rose, M., "Management Information Base for Network Management of TCP/IP-based internets: MIB-II", RFC 1213, Hughes LAN Systems, Performance Systems International, March 1991.

[5] Waldbusser, S., "Remote Network Monitoring Management Information Base", RFC 1271, Carnegie Mellon University, November 1991.

[6] Rose, M., "The Simple Book: An Introduction to Management of TCP/IP-based Internets", Prentice-Hall, Inc., New Jersey, 1991.

[7] Case, Jeffrey D., McCloghrie, K., Rose, M., and Waldbusser, S., "Protocol Operations for the Simple Management Protocol (SMP) Framework", Internet Draft, University of Tennessee, Knoxville, Hughes LAN Systems, Dover Beach Consulting, Inc., Carnegie Mellon University, July 1992.

[8] Case, Jeffrey D., McCloghrie, K., Rose, M., and Waldbusser, S., "Manager to Manager Management Information Base for the Simple Management Protocol (SMP) Framework", Internet Draft, University of Tennessee, Knoxville, Hughes LAN Systems, Dover Beach Consulting, Inc., Carnegie Mellon University, July 1992.

Integrated Network Management, III (C-12)
H.-G. Hegering and Y. Yemini (Editors)
Elsevier Science Publishers B.V. (North-Holland)

Managing the Extended LAN

Anil Rijsinghani[a] and Keith McCloghrie[b]

[a]Networks and Communications, Digital Equipment Corporation, 153 Taylor St., Littleton, MA 01460, USA

[b]Engineering, Hughes LAN Systems, 1225 Charleston Road, Mountain View, CA 94043, USA

Abstract

In recent years, the use of Local Area Networks (LANs) as a means for providing network connectivity has accelerated considerably. In conjunction, there has been a proliferation of transparent bridges, used to interconnect and extend these LANs in a protocol-independent manner. Although such bridges typically do not require any configuration before being put into operation, there is growing interest in the use of network management to retrieve the wealth of information available in these devices about the networks they connect. In this paper, we describe techniques to apply data maintained by transparent bridges in network monitoring and control.

Keyword Codes: C.2.2; C.2.3; C.2.5
Keywords: Network Protocols; Network Operations; Local Networks

1. Introduction

The transparent bridge [1], a device used widely in the interconnection of Local Area Networks (LANs) to form extended LANs, contains valuable data which can be used in the management of the networks it helps to build. This paper explores the application of this data in the monitoring, analysis and control of these extended LANs. Reference is made to two widely deployed standards used to implement and manage bridges.

The IEEE 802.1 committee has specified the standard for MAC bridging in IEEE 802.1D [2]. Transparent bridging is the specified method for interconnection of all 802 LANs. This standard details various aspects of bridge operation, thereby greatly facilitating interoperation of transparent bridges in a multi-vendor environment.

The IETF has specified the Simple Network Management Protocol (SNMP) [4] for management of network devices. In addition, it has specified the Management Information Base (MIB) for bridges [5], a collective definition of variables used to manage bridges based on IEEE 802.1D.

In combination, these two standards allow a vendor-independent, integrated approach to the management of bridges and LANs.

The next three sections of this paper present brief overviews of transparent bridging, SNMP, and the relevant MIBs; the following five sections describe various management applications; and the last section presents a conclusion.

2. Transparent Bridges

This section gives a brief overview of transparent bridging principles. For further detail, the reader is referred to [6].

A transparent bridge extends LANs, while retaining the inherent characteristics of the LANs: low delay, high throughput, low frame loss and no duplication or misordering of frames. End stations communicate across an extended LAN just as if they were attached to the same LAN, unaware of the presence of a bridge.

The bridge operates in "promiscuous" mode, examining every frame sent on the LANs it connects. It learns the location of stations by associating the source addresses in the frames it sees with the ports on which they were they were seen. This information, maintained in a forwarding database, is then used to make forwarding decisions based on the destination address in received frames. Frames from stations that are known to reside on the same port as the destination are filtered, while frames destined to multicast addresses or unknown unicast addresses are sent out on all other active ports.

In Figure 1, frames from station A to station B are filtered by Bridge X. However, frames from station A to either station C or D, both of which are reached via port 2 of Bridge X, are forwarded.

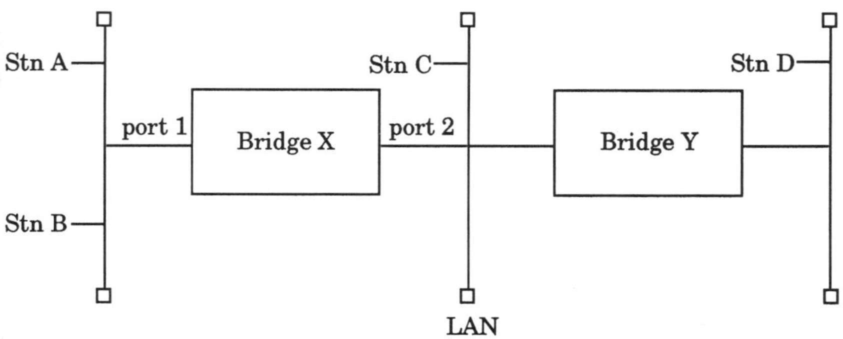

Figure 1. Bridge Filtering

Transparent bridges also participate in a distributed Spanning Tree algorithm [3], running on all the bridges in the extended LAN, to form a loop-free path between any two points in the network. Ports which are not part of the spanning tree are in blocking mode until needed as backup paths for continued connectivity in case of LAN failure.

3. SNMP

This section gives a brief overview of the SNMP architecture. For a more detailed treatment, the reader is referred to [4].

Management information, maintained in a managed entity, and accessed via a SNMP agent, is defined as instances of objects. Such objects are defined in a Management Information Base (MIB) in a format outlined in the Structure of Management Information [7]. Communication between a Network Management Station (NMS) and a SNMP agent occurs using one of five primitives: Get, Get-Next, Set, Get-Response, and Trap.

MIB-II [8], a collection of system and network information independent of the network device, is implemented by all SNMP-compliant agents. Extensions to the MIB-II for management of specific device functions, are defined in separate MIBs which specify objects for managing such functions. *(Note:* When making reference to MIB objects, this paper will often use descriptive names, rather than the complete object names as defined in the MIBs.)

SNMP is used today primarily in the TCP/IP environment, over the connectionless UDP transport layer. The management techniques described here are thus not necessarily limited to extended LANs local to an NMS.

4. Bridge MIB

The IETF Bridge MIB [5] defines objects for the management of bridges based on the IEEE 802.1D standard for MAC Bridges. The MIB is structured as groups of objects corresponding to logically related functions. Of the five defined groups, three must be implemented by all 802.1D-compliant transparent bridges which are SNMP-manageable:

1. *dot1dBase* - Objects applicable to all types of bridges
2. *dot1dStp* - Objects associated with the Spanning Tree process
3. *dot1dTp* - Objects associated with the Forwarding process

This MIB avoids redundant definition of objects applicable to bridges which are already available from the standard MIB-II. As such, it also specifies the correct interpretation of objects in the MIB-II for bridges which implement the Bridge MIB.

5. LAN Monitoring

Since transparent bridges operate in "promiscuous" mode, they attempt to receive all frames sent on the LANs they connect and can effectively be used as monitors for these LANs. This may eliminate the need to place dedicated network monitors at these LANs. The performance of a bridge plays an important part in its utility as a traffic monitor - the greater its ability to receive and examine worst-case, minimum-size frame traffic, the more accurate its counters and thus the more appropriate its use as a monitor.

The *interfaces* group in MIB-II contains several frame and byte counters. On bridge interfaces, the sum of the unicast and multicast frames received, and the unicast and multicast frames transmitted, gives the total traffic on the corresponding LAN. The unicast and multicast traffic rates are useful statistics to monitor individually as well. LAN throughput rates in terms of packets per seconds can be calculated by taking snapshots of these counters at appropriate intervals, in conjunction with the up-time object in the system group. As these are 32-bit counters, care must be taken to ensure that they would not wrap more than once in the interval between which the snapshots are taken. These counters cannot be cleared, being specified as read-only objects, which has the advantage that they are readable without interference by multiple NMSs.

The total receive and transmit byte count may be used to obtain LAN utilization statistics. On high-speed LANs, these may wrap very quickly. As before, it is important to compute the worst-case rate of change of these counters, and use this as a basis for determining polling intervals.

Table 1
LAN Statistics

	MIB object
Total bytes received	*ifInOctets*
Total bytes transmitted	*ifOutOctets*
Unicast frames received	*ifInUcastPkts*
Multicast frames received	*ifInNUcastPkts*
Unicast frames transmitted	*ifOutUcastPkts*
Multicast frames transmitted	*ifOutNUcastPkts*
Frames received on Bridge Port	*dot1dTpPortInFrames*
Frames transmitted on Bridge Port	*dot1dTpPortOutFrames*
Frames filtered on Bridge Port	*dot1dTpPortInDiscards*

It is also useful to gauge the traffic forwarded and filtered by a bridge port in relation to each other. The *dot1dTpPortInFrames* counter in the *dot1dTp* group is the total number of frames received by the bridge entity, while the *dot1dTpPortInDiscards* object counts the frames that were filtered. A constant low ratio of frames filtered to frames received indicates poor isolation of traffic: a fairly large percentage of the traffic is not being kept local, indicating that the bridge placement may be non-optimal. The total number of frames transmitted by the bridge on any given port, also specified in the Bridge MIB, indicates the level of traffic being forwarded on this port. This is also normally much less than the number of frames received. Comparing this counter across multiple ports can help in revealing traffic "funneling" syndromes.

Note that in comparing counters, it is never safe to compare the retrieved values directly, since they may wrap at different rates. Instead, values at the start and end of a period must be retrieved and the differences compared.

6. Topology Mapping

The topology of the extended LAN can be represented by a map of the Spanning Tree formed by all the bridges participating in it, and the LANs that they connect.

It is not a simple task to derive the map from physically tracing cables in anything other than very small networks. Besides, since bridges may be added or removed at any time and LANs may break down, possibly causing standby bridge links to take over, the extended LAN topology is a dynamically changing environment. Network management is the most effective and often the only reliable way to track it.

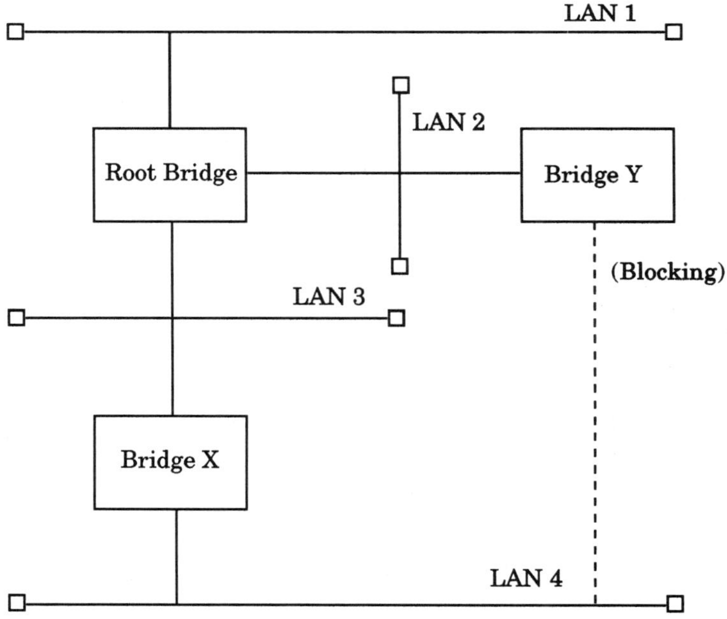

Figure 2. Spanning Tree topology example

The *dot1dStp* group in the Bridge MIB specifies objects relevant to the Spanning Tree process. By examining the contents of certain of these objects in all the bridges in an extended LAN, it is possible to map the topology accurately and quickly.

In a single tree, there is a single Root known by all of the bridges. Bridges immediately adjacent to the Root may be found simply by finding all bridges for which the Root is the Designated Bridge on any port. The particular port of the Root Bridge to which they are connected may be found by looking the Designated Port Identifier.

In a like manner, the rest of the tree can be mapped by looking for the bridges that have so far been placed in the map as Designated Bridges in the spanning tree database of all other known bridges. *(Note:* It is assumed in this procedure that all of the bridges participating in the tree are manually identified, or have been discovered separately.) An iterative invocation of this this procedure will allow the extended LAN topology to be displayed as a single bridged network.

Ports which are not forwarding traffic may be identified by looking at the Port State, which has one of five possible values: Forwarding, Blocking, Listening, Learning and Broken. The tree is formed by all of the paths connected by bridge ports in the Forwarding state.

The type of LAN to which a bridge port is connected may be found by retrieving the corresponding interface type from the *interfaces* group of MIB-II.

The Spanning Tree may change depending on the environment. A new bridge or repeater added into the extended LAN would change its topology, as would the failure of a bridge or a LAN. In Figure 2, if Bridge X failed, the blocking port of Bridge Y would transition into the forwarding state to allow LAN 4 to retain connectivity to the rest of the network. This would result in a different topology.

It is advisable to verify the topology periodically to keep the map current by repeating this process, or at least by verifying that there has not been a change in the Designated Bridge for any LAN. Reception of traps defined in the Bridge MIB may also be helpful in reacting to changes.

7. Station mapping

The logical extension to deriving the LAN map is the mapping of all end stations resident on these LANs.

As explained earlier, a bridge operates in promiscuous mode to learn the location of all active station addresses in order to achieve traffic isolation. Consequently, its forwarding database can be used to locate stations with respect to itself. The forwarding table defined in the *dot1dTp* group reports all learnt addresses, along with the port number on which they were learnt.

However, a bridge cannot distinguish between the location of stations connected to different LANs, if its path to such LANs is through the same port. An intersection of forwarding database information from adjacent bridges must be used to infer the exact location of stations in the bridged topology. The exception to this is addresses learnt on ports which connect to leaf LANs (located at the periphery of the spanning tree), which can be guaranteed to be resident on those LANs without further data reduction.

In this manner, stations may be located with respect to other stations in the extended LAN, including the network management station. In addition, the total number of stations in the extended LAN may be determined.

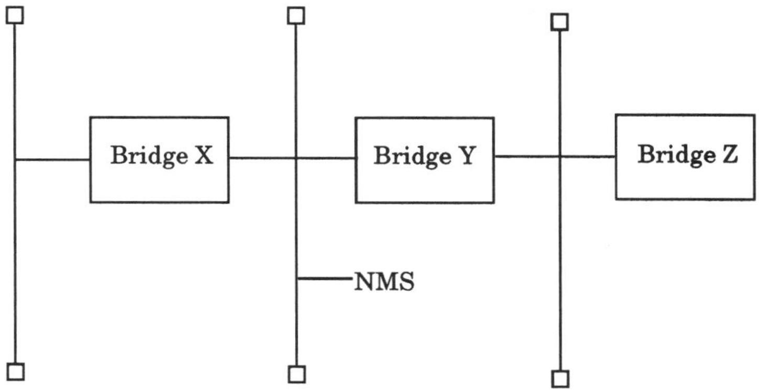

Figure 3. Locating the NMS

In Figure 3, both Bridge Y and Bridge Z point "leftwards" to the address of the NMS, but Bridge X, which is adjacent to Bridge Y, points in the opposite direction. This indicates that the NMS is resident on the LAN to which both Bridge X and Bridge Y are attached.

Addresses in the forwarding database are usually aged out in a small time interval, typically a few minutes, after they have been learnt unless they continue to be active. This time interval is referred to as the *Aging Time* of the bridge. Addresses known by the bridge which have been aged out due to lack of activity may still be reported in its forwarding database, although they are treated identically to unknown addresses for forwarding purposes.

The forwarding database defined in *dot1dTp* contains three attributes for every address learnt: the address itself, the port number on which it was last seen, and a status field. A value of *invalid* in the status field indicates that the address has been aged out. The port number, if one is reported in this case, may be unreliable since the station may since have been relocated, or removed. In such cases, it may be necessary to provoke the station into transmitting a frame in order to get fresh information on its whereabouts.

Sometimes, the learnt information in the forwarding database can be overridden by management-specified filters. One mechanism for doing so is outlined below in the Network Control section. Addresses in the *dot1dTp* Forwarding Database, if also management-specified, will have a status value indicating this fact; for such addresses, the bridge may not use learnt information in forwarding decisions. However, the port number attribute returned for these addresses, if non-zero, will continue to report the actual location of the station.

Hardware addresses of the bridge's own interfaces are reported with a status of *self,* and the port number indicates which bridge port the corresponding address is associated with.

The station map can change rapidly as stations are moved around, removed, or new ones added. In most cases, it is desireable to get the current snapshot by activating this procedure when necessary, rather than have a process continuously track changes via periodic polling. Frequent retrieval of large forwarding tables can have a noticeable impact on network traffic.

8. Alarms and Cautions

Unusual LAN conditions may be detected by monitoring the values of appropriate MIB objects in bridges.

8.1. Congestion

Bridges are required to discard frames that have been queued internally for an excessive period of time, typically on the order of 2 seconds. The *Port Delay Exceeded Discards* object in *dot1dTp* counts the number of frames that were dropped by the bridge due to an excessive lifetime in the bridge. This may indicate outbound congestion at the LAN to which these frames were to be transmitted.

8.2. Bridge Database full

A bridge uses its forwarding database to make forward/filter decisions. If the number of stations on an extended LAN exceeds the capacity of the forwarding database in any of the bridges in the extended LAN, that bridge may have to resort to unnecessary flooding. (A side-effect of this condition may be loss of accuracy in station mapping, described above.)

The *Learned Entry Discards* object in *dot1dTp* is the total number of addresses that could not be learnt due to a full forwarding database. If this counter is increasing, then there is a persistent learning problem in the bridge in its current environment,, indicating that it is unable to cope with the number of stations on the extended LAN. The maximum capacity of the forwarding database of a bridge determines when it will reach this condition. A non-zero but stable value for this counter means that the bridge has run into this condition at least once, but the condition is not necessarily chronic.

8.3. Spanning Tree stability

The stability of the Spanning Tree is crucial in extended LAN connectivity. There are several objects in the MIB that can be used to monitor different aspects of topological stability.

The following guidelines assume that the related counters are accurate, i.e., that bridges can always process received Spanning Tree BPDUs.

dot1dStpTopChanges is the number of topology changes that have been detected by this bridge since it came up. A topology change in the tree is caused by a change in the port state of a bridge into or out of the Forwarding state. This counter should remain constant after the topology stabilizes, usually shortly after network power-up. In addition, *dot1dStpTimeSinceTopologyChange* indicates when the last topology change had occurred.

dot1dStpPortForwardingTransitions indicates the number of times a given port went into the Forwarding state. This counter should stay constant, and is usually 1 in most normally operating networks, since a bridge port which is actively forwarding traffic transitions into the Forwarding state only once after a fixed amount of time (determined by the *forward delay* time in effect for the spanning tree) in which the topology is expected to have stabilized. However, if there is disruption in other parts of the spanning tree, it is possible for the port to transition out of and back into the Forwarding state. A value of other than 1 usually indicates that connectivity has been lost at some time.

9. Network Control

The bridge can be used for network control in several ways. Addresses and protocol types may have status specified by management for security, traffic control, or other purposes. Spanning tree parameters may be tuned if necessary.

9.1. Management-specified filtering

Due to security or other reasons, it is sometimes necessary to control the forwarding decision of a bridge for particular frames via management. There are several fields in the datalink header upon which these decisions can be based. Both 802.1D and the Bridge MIB outline an optional method for specifying filtering based on the destination address of a frame.

dot1dStatic is an optional group, and applies only to bridges which support the model of filtering to which its definition applies. Forwarding information for an address in the destination field may be specified on a per-receive port basis, to a given set of output ports. If the receive port is 0, the information applies to all ports (or ports other than those separately specified).

If a particular type of frame is not being received across several bridges, this table (or other vendor-specified management table) may be retrieved from known bridges to find if the frame is being filtered.

In addition, traffic through the bridge may be controlled by the use of management-specified filters. For example, *dot1dStatic* filtering may be used to prevent a load server multicast from reaching LANs that do not have any load servers on it.

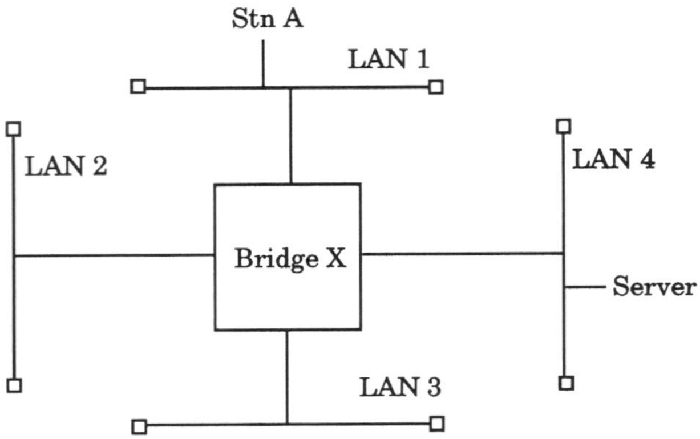

Figure 4. Management-specified filtering example

In the example of Figure 4, assume that Station A is a critical resource restricted to LAN 1. Bridge X could be configured to disallow forwarding of frames destined to Station A to any other port, even if another station "masquerades" as A on another port.

Another use of DA filtering is shown above: if a multicast need be sent only to a particular LAN, as in the case of a multicast address used to find a server, the bridge can be configured to deliver frames destined to this multicast only to the LAN on which that server resides (typically a backbone LAN). In addition to reducing unnecessary multicast traffic on other LANs, this could also allow a manager to set up local servers (eg, one on LAN 3 for use only by clients on LAN 3).

9.2. Spanning Tree tuning

The tree may be pruned to achieve better run-time characteristics and stability, by controlling the appropriate spanning tree parameters.

The Root may be selected to be a bridge connected to the backbone, so that bridges further away from the backbone do not end up forwarding traffic at the expense of bridges connected to the backbone. The Root selection is based on the numerically lowest Bridge-ID. Since the Bridge-ID is composed of a user-settable *root priority* and the unique hardware address of the bridge, and the priority is the more significant field in the ID, it is possible to bias the selection of the Root by controlling the value of this field.

In order to minimize disturbances caused by potential failure of Root, the bridge that takes over in this case should ideally be a bridge adjacent to the Root. This will prevent the tree from changing drastically. This could be done by setting the priority of an adjacent bridge to a value lower than that of the Root, but higher than the default used for all other bridges.

In addition, the cost of bridge ports may be adjusted so as to force the higher speed network links to forward traffic whenever there is a choice. This may be done by assigning lower path costs to higher bandwidth ports.

The global node parameters of the spanning tree - *hello time, max age*, and *forward delay* - determine its timing characteristics. By adjusting these, the switchover time to a backup link may be tuned. However, note that these parameters should be changed only if the topology allows it; they are based on the diameter of the extended LAN. The parameters in effect for the tree are administered by the Root to all bridges in the tree; hence the settable versions of these parameters are significant only for the Root. For recommended values of these parameters, refer to [2].

Finally, the enable switch in dot1dStp may be used to disable ports of a bridge from forwarding traffic and participating in the Spanning Tree. This is one way to isolate ports which are suspected to cause oscillation in the tree.

10. Conclusion

This paper has presented several extended LAN management applications using the SNMP Bridge MIB on transparent bridges based on the IEEE 802.1D standard. The bridge, a basic building block in networks, can be used to manage and monitor the component LANs to which it connects. With these applications, the network manager gains greater visibility and control over the network, and is able to manage as well as resolve network problems in a timely manner.

11. References

1 Hawe, W., Kirby, A., and Lauck, T., "An Architecture for Transparently Interconnecting IEEE 802 Local Area Networks," Digital Equipment Corporation Technical Paper submitted to IEEE 802 Standards Committee, IEEE Document No. IEEE-802.85*1.96, Oct 1984.

2 IEEE Project 802 and Metropolitan Area Network Standards, "ANSI/IEEE Standard 802.1D-1990, MAC Bridges," IEEE, 1990.

3 Perlman, R., "An Algorithm for Distributed Computation of a Spanning Tree," Proc. 9th Data Communications Symposium, Montreal, 1984.

4 Case, J., Fedor, M., Schoffstall, M., and Davin, J., "A Simple Network Management Protocol," RFC 1157, May 1990.

5 Decker, E., Langille, P., Rijsinghani, A., and McCloghrie, K., "Definitions of Managed Objects for Bridges," RFC 1286, December 1991.

6 Backes, F., "Transparent Bridges for Interconnection of IEEE 802 LANs," IEEE Network, January 1988.

7 Rose, M., and McCloghrie, K., "Structure and Identification of Management

Information for TCP/IP-based Internets," RFC 1155, May 1990.
8 McCloghrie, K., and Rose, M., "Management Information Base for Network Management of TCP/IP-based internets: MIB-II," RFC 1213, March 1991.

III.

IMPLEMENTING MANAGEMENT ARCHITECTURES

C.

OSI MANAGEMENT - WORLDWIDE IMPACTS

Integrated Network Management, III (C-12)
H.-G. Hegering and Y. Yemini (Editors)
Elsevier Science Publishers B.V. (North-Holland)
1993 IFIP.

INCORPORATING OSI MANAGEMENT TECHNOLOGY INTO THE MARKETPLACE

L. Phifer[a] and P. Brusil[b]

[a]Bell Communications Research, 331 Newman Springs Road, Red Bank, New Jersey, 07701-7020, USA

[b]The MITRE Corporation (LOA), 35 Brackenbury Lane, Beverly, Massachusetts, 01915-3821, USA

Abstract
Mature ISO/CCITT system management standards and accompanying internationally accepted implementation and testing specifications have set the stage for wide spread introduction of OSI management technology into the marketplace. Procurement recommendations for several organizations and countries are adding impetus to the marketplace. Since its inception, many users and vendors have been committed to OSI systems management technology to provide a major piece of their strategic management solution. Accordingly, such users and vendors should now turn their attention towards orderly deployment of this technology into management environments currently dominated by proprietary management products. This paper examines the strengths of OSI systems management technology and indicates areas in which OSI management should be critically considered as part of the overall management solution. The paper assumes that OSI management technology must coexist with other management solutions, cooperating with and complementing other technologies. This paper identifies the criteria and conditions under which it makes best business sense to introduce OSI management technology.

Keyword Codes: K.6.0; K.6.4
Keywords: Management of Computing and Information Systems, General; System Management;

1. INTRODUCTION

Over the past decade, several industry-agreed management approaches have been developed. All these approaches, while defined fairly independently of each other, were developed to overcome common problems attendant with traditional management solutions. Such problems arose from the functional model approaches adopted by earlier, often proprietary, management solutions. In such earlier functional model approaches, the management protocol typically included a wide and ever growing variety of message types. Managing new resource types, or incorporating new management functionality, often required adding new message types to such a protocol.

Due to this common heritage, more recently-developed management approaches all have somewhat similar conceptual philosophies, and all have a certain amount of functionality and applicability in common. For example, recent management approaches are not based on a functional model, but rather have adopted the fundamental concept of an object model. An object model permits a small, well-defined number of operations to be performed on all managed resources. Protocols which utilize the object-oriented paradigm are designed to foster extensibility.

Today, three "object-oriented" management approaches have emerged with a broad base of industry support. Management protocols and objects developed by the Internet Engineering Task Force have focused primarily on management of TCP/IP-based networks. Object request broker technology developed by the Object Management Group (OMG) provides support for distributed client/server applications in a homogenous environment. OSI systems management developed by the ISO/CCITT community was designed with a broad range of resources and capabilities in mind, and can therefore span management of all kinds of technology resources in both OSI networks and non-OSI environments.

In light of the similarities among alternative contemporary management approaches, this paper focuses on the strengths and appropriate uses of OSI management technology. This discussion may be considered by network administrators and management system vendors as they make their own independent determination of how best to incorporate OSI management technology into their plans, products and procurements.

Section 2 of this paper begins with an overview of the OSI systems management model, its key components, and supporting activities, services, and tools. Some of the strengths of OSI management technology are presented in section 3. Section 4 identifies environments within which deployment and integration of this technology into to overall management solution may be best suited.

2. OVERVIEW OF OSI MANAGEMENT TECHNOLOGY

The OSI Systems Management Model described in CCITT X.701 (ISO/IEC 10040) [1] defines management as interaction between managing and managed systems. System administrators, network operators, and other end users interface with management applications to monitor, manipulate, and control managed resources. Managed resources are made visible to systems management as managed objects (abstract definitions that define the syntax and semantics relevant to management). For example, a network operator might configure a transport connection by using a management application to set values associated with a "transportConnection" managed object. In this example, the management application is taking on a manager role, sending a message to request the management operation (configure a connection). The system that receives this request is said to be taking on the agent role, and is responsible for interfacing with the managed resource to carry out and respond to the management operation. Many managed objects are defined to emit notifications when events occur pertaining to the managed resource (for example, a connection failure). When a notification occurs, the agent forwards event reports to the managing system or logs them locally, depending on configurable parameters set by the manager. This model is shown in Figure 1 below.

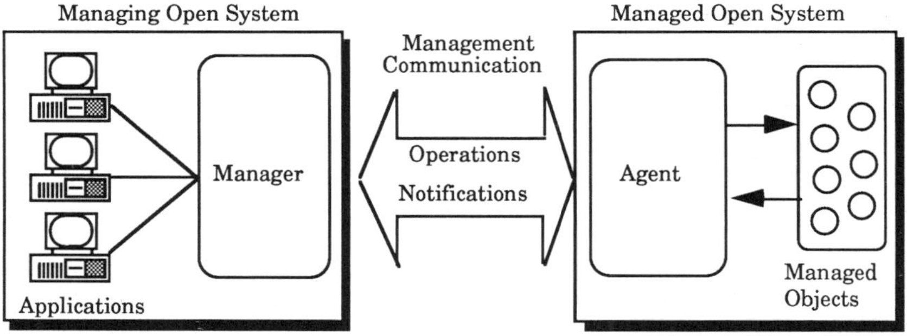

Figure 1. OSI Systems Management Model

There are several key components to this model.

Managed Objects encapsulate the attributes, operations, notifications, and behavior which represent resource properties made visible to management. In the OSI management information model [2], all managed resources that have the same properties are considered members of a common object class. A managed object class is defined using templates and ASN.1 syntax (the abstract data notation commonly used to define OSI protocols). This approach allows management applications to deal with many different resources through their common object class definition. For example, each vendor offers a different (usually proprietary) interface to configure networking software. Using OSI management, a management application can use the same standard managed object to perform this operation, avoiding the need to customize for each vendor interface. Common definitions may be reused in multiple managed object classes, so that common management applications can be developed to perform the same operation on many different kinds of resources which share the same properties.

Management Communication provides the protocol interface between the managing and managed systems. The OSI Common Management Information Service (CMIS) [3] defines the common procedures and parameters that can be used by management applications on all managed objects. CMIS includes the simple services found in other management approaches: a "get" service that can be used to retrieve management information, and a "set" service that can be used to configure managed resources. Unlike other management technologies, CMIS also defines "create" and "delete" services that can be used to invoke or discard instances of managed objects (for example, to update inventory when new equipment is added to or removed from an existing network), and an "action" service that can be customized to request any other type of management operation (for example, requesting an X.25 connection reset). Finally, CMIS includes reliable "event report" services that can be used to notify management applications of asynchronous events occurring about the managed resource. These common CMIS services can be supported by many different protocol stacks, as appropriate for the management environment. In an OSI network, CMIS is supported by the Common Management Information Protocol (CMIP) [4] over two common OSI application layer protocols for application association control and remote request/reply operations. Other defined protocol stacks include CMIP over LLC [17] and CMIP over TCP/IP using RFC 1006 [16] (replaces the earlier, now-deprecated approach identified by RFC 1189).

System Management Functions go beyond basic management communication to provide standard definitions for common management capabilities and control mechanisms that can be used by many management applications. System Management Functions build on services offered by CMIS to provide value-added functionality. For example, the State Management Function [5] defines several status variables that can be included in any managed object definition, allowing a single management application to be developed to monitor and control the status of any managed resource. Other functions define common event report formats, or "support managed objects" which control the management service itself. OSI management technology is unique among contemporary management approaches in defining such common capabilities and control mechanisms, thereby avoiding emergence or proliferation of application- and vendor-specific methods that provide functionally-comparable, but divergent or redundant, features. The set of System Management Functions is open-ended, allowing new, feature-rich functions to be developed over time as management applications become more sophisticated, as new requirements surface, and as the industry gains experience with this technology.

Figure 2 depicts a single management system, showing the relationship between these major components of the OSI management model.

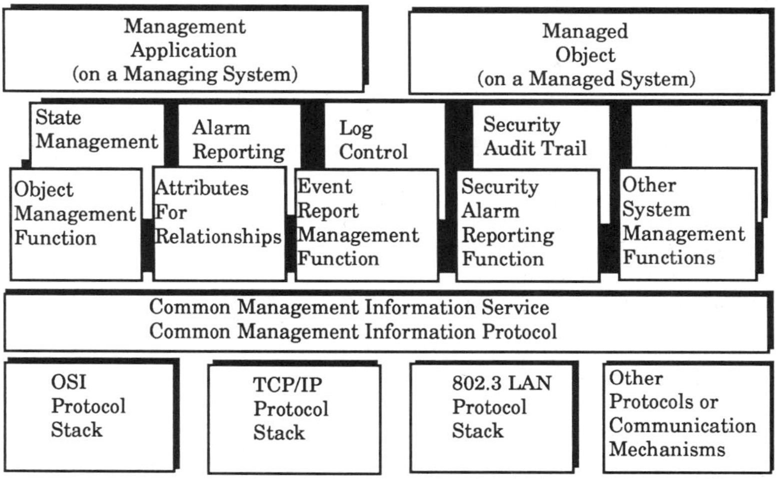

Figure 2. Management System Model

All of the features shown in Figure 2 are defined in existing, stable ISO/CCITT standards and recommendations. Internationally-agreed profiles developed by users and vendors help to ensure that implementations of these standards will interoperate with each other in a multi-vendor management environment. These profiles identify combinations of standards which should be implemented together, select options within those base standards, establish pragmatic limits, and clarify how differences in vendor implementation are to be handled without error. For example, a vendor offering CMIP for an OSI network can choose between two "flavors": basic management communication or enhanced management communication. These two flavors interoperate, at the basic level of functionality, ensuring that all CMIP implementations will be able to communicate with each other. These CMIP profiles were finalized in mid-1992 and provide a highly-stable foundation for OSI management. Additional draft function profiles are sufficiently stable to serve as the basis for implementation.

In order to communicate in a meaningful way, there must be agreement on the managed objects as well as the management protocol. Collections of standardized managed object class definitions are commonly referred to as "MIBs" (Management Information Bases). Many stable, standard MIBs are currently available to manage a wide variety of managed resources, including OSI lower layers [6-9], telecommunications networks [10-12], local area networks [13], security services [18], and even computing system resources [14]. Because the OSI management information model allows existing classes to be refined into new classes, new standard MIBs tend to build on existing MIBs, sharing many common properties and avoiding the wasted effort and redefinition associated with non-object-oriented definitions. This reuse will tend to preserve existing investment in management applications.

In addition to base standards, profiles, and MIBs, OSI management offers a number of infrastructure features which assist in deployment of this technology. These include:

Conformance and Interoperability Test Tools and Services: to help ensure that products faithfully implement the features they claim to support, and to help ensure useful interoperation in multi-vendor management environment.

Application Programming Interfaces: to provide not only a common abstract service interface (CMIS), but also a consistent programming language interface which allows management application portability.

Managed Object Catalog and Registration Services: to facilitate procurement, implementation, and definition of new managed objects.

Open Management Roadmap: represents a working arrangement among numerous standards-making bodies, user groups, and consortia to support specified sets of base standards, profiles, APIs, test tools, and procurement guides which function as a common basis for implementation of OSI management technology. This helps to ensure that the separate piece parts provided by each Roadmap participant will all interwork together to form a composite, total systems management solution intended to meet user requirements. The first of these sets, called OMNIPoint 1, was published in August 1992.

OSI management technology deployment has been slow. Many factors have contributed to this, including the persistence of legacy management systems in a soft economy, competition from other standard and proprietary management technologies on the market, the delayed emergence of consistent stable OSI management standards (no longer a factor), and the limited availability of OSI-based agents to be managed (this factor should diminish over time). As of mid-1992, IBM SNA/NetView continues to be the dominant management technology ($1040 million), trailed by other proprietary technology ($690 million) and standard management technologies like Internet management ($350 million) and OSI management ($20 million) [15]. Insight Research Corporation predicts that protocol converters and parallel stacks will be offered over the next 2-5 years, culminating in a $4.6 billion network management market split 2:1.5:1 between SNA, OSI, and Internet technologies [15]. Growth of the OSI management market can be attributed partly to stimulation provided by major user groups in the US [19, 20], Europe [21], and the Far East [22]. As noted in the Insight report, "it is clear that the vendor that manages a customer's enterprise network will have a competitive advantage and thus maintain a good deal of account control [and] there is a substantial market for vendors to control." Given this, it seems wise to consider carefully strengths that each management technology has to offer.

3. OSI MANAGEMENT STRENGTHS

There are numerous management problems for which OSI management technology may be considered as the technology of choice. The following paragraphs describe the major strengths of OSI management technology, and identify the criteria and conditions under which it is advantageous to introduce and to integrate OSI management technology into the overall management solution.

Aggregate Cost Savings for Application Developers:
OSI management technologies will provide a set of common, sophisticated tools (most notably the system management functions described previously) which go beyond the basic, raw management communication services offered by other management technologies. Since such tools will be implemented and tested once (by the OSI management vendor) rather than by each management applications vendor, there is cumulative downward pressure on the total cost of the OSI-based management system (i.e., the aggregate cost of management application programs and management communication services).

Tailorable Management Operations:
OSI management standards have been bundled into standard packages (called profiles) which allow significant management implementation flexibility. When needed, a basic management communication profile can be used to support only a simple set of management operations (akin to simple

Internet solutions). In other situations, the enhanced management communication profile might be used to offer a richer set of management operations. Additional system management function profiles are available to support broader management capabilities such as alarm reporting, or management controls for event reporting and logging. This small, carefully-chosen set of well-understood profiles strikes a balance between flexibility (scaled solutions to reduce cost) and minimizing the possible permutations (to facilitate interoperation, implementation, and procurement).

Advanced Application Functionality:
Unlike other management approaches, OSI management technology includes a set of standard tools to support common management application needs such as reporting of status and fault information. These basic standard tools enable development of common applications which can manage any resource. For example, OSI management includes an alarm report which uses the same message format for any fault; Internet management requires custom application code to deal with enterprise-specific trap messages; OMG request brokers do not support asynchronous event signaling at all. OSI management technology also includes more advanced standard tools that can be used in multi-vendor environments to support more sophisticated management applications, such as resource performance monitoring, remote diagnostic test execution, and resource usage accounting. Most applications which truly manage (i.e., control rather than simply monitor resources) require this level of advanced functionality.

Support For Management Between Administrative Domains:
OSI management technology includes many features which lend themselves to use between administrative domains, including global X.500-based naming of managed resources, manager to manager communication (through use of dual manager/agent roles), and mechanisms to negotiate and share knowledge about the management capabilities and objects implemented in the different management systems that wish to interoperate. Other management technologies currently rely on locally-unique naming and strict manager/agent or client/server roles, and provide no standardized mechanisms for negotiation or exchange of management capabilities.

Real Time Control of Management:
OSI management is an extremely important technology in situations where management systems must be able to control their own capabilities. Specifically, in addition to being able to manage and control remote resources and services, OSI management includes common, consistent interfaces and controls for enabling/disabling management services such as event distribution, remote test execution, different types of performance monitoring, probes, and security audit trails. By offering real-time management controls as standard tools, devices sold by different vendors can be managed by a single management system without requiring

customization. Without standard tools, control of management in a multi-vendor environment may not be possible at all, or may require costly custom extensions to the management system, the managed device, or both.

Flexible Distribution of Management Functionality:

Other management approaches support only a fixed distribution of management functionality which places the processing burden exclusively on the manager in order to simplify the agent and therefore the managed device. OSI management can be deployed in this simple agent manner, but has been designed with greater flexibility. In some situations, it is useful, perhaps even crucial, to distribute management functionality in other ways. Examples include hierarchical management (where an agent can be an element management system), management of powerful devices (such as computing systems and telephone switching systems), and automated management (where the managed device is capable of self-management to some extent, and remote management is the exception rather than the rule). OSI management technology provides the network designer with the ability to distribute management functionality as best suits his or her business needs, management policies, and/or security policies.

Real Time Problem Detection:

In certain environments and situations, resource monitoring and accompanying control actions are extremely time-critical and are required to stop the spread of faults or security breeches, to prevent or isolate significant resource catastrophes, and/or to constrain or reduce the costs of resource repairs. Powerful OSI management event reporting and control mechanisms are available for real time announcement of resource problems or resource changes. These tools include user-settable criteria to trigger the detection and reporting of problems. Such event reporting mechanisms are especially desirable in situations where low processing or communication bandwidth is available for resource management. Internet management products often use continual or directed polling to identify problems. This approach sometimes results in delayed detection of problems, missed detection of intermittent problems leading to sustained failure, or increased resource usage. OSI management has the flexibility to be deployed in either a polling mode or an event-driven mode, as appropriate for the situation.

Efficient Search and Selection Tools:

In many situations, management applications need to apply operations on ad hoc sets of resources meeting specified criteria (e.g., disconnect from a specified communications service provider all routers that display lengthy queues). In other cases, management applications often need to search through the managed network or system to find management information meeting specified criteria (e.g., a list of all network attachment boards that have been in service for a specified period of time or have been fielded before a specified date, and have had a large number of service calls and are not installed in workstations from a specified manufacturer). OSI management provides very powerful so-called "scoping and filtering" constructs to support such application needs. These constructs allow conditions affecting a management operation to be evaluated locally by the agent system. Other management approaches require the manager to first retrieve all relevant data from the agent, evaluate conditions remotely at the manager system, and then request the operation using a separate message for each selected object. OSI scoping and filtering constructs allow sophisticated management operations to be accomplished in a single manager/agent interaction, relieving the management application from detailed record keeping associated with less powerful tools. Scoping and filtering can also be used to reduce the network bandwidth required to accomplish complex management operations. By having such sorting tools implemented once and making them generally available to all management applications, the costs of many management applications is reduced, and therefore the cost of the total management system is reduced.

Highly Reliable Management Operations:
OSI management technology should be seriously considered when it is necessary to have efficient, reliable management operations with high integrity. OSI management operations can be exchanged via connection-oriented communication services that detect loss of management message exchange, prevent replicated execution of the same management operations, and preserve the order by which management operations were requested to be performed. Such reliability is especially important when management is engaged in critical control activities wherein lost control actions, repeated control actions, or incorrectly ordered control actions can have catastrophic consequences to the system of resources being managed. (When deployed in a local area network environment where the reliability provided by a connection is not needed, OSI management can also be used in a connectionless mode.)

Reusable and Extensible Management Applications:
OSI management technology allows easy MIB expansion without requiring management application code changes. This extensibility is accomplished through a technique known as "allomorphism". New features are simply added to existing MIBs, generating new, refined object classes that are said to be "compatible" with pre-existing classes. This technique allows a new resource to be managed as though it were an existing, known resource. Allomorphism facilitates vendor extension and introduction of new versions in a controlled manner. Management applications which are designed to take advantage of this technique may be able to simultaneously manage both new and old versions of a given device without modification, or may be able to make use of vendor extensions without inhibiting interoperability.

Alternative Configurations:
Different profiles of OSI management capabilities allow different sizes of management implementations to be built to match the management situations at hand. Small (20 KB) implementations of just the simple profile of basic management operations, or of the IEEE 802.1b OSI management stack [17], can be used in situations where extremely limited

computational/memory resources are available to support management. More complex management implementations need only be used in situations where availability of computational/memory resources are not at a premium. In addition, OSI management technology can also be deployed over alternative transports, in order to take advantage of backbone networks already in place.

4. TARGET OSI MANAGEMENT ENVIRONMENTS

The following paragraphs highlight some of the environments in which OSI management technologies may be best suited. This discussion is not intended to preclude other management technologies that may also be applicable, or to imply that OSI management is only applicable to the areas identified here. (In fact, the general approach used by OSI management allows this technology to be used to manage any information technology, resource, service, or environment.) Rather, the discussion is intended to provide guidance regarding areas in which OSI technology is commonly considered a major part of the overall management solution.

Management of OSI Applications and Networks:

It is perhaps obvious that OSI management technology is expected to be used to managed OSI stacks and OSI applications such as X.400 Message Handling, X.500 Directory Service, and the like. An analogy can be drawn to Internet management, which is commonly used to manage TCP/IP stacks and associated applications.

Management of Telecommunications Networks and Services:

The telecommunications industry has strongly embraced OSI management technology. CCITT Recommendations use OSI management protocols for management communication between operations systems and network elements or mediation devices; this is known as the "q3" interface. Most CCITT management information models have been developed using the OSI information model, taking advantage of sophisticated system management functions offered for fault, performance, and accounting to manage transmission technologies such as synchronous digital hierarchy (SDH) [12].

Manager of Managers:

OSI management technology is well-suited for use in hierarchical management environments, where greater complexity and distribution of management functionality is required. For example, OSI management technology might be used by a network control system to interact with element management systems. Element management systems may interact with the actual managed devices using a mixture of management protocols (OSI, Internet, request broker RPC, proprietary) which are native to the local environment.

Management of Heterogeneous Networked Systems:

Information modeling of system resources such as printers, hosts, users, operating systems, and the like is currently underway in several standards bodies (e.g., IEEE POSIX) and consortia (e.g., Open Software Foundation, Unix International). To date, this work has been based on the OSI management information model. There has been a recent shift in emphasis towards the OMG Common Object Request Broker Architecture (CORBA) model for use in systems management, primarily in a homogenous environment where a single object request broker would exist. OSI management is expected to remain a significant factor in management of networked systems in a heterogeneous environment.

Secure Management Environments:

Most industry-agreed management technologies suffer somewhat from lack of security features, and OSI management technology is not exempt from this problem. However, OSI management technology is well-positioned to take advantage of advances made in OSI network security technologies, and may therefore be viewed as a more viable management solution in environments which require a high degree of management security. For example, NATO and the US Department of Defense have chosen OSI technology to manage aspects of security services and mechanisms used in networks (e.g., the DOD Secure Data Network Systems program, work underway in IEEE 802.10).

Enterprise Management:

Many organizations (private industry, public service providers, government entities, branches of the military, etc.) have used OSI management in modeling end-to-end business flows and enterprise-wide management. Typically, these organizations make use of whatever standardized management capabilities exist, supplementing them with enterprise extensions to meet specific business needs. This process represents significant investment in information and process modeling, and may therefore be a business driver for deployment of OSI management technologies.

Management of Large, Highly-Volatile Environments:

Many of the strengths of OSI management technology described previously lend themselves to deployment in management environments which are very large (many resources and devices to be managed, perhaps using hierarchical management domains), especially where control and not just simple monitoring is essential. Additionally, the strengths of OSI management technology lend themselves to deployment in management environments that are very volatile (numerous or frequent changes occurring in management topology, addition or deletion of managed devices, etc.), such as in commercial mobile phone or tactical military environments.

Possible OSI management environments are expanding as MIBs which have been in the development process are now being published at an explosive rate. These new MIBs span a large number of diverse resources such as electronic mail services, computer operating systems, telecommunications transmission equipment, distributed computing environments, network security services, print management services, data communications devices, and enterprise management services. Public MIB availability is key to successful deployment of any open management technology.

5. SUMMARY

Like other contemporary management technologies, OSI management has been designed to solve traditional management problems. However, OSI management's heavy reliance on object orientation positions this technology to provide tremendous flexibility with minimal incremental costs. Delayed completion of OSI management standards can be attributed partially to the time spent architecting a comprehensive management model that is tailorable, simple to extend, provides for flexible distribution of functionality, and maximizes reuse. These characteristics facilitate deployment of products that are easily scaled (up or down) or otherwise adaptable to evolve with changing user needs.

OSI management technology has the flexibility to be used in the simple-agent mode prevalent in other popular approaches, as well as to support more sophisticated needs that are now beginning to emerge as experience with early management solutions reveal additional pressing requirements. Numerous environments can benefit from the strengths of OSI management. Target environments span a broad range of resource domains, but all exhibit a common need for sophisticated, flexible management solutions.

The cost of management flexibility had been the subject of considerable scrutiny during the development of OSI management technology. The solutions which have emerged strike a unique balance between malleability and cost, where cost containment focuses on the cost of the total management system, including the cost of management applications that ride atop OSI management technology. This approach facilitates centralized implementation of many common management tools that don't need to be re-implemented in each and every management application. OSI management technology also preserves investment in management applications by providing a stable environment in which new functions, new resources, and extensions can be added easily and dynamically discovered.

User demand for OSI management technology is increasing world-wide [19-22]. Stable, complete, and consistent standards, profiles, and infrastructure services are now available [1-14]. These enabling factors are likely to stimulate vendor deployment of OSI-based management products to meet market demand, estimated to reach $1.4 billion by 1996 [15]. The strengths and target environments explored in this paper are intended to provide valuable insight for those organizations and individuals who now face the challenge of incorporating OSI management technology into the marketplace.

6. REFERENCES

1. CCITT Recommendation X.701 (1992) | ISO/IEC 10040: 1992, Information Technology - Open Systems Interconnection - Systems management overview.

2. CCITT Recommendation X.720 (1992) | ISO/IEC 10165-1: 1992, Information Technology - Open Systems Interconnection - Structure of management information: Management information model.

3. ISO/IEC 9695: 1991, Information Technology - Open Systems Interconnection - Common Management Information Service Definition. CCITT Recommendation X.710 (1991), Common Management Information Service Definition for CCITT applications - General concepts.

4. CCITT Recommendation X.711 | ISO/IEC 9596-1: 1991 (E), Information Technology - Open Systems Interconnection - Common Management Information Protocol Specification - Part 1: Specification, Edition 2.

5. CCITT Recommendation X.731 | ISO/IEC 10164-2: Information Technology - Open Systems Interconnection - Systems Management - Part 2: State Management Function, ISO/IEC JTC1/SC21 N6356, October 15,1991.

6. ISO/IEC 10737-1: 1992, Information Technology - Telecommunications and Information Exchange between Systems - Elements of Management Information Relating to OSI Transport Layer Standards.

7. ISO/IEC 10733: 1992, Information Technology - Telecommunications and Information Exchange between Systems - Elements of Management Information Relating to OSI Network Layer Standards.

8. ISO/IEC 10589: 1992, Information Technology - Telecommunications and Information Exchange between Systems - IS-IS Routing Protocol Specification.

9. CCITT Recommendation X.721 (1992) | ISO/IEC 10165-2: 1992, Information Technology - Open Systems Interconnection - Structure of management information: Definition of management information.

10. Network Management Forum: Forum 006, Forum Library - Volume 4: OMNI*Point* 1 Definitions, Issue 1.0, August 1992.

11. CCITT Recommendation M.3100, Generic Network Information Model, 1992.

12. CCITT Recommendation G.774, SDH Management Information Model for the Network Element View, 1992.

13. IEEE P802.3K, CSMA/CD Access Method & Physical Layer Specifications, Layer Management for Hub Devices.

14. NIST Special Publication SP 500-202, Stable Implementation Agreements for Open Systems Interconnection Protocols, Version 5, Edition 1, December 1991, including change pages from March 1992 and June 1992 Open Systems Environment Implementors' Workshop (OIW).

15. "Worldwide OSI Network Management Still A Few Years Away", Open Systems Communication, Transmission #130, Phillips Publishing, July 27, 1992.

16. RFC 1006, ISO Transport Services on Top of the TCP: Version 3, Rose, M.T.; Cass, D.E., May 1987.

17. IEEE 802.1B, LAN/MAN Management, January 27, 1992.

18. CCITT Recommendation X.740 (1992) | ISO/IEC 10164-8: 1992, Information Technology - Open Systems Interconnection - Systems Management - Part 8: Security Audit Trail Function, ISO/IEC JTC1/SC21 N7039, June 2,1992.

19. US Department of Commerce, Version 1, Government Network Management Profile (GNMP), Federal Information Processing Standard 179, November 30, 1992.

20. US Department of Defense, Military Standard: Network Management for DoD Communications, MIL-STD-2045-38000, 17 December 1992.

21. UK CCTA, GOSIP OSI Management Subprofile 4.1, Supplier Set and Purchaser Set, January, 1993.

22. Japan Inter-Ministerial Council of Secretary Generals of all Ministries, Standard for Introduction and Utilization of Open Systems Interconnection (OSI) in the Government, December 9, 1991.

OSI Management Technology

Panel Chair: John Pickens, 3com, USA

Panelists: Thomas Rutt, AT&T, USA
Charles Moore, Telecom Australia, AUSTRALIA
Gary Haney, Martin Marietta, USA
Joseph Ghettie, Bellcore, USA
Elizabeth Adams, OSI Network Management Forum, USA
Fran Nielsen, National Institute of Standards and Technology, USA
Yoshimi Teshigawara, NEC Corporation, Japan
Isabelle Valet-Harper, Digital Equipment Corporation, USA
John Westgate, CCTA, United Kingdom
Delbert Brashares, US Airforce, USA
Allen Cressy, The MITRE Corporation, USA

OSI Management Technology is entering the market place from the combined world-wide efforts of several users and vendors working cooperatively toward a highly flexible and powerful management solution. Procurement recommendations of several organizations and countries are adding impetus to the marketplace. Throughout its inception, many users and vendors have been committed to OSI systems management technology to provide a major piece of their strategic management solution. Accordingly, such users and vendors should now turn their attention to orchestrating orderly introduction of this technology into field environments dominated by proprietary and Internet-based management products.

An unprecedented panel of renowned international experts, leaders and consumers representing all phases of the development, introduction and acquisition of this technology will provide insight to its emerging strong market place popularity.

III.

IMPLEMENTING MANAGEMENT ARCHITECTURES

D.

INTEGRATING OSI/INTERNET MANAGEMENT STANDARDS

Integrated Network Management, III (C-12)
H.-G. Hegering and Y. Yemini (Editors)
Elsevier Science Publishers B.V. (North-Holland)

CMIP versus SNMP

Michael Gering

IBM Corporation
P.O. Box 12195
Research Triangle Park, NC 27709

Abstract This paper compares and contrasts the management schemes defined for the TCP/IP and OSI environments, taking into account customer requirements, functional capabilities, and implementation considerations for both TCP/IP and OSI environments.

Keyword Codes: C.2.2, C.2.3, C.2.4
Keywords: Network Protocols, Network Operations, Distributed Systems

1. Introduction

Section 2 discusses some customer network management requirements. Section 3 compares CMIP and SNMP in terms of their information structuring mechanisms, naming and data representation. Section 4 presents a functional comparison, and section 5 compares them from a communications perspective.

In the remainder of this paper, the term *SNMP* will refer, for the most part, to the collection of protocols and services used in managing TCP/IP networks, including SNMP, SMI [1], UDP, and the MIBs. Similarly, the term *CMIP* will refer collectively to CMIP [2], ROSE, ACSE, System Management Functions, SMI [3], and OSI managed objects.

2. Customer Requirements

Today's environment for systems and network management is characterized by these dominant factors:

- Multi-vendor
- Multi-enterprise
- Diversity of types and numbers of resources

The strongest requirements arising from these factors include:

- Integrated management systems with consistent end-user and application interfaces
- Common data definitions and the ability to share data between applications
- Reuse of code to support similar equipment from different vendors
- Communications mechanisms to distribute and coordinate management tasks between different management systems
- Security mechanisms to control the exchange of management information

Consistency is the key to reducing cost and complexity of network management and is the goal for integration of any aspect of a general system.

Although CMIP and SNMP do not address directly the areas of end-user or application programming interfaces, they provide a basis for these areas with their information structure and communications features.

3. Information Structure

This section discusses the differences between the information structures provided for by SNMP and CMIP. Information structuring mechanisms are needed to represent management information about the resources being managed. For example, what information is needed to represent the properties associated with modems, how is that information packaged, how does it relate to other packages, how do we identify them, and how are they communicated between managers and agents?

The following sections discuss information structuring methodology, naming, and data representation.

3.1 Information Structuring Methodology

Network and systems management is a problem of mastering an immense and open-ended amount of diversity and complexity. The motivation behind an information structuring methodology is to bring order to this chaos in ways that:

- Provide tools to model objects in the real world
- Reduce complexity by exploiting natural similarities in the real world
- Provide a clean boundary between the implementation of information from the way it is accessed or used
- Promote reuse of specifications and implementation

CMIP and SNMP use the term "object" to represent something that can be manipulated with their management protocols. Object-oriented methodologies provide these mechanisms to accomplish the above goals: encapsulation, classes, inheritance, and polymorphism. The following sections describe how these object-oriented methodology concepts apply in the CMIP and SNMP worlds.

3.1.1 SNMP Object Oriented Features

The SNMP SMI. specifies the rules and guidelines for names, data syntax, and notation for managed objects. The various MIBs are defined according to the SMI.

The term *variable* is often used interchangeably with *object* in this model because each object represents a single, atomic data element. An example variable is *ifAdminStatus* which is used to represent the up/down/testing status of a physical interface.

3.1.1.1 Encapsulation

SNMP does not provide a formal mechanism for encapsulating related variables. The most serious drawback is that implementers have no formal guidance for how consistency between variables should be maintained. Variables are separately accessible and modifiable. Although the SNMP protocol allows multiple *GET* or *SET* requests to be conveyed in one protocol data unit, it does not require that the requests be processed in order or that they have any relationship with each other.

SMI allows variables to be aggregated in lists and tables. This is not a suitable encapsulation mechanism since the SNMP protocol has no list or table oriented operations, i. e. a manager cannot operate on a list or table as a whole. SMI provides no formal mechanism for specifying that the variables in a list or table are related in any special way, e. g. that they together represent a managed resource.

The SNMP *SET* operation allows side-effects. For example, setting a variable *powerStatus* to zero might have the effect of causing a piece of equipment to power down. This enables a variable to encapsulate behavior. However, the utility of this feature is significantly reduced because the variables are limited to unstructured syntaxes (see 3.3); there is no structured way to set a variable to a list of parameters. An alternative might be to define one variable per parameter and use multiple *SET* operations to effect the desired behavior. However, this introduces problems of consistency, integrity, and synchronization, which are contrary to encapsulation.

3.1.1.2 Classes

The object class concept analog in SNMP is the MIB variable or object *type*. The SMI defines the rules and guidelines for MIB variable types and a formal notation for specifying them. A MIB variable type includes:

- A unique identifier for the type (an ASN.1 *OBJECT IDENTIFIER*)
- A syntax for the type (see 3.3)
- Access - read-only, read-write, not-accessible
- Status - mandatory, optional, obsolete

One of the principle advantages of the class concept in object-orientation is its recognition and expression of similarities between instances of things being modelled. This advantage is negated by SNMP because it intertwines class and instance identification; every MIB variable instance name includes its class name, and the structure of instance names depends on how the data are aggregated in the MIB. This implies a variable type definition cannot be reused in different parts of the MIB to represent similar things. For example, if we define a MIB variable that keeps count of the number of protocol errors, we might want to use different instances of that variable type to count token ring protocol errors, session protocol errors, etc. Unless all these protocol error counters are structured in the same list or table, they cannot share the same class definition. Instead, we would have to specify separate MIB variable types for each of the protocol error counters. The fact that they share many semantics is not formally specified, and the similarities are lost to any managing user or application.

3.1.1.3 Inheritance

SNMP does not support inheritance. The principle advantages of inheritance are specification reuse, and by extension, code reuse. SNMP MIB variable specifications are not reused for other MIB variables. An implementation consequence is that manager application logic depends uniquely on each MIB variable type. The manager application must be changed to specifically accommodate each new MIB variable.

3.1.1.4 Polymorphism

SNMP defines only these polymorphic operations on variables:

- *GET* to retrieve the value(s) for one or more variables
- *GET-NEXT* to retrieve the next value(s) for one or more variables
- *SET* to set the value(s) for one or more variables

SNMP defines another message, *TRAP*, which is not related to any variable, but to the system as a whole.

SNMP handles migration by obsoleting old variables types and defining new ones. SMI provides no specification mechanism for relating a new variable type to one it obsoletes, nor does SNMP provide a protocol mechanism to carry such information.

3.1.2 CMIP Object Oriented Features

OSI management defines an object-oriented paradigm in the Management Information Model standard. The following sections describe the object-oriented concepts as they pertain to OSI management, and contrast them with the SNMP counterparts.

3.1.2.1 Encapsulation

OSI management provides a formal means of encapsulation. OSI managed objects encapsulate behaviors, attributes, operations and notifications at an object boundary. The object boundary is a formal model for specifying aspects of the managed resource which are visible to management processes. Aspects of a managed resource with are inside the object boundary are invisible to the management process. A managed object class specification defines the encapsulated behaviors, attributes, operations, notifications, and integrity and consistency constraints for instances of that class.

One key difference between CMIP and SNMP objects is that a single CMIP object may model a complex resource. For example, CMIP objects may contain attributes to model properties of that resource, whereas an SNMP object models only one property of a resource. From this perspective, SNMP objects are like CMIP attributes, and the CMIP object has no counterpart in SNMP.

3.1.2.2 Classes

OSI management provides for object classes. Each object instance is a member of one class. The specification for an object class uses a formal notation, GDMO. [4]

As with SNMP MIB variables types, OSI managed object classes are assigned globally unique identifiers. However, because OSI managed object instance identification is independent of OSI managed object class identification, instances of the same class may occur in different places of the naming schema.

Class component specifications for attributes, actions, and notifications are assigned unique identifiers which are independent of each other and of the class identifier. This permits reuse of attribute, action and notification specifications in many object classes. A manager application can take advantage of the fact that these identifiers are global and CMIP carries them in the protocol by building a knowledge base of attributes, etc. Presented with an object class it does not understand, the manager may still understand much about the class's behavior because it already knows about the class's parts.

3.1.2.3 Inheritance

OSI management permits inheritance as a means for defining object classes which are extensions or refinements of other object classes. A subclass inherits the properties (attributes, actions, notifications, and behaviors) of its superclasses and may add additional properties. OSI management requires *strict inheritance* which means no inherited properties may be suppressed or redefined. OSI management permits *multiple inheritance* which allows a subclass to have more than one superclass. Because these properties are assigned global identifiers and strict inheritance is required, multiple inheritance does not introduce ambiguity for properties which are inherited from more than one superclass.

3.1.2.4 Polymorphism

OSI management includes provisions for a restricted form of polymorphism, called *allomorphism*, which permits an object instance of one class to be managed as though it were an instance of another class. Usually, allomorphic classes of an object are also superclasses of that object; however this is not a requirement. Allomorphic relationships are modelled in OSI managed objects and may be negotiated and exchanged with CMIP.

The primary benefit of allomorphism is to aid migration and co-existence for managers and agents of different versions and capabilities.

3.1.3 Summary of SNMP and CMIP Information Structuring Methodologies

Table 1 summarizes the features of SNMP and CMIP.

Table 1. CMIP and SNMP object-oriented features		
Feature	**SNMP**	**CMIP**
Encapsulation	single, atomic data only	attributes, actions, notifications
Classes	Not reusable	reusable
Inheritance	none	multiple, strict
Polymorphism	GET, GET-NEXT, SET, as defined by SNMP	GET, SET, ACTION, EVENT, and as specified by the object class

3.2 Object Naming

Object instance names identify the targets and sources, of management operations and notifications. CMIP and SNMP differ markedly in the way they name object instances. The naming techniques are described below, but the most important differences are scope of uniqueness and flexibility.

The scope of name uniqueness is important because it determines where names may be used without the possibility of misinterpretation due to ambiguity. A name is ambiguous if it refers to more than one object instance in a context where the name is used. Note that this does not disallow the possibility of an object possessing more than one name. In management, a manager and agent both must understand and agree on the naming context where ambiguity is not allowed. For enterprise management, global naming is a requirement since the resources to be managed generally cannot be restricted to a predetermined set of contexts. For example, it is unreasonable for an enterprise manager to understand only SNA names and require that all managed resources in the enterprise be assigned SNA names. It is more unreasonable to require that other enterprises assign SNA names to the resources about which they communicate to the enterprise manager.

3.2.1 SNMP Variable Naming

The structure of SNMP variable names are defined in the SMI standard for TCP/IP. SNMP variable names are hierarchical, untyped, and based on the ASN.1 [5] object identifier. Each variable name uses the ASN.1 object identifier syntax of the form $\{x\ y\}$, where x and y are sequences of ASN.1 object identifier components. The first part x (the prefix) is assigned and fixed at the time the variable type is defined. The second part y (the suffix) is assigned and fixed when the variable is instantiated in the MIB.

SNMP variable names are only unique within a system; the same variable name is reused to instantiate variables in different systems. To provide for global naming, variable names must be qualified with globally unique system names. However,

SNMP makes no provision for carrying this additional information. One limitation this imposes is that one SNMP manager has no defined way to direct a request to another SNMP manager to perform an operation on a MIB variable in an agent.

The structure of the name for an SNMP variable is fixed when the variable is architected. The only means for a customer to change this structure is to define a new variable type, with a unique name. Because the class of the variable is incorporated into its name and SNMP does not carry the class information separately, a customer who does this must also change the managing applications to recognize the new variable class.

The X.500 directory standards [6] do not permit directory objects to be named with ASN.1 object identifiers. Therefore, the management system is prevented from accessing a standard directory with SNMP variable names to obtain information about a variable.

3.2.2 CMIP Object Naming

The OSI management information model defines the naming mechanism for managed object instances based on the X.500 *distinguished name*. The names are hierarchical, typed, and global. Because they are compatible with the X.500 Directory, management applications may, but are not required to, use a standard directory service to obtain other useful information about resources. The hierarchical nature of OSI managed object names is defined by a naming tree where each node represents an object instance. The name of each instance is relative to its superior's name. The global name for an instance is formed by concatenating the relative names of each of the object instances in the path from *Root* to the instance.

OSI management uses typed names because the relative distinguished name for an object instance is formed by the combination of one of its attribute types and the associated attribute value. Typed names permit greater flexibility than untyped names for vendors and customers. Because attribute types have global identifiers, vendors can define naming structures for their managed object classes without worrying about ambiguous names for object instances. At the same time, customers are provided flexibility in being allowed to assign attribute values for each instance.

Because object names are typed, hierarchical and allow different ASN.1 syntaxes, they require more octets than SNMP for their encodings.

3.2.3 Naming Summary

Table 2 summarizes the important differences between SNMP and CMIP naming.

Table 2. CMIP and SNMP naming differences	
SNMP	**CMIP**
Unique only within a single system	Globally unique
Based on class name at specification time	Independent of class name, set at instantiation
Untyped	Typed
Incompatible with X.500	Compatible with X.500
Compact encoding	Lengthy encoding

3.3 Data Representation

Distributed management requires exchanging data between managers and agents to perform operations on and to convey information about managed resources. Both CMIP and SNMP use a standard notation, Abstract Syntax Notation One for specifying management data, and a related standard, Basic Encoding Rules, for encoding the data for transmission.

CMIP places no restrictions on which ASN.1 data types may be used to represent the syntax of object attributes, notifications, and actions. The designer of an object class may choose any of the twenty-six appropriate data types; an implementation must accommodate only the data types needed for the object classes it supports.

However, SNMP purposefully limits the ASN.1 data types which a MIB designer may use to four: *INTEGER*, *OCTET STRING*, *NULL*, and *OBJECT IDENTIFIER*. On the positive side, an SNMP agent need only provide the allowed subset of ASN.1 data types. On the negative side, overloading these data types with semantics for more complicated data implies the manager and agent must agree on the encoding; they cannot take advantage of the self-describing data characteristics provided by ASN.1 and BER. This trade-off between simplicity and flexibility is mostly subjective. However, SNMP does not provide for some common requirements.

Some ASN.1 data types are particularly useful in addressing international and inter-enterprise management requirements:

- Character Strings
- Time
- *SEQUENCE*
- *SEQUENCE OF*
- *SET*
- *SET OF*

ASN.1 provides for a variety of character sets, ranging from a minimal set, *PrintableString*, to to most general set, *GeneralString*. In SNMP, character strings would be encoded as *OCTET STRING*. Since *OCTET STRING* carries no explicit identification of the character set it conveys, the manager and agent must understand it *a priori* on a variable by variable basis.

Similarly, ASN.1 defines two representations for time, *GeneralizedTime* and *UTCTime*. SNMP does not have a standard method of encoding time values.

ASN.1 provides data types to structure data. The *SEQUENCE* data type is used to aggregate data which are fixed in number and perhaps of different types. *SEQUENCE OF* is a variation which allows a variable number of elements, each element being of the same type. The *SET* and *SET OF* data types are like their counterparts *SEQUENCE* and *SEQUENCE OF*, except the ordering of the elements is unimportant.

SNMP allows data to be aggregated in the form of tables. In fact, every table is defined as a *SEQUENCE OF Entry*, and *Entry* is defined as a *SEQUENCE*. However, the SNMP protocol does not allow tables or table entries to be transferred as single units. Instead, the manager application must operate on each column or entry separately. Although this limitation simplifies the agent's task of collecting and packaging data, it complicates the manager's logic because the manager must combine multiple, related variables in different operations. In practice, this also prevents the manager and agent from synchronizing the data and ensuring their consistency.

4. Management Functions

A crucial aspect of developing integrated management applications is the creation of functions common to managing a wide range of resource types. CMIP and SNMP differ considerably to the extent that they enable and provide common management functions.

The only functions which SNMP defines common to different resource types are the primitive management operations *GET*, *GET-NEXT*, *SET*, and *TRAP*. In addition to the primitive management operations defined by CMIS, *GET*, *SET*, *ACTION*, *NOTIFICATION*, *ADD*, *REMOVE*, and *CANCEL-GET*, a number of common functions have been defined and more are in development. These common functions build on CMIS and define attributes, actions and notifications which can be incorporated into resource-specific object class definitions.

The common functions, called system management functions, are defined as separate parts to ISO/IEC 10164. Currently, these system management functions have been completed or are in the process of being standardized:

- Object Management
- State Management
- Attributes for Representing Relationships
- Alarm Reporting
- Event Reporting
- Logging
- Security Alarm Reporting
- Security Audit Trail
- Objects and Attributes for Access Control
- Accounting Meter Function
- Workload Monitoring
- Test Management
- Measurement Summarization

5. Underlying Communications

CMIP and SNMP make minimal demands on underlying communications services. One key difference between the two is that CMIP requires a connection-oriented service and SNMP requires a connectionless service. Another key difference is that SNMP assumes a polling paradigm, whereas CMIP assumes an event-driven paradigm. The implications of these differences are discussed below.

5.1 SNMP Communications

The SNMP protocol operates over a connectionless service provided by the user datagram protocol (UDP) and the internet protocol (IP).

The size of a UDP packet is limited by the operational characteristics of the underlying internet routing network. The IP standard requires implementations to be able to relay, under normal conditions, packets of 484 octets or larger. Therefore, a sending system may assume, under normal conditions, packets of 484 or *fewer* octets will be received by the destination system.

As a connectionless service, IP and, by extension, UDP provide no indication to either the sender or receiver of packets lost or discarded by the network. Packets may be lost or discarded for a variety of reasons, including:

- Receiver unavailable
- Packet too large to be relayed by an intermediate system

- Network congestion
- Loss of connectivity

Because traps are unconfirmed events and are sent as datagrams, an agent has no indication that the manager is is unavailable and that it should switch to sending its traps to another manager.

The connectionless model provides no way to control the flow of data between a manager and agent. Managers which are unable to process large volumes of incoming traps have no recourse but to discard them. SNMP partially solves this problem by using traps judiciously to let a manager know that some event has occurred. It is then up to the manager to request more information about the event with subsequent *GET* and *GET-NEXT* operations. The connectionless model does not provide a manager and agent an opportunity to exchange and negotiate information such as capabilities, security parameters, and other operational parameters.

Because the SNMP model relies on polling, event-based management is not feasible. Applications which require periodic samples, such as those for performance management, must periodically poll each agent to collect the samples.

SNMP makes so few demands on underlying services that it easily adapts to operate over other protocols, including connection-oriented protocols.

5.2 CMIP Communications

The CMIP protocol is defined to operate over the connection-oriented OSI presentation service, and it refers to two other protocols: association control service element (ACSE) and remote operations (ROSE). The OSI presentation layer service provides data representation services, including encoding and decoding ASN.1 and BER. The presentation service relies on connection-oriented OSI session services.

Although the OSI session layer is functionally very rich, CMIP, ROSE and ACSE use only a few functions: segmentation/reassembly and connection establishment/termination procedures. CMIP, etc., have no defined maximum data unit size; the session, transport and network layers provide segmentation/reassembly for large data units. CMIP is also readily adaptable to other communications services. CMIP has only one procedure, the multiple reply functional unit, which requires connection-oriented services. CMIP has been adapted to operate over layer two logical link control for Heteregeneous LAN Management and over SNA sessions.

ACSE is used by the manager and agent applications during the connection establishment phase to exchange and negotiate mutually agreeable parameters, including application contexts and functional units. It is also used during the connection termination phase to release the allocated resources in both manager and agent systems. ROSE is a simple protocol machine which is used to structure requests and responses for the CMIP protocol.

5.3 Summary of Underlying Communications

Table 3 summarizes the differences in underlying communications.

Table 3. Summary of Underlying Communications

Feature	SNMP	CMIP
Signalled data loss	no	yes
Signalled data corruption	yes	yes
Negotiation	none	allowed and extensible
Multiple application contexts	not allowed	allowed
Length limitations	484 octets	unlimited
Paradigm	connectionless	connection-oriented
Interaction model	polling	event driven
Adaptable to other stacks	yes	yes

6. References

[1] M. Rose, K. McCloghrie, "Structure and Identification of Management Information for TCP/IP-based Internets," *Internet Working Group Request for Comments 1155* , Network Information Center, SRI International, Menlo Park, California (May 1990)

[2] ISO/IEC 9596, *Information Processing - Open Systems Interconnection - Common management information protocol specification* (November 1990)

[3] ISO/IEC 10165-1, *Information Processing - Open Systems Interconnection - Management Information Services - Structure of Management Information - Part 1: Management Information Model* (August 1991)

[4] ISO/IEC 10165-4, *Information Processing - Open Systems Interconnection - Management Information Services - Structure of Management Information - Part 4: Guidelines for the definition of managed objects* (August 1991)

[5] ISO/IEC 8824, *Information Processing - Open Systems Interconnection - Specification of Abstract Syntax Notation One (ASN.1)* (April 1990)

[6] [CCITT Recommendation *The Directory - Overview of Concepts, Models, and Services*, BLUE Book, Vol. VIII.8 Rec. X.500, ITU, Geneva, 1988

Integrated Network Management, III (C-12)
H.-G. Hegering and Y. Yemini (Editors)
Elsevier Science Publishers B.V. (North-Holland)
1993 IFIP.

Simply Open Network Management: An Approach for the Integration of SNMP into OSI Management Concepts

Sebastian Abeck
Munich Network Management Team
TU München – Institut für Informatik
Arcisstr. 21, Postfach 20 24 20
8000 München 2, Germany
e-mail: abeck@informatik.tu-muenchen.dbp.de

Alexander Clemm *
Munich Network Management Team
Universität München – Institut für Informatik
Theresienstr. 39
8000 München 2, Germany
e-mail: clemm@dhdibm1.bitnet

Ulf Hollberg
IBM European Networking Center
Vangerowstr. 18, Postfach 10 30 68
6900 Heidelberg, Germany
e-mail: hollberg@dhdibm1.bitnet

Abstract

The Internet Activity Board (IAB) and International Organization for Standardization (ISO) are both defining standards for the management of internets and OSI networks, respectively. While IAB's Simple Network Management Protocol (SNMP) has been widely implemented, in the long run a shift to more powerful network management paradigms appears necessary to tackle the increasingly difficult problem of an integrated management of heterogeneous networks. For this, the ISO concepts that can be applied also to non-OSI networks seem to be well suited. Because SNMP resources will still need to be managed in the future, an integration of OSI and SNMP management is needed. In this paper, a deep integration is proposed where an OSI view on SNMP resources as integral part of a heterogeneous network is provided. This allows SNMP resources to be managed from OSI managers. Central to this approach is the transition between the two information models and the integration of the modeled SNMP resources into a global network model. Also some technical aspects involved in the design of a management gateway between SNMP and OSI management concepts to bridge the gap of functional power between the two are discussed.

Keywords: OSI, SNMP, management information models, global network model, network management gateways

---

*A. Clemm is also affiliated with IBM European Networking Center, Heidelberg, Germany

1 Introduction

The ISO is developing standards for the management of OSI networks. Its object oriented approach in identifying management entities ([ISO 10165]) and its management data services and protocol, CMIS/CMIP ([ISO 9596-1]) have become an important conceptual framework for network management solutions. This approach that models network resources as *managed objects* seems to be very well suited to be applied for heterogeneous network management.

Another important set of standards comes from the Internet Activity Board (IAB) for the management of TCP/IP networks, commonly referred to as SNMP (Simple Network Management Protocol). With 'SNMP' we refer to the communication protocol as well as to the underlying management framework. To a large part due to its simplicity, SNMP is today implemented in products of basically every major manufacturer of communication equipment. Thus, Internets are today in fact being managed using SNMP, with almost every management platform commercially available today based on SNMP, e.g. IBM NetView/6000, Cabletron Spectrum or HP OpenView.

However, because SNMP does have shortcomings it is also recognized that more powerful network management paradigms will be needed in the future ([MAZD 91]). The ISO network management standards offer the potential to fulfill this need via OSI-based network management, whose principles are just as applicable to non-OSI networks ([ALSE 92]). Therefore, with standardization efforts being still underway, many platform vendors have already committed to follow OSI and support CMIP in the future, as many have done for other OSI-standardized protocols. However, communication devices which only support SNMP ('SNMP resources') will continue to exist and play an important role in the future. Accordingly, they will still have to be managed with the advent of OSI-based network management stations. For an integrated management of heterogeneous networks that include SNMP resources, some way to integrate SNMP- into OSI-based network management is therefore needed.

In this paper, we will explain our way which is centered on offering a homogeneous OSI view on all resources, including SNMP resources, as part of one integrated global network model. We draw on work currently conducted at IBM European Networking Center, where a management interconnection between OSI and SNMP is being implemented as part of a network management prototype for integrated fault management of heterogeneous networks ([JOPA 93]). Since many complex issues are involved, this paper can discuss only some major topics.

To set the stage for further discussions, in section 2 we will briefly introduce those network management concepts that are relevant to consider for the integration of SNMP into OSI based management. To achieve this, we propose in section 3 as basis the integration of an OSI view on SNMP object types as so-called 'managed object classes' into generic managed object classes as part of a global network model. Section 4 will discuss some of the more technical considerations to be made when building a management gateway between OSI and SNMP based management that result from the different power of the management protocols used. Some conclusions are offered in section 5.

2 Impact of Network Management Concepts on the Integration of Internet- and OSI-Based Management

There are several principles that underly network management independent of the actual network management concept used. In the following, we will briefly discuss the most relevant ones with respect to the integration of Internet and OSI management concepts (see fig. 1).

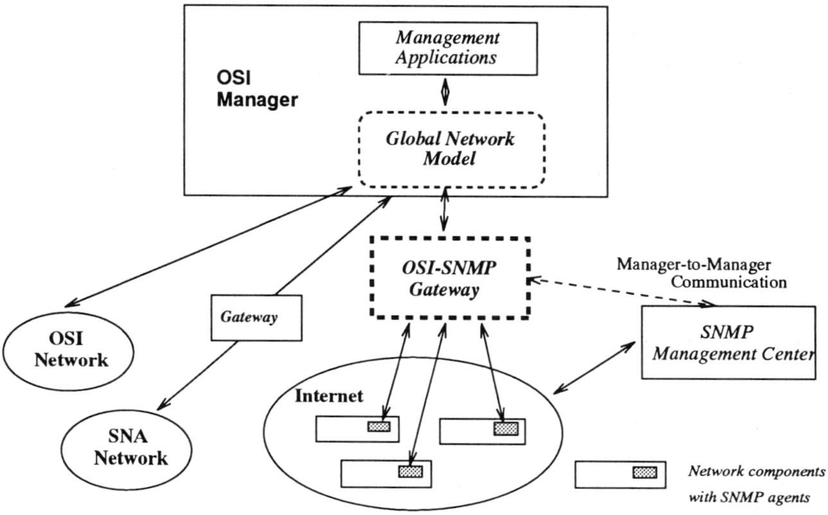

Figure 1: OSI-SNMP-Gateway in the management environment

2.1 Management Hierarchies

Management hierarchies are a means to delegate management functions, such as the accumulation and preprocessing of network management data, to make network management tasks more tractable and to reduce complexity and variety of the network management data collected from the network that is to be processed. Management hierarchies are constructed with 'intermediate agents' (see fig. 2), which act in an agent role towards a manager and act in a manager role towards lower level agents. If the two management protocols that an intermediate agent has to support are not the same, we call the intermediate agent a 'gateway'. Essentially, our approach for the integration of OSI and Internet management concepts is to design and implement such a gateway between CMIP and SNMP.

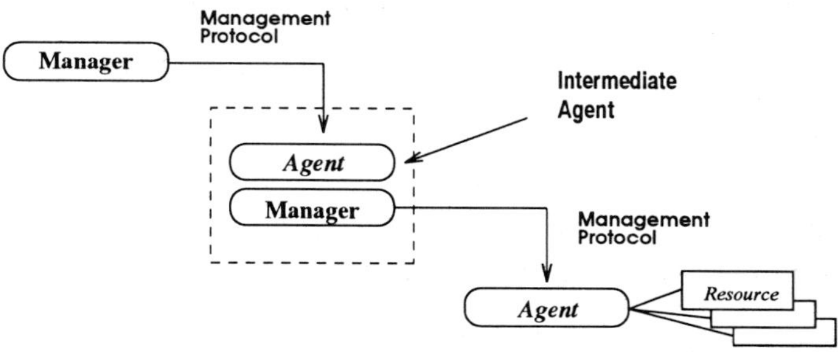

Figure 2: Management Hierarchy

2.2 Management Information Models

While a management protocol provides for communication between distributed components, a shared information model provides a common base for understanding for manager and agents. This model defines every manageable resource with all its relevant properties precisely. Only these resources with their defined properties are 'visible' for management applications.

For homogeneous networks it can be argued that the type of the information model used is of less importance. A small set of communication devices can be defined adequately in nearly every formalism. However, when trying to manage heterogeneous networks, already the sheer variety of devices to be described requires better information models. Here, many variations of entities that basically belong to the same kind of devices need to be defined. Appropriate information models may reduce this complexity effectively if commonalities of otherwise different entities can be captured. For this, object oriented design can serve as a vehicle. Simplified, it can be said that object oriented design is the discipline which concerns itself with the generation of *object models.* It is important to keep in mind that the resulting object models are not completely determined by the subject and the design principles but leave room for human creativity. However, advantageous heuristics and techniques are known which should be respected in the design process ([BOOC 91], [FORUM 90]).

The ISO has decided to use a variation of an object-oriented information model for its network management standards, as object-oriented methodologies are more and more understood as a major advancement. According to the OSI network management standards, resources are represented by *managed objects* which are abstract views of *real* network resources. Managed objects are instances of *managed object classes* which define the visible properties of the particular types of network resources formally in a standardized notation, GDMO (Guidelines for the Definition of Managed Objects). Managed objects are considered part of a *management information base (MIB)* and arranged in a *containment tree* which is also used for naming, while managed object classes are arranged in an *inheritance hierarchy.* The reader unfamiliar with these and related concepts is asked to refer to [ISO 10165], [HEGE 91].

Compared to the OSI 'state of the art' approach, SNMP is based on a much simpler information model. The management information base of SNMP consists of *objects* (not in the object-oriented sense) with *variables.* Both object types and object instances ('SNMP variables') are viewed as parts of the Internet registration tree with instances forming the leaves. Columnar and non-columnar objects are distinguished, depending on whether they are conceptually considered part of a table entry or not. The reader unfamiliar with the concepts of the SNMP information model is asked to refer to [RFC 1155], [ROSE 91].

The very different information models of OSI and Internet management show that a gateway between the two involves much more than merely building a simple PDU mapper. A transition between the information models is required which allows SNMP information to be viewed and accessed as managed objects according to the OSI information modeled. This topic will be covered in more detail in chapter 3.

2.3 Management Protocols

Management protocols are used for the exchange of management data between managers and agents. For details on ISO's and IAB's management protocols, CMIP and SNMP, respectively, the reader is asked to refer to the literature ([ISO 9595], [RFC 1157]).

Because of the gap of functional power between CMIP and SNMP, any functions that go beyond the most basic access and manipulation of managed object attributes have to be provided, or simulated, at the gateway. Especially the scoping, filtering, and synchronization mechanisms offered by CMIP have no counterpart whatsoever in SNMP. Since provision of these mechanisms at an OSI-SNMP gateway adds to its complexity and requires additional effort, an important design decision is which of these features actually to support. We will elaborate more on this topic in chapter 4.

2.4 Management Information Hierarchies

With the information aspect included, management hierarchies can lead also to management information hierarchies. Within network management information hierarchies, the most recent information is collected at the leaves and propagated to some degree into the higher levels. Implementations of the managers, gateways, and agents can be distinguished on how they handle management information.

Stateless components keep no information; instead, information is acquired on demand from the lower management information levels. This keeps the component relatively simple, which as a tradeoff may cause performance problems (see section 4.1). For a management gateway between CMIP and SNMP this means that managed objects are 'simulated' only as they are accessed or manipulated by CMIP management operations, or as SNMP traps arrive that may translate into managed object notifications. However, as a consequence this implies that certain notifications which result from a change of state in the MIB cannot be simulated, since many such changes go unnoticed in SNMP and a comparison of states is not possible.

Stateful components keep some or all information of the lower levels also at higher levels. One benefit of this is that an improvement of performance can be achieved. For an OSI-SNMP gateway, this adds additional complexity because at least parts of the MIB have to be replicated and maintained at the gateway. However, it eases tasks such as scoping and makes a better simulation of managed objects possible because changes in the MIB state can be detected and thus notifications be fully supported.

The similarities to caching or replication problems in distributed file systems are obvious; experience from this area should be built upon. We catch onto this distinction 'stateful - stateless' with the discussion of gateway design principles in chapter 4.

3 Integrating the Internet MIB-II into an OSI Global Network Model

In order to be able to manage heterogeneous networks in an integrated fashion it is desirable to have one global network model in which models of the specific networks are integrated ([VALT 91]). We conceive OSI network management concepts in the way that an OSI network manager should be able to manage not only OSI networks but use the OSI concepts to manage also heterogeneous, non-OSI networks. Therefore we want the OSI network manager to have an integrated, homogeneous view on heterogeneous, also non-OSI networks. One important type of non-OSI-networks is certainly the Internet network. In order to be able to manage SNMP resources from an OSI management station, it is necessary to offer an OSI conformant view on these resources. Therefore, the SNMP variables have to be packaged into managed objects. The managed object classes that result will in the following be referred to as internet managed object classes. The IAB has published several standards which define SNMP object types, the most relevant of which is the definition of MIB-II ([RFC 1213]), which also underlies the following discussion.

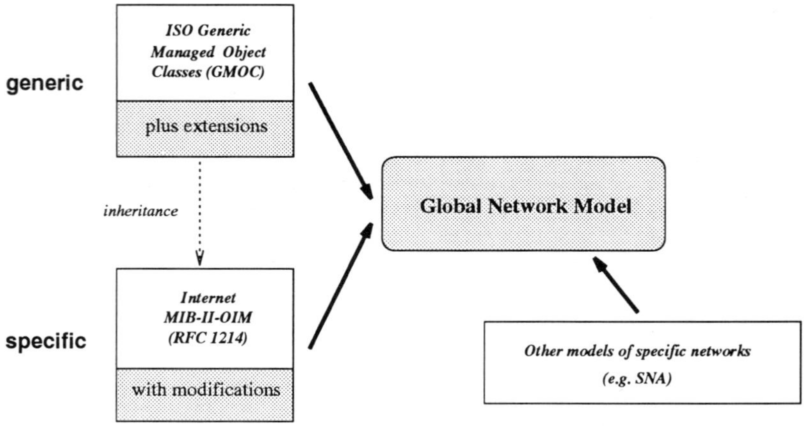

Figure 3: Global Network Model and Internet Information

Fig. 3 gives an overview of our solution. Works from the standardizational bodies we have in particular to take into account are (1) **MIB-II-OIM** (OSI Internet Management) that provide a start for the definition of Internet managed object classes (section 3.1), and (2) ISO's **GMOCs** (Generic Managed Object Classes) that provide the starting point for the Global Network Model that we propose in section 3.2. As we will show, to our opinion some extensions to the GMOCs are helpful or even necessary to integrate IAB's MIB-II-OIM (section 3.3) or managed objects from other network types.

3.1 Internet Managed Object Classes

With the definition of MIB-II-OIM ([RFC 1214]) it has early been shown that it is possible to define internet managed object classes on the basis of SNMP object types. Accordingly, SNMP variables can be interpreted as parts of managed objects. MIB-II-OIM was basically generated from MIB-II according to the following rules:

- MIB groups translate into managed object classes, e.g. `system`, `ip`, ..
- Tables translate into managed object classes, e.g. `ifTable`, ..
- Table entries translate into managed object classes, e.g. `ifEntry`, ..
- All other object types become attributes and are assigned to managed object classes according to their association in the respective MIB.

Additional naming attributes were introduced for each managed object class. Traps can be mapped into notifications of particular managed object classes.

This approach works fine for managing internets independently from other networks from an OSI perspective. However, since all managed object classes are derived directly from the `Top` managed object class and inheritance is not made use of, this model does not fulfill the requirement of being part of an integrated global network model. Additional effort is therefore needed to model

SNMP resources as extensions of generic managed object classes. We will describe in section 3.3 how this can be achieved.

3.2 A Global Network Model Based on Generic Managed Objects from ISO

The ISO is defining a model for OSI networks ([ISO 10165], p.-2,-5). Most of the managed objects classes defined (e.g. `system`, `communicationsEntity`, `clProtocolMachine`, `sap1`, `singlePeerConnection`) currently are rather abstract. Therefore, they might be suited to be used as the generic starting point for the modelling of non-OSI networks. However, unfortunately we see some problems doing this at the current stage of standardization. Additionally, the unstable status makes them a moving target which is hard to keep up with. The following arguments are intended for the knowledgeable reader familiar with the OSI model.

Completeness of the ISO Model

The model is not complete in the sense that all concepts can be modeled with it.

For instance, for *connectionless communication* it may be necessary to model a concept such as routing table entries. There is no corresponding managed object class defined by ISO. In addition, attributes of defined objects need to be adapted. For instance, a `singlePeerConnection` requires an attribute indicating 'underlying connections'. What the value of this attribute would be in the case of a connection relying on an underlying datagram service is not clear.

Also, the *relationship between SAPs* is lost, i.e. information which SAP another SAP depends on to deliver its services. Only what entities use and provide a SAP, respectively, is specified but not, for instance, which underlying SAP-address the selector of an ISO SAP1 refers to. By the same token, also the information from MIB-II-OIM managed object class `ipAddrEntry`, which basically serves to provide a mapping between IP (SAP) address and MAC (SAP) address, cannot be grasped using solely the modeling concepts of the ISO generic managed objects. Only when entities are understood in a sense that they are always attributed to exactly one SAP is this information unambiguous. Accordingly, ISO modeling guidelines suggest that if more than one SAP is offered, entities should be 'split' into several managed objects. However, this contradicts that an entity managed object should contain all aspects of the respective active functional element represented by it.

Because of the problems stated above, we decided to extend the ISO object model and to use this to include internet managed object classes. These extensions and some more are published in [SCHO 92]. For the understanding of this paper, the extensions presented in the following are sufficient. When ISO's generic managed object classes reach IS status, the model will be revisited and any necessary changes for it to be standard conformant incorporated.

- Support for connectionless communication, including a managed object class for routing table entries which can for instance also be referred to by upper-level connections as 'underlying connections'.
- An extended SAP has been introduced which allows the relationship between SAPs on consecutive layers to be directly expressed. For instance, the SAP-id of an (n-1)SAP that a SAP's address depends on can be specified through an attribute.
- As a consequence of these extensions, some attributes and name bindings have been newly defined. In particular, extended SAPs contain connections (or their endpoints, respectively) and have attributes indicating their respective mapping onto other extended SAPs. Extended SAPs are contained by protocol machines.

Documentation of the ISO Model

The model is not well enough documented. For several generic managed object classes it is not clear how they are intended to be used. E.g. a subsystem is intended to model a "part of a system"; should that be a protocol (sub-) stack or a protocol layer? ISO provides no clear answer but suggests 'it may be used' to model the latter. However, if a subsystem is to model a protocol layer, it is not clear where the line should be drawn in modeling between it and a communications entity, or between a communications entity and a protocol machine. To derive internet managed object classes from these managed object classes, one would have to rely on arbitrary interpretation or intuition of which SNMP variables should go into which kind of managed object class. However, such a resulting model is potentially inconsistent with other network models although derived from the same generic managed objects, modelled according to another interpretation. What would be needed is some kind of *GUMO*, 'Guidelines for the Use of Managed Objects'.

For this purpose, we want to briefly summarize the philosophy behind our use of the extended generic model according to which principles also the internet managed object classes shall be modeled, as is elaborated below. It reflects basically the paradigms and terms of the ISO Basic Reference Model [ISO 7498], although managed object classes, such as `subSystem` or `co/clProtocolMachine`, are used in a slightly different manner. However, it should be borne in mind that this model is subjected to the principles of object oriented design, as was previously indicated.

- What relates to the operation of a particular communication protocol is basically modeled by managed object class `protocolMachine` (connectionless or connection oriented). Protocol machines are contained in communications entities that keep all management information relating to the operation of a particular layer. Communications entities are contained in subsystems which are used to structure a system in terms of protocol (sub-) stacks.
- Service Access Points (SAPs) through which services of communication layers are accessed are modeled explicitly. SAPs can be contained by protocol machines. SAPs also contain connection endpoints, and routes pass through SAPs as well.
- System spanning resources which cannot be assigned to any one MIB are modeled separately in each MIB. The respective managed objects represent the local, possibly incomplete view on this system spanning resource. The main example for this are connections and routes.

3.3 The Inclusion of the MIB-II-OIM into the Generic Model

MIB-II-OIM can not be directly integrated into our generic model, i.e. its managed object classes can not be directly derived from the generic managed object classes. One reason for that is that MIB-II-OIM does not model the Internet resources according to OSI philosophies. What distinguishes the model most from MIB-II-OIM is the explicit modeling of SAPs which are also as connection-endpoint containers. Another reason is that attributes to be inherited from generic managed object classes are in part different from those in MIB-II-OIM and would therefore require a redefinition. Therefore, a modified model is needed. However, MIB-II-OIM can be used as a starting point for modeling because it provides a grouping of MIB-II object types and indicates which attributes are available and accessible at SNMP resources and should therefore be part of the model. With some modifications, we have integrated MIB-II-OIM as internet model into the generic model. Some problems remain, however, with attributes from the generic managed object classes which have no direct counterpart in MIB-II. Specifically, this concerns attributes for state management that are rarely supported in MIB-II.

The following table gives a brief overview how, starting from MIB-II-OIM, we achieved an integration of Internet managed object classes into the generic network model, namely which internet managed object classes we introduced, how they relate to MIB-II-OIM, and which generic managed object classes they were derived from. The table shows an excerpt; also, Internet MIB extensions that are vendor specific or geared towards particular network types, such as token ring or FDDI, are not included but will have to be modeled analogously.

MIB-II-OIM Object Class	Global Network Model Managed Object Class	Generic Superclass	Remarks
`system`	`snmpSystem`	ISO:`system`	1
`tcp`	`snmpTcpEntity`	ISO:`coProtocolMachine`	1
`tcpConnTable`	< *not needed* >		2
`tcpConnEntry`	`snmpTcpConnection`	ISO:`singlePeerConnection`	3
< *no equivalent* >	`snmpTcpSap`	ISO/ENC:`extSap1`	4, 5
`ip`	`snmpIpEntity`	ISO:`clProtocolMachine`	1
`ipAddrTable`	< *not needed* >		2
`ipAddrEntry`	< *not needed* >		6
< *no equivalent* >	`snmpIpSap`	ISO/ENC:`extSap2`	5
`ipRoutingTableEntry`	`snmpIpRoutingEntry`	ENC:`routingInfo`	
`interfaces`	< *not needed* >		2
`ifEntry`	`snmpIfEntity`	ISO:`coProtocolMachine`	1, 7
< *no equivalent* >	`snmpIfSap`	ISO/ENC:`extSap2`	5, 8
< *no equivalent* >	`snmpIfPort`	ISO:`port`	9
`udp, icmp, egp, ...` *accordingly*			

Remarks

1. Most MIB-II-OIM objects do not have state attributes. Agents supporting our global model have to infer from other information how to set the inherited state attributes. For example, the `operationalState` of the `snmpSystem` will be set to 'enabled' as long as the SNMP-Agent on this systems replies, similarily, the usageState will be 'active'.

2. Pure container objects that result from MIB-II grouping but serve no particular purpose are not modeled.

3. Local and remote SAP-addresses are obtained through concatenating the respective `ipAddress` and `tcpPort`.
   `underlyingConnectionName` is not supported; the same information can be inferred from mapping of SAPs.

4. SAP-address corresponds to `ipAddress`, concatenated with `tcpConnLocalPort`

5. `userEntityName` not supported because same information already given through information on which SAP a SAP maps onto

6. But provides the mapping between `snmpIpSap` and `snmpIfSap`, and between `snmpIfSap` and `snmpIfPort`.

7. Physical media as indicated by attribute `ifType` in MIB-II-OIM managed object class `ifEntry` not modeled as an own object.

8. `ifIndex` from `ipAddrEntry` serves as SAP-address.
9. Actually part of MIB-II-OIM managed object class `ifEntry`; `ifPhysAddress` serves as port address.

4 Technical Aspects of an OSI-SNMP Gateway Design

The basic management concepts and the global network model provide the basis for a gateway implementation. Based on the distinction 'stateful – stateless' that was introduced in section 2, we point to one of the fundamental design decisions for a gateway implementation. We present a simple scenario to provide a better understanding for the two main problem areas, naming and addressing and service mapping, a gateway has to cope with. The section closes with the discussion of advanced CMIP features which cause considerable additional effort to support by the gateway.

4.1 A Fundamental Decision for the Design of an OSI-SNMP Gateway

An OSI-SNMP gateway must map the management information from the OSI information model underlying the global network model into the Internet information model underlying MIB-II and vice versa. A straightforward solution would be a stateless gateway which maps incoming CMIP indications to SNMP requests. CMIP and SNMP are thus directly coupled. No internal MIB information is kept at the gateway. In the case of a stateful solution, the gateway holds management information about SNMP resources inside a 'gateway MIB'. The gateway honors incoming CMIP indications simply by reading and writing the gateway MIB. Internal mechanisms assure consistency with the actual state of SNMP resources, of course also under the use of SNMP. CMIP and SNMP are therefore decoupled. An update mechanism is necessary to guarantee that the information lying in the gateway MIB corresponds to the represented information lying in the Internet resources. Obviously, this approach can lead to temporary inconsistencies of the gateway MIB with the actual state of the managed SNMP resources.

The advantages and disadvantages of these two solutions with respect to performance are obvious: for more static management information the stateful solution is best suited because updating causes no or only little cost. The more dynamic the information is, the more the stateless solution must be suggested. However, the drawback that a stateless gateway is constrained with respect to the generation of notifications, as was previously indicated in section 2.4, has to be taken into account.

A compromise which combines the advantages of both stateless and stateful solutions is to decide dynamically in the gateway if a CMIP request should be mapped directly onto SNMP services or not. For this, the gateway needs knowledge about the dynamic behavior of the management information (e.g. implemented as flags amended to the managed object description). This idea is further expanded in [HABO 91].

4.2 A Basic Gateway Scenario

To illustrate, consider the process of requesting via a non-scoped CMIP `M-Get` a managed object attribute that corresponds to an SNMP counter value. To fetch the counter value from the SNMP agent, the gateway will in general have to solve the following tasks:

1. **Name mapping**: Upon receipt of the `M-Get-indication`, the gateway has to determine from the specified OSI name which management information in the Internet MIB is referred

to and accordingly perform some kind of name mapping. Here, no scoping is used, therefore the OSI name of the base managed object and its requested attribute can be directly mapped to the respective Internet MIB variable name. With scoping, matters would have been more difficult because now the names of the actually requested managed objects would have to be computed first.

2. **Service mapping**: Another problem which must be treated by the gateway stems from the differences between the management protocols CMIP and SNMP whose services have to be mapped. Here, the CMIP `M-Get` will be mapped to the corresponding SNMP service `get`. The response to the CMIP request will be generated as soon as the corresponding SNMP `get-response` arrives. Since UDP, the mechanism on which SNMP relies, is unreliable, some way has to be devised to cope with outstanding SNMP responses.

We discuss these two problems in the following sections in more detail. It should be kept in mind that a stateful gateway may already keep the requested managed object in internal memory. Therefore, because of the decoupling of CMIP and SNMP the steps described above are not necessarily needed as an immediate result of incoming CMIP indications, although the mechanisms are still needed in general to keep the managed objects consistent with the actual SNMP resources.

4.3 Name Mapping

When access to internet management information is requested from the OSI manager, it is identified according to the naming scheme of the global network model which is based on the containment scheme. The task of the gateway is to reestablish the original Internet identifier which is based on the Internet registration tree.

We illustrate the mapping process of a simple variable with the example of the counter `tcpActiveOpens`. In the global model it is defined as an attribute of the object class `tcpSap`. Figure 4 shows the position of this attribute in the containment tree of the global model. Some of the relative distinguished names of the (full) distinguished name of the managed object of the attribute are used to access the counter via SNMP. We assume that the SNMP agent which holds the requested counter is already selected, i.e. steps (1) and (2) of fig. 4 are already completed.

So the task of the gateway is to execute step (3), which is to identify the object instance in the MIB of the SNMP agent. As the name of the object instance consists of the name of its object type concatenated with a suffix (0 for non-columnar objects; for columnar objects, the index key of the conceptual table entry the object is considered part of), this task can be split into the following two subtasks: (3.1) determining the object identifier of the requested object type, and (3.2) determining the requested instance of the object type, i.e. the suffix which has to be appended to the object identifier.

A straightforward solution for subtask (3.1) is to use an object identifier table in the gateway which holds the object identifier for every tuple (object class name, attribute name).

To cope with subtask (3.2), the identification of object instances, we took the following decision which simplifies this task for the gateway: the relative distinguished name of object classes which are instantiable in MIB-II is the suffix of the object identifier. This means that for MOs with attributes representing noncolumnar objects the RDN value part is '0'; for MOs that contain columnar objects it is the index key of the respective conceptual table row.

So far, only the mapping from ISO to IAB addresses has been discussed. Note that the same scheme could be used to perform mapping also the other way round, to compute for SNMP variables whose containing system is known the respective attribute of a particular managed

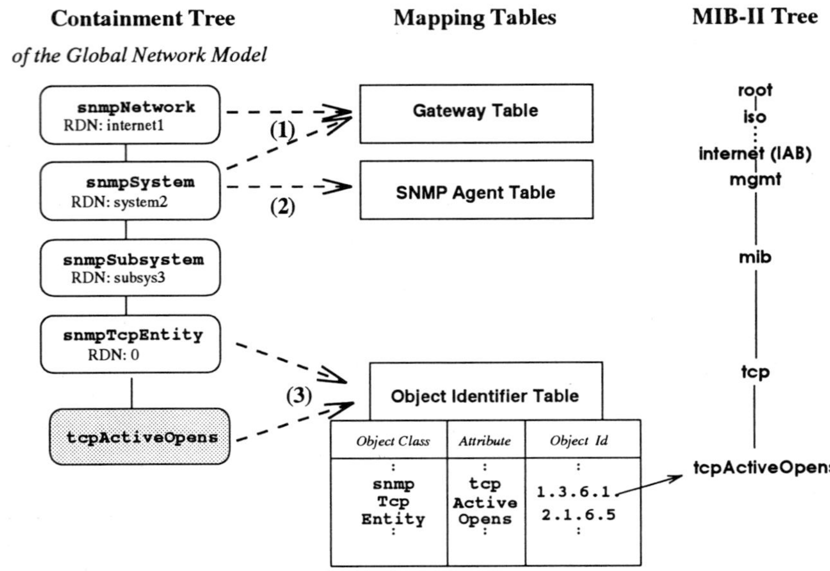

Figure 4: The problem of mapping

object that this variable refers to. However, the need for that does not occur, as an SNMP response always relates to some SNMP request that was issued by the gateway itself before. The only case where some mapping from SNMP to CMIP has to be performed is for traps which refer to no particular SNMP variable and which are attributed to managed objects of a particular class according to a fixed scheme, as was hinted in the discussion of the Internet model.

4.4 Service Mapping

The execution of CMIP services will often require the invocation of SNMP services. For stateless gateways and for frequently changing attributes, this will always be the case. For stateful gateways such as ours, the caching will save many of the these invocations, at least when managed objects are not modified.

This section discusses the mapping of basic service elements without regard to advanced CMIP features like scoping and synchronization. For all interaction with SNMP agents it should be remembered that SNMP services are based on the unreliable datagram services of UDP; some timeout and request-retransmission mechanisms are therefore required to handle outstanding responses.

Attribute Setting and Retrieval: Get and set, the basic management services to read and write attributes in the MIB, have direct counterparts in CMIP and SNMP. When the gateway receives an `M-Get-indication` rsp. `M-Set-indication`, it generates one or more, depending on the number of attributes requested, corresponding `get-requests` rsp. `set-requests` on the SNMP side.

The gateway has to wait for the response(s) of the SNMP agent (i.e. `get-response`) and map it onto CMIP service primitives `M-Get-response` rsp. `M-Set-response`.

Actions: Actions can be simulated in some cases in SNMP by setting variables to particular values. As this feature is at the 'border' of the SNMP standard, it is only available with some implementations of the SNMP agents. Therefore, this will have to implemented in the managed objects for the specific SNMP resource.

Events and Notifications: Events can originate from traps, as was previously indicated and is elaborated in RFC 1214. Traps can be attributed to certain managed objects and are treated as their notifications. Through discriminators they can thus lead to the issue of events.

Another important case to consider is that of other unsolicited notifications, defined for generic and therefore Internet managed object classes that may turn into events with corresponding discriminators. An example for such a notification would be an `objectCreation`- or `attributeValueChange`-notification. Since such 'events' usually do not correspond to traps and therefore go unnoticed in SNMP based management, if they are to be supported they have to be 'simulated' by the gateway, which has to perform to this end periodic polling of Internet MIBs and needs a memory to compare the polling results with to detect any changes. It therefore needs to be stateful; such notifications cannot be supported with a stateless gateway.

Create and Delete: SNMP does not have services for creation and deletion of variables because usually this is not permitted in the course of management operation. SNMP objects become available by the activities of SNMP agents or subagents. However, an exception can be allowed with respect to table entries. Here, an entry can in fact be created by setting columnar objects of a not yet existing table entry. It can be deleted by setting a dedicated columnar object to a value indicating the table entry is invalid. Accordingly, `M-Create` and `M-Delete` on table entries translate into corresponding SNMP `set-requests`. OSI's `objectCreation`- and `objectDeletion`-notifications will be generated in turn.

4.5 Advanced CMIP Features

The issue of advanced CMIP features like scoping, filtering and synchronization which SNMP does not provide have not yet been covered in the discussion. These features are one of the main reasons for the greater power of CMIP over SNMP, as they allow at least in part the delegation of functions to agents. However, because they have no counterpart in SNMP, the gap of functionality, if supported, has to be provided at the gateway. We outline how the features can be implemented:

Scoping: Full scoping is (automatically) supported by a pure stateful gateway implementation, because in that case, conceptually, the complete containment tree of the agent is always available. However, scoping could also be supported by a stateless gateway by applying the SNMP `get-next` operation e.g. for the traversal of MIB-II-OIM tables and table entries.

Filtering: Filtering is not supported by SNMP and therefore has to be performed at the gateway which is uncritical for `M-Get`. For `M-Set` and `M-Action` problems arise because checking of conditions which objects to apply the actions to involves several SNMP operations, i.e. transactions: Before the SNMP `set-request` can be issued, the filter criteria have to be checked using an SNMP `get-request`. The time window between the 'get' and the 'set' can be critical for dynamic attributes.

Synchronization: Best effort synchronization obviously poses no problem. To support atomic synchronization, additional mechanisms are needed, as SNMP offers no transaction concept. We decided not to support atomic synchronization for our gateway; further research on this issue is needed.

5 Conclusions

The integration of Internet management into OSI management certainly is one of the most demanding problems in the field of integrated network management. We have shown that a deep integration can be done in a satisfying manner with our approach of the global information model and the way we have designed the OSI-SNMP gateway.

As a consequence, the definition of a global network model is the fundamental problem which must be solved before any technical gateway problem can be tackled. We believe that the Generic Managed Object Classes defined by ISO are a good starting point, although extensions are necessary to include non-OSI networks, such as the Internet. With the example of the Internet management we have shown how this model can be filled with concrete management information.

The gateway concepts we have presented in this paper can also be applied to integrate other network types like SNA networks or Transdata networks into an overall, 'global' network management. So this paper not only presented the description of OSI-SNMP gateway issues but also defined the central parts of a gateway framework towards OSI.

Acknowledgements

The authors wish to thank the members of the IBM ENC Systems and Network Management Department and the Munich Network Management (MNM) Team for helpful discussions. The MNM Team is a group of researchers of the Munich Universities and the Bavarian Academy of Sciences which is directed by Prof. Hegering.

References

[ALSE 92] Sebastian Abeck, Martin Leischner and Peter Segner, "Applying the Inductive Modeling Technology to Tackle the Problem of Integrated Network Management", Accepted for Distributed Systems: Operations and Management 1992, October 1992.

[BOOC 91] G. Booch, "Object Oriented Design with Applications", In *Benjamin/Cummings Series in Ada and Software Engineering*, Benjamin/Cummings Publishing Company, Inc., November 1991.

[JOPA 93] Derick Jordaan, Martin Paterok, "Event Correlation in Heterogeneous Networks Using the OSI Management Framework", IFIP/IEEE 3rd International Symposium on Integrated Network Management, San Francisco, April 1993.

[FORUM 90] OSI / Network Management Forum, *J-Team Technical Report on Modeling Principles* Network Management Forum, Technical Report 102, Issue 1.0, Draft 6, May 1990.

[HABO 91] Heinz-Gerd Hegering, Sebastian Abeck and Thorsten Böhnke, "Converting MIB Descriptions into MIB Implementations", IFIP/IEEE International Workshop on Distributed Systems: Operations and Management, Santa Barbara, California, October 1991.

[HEGE 91] Heinz-Gerd Hegering, "How to find Significant Management Information in a Heterogeneous Network Environment", Joint SAIEE/CSSA International Symposium on Network Management, University of Pretoria, Republic of South Africa, May 1991.

[ISO 10165] ISO, *Information Processing - Open Systems Interconnection - Management Information Services - Structure of Management Information, Part 1 – 6*, May 1992.

[ISO 7498] ISO, *Information Processing Systems - Open Systems Interconnection - Basic Reference Model*, 1984.

[ISO 9595] ISO, *Information Processing Systems - Open Systems Interconnection - Common Management Information Service Definition*, November 1991.

[ISO 9596-1] ISO, *Information Processing Systems - Open Systems Interconnection - Common Management Information Protocol - Part 1: Specification*, November 1991.

[MAZD 91] F. Mazda, "Convergence or Collision: SNMP and CMIP (Part 2)", In *Telecommunications*, November 1991.

[RFC 1155] IAB, *Structure and Identification of Management Information for TCP/IP-based Internets*, RFC 1155, May 1990.

[RFC 1157] IAB, *Simple Network Management Protocol (SNMP)*, RFC 1157, May 1990.

[RFC 1213] IAB, *Management Information Base for Network Management of TCP/IP-based Internets: MIB II*, RFC 1213, March 1991.

[RFC 1214] IAB, *OSI Internet Management: Management Information Base*, RFC 1214, April 1991.

[ROSE 91] Marshall T. Rose, *The Simple Book: An Introduction to Management of TCP/IP-based internets*, Prentice Hall, 1991.

[SCHO 92] Brigitte Schott, Alexander Clemm and Ulf Hollberg, "An ISO/OSI-based Approach for Modeling Heterogenous Networks", In *INDC 92, Helsinki, Finland*, also published as IBM ENC Technical Report No. 43.9204, IBM ENC, March 1992.

[VALT 91] Robert F. Valta, "Design Concepts for a Global Network Management Database", Proc. 2nd International Symposium on Integrated Network Management, Washington D.C., April 1991.

Integrated Network Management, III (C-12)
H.-G. Hegering and Y. Yemini (Editors)
Elsevier Science Publishers B.V. (North-Holland)

Design of Protocol Independent Management Agent to Support SNMP and CMIP Queries

Subrata Mazumdar, Stephen Brady, and David W. Levine

Network Management Systems
IBM T.J. Watson Research Center
30 Sawmill River Road, Hawthorne, NY 10532

Abstract

Even though CMIP has achieved the level of an international standard for communication between management entities, very few management agents have been written to support CMIP. For the most part, the management agents being built today support SNMP, and a large number of management applications are being written to use this protocol.

This situation adds to the inertia slowing the introduction of CMIP agents, since the existing SNMP-based applications cannot manage CMIP-capable devices. The absence of CMIP-based applications as an alternative leaves network managers with little choice but to stay with SNMP in the search for management interoperability.

In this paper we present an architecture for building management agents that can support both SNMP and CMIP protocols. In particular, we show how to support both protocol using multiple name registration tables from a single protocol independent collection of management objects. We also present a methodology by which one can translate a CMIP MIB specification (with some restrictions) into an SNMP MIB specification.

1. Introduction

As the OSI network management standard committees are focusing on Common Management Information Protocol (CMIP) [ISO90a] based management, Simple Network Management Protocol(SNMP) [RFC90b] based management agents and applications are proliferating. The big advantage of SNMP based management is that the SNMP agent software is publicly available and sufficient to provide monitoring capability for most network devices. The cost of instrumentation of network devices for an SNMP agent is very low and the instrumentation has little impact on the existing software or hardware design. On the other hand, OSI management model imposes a different design philosophy and requires changes in existing software design. The generality of the OSI management model also imposes significant overhead on the MIB implementation.

Another characteristic that distinguishes SNMP and CMIP agents is the amount of processing the agent must perform on the data collected from network devices. OSI management agents collect, process and interpret data at the device level, while SNMP based agents only collects data. In the SNMP case, processing and interpretation of collected data is generally done at the management application level. Thus the processing power required to support an SNMP agent is much less compared to that needed by a CMIP agent, and as a consequence, SNMP is better suited to simple network devices, such as multiplexors, workstations, modems. However, as network devices become more complex (routers, hubs, switches) and contain numerous components and subsystems, special purpose microprocessors are added to the network devices to augment the processing power. Therefore, it is not unreasonable to expect to allocate additional processing power in such devices for the operation of a CMIP agent. In a high-speed network, an SNMP agent will not be adequate because of the volume of the collected data. Raw counter values will be meaningless to an application since they will change very rapidly. A threshold on the counter might be more appropriate, but detection of threshold crossings can only be done effectively if the underlying counter is sampled at a frequency consistent with the rate the counter is changing.

In summary, the reality is that even though CMIP has great potential an SNMP agent costs very little for a vendor to add to their network devices and certain network devices are not equipped to handle the processing requirement of a CMIP agent. For these reasons, some device manufacturers will favor and continue to support only SNMP on specific devices.

Even though a OSI management agent will provide better management capability for complex network devices, the lack of management applications that use CMIP is an impediment to the introduction of CMIP-capable management agents. Because of the proliferation of SNMP agents, the management applications that are being developed or that are planned will work with SNMP only. Since there is a fundamental difference in the way objects are structured and named in the two protocols, SNMP based management applications cannot query the Management Information Base (MIB) of a CMIP capable OSI management agent.

Since a MIB is only a repository of information about a network devices, we feel that an agent implementation that can support both protocols to query its MIB would help to introduce CMIP-capable management agents. Such an implementation will facilitate the introduction of CMIP-based management applications, because it can support the SNMP applications in a live networking environment while the CMIP based management applications are being developed and tested.

In this paper we present an architecture for a protocol independent management agent that can support both SNMP and CMIP on a single MIB implementation. The novelty of our approach is that through multiple name registration, a CMIP-capable OSI management agent can be be extended to support SNMP queries. Note that, our approach is different from the proxy agent approach [ROS91] or the technique of encapsulating SNMP agent by a CMIP agent. Our approach gives equal status to both protocols and allows an agent to support both protocols to access a single MIB.

The paper is organized as follows: in section 2 we present the architecture of the

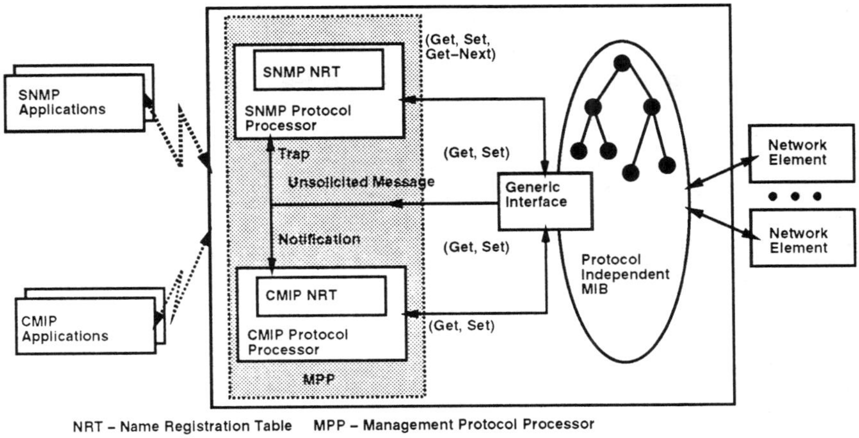

Figure 1: Architecture of Protocol Independent Management Agent.

protocol independent management agent; in section 3 we show to use the name registration tables to support both CMIP queries. in section 4 we show how to translate GDMO [ISO92] style MIB object classes into SNMP style table/table-entries [RFC90a]; and in section 5 we show to use the name registration tables to support both SNMP queries.

2. Architecture of Protocol Independent MIB

Figure 1 illustrates the architecture of our protocol independent management agent (PIMA). The protocol independent management agent has two components: the MIB and the management protocol processor. There is complete separation between management protocol processor and the MIB. The management protocol processors interacts with the MIB through a generic interface that supports Get and Set primitive operations. The MIB emits unsolicited messages through this interface as well. We will ignore the other CMIP primitive operations (e.g., Actions) in order to keep the presentation simple. The MIB collects information from the network devices and updates the corresponding object instances. The object instances in the MIB appear to the management protocol parts as a set of *class-less* objects with *type-less* attributes and values. Thus, we can use generic Get and Set to access/modify the values of the attribute of any object, as long as we can identify the object. For a detailed presentation on the implementation of the MIB see [WU92].

The management protocol processor part has two components: one for SNMP and one for CMIP (see figure 1). Each protocol processor has a name registration table for mapping protocol specific names of objects to the internal names of the corresponding objects in the MIB. The MIB uses the storage pointer of the object instances as the internal name. Thus, when an object instance is created, it registers a protocol-specific name and the internal name with each protocol object. The mapping of the protocol specific attribute names to internal attribute names of the MIB is also handled by an attribute name table outside the MIB.

When the PIMA receives a query from a management application, it identifies the type (SNMP or CMIP) query message. Since SNMP uses connection-less and CMIP connection-oriented protocol, identification of each type of query message is very simple. Once the management protocol type of the query message is identified, the message is sent to the corresponding protocol processor. After decoding the query, the protocol processor uses its object name registration table to find the internal name(s) of the corresponding object instance(s) in the MIB. We will assume that CMIP management protocol processor will implement the appropriate protocol stack for CMIS [ISO90b] and ASCE [ISO88].

The CMIP **Get** and **Set** operations are natively supported without any mapping. The SNMP **Get**, and **Get-Next** operations are mapped to the **Get** operation of the generic interface of the MIB, and the SNMP **Set** is mapped to the **Set** operation of the MIB.

One advantage of our protocol independent management agent is that it can be used in all SNMP, all CMIP or a mixed environment.

3. Registration of OSI MIB Objects for CMIP Query Access

Our implementation of a protocol independent management agent supports the CMIP protocol and GDMO style managed objects natively, that is, the MIB is implemented as if we are implementing the MIB for CMIP only. First we will define a set terms used in CMIP naming scheme and then we will describe the name registration and CMIP operation on managed objects. For a detailed description of naming of OSI managed objects see [EMB91] and [MAZ92].

When the instances of the managed object classes are created, a tree, called **naming tree**, is formed. The root node of the **naming tree**, called ROOT, is the container of all instances. The naming scheme of MIB object instances for CMIP uses the idea of *Distinguished Name* (DN). The components of the DN for a managed object instance are the Relative Distinguished Names (RDNs) of the containing object instances, starting with the ROOT of the naming tree and ending at the instance itself. An RDN of an object instance consists of an *attribute object identifier* and an *attribute value.* The RDN is unique for each managed object instance having the same superior.

The CMIP name registration table (shown in Figure 1) for instances of managed objects is organized in a two level tree and shown in Figure 2. The first level of the tree contains the names of object classes and pointers to the lists of instances of each object class (the second level). The lists in the second level of the tree contain the names of the object instances (DN) and the internal names (storage pointers) of the actual objects in the MIB.

When an instance of an object class is created by the parent object, the parent object passes a list of attribute-value pairs (including the RDN attribute and its value) as well as its own DN. After initializing the new instance with the attribute-value pair list, the new instance forms its RDN by obtaining the attribute OI-value pair using the name binding object of its class. Then the new instance forms its DN by concatenating the DN of the parent with its own RDN. Once the DN of the object instance is formed, the new instance registers with the name registration table by providing its class name, DN and the storage

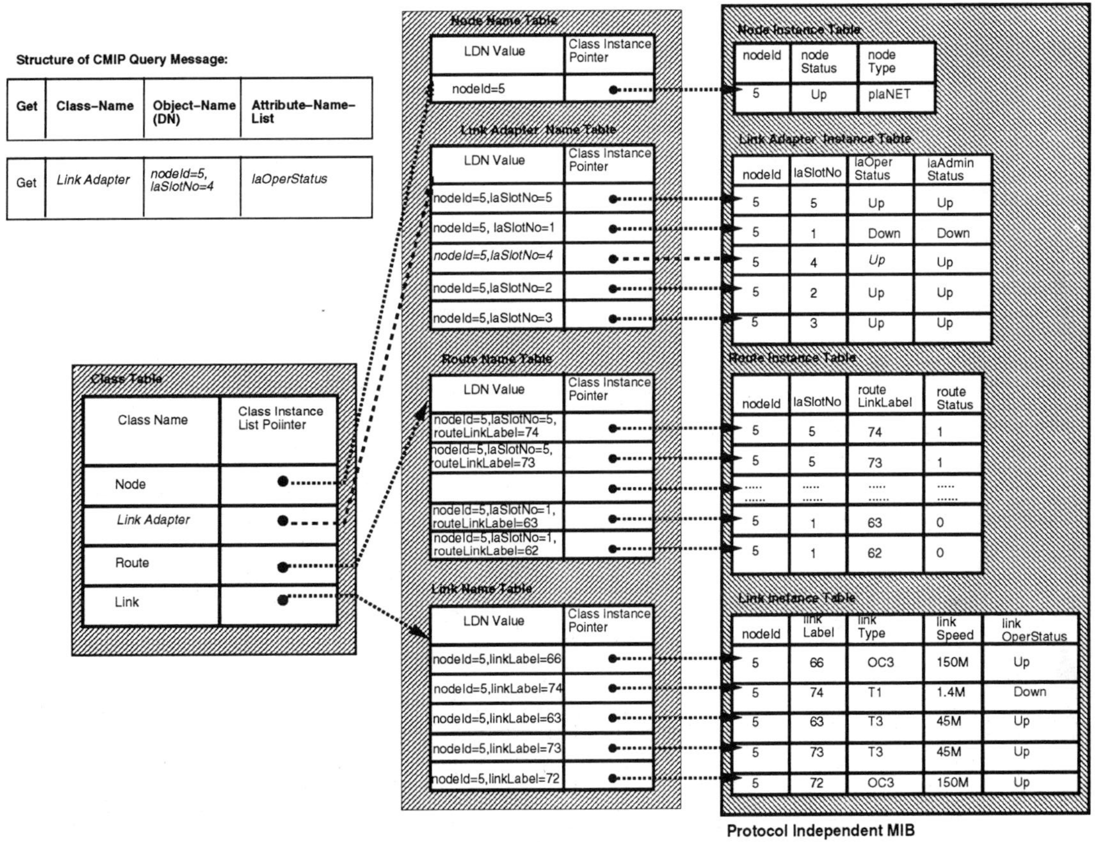

Figure 2: Organization of CMIP Name Registration Table.

pointer of the object instance.

Given a CMIP query with object class name, object instance name (DN), we can easily find the internal storage name for the corresponding managed object. First we search the class table and then we search the name table of the corresponding class. For example, given a Get opertaion on **LinkAdapter** class with DN equal to `nodeId=5,laSlotNo=4` for attribute *laOperStatus* (as shown in Figure 2), first we search the class table for for the corresponding Class Instance List Pointer. Using the Class Instance List Pointer we search the LinkAdapater name table for above DN value. Once we found the match for the DN, we use the Instance Pointer to invoke the Get operation of the generic interface of the MIB and pass the given attribute names as arguments.

4. Translation of OSI Managed Object Classes for SNMP Access

In order to develop a scheme for identifying the objects of the protocol independent MIB for SNMP query access, we will use the concepts of weak entity types of the Entity-Relationship Model [CHE76] to translate the OSI MIB object classes into SNMP Table/Table entries.

Our scheme for mapping the managed object classes of GDMO style MIB into an SNMP addressable MIB is as follows. Since all the objects below the root of containment tree are weak entity, we take the class containment hierarchy (defined by naming object of each classes) of the GDMO style MIB and flatten it to a 1-level deep tree using the following method recursively:

- Identify a node (and call it local root node) in the CMIP containment tree that will be the root node for the subtree in the SNMP MIB object-identifier hierarchy tree
- create a table and table entry for every node (managed object class) below the local root node in the containment tree;
- the attributes of an object class becomes the column variables of the corresponding table entry in the SNMP MIB
- the INDEX parameters of a table entry are formed by concatenating the names of the attributes in the LDN of an instance of the corresponding object class in the CMIP MIB; this scheme is similar to identifying the key attributes of a weak entity type in the ER model (see *ER-to-Relational Mapping* in [ELM89])
- augment the columns of each table entry with the INDEX parameters of the table entry that corresponds to the parent of object class in the CMIP containment tree.

The Figure 4 illustrates the above scheme for mapping objects in the Figure 3 into an SNMP addressable MIB. We have used the **Node** object class in the containment tree as the local root node. First, we represent the **Node** object class by a table, called **NodeTable**, and a table entry, called **NodeTableEntry**. The INDEX parameter of the **NodeTableEntry** is the distinguishing attribute of **Node** object class because the LDN and the RDN of an

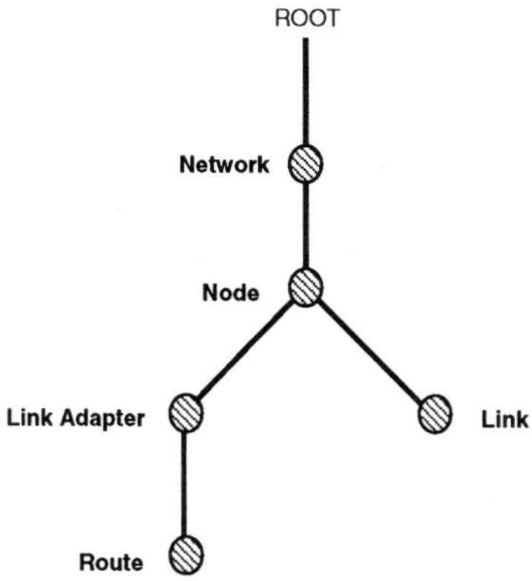

Figure 3: Containment Tree in CMIP MIB.

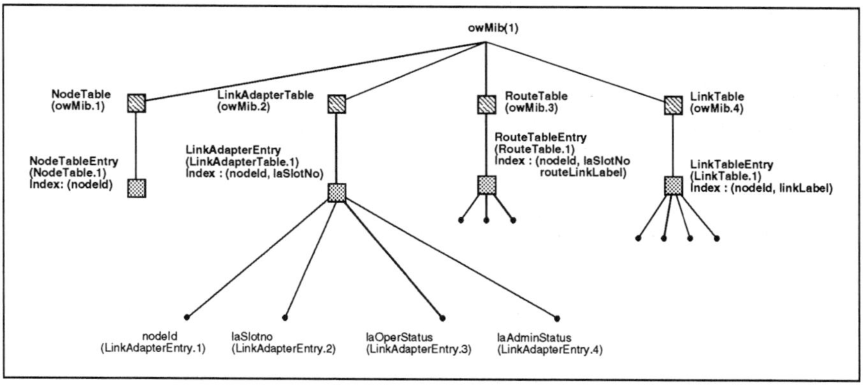

Figure 4: The Translated MIB hierarchy for SNMP.

instance of **Node** are same. The column variables of the **NodeTableEntry** are the attributes of **Node** object class.

Then, we represent the **LinkAdapter** object class by a table, called **LinkAdapterTable**, and a table entry, called **LinkAdapterEntry**. The LDN of an instance of the **LinkAdapter** object class contains the attribute `nodeId` of **Node** object class and `laSlotNo` of **LinkAdapter** (`laSlotNo` is also the distinguishing attribute of **LinkAdapter**). Thus, the INDEX parameter of the **LinkAdapterEntry** are the `nodeId`, and `laSlotNo`. The column variables of **LinkAdapterEntry** are `nodeId` and all the attributes of **LinkAdapter** object class.

Similarly, we transform the **Link** and **Route** objects into tables and table entries for the SNMP MIB, as shown in Figure 4. Figure 5 show scheme for generating SNMP MIB and C++ object classes for the MIB implementation. The SNMP based management application developer will use translated SNMP MIB to write their applications and need not concern them with MIB implementation at the agent. The translated SNMP MIB will also be used during the implementation of the SNMP protocol object shown in Figure 1. See [WU92] for detailed description of implementation of MIB, CoreObject class and their subclasses.

Note that using our scheme to translate a OSI managed object classes to an SNMP objects will not generate any group object. For us, The group objects of an SNMP MIB are special cases of Table objects.

5. Registration of MIB Object Instances for SNMP Query

Like the CMIP name registration table described above, the SNMP name registration table has two levels. At the startup of the agent, we initialize the SNMP name registration table with the names of groups, tables and table entries of the translated SNMP MIB (as shown in Figure 6). In order to facilitate Get-Next operations, the names of table and table entries will be kept in lexicographically sorted order by their object identifier.

The first level of the name registration table consists of the following columns: Table_Name, Total_Attributes, and a storage pointer to the second level (see Figure 6). The column, Table_Name, indicates the name of the table or the name of the group object. The column, Total_Attributes, represents the total number of columns in the table or total number of variables in the group object. The Table_Name and Total_Attributes columns are initialized during the startup phase based on MIB description of objects. Finally, the last column contains the storage pointer to the second level of the registration table. The second level of the registration table represents the unique instance of a group object or the instances of entries of a table object. It consists of two columns. The first column contains the value(s) of INDEX parameter of an entry that uniquely distinguishes the entry in the table from other entries. The second column contains the internal storage pointer of the corresponding entry in the protocol independent MIB. The rows of the table of the individual table entry are also sorted by the values of the INDEX parameters.

When we create an instance of a managed object in the protocol independent MIB, we call a routine called, `DO_SNMP_Registration`, with following arguments:*tableEntryName*, *keyValues*, and *entryPtr*. The function prototype is as follows:

```
DO_SNMP_Registration(char* tableEntryName, char* keyValues, void* entryPtr);
```

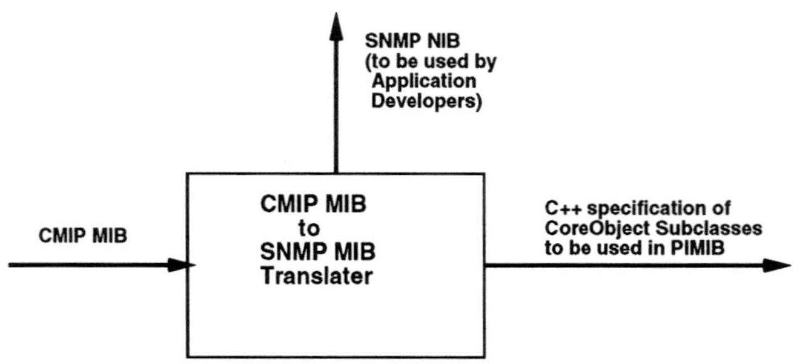

Figure 5: The CMIP MIB translation scheme for SNMP.

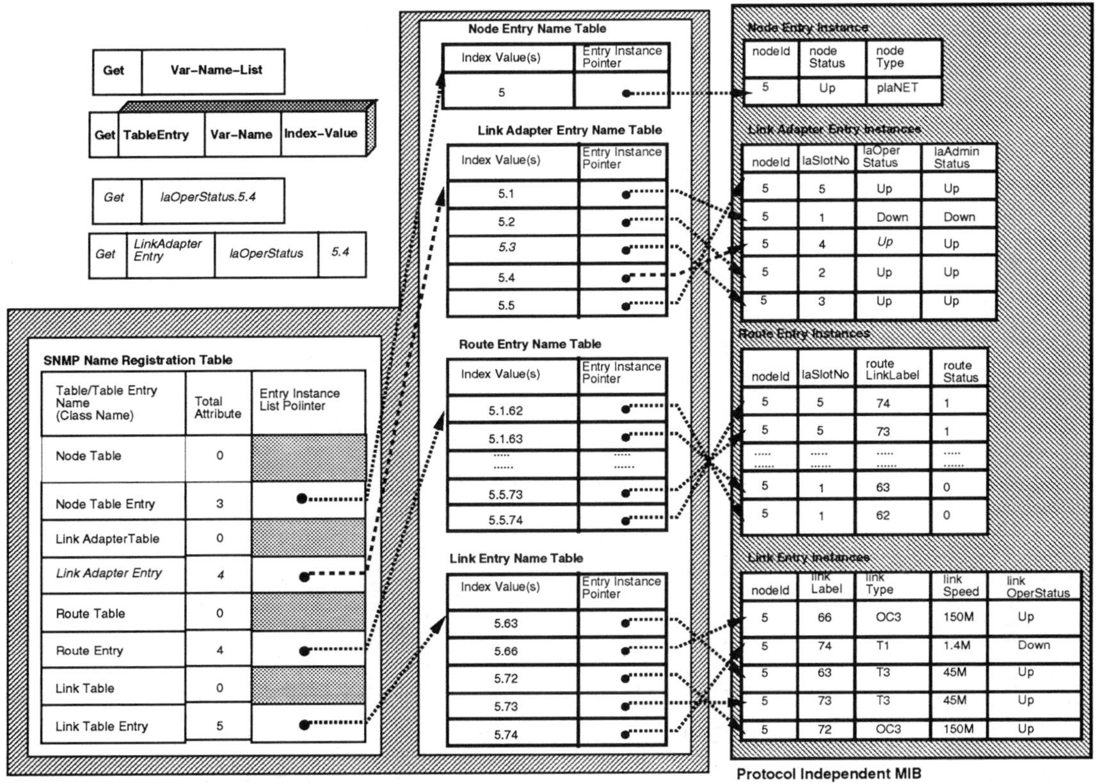

Figure 6: Organization of SNMP Name Registration Table.

The function DO_SNMP_Registration initializes the columns of the first and second level of the registration table. We also make another assumption that when a variable of a table entry is instantiated, all other variables of the same entry will be instantiated. In order to maintain the lexicographic order of the SNMP MIB, rows representing the first level of registration table will be sorted by the first column, *i.e.*, Table_Name. The rows of the second level of the registration table will also be sorted based on the values of the key-values column (as shown in Figure 6). This scheme will provide lexicographic ordering of all instances of the MIB objects.

5.1 How does SNMP Registration Work?

Let us assume that the Link_Adapter_Entry of the Link_Adapter_Table has *four* columns, nodeld, laSlotNo, laAdminStatus, and laOperStatus, and the INDEX parameters are nodeld and laSlotNo. An instance of the Link Adapter with LDN `nodeId=5,laSlotNo=5`, will register using

```
DO_SNMP_Registration(laEntryOid, 4, "5.5", ptrLA1),
```

where laEntryOid is the MIB OI of Link_Adapter_Entry, 4 indicates that there are 4 columns of the entry of the table, "5.5" is the the values of INDEX parameters, `nodeId` and `laSlotNo`, and finally ptrLA1 is the internal storage pointer to the actual object instance. The `DO_SNMP_Registration` first finds the row corresponding to the table entry name specified by laEntryOid and then initializes the columns corresponding to that row. Using the pointer in the last column of the row, it identifies the corresponding name table. Then it adds a row to the table entry name table with the values of INDEX parameters and the internal storage pointer and the orders the rows in the lexicographically based on the values of INDEX parameters.

5.2 How do the SNMP GET and SET operator work?

An SNMP GET request for laOperStatus comes in the following format:
GET laOperStatus.5.4 (or LinkAdapterEntry.3.5.4, since LinkAdapterEntry.3 is laOperStatus) (for the entry with nodeld=5, laSlotNo=4).

After receiving the GET SNMP query, the variable name of the query is parsed to find out the OI of the entry name, OI of the column variable, and the instance identification information. For example, in the above case, laOperStatus.5.4 (which is equivalent to LinkAdapterEntry.3.5.4) will be parsed into {LinkAdapterEntry, laOperStatus, 5.4}. Once the parsing is done, the operation of the GET operator is straightforward. First it finds out the corresponding row in the name registration table that matches the entry class name. Then a specific instance of that entry class name is identified by comparing the instance information (5.4) with the index values of the rows in the second level of the name registration table. After a specific entry is identified, a generic GET is done on the protocol independent MIB using the OI of the requested variable (laOperStatus). Based on the Figure 6, the response to the query would be *laOperStatus.5.4=Up*, where the value of *laOperStatus* is *Up*.

The SET operation of SNMP is similar to the GET operation in accessing a variable in the protocol independent MIB.

5.3 How does the SNMP GET-NEXT operator work?

The GET-NEXT operator is not straightforward and described below using an example. For a detailed description see [MAZ92].

If we send the query GET-NEXT laAdminStatus.5.5 (LinkAdapterTable.1.4.5.5) to an agent for the MIB described in Figure 6, then the variable name will be parsed into "LinkAdapterEntry", laAdminStatus("LinkAdapterTable.1.4") and "5.5". From the laAdminStatus we will get the column number as 4. Once we have identified the instance list pointer from the name registration table, we will find that there are no rows after the index value "5.5" in the LinkAdapterEntry instance name table. So we increment the column number by 1 to 5, but the new column number exceeds the total number of attribute values for LinkAdapterEntry. So we get the next SNMP table name, RouteTable, from the name registration table and column number is initialized 1 and index values to NULL. Since RouteTable has no attributes, we get the next SNMP table name in the name registration table, RouteTableEntry, and again column number is initialized 1 and index values to NULL. Using the NULL index value we will get the first row of the RouteTableEntry instance name table (with index value 5.1.62) and return the corresponding pointer. The new column variable name will be RouteTableEntry.1 (nodeId) and a getValue call on this entry will return value 5. The name of the new variable will be RouteTableEntry.1.5.1.62, obtained by concatenating the new column variable name and new index values. Thus the response to the query would contain: RouteTableEntry.1.5.1.62=5

6. Conclusion

In this paper we have shown how to build a protocol independent management agent that can support both SNMP and CMIP protocols. Lack of CMIP based management applications will slow the process of acceptance of CMIP capable management agents. Through our scheme CMIP capable management agents can support the SNMP protocol with simple extension and without building a separate MIB for SNMP. As we have shown that the cost for SNMP extension is the building of the tool for translation of GDMO style MIB object classes into the SNMP style objects. Based on our scheme, a vendor can build a CMIP agent without worrying about the existence of CMIP based management applications. Since there are large number of SNMP based management applications, they can be used until the application developers can implement CMIP based management applications.

References

[CHE76] P.P. Chen, "The Entity-Relationship Model - Toward a Unified View of Data," *ACM Transactions on Database Systems*, vol. 1, pp. 9–36, March 1976.

[ELM89] R. Elmasri and S.B. Navathe, *Fundamentals of Database Systems.* The Benjamin/Cummings Publishing Company, New York, 1989.

[EMB91] Jock Embry, Peter Manson, and Dave Milham, "Interoperable Network Management: OSI/NM Forum Architecture and Concepts," in *Second International*

Symposium on Integrated Network Management (I. Krishnan and W. Zimmer, editors), pp. 29–44, Washington, D.C., 1991.

[ISO88] International Standards Organization - ISO, "Information technology - Open System Interconnection - Protocol specification for the Association Control Service Element," vol. ISO/IEC 8650, 1988.

[ISO90a] International Standards Organization - ISO, "Information technology - Open System Interconnection - Common Management Information Protocol - Part 1: Specification," vol. ISO/IEC 9596-1, November 24 1990.

[ISO90b] International Standards Organization - ISO, "Information technology - Open System Interconnection - Common Management Information Service Definition," vol. ISO/IEC 9595, November 24 1990.

[ISO92] International Standards Organization - ISO, "Information technology - Open System Interconnection - Structure of Management Information: Guidelines for the Definition of Managed Objects," vol. ISO/IEC 10165-4, June 24 1992.

[MAZ92] S. Mazumdar, S. Brady, and D. W. Levine, "Design of Protocol Independent Management Agent to Support SNMP and CMIP Queries," Technical Report # RC 18246 (Log 80062), IBM T.J. Watson Research Center, Yorktown Heights, NY, August 1992.

[RFC90a] M. Rose and K. McCloghrie, "Structure and Identification of Management Information for TCP/IP-based Internets," May 1990.

[RFC90b] J. Case, M. Fedor, M. Schoffstall, and J. Davin, "A Simple Network Management Protocol (SNMP)," May 1990.

[ROS91] M. T.Rose, *The Simple Book: An Introduction to Management of TCP/IP-based Internets.* Prentice Hall, Englewood Cliffs, NJ, 1991.

[WU92] S. F. Wu, S. Mazumdar, S. Brady, and D. W. Levine, "On Implementing a Protocol Independent MIB," Technical Report # RC 18248 (Log 80064), IBM T.J. Watson Research Center, Yorktown Heights, NY, August 1992.

Integrated Network Management, III (C-12)
H.-G. Hegering and Y. Yemini (Editors)
Elsevier Science Publishers B.V. (North-Holland)

An Application Gateway Design for OSI-Internet Management

Pramod Kalyanasundaram and Adarshpal S. Sethi
(email: kalyanas@cis.udel.edu, sethi@cis.udel.edu)

Department of Computer and Information Sciences
University of Delaware, Newark, DE 19716

Abstract

Network Management is an area of active current research. The problems in managing networks have been compounded by the existence of different standards. These problems can be solved by either forcing just one standard or by achieving unified management of dissimilar networks connected together. The latter solution is the one that is feasible. In this paper, we discuss and contrast two network management models (Internet and OSI) that are prevalent, define problems and issues involved in managing a network belonging to the Internet domain from an OSI domain, develop a paradigm that achieves uniform management of networks (using an application gateway approach) in either domain and present the justification for such a paradigm. Based on the paradigm developed, guidelines for the design of an application gateway are laid out. Future research directions based on this effort are included.

Keyword Codes: C.2.2; C.2.m; K.6.4
Keywords: Computer Communication Networks, Network Protocols; Miscellaneous; Management of Computing and Information Systems, System Management

1 INTRODUCTION

Internetworking has assumed importance due to the following: the existence of multiple standards and realization of the ultimate goal of networks, i.e., ubiquitous connectivity. With the evolution of large networks, network management has become a key issue from the viewpoints of administration, operation, reliability and fault tolerance. The problem of network management is compounded by the fact that more and more large networks with different network management models and schemes are to be interconnected in the process of achieving universal connectivity. This paper addresses the problem of interconnection in the context of network management only.

The scope of this paper in analysing the interconnection problem is restricted to the Internet and the Open Systems Interconnection (OSI) domain. The motivation for such an approach results from the following: these are the models that are prevalent, have current practical significance and allow one to concentrate on solving a subset of the general problem in the framework of the more general problem. An additional motivating factor is that the experience gained from solving this restricted problem will help in addressing the general problem of interoperation of two or more networks supporting dissimilar management models.

Research has been done [WS90] and is ongoing [Kni92] to achieve a unified network management environment. This paper develops a paradigm based on an application gateway approach [Ros90] that helps achieve internetworking of networks supporting the Internet model [RM90] and the OSI model [DIS90a] of network management.

We first describe and contrast these two models in Section 2 of this paper. In Section 3 we discuss different paradigms for internetworking and justify the selection of one paradigm to achieve internetworking namely the use of an application gateway. Section 4 addresses some of the issues involved in internetworking that must be considered in the design of such a gateway. It also presents possible solutions to these issues and outlines the design of an application gateway that achieves internetworking between OSI and Internet domains. The last section presents future research directions based on this effort and summarizes the conclusions drawn from this research.

2 NETWORK MANAGEMENT MODELS

In this section, two common, prevalent models (Internet and OSI) of network management are introduced with a description of their salient features. The main idea is to provide a flavor for these two frameworks and to try and contrast the two models. The reader is referred to [DIS90a, Ros91b, Bla92, RM90] for details.

2.1 The Internet Model

The basic model (shown in Figure 1(a)) comprises the following: at least one Network Management Station (NMS), several managed nodes and a management protocol for communicating management information between an NMS and a managed node. An NMS is a host that runs network management applications and also supports the management protocol. A managed node could be a host, a gateway or a media device. This model aims at having minimum functionality at a majority of nodes and more functionality at a small fraction of the nodes in the network.

Each managed node supports the operational protocols, the management protocol and an interface entity. The interface entity projects the 'real' variables defined and used by operational protocols as managed objects. This entity also defines a set of operations on these variables. The set of managed objects and operations are called the Management Information Base (MIB) [RM90, RM91]. Each Network Management Station performs the role of a manager and supports the management protocol and applications.

The Internet model is based on a request/response protocol. The roles of the manager and agent (managed node) are fixed. The manager by setting the value of a managed object can instruct the agent to perform a particular management operation. The manager can query the values of certain managed objects (which are changed by the managed node), to determine the current status of the node. Thus, the whole communication aspect of this model reduces to just setting and getting the value of managed objects. The Internet model defines objects (Figure 1(c)) [RM90, RM91] which have the following features: access, status, name and syntax. Access controls read/write privileges, status classifies the managed object as optional, mandatory or obsolete vis-a-vis implementation, name denotes the name of the managed object and syntax defines the object specification syntax. The name used to identify an object is called an Object Identifier (OID) which is a unique name in a global name tree.

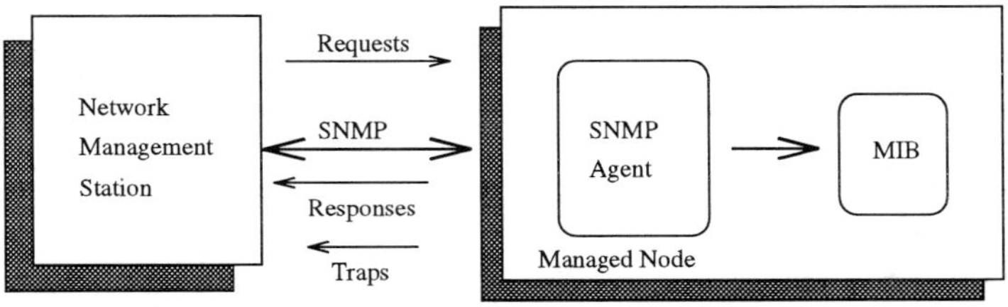

(a) Internet Model for Network Management.

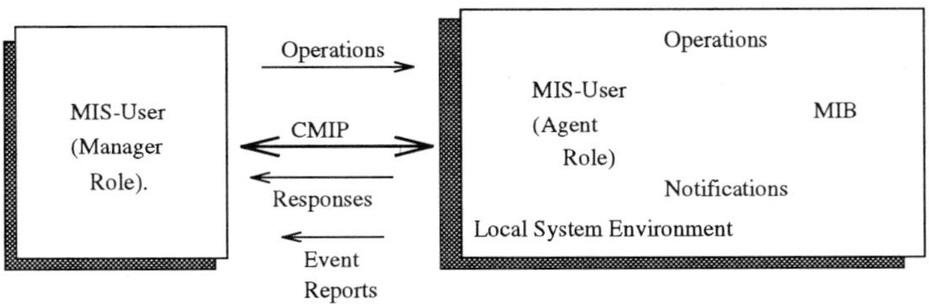

(b) OSI Model for Network Management.

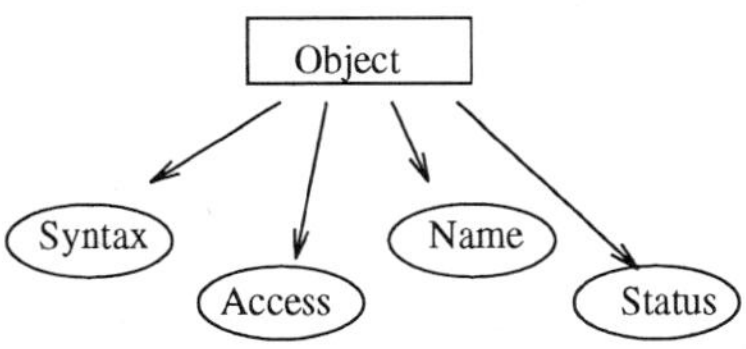

(c) Internet Object Definition.

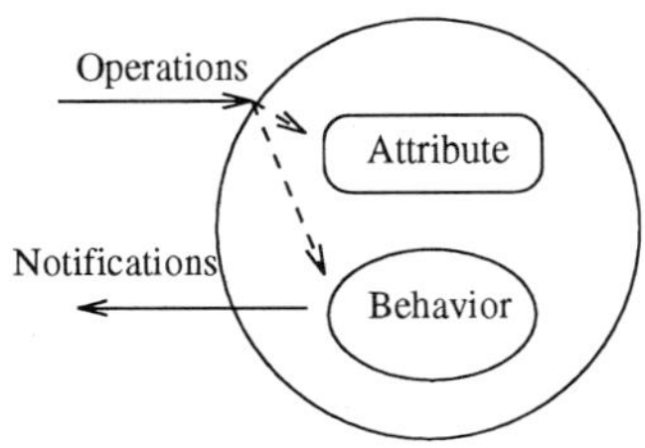

(d) OSI object Definition.

Figure 1: Internet and OSI Management Models and Objects.

The management protocol supported by the Internet Management domain is the Simple Network Management Protocol (SNMP)[1] [JDCD90]. SNMP is built over the well known User Datagram Protocol (UDP) [Pos80]. SNMP supports four basic operations: **Get** allows a manager to query an agent about the value of a particular variable. **GetNext** is a powerful operation that allows MIB traversal and facilitates table entry accesses. A **Set** operation is used by a Manager to set a variable to a specified value. The final operation, **Trap**, allows an agent to asynchronously report the occurrence of an event or fault to the manager.

Another important aspect of this model is the concept of a *proxy* agent. A proxy agent helps in achieving management of nodes that do not implement the complete set of functions defined by the management protocol or nodes that have a different protocol

[1] A revision of SNMP called SNMP Version 2 is currently under discussion in the Internet community.

suite and/or management protocol. It responds to management requests on behalf of the managed agents and can thus act as an application gateway [Ros91b] when there is a need to connect to a network supporting a different management protocol.

2.2 The OSI Model

The OSI model (Figure 1(b)) is primarily a more powerful and generalized model; like all powerful models, the trade-off is in terms of simplicity. Network management is achieved over associations established between application layer processes (called *management processes*). A management process can assume one of two possible roles: manager or agent. The same process may assume different roles over different associations but the roles are fixed for a given association as determined during association establishment. The manager process requests the execution of management operations from the agent which performs them and returns the responses to the manager. The agent may also generate event reports asynchronously to be sent to the manager. The manager-agent interactions are supported by an OSI standard management service known as the Common Management Information Service (CMIS) [DIS90d], implemented by the OSI standard management protocol called the Common Management Information Protocol (CMIP) [IS987].

As in the Internet model, each OSI agent controls a MIB. However, the structures of the MIBs in the two models are very different. The OSI model follows an object-oriented approach. Figure 1(d) highlights the managed object definition principles [DIS90b] used in this model. An OSI managed object is defined by the following: attributes, behavior, notifications and operations. Attributes are the characteristics specific to an object, operations are those that can be performed on the object (eg. setting a value, creating an instance), notifications are emitted by the object to indicate some event, and behavior which primarily dictates the changes in the object due to operations performed on it. This is distinctly different from the way objects are defined in the SNMP framework.

We now list the services provided by CMIS. The **M-GET** service is used by a manager to retrieve attribute values from managed objects. **M-CANCEL-GET** operation allows a manager to cancel a previously requested get operation before such an operation completes. The **M-SET** allows a manager to set the attributes of managed objects. The **M-ACTION** service facilitates operations on managed objects other than those defined as part of the OSI standard. Such a service allows designers to implement object specific actions based on bilateral agreement. **M-CREATE** allows the manager to request the agent to create an instance of a managed object. When the agent receives such a request, it attempts to create the specified object and responds with the result of the operation. To delete an instance of a managed object, a manager uses the services of the **M-DELETE** operation. Finally, **M-EVENT-REPORT** is used by the agent to report to the manager the occurrence of an event associated with a managed object.

The primary source of power in the OSI framework results from two major functions allowed by CMIS, namely *scoping* and *filtering*. Scoping allows a manager to select either a single object or multiple objects within a subtree of the MIB for a specific operation. Filtering allows the manager to specify a set of conditions which have to be met if the specified operation is to be performed on an object. In a typical manager's request both scoping and filtering may be specified. The agent first applies scoping and comes up with a collection of objects suitable for the operation. It then applies the specified filter on the resulting objects. The requested operation is performed on all the objects selected by the filter.

3 PARADIGM DEFINITION

To achieve internetworking it is important to come up with a good paradigm that forms a basis for interoperability. In this section three paradigms that facilitate internetworking between CMIP and SNMP are described and contrasted. Of the three paradigms presented, one is chosen for this research. The reasons for such a choice are also explained. The three paradigms are shown in Figures 2(a), 2(b) and 2(c).

The first paradigm (Figure 2(a)), has most of the complexity built into the Network Management Station [WS90]. This paradigm implements a three tier system with the topmost tier being the management applications with suitable user interfaces, an intermediate layer (Network Management Language System (NMLS)) that provides a standard management command interface to protocol stack machines[2] and finally the stack machines themselves. Users specify commands in NML. These commands pass through the NMLS and are received by the stack machines through a standard command format. The stack machines are then responsible for converting this "standard command" into stack specific command format. These functions add complexity to the manager software and if implemented on a single node would lead to large space and time requirements for the NMS; if the implementation were to be distributed, then a separate interface (or protocol) needs to be designed that implements the distributed interfaces in a consistent fashion. With this paradigm existing managers cannot achieve internetworking without implementing the functions mentioned.

The second paradigm (Figure 2(b)), attempts internetworking using the concept of duplicate MIBs. This approach maps objects in the Internet MIB onto corresponding objects in an OSI MIB (for instance, MIB-II is mapped into OIM-II) [RFC91]. In this paradigm each node has an implementation of an SNMP agent as well as a CMIP agent. The Internet MIB and its mapped OSI counterpart are present at each Internet node. A request from an OSI manager is passed on to the OSI agent whereas a request from an SNMP manager is handled by the SNMP agent. The drawbacks with this paradigm are (a) The MIB translation is manually done. Hence changes in MIB objects involve re-work. Translating all existing MIBs in this manner would be a gigantic task. (b) There needs to be a version of a MIB and an agent that manages the MIB for each type of network being interworked and (c) The presence of more than one agent can lead to multiple reader-writers problems.

The final paradigm being presented (Figure 2(c)) uses the concept of an *application gateway*. In this approach there are special nodes designated as "gateways" that implement the translation between CMIP and SNMP. The rest of the nodes in either network retain their functionality. A scheme that allows easy name and object mapping is used that is independent of MIB changes. The mapping uses the characteristics of the Internet MIB instead of a direct one-to-one mapping of objects. This paradigm does not require more than one agent implementation, since the CMIS agent can achieve translation between protocols and only the translation algorithms need to be added. With such a scheme, a manager in the OSI domain can manage Internet nodes with absolutely no

[2] A stack machine is a term used to denote the software implementing a particular protocol stack/suite of protocols.

change. This definitely is a major advantage on the OSI side since adding more functionality to a manager means that this has to be added to all nodes since any node could function as a manager.

Application gateways interface with the OSI domain on one side and interface

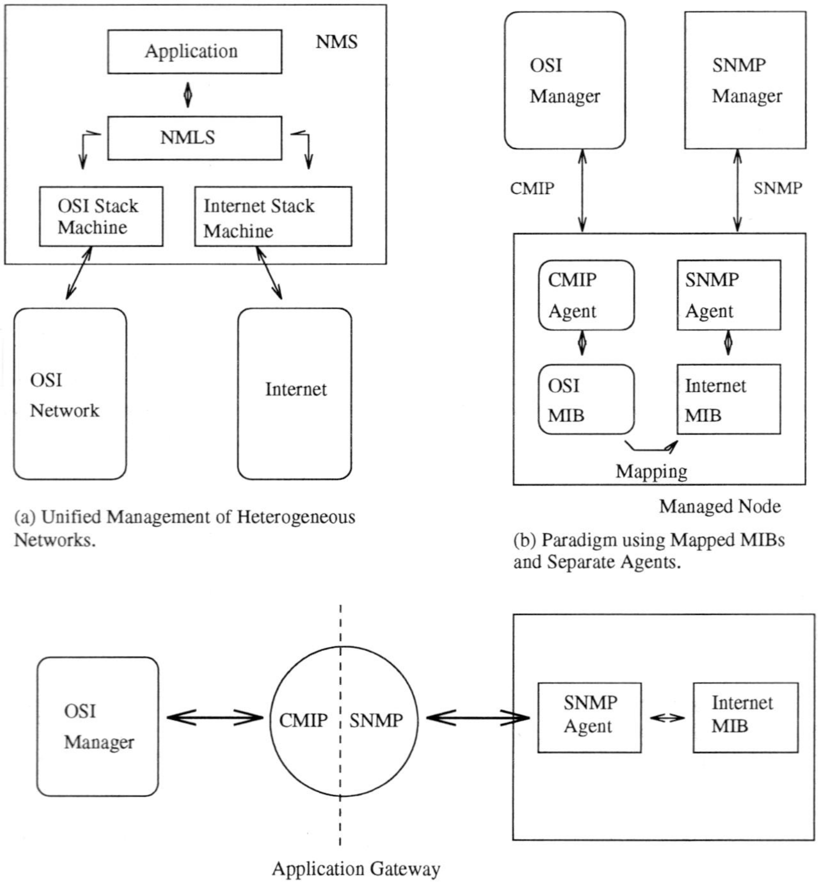

(a) Unified Management of Heterogeneous Networks.

(b) Paradigm using Mapped MIBs and Separate Agents.

(c) Application Gateway Approach.

Figure 2: Paradigms that achieve CMIP-SNMP translation

with the Internet domain on the other (Figure 2(c)). A gateway inputs CMIP requests and outputs SNMP requests; it receives SNMP responses/traps and responds on the OSI side with CMIP responses to achieve internetworking. This paradigm allows a lot of flexibility and power. If a convenient interface can be generated, then several different translations could be achieved by a single agent by just allowing the agent to choose a suitable translation. This paradigm ensures reliability since the loss of a gateway does not result in the manager losing control. By choosing an alternate gateway, the manager can continue unhindered. If an alternate gateway is not available still only that part of the network connected to the failed gateway cannot be managed. Another advantage is that only the required translations need to be present at each of these gateways. All these features have led us to choose this paradigm for our CMIP/SNMP interoperability research.

4 DESIGN OF AN APPLICATION GATEWAY

Based on the paradigm selected, this section presents the design for an application gateway, discusses the problems involved and presents solutions to such problems.

4.1 Name and Instance Mapping

Name and instance mapping assumes importance by virtue of the fact that the two management domains (OSI and Internet) have entirely different schemes for naming and defining objects. The Internet scheme defines two types of managed objects: *scalars*, which have only one instance and are static, and *tables*, which can have one or more dynamically instantiable rows. The scalar objects are named using Object Identifiers (OIDs) that are unique. The instance is identified using a standard suffix to these OIDs [Ros91a]. The table objects are defined in terms of the rows with each row object being made of several column fields. Each column field is named using an OID. Uniqueness of a row entry is achieved by using a sequence of fields called the *index*. The index is usually composed of the values of row entries belonging to specified columns. A column variable in a row is identified uniquely by suffixing the value of the index to the OID of the column field that needs to be operated upon.

The OSI scheme offers an approach where a set of guidelines are provided for object definition [DIS90b] and a user is allowed to specify the definition of object classes based on these guidelines. In the OSI framework, object classes are named using OIDs. Each attribute of an object is identified using an OID. However, instance identification uses the concept of a *containment tree* and is independent of the OIDs used to name the classes and attributes. The containment tree is a dynamic instance tree that captures the containment relationships between managed object instances. There is no equivalent concept on the Internet side. Each object instance in this tree is contained in another instance called its parent. All children of a given parent are identified uniquely with respect to that parent by a *Relative Distinguished Name* (RDN) which is an attribute-value pair. A *Distinguished Name* (DN) is obtained by concatenating the RDNs from the root of the instance tree to the node which represents the object instance being considered. The reader is referred to [DIS90b, Bla92, DIS90c] for further details on OSI object definitions and naming.

Since the two schemes are radically different, finding an efficient and clean method of mapping objects from one scheme to the other is of great importance in achieving internetworking. Such mapping will ensure that either management domain will be able to address objects and object instances that is in keeping with the local scheme by bridging the differences. In essence, name mapping must be simple and easy to automate. It must be generic enough so that changes in managed objects do not lead to the requirement for new mapping schemes. The scheme must capture the essential characteristics of the MIBs.

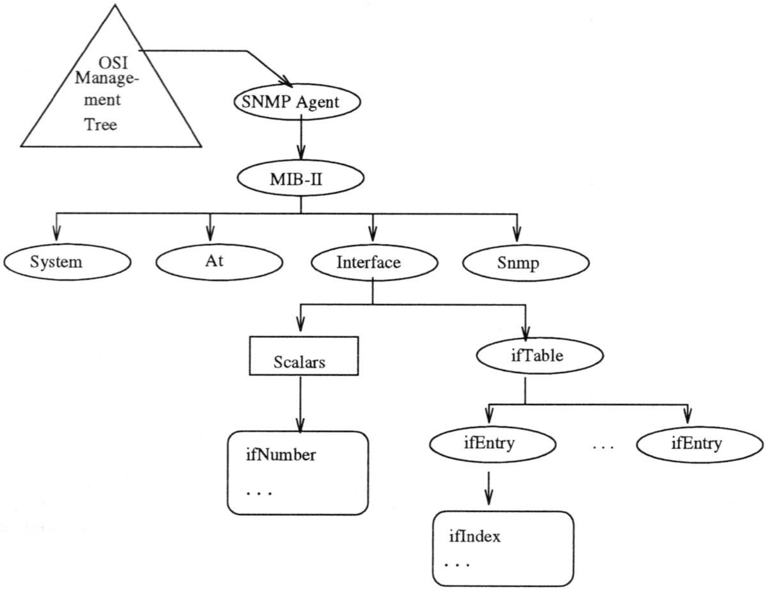

Figure 3: Mapping that preserves Internet MIB heirarchy

The name mapping scheme presented here maps the Internet MIB in such a way that an OSI manager is able to view the Internet MIB as a part of an OSI MIB structure. The name mapping structure has been chosen to preserve the Internet MIB structure (shown in Figure 3). The mapping scheme suggested does not aim at mapping each object on the Internet into an equivalent object or class on the OSI side. Instead it aims at providing the required OSI view by defining new object classes and mapping objects in the SNMP domain to objects and attributes in the OSI domain. This mapping defines two new object classes: *scalars* and *tables.* These classes are defined so that they are consistent with the nature of the objects supported by the Internet domain. These object classes could in future be enhanced to allow better internetworking.

Scalar objects on the SNMP side can have at most one instance. This characteristic maps well with the concept of an object attribute on the OSI side. Hence all scalar objects on the SNMP side can be mapped on to attributes of objects on the OSI side, with their OIDs retained.

Table objects on the SNMP side are made up of rows that can be dynamically created or deleted. This feature of table objects fits in well with the concept of dynamic instantiation or deletion of OSI objects. Hence a *row object class* has been defined. There can be multiple instances of this object since multiple row entries for a table are permitted. A row object has the column variables as its attributes. Each instance of a row object on the OSI side corresponds to one row entry on the Internet side.

Based on the above object class definitions, we present the structure of the mapped MIB (Figure 3). At the highest level, there is an object class for an SNMP agent. Depending on the number of SNMP agents being managed there will be a number of instances of objects of this class. These SNMP agent nodes form the root of the SNMP part of the OSI MIB.

Each SNMP agent node contains a MIB object class. Each SNMP agent node is allowed to have exactly one instance of a MIB object associated with it. This MIB corresponds to either MIB-I or MIB-II. For our examples, MIB-II has been chosen. The MIB class contains the SNMP MIB-II groups (system, interfaces, at, ip etc.) as object classes. There can be exactly one instance of each of the group objects for each MIB. Each of these group classes contain two special object classes: scalar and table object class. Such definition preserves the features of the Internet MIB structure on the OSI side and conform to the OSI MIB environment. Each group object contains an instance of a scalar object and one instance of the table object for each corresponding table defined in the Internet MIB.

The scalar object in a group on the OSI side has scalar variables corresponding to that group in MIB-II as its attributes. Each Internet scalar variable can be operated upon similar to operations permitted on attributes of objects on the OSI side. Powerful operations can be achieved by using filters on these scalar variables. It is also possible to get any combination of these scalar variables. Each table object class consists of row objects that are defined using a row object class. Each table object instance contains row object instances that correspond to the row entries of a table on the Internet side. A row object allows an OSI manager to access a row of a table, and modify, create or delete it. Methods for identifying these row object instances are discussed later.

4.2 Functional Mapping

In order to achieve interoperability of the two management domains providing different services, it is mandatory that the services or functions provided by one domain be translated into the other. Such translation is achieved by functional mapping. Name mapping ensures that Management Trees of two different management domains appear similar to a manager in either domain. Functional mapping allows an operation allowed in one domain to be mapped onto one or more operations in the other domain. Functional mapping is the prime focus of this subsection.

The main problem in functional mapping arises from the difference in complexity of functions offered by the two domains. A powerful function in one domain may map into several simple functions in the other domain. It may not be possible to map all the complex functions provided by one domain due to inherent lack of power in the other domain.

CMIS defines services that allow querying and changing attributes of objects, reporting events, and creation and deletion of object instances. In addition, CMIS defines three other functions namely scoping, filtering and synchronization. These functions add complexity vis-a-vis mapping onto the Internet domain in which there is no counterpart. Hence, any functional mapping scheme between CMIS and SNMP must cover scoping, filtering and synchronization. These mappings must be applied in conjunction with the basic mapping between services in the two domains. For instance, an **M-SET** may map to a **Set** for a single object. However, if scoping and filtering result in more than one managed object selection, then the same **M-SET** service will map to one or more **Sets** if best effort synchronization is specified.

Packet size and bandwidth can present problems during function mapping. Packet size restrictions may force a request on the CMIP side to be broken up into more than one request on the SNMP side. Hence an application gateway will have to maintain state information for such requests. This can add to the complexity of the application gateway.

Functional transformation may result in one of two possible cases: a single operation on one system may result in one or more operations on the other system; several operations on one system may map to a single operation on the other system. In either case there is an uneven distribution of packets resulting from functional transformation. This could lead to bandwidth hogging on the system on which multiple operations are required. Functional transformation schemes must ensure that the resulting transformations are both time and bandwidth efficient.

It is obvious that there are major differences in the schemes adopted by the two domains to access and operate upon managed objects, and traverse the Management Tree. Any functional transformation that is attempted must be simple, clean, efficient and easy to automate. The following paragraphs present the design of functional aspects of an application gateway. The transformations are unidirectional i.e. from OSI domain to the Internet domain.

Distinquished Name (DN) and Relative Distinguished Name (RDN): On the OSI side, generation of DN/RDN is important since a manager has to specify the DN when requesting an operation on a managed object. While accessing a scalar object, the DN is that of the scalar object instance to which the requested attribute belongs. The scalar object instances have fixed RDNs. The gateway maps requests for scalar objects by appending the OID of the attribute requested with a default suffix of zero to request a scalar variable on the Internet side. While accessing a row entry, an OSI manager specifies a DN that has the DN of the table object class suffixed by the values of the index fields. Since the index fields ensure uniqueness, the RDN is also unique. This DN contains sufficient information to enable the application gateway to construct the Internet OID along with the index to be used to access a specific row entry in the request to the SNMP agent.

The DN scheme allows a gateway to extract only the matching entry from the SNMP agent as opposed to getting several entries and then filtering them out at the gateway. This saves bandwidth and time. The gateway has internal tables that allow MIB information to be stored. These tables contain Internet managed object characteristics and are used to interpret and translate the DNs received in the CMIP requests.

Scoping: Scoping can result in multiple object selection. This implies that a specified operation (e.g., **M-GET**) has to be performed on all or a subset (due to filtering) of the objects selected. On the Internet side, scoping can be mapped by repeating the requested operation for each object selected. This repetition is implemented as part of an application gateway functional mapping. The application gateway maintains the mapped MIB information and uses this to decide which objects need to be selected for the requested operation.

The following choices are possible in a scoping request (as defined in CMIP): (1) Base object alone, (2) *n*th level subordinates of the base object, or (3) the base object and all its subordinates.

In case (1), the gateway maps the scoping function as a single get request to retrieve the base object alone. In case (2), the gateway has to traverse the MIB and come up with all managed objects that are at a defined level under the specified object. For instance, if all objects that are at depth 2 from the Interface group object are specified

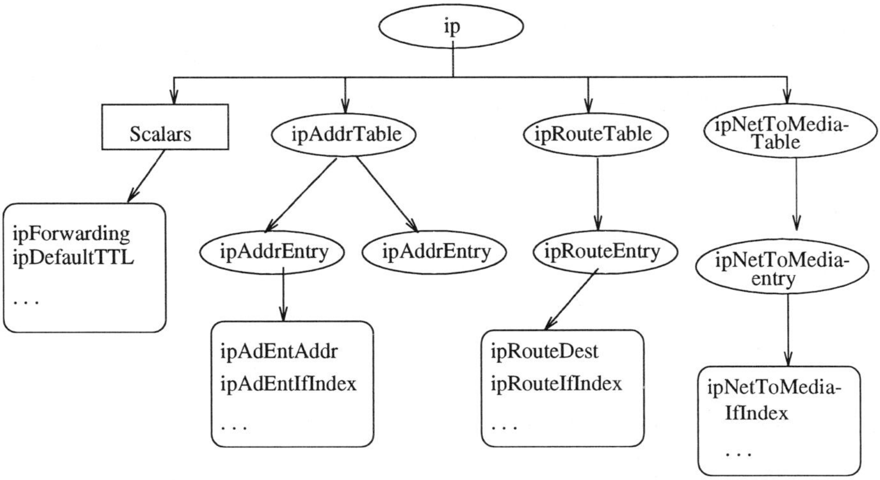

Figure 4: Instance tree of mapped MIB

by scoping , then the gateway will have to come up with all the row objects (Figure 4). Level 1 objects are the scalars and table object instances. The gateway will use the **GetNext** operation to extract information about all the objects at a specified level. In case (3), the function of the gateway is somewhat similar, with the main difference being that the gateway will extract information from all the objects defined in the MIB under the specified object. For instance, if the scoping reflects all objects under group object **Interfaces**, then the gateway will use the **GetNext** operator to extract all the objects defined under group object **Interfaces** using the OID of the Interfaces group.

In effect, the gateway will simulate the actions required by an SNMP manager to operate on several objects and at the same time preserve the interface on the OSI side. Such simulation ensures transparent operation of the OSI manager.

Filtering: A filter specifies a condition that must be true in an object for it to be selected for a specified operation. Filtering imposes further constraints on the objects selected by scoping for an operation. While scoping uses the object hierarchy in an instance tree to select an object, filtering uses the state of an object instance.

In order to implement the filtering function on the Internet side, an application gateway on receiving a request on the OSI side stores the filters associated with the request. The gateway then applies scoping to select objects on which the specified filters need to be applied. The managed objects selected are retrieved from the SNMP agent specified in the operation request. Once the objects have been retrieved, the gateway applies the stored filters and performs the operation only on those objects which test positive for the filter(s) specified.

Packet Size: CMIP has no restriction on packet size. SNMP uses UDP to implement its services. UDP has a packet size restriction which is carried over to SNMP. This packet size restriction adds complexity to the state machine of an application gateway. In cases where a single request on the OSI side translates to several requests on the Internet side the gateway has to maintain the state of each request on the OSI side, collect several replies from the Internet domain and format them into a single reply. In such a case, an application gateway retains the state of the OSI request and collects the responses from the Internet side. Once the complete information is obtained from the Internet side, it is formatted and sent as a response on the OSI side. This adds complexity to the gateway state machine.

Synchronization: CMIS allows two types of synchronization across objects: **atomic** and **best effort**. Atomic synchronization has all-or-none semantics, i.e., the operation succeeds or fails on all objects selected. Best effort aims at performing the operation on as many selected objects as possible. SNMP supports only atomic synchronization[3]. Atomic synchronization on the OSI side translates to all the managed objects being a part of a single request. Best effort synchronization on the OSI side connotes that each request on the OSI side will generate requests that have a single object contained in them.

Based on the above, an application gateway has to allow only atomic synchronization when a manager requests update of multiple objects in a row of a table. Such atomic updates reflect SNMP behaviour and ensure table entry consistency. However, if a manager requests an operation on several scalar objects, then the gateway may either implement them as atomic synchronization or as best effort synchronization.

5 CONCLUSIONS

In conclusion, this paper introduces a paradigm that uses an application gateway approach to achieve internetworking between the OSI and Internet domains. Several key issues in achieving such internetworking are identified and discussed. Based on these issues, the basic design of a gateway that attempts internetworking using the suggested paradigm is suggested. Much work needs to be done in determining the guidelines for reverse translation (i.e from SNMP to CMIP), error mapping and trap to event-report mapping. Future directions include studying the effects of translation between connnection oriented service (CMIS) and connectionless service (SNMP). Extensions or modifications to existing protocols to facilitate better internetworking need to be addressed. This research has helped unearth a myriad of problems, the solutions to which will result in effective internetworking between OSI and Internet management domains. We hope that this work will help in finding a generic solution to the more important problem of achieving internetworking between any two dissimilar network management domains using an application gateway approach.

References

[Bla92] Uyless Black. *Network Management Standards.* McGraw Hill, 1992.

[DIS90a] International Organization for Standardization. *Information processing systems - Open Systems Interconnection - Systems management overview.*, 1990.

[3]Internet Management does not mention synchronization explicitly. It is mentioned as part of the **Set** operation that all or none of the objects specified will be updated.

[DIS90b] International Organization for Standardization. *Information Technology - Open Systems Interconnection - Management Information Services - Structure of Management Information Part 4: Guidelines for the Definition of Managed Objects.*, May 1990.

[DIS90c] International Organization for Standardization. *Information technology - Open Systems Interconnection - Management Information Services - Structure of Management Information - Part 1: Management Information Model*, January 1990.

[DIS90d] International Organization for Standardization. *ISO DIS9595, Information Technology - Open Systems Interconnection - Common Management Information Service Definition*, 1990.

[IS987] International Organization for Standardizaton. *Management Information Protocol Specification - Part 2: Common Management Information Protocol*, June 1987.

[JDCD90] Martin L. Schoffstall Jeffrey D. Case, Mark S. Fedor and C. Davin. *Simple Network Management Protocol. RFC 1157.* DDN Network Information Center, SRI International, May 1990.

[Kni92] Graham Knight, 1992. Private Correspondence.

[Pos80] Jon B. Postel. *User Datagram Protocol. RFC 768.* DDN Network Information Center, SRI International, August 1980.

[RFC91] DDN Network Information Center, SRI International. *OSI Internet Management: Management Information Base. RFC1214*, April 1991.

[RM90] Marshall T. Rose and Keith McCloghrie. *Structure and Identification of Management Information for TCP/IP Based Internets. RFC 1155.* DDN Network Information Center, SRI International, May 1990.

[RM91] Marshall T. Rose and Keith McCloghrie. *Management Information Base for Network Management of TCP/IP Based Internets: MIB-II. RFC 1158.* DDN Network Information Center, SRI International, March 1991.

[Ros90] Marshall T. Rose. Transition and Coexistence Strategies for TCP/IP to OSI. *IEEE Journal on Selected Areas in Communications*, 8(1):57–66, January 1990.

[Ros91a] Marshall T. Rose. *The Open Book.* Prentice Hall, 1991.

[Ros91b] Marshall T. Rose. *The Simple Book.* Prentice Hall, 1991.

[WS90] Unnikrishnan S. Warrier and Carl A. Sunshine. A Platform for Heterogeneous Interconnection Network Management. *IEEE Journal on Selected Areas in Communications*, 8(1):119–126, January 1990.

IV.

MANAGEMENT REQUIREMENTS, EXPERIENCES AND EVALUATION

A.

MEETING USER NEEDS WITH A MIX OF TECHNOLOGIES

Integrated Network Management, III (C-12)
H.-G. Hegering and Y. Yemini (Editors)
Elsevier Science Publishers B.V. (North-Holland)

OMNIPoint: An Implementation Guide to Integrated Networked Information Systems Management

Bruce Murrill, Technical Director
Network Management Forum

Network Management Forum Programme Office, 67 Corder Road, Ipswich, Suffolk, IP4 2XB, United Kingdom

ABSTRACT

This paper describes an industry initiative toward meeting user priority requirements for 'end to end' management of their information environments. The paper outlines the key problems, how the industry structure is changing to solve those problems and the enablers that are allowing this structural change. The paper then goes on to discuss migration issues in moving from proprietary to open management systems, and the applications and support services available today, as part of OMNI*Point*, that will guide the industry toward achieving user objectives.

Keyword Codes: D.4.4; K.6.4; C.2.4;
Keywords: Communications Management, System Management, Distributed Systems.

1. The Challenge of Integrating the Management of Networked Information Systems.

Over the last decade the use of information services has increased dramatically with organisations becoming reliant on widespread availability of information to make them more productive and more competitive. As dependence on information networks has grown, the infrastructure for delivery of information services has become ever more complex and diverse. It is characterised by many components from different vendors interconnected via a range of physical media and communicating using a variety of different protocols. Behind the software applications that the clients see is a complex network of communications and computing equipment, which must work reliably and in concert if the information systems are to meet client requirements.

Operators of complex networks have traditionally relied on a group of proprietary, single-function management systems to control the information network. These proprietary management systems have only two things in common: they each focus on a small part of the network and they do not work together. Such diversity represents a management nightmare.

The complexity of the management task has soared, and as a result the levels of service provided by information networks have been impacted. To make things worse, at a time when equipment costs are falling the overall costs of managing networked systems is increasing because of the need to add extra staff to handle the complexity.

What is needed by operators of networked information systems are management applications that will help them deliver improved service to their clients while reducing the cost of operation. They need applications that integrate management data and that automate management tasks.

For suppliers of equipment and services, the major issue is how best to meet this requirement. Suppliers are faced with a wide range of customer requirements to be met by limited research and development budgets and with an array of potential technical options. Moreover, in many cases, they have limited experience of the management of information networks even though they understand well how to manage their particular system or service within those networks.

The obvious first question for suppliers should be: What do customers actually require from a product or service in order to implement integrated management? Or, rather, what do they want at what price? At a simple level, an analysis of what network components or elements customers want to manage needs to be undertaken, matrixed against what functionality they require. For example, are multiplexers and modems being overtaken by Internet components as a priority for management? Has configuration management overtaken fault or event management as the number one functional priority?

Integrated management is still at a relatively immature stage. However, it is essential to have a keen awareness of what value-for-money perception (or cost-benefit analysis) that customers have. Such analysis can then indicate the marketing and commercial sense in developing interoperable management products - and the bottom line financial justification.

2. Supply-Side Specialisation

For a time, it seemed that every supplier of management systems saw its future as providing the manager of managers, the overall control system for the enterprise network. However, experience and the realities of the complexity of that task have caused most suppliers to concentrate on their own areas of competence. This strategy can result in positive gains for both the supplier and the customer when approached with the intention of working well within a broader interoperable management framework.

Users do not consider systems and network management one large project, any more than they think of, say, computing as one undistinguished project. Instead, they manage specific areas - back-up, access, data, communications, and so on. As has proven to be the case in the personal computer market, when a certain level of standardisation is reached, applications become more and more specific, addressing better-and-better-identified user needs and becoming more and more useful.

The same applies to systems and network management. The market for suppliers of network management systems is evolving into several distinct groups:

* Developers of standard platforms or of platform components.

* Developers of particular network elements and the systems that manage them.

* Developers of management applications that solve certain specific, direct user needs.

* Developers of integrating systems: management systems that integrate the services provided by individual management applications or management systems.

This specialisation comes from a recognition that it is often a more efficient use of limited development budgets to focus on those areas where a supplier can add direct value to the customer's management solution rather than expending effort on developing generic infrastructure code.

3. Technology Choices

The other set of challenges facing suppliers of products deployed in an integrated network environment concerns the technology to be employed in implementing the management system. Choosing the right technology is not easy, simply because there is no single answer to all circumstances. The Internet Simple Network Management Protocol

(SNMP), the Open Software Foundation's Distributed Management Environment (DME), the CCITT Telecommunications Management Network (TMN) model, and OSI Management are all appropriate solutions in some circumstances.

What is required is that choices be made that are appropriate for the environment in which the management systems will operate and that their underlying products can be managed at the level which is necessary in order to employ their capabilities to the full, no matter whether the management system is local or remote, single device type or integrating a range of management systems.

The primary focus of OMNI*Point* is to assist in making those choices concerning both the technology to be employed and the extent to which it needs to be implemented.

4. OMNI*Points*

OMNI*Points*- Open Management Interoperability Points - can be described as providing the best match between user requirements and industry specifications at a given point in time for the integration of the management of multi-vendor, multi-domain networked information systems.

OMNI*Points* were developed through an industry-wide initiative sponsored by the Network Management Forum which incorporates every group or organisation which has an interest in the area of network and systems management, with the aim of encouraging product development. The process is managed by the OMNI*Point* Partners, a collection of user representatives, industry consortia and standards bodies.

OMNI*Point* 1 is a fully implementable and testable profile, agreed by the Partners, for the development of working products. It builds on the Forum's Release 1 specification, and provides smooth migration forward for Release 1 products that are just now beginning to appear. Because it serves as a base for OMNI*Point* 1, and because migration guidelines are provided, the Forum's Release 1 specification is considered an early version of the OMNI*Points* - or OMNI*Point* 0

A key OMNI*Point* concept is the smooth migration path from one to the next. The points are scheduled for completion approximately every two years, fitting in with the time it takes to design, develop, test, produce and update a typical management product. To allow progression and backwards compatibility between products specified according to each OMNI*Point*., a set of mapping specifications from one OMNI*Point* to the next is provided. Thus vendors can follow a single network management scheme, and users will be able to purchase products based on one set of guidelines.

OMNI*Points* include specifications based on the latest standards and industry agreements, development guidelines and rationalised object definitions, to ensure that conforming systems can communicate. As well as the specifications delivered for OMNI*Point,* a guide, 'Discovering OMNI*Point*', offers practical, pragmatic and unbiased help to suppliers and purchasers of management systems.

5. Integration with OMNI*Point*

Because the focus of OMNI*Point* is the enterprise-wide management of networked information systems, in order to deliver cost-effective services to end-users, its main emphasis is on achieving integration of management information through systems interoperability (see Figure 1).

Figure 1: Management Systems Interoperability

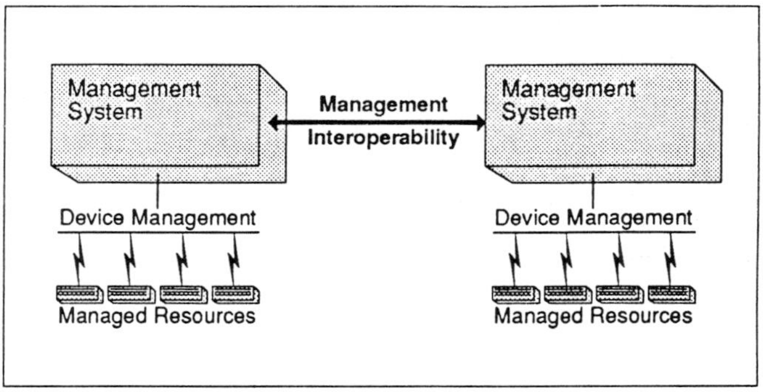

However, OMNI*Point* also starts to address the integration of management applications by specifying API's which should be implemented within Management Systems. This allows the concept of Open Management Platforms (see Figure 2) to be realised, creating a Common Management Applications Environment, and bringing the benefits of applications integration, automation, portability and resultant user choice. Users and vendors are thus able to integrate management solutions in a number of ways; at the applications level, at the information level or a combination of both.

Figure 2: Common Management Applications Environment

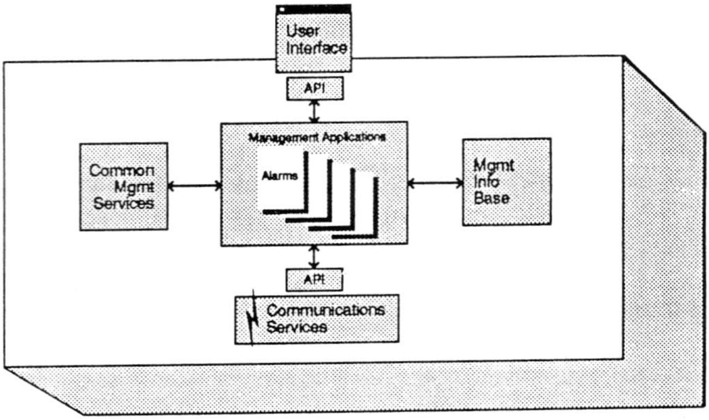

The method for achieving applications integration is through the use of OMNI*Point*-compliant applications. These applications could be run on a single OMNI*Point* compliant management platform, or on a network of similar OMNI*Point*-compliant platforms in an open, distributed, computing environment, (see Figure 3). In this approach, running several management applications in a similar environment, communication between applications can be accomplished either by use of appropriate IPC or RPC mechanisms, or by use of Common Management Information Protocol. For this scenario, the emphasis is on common object services and common, or mapped, management information definitions. This approach is most effective for new systems deployment.

Figure 3: Applications Integration

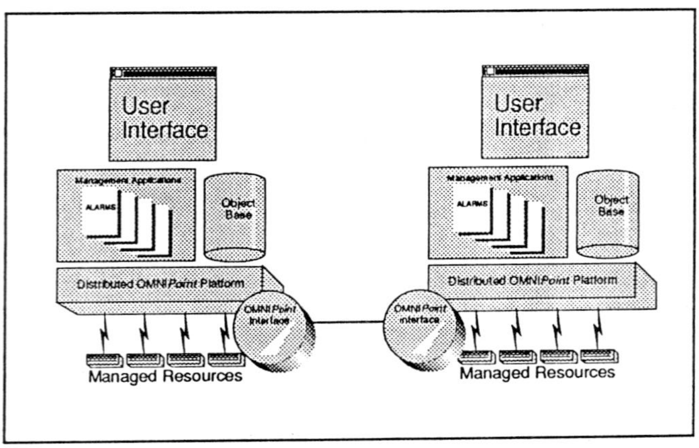

The method to achieve information integration is through the use of a standardised protocol, System Management Functions, and management information data via the OMNI*Point* interface, using the Common Management Information Services (CMIS) over the Common Management Information Protocol (CMIP). The choice of transport network is left to the customer, but with the caution that any two vendors must support the same transport option or support a gateway, if interoperability (and integration) is to occur. (See Figure 4)

Figure 4: Existing Systems Integration

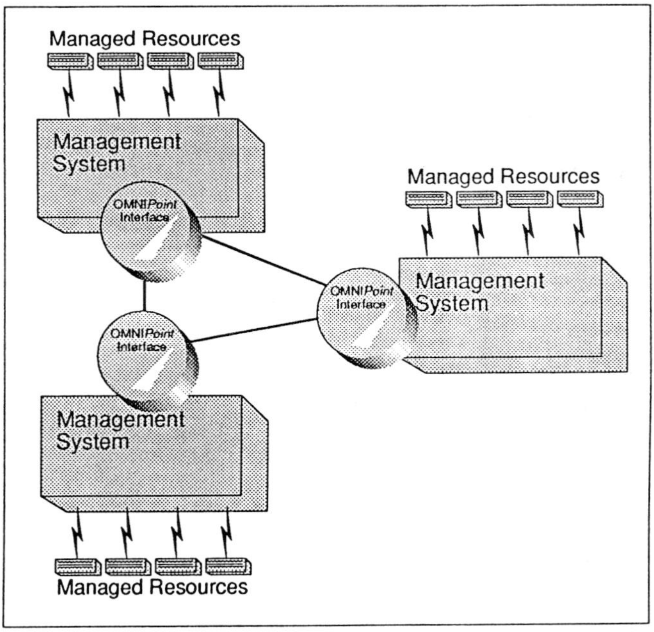

This means that customers can take advantage of OMNI*Point* even though they don't wish to replace their existing management systems, and they may still use a proprietary network to connect their management systems. For companies who have invested heavily in training their management personnel, the ability to gain the advantages of integrated information without disrupting existing systems or installing a whole new management network is critical.

The third way is a hybrid where some management systems are built upon 'open platforms' and others are built in a proprietary way, (see Figure 5). In this scenario, the applications on the open platform could use the Common Management API (X/Open's XMP) and the CMIP protocol to interoperate with other, more internally proprietary, management systems, using the same protocol.

Figure 5: Hybrid Integration

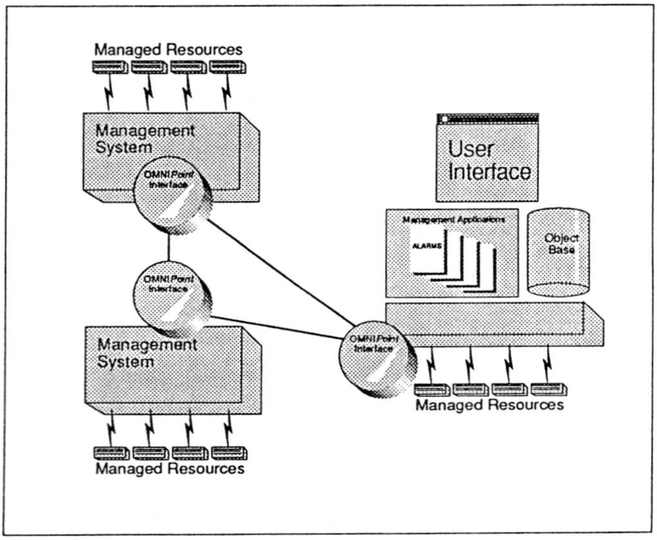

Given that few Management scenarios allow implementation of integration strategies in a 'green field' situation, this approach is likely to prove the most widely adopted as part of a migration plan to open management integrated environments.

6. OMNI*Point* 1 Components

As stated earlier, the components of OMNI*Point* 1 have been selected for their ability to meet top priority user requirements and to provide suppliers with a common tool-kit of industry agreed specifications (and technologies) with which to meet those requirements. OMNI*Point* 1 is not exhaustive; much further work is still needed before it becomes fully comprehensive. It does however start to focus the suppliers on to that set of technologies that the majority of the industry have agreed. Those technologies form the nucleus of ongoing technical strategies to meet user's requirements for functionality and a framework within which it is sensible to provide comprehensive integrated management.

That framework, or Management Model (see figure 6) offers a set of services upon which management systems developers or independent software vendors can build with confidence. It offers a level of stability previously unavailable across the industry which will become more comprehensive, in a controlled way, with time. The plan for enhancement is already firm, thus allowing developers to understand which specific enhancements will become available, and when, and allowing them to make judgements on how to structure products today, which can be more easily enhanced with future capabilities.

The main features of OMNI*Point*1 are identified below although this is not an exhaustive list.

Figure 6: OMNI*Point* 1 Management Model

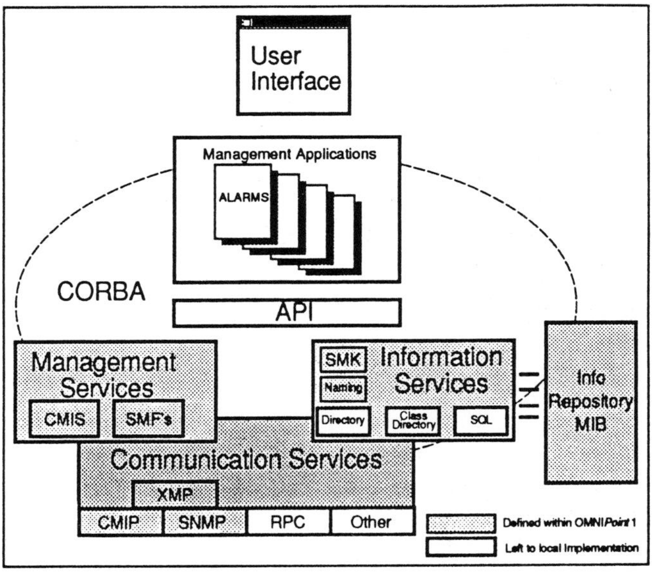

MANAGEMENT SERVICES

Management Services are specialisations of the CMIS Services to support specific functions. These services are modelled in a predefined way with a specific set of "support managed objects" to provide powerful tools which can support specific types of management applications.

System Management Functions -

- General Management Capability (AOM211 DISP) [1]
- State Management & Alarm Report (AOM212 DISP) [2]
- Alarm Reporting (AOM213 DISP) [3]
- General Event Reporting (AOM221 DISP) [4]
- General Log Control (AOM231 DISP) [5]
- Testing Management [6]
- Scheduling Management [7]
- Trouble Management [8]
- Path Tracing [9]
- Security of Management [10]

INFORMATION SERVICES

These include a range of capabilities involved with identifying resources, systems capabilities and locations.

- Managed Object Naming [11]
- Shared Management Knowledge [12]

Object Templates -

- GDMO (IS) [13]

Object Library -

- M3100 Objects [14]
- OIW/NMF Objects [15]
- ISO Transport [16]
- ISO Network [17]
- NMF Objects [18]
- MIB II (GDMO Version) [19]

APPLICATIONS PROGRAMMING INTERFACES (API's)

OMNI*Point* 1 starts to create the Common Management Applications Environment by defining initial API's. Further work is expected during the lifetime of OMNI*Point* 1 to agree further API's, which will be included as 'dot' releases (i.e. OMNI*Point* 1.1)

- X/OPEN
    - XMP [20]
    - XOM [21]
- OMG
    - CORBA [22]

COMMUNICATIONS SERVICES

Commonly supported open protocols which are appropriate to different environments are essential. It is also essential that a common strategy be defined, so that management systems that are required to interwork in mixed environments are able to do so in an agreed way. Again, further capabilities are expected to be defined in the near future.

Protocol - CMIP (ISP's AOM 11 & 12) [23, 24]
SNMP [25]

Coexistence and Interworking Strategies

- SNMP/CMIP Coexistence & Interworking [26]

SUPPORT SERVICES

There are many ancillary services that the industry requires to ensure that management systems can be developed quickly, cost effectively and with maximum opportunity of interoperating. Within OMNI*Point* 1, several such services have been identified and completed. This list is not exhaustive and further requirements have already been identified for future work.

Migration Rules

- OMNI*Point* 0 - OMNI*Point* 1 Migration Strategy and Mapping Rules [27, 28]

Conformance & Testing

Conformance -
- Conformance Requirements [33]
- Conformance Testing Requirements [29]
- Test Specifications [30]
- Reference Implementation [31]

Interoperability -

Test Suite [32]

Procurement/Development Guide (Discovering OMNI*Point*)

Managed Object Library Service

CONCLUSION

The OMNI*Point* programme is the result of unprecedented industry cooperation during the course of the past two years. Its scope and objectives are such that it is the only approach that is likely to succeed in bringing to the user, systems that meet their requirement for the coherent 'end to end' management of their global 'Networked Information Systems'. OMNI*Point* has gained wide support from both the user and vendor community and will continue to be expanded in scope until the management environment is comprehensively specified. OMNI*Point* will guide the developers toward meeting user requirements for Integrated Networked Information Systems Management.

REFERENCES:

[1] ISO/IEC DISP 12060-1, Information Technology - International Standardised Profiles AOM2n OSI Management - Management Functions - Part 1: AOM211 - General Management Capabilities, July 1992.

[2] ISO/IEC DISP 12060-2, Information Technology - International Standardised Profiles AOM2n OSI Management - Management Functions - Part 2: AOM212 - Alarm Reporting and State Management Capabilities, July 1992.

[3] ISO/IEC DISP 12060-3, Information Technology - International Standardised Profiles AOM2n OSI Management - Management Functions - Part 3: AOM213 - Alarm Reporting Capabilities, July 1992.

[4] ISO/IEC DISP 12060-4, Information Technology - International Standardised Profiles AOM2n OSI Management - Management Functions - Part 4: AOM221 - General Event Report Management, July 1992.

[5] ISO/IEC DISP 12060-5, Information Technology - International Standardised Profiles AOM2n OSI Management - Management Functions - Part 5: AOM231 - General Log Control, July 1992.

[6] Network Management Forum: Forum 012, Application Services: Testing Management Function, Issue 1.1, August 1992.

[7] Network Management Forum: Forum 013, Application Services: Scheduling Management Function, Issue 1.1, August 1992.

[8] Network Management Forum: Forum 024, Application Services - Trouble Management Function, Issue 1.0, August 1992.

[9] Network Management Forum: Forum 014, Application Services: Path Tracing Function, Issue 1.0, August 1992.

[10] Network Management Forum: Forum 016, Application Services: Security of Management, Issue 1.0, August 1992.

[11] Network Management Forum: Forum 021, OMNI*Point* 1 Managed Object Naming Specification, Issue 1.0, August 1992.

[12] Network Management Forum: Forum 015, Shared Management Knowledge Requirements, Issue 1.0, August 1992.

[13] CCITT Recommendation X.722 ISO/IEC 10165-4: 1992, Information Technology - Open Systems Interconnection - Structure of Management Information: Guidelines for the Definition of managed objects.

[14] CCITT Recommendation M.3100, Generic Network Information Model, 1992.

[15] NIST Special Publication SP 500-202, Stable Implementation Agreements for Open Systems Interconnection Protocols, Version 5, Edition 1, December 1991, including change pages from March 1992 and June 1992 Open Systems Environment Implementors' Workshop (OIW)

[16] ISO/IEC 10737: 1992, Information Technology - Telecommunications and Information Exchange between Systems - Elements of Management Information Relating to OSI Transport Layer Standards.

[17] ISO/IEC 10733: 1992, Information Technology - Telecommunications and Information Exchange between Systems - Elements of Management Information Relating to OSI Network Layer Standards.

[18] Network Management Forum: Forum 006, Forum Library - Volume 4: OMNI*Point* 1 Definitions, Issue 1.0, August 1992.

[20] X/Open Preliminary Specification, Systems Management: Management Protocols API (XMP), X/Open Document Number: P170, ISBN 1-872630-32-4, July 1992

[21] X/Open CAE Specification, OSI-Abstract-Data Manipulation API (XOM), Document Number: C180, ISBN 1-8720630-17-0, November 1991.

[22] OMG 91.12.1 The Common Object Request Broker: Architecture and Specification.

[23] ISO/IEC ISP 11183-3, Information Technology - International Standardised Profiles AOM1n OSI Management - Management Communications Protocols - Part 3: AOM11 - Basic Management Communications, May 1992.

[24] ISO/IEC ISP 11183-2, Information Technology - International Standardised Profiles AOM1n OSI Management - Management Communications Protocols - Part 2: AOM12 - Enhanced Management Communications, June 1992.

[25] RFC 1157, Simple Network Management Protocol (SNMP), Case, J.D.; Fedor, M.; Schoffstall, M.L.; Davin, C., 1990 May.

[26] Network Management Forum: Forum TR107, ISO/CCITT and Internet Management: Coexistence and Interworking Strategies, Issue 1.0, September 1992.

[27] Network Management Forum: Forum TR110, Migration Strategy, Issue 1.0, August 1992.

[28] Network Management Forum: Forum 022, NM Forum Mapping from Release 1 to OMNI*Point* 1, Issue 1.0, August 1992.

[29] Network Management Forum: Forum TR108, OMNI*Point*1 Testing Requirements, Direction Statement, Issue 1.0, August 1992.

[30] CTS3NM, Conformance Test Specifications, expected Dec. 1992.

[31] CTS3NM, Reference Implementation Description, expected March 1993.

[32] OSINET, Network Management Interoperability Test Specification

[33] Network Management Forum: Forum 020, OMNI*Point* 1 Conformance Requirements, Issue 1.0, August 1992

OMNIPoint

Panel Chair: Jim Warner, OSI Network Management Forum, USA

Panelists: John Payne, DHL Airways, USA
Fran Nielsen, National Institute of Standards and Technology, USA
Gary Francis, IBM, USA

Network managers are responsible for delivering cost effective, quality services to their end users. But none of the current crop of standards, such as SNMP, DME, and CMIP are the complete answer. OMNIPoint is a comprehensive plan for integrating these standards according to real world, user defined requirements. OMNIPoint's object-oriented management model covers interoperable management systems, integrated applications, multiple protocols, installed systems and migration strategies.

This panel will discuss OMNIPoint from both a user and a vendor point of view. Come discover why the OMNIPoint multi-technology model is necessary and why a single technology approach is inadequate. Attendees will also receive a complimentary copy of the guidebook, "Discovering OMNIPoint".

IV.

MANAGEMENT REQUIREMENTS, EXPERIENCES AND EVALUATION

B.

USER REQUIREMENTS

Integrated Network Management, III (C-12)
H.-G. Hegering and Y. Yemini (Editors)
Elsevier Science Publishers B.V. (North-Holland)

Key Technologies for Integrated Network Management

C. Michael Disabato

Network Management Design Group
McDonald's Corporation, 1 McDonald's Plaza, Oak Brook, IL 60521, USA

Abstract

This paper presents a potential network management architecture based on distributed processing and artificial intelligence. The distributed processing component allows a reduction in bandwidth for network management traffic and supports the artificial intelligence component. The artificial intelligence component provides for autonomous diagnosis and trend analysis and takes corrective action, in the event of a network failure.

Keyword Codes: 1.2.11; C.2.3; C.2.4
Keywords: Distributed Art. Intelligence; Network Operations; Distributed Systems

1. Introduction

The common theme in today's computing is a network which attaches geographically separated resources that facilitate computing over a wide variety of transmission media. Applications are no longer centralized with peripheral devices connected to a mainframe. For example, turbulence physics, aerodynamics, chemistry and human vision modeling will challenge the current computing environment. Telepresence is essential in these types of research, in order to keep costs down and maximize use of expensive resources and personnel. Massively parallel systems, while offering the ability to complete the computations in a reasonable amount of time, need reliable supporting networks capable of high speed data transmission[1].

Until now, network management has been focused on hardware failures in static networks, but the real network problem is seldom a hardware failure. It is the constant change induced by software changes, workstation additions and moves, and growth. New releases of software affect the network, often with undesired side effects. Adding servers and workstations changes traffic patterns, requiring bandwidth management. As new devices are introduced to support LAN interconnection, the impact of these devices on the network must be considered.

To support the increasingly critical role of the network, more emphasis must be placed on automation of configuration management and event analysis. Configuration management must learn about changes in device connections on a real time basis. Event analysis must distinguish between normal events (a workstation powering on or off) and an abnormal event (the same workstation losing its power supply). Appropriate action must be taken, and the resolution of the event recorded.

Two key technologies that can reduce operator workload, improve service levels, and reduce operating costs, have received little or no attention until recently. These technologies, when coupled with existing advances, give network managers a powerful suite of tools to address an increasingly complex network environment:

- Artificial Intelligence. Much has been said about using expert systems to automate diagnosis and control. Technologies such as neural networks and case-based reasoning can also contribute to efficient network operation.

- Distributed Management Systems. Spreading the work of managing the network among several cooperating platforms reduces network overhead, and allows intelligence to be placed close to the managed devices.

Good network protocols are robust in the face of partial failures, causing problems in a network to manifest themselves not as overt failures, but as a gradual reduction in performance. Monitoring systems based on these technologies are capable of modeling the network topology, service requirements, and user habits to allow a proactive, rather than a reactive approach to network management [2].

2. Network Management Requirements

In a heterogeneous network, there are useful tools for individual pieces, but not for the whole network. A system is needed that melds these tools into a seamless interface on a single console. Such a system would include:

- A single view of multi-vendor, multi-protocol networks usable by moderately knowledgeable network help desk support staffs.

- Intelligent, self-healing, self-adjusting networks that know how to ease periodic congestion dynamically, automatically route around failures, or suggest permanent changes to optimize the network.

- Automatically updated topologies, configurations, and inventory of all devices on the network, including individual workstations, with mappings to logical networks.

- Predictive "what if" tools to analyze trends, simulate new traffic loads, and plan upgrades.

The goal of network management is to maintain a defined and measurable service level. The management system should assist in this and, itself, be a reflection of the organization's strategic, tactical, and operational processes. Most of the tasks required to do this are routine in nature and can be automated with a good chance of success.

3. An Architecture Model

In order to demonstrate the application of these technologies, the network management architecture developed for McDonald's Corporation will be used[3]. This model divides the network management problem into domains. Each domain, in turn, has three tiers of management, allowing the problem to be decomposed to its basic elements. (See Figure 1.)

3.1 Management Domains

The network management domains encompass the Wide Area Network (WAN), Local Area Networks (LAN), System Services, and Voice Services. Within each of these domains, devices and services exist. (Note that this list is not exhaustive. It is designed to convey a sense of the scope of the problem.)

<u>Wide Area Network</u>

- Fractional T-1 Multiplexors
- CSU/DSU's
- Leased Line & Dial Modems
- Analog and Digital Circuits

<u>Local Area Network</u>

- Network File Servers
- Bridges and Routers
- Workstations
- Intelligent Wiring Hubs

<u>System Services</u>

- SNA Services
- X.25 Packet Switches
- Tandem Systems

<u>Voice</u>

- Private Branch Exchanges
- Software Defined Network
- MegaComm (800) Service
- Integrated Services Digital Network (ISDN)

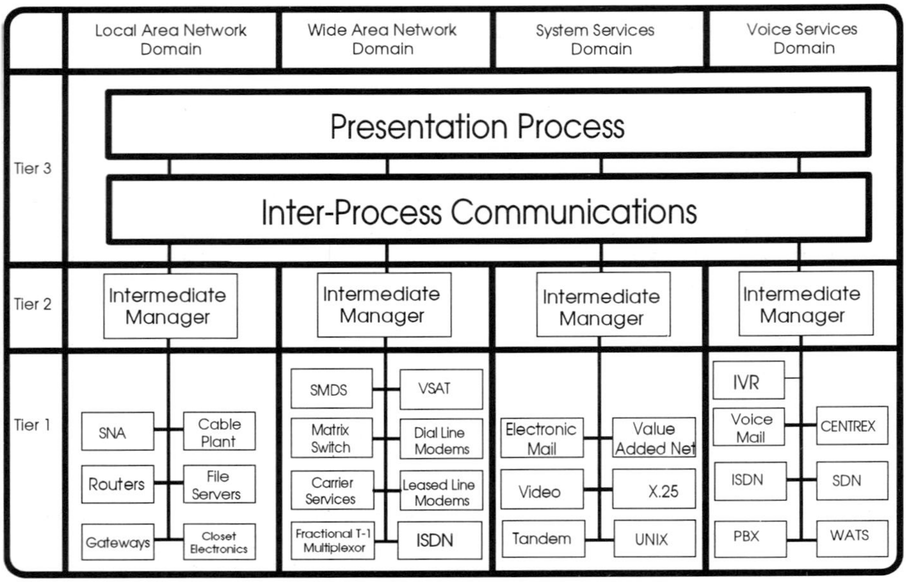

Figure 1.

3.2 Management Tiers

Within each domain, subsystems have been established to distribute the task of managing individual network devices. These are arranged in a hierarchical manner and share information both cross domain and within a single domain.

- Tier 1: On the lowest tier are individual network elements, such as modems, multiplexors, file servers, workstations, etc. These are considered managed objects and are controlled by an element manager dedicated to each class of object.

- Tier 2: The next level is the intermediate or domain manager. This manager controls the subordinate element managers and coordinates their processing within the domain and between domains.

- Tier 3: At the highest level is a sub-system dedicated to coordination of all domain managers, control of the presentation manager, and overall monitoring of the aggregate network. Its primary function is to delegate management tasks to the domain managers.

With the subdivision of work this architecture allows, artificial intelligence and distributed management techniques can be brought to bear on the problem.

4. Artificial Intelligence Components

Fuzziness in fault location exists due to asynchronous alarm reporting and propagation of sub-network failures. Artificial Intelligence (AI) holds great promise for reducing this confusion and providing for the automation of network operations. This can be accomplished by application of classic process control functions and the use of tightly coupled intelligence.

AI processes can enable automatic command submission in response to status changes within the network. This can range from re-routing circuits in the event of failure, to alerting operations personnel of the impending failure of a network element. This results in getting problems diagnosed correctly the first time, reducing down time and the resulting impact on the network customer.

In order to be effective, the scope of each AI element must be limited. The diversity of network elements demands that individual expert tools, tailored to specific devices be developed. This reduces the size of the individual rule bases and eases the maintenance burden. It also allows upgrading or replacing devices with little impact on the other rule bases. Ideally, a standard rule base format would be developed to insure transportability and ease of learning. This format would be developed in the same way as the Management Information Base (MIB) used by the Simple Network Management Protocol (SNMP).

So what exactly should AI do, in terms of network management? Keeping in mind that not all conditions can or will be coded and that human operators will still be needed, a few guidelines should be applied:

- Expert systems should guide the operator through diagnostic procedures.

- Expert systems should react to information generated by the network management sub-systems in a multi-vendor environment.

- Expert systems must be capable of learning from actions taken and documenting the diagnostic process. They must also learn what constitutes normal network operation and inform the operator of any deviations.

- Expert systems must gather whatever information is needed from either the system or the operator, correlate input from multiple systems, and format proper commands for each of them.

4.1 AI Technologies

Traditionally, rule based expert systems have been considered for network management. The "status change-diagnostic response" model has been the paradigm of choice for this type of system. There are two other paradigms that deserve close scrutiny:

- Case-Based Reasoning: This approach examines historical information and develops appropriate rules and strategies. This is a form of machine learning that closely approximates the process an expert uses to solve new problems.

- Neural Networks: These networks model and mimic the operation of the human brain. They are very good at pattern recognition and classification and have application in long-term trend analysis.

4.2 Case-Based Reasoning

Given a previously occurring situation, a rule based system will execute the same statements and reach the same conclusion. In other words, the system does not learn. In addition, if a situation occurs outside the scope of the rule base, the system often cannot provide assistance.

Case-based systems operate in a very different manner. Given an input specification (network event), a case-based system will search for an existing case that matches the input specification. It then provides assistance in the form of a solution to the input specification and stores the actions taken as a new case.

If the system cannot find a matching case, it looks for one that is similar to the conditions in the event. This will not provide a complete solution, but can serve as a starting point. The case-based system, in conjunction with the human operator, modifies the retrieved case to make it fit the event criteria. This is called adaptive reasoning. The result of this is a completed solution, which becomes a new case that is added to the historical database[4].

4.3 Neural Networks

Neural networks are composed of software and hardware equivalencies to the structure of the human brain. These networks are capable of pattern recognition, classification, and of learning their environment. This type of system is capable of detecting slow rates of change in the operation of a network. This is extremely useful in detecting shifts in traffic patterns, slow degradations physical connections, and potential problems in routing configurations[5].

Research continues in the application of neural networks to problems such as network management. The ITRULE algorithm has been developed that speeds the generation of rules from a database. This algorithm has some of the characteristics of a neural network and a case-based reasoning system. It holds promise for development of rule-based systems in the future[6][7].

5. Distributed Management Systems

As the number of managed devices grows, network management traffic will increase, potentially interfering with actual traffic. The polled nature of SNMP requires that the interval between polls be carefully selected to minimize traffic, while maintaining control, since excessive polling can degrade performance through saturation of the network medium[8][9]. Alternatives to polling are the installation of a separate network management backbone, boosting the backbone speed by using FDDI, using traps instead of polling, and distributing the intelligence of the system.

The major problem with traps is they may not trip until the network is just about to crash. As a result, they may never reach the management station to alert the operator. Also, most vendors have not implemented a robust set of traps. A separate network management backbone is costly and must itself be managed. Finally, it is not always cost effective or physically possible to boost the speed of the backbone.

By using a distributed management architecture, the network management system is broken into smaller pieces, suitable for handling the processing of a subnet. This subnet processor collects traps and polls devices, forwarding those statistics, alarms and status information necessary to proactively monitor the system. At night, when network traffic is low, historical data may be sent in a single burst and collected at the central site. This alleviates bandwidth congestion on WAN interconnections and is more conducive to the growth of large networks.

To reduce response time and improve performance, operational decisions should be made as close as possible to the fault. Information concerning these actions should then be distributed wherever necessary. This means that the central management platform should not be directly controlling modems and multiplexors. It also drives a requirement to place the AI components necessary to analyze and direct operations as close to the element managers as possible.

In a geographically dispersed client server environment, this means a configuration for the remote location much like that depicted in Figure 2. Note that there is a network management processor (NMP) with significant hard disk storage in the concentrator. The NMP performs collection and analysis of the network management information. It performs periodic polling of local devices, collects, and analyzes the information. Should a trap be issued, the NMP analyzes the information and makes a determination of the action to be taken.

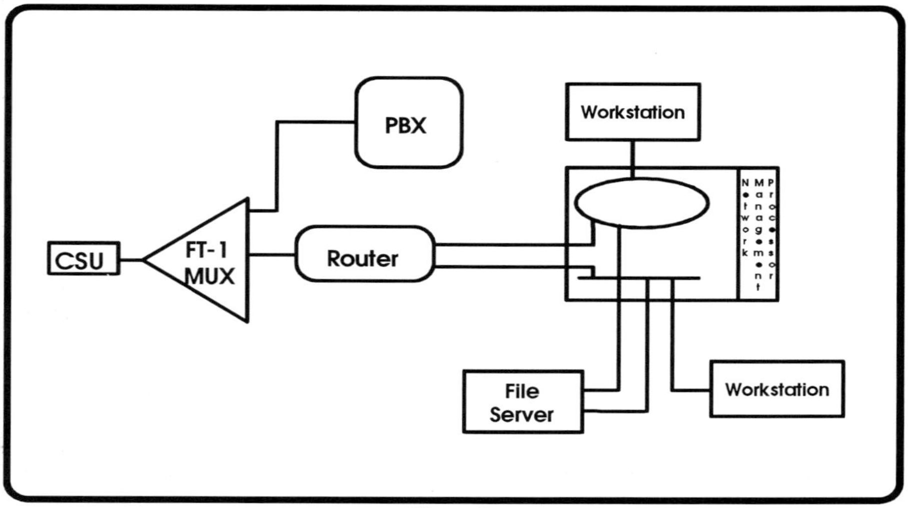

Figure 2.

Note that the NMP also has an expert system and a neural network resident (Figure 3). These AI components perform continuous analysis of the health and status of the network and forward their findings to the central office on a regular basis. Should a failure occur, the contents of the database may be forwarded to the central management station for immediate analysis, if so desired.

Each night, when normal traffic loads are low, the contents of the local database are transmitted to the central management station. This centralized collection of network operations data allows for significant trend analysis to take place with little impact on the routine operation of the network.

6. Policy Generation and Compliance

These technologies come together in an approach which addresses operational, tactical, and strategic operations of the network and can generate operations policies and report compliance[10].

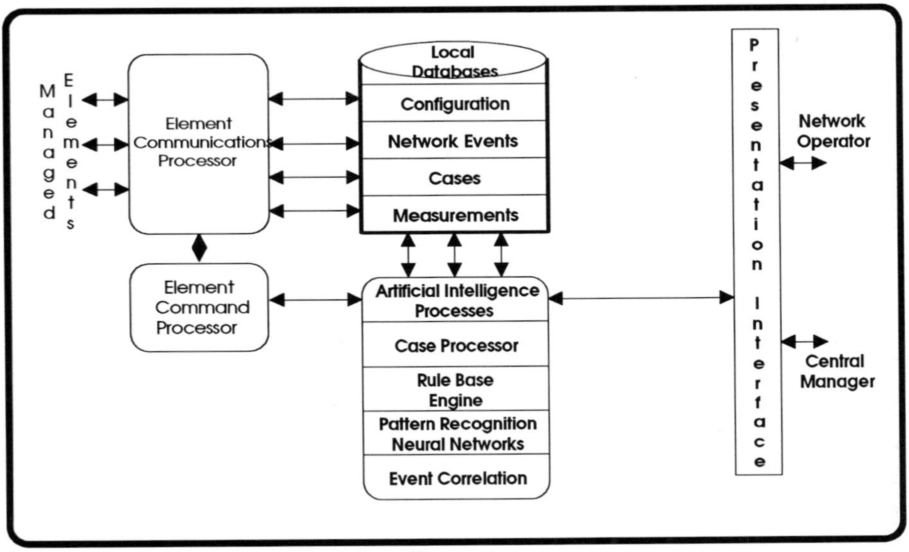

Figure 3.

Analysis of the operating structure begins by breaking the network into its component services, both current and planned. Each element should be identified by application, targeted user groups and service objectives. This enables capacity and performance standards to be established. These internal standards drive network requirements that will be evaluated, managed, and controlled. This was the method used to develop the McDonald's Network Management Architecture[3].

The management system can be decomposed to make functional mapping more visible. The separation of strategic, tactical, and operational aspects allows the integration of a time-sensitive component into the management system. This also allows the controlling infrastructure to be considered separately from the reactionary rule base (Figure 4.) [10].

6.1 Operational Activities

The operational aspects of the network management system is the process control activity, which monitors status and performs real-time measurement of network statistics. As part of this function, the network management system gathers and identifies key network events.

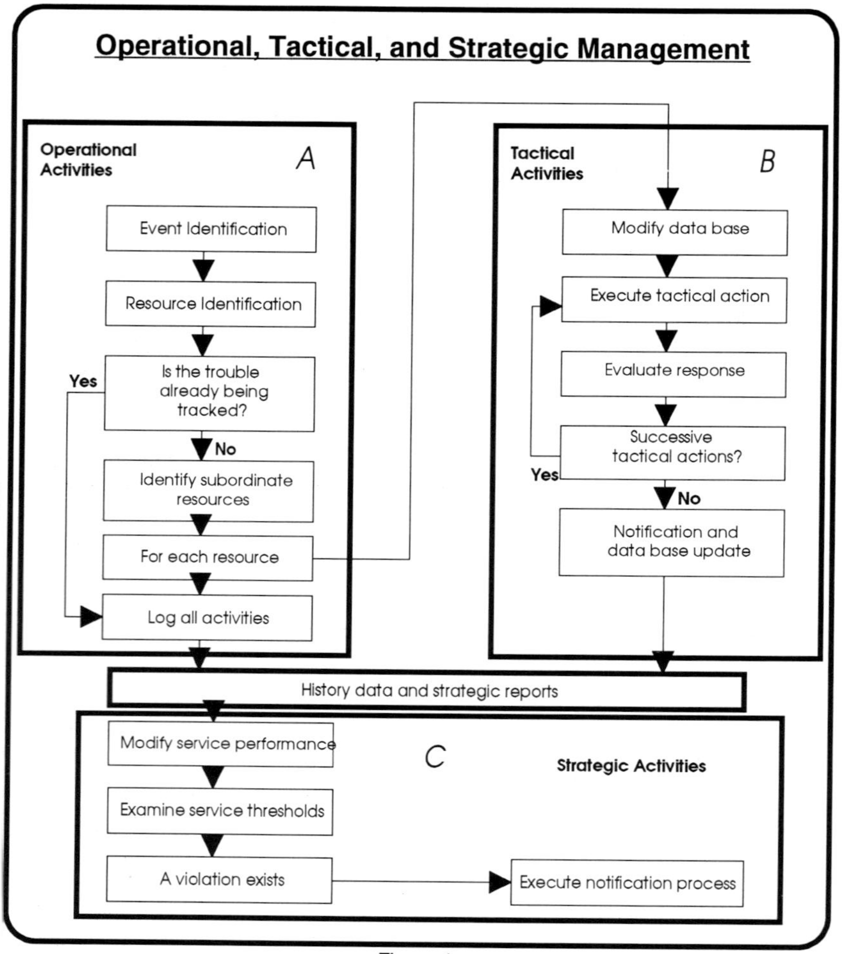

Figure 4.

As individual messages are received by the management system, they are identified and tagged with the source, date and time of origin and classified by affected resources. This association is made using a database representing the network topology and equipment configuration. Every network resource is relationally tied to the network on a vertical and horizontal axis. The vertical axis indicates how facilities are nested within one another (i.e., a tail circuit within a DS-0), while the horizontal axis indicates how facilities interconnect with one another (i.e., how multiple legs of a circuit are related).

At this point, the neural network component learns about the configuration and normal operation of the network based on signal levels, ESF register contents, traffic levels, etc. As traffic increases or lines begin to deteriorate, the neural network notes the change and requests instructions from the expert system. The expert system, in turn, is aware of configuration changes and instructs the neural net to disregard the change, learn about the change and incorporate it into its network model, or takes action on a detected event.

6.2 Tactical Activities

Tactical actions reflect how the management system reacts to specific network conditions. Tactical actions include active fault isolation, automatic fault restoration, maintenance testing, operator notification, alarm queue modification, and direct database modification.

Each event is identified and classified as a network fault or an informational event. If a network fault is detected, the case-based expert system is used to determine the proper action to take. During the restoration and repair processes, the network will generate additional data that is associated with the original fault. The system correlates this with previously received messages to treat this as a single network event. As each network event (input specification) is processed the case history database is updated, improving the system's ability to autonomously respond to future events.

6.3 Strategic Activities

Strategic actions are the management system processes that use network information collected over an extended period of time. Each of the previously discussed operational and tactical activities results in the measurement and reporting of various network parameters. This collected history is compiled into availability and utilization reports. Other types of reports aid in diagnosing chronic fault situations, reviewing resource histories, and performing network engineering.

Each performance measurement results in the generation of a summary value that is compared against baseline values established by the neural networks. Service objectives are compared with actual conditions to predict network failures and warn of potential service impacts.

7. Conclusion

Distributed network management processing and artificial intelligence represent two technologies that offer significant potential for improving reliability and reducing costs. One reduces bandwidth requirements, while the second reduces the skill level necessary to keep the network functioning at acceptable levels. As networks continue to increase in complexity, these technologies will provide cost effective solutions to their management.

Acknowledgments

The author wishes to thank John McConnell of Interworking, Inc. for reviewing this work and making suggestions as to its content.

REFERENCES

1. Peterson, V., Kim, J., Holst, T., Deiwert, G., Cooper, D., Watson, A., Bailey, F., Supercomputers of the Future , NASA Tech Briefs, ARC-12416, 1991

2. Rahalu, I., Gaiti, D., A Multi-Agent system for Network Management, Proceedings of the Second International Symposium on Integrated Network Management, Washington, D.C., pp. 465-479

3. Disabato, M., McDonald's Network Management Architecture, Unpublished Internal Document, February, 1992

4. Barletta, R., An Introduction to Case-Based Reasoning, AI Expert, August, 1991, pp. 43-49

5. Disabato, M., Artificial Intelligence: Strategic Technology for Information Processing, Unpublished Master's Thesis, August, 1987

6. Goodman, R., Latin, H., Automated Knowledge Acquisition from a Network Management Database, Proceedings of the Second International Symposium on Integrated Network Management, Washington, D.C., pp. 541-549

7. Goodman, R., Smyth, P., Algorithm Derives Rules from Data, JPL Invention Report NPO-18114/7631, November, 1991

8. Rose, M., The Simple Book, Prentice Hall, Englewood Cliffs, NJ, 1991

9. Eckerson, Wayne, Net Management Traffic Can Sap Network Performance , Network World, 05/04/92

10. Power, G., Developing a Strategic Network Management System, Networking Management, November, 1990, pp. 62-69

Integrated Network Management, III (C-12)
H.-G. Hegering and Y. Yemini (Editors)
Elsevier Science Publishers B.V. (North-Holland)

X.400 Message Handling Systems Management Functional Requirements

Iyengar Krishnan

The MITRE Corporation
M.S. W558
7525 Colshire Dr
McLean, VA 22102
USA

Abstract

Electronic messaging service based on CCITT X.400 standards is increasingly being deployed and plans are underway, by many organizations, for migration to these standards. One of the issues faced in this context is how to manage a X.400 Message Handling System(MHS). Management of a X.400 MHS is currently based on proprietary technologies and there are very few published papers concerning X.400 MHS management. This paper provides a summary of major functional requirements for managing a X.400 MHS (based on the 1988 CCITT standards). The functional requirements are based on research and are defined in Open Systems Interconnection (OSI) Management framework. In this regard, this paper summarizes the current status of X.400 MHS management standardization efforts at ISO/CCITT.

1. INTRODUCTION

Electronic mail (e-mail) service has traditionally been provided based on proprietary systems such as AT&T's Mail and IBM Professional Office System (PROFS). Generally, e-mail service is based in Host computers which provide the necessary application software for message submission, processing, and delivery. The Host computer checks the format and addressing and, if there is any problem, notifies the sender of the problem via error messages. The sender makes the necessary correction and resends the message. If everything is according to an agreed upon protocol specification, the Host computer forwards the message through the transport network. Proprietary systems such as the ones referenced above require gateways for interoperability between each other.

Simple Mail Transfer Protocol (SMTP) has emerged as a de facto standard for enabling interoperable e-mail service to users on Transmission Control Protocol/ Internet Protocol (TCP/IP)-based networks. CCITT developed the X.400 standards[1], which are international standards for e-mail, with the objective of enabling interoperable worldwide e-mail service. X.400 Message Handling System (MHS) is based on International Standards Organization (ISO) Open Systems Interconnection (OSI) protocols; in particular, it is an application layer service at OSI layer 7. There are many organizations that have already migrated to these standards or are in the process of migration.

There have been very few published papers on how to manage an X.400 MHS[1-4] and these papers are focused on specific aspects of management, such as fault management or routing. This paper aims at providing a set of high level functional requirements to enable a systematic approach to define and develop a X.400 MHS management system. In this regard, the following major areas are discussed:

- Definition of MHS Management Domains (MDs)
- MHS Management Functional Requirements
- Status of Standardization of MHS Management

The following section provides an overview of the X.400 MHS Concept, preceding the discussion of the aforementioned areas of MHS Management.

2. X.400 MHS CONCEPT

A simplified model of a CCITT X.400 MHS is shown in Figure 1 [5]. It has three types of functional entities; User Agent (UA), Message Transfer Agent (MTA), and Message Store (MS). In this model, a user is either a person or a computer process. A user is also referred to as either an originator or a recipient (of a message) depending on the context. A message transfer system (MTS) contains a number of message transfer agents (MTAs). Operating together, in a store-and-forward manner, the MTAs transfer messages and deliver them to the intended recipients. The following is a brief explanation of the functional entities and X.400 protocol architecture [5-6].

2.1 User Agent (UA)

The UA operates on behalf of a user. It interacts directly with the user, performs functions for preparing messages and submitting them for routing to the destination(s). In the process, the source UA interacts with destination UA(s), which performs the delivery function. The UA also assists the user with other message functions, such as filing, replying, retrieving, and forwarding. The UA submits messages to an MTA for transmission across the network. The X.400 series standards specify the interactions of the UA with MTA and other UA entities, but do not specify the interactions between the UA and its user.

Typically, a UA can be implemented as a (set of) computer process(es) in an intelligent terminal.

1. The management concept presented in this paper is intended for a 1988 X.400 standards-based MHS. However, most of the concept is applicable to a 1984 X.400 standards-based MHS also.

2.2 Message Transfer Agent (MTA)

The MTA accepts a message from a UA and delivers it to the recipient, if the recipient is based on the same MTA as the sender. If the recipient is based on a different MTA, the MTA that accepts submission of a message performs routing. In this regard, the message is relayed through one or more succeeding MTA(s), by a store-and-forward mechanism, to the destination UA. For example, if only some MTAs have access to the proper long-distance communications paths, a message addressed to a distant UA might be relayed in several stages. The store-and-forward mechanism also eliminates the need to have all MTAs available on a 24-hour basis.

Typically, an MTA can be implemented as a (set of) computer process(es) in a host computer.

2.3 Message Store (MS)

An MS is an optional general purpose capability of an MHS that acts as an intermediary between the UA and the MTA. The purpose of an MS is to store and permit retrieval of delivered messages. The MS also allows for submission from, and alerting to, the UA.

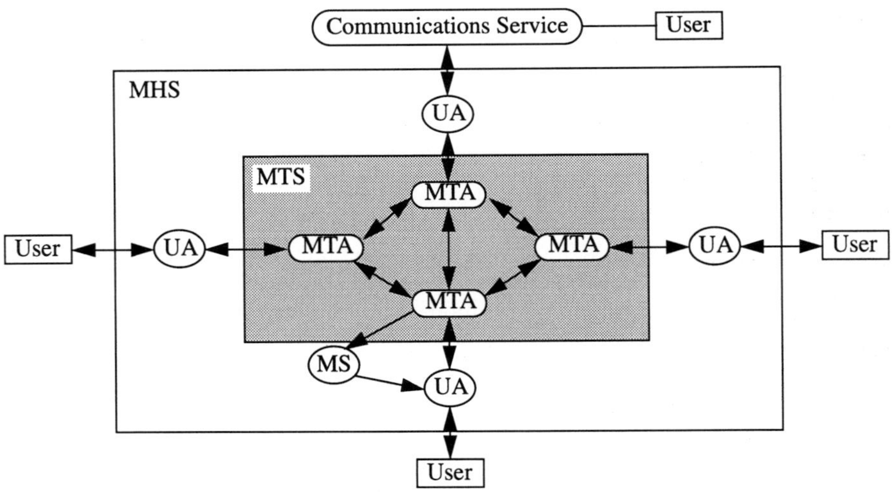

Figure 1. X.400 MHS Model

3 MHS MANAGEMENT DOMAINS

Prior to defining and developing a management system for a MHS, it is essential to identify the domain of authority and responsibility of the management system. X.400 standards define a management domain (MD) as a collection of at least one MTA, zero or more UAs, and zero or more MSs operated by an Administration or organization. An MD managed by an Administration is called an Administration Management Domain (ADMD). Figure 2 illustrates the relationship between ADMDs and Private Management Domains (PRMDs).

3.1 ADMD

An ADMD is characterized by its provision of relaying functions between other MDs and the provision of message transfer service for the applications provided within the ADMD. An Administration can provide access for its users to the ADMD in one of the following ways:

- user(s) to Administration provided UA(s)

- private UA to Administration MTA

- private UA to Administration MS

3.2 PRMD

An organization other than an Administration can have one or more MTA(s), and zero or more UA(s) and MS(s) forming a PRMD which can interact with an ADMD on an MD to MD (MTA to MTA) basis. A PRMD is characterized by the provision of messaging functions within that management domain. A PRMD is considered to exist within one country. Within that country, the PRMD can have access to one or more ADMDs, as shown in figure 2. However, in the case of a specific interaction between a PRMD and an MD (such as when a message is transferred between MDs), the PRMD is considered to be associated only with that ADMD. The name of a PRMD can be either nationally unique or relative to the associated ADMD. The ADMD is responsible for ensuring that the accounting, logging, quality of service, uniqueness of name, and related operations of the PRMD are correctly performed.

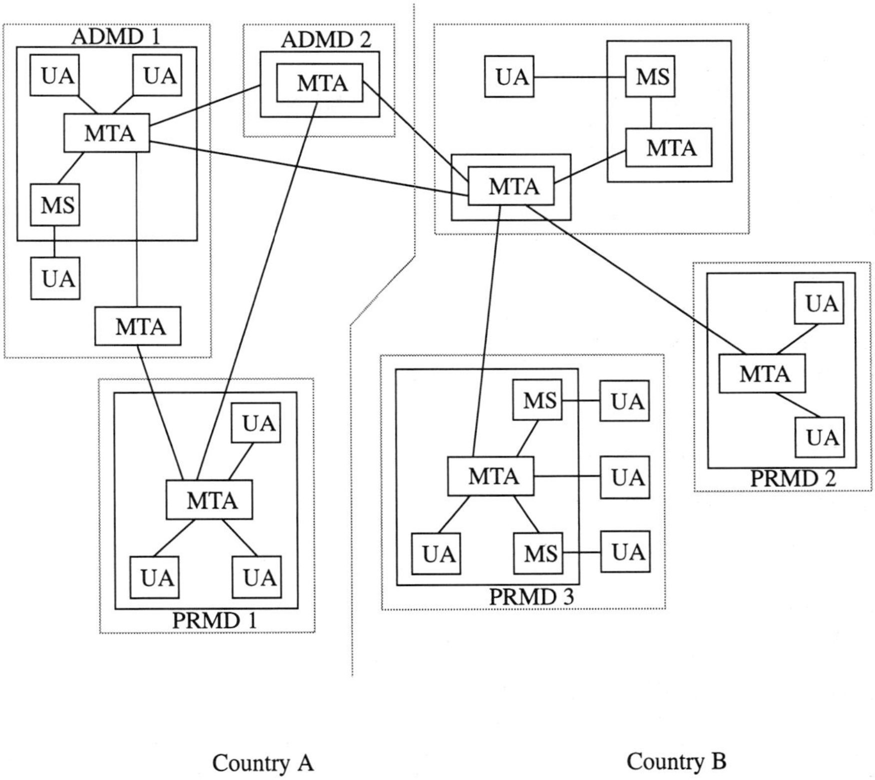

Figure 2. MHS Management Domain

Key:

ADMD - Administrative Management Domain
MTA - Message Transfer Agent
MS - Message Store
PRMD - Private Management Domain
UA - User Agent

4. MHS MANAGEMENT FUNCTIONAL REQUIREMENTS

Management of a X.400 MHS is basically management of UAs, MSs, and MTAs which provide messaging service. It also includes management of messaging service on an end-to-end basis to ensure that submitted messages are delivered to the intended recipients with a high degree of certainty, integrity, and required level of security. Conceptually, management of X.400 MHS can be broadly classsified into two functional areas: **System Operational Management** and **Messaging Service Management. System operational management** is concerned with monitoring the status of MHS components and controlling them to ensure their operational readiness and that they provide uninterrupted and reliable service to the users, in accordance with a set of pre-determined Quality Of Service (QOS) criteria. **Messaging service management** is concerned with resolving user perceived problems such as providing a status of their messages, resolving message delivery problems, and flagging protocol and addressing related errors to the users.

The following sections discuss major management functions pertaining to System Operational Management and Messaging Service Management.

4.1 System Operational Management

The MHS System Operational Management is concerned with maintaining the operational readiness of the MHS as a system, in terms of its hardware and software functionality, to assure continuous and reliable service to the users, and is described in terms of OSI Management functional areas [7-10]: *Fault Management, Configuration Management, Accounting Management, Performance Management, and Security Management.* (Routing Management is basically performance management function, but because of the scope of treatment of this function, it is discussed under a separate topic heading in this sub-section.) As the messaging service relies on the underlying Transport Networks as well as Directory System, overall MHS management also includes interactions between MHS Management System and the Directory and Transport Network Management Systems. It is beyond the scope of this paper to discuss these interactions.

4.1.1 Fault Management

MHS components, such as the UA and the MTA, cooperate to provide the messaging service and, in doing so, make use of the underlying transport services. MHS Fault Management functions should detect faults such as an MTA failure, and initiate recovery action, such as messsage re-routing, so that the failed MTA will not significantly affect the service availability to the users.

An MTA failure will affect all the users connected to it. If the MTA is an intermediate MTA, it could have detrimental effects on the whole network. Hence, MTA faults should be detected by MHS management system and remedial actions taken to restore the failed component or route the traffic around the faulty component. Following are specific examples of fault management considerations.

<u>MHS Communications Related Problems</u>

Communications problems concerning MHS components will, in many circumstances, be detected by OSI lower-layer communications protocols and reported, and X.400 protocols provide for the reporting of many of the faults detected; however, certain categories of failure, for example protocol error in a delivery report, cannot be detected by the MHS protocol. Hence, additional management functions are necessary for Fault Management.

Inter-MHS Domain Considerations

There may occur malfunctions that affect MHS operation in more than one MHS management domain (MD)[2]. For example, a communications link failure between two domains will affect the traffic flow between the two domains. Such malfunctions may require co-operative action to diagnose and recover from the fault. Domains may consider giving other domains access to selected management resources, subject to appropriate access control, in order to efficiently diagnose and recover from faults.

4.1.2 Configuration Management

Configuration management is the maintenance and exchange of information with regards to actual physical and logical placement of the components and the component interfaces. Configuration management information, e.g. connection between two MTAs, is initiated prior to initiating a connection as well as at certain regular intervals during operation, in order to update the configuration information to reflect any changes to the connection. Configuration management functions include:

- Maintaining a database of user equipment, including their hardware/software version numbers, interfaces, location, and other relevant information
- Bringing MHS components, such as MTAs and UAs, into service
- Removal of the above components from service
- Keeping a status of the above components
- Maintaining and updating routing information such as Routing Tables [RT]
- Connection/interface between components and between domains

4.1.3 Accounting Management

Accounting management pertains to logging of information relating to the use of MHS resources to generate the necessary information for billing, settlement of balance of accounts, cost allocation, and account audits. The accounting information captured relates to usage at the application layer. It does not capture resource usage at the lower layers. For example, it does not capture network layer cost between MDs and subscriber systems for the transfer of messages. Accounting management functions include:

- Specifying MHS accounting records to be logged when an MD acts on a message
- Specifying conditions under which accounting records are logged
- Identifying services to be used for the transfer of accounting information

2. As discussed earlier, MHS is divided into a number of MDs for the purpose of delegating the administrative and management responsibilities to concerned authorities.

- Specifying how an accounting log is managed
- Specifying how information for billing or settlement of balance of accounts can be derived from information contained in retrieved records

4.1.4 Performance Management

Performance management includes the following (note that Routing Management and Message Traffic Flow Control are discussed later) :

- Performance measurements
- Identification of performance problems
- Correction/avoidance of performance problems

Performance Measurements

Performance measurements are necessary for input and output ports, MTA processors, route segments, and conversions. Thresholds need to be established for volume of data to be handled by the incoming and outgoing ports and compared against actual data traffic. Also, if messages are delayed due to too much traffic, the waiting time for incoming messages to be processed and queued for outgoing messages should be recorded. For the MTA processor, the volume of incoming and outgoing messages should be set and compared against actual number of messages processed.

Identification of Performance Problems

While the performance measurements to be recorded can be standardized, the evaluation of what constitutes good or bad performance is implementation dependent. Thresholds should be set for various utilization levels and fault management alerts should be generated when the levels are exceeded. Measurements of the throughputs of incoming and outgoing ports may be used to identify performance problems on the route segment between MTAs; it is the receiving MTA's responsibility to detect route segment congestion and initiate fault management (FM) activity.

Correction/Avoidance of Performance Problems

Early warnings of congestion within an MTA or on a route segment between MTAs may be used to initiate measures designed to avoid performance problems. Congestion avoidance measures are taken to selectively restricting the types of messages to be handled or the service request to be honored. Congestion avoidance on the route segment between MTAs requires the sending MTA to apply a filter (of acceptable messages) to determine which should be sent. See section 4.1.6 for an elaboration of routing principles.

4.1.5 Security Management

In order to protect the MHS management system and information, specific security measures are required to control access to management information and processes so that only authorized management operators, or processes acting on their behalf, will have access to read, write, create or delete management information and activate or deactivate control processes.

MHS management facilities, databases, and processes should be protected. Generally, communication of management information is subject to the same or higher level of protection as the network and services it manages. In this respect, facilities should have physical security and management communications links should be encrypted.

Each MHS management domain is responsible for setting its own security policy including the determination of controls on access to the domain's management resources from other MHS management domains.

4.1.6 Routing

The Message Transfer System (MTS) conveys messages in a store and forward manner from the Originator to Recipient through the network of MTAs and spanning MDs. The path taken between an Originator and Recipient may be different for different messages since there may be several possible paths between them, and the MTS may support adaptive routing strategies. The Originator uses an Originator/Recipient (O/R) address related only to the MTS topology at the highest level and not the interconnectivity of its components systems. In the process of conveying a message through the MTS each MD or MTA must make sensible decisions in choosing next MD or MTA on the route to ensure that loops are avoided and efficient use of MTS components and transport media is made.

Route choice decisions for a particular message are based on operational, security, and QOS requirements. A Routing Table (RT) is the source of routing knowledge for routing messages. Configuration information forms the basis for generating the RT. It indicates the type of information required to construct the RTs and various factors that influence routing decisions. Policy information constrains the choice of routes through such factors as security classification or measured QOS. Operational information comprises comparatively short term parameters used for route selection and includes such things as flow control, congestion status, component failures and maintenance schedules that may alter the desirability of using some Route Segments. Operational data are used to qualify the RT on a short term basis only and not to replace longer term configuration and policy information.

Basically the O/R address is used to identify the next MD or MTA, based on the topology at the highest level. In the process of conveying a message through MTS, each MD or MTA must make sensible decisions in choosing the next MD or MTA on the route to ensure optimum routing. Figure 3 illustrates the route selection process based on the RT. The predetermined primary routes and alternate routes are provided by the X.500 Directory Services. Local policy and operational information will influence this process. If there is a problem in the route selection process, default routes can be used. If this also fails, management staff should be alerted to take appropriate action. Message transfer depends on the underlying OSI system for end-to-end transport and, therefore, will involve transformation of names and addresses during the routing process. In particular, it involves the derivation of a Presentation Service Access Point (PSAP) address appropriate to the selected MTA's by means of the directory.

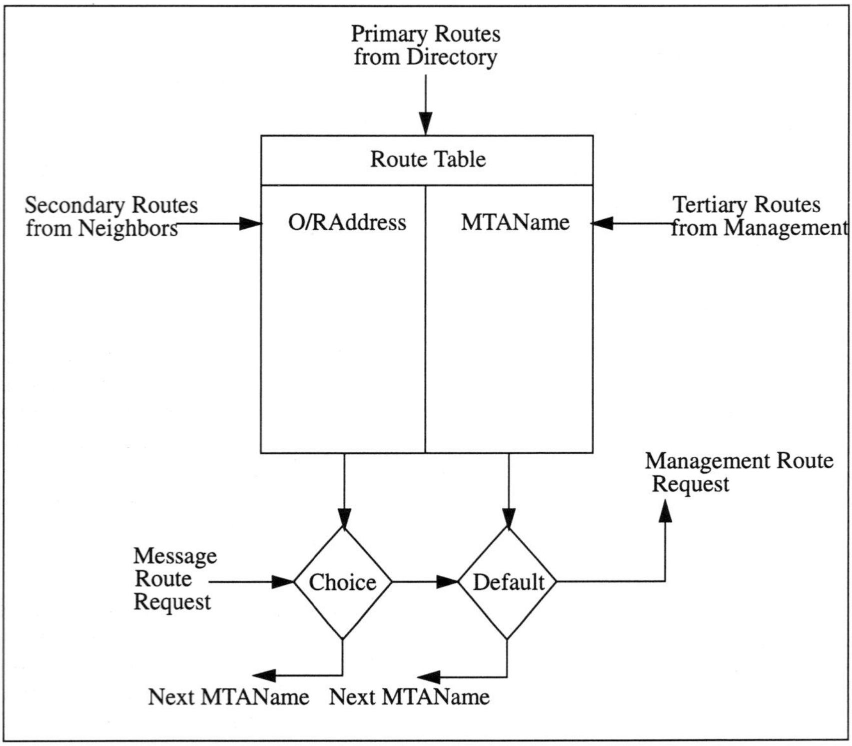

O/R: Originator/Recipient
MTAMessage Transfer Agent

Figure 3. Route Selection Process

4.2 Messaging Service Management Functions

This section summarizes some major management functions that are necessary to resolve user perceptible problems and provide a high quality of service to the users. As was pointed out earlier, messaging service management is concerned with managing the "Messaging Service" at the OSI Application Layer level. *It must be pointed out that even though the MHS management functions are divided into two broad areas of operational and service management, the two areas are closely related and management information pertaining to these two areas is shared in accomplishing management functions.* For example, traffic congestion at an MTA may result in excessive delivery delays. The users perceive the problem through delivery delays and may even report the problem to the MHS management entity. The MHS management entity, acting on user complaint, can correlate the delivery delay with message traffic congestion and take appropriate measures such as traffic re-routing to alleviate the problem.

The management entity may also proactively monitor traffic and control traffic flow to avoid delivery delays. In the former case management function is viewed as service management and the in the latter case it is viewed as operational management.

Major messaging service management functions are described in OSI Management Framework, in the following sub-sections.

4.2.1 Fault Management

The following is a brief description of major generic fault management functions:

- Identification and resolution of non-delivery problems. Non-delivery may be due to incorrect addressing or faulty MTA or due to security classification mismatch. For example, if a specific message cannot be delivered due to incorrect address and if the directory is not up to date, the sender of the message will need assistance from the MHS management entity to find out the correct address. If a delivery cannot take place due to a faulty MTA, the MHS management may take appropriate measures, including delivery through an alternate MTA, to deliver the message.
- Message tracing. There may be situations where a sender of a message needs to know the status of a message. For example, a message may not have been sent to the receiving MTA from the sending MTA due to a fault at the receiving MTA. The sender will not be notified of this problem until the delivery time exceeds a preset limit. In this case, if the use wants to know the status, the MHS management will perform a message tracing by interacting with the management entity associated with the receiving MTA to try to determine the status.
- Protocol and Format Related Faults. Messages that do not conform to predefined protocols and formats will be flagged automatically by the MTA which processes the message through a Delivery Report, as a routine operational procedure. MHS management functions in this context will be to keep logs of the errors, as deemed appropriate, and analyze them later to determine sources of frequent errors. Such analysis can be used to prevent recurrences of such errors by communicating it to the users.
- Interoperability problems. In situations where a message cannot be delivered because of interoperability problems, such as different software versions at the sending and receiving MTAs, MHS management will have to resolve the problem.

4.2.2 Configuration Management

Configuration management will maintain the following information in support of messaging service:

User ID - Directory name and O/R address, Distribution Lists

User attributes - Local/remote, maximum allowable message lengths

Security attributes - Classification level, authentication information

4.2.3. Accounting Management

Users/user organizations need to be provided with accounting/billing information for usage of the MHS resources, in accordance with established policies.

4.2.4 Performance Management

Performance management functions include the following:

- Delivery Delays. Message delivery delays should be determined based on a statistical sampling of delivery delays for some messages. For example, the delay may be due to message queues at an MTA because of overload. Remedial action reroute some traffic thereby alleviating the traffic congestion. Normally, remedial actions will be automatically initiated by the MTAs, based on a pre-determined routing algorithm. Unusual situations, such as a failure of an MTA, may warrant that operational staff be notified to take appropriate measures to resolve a problem.
- Availability/reliability. Availability/ reliability statistics should be analyzed to determine if mean-time-between-failure and mean-time-to-repair (MTBF/MTTR) criteria are met. Specific instances of service deterioration reported by users should be investigated. An example of service unavailability is the MTA outage caused by software problems. Chronic outages due to traffic congestion at an MTA may result in frequent service interruptions and may warrant immediate management action.

4.2.5 Security Management

Security management functions include the following:

- Authentication and access control. Security management is responsible for ensuring that authentication and access control policies are enforced properly through the security mechanisms such as password control.
- Protection of message security. Security Management will ensure the protection of user messages according to their security classification level through security mechanisms dictated by security protocols. Security violations will be investigated through security logs and audit trails.
- Protection of message integrity. Security management will also ensure message integrity; user reported integrity violations will be investigated through security logs and audit trails.

5.0 STATUS OF STANDARDIZATION OF MHS MANAGEMENT

CCITT is in the process of defining various aspects of MHS management. The scope of MHS management is to provide guidelines and develop tools for the management of X.400 MHS. The objectives of this effort are:

- Define Managed Objects (MO's) for MTA, UA, and MS. The MOs are planned to be defined in accordance with the ISO Guidelines for the Definition of Managed Objects [GDMO] [11].
- Functional areas to cover Fault, Configuration, Performance, Accounting, and Security

The initial work in this area was published in ISO/CCITT draft documents pertaining to MHS Management [12-17] in March 1991. The work is in preliminary form and only MTA MO has been defined in these documents. These documents identified many issues pertaining to the MO definitions. ISO/CCITT published a new MHS Management program of work in April 1992 [18], as shown in Table 1.

Table 1: MHS Management Program of Work At ISO/CCITT

Document Title	CD	DIS	IS
MHS Management Model and Architecture	June 1993	April 1994	November 1994
MHS Management Information	June 1993	April 1994	November 1994
MHS Accountability Management Functions	June 1993	April 1994	November 1994
MHS Security Management Functions	February 1994	November 1995	April 1996
MHS Configuration Management Functions	NA	NA	NA
MHS Fault Management Functions	NA	NA	NA
MHS Performance Management Functions	NA	NA	NA

CD - Committee Draft; DIS - Draft International Standard; IS - International Standard, NA - Date of Publication Is Not Announced.

According to this publication, CCITT will produce seven documents for defining standards according to the schedule shown. Even though these documents cover the same MHS management areas as the first set of documents referenced above, it is not clear at this point whether the first set of documents are going to be revised and re-structured or a new set of documents are to be produced. A major issue that implementors are faced with is in the area of standardization of MHS Management. Due to the slow pace of work in the standards bodies, as can be seen from Table 1, the standards will not be completed before 1996. Hence allowing another 3 to 4 years for these standards based products to appear in the market, an OSI standards- based MHS Management system will not be feasible before the year 2000.

6.0 SUMMARY

Many X.400 Standards based MHSs are being implemented and many more are expected to be implemented in the future. A major issue faced by MHS implementors is how to manage the X.400 MHS. This paper summarizes major functional requirements that have to be addressed for managing the X.400 MHS. The functional requirements are based on research and are presented in OSI Management framework. In this regard, this paper points to the current status of standardization of X.400 MHS management standardization at ISO/CCITT.

7.0 REFERENCES

1. Raj Ananthapillai, *Intelligent and Integrated Management of an Electronic Messaging Services Network*, IEEE 1992 Network Operations and Management Symposium (NOMS '92), Volume 2 of 3, page 266, 1992.
2. G. Altomore, P. Pennelli, A. Pepe, and D. Rotondi, MYME: *A Pilot Development For The Distributed Management of Heterogeneous E-mail Systems*, Proceedings of the 1992 IFIP International Conference on Upper Layer Protocols, Architectures and Applications, page 395-408, May 1992.
3. Cristian Constantinof, *MHS Routing Framework*, Proceedings of the 1992 IFIP International Conference on Upper Layer Protocols, Architectures and Applications, page 198-208, May 1992.
4. Christopher J. Lueder, *X.400 Deployment Considerations*, Proceedings of the 1992 IFIP International Conference on Upper Layer Protocols, Architectures and Applications, page 158-171, May 1992.
5. CCITT, *Data Communications and Message Handling Systems*, Volume VIII, Fascicle VIII.7, November 1988.
6. John Henshall and Sandy Shaw, *OSI Explained*, published by John Wiley & Sons, 1988.
7. ISO/IEC 7498-4, *Information Technology - Open Systems Interconnection - Basic Reference Model - Part 4: Management Framework*, 1989.
8. ISO/IEC 10040, *Information Technology - Open Systems Interconnection - Systems Management Overview,* 1991.
9. ISO/IEC IS 9596-1, *Information Technology - Open Systems Interconnection - Management Information Protocol Specification - Common Management Information Protocol,* November 1990.
10. ISO/IEC IS 9595, *Information Technology - Open Systems Interconnection - Management Information Services Definition - Common Management Information Service*, November 1990.
11. ISO/IEC IS 10165-4, I*nformation Technology - Open Systems Interconnection - Structure of Management Information - Part 4: Guidelines for the Definition of Managed Objects*, July 1991.
12. ISO/CCITT, Joint ISO/IEC JTC1/SC18 MOTIS and CCITT SGVII/Q18 Working Document on *MHS Management - Version 3*, December 1990.

13. ISO/CCITT, Working Document on *MHS Accounting Management - Version 1*, CCITT SG VII, Q18/VII, SGI, Q15/I and ISO/IEC JTC1/SC18/WG4 SWG Messaging, February 1991.
14. ISO/CCITT, Working Document on *MHS Service Management - Version 1*, CCITT SG VII, Q18/VII, SGI, Q15/I and ISO/IEC JTC1/SC18/WG4 SWG, February 1991.
15. ISO/CCITT, Working Document on *Security Logs and Audit Trails in MHS - Version 1*, CCITT SG VII, Q18/VII, SGI, Q15/I and ISO/IEC JTC1/SC18/WG4 SWG, February 1991.
16. ISO/CCITT, Working Document on *Proof / Non-repudiation of Transfer - Version 1*, CCITT SG VII, Q18/VII, SGI, Q15/I and ISO/IEC JTC1/SC18/WG4 SWG, February 1991.
17. ISO/CCITT, ISO/IEC JTC1 IST 18 SWG Committee Draft on *MTS Routing - Model and Information Framework*, January 1991.
18. ISO/CCITT JTC 1/SC 18/WG 4, *MHS Management Program of Work*, April 1992

Acknowledgments

The author thanks John Daigle, Walter Lazear, and Rozann Patton of MITRE Corporation - Networking Technical Center, for their critical review of the draft and constructive comments.

Integrated Network Management, III (C-12)
H.-G. Hegering and Y. Yemini (Editors)
Elsevier Science Publishers B.V. (North-Holland)

Process Management: The Future of Integrated Management Systems

James B. Duff[a], James D. Hunter[a], and David C. Matthews[a]

[a]Network and System Management Division, Hewlett-Packard Company, 3404 East Harmony Road, Fort Collins, Colorado 80525, USA

Abstract

Businesses are increasingly dependent on a modern information infrastructure as they strive for competitive advantages amid the rapid rate of change affecting the infrastructure and the organization. The quality of the information infrastructure depends on the processes used to manage it. Process Management is a new capability for Integrated Management Systems (IMS) that focuses attention on the structure and relationships of process knowledge to enable people to operate processes rather than tools. The Management Maturity Model describes the evolutionary levels of an organization's processes. New types of integration are identified to support a process-oriented IMS. The IMS of the future will advance knowledge to continuously improve the effectiveness of organizations that manage the information infrastructure.

Keyword Codes: K.6
Keywords: Management of Computing and Information Systems

1. INTRODUCTION

Many companies are dependent on a modern information infrastructure in their efforts to improve business results. The traditional goal of an infrastructure has been to enable more effective use of information while insulating people from underlying complexities. This goal is important because the effective use of information offers enhanced efficiency, informed decision making, and improved business results. However, the complexity of the modern infrastructure, and the changes affecting it, produce significant management challenges.

This paper focuses on the evolution of today's Integrated Management Systems (IMS), and is organized into three sections. The first section summarizes the challenges facing IMS, and articulates a goal that meets those challenges. The second section introduces a process paradigm based on the use of knowledge. The third section discusses the effects of Process Management on the systems that will address the long-term problems inherent in managing the increasingly complex information infrastructure.

2. THE CHALLENGE OF INTEGRATED MANAGEMENT

Companies in an increasing number of industries rely on their information infrastructure for solutions that enhance their strategic and competitive positions. Continuing innovation and increasing market demand have produced an infrastructure that has never been more powerful, or more pervasive in its influence on business success. While changes in technology have significantly improved the access to information, the resulting increase in size and complexity has made the management of the infrastructure a significant strategic and competitive issue. Rather than enabling results that meet business goals, today's information infrastructure often constrains them.

2.1. The Evolving Information Infrastructure

Traditionally, the information infrastructure was just part of a company's underlying foundation. It facilitated organizational communication (e.g., postal, telegraph, or telephone services) and implemented support functions (e.g., accounting ledgers or personnel records). The "real work" was done with limited technological assistance. In this role, the information infrastructure had an indirect effect on business results. It represented a limited value, an expense to be minimized.

The typical information infrastructure of the 1990's is composed of networks of computer systems and peripheral devices that are connected by various types of network elements. This modern infrastructure is directly involved in delivering business results. In addition to serving in the traditional support role, the modern infrastructure has become the environment in which, and the capability by which, many companies execute their mission-critical processes. Due to this trend, the value that a business delivers is proportional to the value that can be extracted from the infrastructure. Modern information infrastructure expenditures become investments in the business itself.

2.2. Rapid Rate of Change in the Information Infrastructure

Much effort is being spent on improving the management of the information infrastructure. Today's IMS has conquered some of the complexity in the infrastructure. These systems focus on reducing the considerable difficulties experienced by the people and organizations that support today's computer and communications networks.

Current IMS solutions will significantly improve the ability to manage the infrastructure. However, these solutions are unlikely to achieve long-term success, given the most difficult and significant problem: the rapid rate of change that affects the infrastructure, the systems that manage it, and the organization that supports it.

A driving force in the information infrastructure has been the trend to replace a few, large, proprietary, centralized mainframes with many, small, open, geographically dispersed, client-server systems. This technology trend has created many new types of network components, and has caused an explosion in the number of components to be managed. Open architectures and standards for interoperability provide the flexibility to implement needed solutions without undue dependence on a single vendor. Unfortunately, interoperability doesn't automatically enhance manageability. In fact, the trend toward open systems has complicated the management problem by diminishing the scope and effectiveness of proprietary, centralized management systems.

As the proprietary management systems become less effective, the management environment fragments. It becomes inundated with separate management tools. Individually, these tools are often incompatible. Collectively, they often overlap, or leave unmanaged gaps in the infrastructure. Since systems don't exist that can manage the infrastructure as a whole, it is managed in pieces.

Business pressures motivate companies to turn the infrastructure's increased capabilities into competitive advantages. As the use of the infrastructure becomes more ambitious, the pressures and demands on support organizations escalate. Faced with a distributed information infrastructure and fragmented management methods, these organizations struggle to maintain, let alone improve, support levels. In some companies, support organizations have remained centralized; in others they are as distributed as the infrastructures they manage.

Regardless of their structure, few organizations are able or willing to increase support staff in response to the rapid increase in the size and complexity of the infrastructure. Given the limited effectiveness of fragmented management methods, a corresponding increase in staff size would probably not yield an acceptable improvement in results. Not only is the relative size of the support staff shrinking, the absolute size can also decline. As competitive pressures squeeze budgets, support staffs participate in downsizing efforts, and replacement hiring slows down. In response to either relative or absolute reduction in staff size, hidden costs tend to rise as unofficial, ad hoc support efforts expand to fill the void.

2.3. Business Issues

The increasing dependence on the modern information infrastructure makes the management of the infrastructure a key business issue. Many companies have been forced to rely on proprietary management solutions that are augmented by tools developed by internal support organizations. Most of these tools focus on managing particular types of network elements and computer systems. Given the complexity of managing the overall infrastructure, these tools comprise a fragmented, incomplete solution. And given the rapid rate of change in the information infrastructure, long-term success will be elusive.

The quality of available methods and tools makes management a people-intensive activity. In many organizations, the expertise to effectively manage the infrastructure is concentrated in a few experts. Only an expert can deal with the complexity and the pace of change in the infrastructure. The rest of the management staff often depends on the experts for direction. As a result, the experts are often pulled into routine problem solving and tactical details. If they can't apply their expertise to global, strategic management issues, then the organization can become trapped in a reactive mode. Even when proactive management does occur, many decisions must be made with inadequate information. Judgment and intuition give way to guesswork.

As a result, support organizations face a dilemma. Effective management is too expensive to implement, and efficient management is not yet possible. The companies that contain these support organizations face a larger dilemma. Failures in the infrastructure can inflict significant damage on business results. The risk of depending on an unmanageable infrastructure is unacceptably high, as is the cost of coping with it. But the information infrastructure is too valuable and too pervasive to abandon.

These companies are increasingly vulnerable to changes in the infrastructure. In the past, most of the focus has been on increasing the capability and capacity of the infrastructure. In recognition of the widespread vulnerability this focus has caused, the computer and communications industries have turned their attention to the management dilemma.

2.4. The Goal of Integrated Management

Integrated Management must support the evolution to the modern information infrastructure by managing the rapid rate of change that affects it. An IMS must enable a business to make its use of the infrastructure a key competitive advantage. It is unlikely that Integrated Management will achieve its goal with its current focus on providing tools to manage equipment.

Effective management of the infrastructure typically depends on the experience and expertise of the people who use the tools. Management experts have specialized knowledge that tells them which tools to use, how to integrate information across different tools, and how to interpret results. They have the knowledge to penetrate the complexity of the infrastructure, and to understand the vagaries of management processes. These experts are difficult to recruit, expensive to train, and a challenge to retain. They are a scarce resource, and are likely to remain scarce because their expertise is kept locked away inside their brains. If any of their specialized knowledge has been written down, it is likely to be in a form that is inaccessible, or too cryptic for others to use.

The inability to leverage this knowledge makes organizations vulnerable to its loss, and less able to deal with the rapid rate of change. This means that the real problem an IMS must solve is not the management of equipment, rather it is the management of knowledge. The goal of Integrated Management is the advancement of knowledge to improve the effectiveness of organizations that manage the information infrastructure.

3. THE PRINCIPLES OF PROCESS MANAGEMENT

Process Management is the key capability missing in today's IMS. There is already a trend in applications to become more process-oriented. To achieve its goal, the IMS of the future must integrate and manage processes across applications. To understand how Process Management supports the goal of Integrated Management, it is necessary to begin thinking about process in a new way.

Just as Tom DeMarco [1] believed "you can't control what you can't measure", you also can't manage what you can't define. It is necessary to establish a common understanding of what, exactly, is meant by the term "process". The process paradigm links Process Management to Integrated Management by recognizing that knowledge is the essence of process.

3.1. The Process Paradigm - Task and Context Knowledge

A process is a representation of knowledge. It is a design for the action required to meet a recognized goal. Without a goal, there would be no need for action, and so the process would not be needed. Without sufficient knowledge, effective action is impossible, and so a process that meets the goal would not be possible. So, a process is a representation of knowledge organized into the design of an action whose results will meet a goal.

Of course, many goals cannot be achieved in a single step, and so the associated action cannot be executed atomically. In such cases, if an action is partitioned into smaller steps, each step becomes a subprocess partitioned from the original. The subprocesses, each with its part of the original goal and action, can form chains of dependency. Therefore, processes can be related either hierarchically, sequentially, or concurrently, just as their constituent actions can. It is crucial to understand that the same principles apply to any process at any level in a hierarchy, and at any step in a sequence. Every process and subprocess has its own goal, knowledge, and action.

Task Knowledge Some of the process knowledge focuses directly on the required action. It is transformed into a specification of what the action will do, and a design of how the action will do it. It specifies the necessary and sufficient inputs and outputs of the action. For convenience, the action-oriented knowledge component will be called task knowledge, and the action will be called a task.

Context Knowledge The rest of the process knowledge focuses on context knowledge. It specifies the goal, its value, its justification and risks, and the context in which the goal exists. Context knowledge defines the relationships between the goal and other goals, and between the task and other tasks. It also defines a method for validating the quality of the results produced when the task is executed. It is the context knowledge that gives the task its direction and meaning.

A process design is a conceptual representation of knowledge, and cannot produce results until it is implemented. A process implementation is an executable representation of the design. For convenience, the implementation of a process may be called a system. A system is the integration of the process knowledge and the resources capable of executing it. So, it is the system that produces the results to meet the goal of the process. It is the system that puts knowledge to work.

3.2. Process Domains

The process paradigm assumes sufficient knowledge to design a task that will meet a specified goal. The quality of a process design is typically a function of the skill of the designer. However, a process implementation often depends on the complexity of the goal, and the quality of the available knowledge. Figure 1 illustrates three of the domains into which processes can be implemented: physical, computational, and organizational. Some domains inherently encourage and support process quality, others inhibit it. The domains differ according to the amount of task and context knowledge they capture.

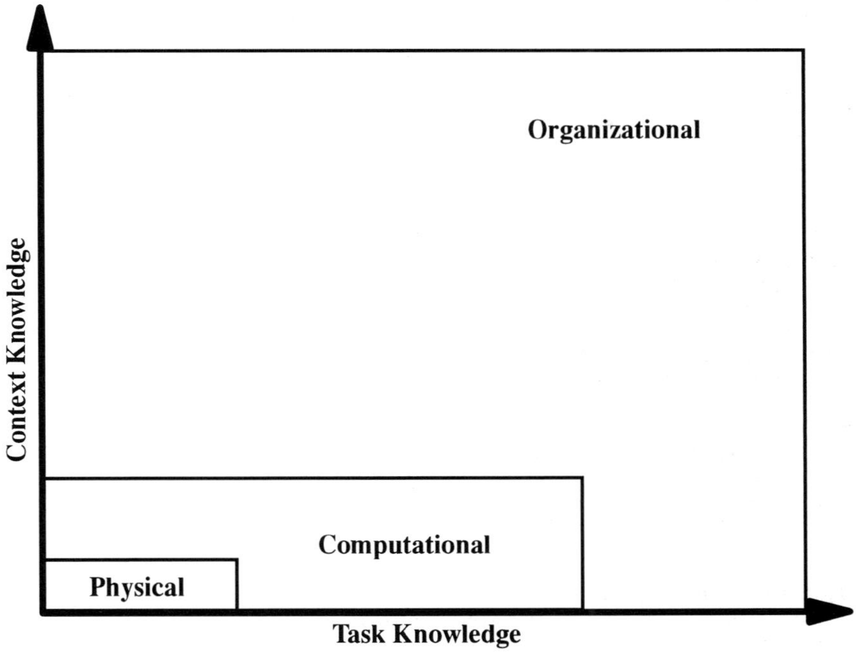

Figure 1. Process Domains

Physical Some goals can be met in the context of a domain in which rigorous technical methods can be applied to map the knowledge into the physical world and express it through the application of physical laws and logical relationships. For instance, the domain of hardware design and manufacture gives rise to physical manifestations of the original knowledge. One can view artifacts such as hammers, televisions, and automobiles as condensed knowledge.

The quality of the processes implemented by physical artifacts is typically quite high, because relationships, dependencies, and priorities are physically established and enforced among the system components. The physical domain is capable of implementing only simple, specific tasks. It cannot handle much context knowledge, which is acceptable given the nature of the goals it must address.

Computational The modern information infrastructure is based, in part, on the computational process domain. Computer hardware and software combine to provide a more general task execution capability than the physical domain can. This domain is also able to capture more context knowledge because of the flexibility of software. The computational domain has seen the constant migration of processes from other domains into the computerized information infrastructure. Robotics is an example of the computational domain absorbing processes from the physical domain. Expert systems are an attempt to capture human reasoning and decision processes. And most significantly, the computational domain has long been the target for the automation of organizational processes.

Organizational The organizational domain implements processes that can't be successfully transformed into any other domain. Typically, these include processes that address goals that can be met only through the direct application of human effort and intelligence. People and organizations can handle task knowledge that is beyond the current capability of the computational domain. More significantly, organizations deal with the rapid rate of change by relying on people to acquire, maintain, and apply context knowledge to successfully execute processes.

Processes that are implemented in the organizational domain tend to be much more defect-prone than those implemented in the physical or computational domain. It is difficult to establish and maintain the visibility and integrity of the relationships, dependencies, and priorities among the people executing organizational processes. Organizations often attempt to address this problem by automating their processes. However, automation often complicates a process, rather than simplifying it.

When an organizational process is translated into the computational domain, only part of the process is actually implemented in the computer. The results that meet the important goals come from a system implemented by the interaction of people and the computer network. In particular, most business processes that are comprehensive enough to deliver real value depend on people to handle the necessary context knowledge. In this sense, people are part of the information infrastructure.

IMS needs to solve the problems in the combined organizational/computational domain. Today's IMS focuses on computational solutions, but fails to recognize people as part of the infrastructure. As a result, current applications don't address the role of context knowledge in organizational processes. Figure 1 indicates that the ability to design an organizational process typically far exceeds the ability to implement it in the computational domain. This leads to the dis-integration of task and context knowledge.

3.3. The Dis-Integration of Task and Context Knowledge

The quality of a process is limited by the degree to which the original task and context knowledge can be embedded in the implementation of the process. Even assuming that a task is initially well-designed, the system could still break down in a variety of ways. It could produce variable results. When the variation exceeds acceptable tolerances, then the process will fail to meet its goal. Or perhaps, changes in the environment may necessitate a change in the goal. A process that produces consistent results, but fails to track environmental changes, will fail to meet a new, possibly unrecognized, goal. One of the root causes for the eventual breakdown in process effectiveness is found in the way in which the two process knowledge components are treated during process implementation.

Task knowledge has always been the easier of the two components to design in a process and implement in a system. Current applications focus on implementing task knowledge because there are fewer technical barriers, and its value is well understood. Many methods exist for transforming this knowledge into an executable task. When the process is executed by a system, it is the task execution that we experience. Because task knowledge draws and holds our attention, the task component of the process is the more visible and persistent of the two.

Context knowledge usually suffers quite a different fate. It has always been difficult to effectively capture and use this knowledge, and its contribution to achieving the goal has long been undervalued. There has been a lack of formal methods for manipulating context knowledge, so it seldom accompanies the task into the system. If it has been captured at all, this knowledge ends up in a different location, having been transformed into a representation that is incompatible with the task implementation. This method of capturing knowledge is sometimes called documentation.

It is the discrepancy in our treatment of the two types of knowledge that leads to inherently unstable and unreliable processes. The conceptual split between task and context knowledge becomes a systematic defect in the process implementation. This is ironic because a given process could never have been designed without the tight integration of the two knowledge components. And yet the process is often implemented in a system characterized by the disintegration of the task and the context knowledge. This disintegration is the reason that organizational processes are so vulnerable to the damaging effects of change. It is the knowledge of context that enables people to recognize, understand, and respond to environmental changes. The systematic gap between task and context knowledge eventually jeopardizes the results produced by the task, and profoundly influences the quality of the process.

4. THE FUTURE OF INTEGRATED MANAGEMENT SYSTEMS

The Process Paradigm asserts that Process Management is primarily concerned with resisting and repairing the disintegration of task and context knowledge. Beyond the technical problems associated with handling knowledge more effectively, organizations have difficulty managing processes because they tend to focus only on the results of process execution. Process Management focuses on results, and also on the location and representation of the knowledge to produce those results. This will enable an IMS to expand the computational domain to incorporate more of the organizational process implementation. Process Management enables an organization to apply TQC[2], or other methodologies[3], to continuously improve the processes used to manage the information infrastructure.

An IMS must be built with the recognition that people are not only part of the system that executes processes, they are also part of the system that manages processes. An IMS must deal with this situation by using organizational process knowledge more effectively. Since each organization and information environment is unique, an IMS will encounter a different set of problems in each.

Perhaps the single most important factor in the success of an IMS is its ability to deal with processes in various stages of disintegration. To succeed on this level, an IMS must improve an organization's ability to manage change. The IMS of the future must be the environment in which processes are executed and managed. As long as processes are executed and managed in separate domains, effective process management will be impossible.

An IMS must adapt to an organization's existing structure, policies, procedures, and constraints. It must recognize the role of context knowledge in the organization, and facilitate collaboration between people and resources. It must be adaptive to changing or evolving processes and technologies. Finally, an IMS must enable people to operate the process by having the process organize the tools.

4.1. The Management Maturity Model

Similar work in software development processes at Software Engineering Institute resulted in the SEI Capability Maturity Model [4,5,6]. The Management Maturity Model in Figure 2 focuses on the representation and location of process knowledge, not just on the use of specific processes. An organization may have processes at all levels simultaneously.

This model enables an organization to evaluate the maturity of its processes. It provides a language and a framework for discussing the organization's current capability, and a direction for evolving the quality of its processes.

It is important to understand that, like all evolutionary models, each stage of the management maturity model depends on capabilities provided by previous stages. This means that a solution aimed at a particular level will fail to deliver its expected improvements if the capabilities of lower levels are absent or ignored.

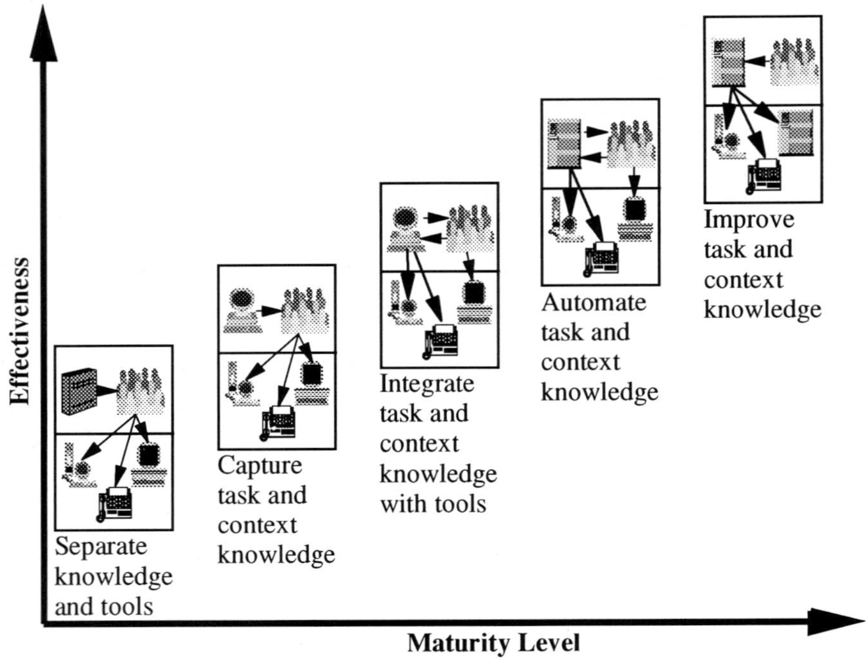

Figure 2. Management Maturity Model

Separate Knowledge and Tools Most organizations are typified by processes that occupy the first level of management maturity. At this level, processes are characterized by the disintegration of task and context knowledge. That is, management tools are separate from the task and context knowledge about the tools.

Since many independent tools exist to support network and system management tasks, each tool requires special expertise to know how or when it should be used. An organization at this level is typically flooded with available tools, and struggles for the knowledge to use them effectively. People must constantly work to acquire this knowledge and apply it to the existing tool set.

Much of this knowledge remains in people's heads. Any that has been recorded is on paper, and any that has been accumulated is in books. This documentation is usually an attempt to rediscover or extend the original process knowledge. It defines the procedures involved in using the tools, and the relationships among the tools. It also defines the relationships among the managed components that the components themselves don't capture.

However, conventional paper technologies are difficult to maintain in the face of the new tools and technologies that are constantly introduced into the environment. Even up-to-date paper documentation provides only limited access to information. The books are few and difficult to share among people.

These conditions lead to a complex, constantly changing management environment. This tends to produce reactive management behaviors, where efforts are inconsistent, ad hoc, and chaotic. Even if proactive management does occur, the organization will remain at this level as long as the tools remain separated from process knowledge. Management processes lack formal structure, so quality, costs, and scheduling are unpredictable. People must perform the integration of knowledge and tools. There is no real automation at this level, and only sporadic success can be expected.

Capture Task and Context Knowledge An organization reaches the second level when it succeeds in capturing its significant process knowledge on-line. The goal is to capture and communicate the sequence and structure inherent in the relationships among tools. Capturing the knowledge on-line makes it much easier to share among people, and to modify when change occurs. This is necessary to enable advancement to future levels.

Moving process knowledge on-line improves repeatability of the process. Different people can easily access the same information when they need to learn about the process. The process will appear to be stable and under control, except to meet evolving needs and changing situations. The process still depends on people to translate the on-line knowledge into the actions required to use the tools. Changes in people, the process, or the environment can easily send the organization back to the first level. This is what happens to many organizations as they move from proprietary to open systems, or from centralized to distributed infrastructures.

Many processes can stop at this level because human interaction speeds are sufficient for their goals. But due to the speed and complexity of the information infrastructure, process knowledge must be on-line so management capability can evolve to the next level.

Integrate Task and Context Knowledge with Tools At the third level, process knowledge is integrated with the management tools in a common execution environment. This will reduce variation in process execution by removing the need for people to translate the on-line process knowledge into a different domain to execute the tools. Emphasis shifts from documenting the tools to defining the process.

This reduces the guesswork that occurs when different people interpret the knowledge and choose to apply different tools. The integration of the knowledge with the tool enables process execution to become independent of the people executing it. It is now possible to train people so that they can consistently follow a well-defined process. The process stabilizes and becomes a qualitative success.

Automate Task and Context Knowledge When an organization reaches the fourth level, people are not the only part of the system capable of executing processes. Once the process knowledge is on-line and integrated into a common execution environment, the next step is to automate the execution of the process. At this level, the individual tools required by a given process will be executed by the computer part of system. The automated process must deal with certain operations that require human intervention. People will interact with the computer only when decisions or manual actions are required. When people operate the process rather than the tools, they will be able to apply their talents to longer term, more strategic activities.

At this level, process measurement is embedded in the automation of the process. Process data becomes an automatic byproduct of process execution. This will provide people with information about the performance of the process. The key improvement is the quantitative understanding of the process. Performance can be predicted based on experience and quantitative analysis.

Improve Task and Context Knowledge When a process is well-defined, automated, and measured, people can manage the process itself, not just the infrastructure. Only at this level can an organization realize TQC's promise of continuous process improvement. The process measurements made at the previous level can be used to iteratively improve the process. Process effectiveness is quantitatively evaluated and the process can be improved experimentally. Since these improvements are made in, and to, the process execution environment, the knowledge base is continuously improved. The improvements are immediately and universally available. This systematic approach continuously improves the effectiveness and productivity of both the process execution and management activities.

4.2. Process Integration Requirements

The Management Maturity Model shows the need for on-line representation and execution of process knowledge. As shown in Figure 3, the real goal of integrating Process Management capability into IMS is to capture a larger proportion of the organization's task and context knowledge in the computational domain. The organization's vulnerability to change is reduced as the management system is able to deal with greater amounts of context knowledge. In addition, a process-oriented IMS will handle manual tasks that continue to resist full automation.

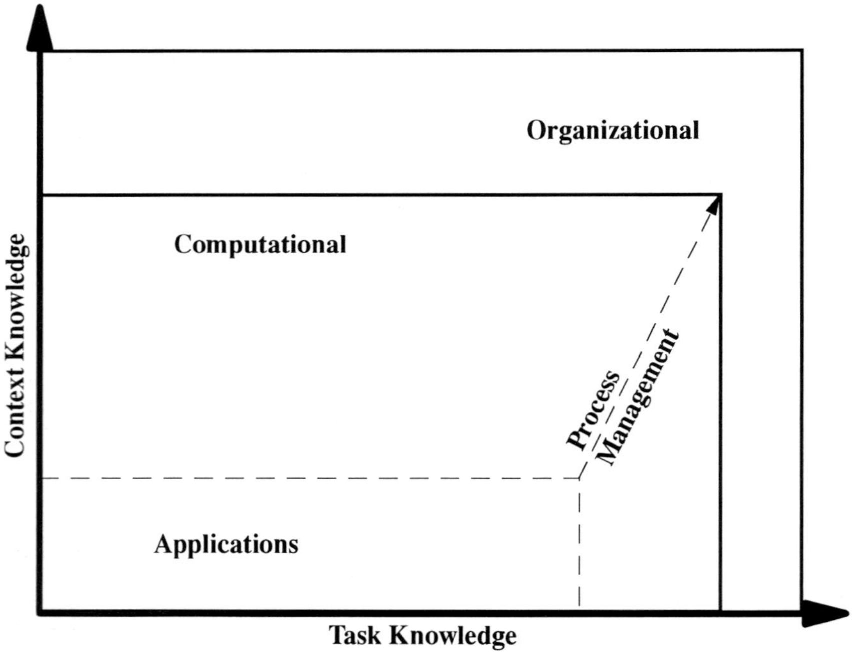

Figure 3. Process Management expands the Computational Domain

Several factors must be considered for the successful integration of Process Management into IMS. The number of peer-to-peer application integrations necessary to provide Process Management at the application level is unrealistically large. Fortunately, the goal of employing particular applications in the implementation of general processes can be achieved by performing the large-scale integration at the process level.

The process-oriented IMS will manage the relationships and capabilities of the many applications available. These applications will be pulled together into an environment that focuses on the process goal. The applications will be hidden when possible, or else streamlined to minimize user intervention. A given process will determine which applications will be used, the order of use, and how the results will be interpreted. The long-term goal will extend beyond the execution and management of process, and will focus on the continuous improvement of the process for individual environments.

Existing technologies for application integration do not sufficiently address these factors, because the problem is not simply application integration. Application integration tends to address only task knowledge, and avoids dealing with context knowledge. In order to repair the disintegration of task and context knowledge, Process Management must expand the computational domain by applying technologies such as knowledge representation, collaboration, and encapsulation.

Representation Fundamental process components and attributes must be defined to enable rigorous representation of process knowledge. It will be impossible to assess the quality of a process or its representation without consistent terminology. The representation chosen for a process-oriented IMS must be based on a general process model, so that the correctness and completeness of any process can be assessed.

Process Management will depend on the ability to accurately and consistently represent processes with arbitrary structure. Processes range from simple sequences of actions to complex flowcharts requiring many decisions. Processes also form hierarchies, so the process representation must be flexible enough to structure and reuse subprocess representations.

Process Management depends on the ability to seamlessly integrate information from a complex, distributed infrastructure. Processes, individually or in hierarchies, may span organizations or geographical locations, so the representation must be scalable.

Collaboration A process-oriented IMS must enable collaboration between people and resources. To successfully handle process execution and management, the IMS must coordinate the efforts of multiple people operating together or independently. Process decision points and parallel efforts must be synchronized. A process-oriented IMS must provide scheduling, synchronizing, tracking, and load-balancing mechanisms.

Encapsulation The process representation must be able to accommodate all types of process implementation and execution. Some processes will be implemented with applications at a high level. Others may require physical interactions, such as flipping a switch or typing an exact sequence of keystrokes. The ability to encapsulate existing applications will be critical.

In the long term, the nature of applications may change to facilitate process automation. Today's applications are driven from their own user interface. In the future, process automation will remove the need for many user interfaces, letting the applications become simple services that plug into processes.

5. CONCLUSION

The key to improving the effectiveness of organizations that manage the information infrastructure is to shift the management burden from people to computer systems. Integrated Management Systems must support this trend by providing the means for process-level integration. This will enable an operations staff to shift its focus from managing equipment to managing processes. This will allow the redirection of human resources to proactive management in the environment.

An organization's long-term business success depends on the quality of its key processes, including the management of the information infrastructure. Process quality can be fundamentally improved only with effective Process Management. Processes are often unreliable and unstable due to the chronic separation of process tasks from relevant context knowledge. Only through the effective use of process knowledge can system results continue to meet their goals. Typically, this knowledge has been lost or obscured due to ineffective implementation of the process. Effective Process Management depends on the ability to recover context knowledge from the infrastructure, and to sustain it over time.

Organizations cannot expect to buy management systems that instantly provide a proactive Process Management capability. Rather, they must evolve this capability. The Management Maturity Model provides a guide for evolutionary improvement. The model describes how knowledge is captured and reintegrated on-line with management tools, automated by computer execution of the process, and measured so that processes can be quantitatively managed and continually improved.

It is crucial to begin building a process perspective in management staffs so they are prepared for the natural evolution of the environment. Organizations can begin operating in a process-oriented fashion even before the environment supports Process Management. Substantial benefits can be achieved even if process documentation and improvement is done only on paper. However, success ultimately depends on the automation that can occur only with process-level integration.

Process-level integration generates new technical requirements, and has long-term architectural impact. Planning must begin now to ensure that Integrated Management Systems will meet these new requirements.

A process perspective is necessary to manage the increasingly complex information infrastructure. There will never be an end to changes in environment, people, and processes. To continuously increase the value of the information infrastructure, Process Management must be the future of Integrated Management Systems.

6. REFERENCES

1. T. DeMarco, "Controlling Software Projects", Yourdon Press, 1982
2. K. Ishikawa, "What is Total Quality Control? The Japanese Way", Prentice-Hall , 1985
3. G.A. Rummler and A.P. Brache, "Improving Performance: How to manage the whitespace on the organization chart", Jossey-Bass, 1990
4. M.C. Paulk, B. Curtis, and M.B. Chrissis, "Capability Maturity Model for Software", Software Engineering Institute Technical Report CMU/SEI-91-TR-24, 1991
5. C.V.Weber, M.C. Paulk, C.J. Wise, and J.V. Wethey, "Key Practices of the Capability Maturity Model", Software Engineering Institute Technical Report CMU/SEI-91-TR-25, 1991
6. W.S. Humphrey, "Managing the Software Process", Addison-Wesley, 1989

IV.

MANAGEMENT REQUIREMENTS, EXPERIENCES AND EVALUATION

C.

EXPERIENCES AND EVALUATION

Integrated Network Management, III (C-12)
H.-G. Hegering and Y. Yemini (Editors)
Elsevier Science Publishers B.V. (North-Holland)

Experiment on and Evaluation of OSI Management Through the INTAP Interoperable Networking Demonstration

Ayako Komatsu* Yoshikazu Maruyama* Tatsuo Tanaka†

Abstract

A great deal of efforts have been made on the standardization of OSI management, but what seems to be needed, however, is verification and evaluation from the viewpoint of practical use. This paper evaluates OSI management by reporting on the Interoperable Networking Event'91(INE '91), held by the Interoperability Technology Association for Information Processing, Japan (INTAP), in November 1991. The demonstration system manages various network resources in the area of configuration and fault management under the assumption of management for a private enterprise network. The managed network included both hardware and software, from modems to application processes. Hierarchical management on three levels was adopted for the demonstration system to satisfy certain requirements. The demonstration specifications were kept as close to available standards as possible. Management protocol and managed information based on the scenario were also decided and the demonstration was carried out successfully. Evaluation of the demonstration has provided three important keys for network management in a heterogeneous environment. The first is that the management scenario has great influence on specification decisions. The second is that agreements on standards in such areas as hierarchical management are needed. Finally, conformance and interoperability testing should be provided as early as possible.

Keyword Code:C.2;C.2.3;D.4.4
Keywords:Computer-Communication Networks,General;Network Operations; Communications Management

1 Introduction

Since its founding in December, 1985, the Interoperability Technology Association for Information Processing, Japan (INTAP) has been eagerly engaged in various projects such as R&D, surveying, promotion and education, and international harmonization , and conformance testing, in order to establish the interoperability of OSI in the information

*First Basic Software Development Division, NEC Corporation
1-10 Nisshin-cho, Fuchu, Tokyo 183 JAPAN E-mail:aya@nw.bs1.fc.nec.co.jp
†INTAP, Sumitomo-Gaien Bldg., 24 Daikyo-cho,Shinjuku,Tokyo 160 JAPAN

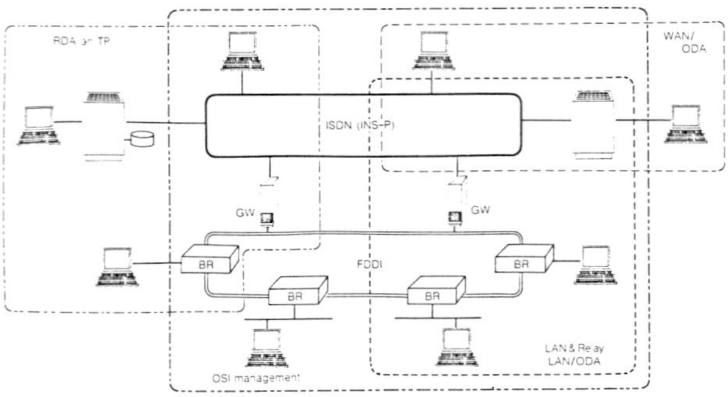

Figure 1: The Overall Demonstration Framework

processing field. The Interoperable Networking Event '91 (INE '91), was held in order to demonstrate some R&D results on this "National Large-Scale Project". Research and development activities were carried out by the Electro-technical Laboratory of the Agency of Industrial Science & Technology (AIST), and INTAP for seven years from 1985 to 1991. This INE '91 presented the final R&D results, mainly on interoperable networking technology.

As a part of the Large-Scale Project activities, INTAP has been testing and evaluating implementation specifications to verify their validity. These results have been presented during INEs since 1988.

In INE '91 the following themes were demonstrated:

- Remote Database Access (RDA) using Transaction Processing (TP),
- Office Document Architectures (ODA),
- LAN & Relay, and
- OSI Management.

The demonstration, consisting of a WAN-LAN connected network, was designed to allocate areas for each theme, and was used to verify the interoperability of each area individually. The majority of the host computers interconnected through ISDN were installed in vendor's offices. Fig.1 shows the overall demonstration framework. The workstations of each vendor installed in this demonstration site were connected to the host computers through LAN, or by direct connection to WAN. ISDN and LAN were interconnected through gateways and bridges. INE'91 offered opportunities to view interoperable networking in a heterogeneous, multi-vendor environment with a variety of advanced applications, which can be realized by OSI.

The demonstration of OSI Management based network management was one of the most attractive and largest demonstrations in the world. Since networks are growing and many resources are involved, OSI management can be regarded as a solution to the network management in a heterogeneous environment. The objectives of this demonstration were to confirm that OSI management is an effective for networks that include multi-vendor network resources, and to clarify many areas which are not dealt with in the standards. Eleven companies participated in this demonstration: Fujitsu, Hitachi, IBM Japan, Matsushita, Mitsubishi, NEC, NTT, Oki, Sharp, Sumitomo Electric, and Toshiba. It was a real time, online demonstration which was carried out after being in preparation for three

years by a special working group established by INTAP.
In this paper, the demonstration of network management is focused describing its configuration, detail management information, testing, and evaluation. Section 2 describes the user requirements and the scenario, including system configuration. Specific demonstration specifications to realize the scenario are stated in Section 3, and the testing process is introduced in Section 4. Finally, in Section 5 some proposed solutions to the issues which arose during implementation are presented.

2 Demonstration Specification

The demonstration system was constructed under the assumption that it would provide a network management system for a private corporate network. The network included distributed offices in which work was performed using the network. At each site, LAN, a proprietary network, or a voice network were installed. The specifications were then decided based on this network scheme.

2.1 Approach

The demonstration was executed according to following approach.

- identification of user requirements,
- decision of management scenario by taking into account user requirements,
- selection of management information and protocol to generate the scenario,
- agreement on the areas outside of the current scope of the standards,
- definition of testing and extraction test cases,
- interoperability testing,
- demonstration, and
- evaluation and feedback to implementation specifications.

2.2 User requirements

The user requirements were identified by considering the market needs of each company, from the viewpoint of fault and configuration management for the corporate network. The following had an effect on the demonstration specifications:

- grasping faults that occur on network resources and confirming fault recovery,
- observing the area affected by a problem,
- observing the administrative state of resources,
- identifying the cause of the trouble and the resource,
- monitoring the connection between resources, and
- monitoring traffic states on specific network resources, especially the LAN equipment.

In addition, requirements based on the network configuration, such as hierarchical management, are stated in the next subsection.

2.3 System configuration

The managed network was configured with network resources as shown in Fig.2. Host computers in various vendor's offices were interconnected by using ISDN (Packet Service), OSI FTAM was executed on the ISDN, and various proprietary networks existed in each vendor's host computer. Workstations installed at the demonstration site were connected through a backbone LAN(FDDI) and branch LANs (CSMA/CD). WAN and LAN were connected by two gateways. Therefore, it can be supposed that a hierarchical management will be needed for such a complex network by following reasons.

1. The management protocol is different in each management domain (e.g. SMT for FDDI),
2. It is effective to manage resources in each location,and
3. The load on the manager must be distributed.

In order to realize the hierarchical management, a proxy agent on the subordinate manager is defined. There are three managers; Integrated Manager, Sub Manager, and LAN equipment Manager. Integrated Manager manages information processing units (both hosts and workstations), terminals and modems. Through Sub Manager, it also manages PBX and LAN equipment (FDDI, Bridges, and Gateways). Sub Manager performs detailed management of specific area, emphasizing locality and operability. LAN equipment Manager monitors the configuration of equipment connected to FDDI and the states and traffic of FDDI by using the layer management protocol SMT[11]. It also monitors the state of connections established on the gateways and the state of traffic on Bridges through LAN equipment agents.

On the other hand agents are classified into four types; system agents for information processing units, modem agents, a PBX agent, and LAN equipment agents for each specific equipment. A proxy agent on a manager relayed its management information to the superior level manager. Fig.3 shows those hierarchical system.

2.4 Management Scenario

A management scenario is the specification that combines the management protocol with management information in order to satisfy certain objectives. In this section, the scenarios which are derived from user requirements are described[3]. The demonstrated scenario was but one example for satisfying the user requirements. Followings are the scenarios for each manager.

1. Integrated Manager
   Integrated Manager monitors entire management domain which includes proprietary networks, modems, a terminal, OSI FTAM, and the domain which is managed by Sub Manager.
   (a) Modems and Terminal
   In order to detect a power failure on the modem during data transfer between a terminal and a host, a notification which indicates alarm is required as management information. Then, Integrated Manager will receive two different alarms, from different agents(a system agent and a modem agent) caused by one trouble. In order to monitor the state

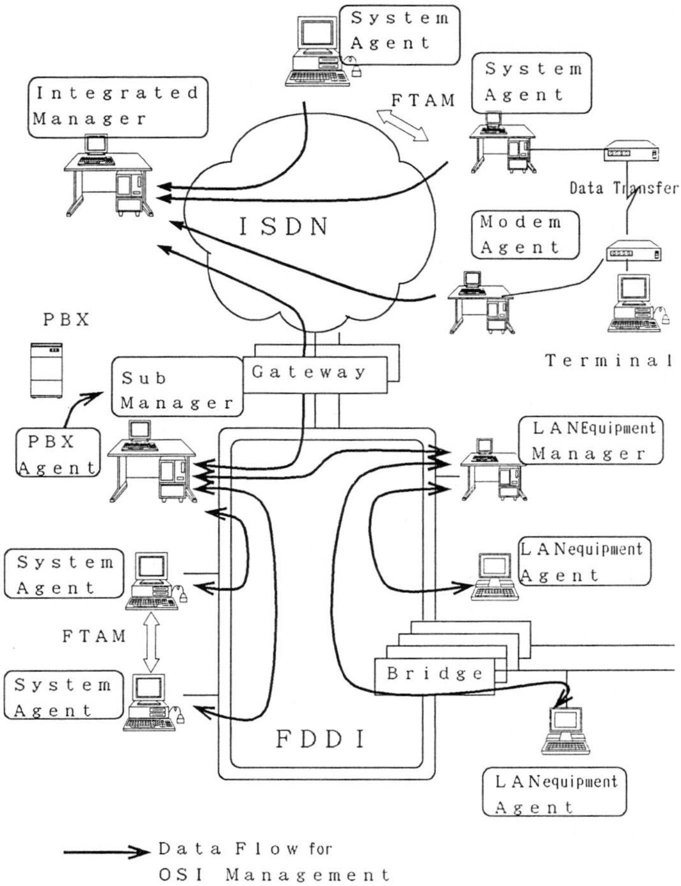

Figure 2: System Configuration

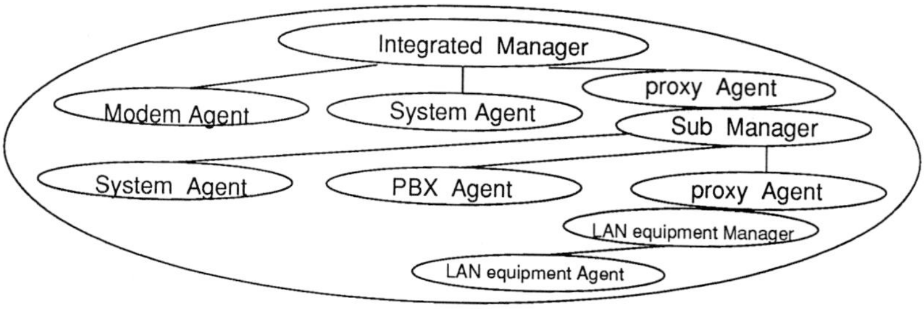

Figure 3: hierarchical Management

of modems and terminals, Integrated Manager retrieves the information representing the state of resources from agents.

(b) FTAM
In order to monitor a connection between remote resources, FTAM is used as a resource because it has been shown to be a stable product for participating companies. The association and application process are defined as Managed Objects(MOs),which represent connections and applications including communication functions,respectively. Alarms are required to report when failures occur on the association and on the application processes. The association should have a information which represents a remote association name in order to observe the area affected by a failure. An association is defined as an MO that can be created and deleted dynamically.

(c) Sub Manager domain
Integrated Manager retrieves the information representing the state of resources in Sub Manager domain from the proxy agent. Integrated Manager should be able to receive any notifications which were reported by the proxy agent.

2. Sub Manager
Sub Manager monitors the site area, including the PBX, FTAM between WSs, and states of LAN equipment through a proxy agent on LAN equipment Manager. The proxy agent on Sub Manager passes the notification, which are received by Sub Manager from agents, to Integrated Manager to report the changes occurred in its domain.

(a) PBX
A notification reporting a change of state is required in order to monitor maintenance work on specific Board of PBX. Sub Manager should distinct it from the alarm occurred by failures. Sub Manager also retrieves necessary information, representing the state of resources, in order to monitor the state of, and statistical information about the telephone.

(b) FTAM
The FTAM is used in the same way as shown above on Integrated Manger, and in addition its reports are observed on the monitors of both Sub Manager and Integrated Manager.

(c) LAN equipment Manager domain
Just as with Integrated Manager, Sub Manager should be able to handle information in LAN equipment Manager domain.

3. LAN equipment Manager
MOs representing equipment, and which have statistical information, are required in order to monitor topological and statistical data. SMT is used to monitor the FDDI infrastructure. Similar to the Sub Manager, the proxy agent on the LAN equipment Manager reports changes which occur in its own domain.

Table 1: Implementation Agreements in the demonstration

item	contents
scope of management	object management state management alarm reporting
CMIS/CMIP	kernel functional unit MultipleObjectSelection functional unit LinkedReply functional unit
Establishment of association	Both manager and agent can request The agent is winner when collision has occurred
Usage of option parameters in standards	-
Usage of error	Identify the error for each operation or reporting
ASN.1 basic encoding rule	5items for confirmation of the rule 4items such as maximum value of INTEGER and LengthOctet value 4items for indefinite form

3 Management Protocol and Information

The Management Protocol and Information are specified by making maximum utilization of the available standards in order to realize the scenarios. The subsetting from the base standards was based on "Implementation Specification" defined by INTAP[2]. Other areas such as agreements related with association were defined for a demonstration referring to the corresponding standards.

3.1 Management Protocol

CMIP[1], Object Management[4], Alarm Reporting[7], State Management[5] and Attributes for representing relationship[6] were used. The outline of the agreement is shown in Table1. Implementation agreements were defined for each MO in a form of table similar to the PICS proforma together with required SMF primitives. The specific determined in the agreement refer to which primitive parameter should be used, and which PPDU parameter should be sent out. The other agreement related to association are also defined.

3.2 Management Information

Management information has been selected by looking at the future practicality of its use, as well as the scenario requirements. Fig.4 and Fig.5 show the inheritance tree and the naming tree, respectively, for the demonstration system. Standard objects require the refinement to suite individual network resources. Therefore, MOs were defined by being inherited from standards, such as DMI[8]and GMI[10]. Twenty four object classes

Table 2: MOs used in the demonstration

reference documents	objectclasses
DMI	top,System
GMI	ApplicationProcess,Entity,CO-Entity,CL-Entity connection,association,sap
none	terminal,modem,switchingsystem,bridge,gateway gatewayProcess,T-entity,fddi-entity T-connection,BridgePort,Cabinet,Unit,Station, Board,Network

were defined, seven by referring to GMI[10]. The other classes were defined for the demonstration by using GDMO[9].

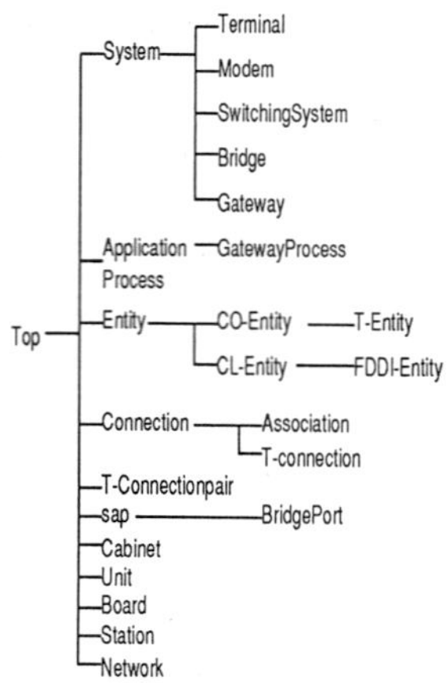

Fig.4 Inheritance Tree

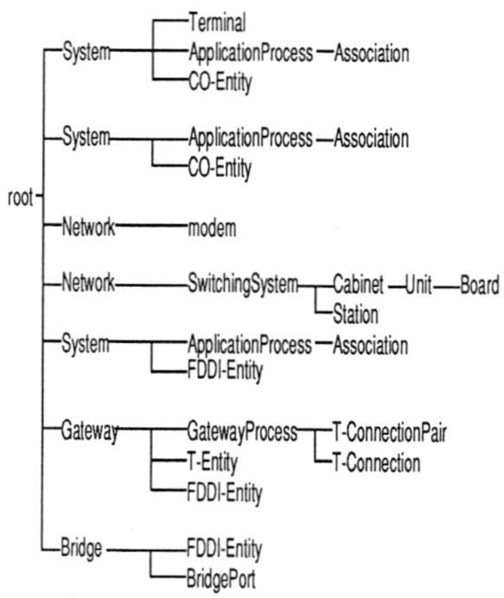

Fig.5 Naming Tree

The MOs used in the demonstration are shown in Table 2. The discussed points related with MOs definition are as follows:

1. It is not easy to map the actual resources which has been implemented by each vendor to MO definition
2. It is necessary for the definition of MOs to know the actual network resources well. Therefore, both experts of management and resources should participate in work of MO definition.

Test items for system agents and momem agents

no		MO and attributes
I1	association establish	
I2	association release	
I3	Get operation	system.state
I4	Get operation	terminal.state
I5	Get operation	ApplicationProcess.state
I6	Get operation	Co-Entity.state.send/recieve data
I7	Object creation reporting	Association
I8	Object Deletion reporting	Association
I9	Alarm(Processing Error)	Association
I10	Alarm(CommunicationErrror)	ApplicationProcess
M1	Equipment Errror	Modem
M2	CommunicationError	Terminal
E1	kill FTAM.restart FTAM	
E2	kill system agent and restart	
E3	kill Integrated Manager and restart	
E4	disconnect between manager and agent	
E5	kill modem agent.restart modem agent	
E6	disconnect between modem agent	

Figure 6: An example of test case sheet for Integrated Manager, a system agent, modem agent

3. The definitions in the BEHAVIOUR clause in the MO class template should be clear.
4. There is no obvious criteria for selecting either the system or equipment class.
5. MOs which do not have other attributes than Relative Distinguish Name(RDN), are effective in representing the configuration.
6. It needs further consideration to determine the timing and the method to get the relationship attributes(e.g. peer attribute).

4 Testing

4.1 Testing Process and Test Items

There had not been any authorized conformance or interoperability testing. That was the first opportunity for the participating companies to connect their equipment to that of the other companies under the OSI management environment. It took approximately six months to carry out testing process, which was divided into several phases. It proceeded as building block from individual to combinations of operations and reporting. The following were all carried out:

1. Checking PPDU dump lists with correct one among all companies.
2. Testing lower layer(ISDN,FDDI) and FTAM
3. Interconnection testing between Integrated Manager and modem agent and system agent, LAN equipment Manager and LAN equipment agent, Sub Manager and system agents and PBX agent, and then Sub Manager and LAN equipment Manager(proxy agent)
4. Confirmation of the arrival of reporting to Sub Manager from LAN equipment agents
5. Interconnection testing between Integrated Manager and Sub Manager
6. Stress testing for Sub Manager
7. Testing based on scenario
8. Evaluating and feed back to Implementation Specification

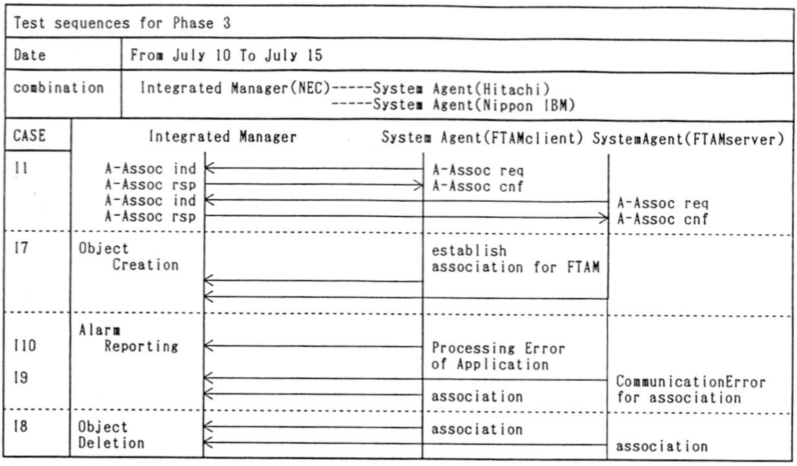

Figure 7: An example of test sequence sheet for Integrated Manager, a system agent, modem agent

Test cases were designed for every phase as shown in Fig.6, and sequencing for testing cases were also designed as shown in Fig.7. Interconnection tests contain common test cases, including those of association establishment, release, program aborts, and recovery after a system failure. The trouble-shooting process was agreed on by the participants due to the test systems being located remotely, except for LAN.

4.2 Testing Result

Under the circumstance of no conformance testing service, it was necessary for interoperability to check the basic operations. The checking of the PPDU dump, done during the first step in testing, was regarded as a phase of conformance testing. Approximately 40% of all troubles were found in this phase, therefore, it is requested that conformance testing service should be realized as early as possible.

5 Discussion

It wasn't sufficient for the demonstration to only use the standards. It was necessary to make such agreements as additional specifications on the standards. In this section, some issues related to implementation are discussed.

5.1 Consistency of the information between manager and agents

The consistency between a manager and its agents is an issue since a manager should monitor the correct information, which is kept in the agent at the same time. This is a problem from the viewpoint of operability. After a failure on the association, the manager must correct its information in the system currently being managed. There are two possible solutions for this issue. One involves only using event reports from the agents. This means that an agent reports the difference between the management information in the trouble time and current management information. The agent is required to maintain information before the trouble happened. This would allow the traffic between a manager and its agents to be kept to a minimum, and would be effective for large scale network management. The second method involves using the "get operation" of CMIP. The manager would issue a "get operation" to all agents regarding all MOs after the association is re-established. This method was adopted because there were not too many MOs, and because the other way requires more implementation for both managers and agents.

5.2 Hierarchical Management

Networks vary in their size, intelligence and their complexity. Therefore, management structure also varies with the networks being managed. Thus, hierarchical management will be required to manage large size,complex network effectively. In this section, issues related to hierarchical management are discussed.

1. Definition the naming tree
   There are two ways to define the naming tree. One method is that each manager has a naming tree of the form shown in Fig.8. This method has the advantage that each manager has less data to be concerned, which requires each proxy agent to summarize its information and transfer this to the manager. It will be necessary to define a new MO representing the information summarized the specific area. Furthermore, a rule between manager and proxy agents need to be determinated. The second method is to give each manager a naming tree as shown in Fig.9. In this tree, each manager shares one naming tree. After undergoing much discussion, this method was adopted for the demonstration. One reason is that there are not so many MOs, and the other is that the solution of the former case will be studied in the area of the management domain. This study of management domain is proceeded in ISO, and therefore, there was no compelling reason to choose the first method as a local rule.
   Peer communication between managers was also discussed. However,it was not implemented due to the lack of appropriate support standards.

2. Consistency of information among managers
   The consistency issue can arise when a superior manager issues an operation to a proxy agent, and when the association between the subordinate manager and a lower proxy agent is lost. During the demonstration, when the Integrated Manager issued an operation to a proxy agent (on Sub Manager) on a certain MO, which are managed by LAN equipment Manager through the LAN equipment Agent, the Sub Manager could not issue an operation to the LAN equipment Manager regarding trouble between Sub Manager and the proxy agent (on LAN equipment Manager). An operation issued by a superior manager does not arrive at the agent which manages the actual resources when some type of trouble

has occurred on the association. There is no appropriate corresponding error in the standards representing this situation. It was decided that the proxy agent should respond to the processing failure when this occurs. This condition will represent merely one state and will be differentiate from errors. It is also difficult for the subordinate manager to identify which of the superior managers have issued the operation , and thus becomes difficult to analyze the trace during testing.

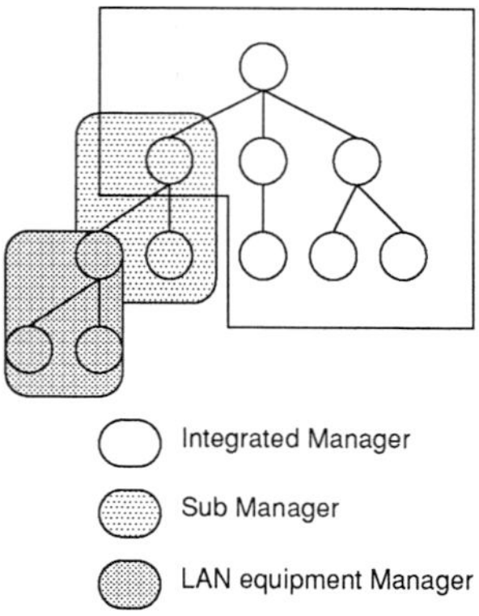

Fig.8 Each manager has a subtree

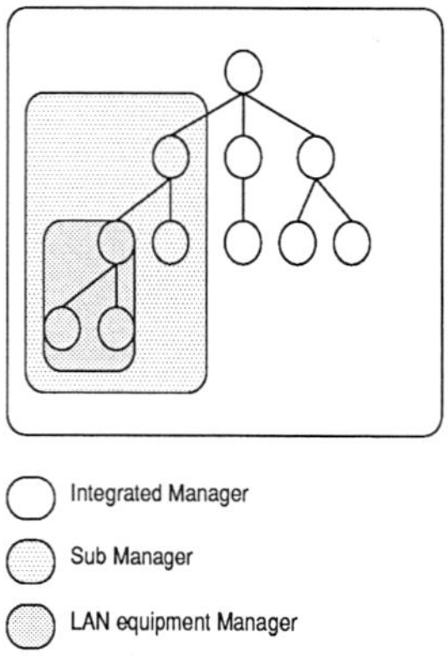

Fig.9 Each manager shares one tree

6 Conclusion

The demonstration was successfully carried out. A majority of visitors were impressed that the interoperability using OSI had become a reality. It should be concluded from overall activities for this experimental demonstration project.

1. It was verified that OSI Management is valid for network management. A heterogeneous network which consisted of various scale equipment, LAN equipment, software, and voice network was shown to be manageable by using the OSI Management.
2. The scenario is an important for the definition of MO and implementation agreements. It is necessary to consider the scenario adding on protocol and managed information. The scenario combines both of them.
3. The rules from aspect of network management system operation ,such as hierarchical management should be needed for interoperability.
4. It can be recognized that conformance testing is important for interoperability.
5. This demonstration contributes the standard specifications , and the scope of the demonstration becomes one of the category of functional standards; pDISP12060-2[12].

It is desirable to demonstrate and evaluate other functional areas such as security management,accounting management. INTAP will contribute development of Implementation Specification , strongly support AOW, with harmonization of the other regional workshops and standardizing organizations. It will also work for conformance testing and interoperability testing in collaboration with other related working bodies and groups.

Acknowledgment

The authors would like to express their appreciation to Dr.Yoshimi Teshigawara of NEC and Dr.Tetsuo Nakakawaji of Mitsubishi for their many valuable comments and refining this paper, and Mr.Akira Date and Mr.Teiji Okumura of NEC for their continuous encouragement and support. Also special recognition is extended to all members of working group for successfully carrying out the demonstration.

References

[1] ISO/IEC 9595/9596: CMIS/CMIP,1989.

[2] S008-01 The implementation specification for OSI Management, INTAP, 1991. (in Japanese)

[3] The specification of demonstration for OSI Management ver.3,INTAP,1991. (in Japanese)

[4] ISO/IEC DIS 10164-1:Object Management function.

[5] ISO/IEC DIS 10164-2:State Management function.

[6] ISO/IEC DIS 10164-3:Attributes for representing relationship.

[7] ISO/IEC DIS 10164-4:Alarm Reporting Function

[8] ISO/IEC DIS 10165-2:Definition of Management Information.

[9] ISO/IEC DIS 10165-4:Guidelines for Definition of Managed Object.

[10] ISO/IEC CD 10165-5:Generic Management Information.

[11] ANSI X3T9.5:FDDI Station management(SMT) May 1990.

[12] ISO/IEC pDISP 12060-2 AOM212 Alarm Reporting and State management Capabilities

Integrated Network Management, III (C-12)
H.-G. Hegering and Y. Yemini (Editors)
Elsevier Science Publishers B.V. (North-Holland)

An Architecture Driven Comparison of Network Management Systems

Kraig Meyer, Joe Betser, Eric Negaard, Dennis Persinger, Steven Wang, Robert Maltese, and Carl Sunshine

The Aerospace Corporation, P.O. Box 92957, Los Angeles, CA, 90009-2957.
Phone +1 310-336-8114. Electronic Mail: kmeyer@aero.org

Abstract
Management of large, multiprotocol, computer networks has become increasingly important and represents a new challenge to many organizations. Many vendors have introduced commercial network management products to meet these growing needs. In this paper, we examine six of these products, including four Network Management Stations and two Passive Network Monitors, or segment monitors. We derived a list of requirements based on the needs of several network architectures of interest. From these requirements, we design an architecture and a set of validation experiments that serve to differentiate among the different products. Finally, we present our conclusions about the abilities of different products to address different program needs.

Key Word Codes: C.2.1, C.2.3
Key Words: Network Architecture and Design, Network Operations

1 Introduction

During the past several years, the need for integrated network management systems has grown very rapidly in universities, industry, and the government. A number of commercially available products have been developed to meet these needs, but they vary widely in functionality and economy. This is largely because the needs of network managers vary somewhat between differing environments, and different product capabilities have evolved to address these needs.

We have developed a set of criteria against which to assess network management products. These criteria are based on the requirements of our own corporate network, as well as the requirements of several government programs with more specialized needs. We then acquired four of the leading, commercially available, Network Management Stations (NMSs) and two Passive Network Monitors (PNMs), and evaluated how well they meet our requirements.

We start this paper by describing the products we evaluated and a framework in which we place those products in Section 2. We then describe the evaluation criteria and the requirements that drove these criteria in Section 3. The environment in which the products were evaluated is described in Section 4. In Sections 5 and 6, we describe the results of specific experiments we performed on the products. Section 7 summarizes the results of all experiments. Finally, we discuss future work and conclusions in Sections 8 and 9.

2 Products

Figure 1 illustrates a framework into which we fit network management products. A specific network management feature typically involves a flow of information through this diagram, and may require the interaction of a number of entities. Consider an example in which a link fails to pass traffic from one end to the other. First, a node at one end of the link must be polled by an NMS—usually via SNMP, the Simple Network Management Protocol. The node must respond to the query, indicating that an attached link is down. The contents of this response (e.g., "link down") must be considered a significant event by the event processing software. The alert screen of the NMS must display this information in a format that can be understood by an operator [MEYE91]. Finally, the alert may be fed into a problem management system that maintains a database of outstanding problems or "trouble tickets." In this paper, we primarily focus on those products that have a role in information collection and display, with limited emphasis on information processing and problem management.

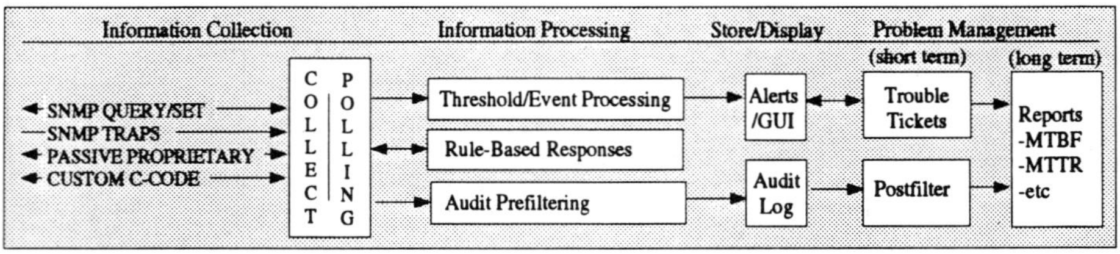

Figure 1: Network Management Framework

We initially looked at a wide variety of commercial network management products in the Fall of 1991. We narrowed this list down to four Network Management Stations and two Passive Network Monitors based on several requirements. First, the vendors had to have proven experience in the networking field and the ability to provide strong customer support. We also considered how long the product had been in the field, and the installed user base. Additionally, we compared the products' claimed capabilities to the required functionality described below in Section 3. Based on these high-level requirements, we decided to evaluate the following products:

Network Management Stations (NMSs)	Passive Network Monitors (PNMs)
SunNet Manager (Sun)	LANtern (Novell)
Spectrum (Cabletron)	LanProbe (Hewlett Packard)
DECmcc Director (Digital Equipment)	
Openview Network Node Manager (Hewlett Packard)	

The software for each NMS was run on a vendor-recommended platform: SunNet Manager 1.2 under SunOS on a Sun SPARCstation 1, Spectrum 1.0.3 under SunOS on a Sun SPARCstation 2, DECmcc 1.2 under Ultrix on a DECstation 5000/125 and HP Network Node Manager 2.0 under HP/UX on an HP/Apollo Series 700. Both of the PNM systems included a proprietary hardware device which is attached directly to the Ethernet segment being monitored. The PNMs also require a PC-based management station that is used to configure and manage the passive monitors.

3 Required Functionality

Based on the requirements of the programs that we are involved with, as well as our experience managing our own corporate (and other production) networks, we developed a list of criteria against which to assess the functionality of Network Management Stations and Passive Network Monitors. These criteria fall into two categories. First are criteria based on generic requirements. This includes functionality that is desirable or necessary in virtually all network management environments. The second group of criteria are based on program-specific requirements. These requirements are driven by a specific network architecture described below, and do not necessarily apply in other environments.

The architecture illustrated in Figure 2 drives the program-specific criteria. This architecture includes a dual network architecture in which two independent networks are connected by an application layer gateway. In this system, users are allowed access to resources on the application network only via the application gateway; general traffic is not permitted to flow between the two networks. Continuous operation of the system is essential, so component outages must be anticipated and avoided whenever possible. In addition, organizational requirements dictate that network status information be made available to multiple operators as well as to higher-level management personnel at all times.

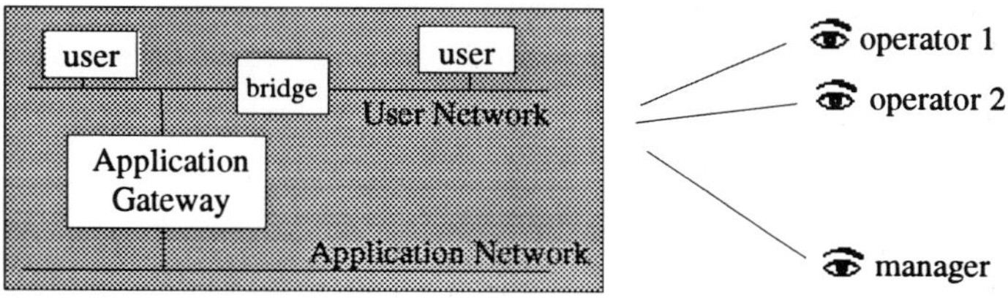

Figure 2: Special Network Architecture to be Managed

Table 1 describes the criteria that we have used to evaluate NMSs. For each capability named in the left-hand column, the right-hand column lists evaluation criteria that were addressed through experimentation with the actual products. The first five requirements are generic in that they apply to an NMS operating in nearly any environment. The last three requirements are primarily program-specific requirements and are driven by the architecture described above as follows: the program's dual network architecture requires that multiple networks be managed by a single station without allowing data flow between the networks. System reliability requires close tracking of workstation health, to avoid outages whenever possible. Organizational requirements dictate that multiple NMSs must simultaneously provide information to multiple personnel.

We also evaluated Passive Network Monitors utilizing the criteria in Table 2. We primarily investigated those features that allow monitoring of the health and activities of a specific Ethernet segment. Although we focused on PNM requirements that were required by the architecture described above, all of the PNM requirements are somewhat generic, in that they apply to a wide variety of network architectures.

In addition to the requirements described in this section, there are additional criteria which will be used to perform future evaluation experiments. These are described in Section 8.

Capabilities	Evaluation Criteria
Topology and Node Discovery	1. Efficiency of algorithm(s) used to discover nodes. 2. Number of types of devices discovered (IP, DECNET, etc). 3. Quantity and utility of Information obtained about nodes (name, MAC address, IP address, etc).
Monitoring Links and Link Devices	1. Ability to determine the status of links and link devices. 2. Effectiveness of method used to display this information.
Alarms, Notifications, Interfaces	1. Number and types of method(s) that can be used to convey information about network status to the user. 2. Ability to group objects in a hierarchy. 3. Ability to propagate alarms between groups. 4. General qualities of the user interface.
Support for Commercial MIBs	1. Ability to use and integrate MIBS from other commercial products.
Reachability of TCP/IP and DECNET nodes	1. Ability to use ICMP echo to determine reachability. 2. Ability to use SNMP. 3. Ability to use DECNET. 4. Ability to use other protocols or methods.
Support for Dual Network Architecture	1. Ability to manage multiple independent networks from a given NMS.
Assessment of Workstation Health	1. Ability to assess health of any generic workstation. 2. Ability to assess health of specific vendor workstations.
Multiple NMS Users	1. Ability for multiple users to simultaneously manage network with single product. 2. Ability for a single database to be shared among multiple users. 3. Ability for a single data collector to drive multiple displays.

Table 1: Required Functionality of Network Management Stations

Capabilities	Evaluation Criteria
Ability to Integrate	1. Ability to communicate with NMS via a standard protocol interface (e.g. SNMP). 2. Ability to communicate with NMS via proprietary interface.
Ethernet Health and Utilization	1. Usefulness and extent of information that can be collected, displayed and stored that indicates the health and utilization of an Ethernet segment.
Host Information and Statistics	1. Usefulness of node information that the PNM has the ability to collect. 2. Ability to map Ethernet addresses to Network addresses. 3. Ability to maintain traffic matrices that show which source-destination pairs are communicating.
Packet Filter and Capture	1. Ability to support remote debugging by providing packet capture features.
Alarms	1. Ability to detect anomalous events and alert the central administrator or management station.

Table 2: Required Functionality of Passive Network Monitors

4 The Testbed

The network we used to perform experiments was initially configured as shown in Figure 3. We attached each NMS to a common subnet, which in turn was attached to the rest of our internal corporate network, Aeronet, via a router. We attached each PNM on one side of the router. From this baseline configuration, we moved PNMs and NMSs to different sides of the router depending on which experiments we were performing. We used the NMSs (and PNMs) to manage (and monitor) devices on Aeronet: roughly 10 multiprotocol routers, 250 IP devices, 100 DECNET devices, and a variety of bridges and hubs.

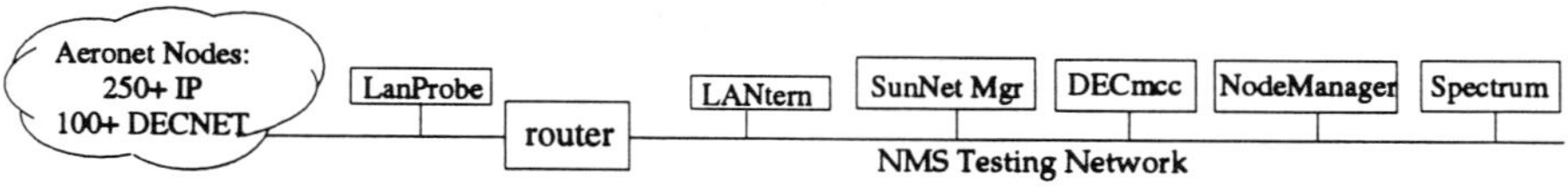

Figure 3: Baseline Network Configuration for NMS Testing

5 NMS Experiment Results

5.1 Topology/Node Discovery

Each of the NMSs we evaluated used one or more of the following approaches to automatically discover IP network nodes and topology: address range scanning, information from ARP tables, and passive monitoring of local segments. The first method, address range scanning, steps through every network address in a range and performs some test on each address—such as an ICMP echo request. Any address that responds is added to a list of nodes. This method can be very time consuming but has the advantage that all nodes on the network that respond to an ICMP echo request are discovered. The second method uses SNMP to collect information from ARP tables, which in turn is used to construct a list of nodes. This allows more information about SNMP-reachable[1] nodes to be collected, but will miss other nodes. In the final discovery method, an NMS listens on an Ethernet segment and creates a list of nodes based on the addresses in the packet headers of each packet that is observed on the network. This method will not discover nodes on remote segments.

SunNet Manager supports address range scanning and passive monitoring of local segments. Each node is tested for ICMP echo, SNMP reachability, and to determine whether it responds to certain Sun remote procedure calls. SunNet Manager builds a limited topology map, grouping nodes by subnet but does not automatically determine which nodes are routers. The discovery process can be repeated as often as necessary, and only newly discovered nodes will be added to the database.

Spectrum supports address range scanning, using ICMP echo requests to determine node existence. In addition, Spectrum will query any user-specified SNMP node and use the contents of that node's ARP table to add to the list of discovered nodes. During discovery, only node names and network addresses are determined, and no topology map is automatically built. Discovery can be run at any time.

[1]A node with a known SNMP community name, or a node that communicates directly with such a node. Note: Unknown community names may trigger SNMP authentication traps.

DECmcc uses address range scanning with ICMP echo requests to determine what nodes exist. When discovery is initiated the user provides a list of SNMP community names used by nodes in the network. Each node that responds to an ICMP echo request is sent an SNMP query, and the ARP table of any node that responds is used to further add to the node list. Node names and IP addresses are determined, and a map is automatically generated. Node discovery can be run at any time. DECmcc was also the only product that automatically discovered DECNET nodes.

HP Network Node Manager uses information from ARP tables to build its node list, and keeps track of node names, network addresses, and Ethernet addresses. The contents of the local ARP table are used to construct an initial node list. As a node is added to this list, that node is sent an SNMP message requesting the contents of its ARP table. This recursive process is repeated until no new nodes are found. In the default configuration, this process repeats itself every few minutes, and newly discovered nodes are added to the topology map according to the internally defined set of rules. HP Network Node Manager generates topological maps in which routers and nodes are automatically attached to appropriate segments.

5.2 Monitoring Links and Link Devices

Network links–such as leased telephone lines–must be monitored using a slightly different approach than network nodes. Because links are typically passive entities, monitoring software must generally obtain information about links from the active entities on either end of the link. In order to communicate with a remote site, both the active and passive entities must be functioning. Some NMSs do not provide a clear way to distinguish between failure of an active entity (such as a router) and a passive entity that is attached to it (such as a T1 line). The usual way to determine the status of a router and its attached links is to examine a MIB variable [MCCL90] that gives the current operational status of each interface on the router.

In order to use SunNet Manager to monitor the status of a serial link on a router, two tests must be set up. One test must check to see if the router is reachable; the other must check the status of each interface on the router. Then, if the router or any of its links go down, the icon representing the router and all of the connections to the router will change color. To determine which specific link is actually down, the event log must be examined.

Spectrum determines which nodes are having link problems based on built-in rules—rules similar to the manually defined tests for SunNet Manager described above. When one of the interfaces is determined to be down, the icon representing the router turns yellow. If the router is unreachable or if all of its interfaces are down, the icon turns red. A window displaying the status of each interface within the router can be obtained with a mouse click on the router icon.

DECmcc uses manually defined tests similar to SunNet Manager's. However, in DECmcc it is possible to represent a link as an independent entity from the node it is attached to. Whenever a link goes down, both the node icon and the link that is down will change color—making it very easy to tell which interface is down. The node icon will also change color if the node itself is unreachable.

HP Network Node Manager also uses built-in rules for determining node status. Routers are automatically identified as any node that, according to SNMP, has more than one network interface and has IP forwarding turned on. A router is shown as up if all of its interfaces are up. It is shown as marginal if at least one interface is down and at least one is up. If all interfaces are down, the router is shown to be down. A window displaying the status of each interface within the router can be obtained by clicking a mouse button on the router icon.

5.3 Alarms, Notifications, and Interface Issues

For each NMS, we determined how the network could be viewed by the user and what methods were available for informing a user of a network alarm. NMSs frequently represent networks in a hierarchic form, and allow the user to specify how alarm information at lower levels of the hierarchy (e.g., a single workstation outage) propagates up to higher levels of the hierarchy (e.g., a subnet icon turning red).

SunNet Manager allows objects in one view to be arbitrarily grouped into a single object in a different view, but each object may appear only in one view in any given database. This allows the administrator flexibility to create any kind of map desired: physical, logical, with or without hierarchy, but makes it difficult to mix different types of views within a database. The highest severity alarm within a view is always propagated up to the top level of any hierarchy. When an alarm occurs, node status can be indicated by automatically changing icon attributes (color, brightness, flashing), electronic mail can be sent, or a program can automatically be executed. Each time an alarm occurs, a line of text is added to an event log that can be viewed in a separate window. In general, SunNet Manager's operations are intuitive and consistent and only minimal reference to documentation is needed.

Spectrum supports three views of networks that it manages: a topological view, a location view, and an organizational view. For certain types of devices, it will also provide a simulated display of the device's front panel in real time. On the screen, a network object's status is always indicated by an icon's attributes. When alarms occur, a line of text is added to an event log that can be viewed in a separate window. It is not possible to use Spectrum to notify users of alarms via electronic mail or automatically invoked programs, although a third party product may be available to support these functions. Propagation of alarms between levels of the hierarchy can be customized on a per-object basis, allowing specific alarms to propagate up while others do not. In general, Spectrum's interface is difficult to work with. Most operations are not intuitive and require training and/or frequent use of a reference manual.

DECmcc allows objects to be arbitrarily grouped into a view or *domain*, and allows devices to belong to any number of domains. An administrator has flexibility to draw any mixture of views desired: geographic map, logical map, and so forth. However, each domain has its own set of rules for alarms and notifications. When an alarm occurs, a notification is added to a textual list of alarms. In addition, node status can be indicated by icon attributes, or through automatic program execution (including mail). Each member of a domain can be individually configured to allow propagation of alarms to the parent domain. DECmcc's user interface is not intuitive in many places, and requires numerous steps to perform some simple actions. Most users will have to read several introductory manuals before they will be able to accomplish basic tasks.

HP Network Node Manager provides a single, three-level topological map interface. The appearance of the map is controlled largely by built-in rules, including what devices can be included at each level of the hierarchy, how objects are initially placed on the map, what conditions cause iconic attributes to change, and how alarms are propagated between levels of the hierarchy. If a user wishes to define a specific alarm condition that is not already included in the built-in rule base, this alarm can either be written to a trap log or may automatically invoke a program. User-defined alarms cannot be displayed on the map. All outstanding notifications, including ones that appear on the map appear in the trap log. HP Network Node Manager's user interface is consistent, intuitive, and easy to use. However, it provides less flexibility in customizing map layout and alarm display.

5.4 Support for Commercial MIBs

In order to manage a wide range of commercially available products, an NMS must be able to utilize MIBs (Management Information Bases) of other commercial products. Most have a MIB compiler that uses the MIB and translates it into a format directly usable by the NMS. We attempted to install the MIB of the Novell LANtern on the various NMSs.

SunNet Manager's MIB compiler was difficult to use. It was necessary to manually remove portions of the LANtern MIB to install the MIB, and to manually modify the resulting schema before using it in SunNet. These problems were attributed to shortcomings in SunNet's MIB compiler.

Spectrum provides very limited support for the use of Commercial MIBs, and provides no MIB compiler. Specific MIB variables may be added by hand, but there is no mechanism for loading an entire product's MIB directly into Spectrum. Instead, Spectrum provides management modules for a limited set of third party products.

We had little difficulty integrating MIBs into DECmcc. In addition to the LANtern MIB, we also installed MIBs for Sun workstations and a Retix bridge.

The LANtern MIB was integrated with HP Network Node Manager with only minor difficulties–some portions of the MIB had to be commented out, but the resulting schema operated smoothly.

5.5 Reachability of TCP/IP and DECNET nodes

All of the evaluated products supported two basic methods for determining the reachability of a TCP/IP node: an ICMP echo request (generally referred to as a *ping*), and an SNMP *get* operation. It is possible to use the DECNET protocols to assess the reachability of DECNET nodes using SunNet with an add-on third party product or by using DECmcc.

SunNet Manager requires the user to create a rule for each node that is to be monitored, specifying how frequently and what method to use to assess node reachability.

Spectrum automatically sends ICMP echo requests periodically to all nodes in its database; if no response is received, the node is marked as down. A third party DECNET product for Spectrum has been announced but was not included in this evaluation.

DECmcc requires the creation of rules to specify which nodes are to be monitored, and how to monitor them. However, a single rule can apply to an entire class of nodes; an individual rule is not necessary for each node being monitored.

HP Network Node Manager automatically sends ICMP echo requests periodically to all of the nodes in its database; if no response is received the node is marked as down. The time interval between queries is a global setting for all nodes in the database and is not individually adjustable. A third party DECNET product for HP Network Node Manager has been announced but was not included in this evaluation.

5.6 Support for Dual Network Architecture

We considered two ways to manage multiple, independent networks from a single NMS. The first design requires multiple Ethernet cards in the NMS, one attached to each network being managed. To ensure separation of the networks, the NMS must not route between the networks. We were able to use SunNet Manager and DECmcc on platforms with two Ethernet cards to successfully manage two independent TCP/IP networks. We were unable to obtain a second Ethernet card for our HP platform and did not attempt to use Spectrum with two Ethernet cards. DECmcc does not support DECNET via multiple Ethernet cards under Ultrix.

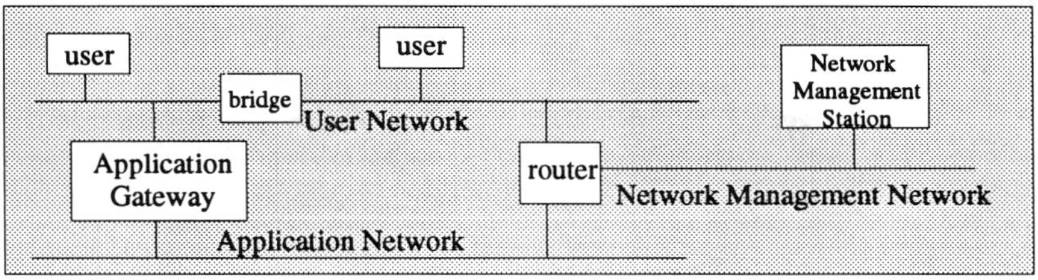

Figure 4: Special Network Management Architecture

The second design for supporting multiple networks is illustrated in Figure 4. This design requires the use of a router with access controls. These controls permit the NMS to talk to devices on each of the attached networks but do not permit traffic to pass between the attached networks. This solution does not require any special capability on the part of an NMS, so any NMS can be used. In addition, this architecture is more modular and expandable, and for this reason, is the preferred solution.

5.7 Assessment of Workstation Health

For workstations and network servers, there are a number of metrics that might be used in assessing node health. Information that is available in many MIBs, and via DECNET protocols, includes network interface statistics and routing information; the number of inbound and outbound packets, and routes to specific destinations. The Sun-4 MIB additionally includes measures of how much time has been used by system and user processes, and how much time the processor has been idle. Most of this information does not appear to be generally useful in determining node healthfulness because it is so narrow in scope. The two most obvious metrics in assessing node health—CPU utilization and free disk space—are not always available via MIB variables.

SunNet Manager can obtain information about CPU utilization and disk usage of other Sun products by using a special remote-procedure-call interface. SunNet Manager can also do threshold checks on MIB values, on the amount a MIB value has changed in a given time period, and on DECNET node statistics.

Spectrum does not provide a general interface to make checks against MIB values. Instead, for each product managed, a "management module" must exist that has predefined rules to determine a node's status. We did not have a management module for any workstation.

DECmcc can use values obtained via SNMP and DECNET protocols to make assessments of node health. It is constrained primarily by the amount of information that is available through the various MIBs and the DECNET protocols.

HP Network Node Manager also can use values obtained via SNMP to assess node health, but is limited in that only basic threshold tests can be performed.

5.8 Multiple NMS Users

In most network management environments, there is a need to provide Network Management displays to more than one user. This can be addressed by having multiple NMSs running completely independently, but there are benefits to a more coordinated approach. If a single data collector can be used to provide status information to multiple displays, the amount of network bandwidth consumed by network management traffic is reduced. If a common configuration database is shared between users, less effort is required to keep the database up-to-date.

Multiple copies of SunNet Manager may execute simultaneously, and each copy collects its own data to drive its displays and to generate alarms. Users may load and make copies of databases defined by other users—provided they can access the other users' files—but no mechanism is provided to maintain consistency between multiple user databases.

Spectrum's data collector runs separately from its display module. Users on any number of nodes can access this data collector and fully share all configuration and status information in a central database. It is possible to give different users different access permissions to the database.

DECmcc allows multiple users to have simultaneous read/write access to the database that includes configuration and status information. In addition to the configuration database, certain configuration options are stored on a per-user basis.

HP Network Node Manager allows multiple simultaneous displays to be running. All copies use the same database, but only one copy has read/write access—all others have read-only access.

6 PNM Experiment Results

6.1 Ability to Integrate

LANtern responds to SNMP queries and generates SNMP traps, but still requires a PC-based manager. LANtern communicates with its manager either via SNMP over the network or via a serial SLIP connection. Some information available on the PC manager is not directly available via SNMP, but may be derived from SNMP-available information.

Data from LanProbe can be collected only via a proprietary protocol that runs on a separate PC. LanProbe communicates with its manager either via an Ethernet or through a serial modem connection.

Both systems can use the serial connection to communicate alarms and other management data to the PC management station when regular network connectivity is not available. There is presently no straightforward way to use lists of nodes collected by either PNM to create or add to a topology map on an NMS.

6.2 Ethernet Health and Utilization

Both LANtern and LanProbe provide instantaneous gauges showing the primary indicators of Ethernet Health: utilization (percentage of time carrier is present), errors/sec, broadcasts/sec, and collisions/sec. As described below, an alarm-triggering threshold can be set for each of these indicators. Both PC systems will maintain a database of trend data that can be reviewed.

LANtern's manager allows the user to display graphs and bar charts using any of the statistics. Statistics can be depicted either as percentages or raw numbers, and there is greater flexibility in choosing scales and sampling intervals. The LANtern can save up to 100 collection points in internal memory and send them to the PC-based manager the next time it connects.

LanProbe's manager displays counts of packets/sec and bytes/sec. Historical data can be viewed, and traffic can be broken down by type (broadcast, multicast, type of error, etc.) and packet size. Trend data is collected only when the PC-based manager can reach the passive monitor.

6.3 Host Information and Statistics

Both PNMs maintain a list of nodes discovered, storing Ethernet and IP addresses. LanProbe will also decode DECNET addresses. When new nodes are discovered, both can send alarms to their PC management stations.

LANtern users may choose any subset of nodes on the network to view as a sorted table. Additionally, the user may see the list of all nodes that a given node has communicated with in the recent past. LanProbe's capabilities are more limited: a user can display only the top 20 nodes according to criteria such as the number of packets, bytes and errors sent.

6.4 Packet Filter and Capture

LANtern has no capability to filter for or capture specific packets.

LanProbe allows users to filter network traffic according to a sophisticated set of criteria, including Ethernet packet type and length of packet. The LanProbe can simply gather statistics based on this filter, or capture the packets themselves for later analysis at the central management station.

6.5 Alarms

An alarm capability is essential to obtain and display manageable alerts from the enormous amount of raw data that is collected. Unfortunately, alarms in both systems were limited to simple thresholds on the primary Ethernet statistics, cable failure, discovery of a new node, duplicate IP address, and power outage or reboot. Improved alarm capabilities could be achieved by using an NMS to perform processing on data obtained from LANtern via SNMP.

7 Matrixed Results

7.1 Network Management Stations

Table 3 shows an overall comparison of how the NMSs performed in experiments. These ratings were determined subjectively by the authors, based on the criteria of Table 2. Some experiments were separated into two entries. No overall ranking of products has been made by the authors, because the weighting given to each capability will be different in different application environments. For example, some environments may require DECNET support while others may not consider it to be important.

7.2 Passive Network Monitors

Both the LanProbe and LANtern systems have areas of strengths and weaknesses, with neither showing clear superiority with respect to our criteria. The LANtern was overall a more flexible monitor, with a larger information base, more customizable reporting and statistics-gathering facilities, and the ability to interact with SNMP-based NMS systems. The LanProbe was more restricted in this regard, but its packet capture facility could allow administrators to debug many network problems remotely, a feature not addressed at all by the LANtern. Table 4 is a summary of how these products performed in our experiments.

8 Future Work

In the near future, we plan to evaluate a number of other NMS and PNM capabilities. A primary area of interest is the integration of NMSs and PNMs into a single integrated architecture

Capabilities	SunNet Manager 1.2	Spectrum 1.0.3	DECmcc 1.2	HP Node Manager 2.0
IP Topology/Node Discovery	FAIR	FAIR	GOOD	GOOD
DECNET Node Discovery	NONE	‡	GOOD	‡
Monitoring Links/Routers	GOOD	GOOD	VERY GOOD	GOOD
Alarms and Notifications	VERY GOOD	FAIR	EXCELLENT	GOOD
User Interface	VERY GOOD	GOOD	FAIR	EXCELLENT
Commercial MIB Support	FAIR	POOR	VERY GOOD	VERY GOOD
TCP/IP Node Reachability	EXCELLENT	EXCELLENT	EXCELLENT	VERY GOOD
DECNET Node Reachability	EXCELLENT	‡	EXCELLENT	‡
Dual Network Architecture	†	†	†	†
Workstation Health	FAIR	POOR	FAIR	FAIR
Multiple NMS Users	FAIR	VERY GOOD	VERY GOOD	GOOD

‡ DECNET support announced but not available at the time of this writing.
† Refer to Section 5.6.

Table 3: Matrixed Results of NMS Evaluation

Capabilities	LANtern	LanProbe
Ability to Integrate	FAIR	POOR
Ethernet Health and Utilization	GOOD	FAIR
Host Information and Statistics	GOOD	FAIR
Packet Filter and Capture	NONE	GOOD

Table 4: Matrixed Results of PNM Evaluation

[YEMI92]. This area is showing promise as products conforming to the Remote Monitoring Management Information Base (RMON MIB) have been introduced in late 1992. Such products have the flexibility and processing power to address not only generic segment health issues, but also may provide a way for administrators to monitor enterprise-specific applications.

We also intend to determine whether custom code can be easily integrated with commercial products. This would allow user-defined alerts to be sent to and displayed by an NMS. Likewise, future generations of PNMs may be able to be customized through modification of filters and alarm rules.

A number of NMS capabilities are tied to the information that is stored in a database, how it is stored, and how it can be accessed. These capabilities include trouble ticketing, configuration management, auditing, and report generation. We plan to assess whether different NMSs support each of these functions, and whether information created performing one of these activities can be easily accessed when necessary to perform a different activity.

Four other areas for future work include SNMP interoperability between different vendors, automatic response or rule-based systems, NMS performance, and SNMP II/SMP (Simple Management Protocol) capabilities. As a candidate for replacing SNMP, SMP will address security shortcomings and provide capabilities to manage at the application level.

9 Conclusion

Network management technologies and products are evolving rapidly. We evaluated six different network management products using a variety of both general and architecture-specific criteria. Several products have some very strong capabilities, but also significant weaknesses. Our laboratory is growing, and we expect to continue to attract state of the art technology to add to our installed base.

There are two key areas of future work that we anticipate. First, we will continue with further experimentation and plan to focus on the customization and integration of Network Management Stations and Passive Network Monitors. Second, we will look for products that implement new standards and utilize new technologies.

10 References

[CASE90] J. Case, M. Fedor, M. Schoffstall, J. Davin, "A Simple Network Management Protocol," Request for Comments 1157, SNMP Research, May 1990.

[MCCL90] K. McCloughrie and M. Rose, "Management Information Base for Network Management of TCP/IP-based internets," Request for Comments 1156, Hughes LAN Systems, May 1990.

[MEYE91] K. Meyer and D. Johnson, "Experience in Network Management: The Merit Network Operations Center," Proceedings of the Second International Symposium on Integrated Network Management, Washington, D.C., April 1991.

[YEMI92] Y. Yemini, "The Data Centered Approach for Network Management," Tutorial at the IEEE 1992 Network Operations and Management Symposium, Memphis, TN, April 1992.

11 Acknowledgments

We would like to express our sincere gratitude to the following vendors for their generous support of this project: Cabletron, Digital Equipment Corporation, Hewlett Packard, Novell, and Sun Microsystems. We would also like to thank Dixie Baker, Joseph Bannister, Phil Schmidt, and Craig Lee who assisted in reviewing this paper and provided helpful comments.

Integrated Network Management, III (C-12)
H.-G. Hegering and Y. Yemini (Editors)
Elsevier Science Publishers B.V. (North-Holland)
1993 IFIP.

Service Management of a Satellite Ground Station Network in Support of Spacecraft Operation

Klaus-Jürgen Schulz
European Space Agency (ESA), European Space Operations Center (ESOC)
Robert-Bosch-Str. 5, D-6140 Darmstadt, Germany
Tel.: +49-6151-90-2472, Fax.: +49-6151-90-3046
e-mail: kschulz@esoc.bitnet

Abstract

This paper describes the Service Management concept of the European Space Agency's (ESA) Ground (Tracking) Station Network ESTRACK that provides communication services in support of spacecraft operation. Starting from a discussion about characteristics of spacecraft operation support networks like ESTRACK, it focuses on the definition of four layers that actually provide the service management capabilities: 1. device management layer 2. subsystem management layer, 3. ground station management layer including the ground stations and the ground communication network, and 4. ESTRACK management layer. The major concepts are based on the application models of Manufacturing Message Specification (MMS) for the device and subsystem management layers and on Common Management Information Services (CMIS) for the other layers. Since the introduction of this complete service management scheme involves upgrades at each level of the management hierarchy the migration strategy is briefly outlined.

Key words: satellite ground station network, integrated network management, service management, MMS, CMIS

1. Introduction

The European Space Agency (ESA) operates a worldwide Ground (Tracking) Station Network called ESTRACK for the support of spacecraft operation. ESTRACK consists of Ground Stations (GS) which are interconnected via a Ground Communication Network (GCN), see figure 1. The Ground Communication Network is a star shaped private network based on leased lines and hybrid switching, i.e. packet switching based on X.25 and circuit switching for operational audio. The ground stations provide Telemetry (TM), Telecommand (TC) and Ranging (RG) services to the centrally located spacecraft control centers (S/C CC) at the European Space Operations Center (ESOC) in Darmstadt, Germany. The Telemetry, Telecommand and Ranging functions in the S/C CC are considered as communication service users from the perspective of ESTRACK, they are applications in the OSI sense. Spacecraft monitoring and control (S/C M&C) makes use of the Telemetry and Ranging information in order to control the spacecraft by sending Telecommands.

The overall Service Management of ESTRACK involves the Ground Stations and the Ground Communication Network. Presently the ground stations are managed by a Ground Station Management System (GS-MS) by operators from a working position either in the ground station or remote at ESOC. The ground communication network is controlled by the Ground Communication Network Management System (GCN-MS) centrally from ESOC.

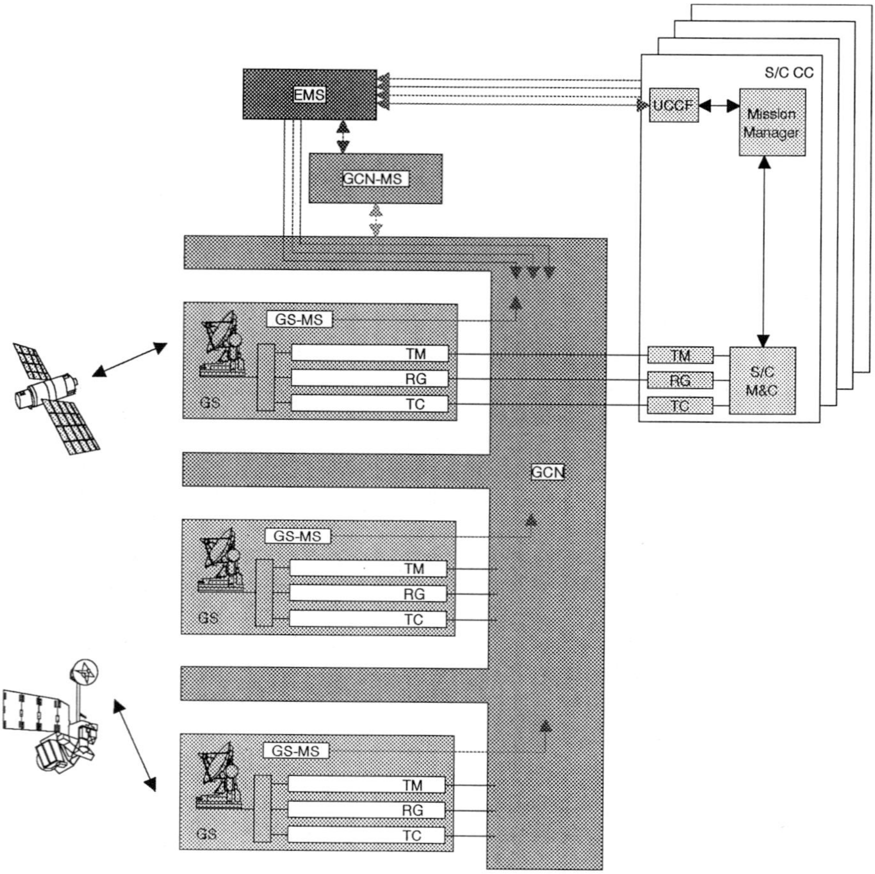

Figure 1: ESTRACK Overview

So far the management systems are not integrated and depend highly on manual intervention by operators. Therefore a service management system called ESTRACK Management System (EMS) is being developed that allows highly automated central management of ESTRACK from ESOC in close cooperation with the User Communication Coordination Function (UCCF) within spacecraft mission operation. This includes major upgrades in all management domains involved starting from elementary processing devices, Subsystem Management Systems (S/S-MS), ground station management systems, ground communication network management system and finally the ESTRACK management system itself, see figure 2.

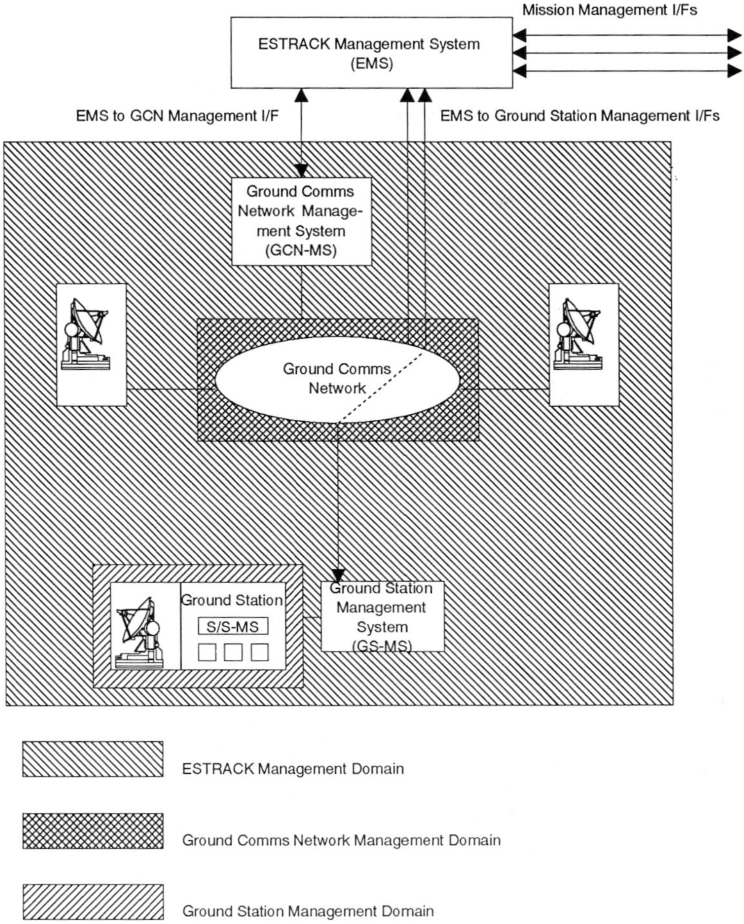

Figure 2: ESTRACK Management Domains

2. Characteristics of Spacecraft Support Networks

Service Management of a network for spacecraft support has to consider some specific characteristics which are not generally applicable to data communication networks. The operational requirements for a spacecraft mission support network can be summarized as follows:

1. It has to work under the constraints imposed by spacecraft mission operation, i.e.
   - physical connectivity given by the visibility of the spacecraft,
   - specific mission phases, e.g. launch, transfer orbit, payload operations requirements, on-board power constraints, etc.

   These constraints lead to the provision of communication services according to a pre-planned schedule according to the mission timeline.

2. Since the spacecraft depends on high reliability communication links, a philosophy of "try later" in case of network congestion is not practical.
   This implies that the allocation of communication resources is pre-planned and implemented before the actual execution of the service instance, e.g. a particular service instance is provided by a dedicated switched circuit that is configured beforehand.
3. As on-board resources for data storage are limited data will generally be transmitted only once, i.e. data not captured by the ground may be lost.
   This implies that a high reliability is required in the ground communication system and appropriate backup capabilities have to be foreseen. Additionally data capture facilities for space-ground data are needed in case of link outage in the terrestrial communication system.

2.1. Schedule Based Operation

The spacecraft orbit requires tracking of the spacecraft position and in all cases, except for geostationary satellites, limits the visibility of the spacecraft from one station. Fortunately the spacecraft orbit is predictable and the support by a specific station may be scheduled for generally well known periods of time.

A normal ground to ground network (e.g. PTT network) may offer all its users the complete set of transmission services it can support and may allocate the available resources by whatever policy best meets its needs, e.g. on statistical basis.

In a support network for spacecraft operation like ESTRACK this offer is constrained by the mission manager in time and in the variety and quality of the services. This fact is modeled by the concept of a "Service Envelope" that allows to describe the communication support to be provided by ESTRACK to a spacecraft mission during a certain visibility period. The Service Envelope has to be negotiated between the spacecraft mission and ESTRACK. This process is called Cooperative Service Management conducted between the ESTRACK Management System and the User Communication Coordination Functions within mission management.

The agreed Service Envelopes are implemented by ESTRACK one after the other. ESTRACK internally derives Resource Allocation Plans and finally configures all equipment such that it actually provides all the service instances as defined in a particular Service Envelope.

Also the operation of the spacecraft and hence its communication requirements in various mission phases as well as constraints on the communication are generally subject to planning and scheduling. The periods allocated to operation of specific payload are determined from the overall mission plan which is updated in "increments" during the mission life-time.

The communication services required by the payload can be described by defining end-to-end Service Instances within an end-to-end Space Data Network [KJS-UC91, KJS91a, KJS91b] which are subject to scheduling. The schedule of such service instances is not necessarily related to visibility considerations, especially when operation of the payload can be supported by on-board recording and later playback of data and by time conditioned on-board instructions.

2.2. Service Management Concept

Today the resources of ESTRACK are generally reserved for a specific mission by human oriented procedures a reasonably long time ahead of their actual use and are often placed under direct control of the spacecraft control center. This approach has some disadvantages :

- the possibility to use resources for multiple missions is limited due to rather long preparation times;

- management of the resources by the spacecraft control center makes automated support nearly impossible as the control center would have to know all the details and, indeed, the approaches seen today are nearly completely human oriented. (Computer support is available only where missions are supported by dedicated stations and by remoting management functions, e.g. from the ground station into the control center).

The future management strategy for ESTRACK assumes an approach whereby ESTRACK provides communication services to its users. The ESTRACK management system manages the resources of ESTRACK in an integrated manner not visible to the mission manager. It is the responsibility of ESTRACK Management System to make sure that the services are provided or to inform the mission manager if this is no longer possible.

3. ESTRACK Service Management

The ESTRACK Service Management concept relies on the definition of 5 management layers, see figure 3:

1. Device Management Layer that is concerned with device specific peculiarities, e.g. representation of monitor and control data in device specific formats.
2. Subsystem Management Layer that transforms the monitoring and control functionality of well identified subsystems in the signal processing chain into a uniform representation. These subsystems may be implemented differently, however the intention is to abstract the Monitoring and Control (M&C) functionality such that these differences disappear.
3. Ground Station Management Layer that transforms the M&C functionality provided by ground stations into a unique view despite the fact that ground stations are build differently and provide slightly different services. All ground station unique detail is hidden, only that M&C information necessary to support a specific spacecraft is visible to the next higher layer.
4. ESTRACK Management Layer combines all information from all ground stations plus the ground communication network into one complete view onto ESTRACK which is provided to mission management .
5. Mission Management Layer to which the ESTRACK services are provided. This management layer is not described further.

Conceptually the Ground Station Layer also covers the abstraction that is provided by the Ground Communication Network management system. Although the Ground Communication Network also consists of devices that are remotely controlled this is not further considered in the model since this functionality is covered by the GCN-MS that will only provide an interface to the ESTRACK Management Layer, but not to individual devices, e.g. modems, switching nodes, routers, etc.

The basic application model that is employed within ESTRACK is based on that of CMIS and CMIP [ISO-9595, ISO-9596], i.e. managed objects that are managed by management agents under the control of managers. However CMIS is not used in a pure Network Management sense, i.e. following the rules defined by the NM Forum and ISO for Structure of Management Information (SMI) [ISO-10165] and System Management Functions [ISO-101064], CMIS is rather used to access a privately defined Structure of Management Information [GS-M&C-92].

In particular the managed objects do not describe the devices or subsystems directly but rather aspects of the monitoring and control process itself. The interface definition makes use of the fact that all managed object instances are pre-defined by "tailoring" (software customization) and are therefore known to both sides.

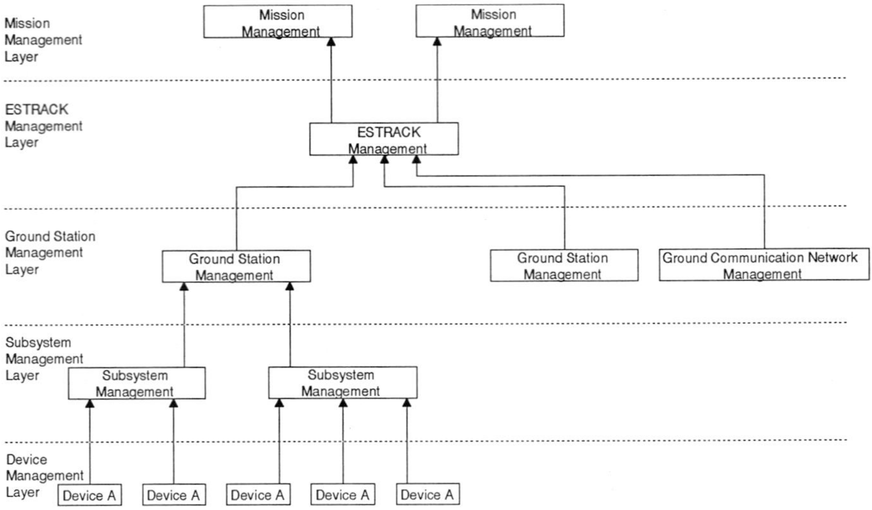

Figure 3: Layered Service Management Model

For the interface between the subsystem layer and the ground station management layer it was initially felt that a richer application model than that of CMIS, e.g. Manufacturing Message Specification (MMS) [ISO-9506-1, ISO-9506-2], would more naturally map to the monitoring and control requirements of complex devices. The main reasons for basing the initial analysis on MMS were the following:

- Context sensitive alarms can be easily based on dynamic event enrollments in MMS, i.e. by issuing an enrollment service primitive a specific event can be mapped to a specific action during run time of the system. This feature would allow the ground station management system, knowing the overall ground station status, to instruct the subsystems accordingly, thereby allowing a high degree of automation in the management system.
- Monitored parameters may be used not only in the M&C hierarchy but also by other subsystems, i.e. MMS allows a fully networked and not only strict hierarchical approach to monitoring and control.

Although the use of MMS for this interface was prototyped and finally found technically feasible, in the actual future implementation MMS will not be used due to limited product availability on VME bus and Motorola hardware platforms. Since all major digital signal processing devices used in the ground stations are based on this platform the limited MMS product availability was considered a major drawback. Instead the necessary functionality will be implemented based on CMIS.

The interface between the ground communication network management system (GCN-MS) and the ESTRACK management system (EMS) should also be based on CMIS. However the initial interface will be based on SNMP due to product availability.

3.1. Device Management Layer

The Device Management Layer consists of elementary processing devices that perform different functions in the signal processing chain, e.g. antenna control, low noise amplification, frequency conversion, modulation / demodulation, frame synchronization, forward error correction, protocol handling. The device layer here is introduced for the purpose of being able to include existing devices in the new service management structure.

New, to be developed, devices will be capable of providing a monitoring and control interface as defined for the subsystem management layer.

3.2. Subsystem Management Layer

The Subsystem Management Layer provides the first level of abstraction such that the identified subsystems provide the same M&C interface irrespective of the actual devices that are used internally.

The definition of the subsystem management layer interface was inspired by the MMS application model for "Virtual Manufacturing Devices". From the manifold of MMS M&C services some were selected which were found useful for ground station subsystem monitoring and control [GSNMF-92]. They were then abstracted such that they could be implemented also by other means, e.g. by using CMIS with a private Structure of Management Information.

The following 3 facilities are defined for a subsystem management system, see figure 4:

1. Elementary M&C Facility
2. Event Facility
3. Configuration Facility

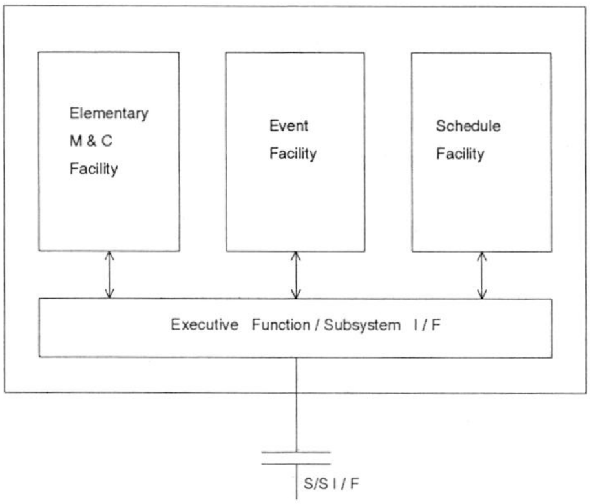

Figure 4: M&C Facilities

Figure 5 defines the service primitives for the 3 facilities of the subsystem management application protocol:

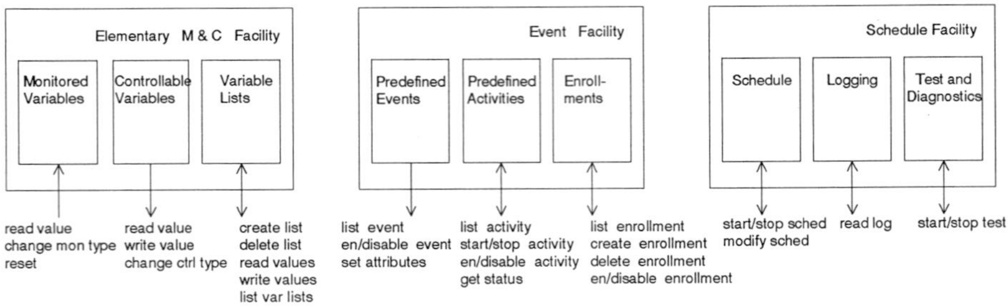

Figure 5: Service Primitives of the M&C Facilities

In the prototype implementation these service primitives of the subsystem management application protocol were mapped to MMS service primitives in a straight forward manner. The prototype subsystem management application acted as a MMS server. A MMS client resembled the ground station management system. Since MMS will not be used in the final implementation due to reasons given above the facilities are mapped to objects of the privately defined structure of management information that can be manipulated via CMIS service primitives.

3.3. Ground Station Management Layer

The Ground Station Layer consists of Ground Station Management and Ground Communication Network (GCN) Management, which are described below.

3.3.1. Ground Station Management

The Ground Station Management Layer within the ground station builds on top of the subsystem management layer. It implements a M&C client application that accesses the M&C server applications in the subsystem management system in order to derive a complete ground station view. This view is represented to the ground station operator in graphical form on a work station and towards the ESTRACK management system as managed objects accessible over CMIP.

To give a rough estimate about the number of parameters visible about a ground station at this interface, this is about 100, whereas the number of parameters from all the devices in the station can be up to 4000. This roughly indicates what the abstraction does with respect to shrinking the number of visible parameters.

The structure of the ground station management to ESTRACK management interface is not purely based on a managed object tree that is accessed. The interface is further structured into the following elements:

- Service Requests to ask the ground station to configure for the support of a particular spacecraft
- Directives to instruct the ground station to take specific actions, e.g. to over-ride a pre-defined schedule or to change a configuration or to simply ask for specific information
- Station Scheduling to transmit ground station schedule information to the ground station or for the ground station to announce down-times
- Information Reports, consisting of the following types of reports:
    - Capability Status Reports periodically list the current capabilities of a station for operational use and thus the availability of the station for service
    - Operational Status Reports periodically provide an overall picture of the current state of the ground station and the services provided

- Event and Alarm Reports are spontaneous reports on pre-defined nominal or non-nominal (alarms) events
- Accounting and Performance Reports are issued after the service envelope has been finished
- Information Reports provide information in the execution phase on demand
- Journal Service and Service Reports provide a list of all events and directives given during the phase of service provisioning for assessment purposes after the service envelope has been finished

Figure 6 shows potential managed objects visible at the ground station management system to ESTRACK management system interface which implement the management capabilities.

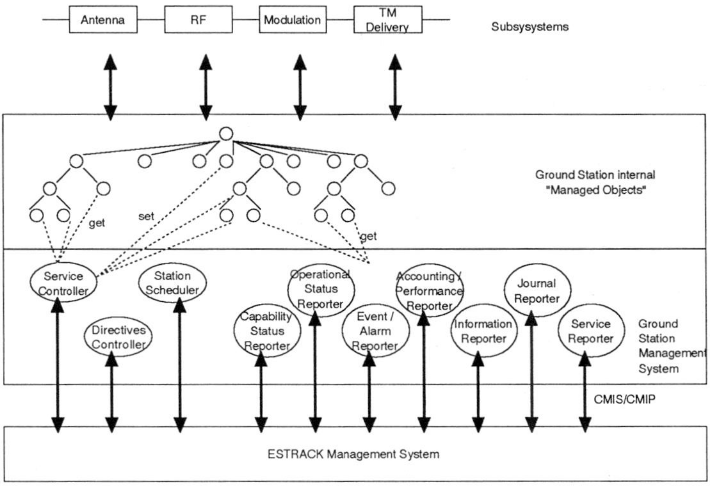

Figure 6: Potential GS-MS Managed Objects

3.3.2. Ground Communication Network Management

The Ground Communication Network (GCN) Management is concerned with the provisioning of network layer connectivity between end systems located in ground stations and in control centers.

The structure of the GCN management system to ESTRACK management system interface is based on the notion of network layer connectivity, i.e. all information about all current connectivities must be accessible such that a connectivity map of all established connections can be drawn thus allowing active access control on network connectivity provided.

The interface basically follows the same structure as that for ground station management defined above. The interface is again based on CMIS with the use of SNMP in a migration period.

3.4. ESTRACK Management Layer

The ESTRACK Management Layer assembles all the information available from all ground stations and the ground communication network to provide a full and complete overview about the operational status of ESTRACK. From this for the various mission managers (Note that multiple mission can be supported simultaneously) all information is provided concerning the provisioning of services to that mission.

The structure of the ESTRACK management system to User Communication Coordination Function interface follows the notion of service provision in support of spacecraft operation. It provides similar management capabilities as the ground station management system, however now gathered from all ground stations that support the particular mission and additionally the ground communication network setup for the particular mission.

The interface is again based on CMIS.

4. Migration Strategy

As already indicated the present operational system is based on human oriented procedures. Monitoring and Control of the ground station is provided directly based on the information provided by devices and therefore rather specific on the individual ground station. The information is provided on a workstation to the ground station operator. In some cases this workstation is remoted and ground station operations are conducted directly from the control center.

By introducing the ESTRACK management structure as outlined above it is expected to highly automate the spacecraft operation support services provided and by the same token make it more robust and reliable.

In the following chapters the migration strategy in the various management layers will be briefly outlined.

4.1. Device Layer Upgrade

There will be no upgrade in the Device Layer for existing devices. It is anticipated that devices to be implemented in the future will directly implement the subsystem management layer interface.

4.2. Subsystem Management Layer Upgrade 1

The present implementation of the Subsystem Management Layer only consists of data gathering devices called Monitoring and Control Modules (MCM), which take the M&C information from the devices attached to a IEEE 488 instrumentation bus and convert it into messages that are sent cyclically every 2 seconds to the ground station management system realized in the present Station Computer (STC-I) via a X.25 direct connection.

Since the hardware and software platform of the MCM is already obsolescent there will be an in-situ replacement on modern hardware and software of the same functionality, which will allow to replace existing MCMs. The same hardware and software platform will then be used when migrating from the current MCM concept to the subsystem management concept realized by use of CMIS over Ethernet.

4.3. Ground Station Management Layer Upgrade 1

The M&C system realized in the present Station Computer (STC-I) will be replaced with a M&C system as specified above on a new hardware and software platform called STC-II. The software architecture of the system is such that it allows a smooth migration from the old interfaces to the new CMIS based interfaces by introduction of a “Mapping Layer”. This is achieved by defining a mapping layer interface which resembles already the future subsystem management like interface which was originally based on MMS services. By this it is possible to build in a first step the in-situ replacement of the present system by keeping the old interfaces to MCMs and to add in a second step the new interface for new subsystem management systems.

The abstraction provided by the ground station M&C system is made visible to the operator and to the ESTRACK Management System. The latter one being completely new and therefore introduced in the second upgrade step.

4.4. Ground Communication Network Management System

The Ground Communication Network Management System is presently a stand-alone closed system and not integrated. It provides however full remote configuration capability of the entire network from a central operator position.

It is anticipated that a high level view of the complete network that is managed from this central work position will be provided. The interface however will be based on SNMP rather than on CMIP. This leads to the requirement that SNMP in addition to CMIP needs to be supported by the ESTRACK Management System.

4.5. ESTRACK Management System

The ESTRACK Management System is a completely new development and is phased with the STC-II development. It will be based on SNMP and CMIP and the integrated management model developed above. It is anticipated that the same hardware and software platform as for STC-II will used, thereby allowing re-use of common protocol stack and man machine interface components.

4.6. Subsystem Management Layer Upgrade 2

The second Subsystem Management Layer Upgrade consists of implementing new M&C servers for the subsystems based on the same hardware as used for the first upgrade.

4.7. Ground Station Management Layer Upgrade 2

The second Ground Station Management Layer Upgrade consists of the use of the new subsystem management interface and replacement of the mapping layer by a real supporting protocol stack based on CMIP in order to support new subsystem management systems and new devices, which will provide this interface directly.

4.8. New Device Layer

New devices M&C interface will be based on the same interface as the subsystem management systems thereby allowing to interface directly to the new STC-II.

5. Conclusion

Starting from special requirements on ground station networks in support of spacecraft operation it was shown how a comprehensive service management hierarchy can be defined by applying appropriate abstractions.

For reasons of commonalty and product availability the management interfaces between the various management layers will be based on CMIS/CMIP. However CMIS is used as a supporting communication service to implement the management abstractions needed. For monitoring and control purposes within ground stations MMS was found rather suitable, major management abstractions were taken from there. However due to limited product availability MMS will not be used, instead the needed management abstractions will be implemented using CMIS based on a privately defined Structure of Management Information.

It is expected that the implementation of the ESTRACK management structure will lead to highly automated procedures in order to conduct service management in support of spacecraft missions more efficiently than in the past.

The migration strategy outlined provides for a smooth migration from the existing system to the target management architecture. This migration will take place in the next 3 to 4 years.

6. Acknowledgment

In particular my thanks go to Dr. H. Uhrig who contributed primarily to a study conducted by Softlab and CRISA, who actually worked out the details of the management structure. Within Softlab I would like to thank in particular Mr. M. Götzelmann and Dr. M. Schedl and within CRISA Mr. A. Laseca and Mr. S. Heredia. For internal discussions I would like to thank the members of the Monitoring and Control Working Group: Dr. G. W. J. Drewes, Dr. E. E. Lijphart, Dr. W. Hell, Mr. Y. Doat and Mr. T. Beck.

7. References

[GS-M&C-92] Ground Station Monitoring and Control - Architectural Baseline, Issue 1, OPS/SCED, September 1992

[GSNMF-92] Study on Ground Station Network Management Facility, Final Study Report, ESOC Contract No 8630/89/D/DK(SC), Softlab, March 1992

[ISO-9506-1] Manufacturing Message Specification (MMS), Part 1: Service Definition, ISO-DIS 9506-1, 1988

[ISO-9506-2] Manufacturing Message Specification (MMS), Part 1: Protocol Specification, ISO-DIS 9506-2, 1988

[ISO-9595] Common Management Information Service (CMIS) Definition, ISO-9595, January 7, 1991

[ISO-9596] Common Management Information Protocol (CMIP) Specification, ISO-9596, January 7, 1991

[ISO-10164] Systems Management, Parts 1 to 15, DIS 1990

[ISO-10165] Management Information Services - Structure of Management Information, Parts 1 to 4, DIS 1990

[KJS-UC91] Klaus-Jürgen Schulz and Uwe Christ: Braodband Networking Requirements for Exploitation of Columbus Space Laboratory Data, Computer Networks and ISDN Systems, Vol 21 (1991), 275-283, North Holland

[KJS91a] Klaus-Jürgen Schulz: Space Data Network — A Challange in Space Communications, Informatik Fachberichte 267, Kommunikation in verteilten Systemen, Grundlagen, Anwendungen, Betrieb, GI/ITG Fachtgung, Mannheim, Februar 1991, Proceedings, 142-156, Springer-Verlag Berlin Heidelberg 1991,

[KJS91b] Klaus-Jürgen Schulz: Space Data Network — Concept and Rationale, Proceedings of the Second European Conference on Satellite Communications (ECSC-2), Liege (Belgium), 22-24 October 1991

Integrated Network Management, III (C-12)
H.-G. Hegering and Y. Yemini (Editors)
Elsevier Science Publishers B.V. (North-Holland)

NetView in a Heterogenous Environment: Implementation of an Experimental Agent Structure

Werner Filip Gautam Kar

IBM European Networking Center, Tiergartenstraße 8, 6900 Heidelberg, Germany

Abstract

Based on the ISO/IEC standards for Open Systems Management it's possible to integrate various types of networks in an OSI management environment and manage it by an OSI systems management application process (SMAE). In this paper the authors describe their experiences with the implementation of a NetView-OSI-Agent prototype to integrate parts of an existing SNA network management system in the OSI management environment. The architecture and implementation described in this paper is of an experimental nature and should not be used to make inferences about IBM's product direction.

Keyword Codes: K.6.4

Keywords: System Management

1. Introduction

A network management project conducted at IBM's European Networking Center (ENC) in Heidelberg, Germany, deals with questions concerning management of heterogeneous networks. The project is based on the standards for Open Systems Management by ISO/ICE [13], [14]. Figure 1 shows one way to depict the management infrastructure that may be used to manage a typical heterogeneous network. This internet consists of three networks, each supporting different protocols (SNA, IDNX, and TCP/IP) as shown. A starting point of our department's activities in network management [17] was to develop the concept of, and infrastructure for, a management system that can interact with each of the three networks in a way that makes it possible to do end to end fault management. In OSI terminology, we defined a Managing Process that interacts with three Agent Processes, shares data about management entities, react to commands, etc. Our goal was to define generic (and specific) structures for the manager and agent processes, protocols needed for the data interchange between the two, and models that enable management entities to be stored in a distributed management information base (MIB). The following sections describe the results of the effort we have undertaken to integrate an SNA network in this heterogeneous environment. Figure 1 also shows that an SNA network is part of a heterogeneous network scenario and NetView acts as an OSI agent. This capability isn't available with the current NetView releases and its complementary set of products. Therefore, for situations where management information has to be forwarded from an SNA network to an OSI management application, NetView with extended functionality could perform this task. Creating a NetView-OSI-Agent can be done using NetView's programming services and NetView's automation facilities available with the current NetView releases. SNA resources can only be managed by a SMAE if managed objects of these resources are modeled and stored in a management information base to which the agent application has access. Part "3. Experimental SNA Managed Objects" describes how this is solved within the NetView-OSI-Agent prototype for a limited number of SNA resources. Details about the agent prototype and how it is integrated in the OSI management scenario follow in section "4. NetView-OSI-Agent". Finally, chapter "5. Concluding Remarks" contains some additional comments about the experiences of implementing the NetView-OSI-Agent prototype. In the following section "2. NetView Facilities" a brief description of the necessary NetView facilities required to build an OSI agent is presented.

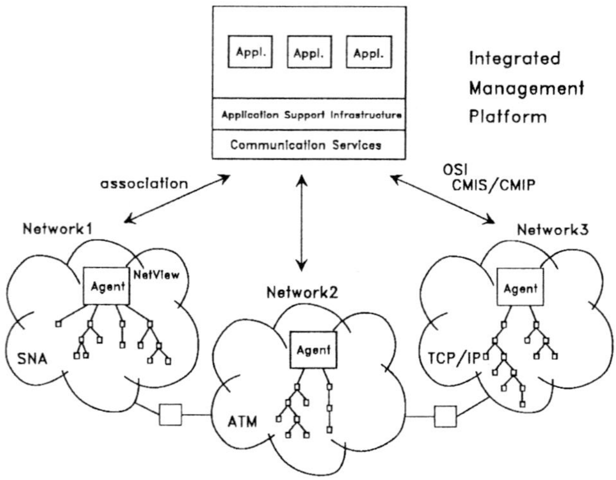

Figure 1. Heterogeneous Network Scenario.

2. NetView Facilities

NetView is IBM's SNA network management system and is used as a central control system. This functionality is known as focal point with it's subordinates, entry point and service point in the SNA network management architecture. An entry point manages a subtree of an SNA network while a service point offers the management gateway capability for controlling Non-SNA networks. Therefore, the service point functionality makes it possible for NetView to manage a heterogeneous network as a focalpoint. To understand how NetView can act like an OSI agent (based on the ISO standards) it is useful to describe some NetView automation and programming facilities first. The following NetView facilities are used to set up the NetView-OSI-Agent concept [1], [2], [3], [4], [5]:

- Automation Tasks (autotasks)
- Message Automation Table (automation table)
- Generic Alerts
- Command Lists (CLISTs)

Automation tasks can be set up to handle messages using the automation table, and they will process messages and commands from the message and command queue. In this way, different automation tasks can run concurrently to control various components of the system (such as the network, applications or the operating system). These tasks will act on messages that concern them immediately, without having to wait for each message in the message queue to be processed sequentially. The *NetView automation table* contains IF-THEN statements that define actions to be taken on the receipt of a message. These IF-THEN statements give the user a control over message processing decisions such as:

- To which operator(s) should the message be displayed?
- Should it be highlighted and held on the NetView screen?
- Should an audible alarm be sounded when the message is displayed?
- What programmed action should be taken, and by which task (including automation tasks)?

The actions that can be specified in the THEN statement include:

- Command or command list execution
- Routing (for either display or command execution)
- Display and logging.

Alerts can be generated from the automation table or a CLIST using the *GENALERT* command. The generated alerts have a generic or non-generic NMVT (Network Management Vector Transport) alert format. A *command list (CLIST)* is a set of commands and special instructions that are grouped together under one name like a computer program. For NetView, a CLIST can be written in either REXX (Restructured Extended Executor) language or the NetView command list language (NCL). There are several ways to run a command list under the control of an operator or automation task. The CLIST name can be entered at a terminal, or a CLIST can be executed at a specified time or time intervals via a timer command. More than one command list can run at the same time under different tasks.

3. Experimental SNA Managed Objects

The NetView-OSI-Agent makes it possible to integrate parts of SNA networks in a heterogeneous management model, as shown by this environment. To give NetView the ability to act like an OSI agent a transformation of the exchanged management information must be made. That means, an OSI-SNA network management gateway has to be defined. This function will be the interface between the real SNA network resources and the SNA managed objects (SNA MOs).

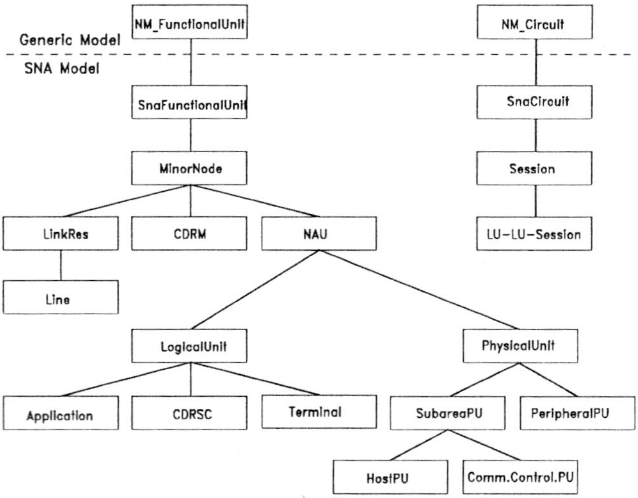

Figure 2. Managed Objects of the Experimental SNA Environment.

For the integration of parts of an existing SNA network it is necessary to create SNA managed object classes [18]. Figure 2 and Figure 3 show a part of the implemented inheritance tree with the related SNA attribute classes. These classes[1] were integrated into an existing generic model, that is the base model of our heterogeneous network management environment[2] [7], [17]. VTAM and NetView internal control block information of SNA resources can only be obtained in a limited way for user-written code. Therefore, for the selection of SNA resources which were modelled as managed objects it was necessary to control these resources through the following VTAM operator commands:

- VARY ACT,ID=*resource name*

---

[1] The managed object tree presented here has not been derived from the structure that will be used in future SNA products. It is an experimental exercise undertaken to solve our specific problem.

[2] Does not correspond to the product-level SNA model.

- VARY INACT,ID=*resource name*
- VARY DISPLAY,ID=*resoure name*

For example, a managing application can send an activate action for a PhysicalUnit MO to the OSI-NetView-Agent. The related MO then causes NetView to send a "VARY ACT,ID=*puname*" to the appropriate node. All necessary information for building this VTAM operator command are transmitted to NetView through the predefined parameter string, "ACTPU*puname****". In NetView a CLIST analyzes the parameter string, builds a VTAM "VARY ACT,ID=*puname*" command and sends it to VTAM. The same technique is used to inactivate SNA resources or display it's current status information. NetView responses and SNA alert information are also carried in parameter strings with predefined structures.

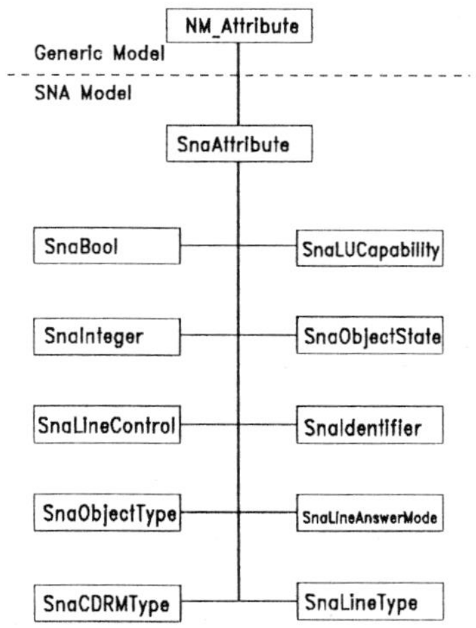

Figure 3. SNA Attribute Classes.

Because of this communication technique, the SnaAttribute class has been created to reflect the specific needs of the SNA MOs (information exchange with the real resource). The attribute classes derived from SnaAttribute of the SNA MOs were introduced to cover two characteristics:

- The SNA attributes encapsulate the generated ASN.1 types and encode/decode functions according to object oriented principles.
- The instances of the SNA MO attribute classes store the information, needed by specific NetView-OSI-Agent functions (see "4. NetView-OSI-Agent") that are responsible for the communication with NetView.

These communication functions ask NetView for attribute values related to SNA resources. Because NetView returns each attribute value in string format, the subclasses of SnaAttribute include a method (SnaStringToOsiAttr), that reads the string generated by NetView and sets the appropriate attribute value of the attribute instance. The mapping of SNA resource states to SNA MO states isn't a trivial task. When, for example, a VTAM "VARY DISPLAY,ID=*resource name*" command is executed in the NetView environment, one of more than hundred different states can be received [6]. The received state information has to be mapped on one or more attribute value of the three possible management states of managed objects. These are [14]:

- operational state
- usage state
- administrative state.

Dependent on the various SNA resource types, a classification of states into major categories aids the mapping functions.

4. NetView-OSI-Agent

NetView's agent support, realized by the NetView-OSI-Agent concept, is provided on VM or MVS systems (see Figure 4). The NetView-OSI-Agent consists of several components. Those incorporated inside NetView are a NetView optional subtask (OPT) and various CLISTs running under the control of automation tasks.

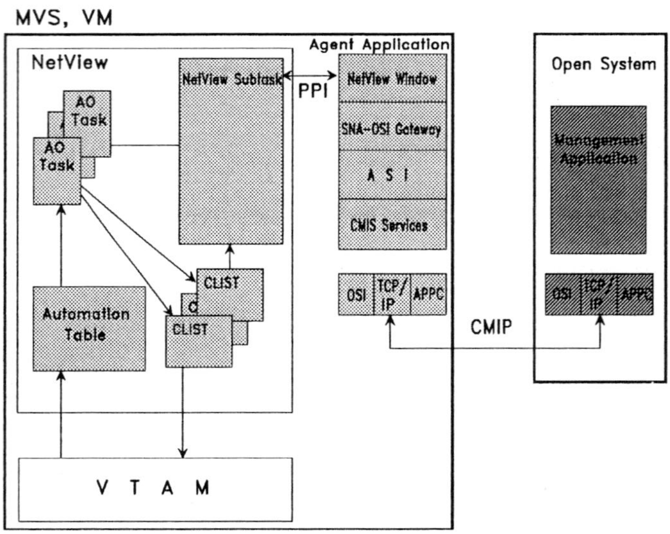

Figure 4. NetNiew-OSI-Agent Structure.

The part of the NetView-OSI-Agent which is outside NetView is an application that communicates with the optional subtask via NetView's program-to-program interface

(PPI). This application is called the "AGENT" program. Data and transfer services allow the AGENT to exchange OSI management information with OSI management applications via different protocols. The implemented protocols used by the AGENT are OSI, TCP/IP, and LU 6.2. Upon receiving requests from a manager with which the AGENT has a management association, the NetView-OSI-Agent can return information about its SNA environment. This includes, for example, status of SNA resources like logical units (LUs), and physical units (PUs). Event reports (SNA alerts) can be sent asynchronously; filtering of alerts is also possible. The management functions that have been integrated into the NetView-OSI-Agent so far are:

- Configuration management functions (parts of object management and state management).
- Fault management functions (alarm reporting).

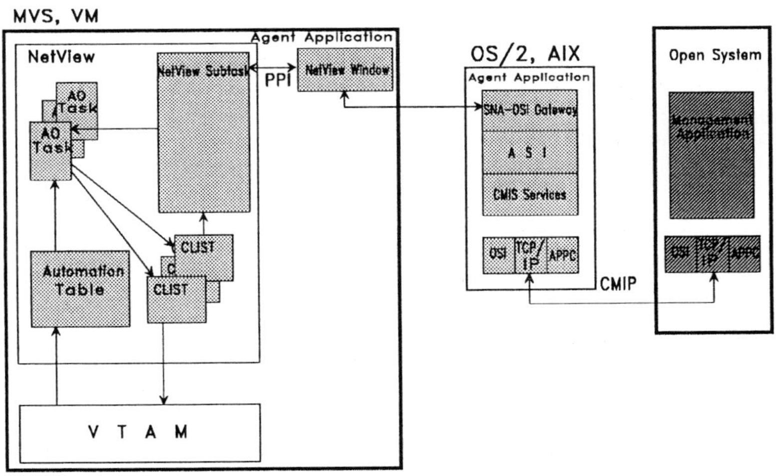

Figure 5. Distributed Agent Application.

4.1. AGENT Application

The AGENT application, the main part of the NetView-OSI-Agent, can run as a single application on a /370 host with a MVS or VM operating system or as a distributed application where parts of it remain on a /370 system while the rest is implemented on an OS/2 or AIX workstation (see Figure 4 and Figure 5). In the case when the AGENT program runs on a workstation (see Figure 5) the AGENT has to be divided into two parts. The first one consists of the NetView Window and remains on the MVS/VM host, the second one includes the rest of the AGENT application. A communication path between the /370 system and the workstation where the two application blocks are running has to be established. TCP/IP or LU 6.2 can be used. Figure 6 shows how the AGENT application is organized. It is divided into the following functions:

- NetView Window
- SNA-OSI Gateway
- Application Support Infrastructure (ASI)
- CMIS Services.

The *NetView Window* is responsible for the communication between the NetView optional subtask and the AGENT application. This is done via NetView's Program-To-Program Interface (PPI). The communication process between the NetView window and the NetView subtask includes, for example, forwarding management requests to an automation task for further processing, and returning the autotask's response to the SNA-OSI Gateway. NetView alerts which are asynchronously received are forwarded to the SNA-OSI gateway for further processing. *SNA-OSI Gateway* functions convert the incoming CMIS requests from an associated management application into VTAM/SNA commands handled by NetView. On the other hand, NetView responses are converted into CMIS data format. This task is supported by the application support infrastructure. The SNA-OSI gateway is also responsible for converting incoming NetView alerts into OSI event reports. The related mapping process is described later in more detail. All services provided by *Application Support Infrastructure (ASI)* and CMIS Services which are implemented in the AGENT program are described in great detail in [8], [9], [15]. We will mention only a few points. *ASI* services currently used in the AGENT offers a portable management communication support that can:

- run in several operating system environments
- communicate over multiple data transport mechanisms among heterogeneous equipment
- provide a basis for applications such that they could run unchanged over both OSI and non-OSI network environments.

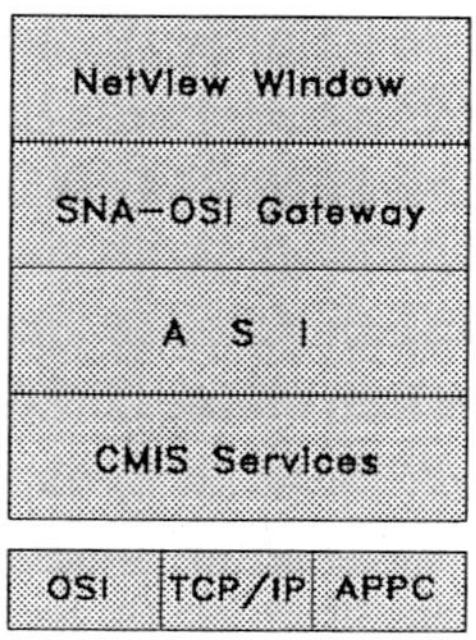

Figure 6. Agent Application.

Services for the application support include, for example, address and name resolution services, data management services, and system tasking and interface models. Furthermore, ASI includes components, such as Association Control Function, a Naming Repository which supports OSI object naming policies. Since the *OSI CMIS Service* has no dependency on complex services from the protocol layers below the OSI presentation layer, CMIS can be provided above any protocol implementation which provides connection and data transfer services. CMIS requires only connection and information transfer services of the presentation layer, and kernel services of the session layer.

4.2. Exchange of OSI Management Information

The implementations used by ASI are:

- OSI
- TCP/IP
- LU 6.2.

The exchange of OSI management information over an OSI stack between the NetView-OSI-Agent and an associated OSI manager can be provided through OSI Communications Subsystem (OSI/CS). Making use of the OSI/CS Application Programming Interface (API) allows the AGENT program to run as an OSI/CS Application Entity (OSI/CS AE) on top of the session layer. The AGENT communicates with OSI/CS over APPC. Either one of the following APPC implementations can be used:

- MVS/VM APPC Bridge
- VTAM APPC
- VTAM's Remote Programming Interface (RPI).

The VTAM OSI Remote Programming Interface allows MVS and VM systems to participate in OSI networks without requiring the installation and support of OSI/Communications Subsystem on every system. The VTAM RPI uses the LU 6.2 protocol to communicate OSI requests between applications (OSI/CS AEs) on an SNA system and OSI/CS on another system.

4.3. Mapping of Alerts to OSI Event Reports

With SNA's Generic Alert architecture it's possible to forward information about an important management event to a central NetView focal point. This can be done from a SNA or non-SNA resource. In the case of non-SNA resources a service point has to convert the event information into the generic alert structure. Also applications, running on MVS/VM hosts, can create alerts and forward them via the PPI interface to NetView where they are stored in a hardware monitor (NPDA) VSAM dataset. It's even possible to create alerts inside NetView using NetView's GENALERT command. An alert provides the following information [20]:

- identity of the alert sender
- identity of the related resource
- indication of the problem's severity
- description of the problem
- list of the probable causes
- list of the recommended actions
- protocol unique error data
- product unique error data
- time at which the problem was detected.

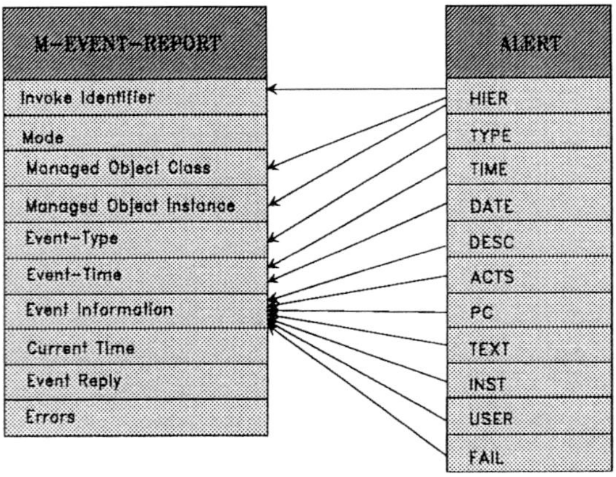

Figure 7. Alert-to-Event Mapping.

The following GENALERT command parameters can be specified using this command to create user specific alerts inside NetView:

- G - Format, generic
- TYPE - Alert type
- PSID - Product set ID
- ALID - Alert ID
- DESC - Alert description
- PC - Probable cause

- ACTS - Actions
- USER - User cause; action
- INST - Installation cause; action
- FAIL - Fail cause; action
- TEXT - Text
- HIER - Hierarchy of names and types.

An agent process of an OSI network forwards similar event information to an associated management process using the M-EVENT-REPORT CMIS service [14], [22]. To transfer alert data from an SNA network to an OSI manager it is necessary to map the Generic Alert information to the structure of an M-EVENT-REPORT. For this to happen the SNA alert has to be considered as a notification of an SNA managed object. The following are necessary to identify an alert as a notification:

- The source of the SNA alert is a SNA MO and included inside the MIB.
- An agreement upon forwarding M-EVENT-REPORTs by managed objects exists between the associated manager and agent process.

Using this information it's possible to identify the SNA MO and the related SNA name of the source that sent the alert. One possible way of mapping alert information to an M-EVENT-REPORT is described in Figure 7. These different event structures rely on the SNA and OSI architecture [14], [20], [21]. Therefore, it's obvious that event information is lost, or has to be described in a different way, during the mapping process. Those parts of the M-EVENT-REPORT without a relationship to parameters of the Generic Alert have to be completed by SNA-OSI Gateway functions. The created event information is not immediately forwarded to the associated management application. A filtering mechanism checks it first. This is done by Event Forwarding Discriminators, separate MO classes [14]. It's also possible to filter event information inside NetView. That allows NetView to forward only specific alerts to the AGENT. In the case, that many alerts of the same type are created in short time intervals, only one alert of that type is forwarded. This prevents NetView from overloading the communication path to the AGENT.

5. Concluding Remarks

Open network management based on ISO standards has become an important aspect for telecommunication strategy worldwide. To get early experience in managing heterogeneous networks and integrating SNA networks in it, we have designed and implemented an experimental NetView-OSI-Agent prototype. The concept was based on OSI management architectures, NetView programming services, and NetView automation facilities. The implementation was done using only those programming interfaces, that are available as part of the standard products, NetView and OSI/CS. As a parallel activity a limited set of SNA managed objects and related attribute classes were developed. This allowed the integration of selected parts of SNA networks. A considerable extension and generalization effort of SNA MOs has to be undertaken if one wants to integrate SNA networks with complicated structures.

Also, an effective object oriented MIB should be part of the agent to satisfy the requirements for managing SNA networks with large numbers of managed resources. In the experimental prototype implementation we used a repository in the agent's storage; that has allowed the handling of only a limited number of MOs. If the communication traffic between manager and agent increases and performance problems appear an extended communication concept can be used to solve the situation. NetView Window (see Figure 6) that is responsible for the communication between the AGENT program and a related NetView subtask can be changed (implementing a multitasking concept) to support multiple communication paths. This would make it possible to exchange management information between several NetView subtasks and related tasks in the NetView Window simultaneously.

6. References

[1] **IBM Corporation** *NetView, Customization Guide Version 2 Release 2* SC31-6048-1, 1991

[2] **IBM Corporation** *NetView, Customization: Writing Command Lists, Version 2 Release 2* SC31-6050-1, 1991

[3] **IBM Corporation** *NetView, Installation and Administration Guide Version 2 Release 2* SC31-6051-1, 1991

[4] **IBM Corporation** *NetView, Customization: Using Assembler Version 2 Release 2* SC31-6090-1, 1991

[5] **IBM Corporation** *NetView, Automation Implementation Version 2 Release 2* LY43-0008-0, 1991

[6] **IBM Corporation** *VTAM, Messages and Codes Version 3 Release 3 for MVA and VM* SC31-6405-0, 1990

[7] **B.Schott et al:** *An ISO/OSI based Approach for Modelling Heterogeneous Networks* Proceedings of the Fourth International Conference on Information Networks and Data Communication, INDC-92, Espoo, Finland, March 16-19, 1992

[8] **J.Dinger, D.Jordaan:** *Experiences with the use of OSI Management Infrastructures* To appear in: Proceedings of IEEE Globecom, Orlando, Florida, U.S.A., Dec. 1992

[9] **D.Jordaan, J.Dinger:** *Requirements for an Infrastructure supporting Management Applications* IFIP/IEEE International Workshop on Distributed Systems: Operations & Management Munich, October 12-13, 1992

[10] **International Standards Organization:** *Information Technology - Open Systems Interconnection - Application Layer Structure.* International Standard ISO/IEC 9545, First Edition 1989-12-15

[11] **International Standards Organization:** *Information Technology - Open Systems Interconnection - Common management information service specification.* International Standard ISO/IEC 9595, First Edition 1990-05-15

[12] **International Standards Organization:** *Information Technology - Open Systems Interconnection - Common management information protocol specification.* International Standard ISO/IEC 9596, First Edition 1990-05-15

[13] **International Standards Organization:** *Information Technology - Open Systems Interconnection - Systems Management Overview.* Draft/International Standards ISO/IEC 10040

[14] **International Standards Organization:** *Information Technology - Open Systems Interconnection - Systems Management Functions.* Draft/International Standards ISO/IEC 10164-1 through 11

[15] **K. Keilmann:** *Eine Infrastruktur zur Unterstuetzung von OSI Netzwerk Management Anwendungen* Universität Mannheim, January 1992

[16] **G. Knight, G. Pavlou and S.Walton:** *Experience of Implementing OSI Management Facilities.* Integrated Network Management, II, Proceedings of the IFIP TC6/WG6.6 Second International Symposium on Integrated Network Management, Washington D.C., U.S.A., 1-5 April 1991, p. 259-270.

[17] **Systems and Network Management Department:** *Heterogeneous Network Management: Initial Project Description:* IBM European Networking Center, Heidelberg, August 1991

[18] **T. Stadtmüller:** *Eine OSI-konforme Management-Informations-Basis fuer SNA* Technische Hochschule Darmstadt, September 1991

[19] **L. Svobodova** *Implementing OSI Systems* IEEE Journal on Selected Areas in Communication, Vol. 7, No. 7, September 1989, p.1115-1130.

[20] **R.E. Moore** *Utilyzing the SNA Alert in the Management of Multivendor Networks* IBM Systems Journal, Vol. 27, No. 1, 1988, p. 15-31.

[21] **K. Garbe** *Management von Rechnernetzen* B.G. Teubner Stuttgart, 1991.

[22] **L. LaBarre** *Management by Exception: OSI Event Generation, Reporting, and Logging* Integrated Network Management, II, Proceedings of the IFIP TC6/WG6.6 Second International Symposium on Integrated Network Management, Washington D.C., U.S.A., 1-5 April 1991, p. 227-242.

V.

TELECOMMUNICATIONS MANAGEMENT

A.

MIGRATION STRATEGIES

Integrated Network Management, III (C-12)
H.-G. Hegering and Y. Yemini (Editors)
Elsevier Science Publishers B.V. (North-Holland)

Uniform Modelling of Management and Telecommunication Services in Future Telecommunication Environments based on the ROSA Approach

Jane Hall, Thomas Magedanz*

GMD-FOKUS / *Technical University Berlin, Department for Open Communication Systems
Hardenbergplatz 2, 1000 Berlin 12, FR Germany
Phone: +49-30-25499200, Fax: +49-30-25499202

Abstract

The development of future integrated multi-service networks, driven by the ongoing deregulation decisions and increasing customer demands, is reflected in a number of different concepts which have to be encompassed for a target telecommunication environment. In this environment a broad range of telecommunication and management services will be offered to the customers by a growing number of competitive public and private service providers based on a common telecommunication infrastructure, which requires considerable interworking between network and service management. The most important concepts in the context of an open telecommunication environment are the Intelligent Network (IN) architecture and CCITT's Telecommunication Management Network (TMN). Focusing on the uniform provision of future telecommunication and management services, this paper discusses the application of the object-oriented paradigm developed in the RACE ROSA project to these concepts for achieving an integrated telecommunication platform.

1. Introduction

In a target Integrated Broadband Communications (IBC) environment, which is best characterized by the major trends in telecommunication service and network evolution, such as broadband capabilities based on ATM, value-added service facilities, multimedia options, user mobility and personal communication services tailored to individual needs, management of both network and services becomes a crucial task, especially when envisaging a deregulated pan-European telecommunication environment. The two most important concepts for the provision and the management of telecommunication services and networks are the *Intelligent Network (IN)* Architecture and *Telecommunications Management Networks (TMNs).*

INs represent an architectural concept for the efficient and uniform creation and provision of new telecommunication services like Freephone, Virtual Private Networks (VPNs) or Universal Personal Telecommunications (UPT). Currently INs are the subject of world wide awareness in the international standardisation bodies, since this network architecture allows a variety of different services to be provided to the customers independent of the underlying network technologies by defining a functional service-oriented network architecture on the one hand and by providing a set of generic service independent building blocks (SIBs) on the other.

In a telecommunications environment, management is realized by means of TMN, which provide a minimum but necessary platform for providing interoperability between different management systems and network components in the short term, and a basis for an integrated management system for all kinds of telecommunication networks and services in the long term.

Additionally, *Open Network Provision (ONP)* sets the scene for future deregulated telecommunication environments driven mainly by IN and TMN concepts. Within these environments users are free to buy services they need from different private and public service providers. The number of services is expected to increase dramatically and there will be a greater variety of more sophisticated services. An important question is whether current concepts are sufficient to handle such an expansion. The introduction of new services, including those previously unforeseen and so not catered for in existing environments, is a very complex issue. There is a need to introduce new services rapidly, and the services must be manageable from within an integrated framework.

An essential attribute of this environment is the facility for allowing users with more demanding requirements to take an active part in tailoring networks and services to fit their specific needs, including access to management functionality and customer control features. For example, users may want to access accounting data, monitor fault statistics, increase the bandwidth subscribed to on a link, etc. Instead of users having to ask operators to do this they will be given direct access to corresponding management facilities. These management services, whether integrated in a service subscribed to or realized as separate services, themselves represent some kind of value-added services. This calls for an analysis of the integration of IN and TMN architectures and the related service models for providing a common implementation platform and a common set of objects and components, as outlined in [1]. So far only a few efforts have been undertaken in IN and TMN standards bodies to investigate the management issues in INs Also, the definition of TMN management services for telecommunication networks (including IN) and services has only just begun.

When considering the possibility of providing a common environment for IN and TMN services, the increasing influence of object-oriented approaches in software and system design has to be recognized. If the RACE ROSA project is taken as a basis for the current state of the art in object-oriented service design for open services architectures, it seems worth investigating the application of the concepts that have been developed so far for their suitability in harmonizing IN and TMN service creation and provision.

After a short presentation of the influencing forces of network evolution towards the target network environment, the key concepts for a future multi-service network are briefly presented. Looking for an object-oriented approach for modelling management and telecommunication services in a uniform way in an integrated service environment, the present paper adopts the ROSA approach and proposes a framework for a target multi-service environment.

2. Key Concepts for Future Multi-service Networks

Focusing on the uniform provision of future telecommunication and management services in an integrated target telecommunication environment, this section discusses the current concepts which have been identified as a suitable basis for such an environment, namely IN and TMN. Important aspects common to both concepts are: reusability of functionality (specifications), network and service independence.

2.1 Intelligent Network

The Intelligent Network (IN) is an evolving service-oriented public network architecture which represents a basis for the uniform provision of services beyond the basic telephone services in the telecommunication environment. Introducing some basic concepts from the data processing world into the telecommunication environment, the IN approach has three major goals: *service independence* through the definition of generic service building blocks, representing generic functions that could be used for the construction of many different services, *network independence* through the definition of functional network elements with specific functionality and *vendor independence* through the definition of unique interfaces and protocols between the defined IN network elements.

Several IN architectures which have the same concepts in common but differ in their scope and their terminology have been defined or are under development. In the following we concentrate on the standardized architecture of intelligent networks as defined in the CCITT Q.1200 recommendations [2]. In order to provide a modelling tool for IN architectures, CCITT developed an *IN Conceptual Model (INCM)* which consists of four planes addressing service design aspects, global and distributed provision functionality and physical aspects of an IN.

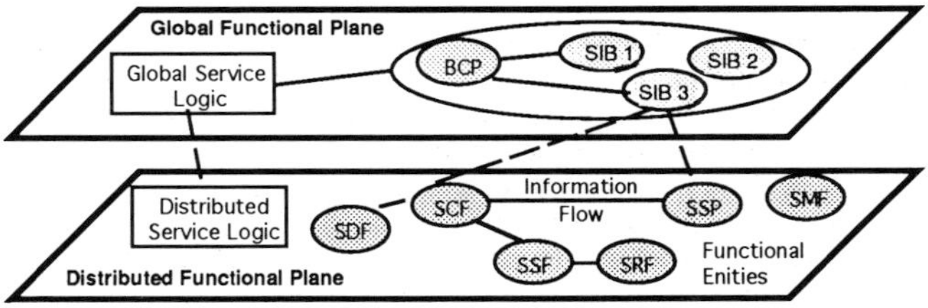

Figure 1: Separation of IN Service Creation and Execution

The *Service Plane (SP)* is the uppermost plane and represents an exclusively service-oriented view where the service implementation and the underlying network technology is transparent to the users. Each service consists of one or more generic blocks called *Service Features (SFs)*, where each SF could be a complete service or part of a whole service, which may or may not have to be present in each of the different variations of the service offering.

The *Global Functional Plane (GFP)* models the network from a global high level perspective as a single programmable entity, hiding the complexity of the distribution of functions. This plane deals with the service creation aspect and contains *Service Independent Building Blocks (SIBs)*, which will be used as standard reusable network-wide capabilities for efficient and fast service and service feature realization.

The *Distributed Functional Plane (DFP)* models a distributed view of an IN in terms of a set of *Functional Entities (FEs)*, such as Service Switching Function (SSF), Service Control Function (SCF), etc. providing transparency to the physical network elements. Each FE may perform a variety of *Functional Entity Actions (FEAs)*. Each SIB is mapped onto one or more FEs in the DFP which means that SIBs are realized by a sequence of particular FEAs performed by specific FEs.

The *Physical Plane* models the physical aspects of an IN and defines different *Physical entities (PEs)* where the relevant FEs from the DFP are located.

In IN a service program is written in the *Service Creation Environment Function (SCEF)*. A particular service can be created by writing a service program (global service logic) which calls various SIBs. In the SIB description the information flows between relevant FEs and the actions to be taken there are being described as a sequence of events. SIBs are defined independently of the underlying network and are only visible in the SCEF. This means that no knowledge about the structure of the underlying network should be required for service creation. Note that there is no notion of SIBs as identifiable entities in the network (DFP). Figure 1 illustrates this separation of IN service creation in the GFP and IN service execution in the DFP.

This means that when an IN service has been deployed in the IN and is actually processed, service processing generally corresponds to the Distributed Functional Plane, where a service consists of distributed service parts (i.e. trigger entries in SSFs, service logic in SCF, etc.) requiring interactions between the related IN functional entities.

2.2 Telecommunication Management Network

The purpose of a TMN as defined in the CCITT M.3010 draft Recommendations [3] is the management (operation, control and maintenance) of all kinds of telecommunication networks and services. The basic concept behind the TMN is the provision of an organized network structure for achieving the interconnection of various types of management systems (which carry out specific management tasks) with the telecommunications equipment in the underlying managed network. In addition to this generic management architecture, the basic ingredients for the TMN are the uniform definition of the necessary management information and the provision of corresponding standardized management services.

Relevant to TMN applications (see figure 2) is the definition of TMN *Management Services*, which are composed of TMN *Management Service Components* and TMN *Management Application Functions*. The management service components and application functions are related to such aspects as *performance, fault (maintenance), configuration, accounting* and *security* management, with the management functions being the smallest parts of the TMN management services as perceived by the user of the services [4]. Each function achieves its objective by using a sequence of actions (management operations) on a defined set of managed objects. Note that TMN has now started work on the definition of management services for INs.

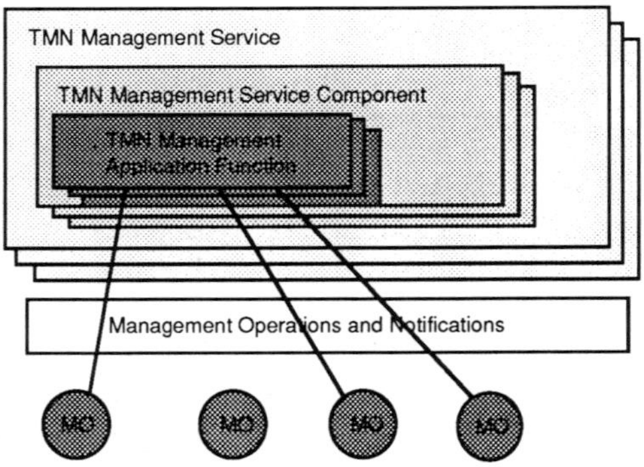

Figure 2: TMN Management Service Components and Management Services

2.3 Important issues for IN and TMN Integration

IN management issues have not yet been considered in IN standardization. In particular, there is no differentiation between service and network management aspects. Additionally, no link between the (IN) telecommunication services and the management services has yet been defined. The IN approach provides more flexibility for what is basically a conventional basic call model but there are expected to be limitations in modelling applications, particularly complex multimedia applications for a more heterogeneous environment.

The TMN functional architecture describes only a functional architecture for management and does not address the aspects of management service creation and processing. No SCEF for management services has yet been defined and the definition of suitable MSCs is progressing very slowly. In addition, the placement of management programs in the TMN function blocks has not yet been defined.

Note that the advantage of fast service creation in the IN concept becomes less powerful if the management of newly created services is not created in the same fast and efficient way. It may therefore be useful to consider the adoption of the SIB concept by TMN for the creation of management services. This would lead to two kinds of SIBs: processing or telecommunication SIBs for the creation of telecommunication services (providing information flow descriptions in the IN manner) and management SIBs for the creation of management services (providing the object-oriented specifications of the relevant object classes) [5]. In the following we investigate an object-oriented approach for the uniform provision of telecommunication and management services.

3. Influences contributing to an Integrated Management Solution

The requirements for managing not just the network but the entire distributed multi-carrier, multi-vendor, multi-service environment are considerable. However, new approaches to management are emerging that are based on integrated management solutions. Influences that have contributed to such solutions and that are significant in the development of an integrated management service environment are introduced below.

3.1 Object-Oriented Techniques

Object-oriented techniques offer several advantages when modelling services for large systems. The details of the individual operations are hidden, which is important in an open environment where openness must be achieved on the basis of well-defined interfaces. An object-oriented design abstracts from the complexity of the system by structuring it as a set of cooperating objects which encapsulate the detailed knowledge of the system and which communicate only via clearly visible interactions. This contrasts with the functional approach where interdependencies between functions make changes to the system more complex and error-prone. The functional approach models functions and the means for grouping them, whereas object-oriented modelling is based on object types and their behaviour. Object-oriented design allows for decomposition and composition and supports refinement through inheritance, both of which encourage reuse and interoperability. Well-defined interfaces, modularity, encapsulation and object interaction via message passing enable the functionality represented by the object interface to be separated from the implementation. This is required for portability and interoperability for if the interface remains the same the service can be implemented on different underlying network technologies. Object-oriented design is therefore a more suitable tool for handling the complexity of future distributed systems when compared with the functional approach.

In the area of management information, ISO and TMN have adopted object orientation for modelling the system resources to be managed as Managed Objects (MOs). This approach is becoming stable with experience gained in defining MOs, for example, in TMN M.3100 [6], the Network Management Forum Library [7], and RACE CFS H550 [8]. However, object orientation here is being used for structuring management information and not for the semantics of object behaviour and object interaction. The object-oriented techniques are based on entity-relationship data modelling techniques and do not fully address the dynamic, behavioural aspects. In addition, the functional approach is still being adopted for modelling the managing side where the managing tasks are defined by management functions. This means that it is more difficult to reuse components and obtain a consistent approach when designing management services. The coexistence of these different modelling approaches will increasingly pose a problem in future as the contrasting ways of modelling together with the greater complexity of the environment will hinder the objective of providing integrated management for an open service environment.

3.2 Management Platforms

When networks were smaller and services less complex, network management focused primarily on monitoring network resources and consisted mainly of a variety of management functionalities tailored for specific hardware and software configurations with no interoperability, coordination or correlation. The concept of service management did not exist and there was little support to help service designers in specifying the management aspects of their services. This resulted in a variety of reference models and concepts which hindered cooperation and prevented a consistent approach to service management from being adopted. It was recognised that an integrated and flexible approach to distributed systems management is required that is responsive to a rapidly changing and increasingly sophisticated open competitive environment. This requirement for integrated management in a heterogeneous environment led to the idea of management platforms.

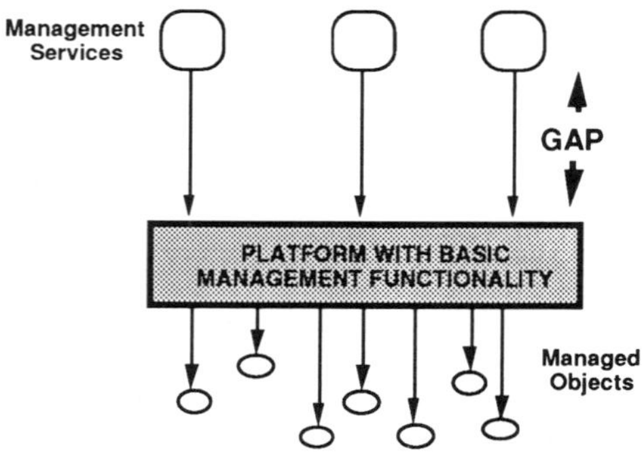

Figure 3: A Management Platform

A management platform allows management services to be developed within an integrated framework independently of the underlying system. The provision of open programming interfaces and the use of standard communication protocols enable a management platform to support a common functionality shared across several management services. The heterogeneity of the underlying hardware and software is hidden so that a variety of resources can be managed via the one platform. A platform with open programming interfaces allows management services, to be considered separately from the resources to be managed, which is required in the era of open provisioning.

A key question is how "high" the services offered by a platform should go and at what level such a platform should be located. Currently, management platforms are located at a fairly low level, offering basic OSI CMIS functionality which then has to be integrated into a management service [9]. This is certainly a step forward but the gap between the basic functionality and the management services is wide (see Figure 3). To design and write management services using such basic functionality is difficult and error-prone, and large distributed systems cannot be managed with the primitive operations offered by platforms alone. Effort must now be concentrated on developing services for managing a wide variety of services. In future, management platforms should be able to offer a higher-level functionality in the form of generic components providing, for example, accounting, monitoring or configuration functionality. More attention can be paid to the management services themselves by combining the generic components offered by the platform so that the managing components required for the particular management service are available, reusable, and can be combined to provide the management services required. For example, components can be selected and grouped to provide quality of service management, customer control features, or security services for a variety of different types of service and network. In this way, platforms can enable management services to be introduced rapidly and efficiently.

3.3 Management Services

Management services are considered here as services that need to be specified and designed in the same way as any other service. They provide management services as part of their "core" functionality rather than some other service, but they still need to be designed, specified, implemented and even managed in the same way as these other services. The functional approach has been widely used in specifying management services, but the functions are not grouped or coordinated according to any particular method, and in themsevelves do not contribute to open, modular, interoperable services. There is not an accumulated body of knowledge or experience in defining management services as there is with specifying Managed Objects and so there is no common approach. Both IN and TMN provide a means of separating services from the underlying network, but they both regard services through the capabilities of the network and not from the service aspect, i.e. a bottom-up approach. They do not provide an integrated model and have different views of what a service is.

Work has been started in adopting an object-oriented approach to specifying management services. Although this work is relatively new and still immature it promises a more consistent approach. The management aspects of services are becoming more open in that not only "managers" but also end-users want to be able to control the services they use and so the control element of a service must not only be modelled at the same time as the service is being specified, its control and management functionality must be offered in the same way as the core functionality of the service. It is therefore important to have a model that enables both general services together with their management as well as the management services themselves to be modelled in the same way. As management services may in turn need to be managed, the relationship between managed and managing is relative and can depend on the context, which is another argument in favour of adopting the same modelling approach for both the management services and the services to be managed.

4. An Integrated Management Service Environment

An integrated management service environment should provide support for service management so that all services can be managed according to a common model. The developments outlined so far do not in themselves provide an adequately integrated approach when a far higher number of services involving a wide range of media in a heterogeneous environment will need to be developed, configured and managed. An orientation towards services, including their requirements, their relationships to the underlying networks and their specific management needs is essential. This is equally true of the services required to manage such services. An integrated service environment for management requires an architecture which is independent of any one specific network configuration and able to support a very wide range of services, including management services. Using an approach such as ROSA would enable service designers to use generic building blocks for specifying and constructing management services. An integrated management service model based on this approach is therefore being proposed.

4.1 The ROSA Approach

ROSA *(RACE Open Services Architecture)* was a project in the European RACE programme responsible for defining an architecture and methodology for modelling and specifying open telecommunication services. The aim is to facilitate the rapid, cost-effective and smooth development and deployment of new and complex Integrated Broadband Communications services in a heterogeneous, multi-vendor, multi-service environment independent of any specific network technology. Important is both the openness of a service as well as the integration of the management aspects of a service with its core functionality so that the service´s management requirements are considered, designed and specified at the same time as the service specification itself [10]. ROSA is representative of similar approaches focusing on an open service environment, such as Bellcore´s Information Networking Architecture (INA), which is addressing the same problems but emphasising more the infrastructure support for open services [11]. RACE Project R2049, CASSIOPEIA, is continuing the work started by ROSA.

The ROSA architectural approach provides support for designing, specifying and modifying open services for implementation in an IBC environment where both the services themselves and the underlying network technologies will be constantly evolving. The emphasis is on services and service requirements, so that service designers can focus on what the service does using a top-down approach in contrast to the usual bottom-up approaches which emphasise the network technology and its capabilities. It thus decouples the service design and specification from the technological infrastructure and its evolution. By abstracting away from the specific network models, ROSA is providing the service designer with a common platform for all services, including management services.

A single object-oriented approach has been adopted as a basic principle in ROSA with object definitions providing the framework for specifying services. The dynamic aspects of object orientation are included so that a computational model is obtained. Telecommunication services are described as a composition of interacting component object types modelling various aspects of the complete service. Objects are the only entities that provide services and one object model for information modelling and expressing service semantics is used throughout the architecture - the ROOM *(ROsa Object Model)* [12]. The object model has a formal basis and a notation has been developed for describing both the behaviour and the data structure aspects of an object type in accordance with ROOM [13].

The architecture defines concepts, rules and guidelines for using the component object types to specify a service [14]. It is organised into two frameworks. The *Service Specification Framework* is of most relevance to the service designer as it is concerned more with developing the service itself than with how the service will be implemented on the underlying network. It can be aligned to the ODP computational viewpoint [15] as it provides a high-level view of the logical infrastructure required by services and defines object types that can be used as building blocks when modelling an open service. The *Resource Specification Framework* reflects the ODP engineering viewpoint. It is of more relevance to a system designer as it provides object types concerned with modelling the underlying infrastructure requirements of the service. The Resource Specification Framework is intended to aid the mapping from the Service Specification Framework to the specific network technologies that the service will be implemented on and the object types that it provides are logical representations of physical network resources [16].

When using the ROSA architecture, the service designer is supported by the ROSA *service specification methodology* which provides a structured approach to service specifications and their refinement using object types in order to create realisable service specifications. In accordance with the ROSA focus on services, it is a top-down approach that first analyses the requirements on the service of stakeholders, i.e. the customer, the service provider, the network provider and the intended end user, and describes the behaviour of the service from the users´ perspective. Service specifications are refined iteratively so that the relevant object types are described at different levels of abstraction [17].

Telecommunication services are specified by applying the architectural concepts at various levels of abstraction using the ROOM and service specification methodology. The concepts of the architectural frameworks, the Service Specification Framework and the Resource Specification Framework, provide guidance and direction in this process and make available object types that can be reused in the specification. As the architecture matures it is expected that object type libraries will be created and that increasingly object types will be re-used rather than designed anew for each service specified. A sample set of object types - Reference Object Types - that embody the basic architectural concepts and conform to the ROOM has been specified, for example, Directory, Clock, ChargingAgent, UserProfile, Trail, and Connector [14]. The specification procedure is quite complex for a complete service and so it is only briefly described here. However, in [18] examples are given of how the ROSA approach can be applied to specifying, for example, a Multimedia Conference Service, a Virtual Private Network Service and an Alarm Surveillance Service.

The ROSA approach to an open services architecture is characterised by its adoption of an object orientation that includes the dynamic aspects and so provides a computational viewpoint of the services being designed. The service model matches more closely the computational model and the service specification can be more easily mapped onto an existing software platform. In addition, a close linkage has been established between general telecommunication services and management services.

4.2 Integrated Modelling of Telecommunication and Management Services

A unifying paradigm is required for modelling telecommunication and management services. It should be service-oriented and able to integrate current approaches into a single model in order to better meet the requirements of the many types and varieties of services offered in an open telecommuncations environment. It is important that the separation of the service functionality from implementation details is applied to management services as well as to the services to be managed. A formalisation of concepts would contribute to the design and specification of management services and would also enable them to be developed in the same way as other services, thus allowing the same model to be used both for the services to be managed and their management. A model based on object types will enable future technological developments and new types of services to be more easily incorporated than models that are not object-based or which are based on a variety of approaches.

Reusability is a common thread that links ROSA with many other approaches. They all have some concept of reusable components with open interfaces that can be used to construct new services by means of composition, for example the SIBs in IN, or the management service components in TMN. Management functionality can be included when applying (de)composition and resusability techniques in the design of services that are to be managed as well as when designing the management services themeselves. As various approaches have this common thread, it can form the basis for integrating them in order to suppport the development of management services using an object-oriented architectural framework. If services are to be designed and specified as a set of interacting objects, it follows that this should also be true of management services. A library of generic object types for use in specifying management services will enable management services to be developed using existing components, possibly by refinement for specific services. Standard interfaces are provided by these generic object types which carry out the management tasks required. Not only does this provide an integrated model, but it also is more cost-effective in the longer term. Such generic object types reduce the effort and complexity involved in developing management services as they are already provided and service developers can concentrate on the service-specific aspects of the new service itself.

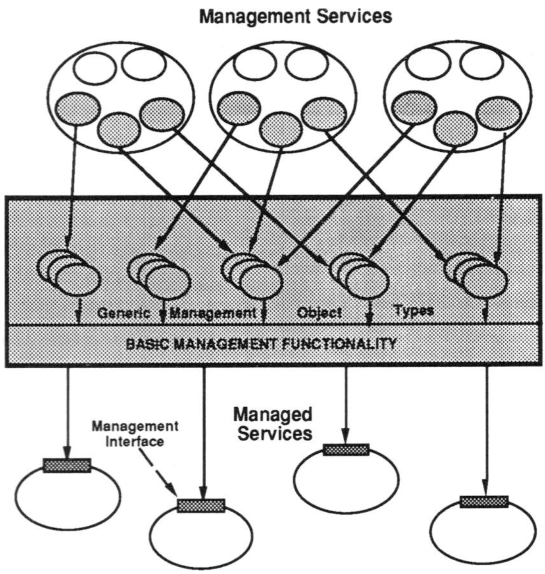

Figure 4: A Service-Oriented Management Support Platform

Using the recipes, rules and guidelines provided by the ROSA architecture a variety of intermediate-level components can be composed for use in management services. These components represent the management functionalities required for different types of service and different kinds of physical network models. Figure 4 shows how the gap between basic management functionality and high-level complex management services can be reduced by offering such generic components to be used in composing management services. A consistent object-oriented approach is used throughout and computational object types are used for designing both the management services as well as the services being managed.

These generic object types incorporate aspects of management, and can be grouped according to particular management tasks and for particular types of services and networks and in order to be compatible with concepts such as those from TMN. For example, an accounting group would provide object types enabling billing and charging to be carried out, and a service designer would choose the object types relevant to the service being designed and the type of charging intended. Service provisioning object types can provide support for provisioning and configuring object types for different network technologies. Object types representing various security policies can be made available, and object types for managing, for example, various types of VPNs. Sets of object types can be established along the same lines for other aspects of service management, such as security, customer control, or performance. Many of the object types being provided for services generally, such as the communication or access object types, would be also required for the management services and so the management services can share many of the generic object types available to all services.

The ability of services to manage other services requires a corresponding component in the services being managed and object types corresponding to those being used to specify management services must be available for those services being managed. The services being managed need to be designed with the corresponding management interface to allow them to be managed. If all services are specified using the same architectural concepts and object model this task will be facilitated as the appropriate "managed" object types to correspond to the matching "managing" object types will be available to the designer of the services to be managed.

5. Conclusions

This paper focuses on the key concepts for future telecommunication networks and services, namely the IN architecture and TMN, and proposes a unified object-oriented modelling of telecommunication and management services in a future telecommunication environment. Networks have become too complex and unwieldy for ad hoc management. Greater flexibility with reusability within an integrated framework is required and this is what the ROSA approach of an Open Services Architecture offers. An increase in the number and type of services available and their need to be managed means that service requirements as well as the network technologies must be considered.

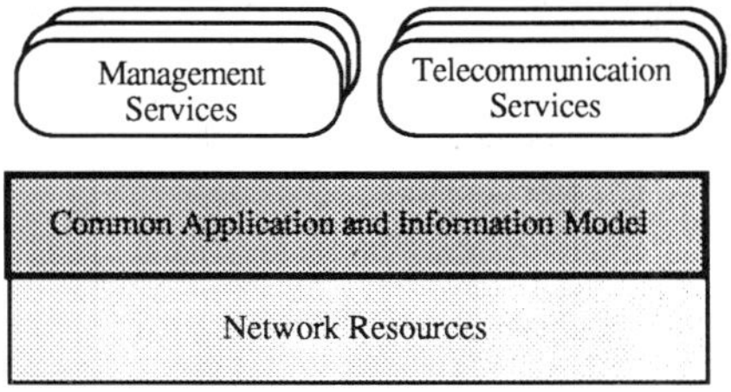

Figure 5: Integrated Platform for Management and Telecommunication Services

As systems become more complex and a wider range of services are introduced and supported, management will also become more difficult. The real challenge will be to master the complexity of these systems. It is vital that systems can be structured in a robust and comprehensible way and the trend towards adopting an object-based approach for designing and implementing services suggests that the approach adopted by ROSA is a step towards an object-oriented architectural framework for future systems. It enables management services and the services being managed to be designed and specified for a broadband environment using the same modelling techniques (see Figure 5).

An integrated approach based on the ideas of ROSA together with existing developments in IN and TMN can provide a structuring for all services, both management services and other telecommunication services, together with their management aspects. In this way a single paradigm can be adopted for the future open environment which offers a more powerful and flexible approach to the many management issues that will arise.

Acknowledgements

The ROSA project was partly funded by the Commission of the European Communities through the RACE programme. The consortium consisted of thirteen partners led by BT Laboratories. The authors would like to acknowledge the work of the ROSA project members and to thank those who commented on earlier drafts of this paper.

References

[1] T. Magedanz: "Integrating Intelligent Networks and TMN", 3rd TINA Workshop, Narita, Japan, January 1992

[2] CCITT Draft Recommendations Q.1200 series: "Intelligent Networks", Geneva, September 1991

[3] CCITT Draft Recommendation M.3010: "Principles for a Telecommunications Management Network", Geneva, September 1991

[4] CCITT Draft Recommendation M.3020: "TMN Interface Specification Methodology", Geneva, September 1991

[5] ETSI DTR/NA-43308: "Baseline Document on the Integration of TMN and IN", Version 3, September 1992

[6] CCITT Draft Recommendation M.gnm (M.3100) : "Generic Network Information Model", December 1991

[7] "Forum 006, Forum Library - Volume 4: OMNI*Point* 1 Definitions ", Network Management Forum, Issue 1.0, August 1992

[8] "Telecommunications Management Objects", CFS H550, in: RACE Common Functional Specifications, Issue B, December 1991, Document 6, Network Management, General Aspects, published by RACE 1045 Project Consensus Management

[9] "Information Processing Systems - Open Systems Interconnection - Common Management Information Service Definition", ISO/IEC IS 9595, January 1990

[10] M. Key, M. Chapman, A. Oshisanwo, A. Mullery, J. Saint-Blancat: "The RACE Open Services Architecture Project", to appear in IBM Systems Journal, volume 31, no. 4

[11] N. Natarajan, G. Slawsky, "A Framework Architecture for Information Networks", IEEE Communications Magazine, 32(4), April 1992, pp. 102-109.

[12] M. Chapman, I. Tönnby, P. Schoo: "Suggestions for Object-Oriented Modelling from ROSA", 3rd TINA Workshop, Narita, Japan, January 1992

[13] P. Schoo, I. Tönnby: "The ROSA Object Model", in Open Distributed Processing. Proceedings of the IFIP TC6/WG6.4 International Workshop, Berlin, Germany, 8-11 October 1991, ed. J. de Meer, V. Heymer and R. Roth, North-Holland, Amsterdam, 1992, pp. 291-300

[14] "ROSA Architecture, Release Two", ROSA deliverable 93/BTL/DNR/DS/A/005/b1, May 1992

[15] "Working Draft for the Basic Reference Model of Open Distributed Processing - Part 3: Prescriptive Model", ISO/IEC JTC1/SC21/N7055, August 1992

[16] G.Bregant: "ROSA: from the Services to the Architecture", 2nd TINA Workshop, Chantilly, France, March 1991

[17] "The ROSA Handbook, Release One", ROSA deliverable 93/BTL/DNS/DS/B/012/b1, September 1991

[18] "Specifying IBC Services using the ROSA Architecture", ROSA deliverable 93/BTL/DNR/DS/A/003/b1, May 1992

Integrated Network Management, III (C-12)
H.-G. Hegering and Y. Yemini (Editors)
Elsevier Science Publishers B.V. (North-Holland)

A TRANSITION ARCHITECTURE FOR BREAKTHROUGH SPECIAL SERVICES AUTOMATION

Donna Bernhardson[a], Susan K. Wilson[b] and Douglas N. Zuckerman[a]

[a]Systems Technology Center, AT&T Bell Laboratories, 200 Laurel Ave., Middletown, NJ 07748, USA

[b]Systems Technology Center, AT&T, 300 Eastside Dr., Alpharetta, GA 30202, USA

1. ABSTRACT

AT&T and other large telecommunications service providers, in response to a highly customer-focused, competitive business environment, have sought "breakthroughs" in provisioning, maintaining, and responding to customer calls, for private-line special services. However, the pre-existing system and database infrastructure based on older views of the business presents a challenge to significantly reducing cycle times, lowering defect rates, and cutting costs.

To address modern business needs, a long-term target architecture has been established where applications share a centralized database (i.e., client-server). AT&T has established a transition architecture which takes into account the pre-existing systems environment, while building towards the desired target. The architecture is centered around a new system which integrates and mechanizes interactions with existing systems and results in significant customer servicing improvements. These improvements are achieved mainly through increased automation of operations from receipt of a service order, through inter-office installation testing, service turn-up, proactive network service and alarm correlation, and finally call-receipt and maintenance tracking activities.

Keyword Codes: H.4.0; H.4.3
Keywords: Information Systems Applications, General; Communications Applications

2. SCOPE

Earlier publications have presented overall architectures for customer servicing across a wide range of network services and premises products.[1] Due to business structure directions, the scope of services supported by the new system is Private Line, though many of the same concepts and functionalities could be carried over to other types of services such as circuit- or packet- switched. The basic network element for the services supported is the AT&T Digital Access and Cross-connect System (DACS). Approximately one-half million private line circuits are included, with the vast majority being digital private-line services (such as AT&T ACCUNET Spectrum Digital Services).

3. OPERATIONS SUPPORT ENVIRONMENT

Many telecommunications carriers have had Operations Support Systems (OSS) architectures resulting from earlier structures for our business, and have not benefited from integration and automation opportunities, enabled by more modern technologies (see Figure 1).

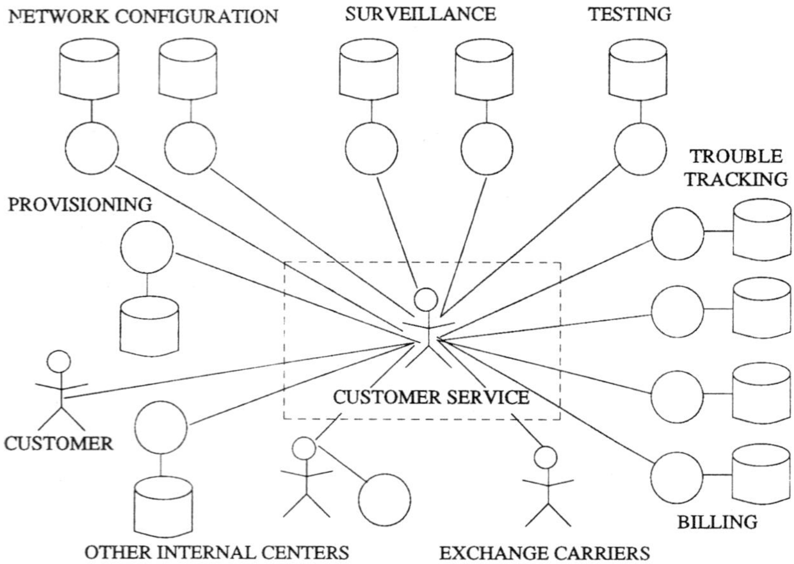

Figure 1 - Operations Support Systems Complexity

The processes that are built up around these earlier architectures containing specialized, non-cooperating OSSs are labor intensive, and mainly aim at: taking phone calls from customers who are reporting troubles, doing manual look-up of information, using separate systems for testing, verification of service or circuit status (often using inaccurate or hard-to-obtain data), referrals via telephone call to providers such as the core network centers or LECs, and, eventually, manual close-out of a trouble, along with a telephone call to the customer. The difficulties associated with using these older architectures in the newer business environment have often led to numerous specialized "work-around" systems that try to fix particular pieces of the problem, but with limited attention to achieving cooperation and synergy among the OSSs.[2]

4. BREAKTHROUGH SOLUTION

The main functions supported by the initial phase of the new system focus on: populating a single, shared database with quality data; automating pre-service and maintenance testing; providing an integrated user interface, correlating network failure alarms with affected circuits and services, and implementing call-receipt capabilities which meet business-driven work-center needs. The breakthrough characteristics translate into significant operations cost reductions and servicing improvements when compared with the present characteristics. In addition, the use of new technology (such as client-server, built on AT&T System-7000 computers, using a high-performance relational database), may translate to significant reductions in systems development and production support, database administration, and communications costs (with a slight increase in hardware costs).[3] For example, moving an older OSS (distributed across about 20 minicomputers nationwide) to the newer client-server architecture, can result in an annual "systems" cost reduction of approximately 75 percent. The following table compares the present and breakthrough functionalities.

Functions	Present Mode	Breakthrough Mode
Populate Database with Quality Data	Manual, Many Database	Automate, with Rules, One Database
Pre-Service and Maintenance Testing	Manually Initiated, Multiple Screens	Automated, Exceptions Handled Manually
User Interface	Multiple Document-Based	Single, Information-Based
Correlate Alarms with Service	Slow, Manual, Reactive	Automated, Status Available When Needed
Call Receipt	Center-Dependent, Paper Tickets For Backup	Not Center Dependent, Full System Availability

4.1 Near-Term Function Architecture

The near-term systems architecture shown in Figure 2 integrates existing systems with newer functionalities, to streamline provisioning, call receipt and maintenance activities. The new system has four main processes (i.e., applications):

- Service Implementation and Quality Control (SIQC) - includes automated provisioning of the customer servicing database with accurate data, and control of network and pre-service installation testing.
- Service Verification and Analysis (SVA) - includes initiation of automated circuit testing and correlation.
- Network Service and Alarm Correlation (NSAC) - includes real-time receipt of network alarm and performance information, and mapping to affected customer services and circuits.
- Customer Service Tracking (CST) - includes call receipt automation, and trouble reporting and tracking.

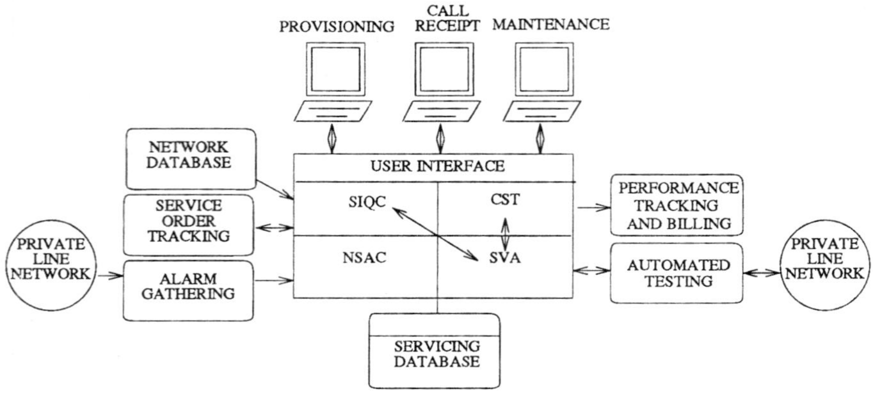

Figure 2 - Near-Term Functional Architecture

A major characteristic of the near-term environment is a central database shared among customer servicing applications. This significantly reduces database administration costs, greatly improves the quality of customer servicing data, and enables customer servicing applications to be implemented on a wide range of processor configurations (distributed, or centralized). Another major characteristic is that the new system effectively integrates use of some existing systems, and supplants the functionality of other existing systems, resulting in significant cost benefits starting almost immediately.

4.2 Service Implementation and Quality Control

SIQC is responsible for integrating essential automated support for the pre-service technician during service implementation and for supplying downstream processes, both internal and external, with necessary data for continued customer servicing.

SIQC receives data from two external sources: A corporate service order tracking system and a corporate network database. It performs format and syntax checks on the received data and then validates it by using a pre-defined set of rule-based criteria. Then service order information along with circuit design data, technician work authorization information, engineering design notes, LEC access information and customer information are placed in the customer servicing database.

SIQC has a user interface that provides technicians with access to the customer servicing database service order and circuit information. SIQC interfaces with SVA to automate pre-service inter-office installation circuit testing. It also allows manual triggering of pre-service inter-office installation circuit testing via the SVA process, as well as viewing of results in the customer servicing database. SIQC and CST share user-interface processes for displaying data needed by both pre-service technicians and maintenance technicians.

SIQC automatically transfers required inventory data to an external system responsible for customer rebates and operations performance tracking. SIQC is also responsible for managing requests for temporary circuits and purging data for disconnected circuits.

By providing the overall interface with the external corporate service order tracking system to manage automated pre-service testing triggers/results, data quality jeopardies and in-service completion status, SIQC is a major step forward in consolidating new work center functions and existing systems.

4.3 Service Verification Analysis

SVA is responsible for providing automated and manual circuit testing support for pre-service technicians during service implementation, for call receipt agents during initial customer trouble reporting and for maintenance technicians during trouble shooting. SVA has an external interface with AT&T's private line testing system, SARTS (Switched Access Remote Test System). SVA sends test requests to SARTS and receives test results back. SIQC and CST provide the user interface screens that are used to request manual circuit tests and view test results.

For provisioning, SVA automates pre-service inter-office installation circuit testing for new circuits. SVA determines the test capability for circuits and initiates the appropriate automated tests in SARTS which then returns results to SIQC. Pre-service technicians using the SIQC interface can view test results and, as needed, manually re-initiate certain tests.

For maintenance, when a call receipt agent receives a circuit trouble report and opens a maintenance request on a circuit, SVA first determines valid tests for the circuit reported in trouble. Then SVA reserves an automated testing time slot in a SARTS system and downloads test data to support later tests initiated with customer authorization. SVA returns test results from SARTS to the user via CST's user interface. A maintenance technician can view test results and initiate manual circuit tests while trouble shooting circuit failures. Some conditions will require that the technician execute the test directly from SARTS. Since SARTS private line testing consists of many SARTS Process Controllers (PC), SVA also balances the test request load among SARTS PCs.

SVA improves private-line customer servicing by performing inter-office circuit tests automatically for new circuit orders and by reducing the overall time it takes to respond to customer-reported troubles.

4.4 Network Service Alarm Correlation

NSAC is responsible for receiving DACS carrier group alarm messages via the DACS alarm monitoring systems. NSAC uses the circuit and facilities information contained in the customer servicing database, along with algorithms aimed at consolidating multiple alarms, and then mapping relevant facility alarms to associated customer circuits. When the DACS alarm monitoring systems transmit alarms "clears", NSAC correlates the clear status with the original alarm and maintains historical alarm statistics for later viewing.

Using the CST user interface, call receipt agents and maintenance technicians can view current-hour alarms, 24-hour alarms and alarm history for a selected circuit. This means that while talking with a customer reporting a possible circuit outage, the call receipt agent can verify the existence of alarm conditions on the circuit.

The alarm history is a valuable maintenance tool for working with recurring circuit problems. NSAC's alarm correlation capabilities speed up our response to customer calls by reducing the need for testing, helping call receipt agents determine when to initiate a SARTS test and helping maintenance technicians analyze and refer the trouble to the correct work center.

4.5 Customer Service Tracking

CST provides highly-responsive quality customer servicing, through its customer call receipt and technician maintenance functions. This is accomplished through automated diagnostic tools, and access to accurate current network, service and customer information in the servicing database.

For call receipt, CST enables easy entry of maintenance request information by a call receipt agent. In most cases, while a customer is still on the phone, an initial sectionalization of the trouble takes place, and the trouble is referred out to the appropriate entity for repair. The NSAC alarm functionality, as well as the SVA testing functionality, enable this speedy sectionalization. Where necessary, maintenance technicians can conduct more specific sectionalization through use of additional SVA and NSAC capabilities. CST also supports the call receipt agent in providing customers with updated maintenance request status.

CST automates the processing of maintenance requests from the time a trouble is reported, through referrals, testing, repairs, and closure.

For closed troubles, CST automatically transfers associated maintenance request data to the performance tracking system. This information, together with SIQC's, allows the system to generate work center quality assurance reports, as well inputs to downstream billing systems involved with rebating customers.

CST improves private-line customer servicing by making customer interactions more efficient, and taking advantage of reliable data appearing in the shared, centralized database. In addition, CST realizes a significant systems-operations cost benefit by consolidating onto new technology the trouble reporting and tracking functionality that was previously distributed across multiple mini-computers built on older technology.

5. MAIN ARCHITECTURE FEATURES

Figure 3 shows the physical architecture for the four main applications, servicing database, user interface and external system interfaces.

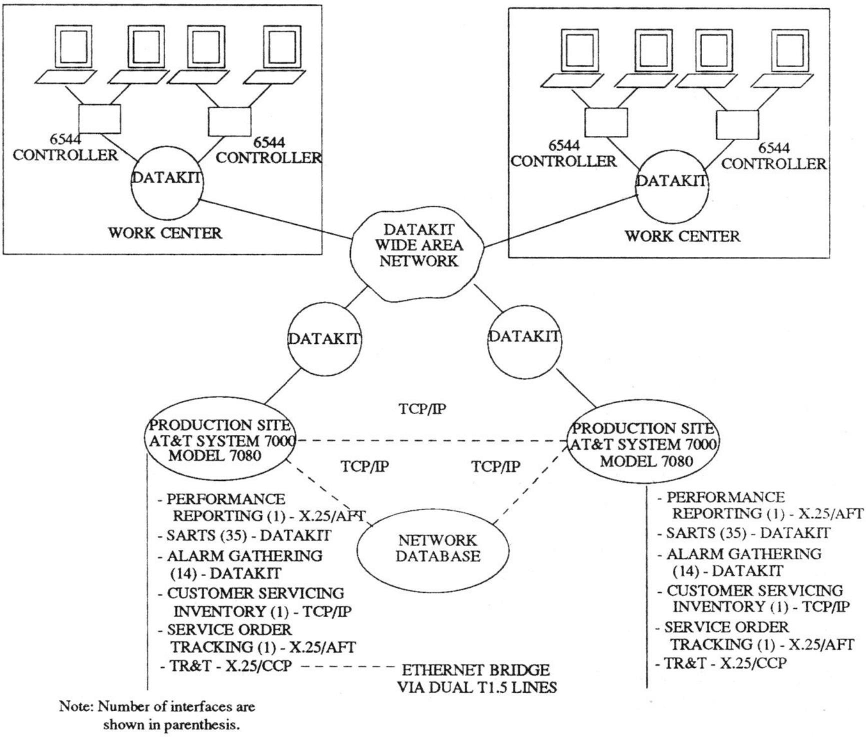

Figure 3 - Physical Architecture for Main Applications

Due to the high communications connectivity required and the expected load on the database, the hardware architecture has four AT&T-7080s as the processing core. The main functions distributed via Ethernet across these processors are database, communications and user-interface applications.

This architecture supports the system requirements of 800 megabytes of main memory, 32 gigabytes in the main database, an additional 12 gigabytes for historical data and about 370 MIPS for overall system operation. To meet stringent user requirements, the system will operate 24 hours a day, 7 days a week. The system is not expected to have any scheduled downtime.

Production and disaster recovery sites are interconnected using diverse T1.5 lines and Vitalink "TransLAN-350" smart bridges. The data center where the shared network database resides is similarly connected. The user community gains access to the breakthrough system using asynchronous terminals via a Datakit Wide Area Network (DWAN). External systems interface using three different communications protocols: X.25, TCP/IP and Datakit.

The database is manipulated through use of ORACLE relational database software. ORACLE Table replication software will provide a near "shadow" copy of the production site database. The TUX/T (TUXEDO) system is used to provide the interface with the RAPID/RCT package to interconnect the computers.[4] The Operating System is UNIX SVR3 (OSX/5.1a for the AT&T 7080s) and C and C++ programming language will be used. Datakit Build 35, X.25 and AFT (X.25) Release 2.0 is used for network connectivity.

6. FUTURE DIRECTIONS

The transition architecture described in this paper is currently deployed and being tested with users. It is a major step towards a target architecture (Figure 4) with even more automation, as well as mechanized interfaces to private-line customers, internal hands-on work groups, and exchange carriers responsible for providing access service.

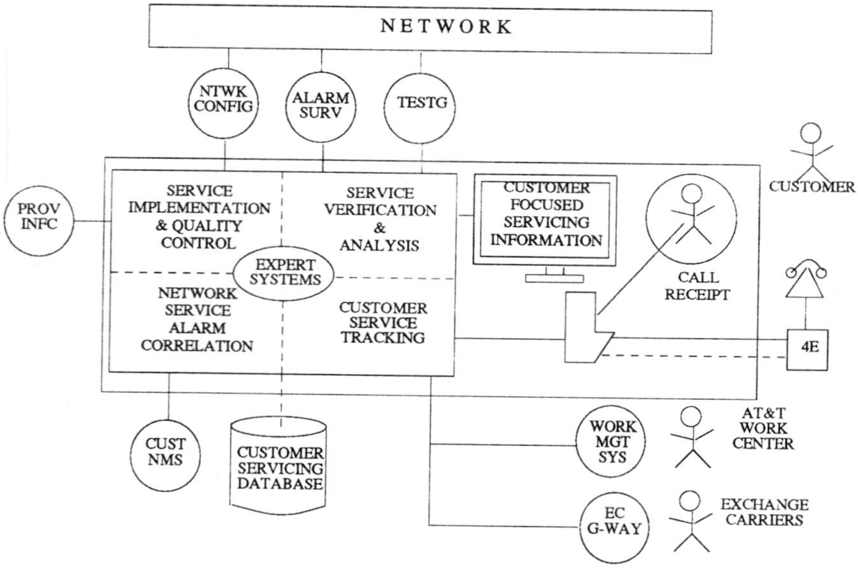

Figure 4 - Simplified Target Architecture

The transition architecture focused on simplifying the systems environment by building a central database shared among customer servicing applications. This significantly reduces database administration costs, greatly improves the quality of customer servicing data, and enables customer servicing to be implemented on a wide range of processor configurations (distributed or centralized). Future expansions of this architecture will aim at providing expert interworking of customer servicing applications to reduce or eliminate the need for human involvement in coordinating the activities of the various customer servicing applications (such as trouble management, testing, surveillance, referrals, etc.). The expansions will also aim at providing data quality maintenance feedback loops with "the network", as a highly-effective means to assure the accuracy of data from the provisioning process, and to make sure the customer servicing database and actual network configurations are in synch at all times, capturing all changes in network equipment.

To accommodate future functionality, and facilitate retirement of older systems or remaining redundant databases, we will build on the logical data model (LDM) currently designed to support *all* four applications.

7. ACKNOWLEDGMENTS

This project involved many individuals. The authors especially would like to acknowledge work which most directly relates to this presentation by the following individuals: Al Baldwin, Pari Bajpay, Jim Carleton, Mark Clay, Chris Kolmodin, Tum Lee, Ronald Slosberg and Harry Stovall. From the "user" community, the authors acknowledge and appreciate the perspective and guidance provided by: Mary Ann Ketterer, John Pfister and Joe Wisniewski.

REFERENCES

1. W. S. Caldwell et al, "*A Unified Operating Environment for Customer Servicing*," Paper No. 802.1, IEEE GLOBECOM '90, December 1990, San Diego.

2. R. R. Roscitt and D. N. Zuckerman, "*The Future of Operations Support Systems: High Tech and High Touch*," National Communications Forum, October 1990, Chicago.

3. P. E. Prozeller, "*Applying Distributed Systems Principles and Technology to Downsize and Improve Common Carrier Operational Infrastructures*," Paper No. 21.3, NOMS '92, April 1992, Memphis.

4. Y. S. Bagg, S. R. Desai, D. V. Kanwar, G. Sinha and R. W. Tower, "*AT&T RAPID Application Platform (AP) for Operations Support Systems*," Paper No. 4.2, NOMS '92, April 1992, Memphis.

Integrated Network Management, III (C-12)
H.-G. Hegering and Y. Yemini (Editors)
Elsevier Science Publishers B.V. (North-Holland)

Potential Network Management Services And Capabilities Associated With SMDS

Patrick Sher and Kaj Tesink

Bell Communications Research, 331 Newman Springs Road, Red Bank, NJ 07701, USA

Abstract

Switched Multi-megabit Data Service (SMDS) is a high-performance, public, packet-switched data service. In conjunction with SMDS, a network management service offered to end-users has been defined. This service is called Customer Network Management (CNM) service. The specification of generic requirements that allow for Interexchange SMDS impacts the CNM service. It further implies the possibility of additional network management services and capabilities, for example, between local exchange carriers and interexchange carriers, and between local exchange carriers mutually, to shepherd smooth traffic flow for SMDS. This paper reviews all these management services and capabilities, defines their relationship, and lists example network management functions that can be part of these respective network management services and capabilities.

Keyword Codes: B.4.0; C.2.1; C.2.3
Keywords: Input/Output and Data Communications, General; Computer-Communication Networks, Network Architecture and Design; Computer-Communication Networks, Network Operations

1. INTRODUCTION

Network Management refers to a set of activities that are designed to shepherd smooth traffic flow, resolve problems efficiently, and allow for easy addition (or deletion) of network resources (e.g. introduction of new subscribers). The primary interested party in these activities is the network provider itself, motivated by the desire to provide an attractive and viable service. However, since a Local Exchange Carrier (LEC) provided network service like SMDS may be used as part of larger networks, providers of connected networks may also be interested in these activities, i.e., the managers of customer premises based networks, managers of interexchange carrier networks, and managers of other LEC networks. But while the manager of the LEC network necessarily must have a complete view of all intricate NM details of the LEC network, managers of other interconnected networks, e.g., Customer Premises Network (CPN), Interexchange Carrier (IC), LEC, Independent LEC (ILEC), are unlikely to be interested in all these details. They will probably only be interested in those NM aspects of LEC network(s) that affect the operation and users of their own network. In other words, they are only interested in a portion of the complete view that the LEC manager has: a portion that is abstracted and tailored to their needs. Thus, the art of a good NM service is to tailor and abstract an appropriate view of a network's NM capabilities to the needs of the prospective user.

Section 2 of this paper summarizes the relationship of XA-SMDS with Exchange SMDS and Interexchange SMDS. The introduction of XA-SMDS and Interexchange SMDS imply that end-to-end traffic, from the perspective of the end-user, will often traverse a number of network management domains: customer premises based networks (e.g., one or more LANs), source LEC network, IC network(s), destination LEC network, and remote customer premises based networks. As discussed above, these management domains may provide NM services to each other.

Note that these network management information exchanges technically follow the client-server model. In that sense, these exchanges are Services. However, from a business perspective these exchanges may be treated differently, and may not always be labeled as services. Some of these exchanges may be billable. Others may be offered via a mutual agreement (e.g., Inter-LEC OM, defined below), and again others may be provided as integral part of the SMDS offering. This paper describes all exchanges as services in the technical sense.

Section 3 of this paper identifies relevant potential NM services, and summarizes their functions:

1. Customer Network Management (CNM) service - NM services provided by a LEC to End-customers (see Section 3.1),
2. EXchange Access Operations Management (XA-OM) service - NM services provided by a LEC to Interexchange Carriers (see Section 3.2),
3. Interexchange Carrier Operations Management (ICOM) service- NM services provided by an IC to a LEC (see Section 3.3),
4. Interexchange carrier Customer Network Management (ICNM) service - NM services provided by ICs to End-customers (see Section 3.4),
5. InterLEC Operations Management (InterLEC OM) service - NM services provided by one LEC to another LEC (see Section 3.5).

For completeness, one should also mention the use of the LEC network capabilities by the LEC manager. This capability is likely the most comprehensive NM use of the LEC network capabilities and includes all network operations, and most likely, all NM services and capabilities listed above.

6. Internal LEC NM Information Retrieval - NM information provided by a LEC for its own internal use.

This NM service is sufficiently described elsewhere[Bell4, Bell5], and is not further discussed in this paper.

SMDS CNM Service has been specified by Bellcore[Bell3], and has been demonstrated[JLT][Brunner] and is being deployed. An initial Bellcore specification of SMDS XA-OM Service has been drawn up[Bell6], and consultation is ongoing on that subject. InterLEC OM Service is subject of continuous study by Bellcore and its clients. ICOM Service and ICNM Service both would be provided by ICs. Some ICs have already planned in particular ICNM Service for their fast packet services.

2. SMDS AND EXCHANGE ACCESS-SMDS

Bellcore has published TR-TSV-000772[Bell1] which describes generic systems requirements in support of SMDS. SMDS provides a packet switched service to End-customers of a LEC. TR-TSV-001060[Bell2] describes generic requirements for Exchange Access SMDS (XA-SMDS), which is intended for customers that are Carriers (i.e., Interexchange Carriers) that use XA-SMDS to provide their end-customers access to their service (e.g., Interexchange SMDS). The relationship among Exchange SMDS, Interexchange SMDS and XA-SMDS is illustrated below.

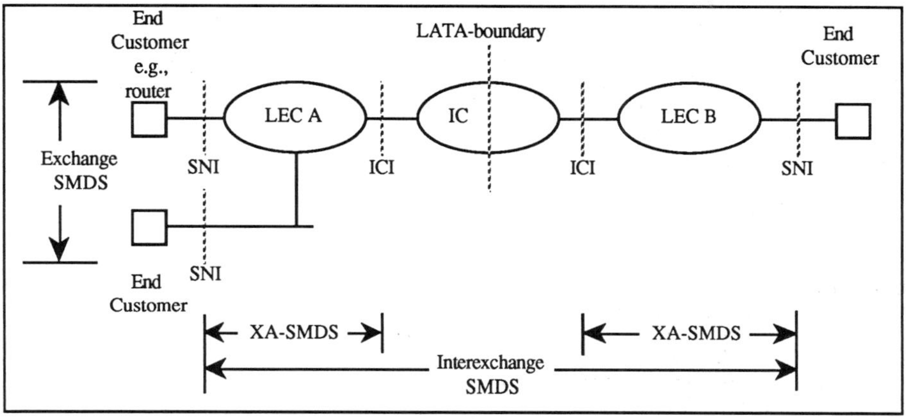

Figure 1: Relationship of XA-SMDS, Exchange SMDS and Interexchange SMDS

In Figure 1, either one of the two Local Exchange Carriers (LECs) may offer originating and/or terminating XA-SMDS to the Interexchange Carrier, which in turn employs the XA-SMDS service to provide end customers directly connected to the LEC's network with access to their Interexchange service.

Note that in Figure 1, the end customers that employ the IC's Interexchange SMDS offering are located in different LATAs. In some jurisdictions, the IC may also offer Interexchange SMDS to connect end customers in the same LATA. Each LEC may also interconnect to other LECs in the same LATA through BCC-Independent LEC (ILEC) serving arrangements. Through these serving arrangements, each LEC may reach destinations served by other LECs in the same LATA. The two LECs are interconnected via ICIs as for example described in the Separate Networks Architecture Model of the BCC-ILEC serving arrangements. Aspects of these serving arrangements are described in TR-TSV-001060[Bell2].

3. GENERAL IDENTIFICATION OF NM SERVICES FOR SMDS

Network Management services associated with Exchange SMDS are limited to a single case, i.e., a LEC offering network management services to the SMDS customer[1]. A scenario of this service, generally referred to as CNM Service, is shown in Figure 2.

A CNM service generally parallels the client-server paradigm, where the customer acts as requester, and the LEC network acts as responder. While Figure 2 represents this

---

[1]This discounts the theoretical though less realistic possibility of the customer offering NM services to a LEC.

communication as terminal/workstation to host/database communication, the paradigm applies equally well to customers calling LEC operators, sending email to LEC operators, and customers sending a test packet to a designated point in the network to test reachability (e.g., use of *ping()* in Unix®). Selection of the appropriate way to communicate between customer and a LEC for CNM is tailored to customer needs, and may differ among different CNM functions.

Note further that the LEC NM capabilities to run the internal LEC network operations typically go beyond those necessary to provide CNM service to customers; the CNM service offers customers a particular *view* of the LEC NM capabilities that is tailored (and restricted!) to the customer's needs. For example, the exact LEC network topology is critical knowledge for LEC operations personnel, while customers may prefer to not be dependent on or familiar with the evolving LEC network topology.

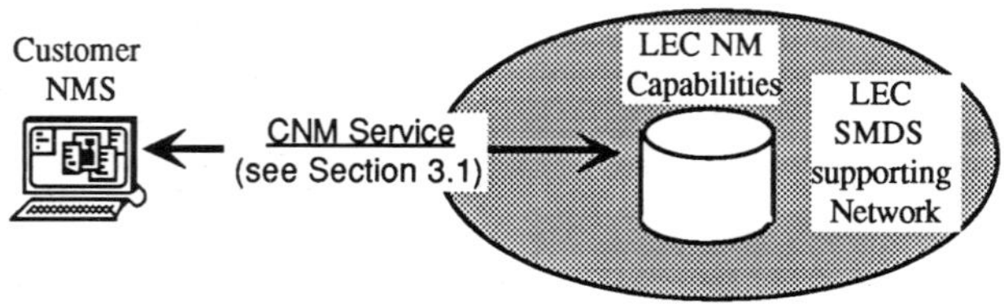

Figure 2: CNM Service associated with Exchange SMDS

The introduction of XA-SMDS fosters Interexchange SMDS, and thus introduces additional parties involved in SMDS traffic, i.e., interexchange carriers (ICs), ILECs, and remote LECs. As in other telecommunications services, involved parties may have to exchange some management information in some form, to ensure smooth traffic flow for SMDS. Each of these different management information exchanges represents a separate NM service. This is illustrated in Figure 3.

Note that the client-server paradigm explained for CNM service applies equally well to these additional services and capabilities; for each service, one party acts as server, offering NM information to parties taking the role as client. Note further that while some of these services may be based on common NM capabilities, different *views* are offered of these NM capabilities.

In practice, there may often be a relationship among these services; some of them may for practical reasons be packaged as a single service, while in other cases one service may be a precondition for the offering of another one (e.g., chained databases). Figure 3 also shows the Sections in this paper that discuss these respective services and capabilities; their mutual relationship will also be discussed in those Sections.

® Unix is a registered trademark of Unix System Laboratories, Inc.

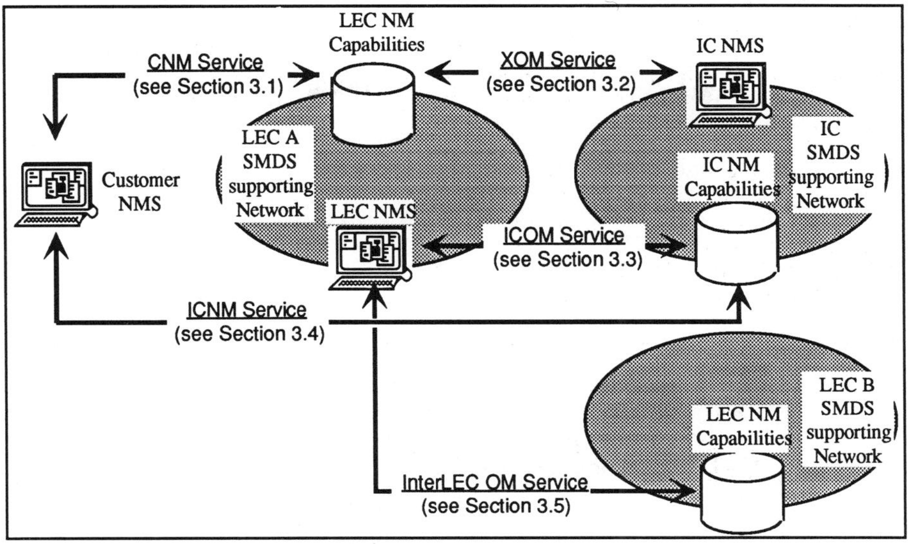

Figure 3: Potential NM Services and Capabilities Associated with SMDS

3.1 CNM Service - NM services provided by the LEC to SMDS End-customers

To assist managers of CPNs in managing their use of LEC-provided SMDS, Bellcore has planned an SMDS Customer Network Management (CNM) service to provide these managers with capabilities to directly manage many aspects of SMDS offered by LECs. Bellcore has published TA-TSV-001062[Bell3] which provides generic requirements for SMDS Customer Network Management (CNM) Service. This TA approaches the service from a platform independent perspective. For example, several access protocols could be used to implement the service. At this point, access to most functions is provided using SNMP. This selection was based on current customer market conditions. Other protocols may be considered in the future as warranted by the market.

Managers of customer premises networks, using a customer premises based[2] Network Management Station (NMS), could use the following capabilities through the CNM service:

- Receive event notifications. Of particular interest are SNI related events (from the LEC perspective). Examples are link Up/Down.
- Retrieve subscription profile information. This includes the identification of SNIs, and information like the address screens, group address list, End-customer Blocking, Carrier Preselection, etc. (see the TA-TSV-001062 Section 4, and the SNMP based Subscription MIB[CT3]).

---

[2]Note that the customer is free to authorize anybody or any organization to manage the network on the customer's behalf, including organizations like ICs. The CNM service is not concerned with this. Outsourcing of a corporation's NM needs is a good business example of this possibility, and complies with TA-TSV-001062.

- Modify subscription profile information. For example, the LEC may allow the end-customer to modify the address screening list.
- Retrieve performance information. In particular the LEC perspective of the SNI behavior (performance, up/down status of the SNI) is of interest (this information is described in TA-TSV-001062 Section 4; monitoring via SNMP is defined in the MIB modules for SIP[CT2], DS1[KB], DS3[CT1]. This information is useful for trouble isolation, for example by comparing it with the peer information kept by CPE.
- Retrieve usage information.

A number of additional service capabilities are under study:

- Request network intervention
- Request event notifications: turning on/off LEC generated events relevant for the end-customer .
- Initiate tests. An example may be loopback on the ICI.
- Trouble ticketing: the end-customer obtains trouble ticket services from the LEC.

The introduction of Interexchange SMDS required some modest extensions to CNM Service. These extensions are already accommodated in TA-TSV-001062 Issue 2. The extensions allow customers to read their carrier selection profile, including their presubscribed carrier, carrier selection errors and End-Customer Blocking situations. Furthermore, customers can obtain status information on the reachability of carriers from the LEC perspective (linkUp/Down)[3].

3.2 XA-OM Service - NM services provided by the LEC to directly connected ICs

LECs, Interexchange Carriers and premises based networks fulfill different roles in the end-to-end chain of networks. For example, a LEC may provide certain services on behalf of the IC (Carrier Preselection, End-Customer Blocking, billing of an IC's Interexchange SMDS offering to end customers). Also, LECs and ICs are connected via one or more ICIs, while LECs and premises based networks are connected via SNIs. The NM view on the LEC network provided by the SMDS CNM service is therefore less appropriate as a service provided to ICs. ICs would likely be interested in NM aspects of Carrier Preselection, End-Customer Blocking, billing, and the detailed status of the ICIs. It is quite possible that ICs may also be interested in certain information relevant to end-customer's subscription to Exchange SMDS, for example, the aggregated sustained information rate subscribed by SNIs that have preselected a given IC. This information assists the IC in forecasting interexchange traffic that may be expected from this LEC network.

The NM service provided by the LEC to Carriers is named XA-OM service - eXchange Access Operations Management service [Bell6]. The XA-OM service may be offered by a LEC to ICs. The LEC will be connected to the ICs via one or more ICIs.

Presently Bellcore is investigating the XA-OM network management needs of carriers that utilize XA-SMDS. These carriers will likely have in place sophisticated network management systems and strategies that monitor, measure, and analyze the performance of their network equipment and transmission facilities. As users of XA-SMDS, these carriers

---

[3]Here is an example where a detailed picture, as relevant for the LEC and IC, is translated into a simple view for the customer. The linkUp/Down refers to whether connectivity exists to a particualr IC via ANY of the existing trunks. In other words, if any of the trunks is up, the customer will read linkUp, and if all trunks are down, the customer will read linkDown.

may also require capabilities for managing their use of XA-SMDS. LEC offerings of a network management service associated with XA-SMDS will make XA-SMDS more attractive to carrier-customers, thus increasing the likelihood that carrier-customers will employ XA-SMDS for accessing their end-customers rather than connecting directly to them. The XA-OM service is intended to provide carrier-customers with capabilities they desire and is meant to reduce the measure of distinction between a carrier's alternatives for accessing end-customers. An high level service description has been specified by Bellcore[Bell6].

Through the XA-OM service, carrier-customers would gain the capability to directly manage information about their use of the XA-SMDS service. The scope of managed information under XA-OM includes subscription profile information, ICI traffic information, maintenance information, and usage information. Some information may be modified directly by carrier-customers, while other information may only be queried without any direct changes.

Managers of IC networks, using a IC network based Network Management Station (NMS), could use the following capabilities through the XA-OM service:

- Receive event notifications. Of particular interest may be ICI-related events (from the LEC perspective). Examples are the ICI counterparts of the SNMP based events defined for CNM (e.g., linkUp/Down).
- Retrieve Carrier subscription profile information. This may include the identification of ICIs, and information like the End-customer Blocking list, the list of end-customers that have preselected that Carrier (for CNM counterparts, see the TA-TSV-001062 Section 4, and the SNMP based Subscription MIB).
- Modify Carrier-subscription profile information. For example, a LEC may allow a Carrier to make certain types of changes to the End-customer Blocking list.
- Retrieve performance information. In particular the LEC perspective of the ICI behavior (performance, up/down status of the ICI) may be of interest. Protocol monitors for network management similar to the ones used for the SNIs for CNM would have to be developed for ICIs (for the SNIs, this information is described in TA-TSV-001062, Section 4; monitoring via SNMP is defined in the MIB modules for SIP, DS1, DS3. The DS3 ICI could use the DS3 MIB and would likely employ the SIP MIB with minor variations. This information is useful for trouble isolation, for example by comparing it with similar information kept by IC.
- Retrieve usage information. Information of interest may be to provide aggregate usage information on the originating and terminating services employed by the requesting IC. Note that this may not be an "on-line" service.
- Request event notifications. One example is the ability to activate notification of a change in the IC preselection of an end user subscription profile, i.e., the end-user subscribing/unsubscribing to that IC.
- Initiate tests. An example may be requesting a Layer 1 loopback on the ICI, or initiating internetwork test packets.
- Trouble ticketing; the IC obtains trouble ticket services from the LEC.

Note that the XA-OM service described here focuses on potential on-line services offered by LECs to ICs. This is just one piece of the much broader operations interaction that will take place between LECs and their carrier-customers[4].

---

[4]For example, consultation with LECs and ICs showed that the ordering of ICIs is likely to be done with the Access Service Request (ASR) form. Thus ordering of ICIs will not be part of the initial XOM service.

NM information on the status and monitoring of SNIs are not part of XA-OM. As noted in Section 3.1, if ICs require access to such information as part of certain outsourcing arrangements with end customers, they would use CNM service as a regular CNM service user acting on behalf of the end-customer.

Both the SMDS CNM service and the XA-SMDS XA-OM service are offered by the LEC, and may share a common OS that will process information collected in various SS NEs and present the information to customers of each network management service in an appropriate form. The CNM service is offered to end-customers of the SMDS service via the SNI or other interfaces (e.g., dedicated link between a customer's NMS and the BCC's NMS), while the XA-OM service is offered to carrier-customers of the XA-SMDS service via a XA-OM interface.

3.3 ICOM - NM service provided by ICs to the LEC

As a counterpart of XA-OM, the LECs may wish to receive similar information from their carrier-customers to assist the LECs in identifying and sectionalizing internetwork troubles. From the service provider's viewpoint, the end user is the end-customer of the LEC's Exchange SMDS service and is at the same time the end-customer of the IC's Interexchange SMDS service. In most cases, the end user will have the need to use both Exchange SMDS and Interexchange SMDS. Thus it is beneficial to both LEC and IC to provide services that appear as an integral service to the end-users.

When an end user's use of Exchange SMDS is interrupted, the end user will typically report the trouble to the LEC offering the service. However, when an end user's use of the Interexchange SMDS is interrupted, the end user may contact either an IC or a LEC. If an IC is contacted first, the IC may use the proposed XA-OM service to obtain information necessary to form an end-to-end perspective of Interexchange SMDS to diagnose the trouble properly. If a LEC is contacted first, the LEC may need to gather information from other networks traversed by the end-user's Interexchange SMDS communication (which may include one or more IC and LEC networks) to sectionalize the trouble properly. This type of information can be offered to the contacted LEC by other carriers as a service equivalent to XA-OM.

The NM service provided by an IC to a LEC is called the Interexchange Carrier Operations Management service. Managers of LEC networks, using a LEC network based Network Management Station (NMS), could use the following capabilities from ICs through such a service:

- Receive event notifications. Of particular interest may be ICI related events (from the IC perspective). Examples are the ICI counterparts of the SNMP based events defined for CNM (e.g., linkUp/Down).
- Retrieve performance information. In particular the IC perspective of ICI behavior (performance, up/down status of the ICI) may be of interest. Protocol monitors for network management similar to the ones used for the SNIs for CNM would have to be developed for ICIs (for the SNIs, this information is described in TA-TSV-001062, Section 4; monitoring via SNMP is defined in the MIB modules for SIP, DS1, DS3. The ICI could employ those MIBs with a relatively small variation). This information is useful for trouble isolation, for example by comparing it with the peer information kept by LEC (each shows the interface behavior from their local end).
- Request event notifications. One example is the ability to activate notification of a change in the operational status of the IC network.

- Initiate tests. An example may be requesting a Level 1 loopback on the ICI, and test packets initiated by a LEC and returned by an IC. This could for example be needed after an event notification of a malfunctioning ICI.
- Status information about connectivity with remote LECs.

These management capabilities mirror those of the XA-OM capabilities, but provide the carrier perspective on issues like ICI performance data. Also, features such as Retrieve Subscription Information, Retrieve End-customer Blocking List, and Retrieve Usage Information would not be applicable in this service[5].

3.4 ICNM - NM services provided by Carriers to End-customers

Carriers other than the LECs may also offer network management services to their end-users. In Figure 3, for example, the end-users are customers of a LEC for Exchange SMDS, but are also customers of the IC for Interexchange SMDS (several network access functions are still performed by the LEC). Consequently, the IC may offer end-users a network management service that assists them in managing their use of Interexchange SMDS.

The NM service provided by an IC to end customers is called ICNM Service - Interexchange carrier Customer Network Management Service. The exact contents of such service is of course speculative here. However, it would be in the interest of end customers for this service to complement CNM service. In that case, managers of customer premises networks, using a customer premises based Network Management Station (NMS), could use the following capabilities from ICs through the ICNM service:

- Receive event notifications. Of particular interest may be LEC-IC connectivity related events (from the IC perspective); the LEC-IC connectivity[6] counterparts of the SNMP based events defined for CNM (e.g., linkUp/Down). Note that this may include the IC connectivity events with a variety of LECs.
- Retrieve usage information. Information of interest may be to provide end-customers with aggregate interexchange usage information for a particular IC.
- Request event notifications. One example is the ability to activate notification of a change in the operational status of the IC network.
- Initiate tests. An example may be test packets initiated by the end-customer and returned by the IC.
- Trouble ticketing; the end-customer obtains trouble ticket services from the IC.
- Status information about connectivity with remote LECs.

3.5 InterLEC OM service - NM services provided by a LEC to another LEC

A LEC may wish to receive network management information from another LEC (including ILECS), for example, to assist the LEC in identifying and sectionalizing

---

[5]In practice, the scope of such a NM service always depends on the business interests of the service provider to provide network management information to the LEC. When the LEC offers XOM to the IC, the network management service may be offered in conjunction with the LEC's offering of XA-SMDS to the IC. On the other hand, the IC is offering the Interexchange SMDS service to the end user, not directly to the LEC. If an IC has a business interest to offer such a NM service to the LEC, the desirable capabilities would be similar to those available through the XOM service, except for features specifically related to the offering of XA-SMDS. The winner would be the end customer, who would need only a single call for trouble resolution.

[6]I.e., the combined number of applicable ICIs.

internetwork troubles. For example, when an end-user contacts a LEC for service interruption in the Interexchange SMDS service, the contacted LEC may use ICOM service to obtain NM information from the IC and a similar NM service to obtain NM information from the other LEC involved in the end-user's Interexchange communication. The contacted LEC may use this information to form an integrated perspective of the end user's end-to-end communication and diagnose the trouble properly[7].

One example of the use for such a capability is related to the trouble sectionalization capability in internetwork testing (see Appendix A in TR-TSV-001060[Bell2]). When an end user's Interexchange service is interrupted, one of the involved end users may contact a LEC for trouble shooting. That LEC would check its own network for trouble (depending on the trouble symptoms described by the end-user). If no troubles are found, the LEC may use the InterLEC OM service to verify conditions in the remote LEC.

The NM service provided by the LEC to another LECs is named InterLEC OM service. Managers of LEC networks, using a LEC network based Network Management Station (NMS), could use the following capabilities from LECs through such a NM service:

- Receive event notifications. Of particular interest may be remote LEC-IC connectivity related events; the LEC-IC connectivity[8] counterparts of the SNMP based events defined for CNM (e.g., linkUp/Down). Note that this may include the remote LEC connectivity events with a variety of ICs.
- Request event notifications. One example is the ability to activate notification of a change in the operational status of the remote LEC network.
- Initiate tests. An example may be the sectionalization of customer reported troubles.
- Trouble ticketing; the LEC obtains trouble ticket services from the remote LEC. For example, the remote LEC may have received a Trouble Ticket request from its customer, and after trouble sectionalization it is determined that the trouble is in the other LEC. The Trouble Ticket may then be handed off to the other LEC.
- Status information about connectivity with ICs.

4. SUMMARY

This paper identifies and discusses five potential network management services and capabilities that may be important for the offering of SMDS. The CNM Service is the most relevant for SMDS end customers, and therefore has already been defined in TA-TSV-001062, and is on its way to full deployment. However, both the end-customer and the LECs would still gain by the offering of the other NM services (assuming that the costs are not excessive), that could be used to complement the CNM service, and would make SMDS even

---

[7]Since all the LECs involved in the Interexchange communication serve one or more end-users, these LECs play equal roles to one another and will likely have different incentives to provide InterLEC OM information exchange capabilities as compared with the IC's willingness to provide IC-to-LEC information exchange capabilities to the LEC (Section 3.3). Although a LEC does not offer communication service to other LECs directly, each LEC offers both originating XA-SMDS and terminating XA-SMDS to the IC in its Interexchange SMDS offering. At the same time, each LEC offers Exchange SMDS to its end-customers. In order to provide customer satisfaction, each LEC will be motivated to offer an InterLEC NM information exchange capability to other LECs. The scope of the InterLEC OM information exchange capability may very well be similar to that of the XOM service offered by the LEC to ICs, with appropriate modifications of service features and service interface.

[8]I.e., the combined number of applicable ICIs.

more attractive. LEC offerings of a network management service associated with XA-SMDS will make XA-SMDS more attractive to carrier-customers, thus increasing the likelihood that carrier-customers will employ XA-SMDS for accessing their end-customers. Also, the LEC's offering of the XA-OM service to the IC would enable the IC to gain end-to-end perspective of its offering of Interexchange SMDS and provide end users with more efficient service. Likewise a LEC's offering of InterLEC OM service to other LECs would strengthen the LEC's reputation as a service provider to end users.

It is important for technical reasons to make a distinction between each of the five NM services described in this document. This paper describes some possible functions for each. It may, in individual cases, be desirable to appropriately package combinations of these NM services.

CNM has been well documented in TA-TSV-001062. XA-OM (LEC-to-Carrier) has been discussed with interested parties. ICOM service and InterLEC OM service are more speculative at this point. The specification of the ICNM service is completely determined by the ICs.

5. REFERENCES

[Bell1] Generic System Requirements in Support of Switched Multi-megabit Data Service, Bellcore Technical Reference TR-TSV-000772, Issue 1, May 1991.

[Bell2] SMDS Generic Requirements for Exchange Access and Intercompany Serving Arrangements - Bellcore TR-TSV-001060, Issue 1, December 1991

[Bell3] Generic Requirements for SMDS Customer Network Management Service, Bellcore Technical Reference TA-TSV-001062, Issue 2, February 1992.

[Bell4] Local Access System Generic Requirements, Objectives, and Interfaces in Support of Switched Multi-megabit Data Service, Bellcore Technical Reference TR-TSV-000773, Issue 1, June 1991.

[Bell5] SMDS Operations Technology Network Element Generic Requirements, Bellcore Technical Reference TR-TSV-000774, Issue 1, March 1992.

[Bell6] A Framework of High-Level generic requirements for SMDS Exchange Access operations Management (XA-OM) Service - Bellcore Framework Advisory FA-TSV-001237 Issue 1, October 1992.

[Brunner] T. O. Brunner, A demonstration of SMDS Customer Network management Features, IEEE Proceedings 17th Conference on Local Computer networks, Minneapolis, September 13-16, 1992.

[CFSD] J. Case, M. Fedor, M. Schoffstall, and C. Davin, Simple Network Management Protocol, RFC1157, Network Information Center, SRI International, Menlo Park, CA, May 1990.

[CPT] SNMP-based Customer Network Management (CNM) for SMDS - Tracy Cox, David M. Piscitello, Kaj Tesink - IEEE Network Management Magazine, September 1991

[CT1] T.A. Cox & K. Tesink (editors), Experimental Definitions of Managed Objects for the DS3 Interface Type, RFC1233, Network Information Center (NIC), SRI International, Menlo Park, CA, May 1992.

[CT2] T.A. Cox & K. Tesink (editors), Experimental Definitions of Managed Objects for the SIP Interface Type, RFC1304, Internet Engineering Task Force, Network Information Center (NIC), SRI International, Menlo Park, CA, March 1991.

[CT3] T.A. Cox & K. Tesink (editors), Definitions of Managed Objects for SMDS Subscription, Version 2.0, Enterprise-specific MIB, Bellcore posting via the Internet Assigned Numbers Authority on host venera.isi.edu, February 1992.

[KB] C. Kolb & F. Baker (editors), Experimental Definitions of Managed Objects for the DS1 Interface Type, Internet Draft, RFC1232, Network Information Center (NIC), SRI International, Menlo Park, CA, May 1991.

[JLT] T. B. Jennings, J. C. Lawrence, K. Tesink, practical SMDS demonstrations, IEEE Proceedings 17th Conference on Local Computer networks, Minneapolis, September 13-16, 1992.

[RM] Management Information Base for Network Management of TCP/IP-based Internets: MIB-II - Marshall Rose, Keith McCloghrie (editors), RFC1213, Network Information Center (NIC), SRI International, Menlo park, CA, March 1991

6. ACCRONYMS

CNM	Customer Network Management
CPE	Customer Premises Equipment
CPN	Customer Premises Network
IC	Interexchange Carrier
ICI	Interexchange Carrier Interface
ICNM	Interexchange carrier Customer Network Management
ICOM	Interexchange Carrier Operations Management
ILEC	Independent Local Exchange Carrier
LATA	Local Access and Transport Area
LEC	Local Exchange Carrier
MIB	Management Information Base
SIP	SMDS Interface Protocol
SMDS	Switched Multi-megabit Data Service
SNI	Subscriber Network Interface
SNMP	Simple Network Management Protocol
XA-SMDS	Exchange Access SMDS
XA-OM	eXchange access Operations Management

V.

TELECOMMUNICATIONS MANAGEMENT

B.

IMPLEMENTATION DIRECTIONS

Integrated Network Management, III (C-12)
H.-G. Hegering and Y. Yemini (Editors)
Elsevier Science Publishers B.V. (North-Holland)

Object-Oriented Intelligent Telecommunications Network Management System

Jason L. Tsay

AT&T Bell Laboratories
101 Crawfords Corner Rd., Holmdel, NJ 07733, USA

Abstract

The design and development of an intelligent network management system (INMS) for an advanced telecommunications network involves several teams, of which each has its own perception of a problem domain and its own discipline regarding solutions. A domain model, based on the simple and powerful Entity-Relationship (E-R) diagram, provides everyone involved in the system with a common abstraction of the problem domain. When one extends the domain model into the subsystem specific-design models, (also based on the E-R diagram), all software subsystems achieve integrity of design. With these two models, the fundamental concepts in the problem domain are mapped directly into the design, enabling everyone involved in the system to discuss the domain with a common vocabulary. Furthermore, the object-oriented (OO) design is mapped directly into the OO programming language C++; as a result the intelligent network management system is better and is developed more quickly than other approaches. The object-oriented approach improves software productivity and quality; it also supports and enhances rapid prototyping, incremental design/development, and user programmability.

Keyword Codes: D.1.5; D.2.10; I.2.1
Keywords: Object-Oriented Programming; Design; Applications and Expert Systems

1. Introduction

A telecommunications network is a complex aggregate of electronic gear (switches) and transmission systems, which together provide a multiplicity of channels over which many customers' messages (voice, data, and video images) and associated control signals can be transmitted. The goals of network management are to (1) maintain efficient, reliable operations, and (2) increase the performance of the network in terms of the quality and quantity of service provided to its customers.

Rapid technological changes in computing and telecommunications industries are expanding customer expectations and service requirements for the public network. Meeting these needs will require several major changes in the network, notably digitization, ISDN (Integrated Services Digital Network), SDH (Synchronous Digital Hierarchy) facilities, and high-bandwidth networking supported by fiber optics. Figure 1 shows advanced public telecommunications network architectural components connected by optical fibers.

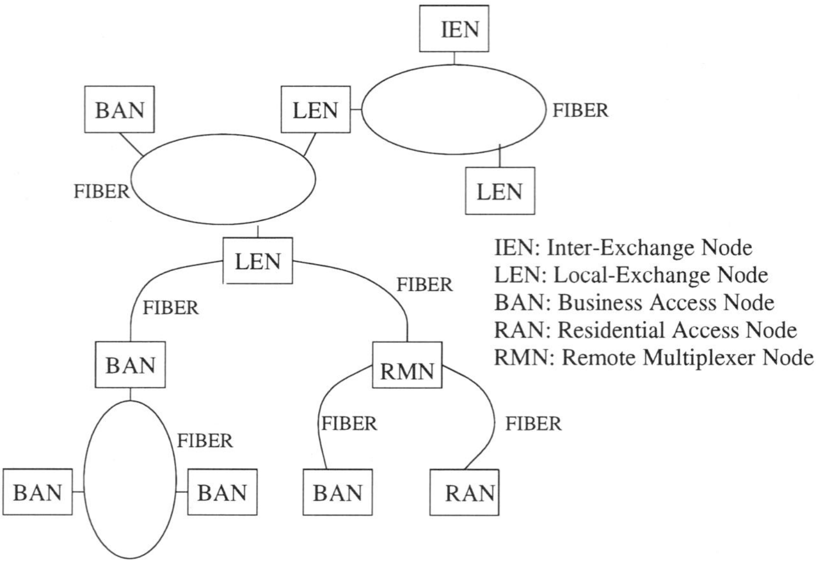

Figure 1. Advanced public network's architecture components connected by high-bandwidth optical fibers.

2. Network Management

Although there are several definitions of network management, the International Standards Organization's (ISO) network management standards consider the following five functional areas:

- Configuration Management: provisioning functions, status and control functions (including restoration), installation functions.
- Fault Management: alarm surveillance, failure localization functions, testing functions.
- Performance Management: performance monitoring functions, traffic and network management functions, quality-of-service-observation functions.
- Accounting Management: billing functions.
- Security Management.

All network management performed by a telecommunications operating company falls under the general classification of Operations, Administration, Maintenance and Provisioning (OAM&P). The major components of an OAM&P activity are service provisioning, network provisioning, performance monitoring, maintenance, and planning.

The ANSI (American National Standard Institute) T1 Committee has published a standard that defines fault management messages for interfaces between Operations Systems (OS) and Network Elements [2]. This article addresses the management of telecommunications networks through advanced OAM&P. Specifically, it presents an intelligent network management system (INMS) or an intelligent OS (IOS) that integrates expert system technology into conventional OSs.

3. Expert System Technology

Traditional approaches to network management are exhibiting great difficulties and may not be able to work in the future [14]. The complexity and diversity of telecommunications systems providers demand experimentation with new approaches to solve these problems--using methods that will extend the capability of the network operations involved. One of these methods involves the use of AI (Artificial Intelligence) techniques. The techniques focus on the use of declarative (factual) knowledge and relatively simple rules of inference in order to put that knowledge to work on a specific problem. AI is a powerful method when it is applied to the right domain. One of the AI techniques, expert systems, captures domain expert problem-solving strategies and selectively applies them to that domain.

The advantages of expert systems -- twenty-four-hour availability, automation of diagnostic reasoning, distribution of expertise -- have been demonstrated in the day-to-day network management domain [8, 15]. The success of this expert system technique has not led to the integration of the technique into the conventional network management systems; most expert systems are still hosted in special hardware platforms that are different from those used by network management systems. The separation is mainly for the following reasons:

- Platform: Some expert systems require special hardware and software platforms that are different from those used by conventional software systems; integration may be difficult to achieve.
- Perception: AI is perceived as a new and untested technology--a risky technology.
- Methodology: Most expert system development does not follow conventional, understood software development and management methodology.

The gradual convergence of AI and software engineering will pave the way to the development of intelligent network management systems. Notably, object-oriented technology and rapid prototyping methodology are more accepted and currently are being applied by both AI and the software engineering communities.

4. Entity-Relationship Diagram-Based Models

There are several driving forces for using E-R diagrams to describe domain and design models.

4.1 ANSI Standards

The ANSI T1 Committees have applied OO data modeling techniques to the development of standards for the OAM&P of telecommunications networks [2]. The use of object-oriented data modeling techniques is launching the North American telecommunications network into an open, standard, multi-vendor communications environment.

In the 1990 version of Generic Network Model [2], Entity-Relationship (E-R) diagram techniques were used to represent the way resources or entities can be combined or interrelated, as shown in Figure 2.

4.2 Knowledge Acquisition

Expert system design needs domain experts and knowledge engineers to start with knowledge acquisition. The knowledge acquisition process is in fact like the domain analysis and system design in software engineering methodology.

Though model-based reasoning (an OO approach) was applied to the development of expert systems [11], function-oriented prospective reasoning (what shall I do next?) dominates most

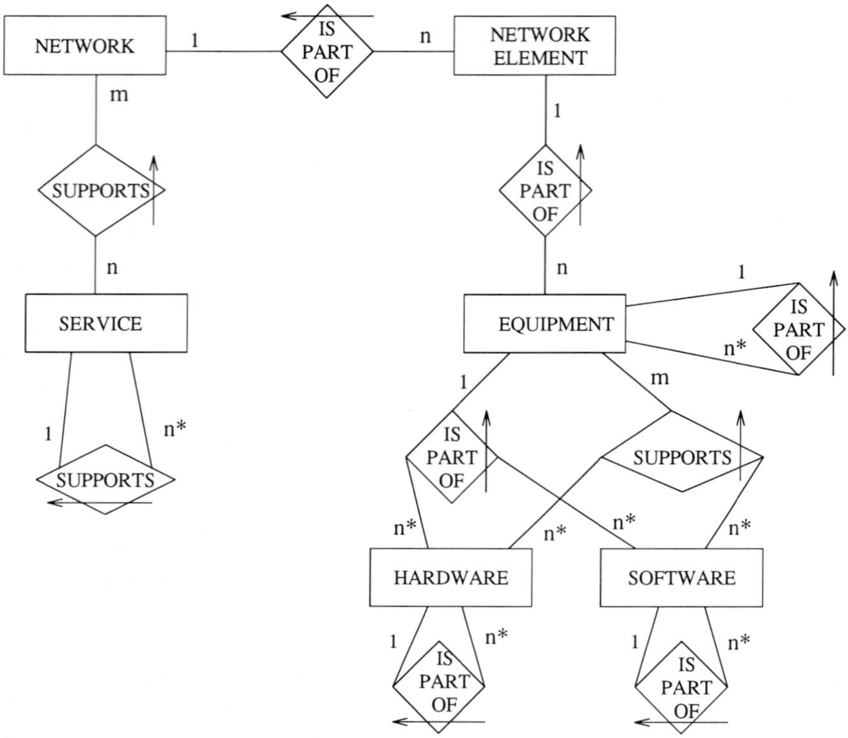

Figure 2. Entity-Relationship diagram of the ANSI T1' Generic Network Model--a partial figure.

expert system applications [14, 16]. Arising originally through AI research, the OO paradigm is currently one of popular AI paradigms and is penetrating the commercial programming market. For example, the KEE^{TM1} (Knowledge Engineering Environment) is one of the largest and most advanced OO expert system development tools currently available commercially [19]. For each object, the system has rules defined and stored in that object's rule attribute.

Knowledge acquisition emphasizes real-work things or objects and their attributes, instead of just domain-problem-solving strategies, procedures, or rules of thumb. Ideally, knowledge acquisition should have the following steps [3]:

- Identify the physical and conceptual objects that characterize the decision-making process. Data-modeling techniques can accomplish this goal. They identify the physical entities,

1. KEE and Knowledge Engineering Environment are trademarks of Intellicorp.

conceptual entities, and event or state entities and their relationship to one another. The entities and their relationships can be captured and documented graphically by the E-R diagram.

Ultimately, all or part of the data model is used for logical database design. Because almost all expert systems in telecommunications reason about existing network management databases [15], it is easy to see the benefits of using the same data modeling technique in knowledge acquisition activity.

- Uncover the reasoning process used by the experts for decision making: E-R models incorporate important semantic information about the applications. Thus, an E-R diagram not only identifies all relevant entities and their attributes but also uncovers many of the rules, procedures, and constraints used in decision making.
- Identify the types and sources of knowledge the experts use: An expert system incorporates three basic types of knowledge: factual (data), judgemental (rules), and procedures or control knowledge. Data modeling identifies the factual knowledge and describes the relationships that govern the data. These relationships are used to identify some of the expert-system rules and constraints and determine data supertypes and subtypes (class and object inheritance hierarchies).

However, when knowledge engineers develop rules, they tend to examine the goals to be achieved and then design rules to carry out the necessary actions to satisfy the goals or to set up subgoals [5]. For each action related to a goal, a set of rules, termed a rule cluster, is developed. The goal-driven or task-oriented knowledge acquisition and construction causes "three C" problems--consistency, correctness, and completeness--in the expert system.

4.3 Database Management

Database designers tend to see the world in terms of persistent, monolithic tables of information, whereas application developers tend to see the world in terms of its flow of control [4]. The database designer/developers are concerned more about the data model--how information can be represented and manipulated within the formal framework of a database system [9]. The data model consists of a collection of object types, operators, and general integrity rules. In the relational model, the objects are relations and domains, and the operators are relational algebra.

Database technology has contributed to the object model primarily through the ideas of the Entity-Relationship approach to data modeling [4, 6, 17]. The E-R diagram succinctly depicts the basic data elements in the system and describes the relationships between them; if the diagram is complete, the database structure can be derived directly from it.

5. Information Models

To provide a common conceptual view of the telecommunications network management domain in its entirety, an E-R-diagram-based information model is clearly required. The previous section has offered several reasons for the choice of the E-R diagram. That diagram is the representation of a semantic data model of the problem domain [12]. A semantic data model is so named because it allows the representation of a broader range of information than other data models. Furthermore, each subsystem's design specific models, based on the common conceptual model, are needed by software designers/developers to describe a more detailed data modeling information for each software subsystem.

We use two levels of abstraction: the low level is the domain model and the another level, expanded from it, is the design model. Although the domain model is modified and expanded to subsystems' design models, the basic object structure is retained. Collectively, the E-R-diagram-based domain and design models can adequately describe the existence and meaning of the key abstractions that form them.

The focus here is on the information models that document classes--attributes and operations (or services as used in [7])--and their associations with each other. These models provide a static view of the classes of objects within the domain and their associations with one another. The dynamic view of the interaction of the classes of objects over time can be provided by Object State Models and Object Communication Models [4, 18].

The domain models will get more elaborate as the design progresses. The same generalization/specialization process of the object class design [12] is also applied to the assignment of OO data to either domain or design models. Generalization supports the exploitation of commonalities between the subsystems' design models. When two or more design models have the same submodels, the submodels will be moved down to the domain model to represent some common conceptual view of the domain. Thereafter, the newly expanded domain model is available for other subsystems to reuse.

5.1 Domain Models

The domain model furnishes a common view or data model for all the teams mentioned in this article. System engineers define the domain model, using the E-R model, in the system requirement documents. The data model, providing the same common abstractions, is not only for software developers and testers but also for end users (e.g., network managers and technicians).

5.2 Design Models

Design models are extensions of the domain model, which provides the common conceptual data model of the system domain. The extensions are based on the design of each software subsystem. For instance, Figure 3 shows four disjointed subsystem specific data models for four software subsystems. More specifically, the database design model consists of both the domain model and the database-specific design model. Since the design models are extensions of the domain models, the common E-R diagram or notation is applied to all design models for consistency and uniformity.

5.2.1 User Interface Design Model For the graphical user interface (GUI) subsystem, the following are some examples of object classes that are unique to GUI design models: window, menu, colors, keyboard/text, mouse/pointer, help, shape, button, etc. These GUI-specific object classes are not concerns of other subsystems. Thus both the domain- and GUI-specific design models can describe the complete design model for the GUI.

GUI provides layered multidimensional network views offering a range of network relationships and perspectives. The highest layer network view of INMS resembles the figure shown in Figure 1. Each object represents one fiber center office, a network element, or a circuit pack. The shapes, colors, and links reveal many important attributes, behaviors, and relationships of the visible objects. For example, in network surveillance, end users can view the total network topology, performance, health, and status. If needed, the users can zoom in to obtain a more detailed view of the interested object or cluster of objects. Furthermore, equipping the point-and-click device with on-screen menus that appear and disappear under a pointing-device control, users can quickly select and interrogate a concerned individual or cluster of objects.

The GUI's object-oriented presentations are essential and necessary to effectively manage a complex telecommunications network. In contrast to the character- or command-line-based user interface, end users are constrained to perceive the network management system as function-

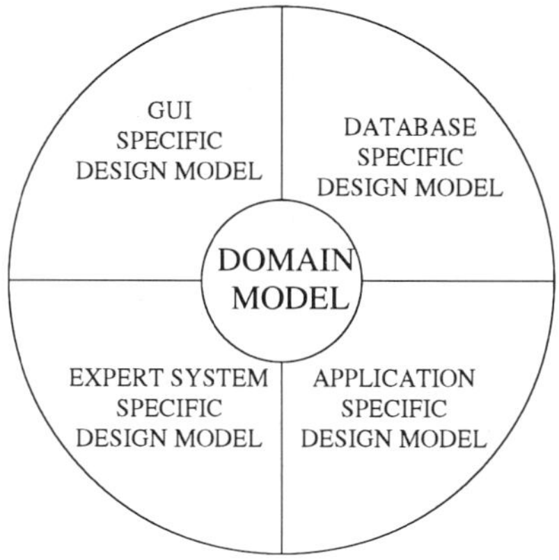

Figure 3. Four design models are extension of INMS' domain model, the common conceptual data model.

oriented: the system performs certain functions and users use it to "manage" or "operate" the network. The limitations of the display and input devices present no or blank network views to the users. Since no associations can be established between what the display device reveals and the users' conceptual view, it is difficult for end users to effectively manage a complex network.

5.2.2 Expert System Design Model Like the other subsystems, the expert system design model is an extension of the domain model. Each object class has rule attributes for storing a set of rules or procedures for intelligent processing, such as service provisioning, alarm correlation, and trouble sectionalization. The design model also contains special rule objects for defining and storing control strategies/knowledge and problem-solving rules.

Rule-based knowledge representation is commonly adopted in the telecommunications area, especially in network maintenance and service provisioning. The main reason is that most domain experts and experienced technicians perceive that the telecommunications domain consists of many objects and each has special attributes, behavior, and rules/procedures for operations.

The rule-based representation using an object-oriented approach provides common conceptual frameworks for both domain experts and knowledge engineers. As long as domain experts and knowledge engineers have a common understanding of the domain and its rule attributes, "three C" problems can be avoided.

5.2.3 Application Processing Design Model To perform enhanced and intelligent OAM&P, the INMS's application processing subsystem needs to expand the common domain model to construct the subsystem specific design model. For instance, in fault management for processing network alarms, the design model should include more object classes than the domain model, such as alarm message log, message suppression, alerting and display.

Some data submodels of the design model, such as message log and report, initially defined in the application design model, will be generalized and move down to the domain model in order to be possibly applied to the database and the GUI subsystems.

5.2.4 Database Design Model Identifying the key abstractions of a database is much like the process of identifying classes and objects in object-oriented design [4]. After an E-R data model is used in the domain model, the foundation for the relational database model is almost completed. The database design model will get more elaborate as the design progresses. New entities will be added, and the description will gain precision as the E-R model starts to take shape.

6. Software Platform

Figure 4 shows the layered software architecture with UNIX[TM2] as the kernel of the software platform. The architecture consists of the following subsystems for the intelligent network management system:

- GUI: X-window and C++ based.
- Application processing subsystem: C++ based.
- Expert system subsystem: based on OPS83[TM3] and C++
- Database management subsystem: based on commercial products such as ORACLE[TM4] , INFORMIX[TM5] , or other relational DBMS.
- Network Element Interface Subsystem: based on X.25 PDN (Public Data Network) for synchronous communication and DATAKIT[TM6] for asynchronous communication. The system will evolve into an OSI (Open Systems Interconnection) 7-layer protocol stack whenever the NE supports it.
- Distributed Networking Subsystem: The system can be centralized or distributed, depending on the network configuration and customer requirements. The distributed architecture, based on IEEE 802.3 standard local area network (LAN), supports distributed servers or subsystems on different hosts over LAN network connections. The communications platform supports transparent and explicit routing between clients and servers, encoding/decoding routines, and connection resources management.

The client/server model of these subsystems is also revealed in the figure; the outer layer subsystems are clients of the servers that reside in the inner layers. This structure not only provides the expandability and transparency of the hardware and software platform but serves to allocate and manage the resources of the development teams.

7. Advantages

The object-oriented approach, using the E-R diagram to define the domain and design models, has many significant advantages over the conventional function-oriented approach.

---

2. UNIX is a trademark of UNIX Systems Laboratories.
3. OPS83 is a trademark of Production Systems Technologies, Inc.
4. ORACLE is a trademark of Oracle Corporation.
5. INFORMIX is a trademark of Informix Corporation.
6. DATAKIT is a trademark of AT&T.

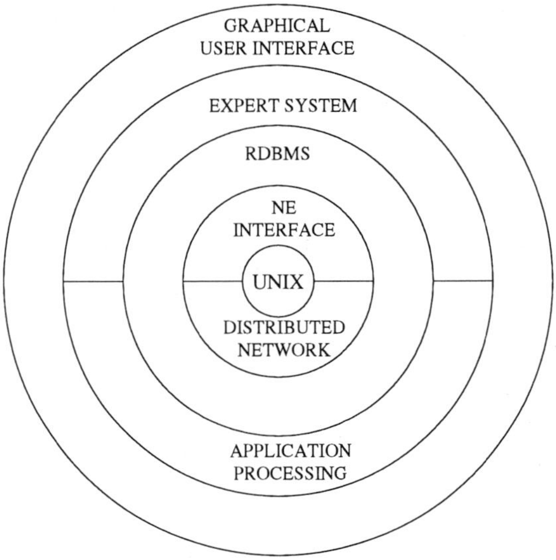

Figure 4. Layered software architecture; the outer layers are client and the inner layers are servers.

7.1 Consistent and Common Conceptual Model

The E-R-diagram-based domain model, defined by system engineers in the requirements, provides a consistent and common conceptual model for the network management end users, domain experts, and all software subsystem designer/developers/testers. Everyone involved in the system can access and understand the common high-level view of the abstractions relevant to the problem domain and solutions. The common models expedite the information exchange and knowledge transfer in all the system development phases.

7.2 Rapid Prototyping and Incremental Design/Development

To shorten the new system delivery cycle, the conventional waterfall development model is no longer appropriate. INMS is created through the use of a "round-trip gestalt design" [4]. This style of design emphasizes the incremental and iterative development of a system through the refinement of different yet consistent logical and physical views of the system as a whole.

Rapid prototyping is also used extensively to develop and define new requirements. Without using rapid prototyping as a tool to study the effectness and feasibility of new features, system requirements in some areas of INMS are difficult to specify, especially in areas such as alarm correlations and automated service provisioning.

The OO approach to rapid prototyping results in an evolutionary prototyping, which can provide the framework and basis for subsequent design and development. By contrast, the function-oriented throw-away prototyping starts and ends with no reusable components/facilities. The reason that both inheritance and polymorphism, supported by the OO design and programming, enable the rapid prototyping to occur is that the derivation of objects and operations from already developed objects and operations often takes no more than a few lines of program code. The prototyping starts scanning and analyzing the existing database or library of objects and operations defined in the domain model and all four design models.

The OO approach furnishes not only a good paradigm for incremental design and development, but also an effective tool to conduct requirement analysis and specification via rapid prototyping. Moreover, the prototype can then be evolved and integrated into the final product.

7.3 User Programmability

As more complex telecommunications networks evolve, users will request more control and feasibility of the management system. User programmability is becoming an important feature of the system; it must have the strengths of flexibility and adaptability to meet the end users' needs, especially in the following three areas:

- User interface: The OO presentation of GUI reflects the end users' OO perception of the domain. Therefore, the definitions and control of all visible objects and their attributes, behavior and operations are easily understood by the end users. In addition, users not only are given control on user-modifiable objects but are allowed to create and define new objects from existing or sample sets of objects.
- Knowledge bases: Separating rules/procedures from the control rules and storing rules/procedures in the domain's object class make the modification and customization possible and easy.
- User defined reports: Most off-the-shelf relational database management systems provide report generation utilities. Since the OO domain model reveals the needed information for the internal data model and tables, it makes report writing easier.

7.4 Productivity and Reliability

Like other OO systems, code reuse improves the system's development productivity and the product's reliability [13]. This is extremely important for INMS to support multi-vendor network elements. Code reuse is supported by two C++ main language concepts: polymorphism and inheritance [1, 10, 12].

System testing is also enhanced with the OO approach. Without the OO domain and design models, most system testing cases are derived from the requirements and regard the system under test as a black box. Since the domain and design models are available to the system testers, more thorough and complete test cases can be derived and implemented in a shorter interval.

REFERENCES

[1] Al-Haddad, H. M., George, K. M., and Samadzadeh, M. H., Approach to Reusability in C++ and Eiffel, *Journal of Object-Oriented Programming*, Sept. 1991.

[2] ANSI T1.214-1990, American National Standard for Telecommunications -- Operations, Administration, Maintenance and Provisioning (OAM&P) -- Generic Network Model for Interfaces between Operations Systems and Network Elements.

[3] Blackman, M., "CASE for Expert Systems," *AI EXPERT*, Feb. 1990.

[4] Booch, G., *Object Oriented Design*, The Benjamin/Cummings, 1991.

[5] Brownston, L., Farrell, R., Kant, E., and Martin, N., *Programming Expert Systems in OPS5*, Addison-Wesley, 1985.

[6] Chen, P., *The Entity-Relationship Approach to Logical Data Base Design*, Q.E.D. Information Science, Wellesley, Massachusetts, 1977.

[7] Coad, P., and Yourdon, E., *Object-Oriented Analysis*, Second Edition, Yourdon Press, 1991.

[8] Corn, P., Dube, R., McMichael, A., and Tsay, J., "An Autonomous Distributed Expert System for Switched Network Maintenance," *Proceedings of the IEEE Global Telecommunications Conference, 1988*.

[9] Date C. J., *An Introduction to Database Systems*, VOLUME I, Fifth Edition, Addison-Wesley, 1991.

[10] Jordan D., "Implementation Benefits of C++ Language Mechanisms," *Communications of ACM*, Vol. 33, Number 9, Sept. 1990.

[11] Khan, N, Callahan, P., Dube, R., Tsay, J., and Van Dusen, W, "An Engineering Approach to Model-Based Troubleshooting in Communication Networks", *IEEE JOURNAL ON SELECTED AREAS IN COMMUNICATIONS*, VOL 6, NO 5, JUNE 1988.

[12] Korson, T. and McGregor, J., Understanding Object-Oriented: A Unifying Paradigm, *Communications of ACM*, Vol. 33, Number 9, Sept. 1990.

[13] Lewis, J. A., Henry, S. M., et al, An Empirical-Study of the Object-Oriented Paradigm and Software Reuse, *SIGPLAN Notices*, 1991.

[14] Liebowitz, J. (Ed.), *Expert Systems Applications to Telecommunications*, John Wiley & Sons, 1988.

[15] McMichael, A., Thien, J., and Tsay, J., "TERESA: An Integrated System for Network Diagnoses," *Proceedings of the International Conference on Data Base and Expert Systems Applications (DEXA 90)*, August 1990, Vienna, Austria.

[16] Prerau, D., *Developing and Managing Expert Systems*, Addison-Wesley, 1990.

[17] Shlaer, S., and Mellor, S., *Object-Oriented Systems Analysis: Modeling the World in Data*, Englewood Cliffs, NJ, Prentice-Hall, Inc., 1988.

[18] Shlaer, S., and Mellor, S., *Object Lifecycles: Modeling the World in States*, Englewood Cliffs, NJ, Prentice-Hall, Inc., 1992.

[19] Tello, E. R., *Object-Oriented Programming for Artificial Intelligence: A Guide to Tools and System Design*, Addison-Wesley, 1989.

Integrated Network Management, III (C-12)
H.-G. Hegering and Y. Yemini (Editors)
Elsevier Science Publishers B.V. (North-Holland)

Managed Object definition for Customer Network Management in public data networks

Hiroshi Sunaga, Shuji Tomita

NTT Communication Switching Laboratories, 3-9-11 Midori-cho, Musashino-shi, Tokyo 180, Japan

Abstract

This paper presents the basic concepts of defining managed objects (MOs) for the Customer Network Management Service currently studied in CCITT SG VII. First, we show the Customer Network Management architecture and service menus. Based on this, principles of MO definition are described. MOs are defined for Customer Network Management services such as the management of fault occurrence, fault notification, tests, subscription, permanent virtual circuits, and accounting. These MOs are defined as generic MOs so that they can be used and extended for providing specific Customer Network Management services of a network operating company.

Keyword Cords: C.2.2: C.2.3; H.5.2
Keywords: Network Protocols; Network Operations; User Interfaces

1. Introduction

CCITT SG VII is currently studying a new type of management service for customers of a public data network (PDN). This service is called the Customer Network Management (CNM) service, and it allows customers to access management information or functions of the network services to facilitate each customer's own management. This service could offer higher level functions for customers than the usual communication between them and the network operator, where, for example, complaints are submitted by telephone and bills are delivered by mail. The architecture and services to be provided are being defined and will be approved as new Recommendations [1], [2], in the early stages of the CCITT tenth study period (1993-1996).

The definition of managed objects (MOs) is closely linked to implementation of management services based on the OSI Systems Management, including CNM service. We must consider this point when defining MOs and management information, because they specify the service behaviour, information elements, and their data structure (ASN.1 base), all of which are essential for the implementation. CCITT SG VII is currently discussing the definition of management information for public data networks, focussing on packet switched networks, and this will become a Recommendation after the architecture and service Recommendations are completed [3].

This paper presents the basic architecture and services for CNM, and the basic concepts for defining managed objects for the CNM services. Then, we show several MOs and items of management information. Since the MO definition for the CNM service is currently under discussion in SG VII and has not yet been fully approved, we will present mainly MOs that we have proposed, and some additional MOs for enhancing CNM services.

2. Outline of the Architecture and Services

2.1 CNM-related Recommendations

Three CNM-related Recommendations have been drafted, for its architecture, services, and the definition of MO and management information:

(1) The architecture Recommendation (X.cnma) defines the architecture and models for supporting CNM services making use of the concept of the systems management, layer management, and TMN. It also outlines the relationship between the network operator, network management, service management, and the customer. It gives the protocol structure and protocol stack for these services, and a security mechanism to prevent illegal access.

(2) The service Recommendation (X.cnms) defines management functions which may be used by customers of PDNs to perform certain customer network management functions. Examples are the utilization of a line test loop, the retrieval of management information, and the provision of a service inventory.

(3) The MO / management information Recommendation (X.cnmi) defines management information in accordance with Recommendation X.722 (the MO classes, attribute types, action types, notification types, etc.) for CNM services specified in X.cnms. This Recommendation provides generic definitions for developing CNM services. Currently, it only considers the definition of management information for X.25 packet-mode access to PSPDNs. The definition of MOs for other types of access to PDNs, e.g., PADs, X.32, ISDN packet, and CSPDN is for further study.

2.2 CNM architecture

2.2.1 CNM Service provision model

The CNM service model is based on the mechanism by which MO notification and operations are communicated between open systems, as shown in Figure 1. Applications with the manager and agent roles, which are in most cases located in a customer terminal and in the network respectively, interact using the underlying transfer protocol, i.e., CMIP or MHS. The information elements at the service boundary are defined by the management information definition, i.e., Recommendation X.cnmi. Note that the usage of MHS is optional and for further study.

The management information and functions necessary for providing the CNM services are logically encapsulated in an MO. An agent which receives a request from the manager accesses the MO, and then emits responses or notifications, in accordance with the MO definition. The manager or agent uses the PT service primitives defined in Recommendation X.730, which are equal to the CMIS primitives, to invoke or process a CNM operation.

2.3 CNM Service Menus

An MO contains all the necessary information (e.g., attributes, behaviour, notification, etc.) for implementing services. Thus, service definition and MO definition are closely related. Before defining MOs, which will be presented in section 3, we should clarify what services are to be provided.

CNM services are classified into five categories based on the OSI Systems Management:

-Fault management
-Configuration management
-Accounting management
-Performance management
-Security management

In the following sections, we will discuss main services which CCITT SG VII has already specified clearly enough to define their management information.

2.3.1 Fault Management

(1)Line alarm

When a fault occurs on a subscriber line, information about the location, time of occurrence, probable cause, etc., is reported to the customer.

(2)Network problem alarms

Information on problems regarding a common network, e.g., the location (if possible), time of occurrence, probable cause, etc., is provided. However, this information is sent to the customer only when the problem affects the customer.

(3)Fault log access

By log access, customers can obtain information on problems having affected the customer, e.g., fault notification log records related to the customer's access line or network.

(4)Line test

A line test with loop back test capability can be performed on a customer's request. The loop back point can be set at several places in the network including the remote end terminal.

(5)Protocol test

A protocol integrity test and protocol monitoring are performed on a customer's request. For this purpose, a test host function may be provided.

2.3.2 Configuration Management

(1)activation/deactivation of a line

A specified subscriber line can be activated or deactivated by the customer on demand or on schedule.

(2)Subscription management

Customers can subscribe to new lines, cancel an existing line, or change the subscription (facilities/services).

(3)PVC management

Customers can create, delete, or change PVC subscription parameters (logical channel number, packet size, window size, throughput class, etc.).

2.3.3 Accounting Management

(1)Charging information

The history of charging information, e.g., usage, tariffs, third-party addresses, is provided on demand or on schedule.

(2)Real-time charging information

The charging information is reported just after each call is finished.

2.3.4 Performance Management

(1)Traffic information

Traffic information is observed and accumulated, based on Recommendation X.738. The information can be either reported by notifications or accessed by log management.

(2)Quality of service information

Customers can retrieve or be notified of the QOS information recorded in a log record.

2.3.5 Security Management

(1)Security audit trail record access

Customers can retrieve their own security audit trail record containing the record of illegal access to the user-network interface.

3. Basic concepts for defining managed objects

3.1 Aspects of CNM MOs and an Approach to the Definition of CNM MOs

In defining CNM MOs, we should consider two types of MOs: those which can also be used for network operation within a network and those dedicated to service management. The former type of MO is basically of the same structure as that of network operation MO. By inheriting a network operation MO, this type of CNM MO is defined. Customers are allowed to monitor and control a limited set of the parameters of the resources that they are entitled to access. Typically customers have partial control/view of these objects. For example, a remote customer's line MO can be defined so that both the customer and network operator can retrieve its communication status, while the customer is not allowed to deactivate the line MO. On the other hand, MOs for service management are owned by a particular customer. Examples are logs or local memory over which the customer has total control (create, delete, etc.). Which type of MO should be used depends on the definition of each service menu.

However, some MOs have both aspects. For example, an MO dedicated to subscription management can be defined, while subscription data can also be controlled by a protocol entity MO used by the network operator. Since the network operation can share an MO with the CNM service, the former approach can make it easier to implement the management software, on condition that security is guaranteed.

Also, it is necessary to examine how the object accessed is seen by a customer, and what concerns customers have, for defining MOs and specific information elements for CNM services. At the same time, from the standpoint of the network operating company, they may want to limit information elements provided to customers for security reasons. Taking these situations into account, we should model CNM MOs and should define common elements or characteristics of CNM services in them.

These CNM generic MOs may be subclassed by adding specific features to extend CNM services by service provider (a network operating company). On the other hand, it is desirable to inherit and subclass MOs of the Systems Management and Layer Management Standards / Recommendations for CNM MO definition because the portability of MO definition and products is expected to increase. In the following section, we will describe the definition of CNM MOs based on these concepts.

3.2 MO definition

MOs for CNM services are defined in accordance with the classification stated in section 2.3 and detail service requirements from the customer and network point of view. User requirements for these services are modeled as shown in figure 2. Communication-protocol-dependent-parts are encapsulated in accordance with the OSI communication model, and the other parts are based on the SMASE model[4], [5].

3.2.1 Fault Management

(1) Finding and Locating faults

The customer wants to know the status of his communication, and any faults or congestion. The location of the problem is also important. The customer also needs to know the conditions of the network, his access line, datalink, and virtual circuit. In addition, he might be concerned with the state of the remote customer terminal or, in the internetwork communication, the condition of other networks.

Thus, the following items should be defined as MOs for CNM services:

-the network to which the customer is directly attached

-other networks involved in an interworking call

-the customer's access line (expressed as a physical connection) and its layer 1 interface circuits (expressed as physical entities)

-the data link (expressed as a data link layer connection) by which the customer accesses the network, and the datalink entity in the switch

-X.25 virtual circuits (expressed as VC, PVC), and the packet layer entity in the switch.

-the interface circuit, data link, and virtual circuit of the remote customer

Here, note that the internal trunks and nodes are invisible to customers. For customers, the status of the network is more important, and other information is concealed. Also note that the remote customer's security must be guarded. When a customer tries to retrieve the other party's information, the network should check his access privilege.

(2) Notification of a problem

When problems are reported to a customer, the criterion for notification generally differs for each customer, e.g., the type of alarm, and the threshold for the number of alarms. A criterion and threshold handling function is hence required by the customer. The Event Forwarding Discriminator defined in Recommendation X.734 should be used for this CNM service, and should be installed for each customer.

(3) Logging and Retrieval of Fault Information

Problem information is not only output as an alarm, but also needs to be accumulated in a log that can be accessed freely by the customer. MOs concerning logs are defined in Recommendation X.735, and they should be applied to the CNM service.

Log records, which are also managed objects, include many types of records. The customer should create the records required for his management jobs, e.g., traffic information log, alarm log, and audit trail log. Hence, the following MOs should be used for this CNM service:

-log --installed for each customer
-Event Log Record
-user defined Log Record

(4) Test and Diagnostic Functions

The test functions of Recommendations X.737 and X.745 should be used for fault location or preventive maintenance. The TO (test object) and test result record MO class definitions specified in them are imported for this CNM Recommendation. The MOT (MO on test), which represents a resource to be tested in the system, and associated object(AO), which supports a test action, are specified for a test action. Also the TARR (test action request receiver) functionality should be contained in one or more managed objects related to the test. The actions for the test, such as making a loop back point or protocol monitoring point, must also be defined. The following MOs are necessary for the CNM test management:

-TO, defined in Recommendation X.745 --installed for each customer
-MO with the TARR functionality
-MOT, which represents the MO to be tested
-AO (associated object), with a loop back function and/or a monitoring function

3.2.2 Configuration management

(1) Management of customer subscription data

To manage customer subscription data, the X.25 packet layer entity and data link entity are used for setting X.25 parameters. Their Initial Value Managed Objects (IVMO) may also be used, when the data is to be set in a way other than that in normal operation. We can import layer management object definitions such as ISO/IEC 10733 and ISO/IEC 10742, which include X.25 service/facility attributes, e.g., Closed User Group, the D-bit modification facility, etc. The MOs are as follows:

-X.25 data link layer entity MO for the CNM services
-X.25 packet layer entity DCE MO for the CNM services
-X.25 data link layer entity IVMO for the CNM services
-X.25 packet layer entity DCE IVMO for the CNM services

(2) Management of PVCs

The PVC DCE MO is used for PVC subscription. To change PVC parameters, the successive operation of *DELETE* and *CREATE* (with new parameters) is used.

(3) Retrieval and Control of Customer Information

If a customer wishes to retrieve or change the information concerning his terminal, he only has to access the X.25 packet layer entity, data link entity (also their IVMOs) or the PVC MOs for subscription, VC or PVC MO for the retrieval of status information or statistical data, and eventLogRecord managed objects for collection of system alarms.

3.2.3 Performance Management

There is also a need for performance management. The model for this in the customer network management is based on that of draft Recommendation X.738. Three kinds of scanner (e.g., HomogeneousScanner) will be used. The object under observation is the MO that represents the resource in the network from which performance information is obtained, such as a physical entity, data link, or virtual call. The contents of the log record are based on the definition of Layer Entity MOs. The QOS alarm (defined in Recommendation X.733), etc., is used as the log record.

3.2.4 Accounting Management

Two MO classes, the Accounting Meter Data and Accounting Meter Record, defined in draft Recommendation X.742, are used for immediate retrieval of charging information, and non real-time retrieval, respectively. The Accounting Meter Control managed object class is handled by the network itself and is invisible to customers.

3.2.5 Security Management

The security Audit Trail Record MO in Recommendation X.740 is used for retrieving a record of all potentially security-related events in the management domain (the CNM domain). The other security management related MOs defined in Recommendations X.712, X.736 and X.740 are for network functions and hence should not be handled by customers.

3.3 Examples of Managed Object definitions

The above-mentioned MOs are shown in Figure 3. This section shows some of the MO definitions focussing on test management and subscription / PVC management, which would become the main functions at the implementation time.

3.3.1 Test Management Information Definition

As described in section 3.2.1(4), an MO with the TARR package, a MOT, TO(s), and AO(s) have to be defined to implement the test function.

When the network performs a loop back test on a customer access line, the Physical Connection MO, namely the access line, should be the MOT, i.e., the object to be tested. This MO includes the TARR functionality so that the implementation would be simpler. In addition, the Physical Entity of the customer terminal should be defined as an AO to set a loop back point. To control the test, a TO has to be defined which executes a loop back test, and sends its result to the manager (customer) by a notification. The input and output parameters are based on those defined in Recommendation X.737. An example of the definition is shown in ANNEX.

3.3.2 Subscription / PVC Management Information Definition

Subscription management related MOs are based on the service definitions described in section 2.3.2. The DCE Layer Entity MO (here, especially for X.25 Packet Layer) includes attributes on X.25 (X.2) facilities, which are used as subscription data. For PVC management, the DCE PVC MO, which include attributes for logical channel number, packet size, window size, throughput class, etc., is handled by each customer (see ANNEX).

4. Conclusion

In this paper, we have discussed principles of MO definition for CNM services. First, we have clarified customers' requirements for CNM services in order to specify functions and information elements necessary for the services. Then, CNM generic MOs are defined based on the OSI protocol model and OSI Systems Management model, by encapsulating necessary information elements and functions as attributes and behaviour definitions in an MO. Considering these points, the following MOs are defined for CNM services:

-MOs representing each layer entity, test related MOs, and log control MOs are defined for fault and test management, which can be used for reporting trouble, accumulating alarms, and performing loop back and protocol tests, etc.

-layer entity MOs and PVC MOs are used for the subscription and PVC management, which have attributes for indicating subscription of services.

-other MOs to enhance the CNM services are currently under discussion in CCITT.

These CNM generic MOs can be used and customized for providing specific CNM services of a network operating company which offers the services.

Acknowledgement

The authors would like to express their appreciation to Mr. Kazuyoshi Morino, Executive engineer of Nippon Telegraph and Telephone Corporation (NTT), Mr. Tatsunari Kikkawa, Mr. Yasunori Hataya and Mr. Tsuyoshi Snimazaki, engineers of NTT for their useful advice.

References

1. CCITT SGVII Q9: Draft Recommendation X.cnma "Architecture for Customer Network Management Service for Public Data Networks", COM VII-224 (1989-1992)
2. CCITT SGVII Q9: Draft Recommendation X.cnms "Definition of Customer Network Management Services for Public Data Networks", COM VII-224 (1989-1992)
3. CCITT SGVII Q9: Draft Recommendation X.cnms "Definition of Customer Network Management Services for Public Data Networks", D442 (1989-1992)
4. CClTT SGVIl: Recommendation X.701 "Systems Management Overview"
5. ISO/IEC JTCl SC21: International Standard 10164 "Information Technology - Open Systems Interconnection - Systems Management

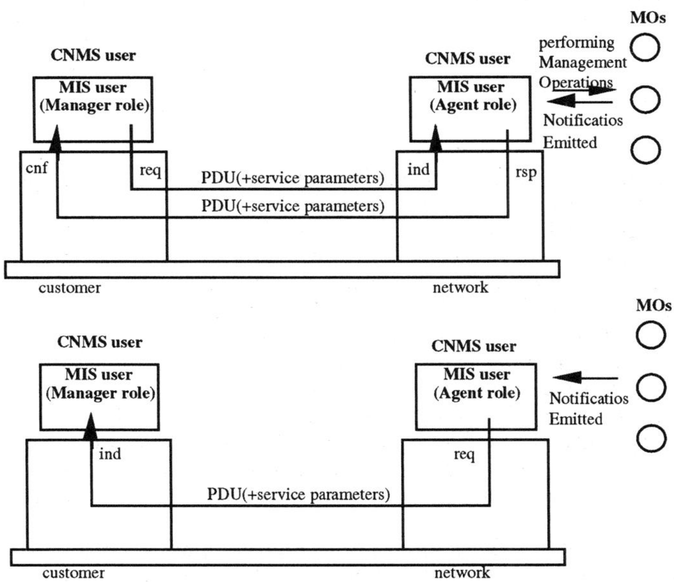

<As the transfer protocol, either CMIP or MHS may be used.>

Figure 1. CNM service model

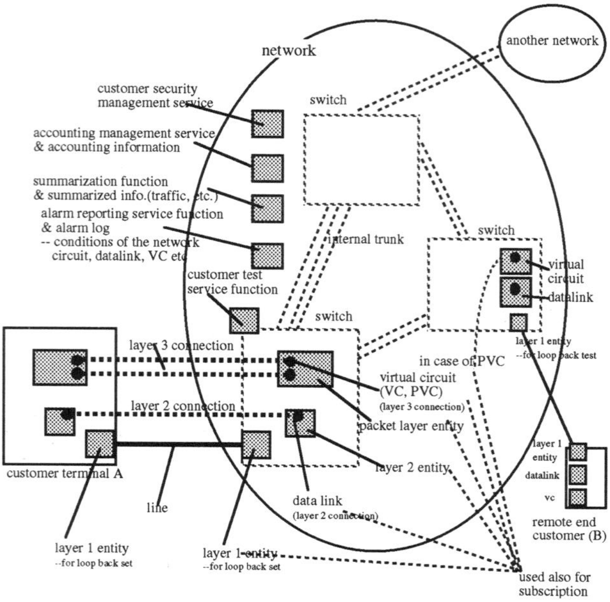

Figure 2 Physical view of the CNM service (viewed from Customer A)

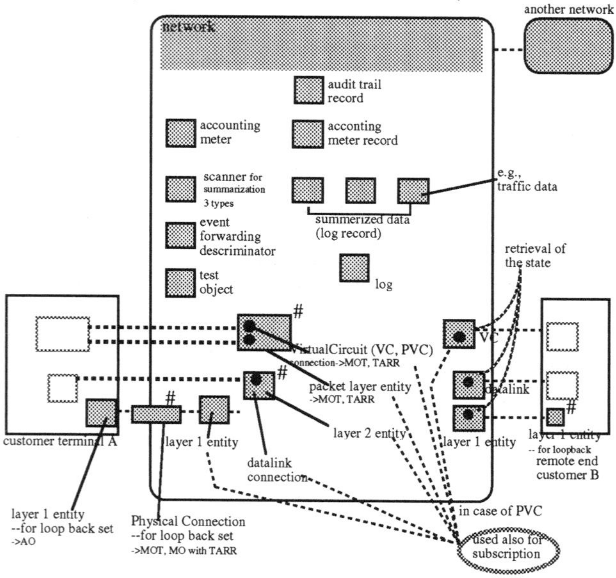

<note #> The MOs marked with # may be a MOT.
In this case, the MOT must indicate test behaviour in the MO definition.

Figure 3 Management view of the CNM service (viewed from Customer A)

ANNEX Management Information Definition for CNMS

[MOT and MO with TARR]
cnmsPhysicalConnection MANAGED OBJECT CLASS
DERIVED FROM top;
CHARACTERIZED BY cnmsPysicalConnectionPackage PACKAGE
ATTRIBUTES
X.721:usageState GET, etc. ;;;
CONDITIONAL PACKAGES
cnmsLoopbackTestRqstAsyncPKG PRESENT IF a protocol test is executed in this MO,
cnmsLoopbackTestPKG PRESENT IF a protocol test is executed in this MO,
"X.721/745":testSuspendedResumePKG PRESENT IF a protocol test is executed in this MO,
"X.721/745":testTerminationPKG PRESENT IF a protocol test is executed in this MO;
REGISTERED AS { ? };;

[AO]
customerPhysicalEntity MANAGED OBJECT CLASS
DERIVED FROM top;
CHARACTERIZED BY cnmsLoopbackPKG
BEHAVIOUR DEFINED AS When a loopback point is required to be set from the TO, this object makes a loopback point as defined in Recommendation X.745. This MO acts as an Associated Object.
loopbackState GET-REPLACE, -- to set the loopback point and retrieve the status of the loopback point.
loopbackType GET-REPLACE; -- indicates which type of loopback point is set
-- other items are for further study;;
REGISTERED AS { ? };;

[TO]
cnmsLoopbackTestObject MANAGED OBJECT CLASS
DERIVED FROM "X.745":testObject; --TO
CHARACTERIZED BY
cnmsLoopbackTestObjectPKG,
"X.746": schedulerObject,
"X.746": schedulingMode;
REGISTERED AS { ? };;

[Related Definitions]
cnmsLoopbackTestRqstAsyncPKG PACKAGE
BEHAVIOUR cnmsLoopbackTestRqstAsync-B BEHAVIOUR
DEFINED AS This package defines the TARR functionality to receive an action of type testRequestAsync. It is to be imported in an object class when such functionality is required and allows creation of TO(s), corresponding to the testObjects parameter as a side effect.;
ACTIONS cnmsLoopbackTestRqstAsyncAction "Rec. X.737": loopbackTestParam;;
REGISTERED AS { ? };;

cnmsLoopbackTestPKG PACKAGE
BEHAVIOUR cnmsLoopbackTest-B BEHAVIOUR
DEFINED AS Controlled by a TO, a MO with this package cooperate with the TO for execution of the loopback test.;;
REGISTERED AS { ? };;

cnmsLoopbackTestControlPKG PACKAGE
BEHAVIOUR cnmsLoopbackTestControl-B BEHAVIOUR
DEFINED AS a TO with this package executes a loopback test, and inform its result of the manager by the following notification.;
NOTIFICATIONS
cnmsLoopbackTestRsltNotification "Rec. X.737": loopbackTestResultParam;;
REGISTERED AS { ? };;

cnmsLoopbackTestRsltNotification NOTIFICATION
BEHAVIOUR
cnmsLoopbackTestRsltNotification-B BEHAVIOUR
DEFINED AS When the specified test execution has been completed, the TO shall reply with result information.
MODE NONCONFIRMED;
WITH INFORMATION SYNTAX "Rec. X.721/745": NotificationASN1Module.TestResultInfo;
REGISTERED AS { ? };

cnmsLoopbackTestRqstAsyncAction ACTION
BEHAVIOUR
cnmsLoopbackTestRqstAsyncAction-B BEHAVIOUR
DEFINED AS When an action request for the protocol test is received, a protocol test TO shall be instantiated. The information associated with the action request is handed to the TO, which in turn executes the test.;
MODE CONFIRMED;
WITH INFORMATION SYNTAX "X.721/745": ActionASN1Module.TestRequestAsyncInfo;
WITH REPLY SYNTAX "X.721/745": ActionASN1Module.TestRequestAsyncResult;
REGISTERED AS { ? };;

<note>The above definitions have not yet been approved in CCITT. These are based on the authors' proposal.

[THE X.25 PLE packet layer entity Managed Object (Subscription related parts only)]
cnmsX25PLE MANAGED OBJECT CLASS
DERIVED FROM "ISO/IEC10733":x25PLE;
CHARACTERIZED BY cnmsX25PLE PACKAGE
ATTRIBUTES
"ISO/IEC10733":incomingCallBarred GET-REPLACE,
"ISO/IEC10733":outgoingCallsBarred GET-REPLACE,
"ISO/IEC10733":oneWayLogicalChannelOutgoing GET-REPLACE,
"ISO/IEC10733":cUG GET-REPLACE,
"ISO/IEC10733":fastSelectAcceptance GET-REPLACE,
......;
REGISTERED AS { ? };

[PVC Managed Object (Subscription related parts only)]
cnmsPVC MANAGED OBJECT CLASS
DERIVED FROM "ISO/IEC10733":permantVirtualCircuit-DCE ;
--this MO contains chargingDirection, logicalChannel, packetSize, throughputClass, virtualCircuitId, windowSize.
CHARACTERIZED BY cnmsPVCMO-P PACKAGE
......;
REGISTERED AS { ? };

Integrated Network Management, III (C-12)
H.-G. Hegering and Y. Yemini (Editors)
Elsevier Science Publishers B.V. (North-Holland)

A State Dependent Reliability Model and Its Applications

T. Lin[a], R.E. Mallon[a], C-W Chao[a] and H. Eslambolchi[b]

[a]AT&T Bell Laboratories, 480 Red Hill Road, Middletown, NJ 07748, U.S.A.

[b]AT&T, Network Service Division, Rtes 202/206, Bedminister, NJ 07921, U.S.A.

ABSTRACT

FASTAR[SM] is a fast automatic restoration system for restoring AT&T DS3 services (in the event of a cable failure). Its deployment has helped to reduce restoration times significantly. Reliable operations throughout a cable failure event is the key to the success of the restoration process. Since the hardware components involved during restoration vary, the traditional reliability block diagram is too simplistic to capture the state dependent system availability. In this paper we propose a state dependent reliability model that incorporates both state transition and reliability block diagrams. This model is designed to handle a general cable cut case which usually results in multiple reroute paths. We demonstrate how the model works using a typical example. The primary application of the model is to support the AT&T FASTAR program in meeting its very high availability design criterion.

Keyword Codes: B.4.5; C.2.3; I.6.4
Keywords: Reliability, Testing, and Fault-Tolerance; Network Operations; Model Validation and Analysis

1. INTRODUCTION

The FASTAR (FAST Automatic Restoration) system employs a great many components deployed over a large geographic area. Communications and interdependency among components are essential to the operation of the system. The availability of each component and its configuration in the system have a great impact on system performance.

A cable failure may affect hundreds of T3 facilities, resulting in multiple **reroute paths** (a restoration path used to restore service for a T3). The successful implementation of each reroute path depends on its hardware components, software availability and database accuracy. The traditional Reliability Block Diagram [1,4] is not adequate for a multiple reroute paths since each has a different number of hardware components. This changing hardware configuration calls for a Markov model [2,3] to monitor system transition in

---

[SM] FASTAR is a service mark of AT&T

different configurations. We propose a model which combines a state transition and a reliability block diagram to evaluate FASTAR availability based on hardware components.

A number of reliability modeling techniques are widely used today [2,4]. Each technique provides specific advantages and disadvantages. The concept of combining different techniques has been explored and various software tools which utilize one or a combination of the techniques has been compared and discussed [5,6]. Other hybrid model has been proposed as well. A Markov model and fault tree model were used to evaluate the sequence dependencies [7]. Contingence enumeration and Monte Carlo simulation were used to evaluate different operating states of power systems [8]. One software tool [9] uses Markov model for subsystem modeling and fault tree/reliability block diagram for modeling the resulting subsystems. Our model uses reliability block diagram to evaluate the successful probability of each reroute path, then uses Markov model to model the multiple reroute path restoration.

This paper first briefly describes the architecture of the FASTAR system and the restoration procedures. The state dependent reliability model that is used to evaluate FASTAR availability is proposed in section 3. The results and sensitivity analysis are presented in section 4.

2. FASTAR ARCHITECTURE AND RESTORATION PROCESS

2.1. Architecture

FASTAR is a centralized processing system with distributed monitoring functionalities [10]. The system architecture of FASTAR is shown in Figure 1. The components and their functionalities are explained as follows:

- **RAPID** - Restoration and Provisioning Integrated Design (RAPID) is the central processing unit that receives alarm reports, calculates reroute paths and implements the restoration paths. It also rolls the system back to the normal operating configuration after the failure is repaired.
- **SKYNET™ and PINet** - SKYNET is a satellite-based data communication network and PINet is a terrestrial Packet Internal Network. They support diverse communications routes between the central processor (RAPID) and the distributed nodes (described below). The redundancy increases the survivability of the communications network if one of them is down.
- **FT Series G LTE** - The Line Terminating Equipment (LTE) of the AT&T FT Series G lightwave digital transmission system monitors the fiber status and reports any changes to the local restoration node controller (RNC).
- **DACS III** - The Digital Access and Cross-connect System (DACS) provides DS3-level cross-connect capability, and monitors T3s for intra-office Loss of Signal (LOS) failures.

™ SKYNET is a trademark of AT&T

- **RNC** - The Restoration Node Controller (RNC) is the distributed intelligent processor which gathers alarm information from LTEs and DACS IIIs, ages the alarms and finally reports alarm status to RAPID.
- **RTE** - The Restoration Test Equipment (RTE) provides DS3 continuity check signal generation and detection.
- **TMAS** - The Transport Maintenance and Administration System (TMAS) provides the protection switch lock-out of FT Series G fibers, under RAPID control.

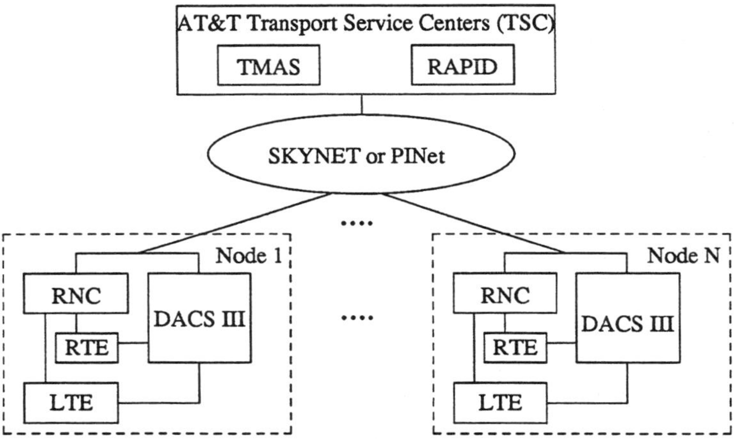

Figure 1. FASTAR System Architecture

2.2. Restoration Process

The operation of the FASTAR system is divided into three stages: alarm gathering and reporting, path implementation and path assurance.

- **Alarm Gathering and Reporting**

  A T3 signal failure is detected by the fiber link LTE or downstream DACS III. These failures are reported to the local RNC and are aged[1] by the RNC before being reported to RAPID. A failure detected can be either a T3 channel failure or a cable failure. A channel failure normally is recovered by switching the service to its protection channel quickly. A cable failure affects all the T3 channels in the cable, and results in an automatic restoration via the FASTAR system.

1. Aging is a process for RNC to filter out transient faults.

- **Path Implementation**
  When RAPID receives a failure report from the RNC, it waits a time window to assure that the entire event is reported before initiating restoration for the failed T3s. When the window expires, RAPID first calculates the reroute path. RAPID then implements the path by sending (via SKYNET/PINet) cross-connect commands to all DACS IIIs along the reroute path except those at the ends. Lockout commands are sent to TMAS for relay to the LTEs of protection channels used to provision the reroute path.

- **Path Assurance**
  RAPID then issues Service Failure Verification (SFV) requests to RNCs at both ends of the path. After the RNCs verify and report service failure, RAPID transfers service to DACS IIIs at the ends of the path. RAPID completes the restoration by requesting Service Verification (SV) by the RNCs.

2.3. Exception Handling

The procedures described in section 2.2 constitute a successful restoration. However, exceptions may occur. Some T3s may not be successfully restored due to failures of components (of that restoration process) during restoration. If both RNCs of the cut link fail during the aging period and the failure persists, the path cannot be restored automatically by FASTAR. If RAPID fails during the restoration process, the remaining restorable T3s are handled manually. Any component failure except RAPID will only interrupt the restoration process for a single path. After RAPID visits every T3 on the priority list, those T3s not restored are handled manually. When the failed link is finally repaired, RAPID rolls back the network to the normal operation status.

3. THE STATE DEPENDENT MODEL

We use a state transition diagram to monitor the system states throughout the restoration cycle. For each state, a reliability block diagram is used to evaluate the probability of successful restoration. The combination of state transition and availability block diagrams offers a solution to the state dependent system availability evaluation.

3.1. Model Assumptions

The model we describe here handles cable cuts with multiple reroute paths. There are cases where restoration of a T3 may not be successful due to software or database error, or lack of continuity. We neglect these cases in this paper. An extended model which includes the database availabilities is being studied. In addition to the error-free software assumption, the hardware component availabilities were taken from manufactures' data. Therefore, the model represents the upper bound of FASTAR availability. Currently there may not be sufficient spare capacity for every T3 affected. However, a capacity shortage does not constitute a FASTAR failure. Thus for this study, we only focus on those T3s that are restorable.

3.2. Definition of Terminology

Here, we define the terminology used in the model.

- **Bold** face characters represent a state.
- *Italics* are used for transition rates.
- Availability of a component is denoted as $A_{component}$.
- The steady state probability is denoted as P_{state_name}.
- r_{lf} is the average link failure rate.
- r_{cw} is the average RNC window aging rate.
- r_{rw} is the average RAPID window aging rate.
- r_{res} is the average FASTAR restoration rate per reroute path.
- r_{repair} is the average cable repair rate by maintenance crew.
- **S(i,j)** is the system state where *j* of the *i* paths which FASTAR attempted failed.
- Each cable cut requires *m* reroute paths. Each reroute path contains *l* DACS III offices.
- For every reroute path, components involved in the restoration effort may suffer from failure. We define the probability that no component failure occurs during restoration to be **component coverage, C**.

3.3. State Transition Diagram

State transition diagram of a two-reroute-path case (i.e., m=2) is shown in Figure 2. Figure 4 in the Appendix provides a generalized state diagram. State **NORM** is the normal state where there is no failure in the system. When a failure occurs, with rate r_{lf}, the system transits to the RNC aging state, **RNC_AGE**. If both RNCs fail (with rate $(1-A_{rnc})^2 r_{cw}$), the failure cannot be restored by FASTAR (instead they are recovered manually). The system then moves to **S(m,m)** where not a single T3 restoration is successful. Otherwise when the aging window expires, RNC reports to RAPID. Therefore the system moves from state **RNC_AGE** to state **RAPID_AGE** with rate $(1-(1-A_{rnc})^2)\cdot r_{cw}$. The system state **RAPID_AGE** has similar transitions; it either transits to state **S(0,0)** with rate $A_{rapid}\cdot r_{rw}$ or transits to state **S(m,m)** with rate $(1-A_{rapid})\cdot r_{rw}$.

When the system enters state **S(0,0)**, it begins the path implementation and path assurance phases. From this point on, each transition involves all the hardware components along a reroute path. Each transition is either a restoration success or a failure of a reroute path. The component coverage factor C_i and RAPID availability A_{rapid} are used to determine the outcome of the restoration. From state **S(i,j)**, $0 \le i,j < m$, the system has three transitions. When RAPID successfully restores one reroute path, the system moves from **S(i,j)** to **S(i+1,j)** with rate $r_{res}\cdot A_{rapid}\cdot C_{i+1}$. When the restoration attempt fails due to component failures, the system moves to **S(i+1,j+1)** with rate $r_{res}\cdot A_{rapid}\cdot(1-C_{i+1})$. When RAPID fails (with rate $r_{res}\cdot(1-A_{rapid})$) before the restoration completes, the system moves to **S(m,m-i+j)**. (Note: In Figure 2, the failures rates of the transitions **S(1,0)** -> **S(2,1)** and **S(1,1)** -> **S(2,2)** are combinations of RAPID failure and components failure events to give a failure rate of $r_{res}\cdot(1-C_2 A_{rapid})$.)

When the system reaches state **S(m,j)**, $0 \leq j \leq m$, the restoration cycle has been completed. However, the system is not at normal operation mode yet. Therefore, with rate r_{repair}, maintenance crew repairs the cut link and the system returns back to normal state, **NORM**.

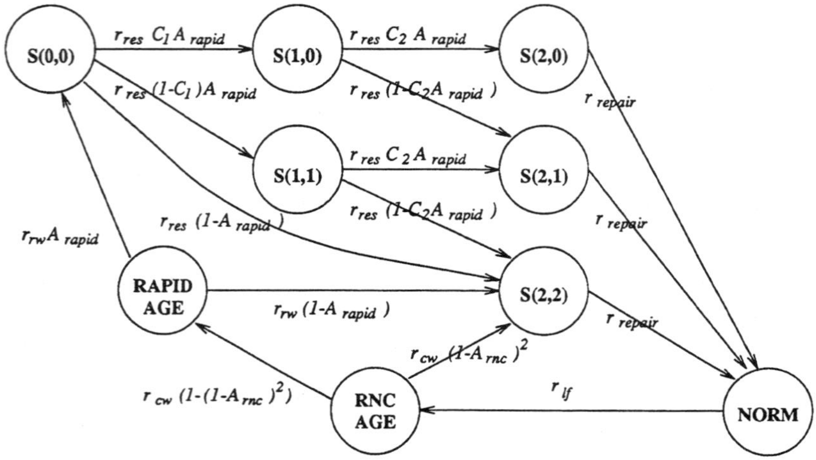

Figure 2. State Transition Diagram of the FASTAR System with two reroute paths case

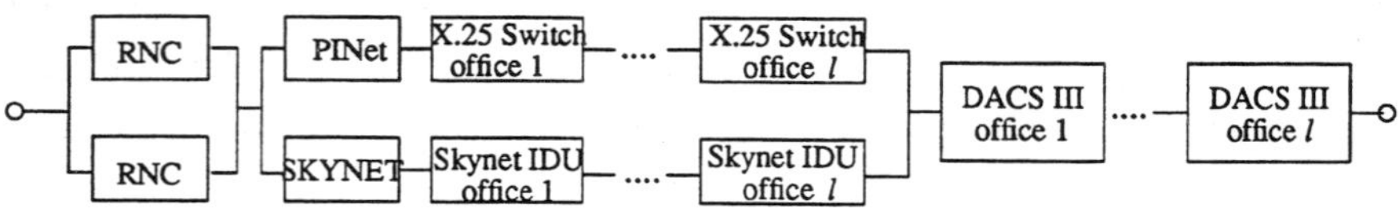

Figure 3. FASTAR Components Reliability Block Diagram for a path length l

3.4. Component Coverage C - The Reliability Block Diagram

Component coverage C is the probability that all components involved in the restoration (of a reroute path) do not experience failures that endanger the restoration. Some components are designed with redundancy that they can tolerate certain failures without disrupting the restoration. A reliability block diagram of the FASTAR components involved in a restoration path is shown in Figure 3. The component coverage C can be found as a function of the reroute path length, l :

$$C=\left[1-(1-A_{rnc})^2\right]\cdot\left[1-(1-A_{sky\_at\_rapid}\cdot(A_{sky\_at\_dacs})^l)(1-A_{pinet}(A_{x25})^l)\right](A_{dacs})^l \quad (1)$$

In addition to hardware failure, DACS III can become unavailable for restoration due to daily scheduled maintenance. Since this factor is controllable by maintenance crews, we separate it from the DACS III hardware failure and denote the availability of a DACS III in terms of maintenance as S_dacs. Then C_{S_dacs}, component coverage including DACS III scheduled maintenance outage, is as follows:

$$C_{S\_dacs}=\left[1-(1-A_{rnc})^2\right]\cdot\left[1-(1-A_{sky\_at\_rapid}\cdot(A_{sky\_at\_dacs})^l)(1-A_{pinet}(A_{x25})^l)\right](A_{dacs}\cdot S\_dacs)^l \quad (2)$$

3.5. Consideration of DACS III Scheduled Maintenance

In the real system, when one or more DACS III's do not respond to RAPID during restoration, DACS III can be either down or at scheduled maintenance. If DACS III is at scheduled maintenance, it will become available again when the maintenance is done. For those DACS III not responding, RAPID would try to restore that reroute path again after a retry window expires. If the DACS III is still not answering the second RAPID attempt, RAPID labels it as a Block Fail and resorts to manual restoration.

The diagram becomes complicated if we incorporate the second attempt exactly into the state transition diagram. Instead, we make the following approximation: we accumulate the success probabilities of both the first attempt and the second attempt as though they "happen" simultaneously. Neglecting the re-try window will not affect the model accuracy since it's the relative state probabilities in which we are interested.

We modify the component coverage C to account for the second RAPID attempt as follows. When the restoration failure is due to DACS IIIs in scheduled maintenance, RAPID will try to restore the reroute path after the retry window expires. Let us assume the retry window is k minutes. The probability, q, that a DACS III under maintenance becomes available again during the k minutes time frame (if the maintenance time is greater than k minutes) is $q=\frac{k}{DACS\ maintenance\ time}$. If the maintenance time is less than k minutes, then the RAPID's second attempt will always be successful ($q=1$). Therefore, the modified component coverage $C_{modified}$ can be expressed as :

$$C_{modified}=C\cdot\left[q\cdot(1-S\_dacs)+S\_dacs\right]^l \quad (3)$$

3.6. System Availability

To solve for system availability, we need a set of link failure information about the number of reroute paths involved, the number of T3s restored on each reroute path and the number of offices (nodes) on each reroute path. This information is input to the model we have described. For each path, we calculate the component coverage C_i. The value of the path length l can be either calculated from a probability distribution function or taken from a representative cut information. We then solve for each steady state probability using the general solution described in the Appendix.

We define the system availability to be the ratio of the average number of successfully restored paths to the total number of restorable paths as follows:

$$A_{system} = \frac{E[number\ of\ successful\ reroute\ paths]}{total\ number\ of\ usable\ reroute\ paths} = \frac{\sum_{i=0}^{m}\left[P_{S(m,i)}\cdot(m-i)\right]}{m\cdot\sum_{i=0}^{m}P_{S(m,i)}} \tag{4}$$

However the above formula is correct only when each reroute path contains the same number of T3s. In reality, where each reroute path contains a different number of T3s, the system availability can be more accurately defined as follows:

$$A_{system} = \frac{E[number\ of\ successfully\ restored\ T3s]}{total\ number\ of\ restorable\ T3s} \tag{5}$$

The state probabilities do not provide enough information to solve for the number of successfully restored T3s. We need to enumerate all possible combinations of successful restoration to calculate the number of successfully restored T3s. Let $t3[i]$ be the number of T3s restored by the *ith* reroute path. Let C_i be the component coverage of *ith* reroute path in the diagram. We propose a recursive algorithm to calculate the number of successfully restored T3s:

```
depth_search(i, eventprob, succt3)
    if (i < m)
            depth_search(i+1,eventprob·C_i,succt3+t3[i]);
            depth_search(i+1,eventprob·(1−C_i),succt3);
    else
            total_succ_t3=total_succ_t3+eventprob·(succt3·(1−C_i)+(succt3+t3[i])·C_i);
```

By calling the recursive function, **depth_search**(1,1,0), the algorithm performs a depth-first search of the state diagram and accumulates the average successfully restored T3s. The recursive function may become too time-consuming when there are many reroute paths. At this point, we need to confine the computation time to a reasonable range. For example, we can accumulate the ending state probabilities from **S(m,0)** until a state **S(m,t)** when the summed value exceeds a reasonable portion (e.g. 0.999) of the total of all the ending state probabilities. We use t to confine the search space in the recursive function by searching cases with the number of failed reroute paths less than or equal to the threshold value t. The contribution from those cases where more than t reroute paths failed is assumed to be negligible. This approximation can greatly reduce the computation time without losing the model accuracy.

4. APPLICATIONS

In this section, we first demonstrate how to apply the model using a typical example. We then illustrate sensitivity analysis of key parameters. Finally, we describe how FASTAR benefits from this study.

TABLE 1. Representative Cuts for an Example Network

Example Cuts					
Cut #1		**Cut #2**		**Cut #3**	
# of T3s	# of nodes	# of T3s	# of nodes	# of T3s	# of nodes
8	5	3	4	29	4
19	6	15	7	20	5
21	8	13	8	23	10
18	10	10	13	24	11
18	13	10	15	18	12
11	16	13	16	13	14
7	19	14	18	12	16
6	23	3	28	13	17
3	31			10	21
				6	26

TABLE 2. System Availability for Three Example Cuts

Cable Cut	**Total restorable T3s**	**Average path length**	**System Availability (Percent)**
Cut #1	111	11	93.78
Cut #2	81	12	93.03
Cut #3	168	11	93.92
Mean	120	11	93.58

4.1. A Typical Example

Table 1 shows the reroute path information in terms of the number of T3s and number of nodes per reroute path for three representative cuts of a sample network. For example, cut #1 requires 9 reroute paths. Eight T3s are restored on the first reroute path with 5 DACS III offices involved (thus, path length is 4).

The system parameters used in the example are assumed as follows: r_{lf}=0.0011/hr., r_{cw}=360/hr., r_{rw}=120/hr., r_{res}=720/hr., r_{repair}=0.1/hr.. A_{rnc}=99.97%, A_{x25}=99.98%, A_{pinet}=98.00%, A_{dacs}=99.99%, $A_{sky\_at\_rapid}$=99.94%, $A_{sky\_at\_dacs}$=99.92% and A_{rapid}=99.99%. We also assume a 10-minute scheduled maintenance time for DACS III per day (S_dacs = 99.31%) and a 2-minute retry window for RAPID. Applying these data to the model proposed in section 3, the results are given in Table 2. To calculate the average path length, we divide the sum of path length by the total number of T3s. System availability, as calculated here, refers to A_{system} which was defined in equation (5), section 3.6.

4.2. Model Analysis

Using the cut #2 data described in the previous section, we analyze the key system components to illustrate how to apply the model to locate any bottleneck and to improve the system availability.

- **Sensitivity Analysis of Average Path Length**
  From equations (1)-(3), component coverage (hence the system availability) decreases as the average path length, *l*, increases. With an increment of 4 offices on every path for the cut #2, the average path length increases from 12 to 16 and 20. The corresponding system availabilities are 0.9303, 0.9096 and 0.8893. In order to achieve higher system availability, an algorithm to find a shortest reroute path (such that fewer and shorter restoration paths are used in the event of cable failure) is important.

- **Sensitivity Analysis of RAPID Retry Window**
  When DACS III scheduled maintenance is eliminated, the system availability is 99% and up. Therefore, when the RAPID retry window is greater than the scheduled maintenance time, the DACS III maintenance outage will not be of concern and the system will be very reliable (i.e., at least 99%). However, due to performance considerations, a long re-try window is not realistic. Let us vary the retry window from 1 to 5 minutes by increments of 1 minute. The system availability for cut #2 becomes 0.922, 0.9303, 0.9387, 0.9471 and 0.9555 respectively. It confirms that one can improve the availability by increasing the retry window. However, there is a trade-off between availability and performance (restoration time). As long as the DACS maintenance time is significant, most likely improving availability using a retry window will be limited due to minimum performance considerations.

- **Sensitivity Analysis of DACS Maintenance Time**
  As just discussed, the DACS maintenance time is an important parameter in the model. Let us explore the relationship between system availability and DACS maintenance outage. We assume a 2-minute RAPID retry window. With the maintenance time reduced from 10 minute to 5 minutes or to 1 minute, the corresponding system availabilities are: 0.9303, 0.9728, and 0.9992. The result indicates a significant availability improvement can be achieved if the DACS III maintenance time can be reduced. If the maintenance time is less than the retry window, the system availability is less sensitive to the average path length as well. It only decreases from 0.9992 for the l=12 case to 0.9982 for the l=30 case.

By applying the example cuts data to the model, we show that the model enables us to analyze several key system parameters and locate the dominant availability bottleneck. We conclude from the study of the sample cuts that reducing the DACS III maintenance time to be less than a reasonably small RAPID retry window (to maintain expected restoration performance) can achieve higher system availability and less sensitivity to the reroute path length.

4.3. Benefits to FASTAR

As FASTAR is strategic to AT&T's ability to serve its customer base with minimal impact from intrusions on its digital network, very high availability design is essential to FASTAR. We have applied the model described in section 3 to the AT&T digital network during the FASTAR design phases. A sensitivity analysis with respect to key system parameters was also performed. As a result, several enhancements (such as redundant RAPID, dual processors/mirror disks of DACS III, an optimum rerouting algorithm, dual communication media between RAPID and RNC/DACS III, etc.) have been incorporated into the FASTAR design to meet the vitally important need for very high availability.

5. SUMMARY

In this paper a state dependent reliability model, which can handle a general cable cut requiring multiple rerouting paths, has been introduced. This model expands a traditional reliability block diagram in combination with a state transition diagram. We have also demonstrated how the model worked using a typical example. The key application of the model is to evaluate the AT&T FASTAR system via a sensitivity analysis of some key system parameters. Therefore, one of FASTAR's system design criteria, very high availability, can be met and thus help ensure the success of this service.

6. REFERENCES

[1] US MIL-HDBL-217: *Reliability Stress and Failure Rate Data for Electronic Equipment*, Military Handbook

[2] R. Billinton, R.N. Allan, *Reliability Evaluation of Engineering Systems: Concepts and Techniques*, Plenum Press, New York and London, 1983, 349p.

[3] Y.W. Ng, A.A. Avizienis, "A Unified Reliability Model for Fault-Tolerant Computers", *IEEE Transactions on Computers*, Col. C-29, No. 11, November 1980, pp. 1002-1011.

[4] P.D. O'Connor, *Practical Reliability Engineering*, Wiley, New York, 1989

[5] M. Mulazzani, K. Trivedi, "Dependability Prediction: Comparison of Tools and Techniques", IFAC SAFECOMP 1986

[6] R. Geist, K. Trivedi, "Reliability Estimation of Fault-Tolerant Systems: Tools and Techniques", *Computer*, Vol.24, No.7, July 1990

[7] J.B. Dugan, S.J. Bavuso, M.A. Boyd, "Fault Trees and Sequence Dependencies", *Reliability and Maintainability Symposium Proceedings: Annual Reliability and Maintainability Symposium*, NSYM, 1990, p286-293

[8] M.E. Khan, R. Billinton, "A Hybrid Model for Quantifying Different Operating States of Composite Power Systems", *IEEE Transactions on Power Systems*, V7, N1, 1992, p187-193

[9] R. Sahner, K. Trivedi, "A Hierarchical, Combinatorial-Markov Method for Solving Complex Reliability Models", *Proc. ACM/IEEE Fall Joint Computer Conference*, Dallas, Texas, Nov. 1986.

[10] C-W Chao, *et al*, "FASTAR - A Robust System for Test DS3 Restoration", *GLOBECOM '91*, pp. 39.1.1 - 39.1.5

Appendix : General Solution to the Multiple Reroute Path State Transition Diagram

Here, we solve for the solution for the general state diagram with *m* path. The state diagram is shown in Figure 4. The steady state probability is denoted as P_{state_name}. The general solution to the steady state probabilities of a cable cut with *m* reroute paths is as follows :

$$P_{RNC\_AGE} = \frac{r_{lf}}{r_{cw}} \cdot P_{NORM} \qquad P_{RAPID\_AGE} = \frac{r_{lf} \cdot (1-(1-A_{rnc})^2)}{r_{rw}} \cdot P_{NORM}$$

$$P_{S(0,0)} = \frac{r_{lf} \cdot (1-(1-A_{rnc})^2) \cdot A_{rapid}}{r_{res}} \cdot P_{NORM}$$

$$P_{S(i,0)}=C_{i-1}\cdot A_{rapid}\cdot P_{S(i-1,0)} \qquad 1\le i\le m-1$$

$$P_{S(i,i)}=(1-C_{i-1})\cdot A_{rapid}\cdot P_{S(i-1,i-1)} \qquad 1\le i\le m-1$$

$$P_{S(i,j)}=A_{rapid}\cdot\left[(1-C_{i-1})\cdot P_{S(i-1,j-1)}+C_{i-1}\cdot P_{S(i-1,j)}\right] \qquad 2\le i\le m-1,\ 1\le j<i$$

$$P_{S(m,0)}=\frac{C_{m-1}\cdot A_{rapid}\cdot r_{res}}{r_{repair}}\cdot P_{S(m-1,0)}$$

$$P_{S(m,1)}=\frac{r_{res}}{r_{repair}}\cdot\left[(1-C_{m-1}\cdot A_{rapid})\cdot P_{S(m-1,0)}+C_{m-1}\cdot A_{rapid}\cdot P_{S(m-1,1)}\right]$$

$$P_{S(m,m)}=\frac{r_{res}}{r_{repair}}\cdot\left[(1-A_{rapid})\cdot\sum_{k=2}^{m}P_{S(m-k,m-k)}+(1-C_{m-1}\cdot A_{rapid})\cdot P_{S(m-1,m-1)}\right]$$
$$+\frac{r_{rw(1-A_{rapid})}\cdot P_{RAPID\_AGE}+r_{rw(1-A_{rnc})}\cdot P_{RNC\_AGE}}{r_{repair}}$$

$$P_{S(m,j)}=\frac{r_{res}}{r_{repair}}\cdot\left[(1-C_{m-1}\cdot A_{rapid})\cdot P_{S(m-1,j-1)}+C_{m-1}\cdot A_{rapid}\cdot P_{S(m-1,j)}\right]$$
$$+\frac{r_{res}}{r_{repair}}\cdot(1-A_{rapid})\cdot\sum_{k=2}^{j}P_{S(m-k,j-k)} \qquad 2\le j<m$$

$$P_{NORM}+P_{RNC\_AGE}+P_{RAPID\_AGE}+\sum_{i=0}^{m}\sum_{j=0}^{i}P_{S(i,j)}=1$$

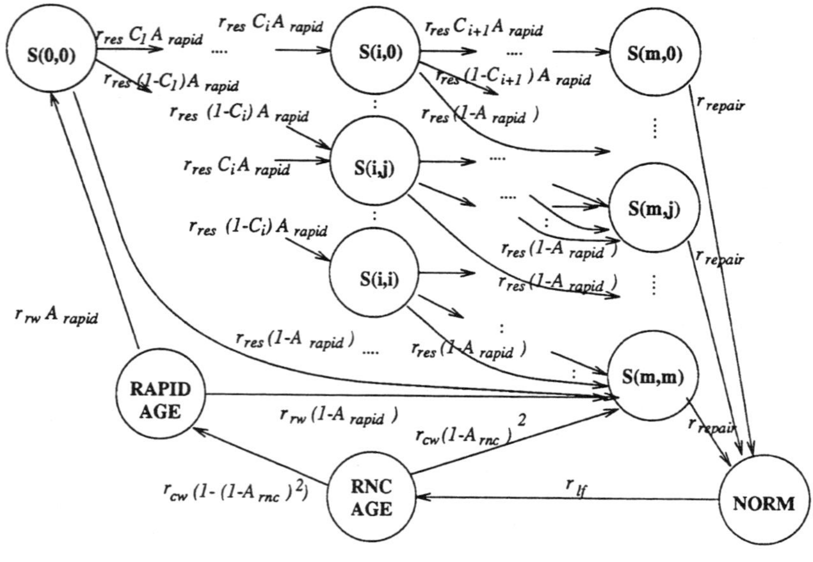

Figure 4. A general state transition diagram for a cable cut with *m* reroute paths

Integrated Network Management, III (C-12)
H.-G. Hegering and Y. Yemini (Editors)
Elsevier Science Publishers B.V. (North-Holland)

A software architecture for Management Information Model definition, implementation and validation

Florence Dossogne and Marie–Pascale Dupont

Alcatel Alsthom Recherche, DOM,
Route de Nozay, 91460 Marcoussis, FRANCE

Abstract

This paper proposes software tools and techniques for the definition (together with static validation), implementation, and dynamic validation of systems management information models. Definition of information models can be done with the help of the *ASN.1/GDMO editor*, a software tool which provides dedicated edition and visualization support, document library management, and a whole range of static validation facilities, of prime importance due to the amount of information to manipulate and to the complexity of the syntactic and semantic rules to be enforced. Implementation can be done with the help of the *ASN.1/GDMO translator*, a software tool which automatically generates C++ source files from ASN.1/GDMO information model specifications, on the basis of a well–defined generic implementation model and of a library of C++ basic classes. Dynamic validation can be done with the help of a prototype–based architecture, which provides support for (a) customizing an implemented information model, (b) integrating it in a prototype agent/manager architecture, (c) creating a prototype *management information base (MIB)*, (d) selecting prototype TMN management functions from a library, and finally (e) integrating and running the resulting prototype application. The dynamic validation architecture is flexible enough, so that its use in wider contexts can be envisaged.

1. INTRODUCTION

One of the main Alcatel activities concerns the development of Operation, Administration, Maintenance and Provisioning (OAM&P) software for Telecommunication Systems. Developing such applications is largely driven by existing international standards, the most relevant of which describes the notion of Telecommunication Management Network (TMN) [1]. Amongst other things, the TMN specifies an information model using object oriented concepts. This information model is conceptual, and therefore does not at all address implementation aspects, which are left to the application developers.

In the context of an Alcatel N.V. research project, Alcatel Alsthom Recherche is developing software tools and techniques, presented in this paper, which target those aspects that are not addressed in the standards.

Section 2 justified the Alcatel Alsthom Recherche approach to information model development, which identifies three major stages: definition (and static validation), implementation, and dynamic validation.

The following sections 3 to 5 then detail the software tools and architecture which support each stage.

Finally section 6 addresses possible re–use of the dynamic validation architecture in wider contexts.

2. RATIONALE

The application domain that is targeted in this study is the development of Operation, Administration, Maintenance and Provisioning (OAM&P) software for Telecommunication Systems.

In the standardization context, TMN provides a telecommunication–specific understanding of *systems management* principles, specified by joint work of CCITT and ISO/IEC organizations (see [2] for a synthesis of this work, or [3], [4], [5], [6] for detailed information).

Systems management standardization provides a conceptual model for the distributed management of open systems. Central to this is the support for expressing the information that is exchanged between managing and managed systems. The information is described according to object–oriented concepts, and is based on *managed objects*, which are capable of emitting *notifications*, and on which *operations* can be performed. The description of the information model is achieved in accordance with the generic *management information model* [4]. Pragmatic standards have also provided libraries of information models, the most general of which is specified in [5].

When a systems management OAM&P application is being developed, a major part of its specification consists in the *information model* it manipulates. This information model is usually composed of the most general managed object classes (e.g. *top*, *system*, both defined in [5]), with the addition of more specific ones, dedicated to the particular kind of equipment that is being managed (e.g. *indirect adaptor sink, regenerator section orderwire*, from the SDH model [7]) and/or to the particular functional area that is being applied (e.g. *alarm event record* in the case of an alarm reporting application).

Telecommunication manufacturers like Alcatel also develop new generation equipments (e.g. cross–connects for ATM networks), for which a standardized information model does not always exist. In such case, it is of prime importance to define new managed object classes and their associated characteristics, which at a later stage will be submitted to the standardization bodies for conformance and inclusion in standardization documents.

As a consequence the number of definitions continuously increases, leading to growing libraries of managed object classes and characteristics, the management of which is not simple and needs to be supported. In addition the definition of new information models is driven by quite complex syntactic and semantic rules (corresponding to ASN.1 and the Guidelines for the Definition of Managed Objects, or GDMOs, described in [8], [4] and [6]), which involves a requirement for *static validation* support. Software tools can be very valuable in answer to these needs. Section 3 describes such a tool: the *ASN.1/GDMO editor.*

Following this task of information model *definition*, providers of TMN OAM&P solutions for telecommunication systems must develop what we can call the TMN interface of new or existing equipments, which consists in an *implementation* of the various information models supported by the equipments, in the perspective of a standardized management.

The standardized syntax and semantics of information model definitions being largely specified, a very valuable assistance for the implementation of information model consists in supporting the automation of this implementation (in all or part). For this purpose it is necessary to first produce an implementation model on the basis of the standardized specifications, then construct a translation tool from the specification to the implementation models (see Figure 2).

It is worth recalling that the information models are specified according to object–oriented concepts like class, inheritance, message passing, etc. This is the reason why it is thought of prime interest to develop an object–oriented implementation model, and later to choose object–oriented technologies in order to support the implementation. The C++ language has been chosen, because it conveys an adequate trade–off between the advantages of object–oriented languages and the qualities of an efficient and powerful language. Additionally C++ permits a continuity from the procedural C–language, previously widely used for such applications in the Alcatel context.

Section 4 describes the chosen implementation model and a corresponding automatic translation tool: the *ASN.1/GDMO translator.*

The static validation of an information model definition consists in verifying that its fulfills the syntactic and semantic rules defined on ASN.1 and GDMO. Likewise the static validation of an information model *implementation* consists in verifying the adequacy of the implementation model with respect to the syntactic and semantic rules of the chosen implementation language: therefore this validation is taken care of during the C++ compilation of the generated source code.

In such a complex context, static validation is not sufficient in order to ensure that the definition and implementation of the information model fulfills the requirements that are demanded of it. In addition, *dynamic validation* is also required, for which the proposed solution follows a prototyping approach.

A detailed state–of–the–art study on software prototyping [9] has shown that, in numerous cases, validation of the requirements (in TMN contexts, the information model is part of the requirements) can receive a significant amount of help through prototyping. Following this, the proposed approach consists in developing rough implementations of some *TMN management functions* (as specified in CCITT recommendation M.3400 [10]), and running them on the implemented information model. This approach, detailed in section 5, participates in the dynamic validation of the information model *definition* and its *implementation.* It consists in several steps: (a) a customization of the implemented information model, (b) the integration of the customized information model in a prototype agent/manager architecture, (c) the creation of a prototype *management information base (MIB)*, (d) the selection in a library (or development) of prototype TMN management functions and facilities to pilot the prototype, and finally (e) the integration and running of the resulting prototype application, which permits the validation.

The following sections detail the software tools and techniques for the three steps of: definition and static validation, implementation, and dynamic validation of information models.

3. DEFINITION AND STATIC VALIDATION OF AN INFORMATION MODEL

The standardized specification of an information model consists in defining managed object classes and their associated characteristics using a set of GDMO templates [6], themselves pointing on ASN.1 type and value definitions [8]. The class characteristics relate to attributes, management operations that can be performed upon a managed object, the behaviour exhibited by the managed object in response to these operations, and the notifications it may emit. The various templates and ASN.1 modules are grouped within documents.

Following this, the purpose of the *ASN.1/GDMO editor* is to help the specifier define statically valid sets of managed object classes with their associated characteristics and ASN.1 definitions. Help consists in supporting the syntactic and semantic rules, checking the consistency rules and managing the various definitions, as it is detailed in [11].

First of all, the *ASN.1/GDMO editor* provides facilities to consult the content and the organization of an information model: lists of existing entities (documents, templates or ASN.1 modules); definitions of templates or ASN.1 modules; inheritance tree and naming schema [4].

Then, it supports edition facilities concerning the addition/suppression, of the various document entities on the one hand and of the information contained within their definitions on the other hand. This information concerns references to other definitions as well as specific possibilities according to the type of definition currently edited. The associated support consists in providing a means to select an entity among a list of defined ones (in order to use it as a reference in the definition of another entity), and customized menus.

Static validation of the information model is ensured by the *ASN.1/GDMO editor* in the following way:

Forward syntactic and semantic checking is enabled during edition, thanks to three main features. First, the visualization/selection/reference facilities ensure that all references exist and are valid with respect to the types (of templates or ASN.1 definitions) requested by the standards. Then, the systematic checks of any user–input value ensure that all names and object identifiers (OIDs) that are entered, are valid and unique. Finally, the customized menus ensure that all specific elements contained in a definition are allowed (e.g. a *confirmed mode* indication can only be added to an action definition).

Additionally commitment–based checking is proposed, which carries out a wide range of verifications on an existing document. They ensure that all the definitions are complete, and perform consistency verifications according to the rules and constraints defined in the standards [4][6]. For example an attribute group to which attributes have been added must be extensible, a parameter must be used in a compatible way with respect to the context it has been defined for (e.g. specific–error), etc. If valid, the document may be transformed into a library, in order to allow other documents to reference its entities for the definition of an information model, and to be translated into a programming language.

Finally, information model documentation publication is an additional facility which links the tool with the outside world. Documentation can be a textual document for the purpose of producing a report, but it can also be a set of files which contain the ASN.1 and GDMO description of the information model, to be further treated as input for the *ASN.1/GDMO translator* (see section 4.2).

The *ASN.1/GDMO editor* proposes a graphical, window–based, user interface. The various windows are summarized on Figure 1 which also shows the navigation possibilities between windows.

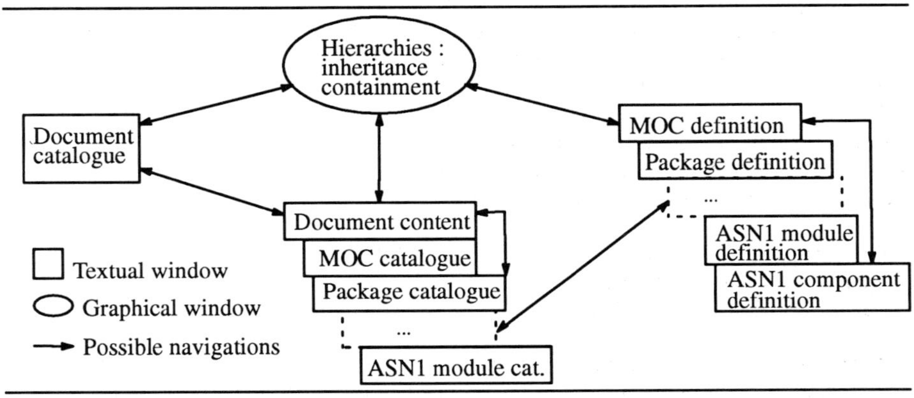

Figure 1. GDMO Editor user interface windows

4. IMPLEMENTATION OF AN INFORMATION MODEL

The proposed approach is that of automatic generation of part of the information model implementation from its specification. As shown on Figure 2, this involves the definition of an *implementation model* [12], which maps the ASN.1/GDMO syntax and semantics onto a C++ equivalent, and the development of the *automatic translator* [13] producing C++ source code from ASN.1/GDMO specifications according to the implementation model.

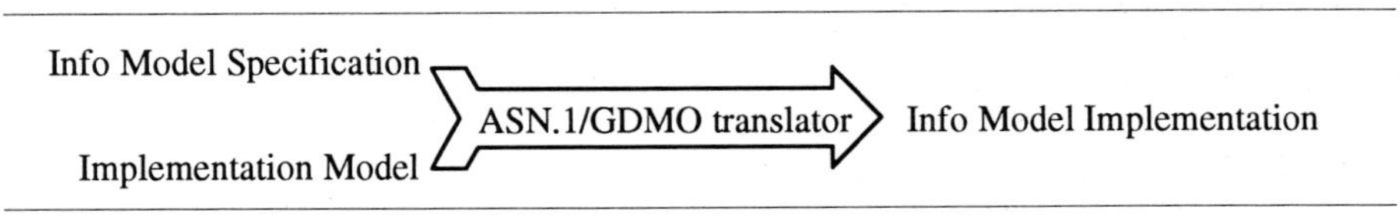

Figure 2. Information model implementation

4.1. The chosen implementation model

The chosen C++ implementation model was developed with the main constraint of *automatization*, since it is automatically produced by the *ASN.1/GDMO translator*.

In order to implement the *shared management knowledge* [3] as distributed information, specific C++ classes are defined, which encapsulate the information specific to the access procedures (i.e. local or remote access) to the managed object classes. These classes are clearly separated from the C++ classes which implement the managed object classes and their associated definitions.

The management information model is mapped onto the C++ object model: each managed object class is implemented as a C++ class, the standardized inheritance is implemented by the C++ inheritance, the various templates are implemented as C++ classes. The packages are represented as a C++ class hierarchy (one C++ class per package). The instances of these classes are exclusively used during the construction of managed objects.

All standardized rules are implemented in C++: they include the mechanism of name binding; object destruction restrictions; *scoping* and *filtering* for CMIS [14] messages.

Some of the information described in ASN.1/GDMO is in natural language, which cannot be automatically translated into a C++ equivalent (e.g. the natural language description of managed object class *behaviour*). Therefore it is not addressed by the model. So that the resulting implementation can be completed, the implementation model includes the notion of *C++ class doubling*: every incomplete aspect is described in an abstract C++ class (the "automatic" class), and another C++ class is also generated (the "user" class), which directly derives from the abstract one, but does not contain additional information, and whose empty implementation can be customized through a process other than the *ASN.1/GDMO translator*.

The implementation model actually identifies two types of C++ classes: (a) *basic* classes, which are common for every information model implementation (e.g. the abstract C++ classes corresponding to the GDMO templates: *managedObjectClass*, *package*, etc), are developed once and for all, and kept in a *basic class library*, (b) *generated* classes, which depend on the information model being treated (e.g. a specific *alarmRecord* C++ class).

4.2. The *ASN.1/GDMO translator*

The input of the *ASN.1/GDMO translator* is a set of ASCII files containing ASN.1 module or GDMO document specifications.

The translation process is decomposed in three steps, according to a well–known compilation schema (see Figure 3): (a) the input documents are processed by a *parser*, which performs lexical, syntactic, and semantic analysis of each document independently; (b) the parsed documents are processed by a *linker*, which performs the resolution of external references between documents, and generates archive files; and (c) the linked documents are processed by a *C++ code generator*, which produces C++ source files.

ASN.1 type definitions are actually translated into C language type definitions and access functions by the ISODE *pepsy* tool [15]. These elements are encapsulated behind C++ classes, therefore the call to the *pepsy* tool is transparent when the ASN.1/GDMO translator is invoked.

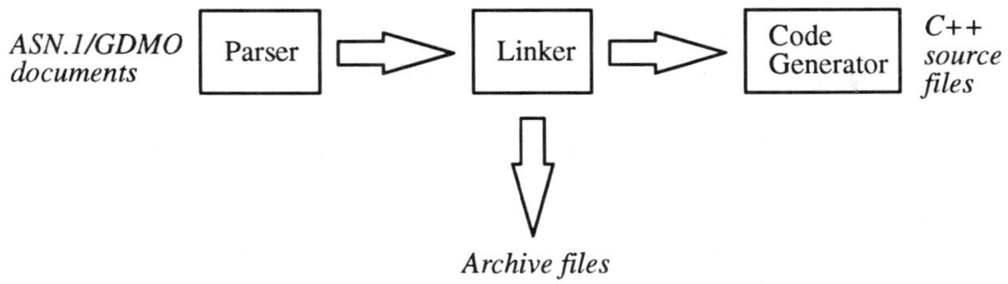

Figure 3. ASN.1/GDMO translator

5. DYNAMIC VALIDATION OF AN INFORMATION MODEL

The dynamic validation concerns both the information model *definition* and *implementation*, and is supported by a prototyping approach, which consists in quickly running a prototype systems management application on an information model implementation, in order to assess the information model's quality and adequacy.

A prototype systems management application is made of several architectural elements, whose contents originate from various sources, as shown on Figure 4: some are available as libraries, developed once as for all; some are automatically generated with the assistance of the information model tools (see section 4.2); some are specific to a given prototype. The dynamic validation activity consists in linking those elements together, running the resulting C++ `main()` program, assessing the execution of the prototype, consequently modifying the information model, and looping the link–evaluate–modify process until a satisfactory solution is found.

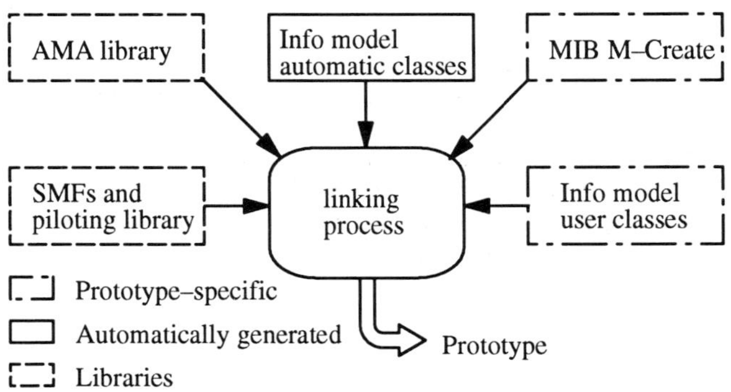

Figure 4. Architecture of a prototype systems management application

The following sections detail the creation and use of the various components of this architecture.

5.1. The *AMA library*

Following the Systems Management standards, it is possible to identify common entities for every open system, which fulfill the roles of agent/manager/communication [3]. This can be implemented under the form of a C++ library, called the *agent/manager architecture* (AMA) *library* [16], and provides basic support for implementation of open system prototypes.

The *AMA manager classes* are used to convey the information required to access managed object information, in response to a management application call. These classes implement the shared management knowledge in providing a uniform access to managed objects, be they locally or remotely located.

The *AMA agent classes* implement some basic *systems management functions*, corresponding to the most generic functionalities required from any open system accessible from a CMIS–like application programming interface (based on a limited number of requests as in CMIS, like `get()`, `set()`, `create()`): the reception, forwarding and management of systems management operations (*operation handler*), the reception of managed object notifications and their forwarding towards adequate open systems (*event handler*), the subscription, reception and logging of event reports from local or remote open systems (*log handler*). Additionally, AMA agent classes provide an adequate application programming interface to a managed object repository mechanism. Figure 5 below sums this up.

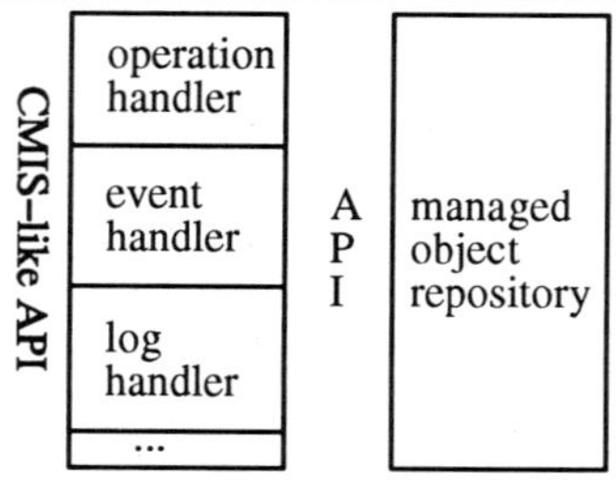

Figure 5. AMA agent classes support

The *AMA communication classes* support the inter–working between several open systems, and provide a CMIS–conformant application programming interface, interfaced with a communication protocol stack.

5.2. TMN management functions and prototype piloting facilities

The dynamic validation architecture shown on Figure 4 includes a library which contains C++ classes corresponding to the implementation of CCITT–standardized TMN management functions. Additionally, it also contains classes providing facilities to pilot a systems management application prototype.

TMN management functions

The standardization in the field of management application description is led by the CCITT work: Recommendation M.3020 (M.meth, [17]) articulates the TMN functional specification in the form of *TMN management services* (e.g switching management, customer administration), which can be decomposed in *TMN management service components* (e.g alarm surveillance), themselves decomposed in *TMN management functions* (e.g report alarm, allow/inhibit logging). In this decomposition, TMN management functions (described in M.3400 [10]) represent the smallest part of the TMN management service as perceived by the user of this service.

The chosen approach consists in selecting adequate TMN management functions for a given prototype. Since the function specification is standardized, once they are implemented, they are integrated in the library, where they can be retrieved and re–used in another prototype.

Prototype piloting facilities

A systems management application prototype is made of certain functionalities and associated semantics, but also of what we can call a *piloting interface*, which include such facilities as graphical visualization of a management information base, human–machine interface commands to initialize, run, step–by–step run, stop the prototype executions. These facilities are contained in the library, accessible by each prototype.

5.3. Management information base initialization: the *MIB editor*

A systems management application prototype not only manipulates an information model, but generally an instance of it, under the form of *management information bases (MIBs)*, made of managed object class instances and their instantiated characteristics: *values* of attributes, *selection* of conditional packages, etc.

The purpose of the *MIB editor* is to support a user–friendly instantiation of managed objects. Its graphical user interface provides the same kind of visualization/selection/reference facilities as the *ASN.1/GDMO editor* (e.g. visualization of the containment tree). Furthermore the edition of both the information model and its instantiation can even be done in a single session. The *MIB editor* provides additional facilities, like internal maintenance of the mapping between *labels* and corresponding *object identifiers.*

MIB edition results in generating as many CMIS M_CREATE indications as there are instances to create, which can be "plugged" in the prototype at the *AMA agent's operation handler.* This choice has several advantages: (a) M_CREATE parameters provide for all necessary information on a new managed object class instance (see Table 1); (b) M_CREATE indications can be used in a wider context as the one presented here, in particular it is independent of the *AMA library.*

Table 1
CMIS M_CREATE indication parameters

Parameter name	**Comments**
Invoke identifier	uniquely identifies the indication
Managed object class	which is to be instantiated
Managed object instance	the name of the new instance
Superior object instance	the name of the containing MO
Reference object instance	to create a "clone" of another MO (optional)
Attribute list	indicates all necessary attribute values

5.4. Information model *user* classes

As is described in section 4.1, the implementation model includes a principle of C++ class doubling supporting the customization of all incomplete aspects of the information model definition.

This customization consists in implementing these empty classes according to the prototype, and mainly concerns action mock–ups, behaviours and conditions of notification emission. It is assisted by the file modularity produced by the *ASN.1/GDMO translator:* it generates one file (two actually, header plus implementation) for all automatic classes, and as many files as user classes. Furthermore, the user classes are not overwritten when the *ASN.1/GDMO translator* is run again.

6. FUTURE DIRECTIONS

The prototype–based architecture proposed in section 5 was developed in order to fulfill the requirements of dynamic validation of information model definitions and implementations. The purpose of this section is to mention alternative uses of this architecture, in order to illustrate its flexibility.

The *informal* assessment proposed in prototyping could be formalized. In the described architecture, the quality and adequacy of the information model with respect to the requirements is assessed through several trial–and–error loops, and it would be of great interest to quantify the "distance to the ideal solution", with the use of metrics (integrated in the processing of the prototype management functions).

When an OAM&P application developer faces the implementation of management information bases (MIBs), and not only information models, he/she has to raise the issue of *managed object storage*. In that context, several possibilities exist as regards the information repository used to store the managed objects and their characteristics: using a relational database system, using an object–oriented database system, using a file system, using a hybrid approach (i.e. choosing a different system for different managed objects, or different management applications), etc.

The architecture can be used as a starting point for the validation of a choice among management information base storage solutions. The required effort consists, of course, in adding the various repository solutions to the environment, in extending the *managed object editor* with the ability to select a repository among several, and also in extending the *AMA library* with access to the repositories.

The final purpose of OAM&P development is to provide solutions for network management systems, including real implementation of management applications which manipulate an information model. The architecture can be used as a starting point in this context.

The required effort to arrive at such a solution involves the enhancement of the *AMA* and *management function libraries*. The *AMA library* can be largely re–visited, with the inclusion of distributed application management, since it is advisable to cater for the development of an operations system as a distributed application. Notions of efficiency, additional functionalities like support for *de facto* standards (e.g. Open Software Foundation's Distributed Management Environment) can be introduced. The implementation of the TMN management functions themselves can be improved, with a better efficiency, quality insurance, user–documentation, configuration and version management, etc.

7. CONCLUSION

This paper proposed software tools and techniques which support the definition, implementation, and dynamic validation of systems management information models, as a possible contribution to a development platform strategy.

Initial versions of the *ASN.1/GDMO editor* and *translator* were developed at Alcatel Alsthom Recherche in 1992, and they are being evaluated in Alcatel real contexts, with a view to specifying tool upgrades, to be provided in 1993.

The dynamic validation architecture is currently in an advanced specification phase, and an initial running prototype systems management application is targeted in 1993.

8. REFERENCES

[1] CCITT, Recommendation M.3010 (formerly M.30), *Principles for a telecommunications management network*, Draft recommendation, Study Group IV, Report R28, Dec. 1991.

[2] MP. Dupont, A. Plainfossé and F. Dossogne, *Survey of the standardization work in the TMN field*, TOOPIE deliverable D2.1.1, Alcatel Alsthom Recherche, 91460 Marcoussis, France, 1992.

[3] ISO/IEC, DIS 10040, *Information Technology – Open Systems Interconnection – Systems Management Overview*, N6353, Jun. 1991.

[4] ISO/IEC, DIS 10165–1, *Information Technology – Open Systems Interconnection – Structure of Management Information – Part 1: Management Information Model*, N6351, Nov. 1991.

[5] ISO/IEC, DIS 10165–2, *Information Technology – Open Systems Interconnection – Structure of Management Information – Part 2: Definition of Management Information*, N6363, Sep. 1991.

[6] ISO/IEC, DIS 10165–4, *Information Technology – Open Systems Interconnection – Structure of Management Information – Part 4: Guidelines for the Definition of Managed Objects*, N6309, Aug. 1991.

[7] CCITT, *Draft SDH Information Model*, Temporary document 17/XV/6, Nov.1991.

[8] ISO/IEC, IS 8824, *Open Systems Interconnection – Specification of Abstract Syntax Notation One (ASN.1) – Second Edition*, Dec. 1990.

[9] F. Dossogne, *A State of the Art on Prototyping*, TOOPIE project deliverable D5.1.1, Alcatel Alsthom Recherche, Marcoussis, France, Nov. 91.

[10] CCITT, Recommendation M.3400 (formerly M.func), *TMN Management Functions*, Study Group IV, Report R28, Dec. 1991 (first published 1992).

[11] MP. Dupont, *OCE: An Object Class Editor*, working document TOOPIE/OCE/SP/002, Alcatel Alsthom Recherche, 91460 Marcoussis, France, June 1992.

[12] F. Dossogne, MP. Dupont and Y. Berteaud, *REP: A representation of the MIM in C++*, working document TOOPIE/REP/SP/001, Alcatel Alsthom Recherche, 91460 Marcoussis, France, Jul.1992.

[13] F. Dossogne and C. Santellani, *ACC: An ASN.1/GDMO compiler to C++*, working document TOOPIE/ACC/SP/003, Alcatel Alsthom Recherche, 91460 Marcoussis, France, Jul. 1992.

[14] ISO/IEC, DIS 9595, *Information Technology – Open Systems Interconnection – Common Management Information Service Definition*, N5302, Nov. 1990.

[15] M.T. Rose et al, *The ISO Development Environment: User's manual, Volume 4: The Applications Cookbook*, Version 7.0, Jul. 91.

[16] F. Dossogne, *AMA: Agent/Manager Architecture, a functional architecture for TMN–compliant Open Systems*, working document TOOPIE/AMA/SP/005, Alcatel Alsthom Recherche, 91460 Marcoussis, France, Oct.1992.

[17] CCITT, Recommendation M.3020 (formerly M.meth), *TMN Interface Specification Methodology*, Study Group IV, Report R28, Dec. 1991 (first published 1992).

V.

TELECOMMUNICATIONS MANAGEMENT

C.

ASSURING NETWORK QUALITY

Integrated Network Management, III (C-12)
H.-G. Hegering and Y. Yemini (Editors)
Elsevier Science Publishers B.V. (North-Holland)

A Hybrid Expert System / Neural Network Traffic Advice System

Rodney M. Goodman[1], Barry Ambrose[1], Hayes Latin[2] and Sandee Finnell[3]

[1] Electrical Engineering Dept., California Institute of Technology, Pasadena, CA 91125.

[2] Systems Technology, Pacific Bell.

[3] Southern California Network Management Center, Pacific Bell.

Abstract

A hybrid Expert System / Neural Network Traffic Advice system has been developed to aid traffic management in the Southern Californian telephone network. The system has been working on live data since September 1991 and generates rerouting advice that agrees with that generated by the present network management procedures. A modular software design was adopted to allow for evolution. A graphics interface allows the user to easily navigate through the display of exception conditions and advice. Exceptions are shown highlighted on a map of Southern California. A neural network telephone traffic predictor was tested prior to incorporation in the system.

Keyword Codes: I.2.1, I.2.6, C.2.3

Keywords: Applications and Expert Systems, Learning, Network Operations

Introduction

In September 1990, a project was started by CalTech and Pacific Bell to investigate the implementation of a real-time autonomous expert system which would aid network management of the Pacific Bell Southern Californian telephone network [GOOD92]. The ultimate aim of the project was to produce and automatically implement recommendations for network controls. To do this, the knowledge in everyday use by the network management staff had to be written down, and coded for use in a computer system while retaining flexibility and openness to change.

The situations that the network management staff are exposed to are complex. Every day new problems arise. Examples are single random overloads of traffic on a given route, high volume phone-in conditions, disasters such as earthquakes, unusual calling patterns such as occur on Mother's Day, killer trunks, which are trunks that seize a call attempt and immediately release and other events. The introduction of new facilities, such as advanced signalling systems, adds additional problems that have to be diagnosed.

While many implementations exist that embed Expert Systems in existing operations software

[CALL88, CRON88, ERIC89, FRON91, MARQ88], the constraint of having to operate in real time and therefore prioritize exception conditions made this an interesting project. In addition, to our knowledge, the incorporation of neural networks to aid in time series prediction of telephone traffic statistics had not been previously attempted.

To date, a number of software modules have been developed which use data gathered in real time from the network. The software can produce a graphic analysis of the network, describe how the network is functioning in terms of exception conditions and recommend controls. The controls are given in the format that the network manager is accustomed to. Each module has been extensively tested.

The rest of the paper gives a description of the Pacific Bell telephone network and the architecture of the Network Operators Advice and Assistance (NOAA) system. This is followed by the details of how NOAA works presented logically in the order (i) input information available to NOAA (ii) decision making and decision support for NOAA (iii) controls recommended by NOAA. The map based Graphics Interface developed for NOAA is described separately followed by the neural network experiments and finally an indication is given of future work.

The Pacific Bell Network

In order to gain some appreciation of the network management tasks, one must have a description of the network to be managed. The Pacific Bell Southern Californian telephone network provides service to at least 1 million subscribers and has a hierarchical structure. For administrative and legal purposes it is divided into Local Access and Transport Areas (LATAs) [BELL86a] and the network management center in Pasadena manages LATA's 5 and 6. Geographically LATA's 5 and 6 extend to San Diego in the south and include all the Los Angeles urban area.

The network is hierarchical. <u>End offices</u> are the exchanges that serve customers, and <u>tandems</u> are the exchanges used for traffic between end offices that are not directly connected [BELL86a]. There are 6 tandems to be managed and over 200 end offices.

There are two types of trunk groups. <u>High usage</u> trunk groups exist between offices that are sufficiently close and have sufficient traffic flow between them. For economic reasons, these high usage trunk groups, which are the most common type of trunk group, are dimensioned to be lossy, i.e. during the busy hour they are not guaranteed to have enough capacity to carry all offered traffic. Traffic (call attempts) will therefore overflow onto a smaller number of <u>Final</u> trunk groups which are dimensioned to provide a good Grade of Service. In general there will be a final route between each end office and its parent tandem and between the tandems. It is these final routes that provide the backbone of the network. The final routes are therefore closely monitored by the network managers.

Architecture of NOAA

The Architecture of NOAA is shown in Figure 1. The Pacific Bell network management system is called EADAS [BELL86b]. NOAA is connected over a 4800 baud data link and appears as an ordinary operators terminal to EADAS. NOAA then runs on a Sun workstation.

There are four software modules that make up NOAA, each made up of about 3000 lines of C code. These are:

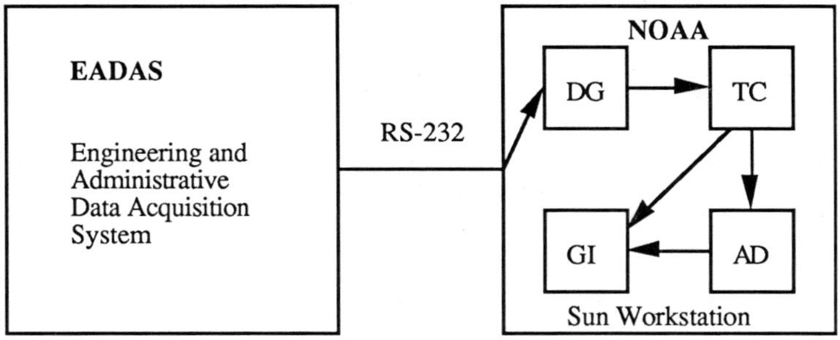

Figure 1 Architecture of NOAA

- data gathering (DG).

  This module is responsible for logging into the EADAS system and maintaining the basic dialog with EADAS. The UNIX tip command is used for this purpose. A significant effort was put into making this module robust in the face of communication difficulties, such as loss of synchronization in the EADAS-NOAA dialogue.

- tandem calculations (TC).

  This modules takes the data which has been downloaded from EADAS and carries out the calculations needed to transform the data into usable statistics. The data is downloaded in register format. These registers, also known as peg counts, are integer counts of events such as trunk seizures, that have happened in a five minute interval. The calculations that are carried out are well defined and produce more useful statistics such as seizures per circuit per hour. In the next version of NOAA, neural networks will be used to predict future traffic statistics given the previous readings. This should improve the accuracy of the advice.

- Advice (AD)

  Given the statistics on the various trunk groups, exception situations are noted and a severity measure associated with each exception. A diagnosis based on observed symptoms is carried out according to rules supplied by network management staff. Finally advice is given about the response to the exception. If a control is advised, it is given in the format expected by the network operator. This can be then approved by the operator and implemented through EADAS.

- Graphics Interface (GI)

  The graphics interface uses a map of Southern California. Exceptions are indicated by red, suggested controls by green and actual controls already in place by blue. Additional information is given when the operator clicks on the map. This information includes calculated statistics on the route and any advice generated for that route.

Statistics

NOAA makes its diagnosis on the basis of available information about the trunk groups. This information is available to NOAA in the form of counts of certain events for each trunk group during a 5 minute period. These events are for instance (i) a trunk being seized (ii) a seizure that is not successful i.e. overflow (iii) a survey of trunk occupancy at a given instant. On the basis of these counts, the following standard statistics are calculated:

ACH	Attempts per circuit per hour
CCH	Connections per circuit per hour
OFL	Percentage of attempts that overflowed
USG	Percentage of trunks occupied on average
HT	Holding Time of Calls

The 5 minute period is a compromise. The shorter the period between data collections, the faster the response to any potential problems, but the larger the amount of information to be processed, requiring more expensive hardware. The 5 minute period means that problems can be cleared relatively rapidly. It is known that mean traffic increases do occur on a time scale of the holding time whereas traffic fluctuations occur on a time scale of the inter-arrival rate of calls [FILI90].

Since a large amount of information is available, filtering is carried out. Firstly exceptions on high usage routes are ignored and only final route data are examined. Secondly exceptions on special purpose final routes are ignored. For example, we expect a trunk group for the time announcement to have a short holding time. Thirdly overflows are treated as having higher priority than other exceptions, since an overflow on a final may represent lost traffic.

Decisions

When an exception condition has been noted on a trunk route, there could be many possible explanations for it. Typically phone-ins to radio stations and TV stations may generate excess call attempts. Facilities failures may mean that overflow shoots up on related trunk groups. Occasionally the data gathering may interfere with maintenance operations and unreliable data is returned. Finally random overflows can occur on individual trunk groups.

For any reroutes, a search for spare capacity is carried out. If there is an overflow problem between A and B where A is an end office and B is a tandem, then the program first looks for capacity on other trunk groups going from A to B, and then carries out a search for A to B via C possibilities. Here C has to be one of the 6 tandem exchanges, because of the hierarchical nature of the network. The candidate reroutes are then sorted according to available capacity and controls are suggested that make use of the minimum number of reroutes that achieves the required capacity.

A severity measure is attached to each exception. This measure is biased to give overflow exceptions a high severity figure, since overflows can result in lost traffic. The severity measure is also made proportional to the size of the trunk group and the amount by which the statistic is above threshold. Thus a 10% overflow on 10 circuits is considered equivalent to 1% overflow on 100 circuits. A log scale is then used to scale the figure between 0 and 100. This severity measure has proved useful for sorting exceptions prior to displaying them. In a full scale emergency with the loss of a lot of capacity, a sort by severity would indicate which trunk groups to deal with first. Note that 911 (emergency) calls have their own reserved trunks from end offices and are not supervised by NOAA.

The rules used in the program are of three separate types:

- rules which indicate which exceptions can be safely ignored. For example overflow on high usage routes is ignored;
- rules which indicate which routes can be use as candidate reroutes;
- rules that map a suggested reroute into a list of controls to effect the reroutes. E.g. certain other routes may have to be "finalized" first to prevent a round-robin situation.

```
DISREGARD ANY EXCEPTIONS ON TRUNK GROUP COMMON LANGUAGES (CLLIS)
ENDING WITH "MD" (EX: LSANCA02AMD; LSANFDRCCMD). EXCLUDE SAME WHEN
SEARCHING FOR VIA ROUTE CANDIDATES.

DISREGARD ANY EXCEPTIONS ON CLLIS INDICATING "PB" IN THE STATE
DESIGNATION (EX: OKLDPB0349T). EXCLUDE SAME WHEN SEARCHING FOR VIA
ROUTE CANDIDATES.

DISREGARD ANY EXCEPTIONS ON THE FOLLOWING HIGH VOLUME CALL-IN
CLLIS:

  HLWDCA01520 SNANCA01977 COTNCA1143A
  SIMICA11629 SNDGCA0157X
```

Table 1 Typical Network Management Rules

Some of the above rules were already written down in operators handbooks. Others were supplied by the network management staff. Examples of the rules are given in Table 1.

As with any rule based system, including the rules in a computer program has the advantage that any special cases that only occur rarely are immediately recognized and the operators handbook does not have to be referred to. However for complete trust in the system, the rule base has to be extensively tested and compared with the experts analysis in a wide range of cases.

Controls

It was mentioned above that there are many causes of exceptions on the trunk groups. Similarly there are many possible controls that can be put in to improve the network throughput. In general, controls can be divided into protective controls and expansive controls [FILI90, BELL88].

Protective controls are used to cut down the amount of call attempts entering the network. For example in the event of a phone-in, controls can be put into each end office to block callattempts to that telephone number. Alternatively a certain percentage can be let through. This limits the amount of congestion in the network. Clearly the calls destined for a phone-in are "bad" traffic as there is a high possibility of non completion, and therefore they should be blocked at source and not rerouted.

On the other hand, expansive controls allow trunk groups that are normally separate to share their capacity. If there is a single random overload, then expansive controls are called for. Two alternatives here are overflow rerouting, which reroutes overflow traffic only, and an immediate reroute which puts all call attempts onto the alternative route first. Each is appropriate for certain cases. Finally spray rerouting is possible, whereby a number of alternatives are given and each is tried for a given call. Clearly todays telephone exchanges have a lot more flexibility than the hardwired mechanical exchanges that were common thirty years ago.

Graphics Interface

It was found that this application was eminently suitable for graphical interfacing. Initially a scrolling list of exception conditions was presented to the operator in a window. Buttons allowed the operator to pull up information about the statistics on a trunk group and the recommended advice.

This was then modified to allow the operator to choose from a number of lists of exceptions, each list corresponding to a different statistic over threshold.

Finally a map interface was developed (see Figure 2). Without doubt, this was the easiest way to (i) display an analysis of the network, showing the exceptions clearly (ii) pull up information about individual exceptions. The map, with its translation and zoom features, could be mastered by any user in minutes and added a "fun" element to navigation through the program. Coordinates of most of the cities in Southern California were available from a public database.

Neural Networks

Neural networks have been used in applications ranging from pattern classification to associative memories. One of their main features is the ability to learn a mapping between the network inputs and the outputs. In contrast to artificial intelligence algorithms, the learning is based on memorizing examples by the process of *adjusting weights* in the network, rather than looking up rules. Much progress has been made on the algorithms used to train neural networks [HKP91].

Figure 3 gives an example of a two input single output neural network. The two input single output linear predictor can be considered as a simple form of neural network.

In our case, to aid in traffic management, we tried to use the neural network to predict a future value of trunk occupancy on a route, based on six previous readings over a thirty minute period. The prediction would provide a better indication of spare capacity for rerouting purposes and could also be used for extrapolation in the event of data not being available. The advantage of using a neural network for this application is that it can implement non-linear mappings between the inputs (in this case the previous occupancy readings) and the output (the predicted occupancy reading). Figure 4 shows a graph of occupancy on a trunk route over the course of a day.

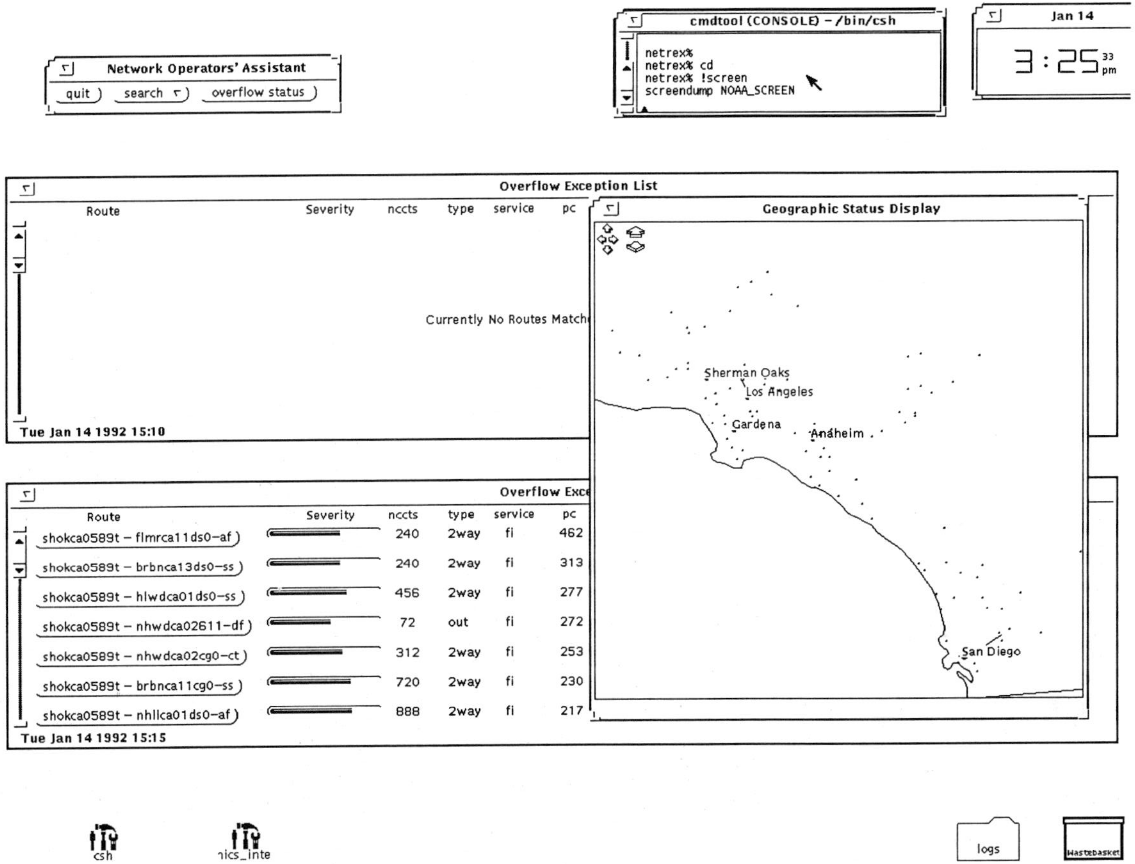

Figure 2 Graphical Interface for NOAA

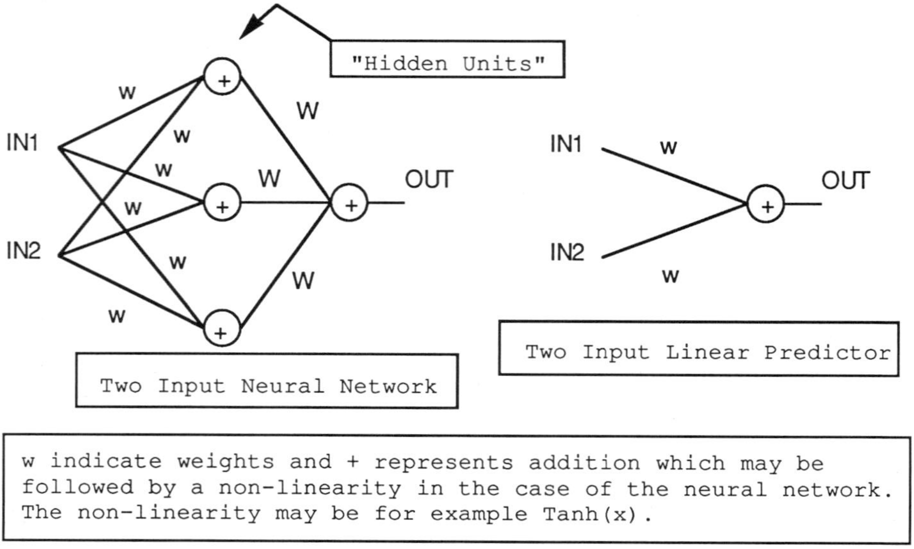

Figure 3 Neural Network and Linear Predictor

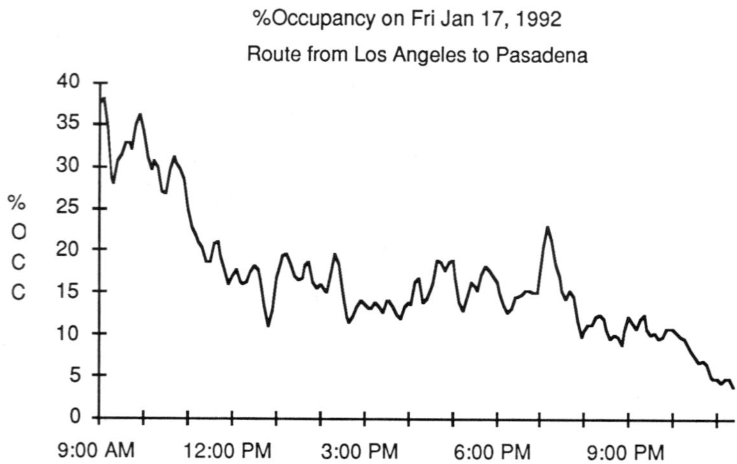

Figure 4 Traffic on a Trunk Route

After trying a number of simple examples, we choose a network with a small number of hidden units (3-5) for faster learning. We scaled the inputs to be between 0 and 1, although this was not essential. We used non-linear hidden units but linear output units. We used a training set consisting of 893 examples and a test set of 609 examples. We used a linear predictor (also known as a perceptron in the neural network literature) as a standard for comparison. The results are shown in Table 2.

	Training Set	Test Set
Linear Predictor	2.39×10^{-4}	3.63×10^{-4}
Neural Network	2.18×10^{-4}	8.20×10^{-4}

Table 2 Mean Square Error for Two Prediction Methods

The table indicates that the neural network was better able to memorize the training examples but less well able to generalize on the test set. Given the added computation required to learn the non-linear mapping, it would seem that a linear predictor is the most suitable method for prediction of telephone traffic. This can also be justified theoretically. It is well known that a linear predictor works well for stationary processes whose probability densities are approximately normally distributed as in the case of telephone traffic [CHRIS91].

Future Work

Over the past 18 months, much work has been done in interfacing NOAA to the Pacific Bell network management computer and building the infrastructure for an expert system. Over the coming months the rules implemented in the program will be further tested by continuing to run the program on live data. When sufficient cases have been tested, it is intended to close the loop and allow NOAA to carry out controls autonomously. Clearly considerations of reliability and robustness have also to be taken into account when this step is to be carried out.

The ability of NOAA to diagnose problems correctly and to take the correct actions will be enhanced if the system has other information sources besides EADAS. Two other sources being considered at present are NSCS which provides information about seizures of trunks, and a separate system which provides information about the SS#7 signalling network.

Experiments with neural networks have shown that at little cost, linear predictors can be used to extrapolate statistics and generate more accurate assessments of available route capacity. A short term goal is to build the linear prediction capability into the TC module.

Looking at the long term future for NOAA, the definition of a standard data format for exceptions and for statistical information about trunk group performance would help in minimizing the cost of upgrade of NOAA, as new versions of EADAS become available. As in any network management application, standardization of data formats between applications that share the data is an important requirement. The ISO standards work on fault management and performance management should go some way to filling this gap.

To summarize, network management advice is currently being generated for the Southern California telephone network. As the rules that generate this advice are tuned, a robust network management application will have been developed that relieves network management staff of some of the need to supervise the day to day running of the telephone network.

References

BELL86a Bellcore, Network Management Intra-LATA Network Fundamentals, BR 780-150-122, Issue 1, December 1986.

BELL86b Bellcore, Network Management Technical Overview, BR 780-150-130, Issue 2, December 1986.

BELL88 Bellcore, Network Management Control Descriptions, BR 780-150-128, Issue 3, June 1988.

CALL88 Callahan, P., Expert Systems for AT&T Switched Network Maintenance, pp. 93-103, AT&T Technical Journal, Vol. 67, No. 1, January 1988.

CHRIS91 Christensen, R., Linear Models for Multivariate, Time Series, and Spatial Data, Springer-Verlag, New York, 1991.

COOP81 Cooper, R. B., Introduction to Queueing Theory, 2nd ed., North Holland, New York, 1981.

CRON88 Cronk, R., Callahan, P., Bernstein, L., Rule Based Expert Systems for Network Management and Operations: An introduction, pp. 7-21, IEEE Network Magazine, Vol. 4, No. 5, September 1988.

ERIC89 Ericson, E., Ericson, L., Minoli, D., (eds), Expert Systems Applications in Integrated Network Management, Norwood MA, Archtech House, 1989.

FELL50 Feller, W., An Introduction to Probability Theory and its Applications, Volume I, ed. 3, Wiley and Sons, New York, 1950.

FILI90 Filipiak, J., Real Time Network Management, Draft Copy (not published), April, 1990.

FRON91 Frontini, M., Griffin, J., Towers, S., A Knowledge-Based System for Fault Localization in Wide Area Networks, pp. 519-530, IFIP Conf. on Integrated Network Management II, Washington, 1991.

GOOD92 Goodman, R. M. et al, Network Operators Advice and Assistance (NOAA): A Real-Time Traffic Rerouting Expert System, Globecom, Orlando, December 1992.

HKP91 Hertz, J., Krogh, A., Palmer, R.G., Introduction to the theory of neural computation, Lecture Notes Volume I, Santa Fe Institute Studies in the Sciences of Complexity, Addison Wesley, 1991.

MARQ88 Marques, T., A symptom driven Expert System for Isolating and Correcting Network Faults, pp. 6-14, IEEE Communications, Vol. 26, No. 3, March 1988.

Integrated Network Management, III (C-12)
H.-G. Hegering and Y. Yemini (Editors)
Elsevier Science Publishers B.V. (North-Holland)

AN EXPERT SYSTEM FOR REAL TIME FAULT DIAGNOSIS OF THE ITALIAN TELECOMMUNICATIONS NETWORK

Simona Brugnoni[a], Guido Bruno[a]
Roberto Manione[b], Enrico Montariolo[b], Elio Paschetta[b], Luisella Sisto[b]

[a]Operation and Maintenance System Department

[b]Artificial Intelligence Department

CSELT, Centro Studi E Laboratori Telecomunicazioni S.p.A.
Via G. Reiss Romoli 274
10148 Torino (ITALY)
Tel. +39 - 11 - 2285111
Fax. +39 - 11 - 2285520
Telex 220539 CSELTI

Abstract

Sinergia, an Expert System for the isolation and diagnosis of faults in the Italian Telecommunications Network, is presented.

Unlike most diagnostic Expert Systems, Sinergia works in real time providing fault diagnosis in a dynamic context, while the anomaly signals are continuously received.

Sinergia interfaces to various Operation Support Systems on the Italian Telecommunications network which are treated as Agents, and loads the network configuration incrementally on a demand basis from the corporate Data Base.

At present the fault diagnosis activity is carried out by maintenance experts faced with huge amount of data to analyze; the introduction of Sinergia has cut the data presented to the operator of more than one order of magnitude.

Keyword Codes: I.2.1; I.2.8; J.7
Keywords: Artificial Intelligence, Applications and Expert Systems; Problem Solving, Control Methods, and Search; Computers in Other Systems

1. INTRODUCTION

The Expert System (ES) technology has been largely and successfully applied in different telecommunications areas which traditionally include maintenance, provisioning and network administration [1, 2]. ESs are usually built up using rule-based languages that are especially designed to heuristically solve problems [3]. In fact, ESs rely on productions, or rules, that allow the specification of non-deterministic procedures that human experts use to solve real problems. Rules are coded with the traditional *If-Then* structure having conditions in the Left-Hand-Side (LHS) and actions in the Right-Hand-Side (RHS). The execution of rules is headed by the Recognize-and-Act cycle where the LHSs of rules are matched against the content of the Working Memory in order to determine the appliable rules by means of the Inference Engine. The rule-based techniques have proved to be a natural way to express the necessary relationships among messages needed by monitoring systems.

This paper presents an enhanced version of Sinergia, an Expert System for real time fault diagnosis of the Italian Telecommunications network. The first prototype version [4], completed in 1991, is soon expected to be operating in the field. The first version was aimed at the correlation of mulfunction signals (events) from the transmission network in order to deduce isolated and simple faults.

The actual version, planned for the end of 1992, is aimed at the correlation of alarms from transmission and switching networks and at a more sophisticated diagnosis, such as multiple faults handling; futhermore, due to its enhanced real-time capabilities, it is able to diagnose intermitted faults.

The paper is organized in the following way. In section 2 the typical scenario of network management is described, and the main problems of this task are outlined. Section 3 presents the Expert System approach of Sinergia while Section 4 deals with the overall system architecture.

2. FAULT MANAGEMENT IN THE ITALIAN NETWORK

In this section some basic concepts of the fault management in the Italian Telecommunications network will be sketched in order to show the context of Sinergia and the data used in the fault diagnosis.

The goal of Sinergia is to locate faults in the *digital paths* along the transmission network, which are caused by failures of Lines, Line Terminations (T), Multiplexers/Demultiplexers (M) as well as of the trunk interfaces of Exchanges, Digital Cross Connects (DXCs) and other network devices (internal failures occurring in exchanges and DXCs are not considered at present since redundancy is provided at that level).

The CCITT Recommendations G.704 [5], provide a mechanism for the propagation of fault alarms in the digital transmission paths to allow an easy identification of the faulty equipment; however, almost always, the occurrence of a trouble in one equipment originates different signals from a number of equipments somehow related to it. The point is to separate the signals which lead to the cause of the failure from the ones that make the secondary effects to arise.

At present this process is not straightforward and is usually carried out by maintenance experts.

In the following, the region of the topological network containing all the equipments affected by a fault (f_i) occurring at the equipment (e_j), will be referred to as the *Fault Influence Area* (FIA) $A'(f_i,e_j)$.

More formally, the telecommunications network can be seen as a graph N(D,E) where E is the set of Network Equipments (nodes) and D is the set of the digital paths (arcs) that connects the nodes; N is not necessarily connected.

In the following it will be described how the graph N is built up from the telecommunications network data base.

Given a fault f_i occurring at the equipment e_j the fault influence area A' is defined as:

$$A'(f_i,e_j) = \{e_1, \dots e_n\} \mid \text{every } e_k \text{ is affected by fault } f_i \subseteq E$$

$$N(D,A') \subseteq N(D,E) \text{ is not necessarily a connected subgraph.}$$

The signals coming from the FIA will be analyzed by the Expert System in order to isolate the faulty devices and to provide a diagnosis of the fault.

The fault propagation mechanism provided by the CCITT recommendations G.704 leads to large fault influence areas, since the Alarm Indication Signal (AIS), is propagated downstream and, to some extent, upstream.

During the analysis phase of the problem it was found out that, as far as the Fault diagnosis is concerned, a *Reduced Fault Influence Area* (RIA) could instead be used:

$A(f_i,e_j) = \{e_1, ...e_m\} \mid$ every e_k is affected by fault f_i without considering AIS

$A(f_i,e_j) \subseteq A'(f_i,e_j)$

$A(f_i,e_j)$ is a connected subgraph.

The A area instead of A' will be used for sake of efficiency.

With the above definition of reduced fault influence area it turns out that: "*for any kind of fault in the digital network and for any topological connection of the network equipments, the shape and topology of the relative Reduced Fault Influence Area falls into a limited number of cases*": in the following this concept will be referred to as the *Fault Diagonalization Principle*; this is the basic consideration behind the idea of implementing an Expert System for such a diagnosis and is the theoretical foundation for building the Knowledge Base of the Expert System.

The following picture (Fig. 1) reports the situation where a fault occurs within a multiplexer located in the middle of a multiplexing hierarchy.

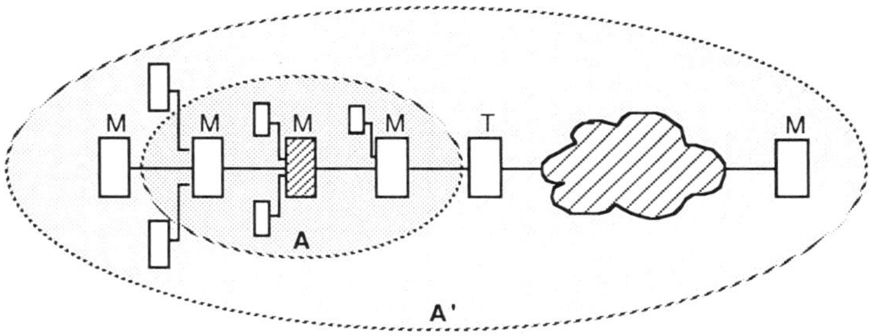

Figure 1. Representation of the Fault Influence Area (A') and the Reduced Influence Area (A).

Different Operation Support Systems (OSSs) are available in the Italian management network that allow the management of the aforesaid equipments, namely FAMAs that monitor the transmission equipments and CEMs that deal with the exchanges; in particular, while FAMAs have a management domain as large as an Italian province, CEMs have a management domain as large as an Italian region.

With reference to the OSI Management Architecture [6], Sinergia acts as a Manager of the previously described OSSs (which play the Agent role) that send the fault signals to it; Sinergia correlates the various signals isolating and highlighting the faulty device; furthermore, depending on the particular configuration of the signals, the particular type of fault is also inferred.

In the following, some peculiarities of the behaviour of the Network Equipments and of the respective OSSs will be analized in order to motivate the key choices made in the development of the Expert System.

The transmission equipments are monitored by FAMA, which makes the status of each monitored equipment available to Sinergia; the status of the equipment is a discrete variable that can assume one out of 42 values: Es = {OK, IND, INT, EXT, ...} (that stand for: Working, Alarm Indication Signal, Internal Fault, External Fault, ...). Such values are in turn set by the operating status of the equipment and of the digital paths connected to it (as specified in the CCITT recommendations G.704 [5]).

We are interested in faults in a multiple faults context, even in the case of $A(f_1,e_1) \cap A(f_2,e_2) \cap ... \neq \emptyset$, where multiple anomaly conditions (theoretically one for each fault) are (partly) on same equipments at the same time.

In the above situation the anomaly signals generation logic, built in the equipment, does not always generate all of the expected signals; in fact two different masking conditions can take place:

- a *stronger* signal masks a *weaker* one: (if the stronger one is already active when the weaker should be sent, the latter is not sent);
- if a signal is on, the same signal deriving from a different anomaly is not sent a second time (*auto-masking*).

In Fig. 2 there are shown two overlapping RIAs with the signal INT masking the signal EXT on the same device (d_3).

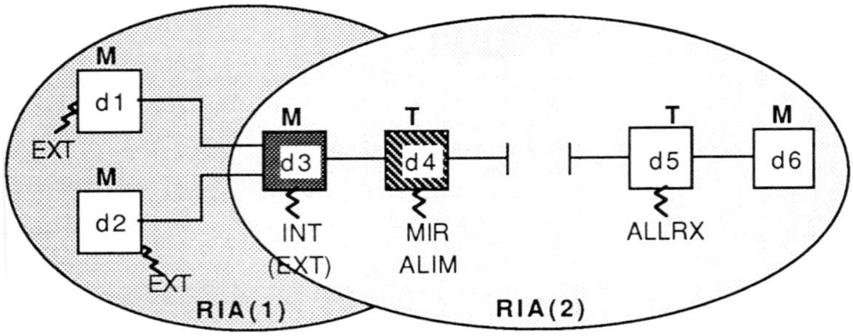

Figure 2. Example of two overlapping RIAs with INT masking EXT on d_3.

The signals belonging to the RIA are close in time, since the signals generated upon the occurrence of a fault occur within few milliseconds while the notification to the Manager takes place within seconds; it follows that all the signals belonging to one Influence Area are *Temporally Local.*

It will be shown later that the temporal locality principle is the main theoretical support of the real time diagnostic technique implemented in Sinergia (see § 3.7).

As it has been already mentioned, the graph N(D,E) is built on the basis of the information on devices and their connections in the network data base; the data base for the Italian telecommunications network is a relational one, named GAT.

Thanks to the Diagonalization principle the whole network does not need to be loaded into the Expert System; instead, for each fault to be isolated only the RIA has to be available in the Working Memory; in the following how the $A(f_i,e_j)$ are loaded from the GAT data base (without actually knowing which fault f_i represents on which e_i equipment) is shown (see § 4.2). The process of faults isolation and diagnosis in a transmission network cannot be carried out in a batch mode: it has to take place on line in order to provide real time advice to the operators; this goal has been achieved by Sinergia by means of both theoretical considerations like the temporal locality principle and via implementation choices, explained in the following sections.

3. THE EXPERT SYSTEM APPROACH

3.1. Knowledge acquisition

According to the "Fault Diagonalization Principle", the occurrence of a single fault in the network may be detected as an anomaly by many different types of equipment. This phenomenon leads to the so called "fault propagation", i.e. a typical sequence of signals emitted by the devices in the relative RIA, when a particular fault occurs.

Once identified all possible RIAs the Domain Experts listed all the possible fault propagations: in this way the Domain Knowledge was then easily encoded into the so called "Data Sheet Charts" (special modules that the Domain Experts were asked to fill in: see Fig. 3).

DataSheetChart (MTTM-2Mbit/s)

MPX	TL	TL	MPX	Technique 2Mbit/s CC(N2)	CC(N2C)	Interpretation
INT	INT	EXT	EXT	X	X	Power supply or SUM for Multiplex N2
EXT	**EXT**	**EXT**	EXT	X		Line Trouble or both regenerators
EXT	**EXT**	-	(-)	X		Regenerator (one direction) or CP RX
(-)	**INT**	EXT	(-)		X	Interface TX or Power supply

Figure 3. Example of a Data Sheet

Each fault propagation included in a Data Sheet represents two different kinds of knowledge: a *Topological Knowledge* of the involved devices, their hierarchical interconnections and physical characteristics (such as the type of equipment, its bit-rate and its manufacture technique), and an *Expert Knowledge*, derived from the maintenance Experts' skill and regarding the signals expected when a specific fault occurs. Moreover each propagation embodies the fault, i.e. an indication of the faulty device together with the description of the trouble occurred. We can therefore say that a single propagation is the fundamental piece of knowledge on which the reasoning process can be built.

The step between a given fault propagation and a forward rule is short; an IF-THEN rule of the form:

IF <conditions> THEN <actions>

is in fact its most natural translation. The conditions will test the topological constraints among the involved devices and the presence of the expected sequence of signals, while the actions will generate the corresponding fault hypothesis.

3.2. The methodology

The overall methodology, exploited by our ES, is built up of two main reasoning steps that implement a sort of *generate and test* paradigm as is depicted in Fig. 4.

The first step is based on a set of rules which instantiate fault hypotheses, whilst the second is a classical AI heuristic search to determine the best solution among the hypotheses (the fault diagnosis result). In the Fig. 4 are also depicted the fundamental blocks of the first step. In fact the execution of the rule component relies mainly upon the Working Memory (WM), useful to determine what rules are executable, the conflict set (CS), which contains the executable rules, and the Inference Engine (IE) which governs the whole process.

3.3. Rules

The rules block works mainly on the signals collected from the telecommunications network (also called real events) to produce the Fault Hypotheses (FHs). The real events are not generally sufficient to directly deduce which faults caused those network signallings. If the system is facing an isolated fault, all the events indicated in the Data-Sheets will be received triggering the application of the appropriate rule (the one corresponding to that fault in the Data-Sheet), and will deduce and highlight the diagnoses to the operator. However, very often, some needed elementary events cannot reach the ES because of two, or more, overlapping RIAs.

The rules block is made up of three parts: the Mask Rules, the Base Level Rules and the Control Rules.

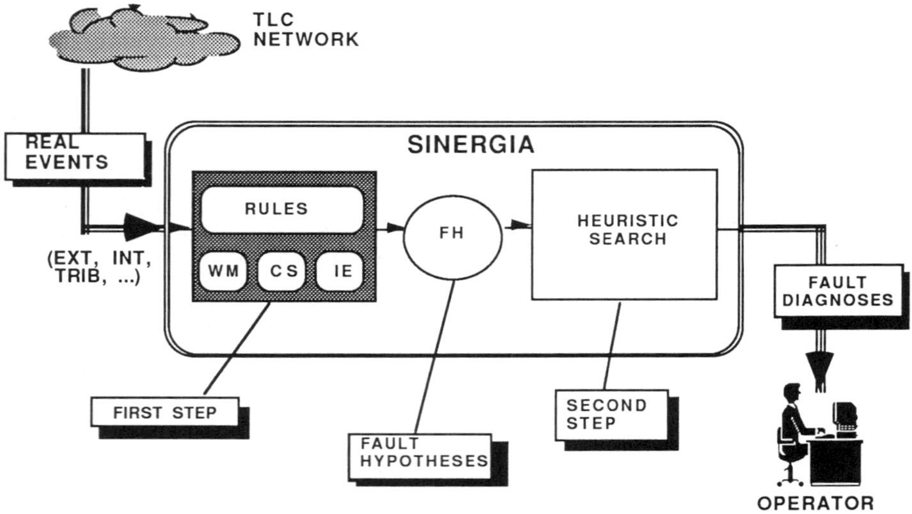

Figure 4. The Expert System Approach

3.4. Hypothetical Events

The Mask Rules provide a solution to the missing (e.g. masked) events by automatically instantiating the *Hypothetical Events*. Such rules fire upon receiving a real event in order to create all the events that could be masked in the future by the received event on the same device. Those events are classified as hypothetical. In the following there is an example of two mask rules:

```
rule INT_masks_EXT
{
  (RealEvent  INT  on device <D>)
  -->
  make (HPEvent  EXT  on device <D>)
}
```

```
rule AUTOMASK
{
  (RealEvent  <TYPE> on device <D>)
  -->
  make(HPEvent <TYPE> on device <D>)
}
```

3.5. Fault Hypotheses

The Basic Level Rules are derived from the Data Sheets and their LHSs match against elementary events (whether they are real or hypothetical) and against topological portions (devices and their connections) of the transmission network. By executing their RHSs, these rules instantiate hypotheses in the Working Memory. Every fault hypothesis is strictly bound to the component devices and events and include at least one real event. The information about the involved devices is taken from the external Data-Base GAT. Every event contains the name of the device from which it was issued. Due to the creation of the hypothetical events, at every instant many FHs may be active. Besides, only some of them are *compatible*. We say that two, or more, FHs are compatible when every real event is at most bound to one of them. For example, let be E_1, E_2, E_3 three real events and E_1^*, E_2^*, E_3^* three hypothetical events. If we have the following fault hypotheses: $H_1(E_1, E_2, E_1^*)$, $H_2(E_3, E_1^*, E_2^*)$ and $H_3(E_1, E_2^*, E_3^*)$ then H_1 and H_2 are compatible because they do not share any real events but only (one) hypothetical event. For the same reason H_2 and H_3 are compatible, but H_1 and H_3 are not.

So the intersection of the events of two compatible FHs gives as a result at most a set of hypothetical events and no real events. In fact a real event represents only one fault. A hypothetical event is artificially created to indicate the hypothetical presence of any number of such events, and so it could be associated to more than one compatible FHs. Compatible FHs represents a set of FHs that can explain the received real events by means of multiple faults.

A separated set of rules, the Control Rules, have the responsibility to govern the activation of the other rules and to chain the execution of the heuristic search algorithm.

3.6. The Goal

The second component of the Sinergia ES methodology is applied to the FHs generated by the previously described rules in order to determine the optimal (or near optimal) solution. Before talking about the method used searching the solution, it is better to stop for a while and clearly assert what the goal is.

Let be

$$H = \{H_1, H_2, \dots, H_N\}$$

the set of all FHs active at a certain time, triggered by the received events.
We indicate with

$$C \subseteq H^*$$

the set of all candidate solutions, where:

$C = \{C_1, C_2, \dots, C_n\}$
and
$Ci = \{H_{i_1}, H_{i_2}, \dots, H_{i_{n_i}}\}$ and $\forall\ k,j \in [1, n_i],\ k \neq j \implies \text{compatible}(H_{i_k}, H_{i_j})$

Every C_i is characterized by a *score* (r_i, n_i, h_i) where:

r_i = number of real events considering all the components H_{i_j};
n_i = number of H_{i_j} in C_i;
h_i = number of hypothetical events in the H_{i_j} of C_i.

The goal is to determine the best solution:
$S \in C$
and $S(r_s, n_s, h_s)$ such that $\forall\ C_i \in C$ and $C_i \neq S$ and score(S) is better than score(C_i):

$(r_s > r_i)$
or $(r_s = r_i$ and $n_s < n_i)$
or $(r_s = r_i$ and $n_s = n_i$ and $h_s \leq h_i)$.

The first condition expresses the *Maximum Events Coverage* in the sense that the best solution is the one that covers (explains) the greatest number of real events without any regard to other conditions. If two solutions have the same r $(r_s = r_i)$ it is chosen the one with the least number of faults. This is the *Minimum Faults Principle* based on the physical consideration that the probability of a multiple fault decreases when the cardinality of faults increases. If the previous conditions are equivalent (i.e. same r and n) then the preferred solution is the one with a minimum number of hypothetical events (*Minimum Hypothetical Events*) because it is more heavily based on real signals.

3.7. The time evolutive tree structure

The set C may be regarded as the solution space to be searched. An exaustive representation of the set C, given a certain H, would lead to an exponential algorithm (in time and space). In our research we found out that still a correct and complete representation of the space C may rely on a *tree structure* with a polynomial algorithmic behaviour.

The chosen tree structure has the following properties:

- it represents the whole set of candidate solutions C;
- at every time it keeps the best solution;
- the computational effort is polynomial in time and space.

The tree is built incrementally by inserting one FH at a time. As soon as a new FH is made by the rules, it is inserted in the tree. Every node of the tree, except the root, contains a FH. The insertion operation starts from the root and recursively inserts the new node (the new FH) as a son of every compatible existing node (the root is always compatible by definition because it has no FHs associated with it). Fig. 5 shows a simple example of H and the relative tree.

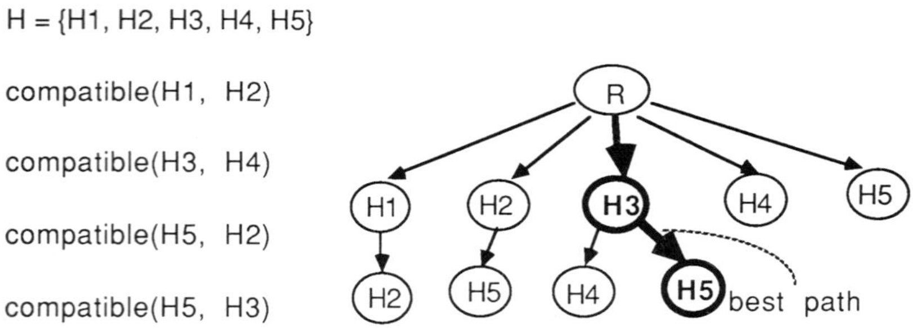

Figure 5. Example of tree

Every path from the root to a given node (not necessarily a leaf node) is a candidate solution of C. Due to the conditions stated for the best solution we are considering only the subset of C including the paths on the tree from the root to the leaves because they are better than any subtree (in fact a complete path will cover a greater number of real events than a subpart of itself). Every node of the tree contains the score information regarding the best path from itself down. When a new FH is inserted in the tree the information of the best path and its score is updated accordingly.

During the processing of the tree it is difficult to establish the precise time when it's better to extract a node. In fact, at run-time, real events keep arriving asynchronously and new FHs are consequently added to the tree with possible modification of the best path. Anyway, because of the Temporal Locality it is very likely that at least part of the best path will remain unchanged. This brings us into the *Time-out Concept.* When a FH joins the best path (it becomes part of the best solution) a time-out counter is set indipendently of other FHs that were already in the best path. When a time-out expires (i.e. a FH has been part of the best solution for more than a given time - a few seconds) the FH will be released to the operator as a diagnosed fault. The tree structure will then be reorganized accordingly, eliminating the FH just released and all those that are incompatible with it (because an incompatible FH relies upon some of the real events already covered by the diagnoses).

3.8. Heuristics

In the tree representation of the space C the dimension of the tree is a crucial point.
Le be
NT = number of nodes on the tree;
N = number of FHs in H
then in the worst case (when every FH is compatible with each other):

$$NT = 2^N - 1$$

and in the best case (every FH is incompatible with each other):

$$NT = N;$$

but in the average we have heuristically found out that:

$$NT = N\left[1 + \sum_{i=1}^{L} \frac{F^i}{2^{i(i-1)/2}}\right]$$

where:
F = average number of sons to a node of the first level (immediately under the root);
L = total (max) number of levels in the tree.

The distribution of nodes per level is depicted in the Fig. 6 where the area beneath the curve is approximately the dimension of the tree.

Practically the tree path with the maximum number of FHs has length L. This is usually much beyond the dimension of the best path. In fact, due to the Minimum Faults Principle we are looking for the shorter path with higher real events coverage. Although the dimension of the tree is polynomial, its value may anyway become very high and consequently algorithmically intractable. The first heuristics we used to cut the search space is the *Tree Depth.* The tree-depth heuristics limits the tree expansion (and searching) at M levels, $M<L$ (actually $M=5$). As we can see in the Fig. 6 the search space is much reduced (the highlighted area is the new dimension).

Another very useful heuristics we adopted is the *Parallel Deduction.* When two faults arise close in time but belong to non-overlapping RIAs it turns out that every FH around the first fault is compatible with the FHs of the second fault. This high compatibility brings to an explosion in the tree structure. Besides it is also useless to put the two sets of FHs in the same structure. In this case it is more profitable to split the tree in 2 parts corresponding to the two portions of the network. Generally speaking the parallel deduction heuristics build a tree forest of independent and parallel reasoning processes.

Some other minor heuristics are under evaluation and are devoted to further reducing the search space and the reasoning time.

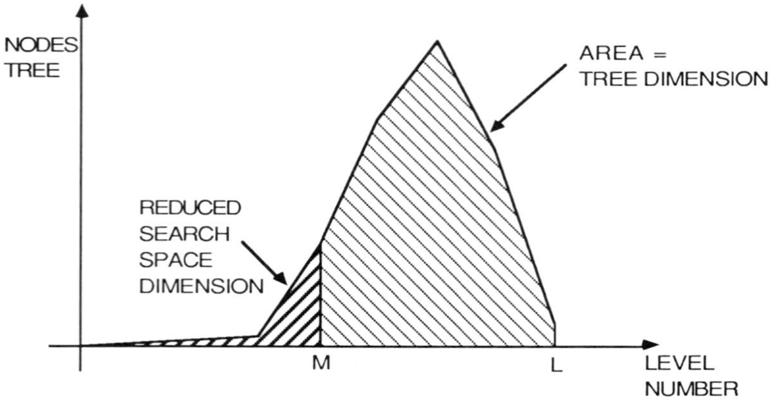

Figure 6. Distribution of nodes per tree level.

4. OVERALL SYSTEM ARCHITECTURE

Sinergia has been developed on a Sun 4 Workstation (RISC architecture) with a graphical display. The Operating System is SUN OS 4.X (a UNIX dialect) and the base language for the system is the C language. The various modules are internally implemented as Processes which communicate by means of sockets [7].

The overall system Architecture consists of the following modules, as sketched in Fig. 7.

4.1. The Reasoning Module

The Reasoning Module (RM) is the heart of Sinergia. It instantiates the Fault Hypotheses applying the Base Level Rules and, by means of a Heuristic Search Algorithm, driven by the Control Rules, it determines the subset of compatible Fault Hypotheses that best explains the received events in terms of fault minimality and best events coverage.

Both Base Level Rules, Control Rules and Mask Rules have been written in the C-Rule language [8], a proprietary tool which allows to build Rule Based Expert Systems. C-Rule is an extension to the C language, from which it inherits efficiency, portability and integrability, which are highly desirable features for a system that has to cope with the real-time aspects originated by the telecommunications environment. The Reasoning module receives the event stream from the external world, reads the topological data from a local Data Base and deduces the fault diagnoses. The system deductions are also stored in a LOG Data Base, according to CCITT recommendations [9, 10].

In the current system implementation the total number of Rules encoded in the system is about 400.

4.2. The DB Interface Module

The DB Interface Module allows to retrieve, according to the Federated Data Base Approach [11], the information relevant to topological aspects, such as the interconnections of involved equipments, their physical localization, their type (MPX, TL, DS, ...), bit-rate, and technique. These data are retrieved on demand by the RM and are internally stored as C-Rule facts, to be matched against the LHSs of the Base Level Rules. When an event is received from the network, the system queries the corporate Data Base in order to collect information about the device that originated the event and the adjacent ones, provided that their records are not already into the RM. Furthermore such information is removed when no longer useful.

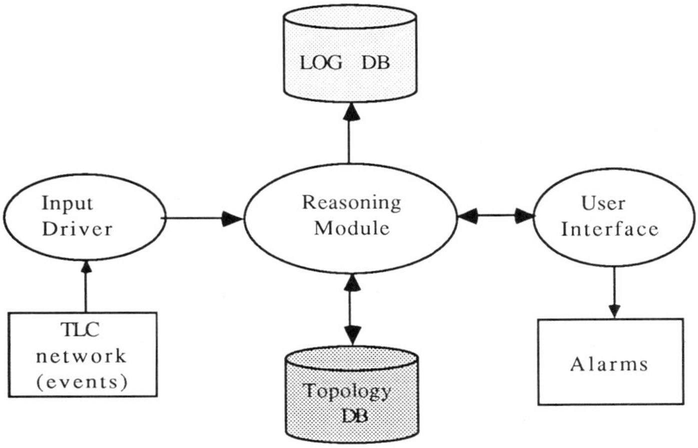

Figure 7. The System Architecture

4.3. The Input Driver

The asynchronous sequence of signals emitted by the equipments in trouble are sent to the Reasoning module by means of an Input Driver, whose purpose is to collect them from the lower level OSSs. The RM stores the received events as C-Rule facts, the so called "Real Events", which will be matched against the LHSs of the Base Level Rules. Moreover, for each real event, the Reasoning module automatically generates, by means of the Mask Rules, all the Events that could be masked by it, the Hypothetical Events.

4.4. The User Interface module:

Sinergia is provided with a sophisticated and user-friendly graphical Interface based on the X Windows Graphical System [12], which, by means of textual and graphical scrolled windows and pop-up menues, allows to:

- graphically display the considered network, in term of devices and interconnections;
- select a device and display the fault history associated to it;
- highlight with different colours the devices that sent signals;
- highlight the faulty devices;
- list the Elementary events received, and the Fault Diagnoses released by the reasoning process;
- show, by means of numerical indicators, the actual level of search complexity.

5. CONCLUSIONS AND THE FUTURE EVOLUTION

The Expert System technology used to develop Sinergia has proved to be worthwhile because of the straightforward mapping between rules and the Data-Sheets. Rules themselves represent a declarative model of the knowledge which can be easily entered into the system and furthermore allow simple updating operations of the knowledge base itself.

The proprietary C-Rule language has helped the coding of Sinergia for both easiness of the formalism (very close to the traditional C language) and the integrability with other programming paradigms. The efficiency of C-Rule makes it suitable for building ESs targeted to the field and not only for prototypes.

Besides, the multiple fault diagnosis would have been hard to treat without the heuristics search algorithms that reduced the diagnosis effort of about 2 order of magnitude. Anyway the heuristics we adopted did not modify the correctness of the diagnosis process as came out from a large number of tests run. The examples included real networks up to 5,000 devices with as much as 15 multiple faults on overlapping RIAs. To evaluate the performance of the system in very demanding conditions we used an Events Generator whose purpose was to simulate timed signals emission from devices.

At present the fault diagnosis activity is carried out by maintenance experts faced with huge amount of data to analyze; the introduction of Sinergia has cut the data presented to the operator of more than one order of magnitude.

The future evolution of Sinergia includes the study and integration of Machine Learning techniques in order to allow the induction of new Data-Sheets and their encoding as forward rules.

Other main topics of interest for the near future are the extension of the range of the networks monitored by Sinergia including, for instance, more complex and technologically advanced networks such as IN, Integrated Services Digital Network (ISDN) and Synchronous Data Hierarchy Network (SDH).

6. REFERENCES

1 J.R. Wright, G.T. Vesonder, Expert Systems in Telecommunications, Expert Systems With Applications, Vol. 1 pp. 127-136, 1990.

2 J. Liebowitz, Expert System Applications to Network Management, Jhon Wiley and Sons, New York, 1988.

3 F. Hayes-Roth, D. A. Waterman, D. B. Lenat, Building Expert Systems, Addison-Wesley, Reading, 1983.

4 E. Ronco, A Real-Time Expert System for Monitoring a Domestic Network, Proc. of the 12th Int. Conference on AI, ES & NL, Specialized Conference: Computer Systems & Telecommunications, Avignon, FR, June 1992.

5 CCITT, Digital Networks Transmission Systems and Multiplexing Equipment, Recommendations G.701-G.941, Yellow Book, Vol. III - Fasc. III.3, Geneva 1981.

6 CCITT, Telecommunications Management Network, M3000.

7 W. R. Stevens, UNIX Network Programming, Prentice Hall Software Series, 1990.

8 E. Paschetta, S. Ponsero, C-Rule: Providing the C Language With a Rule-Based Technique, Proc. of the 2nd Congress of AI*IA Applications, Palermo, IT, October 1991.

9 Recommendations X.721; ISO/IEC DIS 10165-2.

10 Recommendations X.730; ISO/IEC DIS 10164-1.

11 A. P. Sheth, J. A. Larson, Federated Database Systems for Managing Distributed Heterogeneous, and Autonomous Databases, ACM Computing Surveys, Vol. 22, N° 3, September 1990.

12 A. Nye, T. O'Reilly, X Toolkit Intrinsics Programming Manual - OSF/Motif Edition, Vol. 1- 5, O'Reilly & Associates, Inc, 1990.

Integrated Network Management, III (C-12)
H.-G. Hegering and Y. Yemini (Editors)
Elsevier Science Publishers B.V. (North-Holland)

Applying OSI systems management standards to virtual path testing in ATM networks

Satoru Ohta[a] and Nobuo Fujii[a]

[a]NTT Transmission Systems Laboratories, 1-2356 Take, Yokosuka, Kanagawa 238-03, Japan

Abstract

Asynchronous Transfer Mode (ATM) is an information transport technique that comprehensively supports Broadband ISDN (B-ISDN). One of the unsolved problems in ATM network development is that of virtual path testing through the standardized Telecommunications Management Network (TMN) model. The TMN model employs the OSI systems management concept as a basis for information exchange. This paper, in addressing the applicability of OSI systems management standards to virtual path testing, clarifies requirements for remote control of test resources and proposes a method for defining information models. The method proposed ensures easy storage and retrieval of intermediate test results as well as a consistent test environment assured by concurrency control on managed objects. In addition, an application scenario is derived showing that a test can be executed using standardized communication services. Results suggest that the OSI systems management concept is useful for virtual path testing in ATM networks.

Keyword Codes: C.2.3, C.2.2
Keywords: Network Operations; Network Protocols

1. INTRODUCTION

Asynchronous Transfer Mode (ATM) is an information transport technique especially designed to support Broadband ISDN (B-ISDN). B-ISDN is the long anticipated advanced telecommunication network that will flexibly provide a wide variety of services and applications [1]. The importance of ATM is shown in the enormous number of studies [2]-[10] that have recently tackled various problems associated with it. Technical features of ATM include its use of fixed length packets, called cells, and a hardware switching technique.

The virtual path concept emerged from studies aimed at the construction of an efficient ATM-based network [4]-[6]. In this concept, a virtual path is a bundle of virtual channels to which a route is assigned. At a node in the network, a cross-connect system forwards a cell along a virtual path route by referring to a virtual path identifier (VPI) field in the cell's header. A virtual path is also assigned a capacity to provide a basis for traffic control. Various aspects of the virtues and effectiveness of the virtual path concept have been reported [2], [4], [6].

Undoubtedly, testing of virtual path connections is a critical subject in the realization of ATM networks. Recently, principles of performance measurement for testing and monitoring have been proposed [7]-[9], and performance parameters for virtual path/channel connections are now being discussed by standardization organizations. Moreover, there are currently some experimental ATM systems that employ testing mechanisms for virtual path/channel connections [10].

To maintain an ATM network efficiently, a network provider must be able to remotely execute virtual path testing. This should be done through the Telecommunications Management

Network (TMN) [11] and a standardized interface that conforms to OSI systems management [12] standards. The OSI standards specify, for the purpose of management, the exchange of information between communicating processes. Here, one communicating process is a manager, while the other is an agent that performs management operations on managed objects representing network resources. To apply these standards to a particular management problem, all associated information models must be defined, and control of resources must be achieved through standardized communication services/protocols.

OSI systems management standards are generic specifications covering a wide range of communications management problems. Moreover, for testing functions, a draft standard was recently proposed [13] that provides a basis for defining information models. However, it is not clear whether these standards are applicable to virtual path testing.

The purpose of this paper is to show that OSI systems management standards are applicable to remotely controlled virtual path testing. To investigate this point, requirements for a control scheme for virtual path testing are clarified. One of these requirements is the provision of concurrency control. Concurrency control accommodates concurrent test invocations that access the same managed object and that be executed by the multiple objects involved in each test invocation. This control is indispensable for consistent testing, thus, alternative control algorithms are identified and compared with the result that a method based on the wound-wait algorithm is found to be the more advantageous. Finally, a method of defining relevant information models and application scenarios are deduced from the requirements. The method proposed enables easy storage and retrieval of intermediate test results as well as the effective implementation of the concurrency control scheme. The method satisfies not only the clarified requirements, but also the restrictions imposed by the standards. This suggests that the OSI systems management standards are applicable to virtual path testing.

2. REQUIREMENTS FOR VIRTUAL PATH TESTING

Virtual path testing can be initiated in response to an observed performance degradation, as a routine maintenance action or as a check prior to bringing a system into service. Testing must be able to distinguish degradation types and to find exact causes. For these purposes, it may need to examine performance parameters such as connectivity, bit error ratio, aborted cell ratio, errored cell ratio and misinserted cell ratio. Several measurement schemes for obtaining these parameters have been reported [7]-[9] for the purpose of testing or monitoring virtual path/channel connections.

In measurement schemes proposed to date, virtual path performance is measured either in service or out of service [7]. In both schemes, performance is measured using test cells. These are operation, administration and maintenance (OAM) cells labeled with the same VPI as that of the virtual path to be tested. The cells can also carry other information used to examine performance parameters. Test cells are inserted into the virtual path to be tested at a specified insertion point and leave the path at a specified drop point. Path performance is then evaluated based on information from the test cells. Every drop point, therefore, must have the functions necessary to summarize performance parameters. To localize failures effectively, these summarization functions can be placed at points along the path between the insertion and

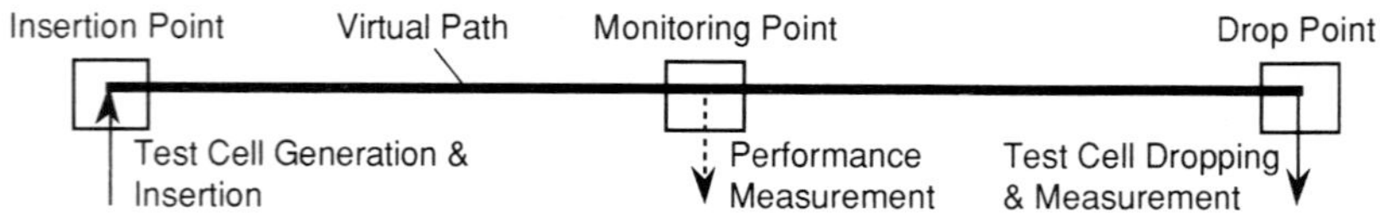

Fig. 1 The principle of virtual path testing.

drop points. Such intermediate points are referred to as monitoring points. At a monitoring point, test cells are not dropped but measurements are executed in the same way as at the drop point. This measurement principle is illustrated in Fig. 1.

For remote control execution of the testing scheme described above, several technical requirements should be satisfied. Some of these originate from characteristics specific to ATM or B-ISDN. A significant characteristic is the effect of increased traffic load from test cells on network performance. Test traffic must be specified so that it does not induce congestion and degrade network performance. Note that some B-ISDN services, such as video applications, are basically asymmetrical. In paths that transfer such services, a loop-back test [8] is not applicable because the required test traffic load cannot be supported in both directions. This means that one-way testing is mandatory for virtual paths, although loop-back testing or echo-back testing may be useful for simple connectivity checks.

Another issue that arises concerns two performance parameters specific to ATM. Measurements of the aborted cell ratio and inserted cell ratio, although important performance parameters, may have to be executed over an extremely long time period to obtain accurate values. This is because cells are, in fact, very seldom aborted or misinserted. For example, in a path whose cell loss probability is 10^{-7}, it will take an hour, on average, to observe just 10 aborted cells with a test traffic rate of 10Mb/s.

Additional requirements are derived from the general characteristics of transmission paths. First of all, test signals must not be received by the user. Furthermore, the ability to insert test signals during testing must be guaranteed. To satisfy these requirements, control of network resources must be accomplished in a specified order. Finally, since resources associated with a test are often shared among multiple paths, concurrency control, discussed in detail later, is necessary.

From the above considerations, the requirements for virtual path testing can be summarized as follows.

- One-way testing must be provided. This leads to a testing environment comprising two or more manager-agent models (see Section 3).
- Long term measurements must be supported. This demands several test control functions such as suspension, resumption, monitoring and forcible termination.
- Test resources should be established in an order such that the test cell drop is set before test cell insertion begins, and that measurement starts after test cell insertion begins.
- Concurrency control is necessary since test resources are shared.

3. INTERNATIONAL STANDARDS

CCITT has recommended the TMN concept as the basis of telecommunications network management [11]. Figure 2 illustrates how the TMN concept accommodates an ATM network. In this figure, the Operations System (OS) performs various network management processes by remotely monitoring or controlling Network Elements (NEs), which include resources associated with virtual path testing. NEs and the OS are connected to the Data Communications Network (DCN) that transports information between the OS and the NEs. Standardized interfaces link the OS and NEs to the DCN. Information is exchanged across these interfaces according to the OSI systems management concept. Remote virtual path testing must be also executed through this information exchange scheme.

Figure 3 shows the OSI systems management model [12]. In Fig. 2, suppose that the OS is communicating with an NE to perform network management. Then, the OS corresponds to a managing open system, and the NE represents a managed open system in the context of OSI systems management. Here, the process in the managing system takes a manager role, while that in the managed system takes an agent role. The network resources being controlled or monitored are represented as managed objects that are managed by the agent. The managing system can remotely control or monitor the resource by sending operations to, or receiving notifications from, the managed objects.

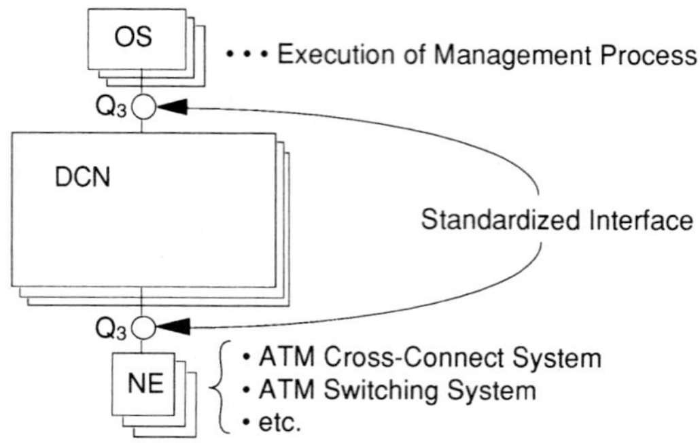

Fig. 2 The TMN model applied to an ATM network.

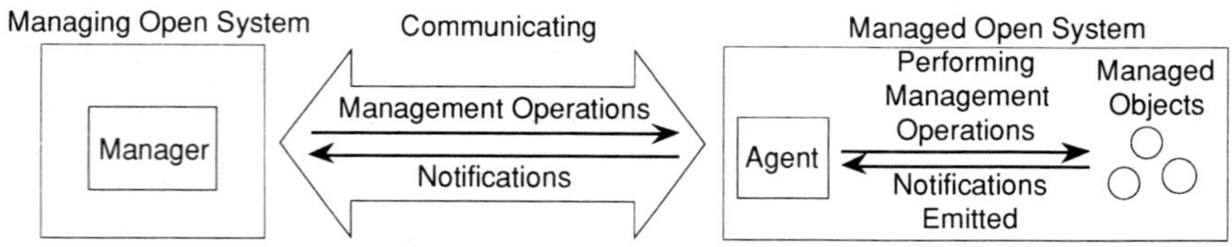

Fig. 3 The OSI systems management model.

Reference [13] specifies the test management function in the OSI systems management model. In this specification, the manager process issuing test operations is called the test conductor, while the agent process receiving test operations is termed the test performer. Test operations issued by the test conductor and directed to a managed object must comply with the test action request receiver (TARR) functionality. The types of tests supported by the test performer are described as 'test categories.' A test is defined as either an asynchronous test or a synchronous test. The former can be monitored and controlled, while the latter cannot. For an asynchronous test, a Test Object (TO) is defined to hold information pertaining to the test. A test invocation is a specific test instance, while a test session is defined as a set of test invocations. In this paper, a test session means the set of all test invocations concurrently issued to the drop point and to monitoring points for the purpose of examining a single virtual path.

In one-way virtual path testing, the insertion point is geographically apart from the drop or monitoring points. Therefore, it is natural to assume that the insertion point is managed by a managed system that is different from the system managing the drop or monitoring points. In addition, the managing system that invokes a test is assumed to know the following before test invocation:

- traffic load of test cells,
- destination addresses of managed systems including insertion, monitoring, and drop points,
- identifiers of links or equipment to be accessed, and
- VPIs of the virtual path to be tested.

Since two or more managed systems are concerned with virtual path testing, there are several alternatives for applying manager-agent relations, as follows.

(1) Manager-agent relations are applied in parallel. That is, the test conductor exchanges information directly with each managed system as shown in Fig. 4 (a).

(2) Manager-agent relations are applied recursively. The test conductor exchanges information with only one managed system. The managed system then generates another manager-agent model, and so on. Figure 4 (b) illustrates this scheme.

(3) The test conductor exchanges information with only one managed system, which in turn exchanges information with all other managed systems as shown in Fig. 4 (c).

(4) One agent manages all resources through nonstandard communication paths as depicted in Fig. 4 (d). Namely, only one manager-agent relation is employed.

The definitions of managed objects and the application scenario depend greatly on which of the alternative methods is employed. For example, alternative (2) requires that every managed system have an application supporting both manager and agent, while alternative (1) does not require this of any managed system. Thus, it is necessary to choose one of these alternatives for studying the objects and scenario. The system processing load may be a criterion for such a choice. From this viewpoint, alternative (1) is characterized by a heavier load on the managing system than on the managed systems. On the other hand, in alternatives (3) and (4), one test performer is more heavily loaded than the others. In alternative (2), processing load is not concentrated on any particular system. Thus, if test performer load is to be decreased, alternative (1) should be chosen. Otherwise, one of alternatives (2) through (4) should be selected.

Presently, it is difficult to determine a best choice among the alternatives using the above criteria because we have no information concerning which system is most critical in terms of processing load. Furthermore, it is not known whether virtual path testing generates serious processing loads on systems or not. Therefore, this paper concludes that further studies are before a best alternative can be determined. Nevertheless, since a method of applying

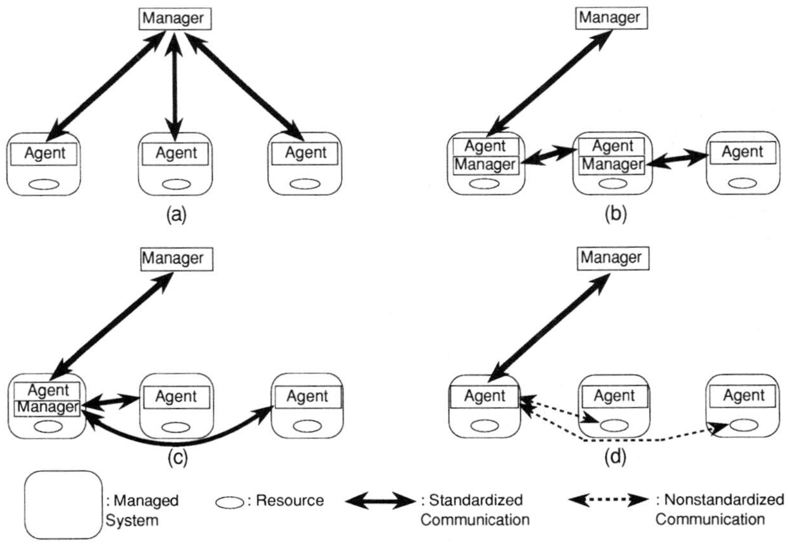

Fig. 4 Alternative ways of applying manager-agent relations.

needed manager-agent models is required as a foundation of this study, only alternative (1), shown in Fig. 4 (a), is considered hereafter. Note that this does not necessarily mean that alternative (1) is the superior method.

4. CONCURRENCY CONTROL

Concurrency control is an indispensable technique for consistency in test execution. The definition of managed objects and the application scenario are affected by the concurrency control mechanism. The following describes this problem.

In general, a resource involved in testing is shared by virtual paths that are multiplexed on the link or that pass through common equipment such as a cross-connect system. Thus, two or more test performers may initiate test sessions using the same test resource concurrently. This leads to inconsistent settings in test environments and to faulty test results. Therefore, any object associated with a testing environment should have a "write lock" mechanism to prohibit other test sessions from resetting its attributes, thus fixing the environment for the duration of a test. Furthermore, since there are multiple test sessions and multiple resources associated with a test session, deadlock can occur. Figure 5 shows an example of deadlock. Here, test session A cannot start because resource #2 is locked by test session B while test session B is waiting for test session A to complete the test and unlock resource #1. Thus, both test sessions will wait permanently. Therefore, managed objects must have a concurrency control mechanism that supports "write locking" as well as a deadlock prevention method.

The concurrency control mechanism selected must work correctly among distributed objects managed by different agents in the model shown in Fig. 4 (a). Additionally, the mechanism must be implemented according to the OSI systems management standards. That is, it should be described using managed objects, an application scenario of the managing system and standardized communication services.

Fortunately, these requirements are satisfied with the method proposed in Ref. [14]. This method defines a concurrency control package for inclusion in all managed objects that require concurrency control. The package achieves the write locking mechanism by applying the CMIS filtering function to the identifiers (IDs) of transactions, i.e., test sessions. A transaction ID contains the transaction invocation time, i.e., a timestamp. To avoid deadlocks, the method exploits the "wound-wait" algorithm [15], implemented as follows. When a lock request from transaction T_i conflicts with another request from transaction T_j at a managed

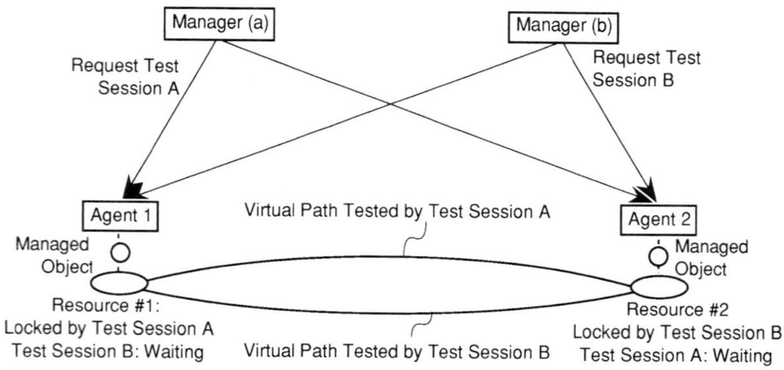

Fig. 5 An example of deadlocked test sessions.

object, the object compares their timestamps. If T_j is younger than T_i then the object emits a "wound" notification to the manager executing T_j. Otherwise, T_i waits for the object to be unlocked by an unlock operation from the manager executing T_j. These steps are described in the behavior of the concurrency control package. A manager receiving a "wound" notification aborts its transaction if it is in the locking phase. In this way, wound-wait is incorporated into the OSI systems management environment. The scheme is summarized in Fig. 6.

By observing Fig. 6 carefully, it can be seen that other known deadlock prevention algorithms are also applicable with slight modifications. Figure 7 shows two such alternatives: Fig. 7 (a) shows the "running priority" algorithm [16] and the "wait-die" algorithm [15] is shown in Fig. 7 (b). Other deadlock prevention techniques may also be applicable. The resource preordering technique (for example, see [17]) can be achieved by locking every object in a specified order. Time-out [18] is another available technique.

The important issue is to choose the best strategy from among the alternatives. Among the alternatives, resource preordering has the disadvantage of increased response delay [17]. The time-out technique is also disadvantageous in its difficulty of finding an appropriate time-out period [18]. For the remaining three alternatives, the number of transactions completed per unit time have been evaluated and compared [16], [18]. However, it is still difficult to compare alternatives because the difference between them is small at the low conflict levels commonly experienced in testing. For example, suppose that a test resource accommodates 1000 virtual paths. For each virtual path, test conductors request a simple check once a month as well as a thorough performance measurement once a year. Assuming that it takes 1 minute to execute the former and 1 hour to execute the latter, the test resource is occupied for 3.3 hours per day on average. This implies that the probability of resource occupancy is not very high, and thus conflicts do not occur frequently.

Stability is a more useful measure for ranking the alternatives. From this viewpoint, the wait-die and running priority methods both have drawbacks. In wait-die, a transaction can be restarted repeatedly [18], and the older transaction tends to slow down [15]. The running priority scheme can also repeatedly restart a transaction and does not guarantee that a transaction will finish safely [18]. These characteristics are not acceptable in network management. By contrast, wound-wait has good stability. With wound-wait, a restarted transaction is not restarted by the same conflict, and the oldest transaction is assured of finishing [18]. This analysis indicates that wound-wait is more advantageous for virtual path testing.

Another benefit of the wound-wait algorithm is the availability of priority control corresponding to test significance or urgency. This is done by representing priority with a number (the smaller the number the higher the priority) then defining a timestamp for each transaction by adding the priority number to the test invocation time. The result is that transactions with higher priority tend to be executed before those with lower priority because the algorithm runs 'older' transactions first. Since test priority is a practically significant technique, the availability of this mechanism also favors employment of the wound-wait algorithm.

5. DEFINING INFORMATION MODELS

This section describes how to define the test category, the managed objects and the application scenario for virtual path testing. The definitions must satisfy all requirements stated above and must also conform to the OSI systems management standards.

In defining the test category, one issue is to determine whether the test is asynchronous or synchronous. Virtual path testing should be asynchronous because monitoring and control, such as suspending/resuming, are necessary to flexibly accomplish long-term tests. Obviously, the test category should include suspending and resuming operations. Additionally, the test conductor must be able to set resources in a consistent order, and to perform a forcible termination. Therefore, the test termination type of the test category must be defined as a 'solicited' termination, in which a test is terminated by a request from the test conductor.

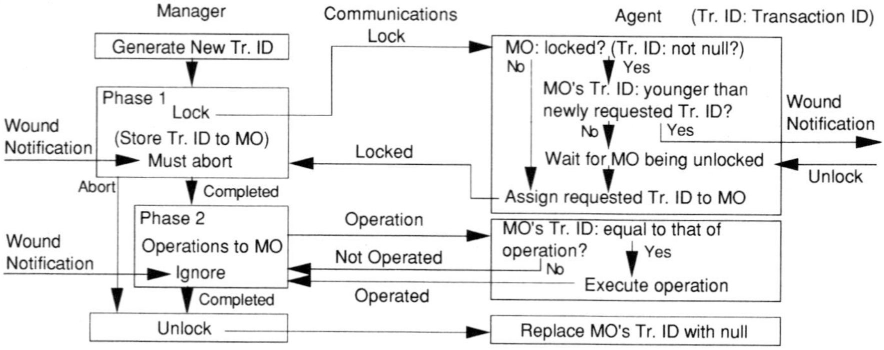

Fig. 6 The concurrency control scheme of Ref. [14].

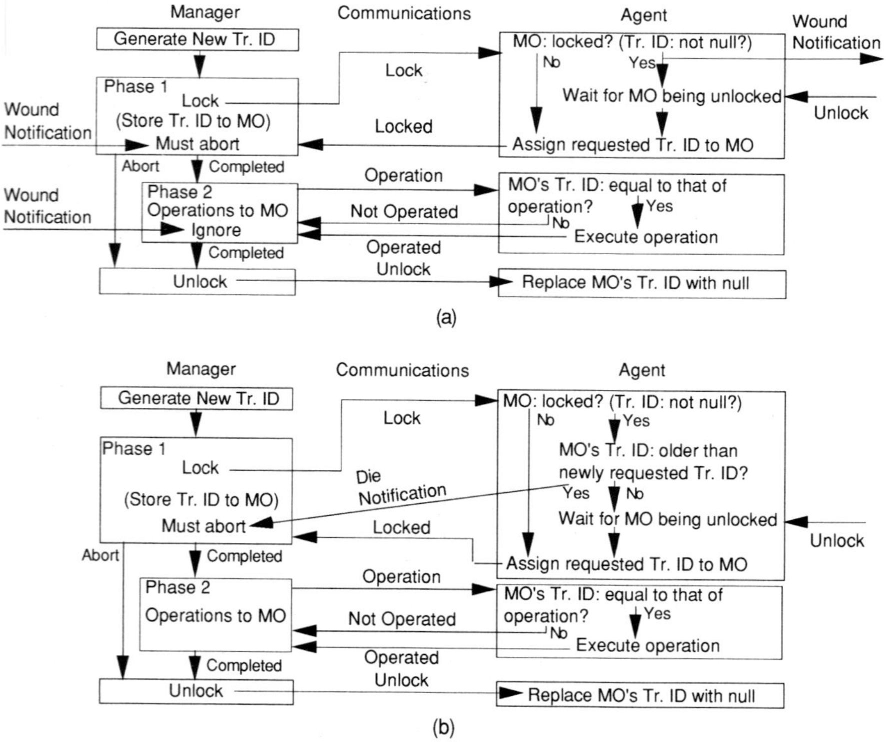

Fig. 7 Alternatives for concurrency control. (a) running-priority. (b) wait-die.

Network resources involved in testing include the virtual path to be tested, the resource generating test cells at the insertion point and the resource examining the received cells at the drop/monitoring point. Objects representing these resources are a set of managed objects under test (MOTs), one or more Test Objects (TOs), and Associated Objects (AOs) [13]. As described in Section 3, a TO is defined for asynchronous tests. AOs are managed objects involved in a test and other than the MOT or the TO. It is trivial that the MOTs represent the virtual path and its components. Thus, the resources at the insertion, monitoring, and dropping points should be represented by appropriately defining the TO and AOs. The difference between TOs and AOs derive from the TO's characteristics as specified in the standard [13]. For example, a TO is created at a test invocation, and test results must be reported by a TO or be available from TO attributes. The TO should be defined taking these characteristics into account.

The TO for virtual path testing must be placed at every drop or monitoring point. A TO is not necessary at the insertion point because no measurement is executed and no test results are obtained. Thus, the TO must be determined in relation to the drop or monitoring point resource.

A TO definition is effective when its attributes represent data measured by the resource. This enables a test conductor to retrieve information about the measurement results, or the progress of the measurement process, using a 'get attribute value' operation. It also makes it easy to record intermediate test results contained in the TO when a test is suspended.

From the viewpoint of concurrency control, however, it is difficult to define the TO as representing the whole resource. Since a TO is created at test invocation, it does not necessarily correspond to the resource in a one-to-one manner. To illustrate this point, suppose that a test is suspended and that the resource is used by a newly invoked test. Two TOs will then exist that are associated with one resource. Namely, a TO is related to a test invocation more tightly than to a resource. This characteristic makes it difficult to provide comprehensive concurrency control on resources through TOs. This is because a 'lock' operation on a TO does not necessarily mean that there is a lock on the resource if two TOs are associated it; a 'replace value' operation on the other TO may change the resource setting.

Considering the above discussion, the TO is defined as representing only one of the drop/monitoring point resource functions, which are categorized into a connecting function and a summarization function. The connecting function supports the selection of cells to be examined, the execution of cell dropping/monitoring and the insertion of alarm indication signal cells for out-of-service testing. For this purpose, the function stores the VPI of the path, the point to be accessed and a monitoring/dropping indicator. That is, the connecting function determines the test environment. The summarization function observes and calculates performance parameters using the cells selected by the connecting function. The TO is defined as describing the summarization function. Therefore, the connecting function is expressed by another managed object, i.e., an AO. This object contains the concurrency control package mentioned in Section 4. These functional object definitions are reasonable considering the following points.

- Since the TO holds data obtained by the test, it is easy to retrieve measurement information, or maintain intermediate test results, while the test is suspended.
- TO attributes related to the resource do not need to be "write locked" because they are essentially read-only and are updated only through measurement.
- Attributes requiring a locking mechanism are all contained in the AO, which is defined as corresponding to the resource in a one-to-one manner and as offering effective write locking.

Note that measurement can only be started after confirming the establishment of the test environment. Therefore, measurement must be initiated by a request from the test performer. This requires that the TO have an action that initiates measurement.

Another AO is defined to represent the resource at the insertion point. This object also contains the concurrency control package in order to hold the test traffic load and VPI to the specified values during the test. Figure 8 summarizes the above manner of defining objects.

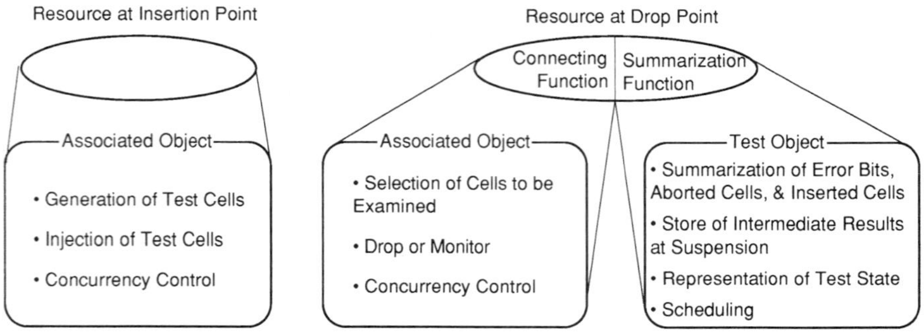

Fig. 8 Test resource functions and object definitions.

Since the manager-agent relationship shown in Fig.4 (a) is employed, the application scenario performed in the managing system must be defined. The scenario must incorporate functions such as initialization, termination, suspension and resumption using standardized management information services. Also, the scenario must be defined so as to set resources in a consistent order and to accomplish the concurrency control mechanism. Figure 9 illustrates such a scenario involving test initialization and test termination.

As shown in Fig. 9, the scenario consists of the locking phase, the environment setting phase and the test execution phase. In the locking phase, the manager requests locking of the AOs using the PT-SET service [19]. This is done prior to other operations to achieve the concurrency control mechanism described in Fig. 6. The test environment is established after confirming that all AOs have been locked by the test session. During the environment setting phase, the manager requests insertion, monitoring, and dropping of cells using the PT-SET service. This phase is executed prior to test execution. The insertion of cells is requested after confirming completion of the drop point setting to not send test cells to the user. Measurement is started by request from the test conductor through the PT-ACTION service [19], and the

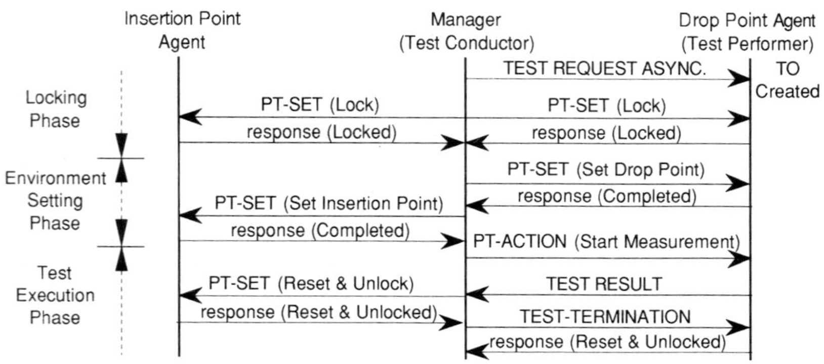

Fig. 9 An example of an application scenario (1).

result is reported through the RESULT REPORT service [13]. The resource at the drop point must be reset after termination of cell insertion. Thus, the test is terminated by request from the test conductor, using the TEST TERMINATION service [13], after confirming reset of the insertion point. Similarly, Fig. 10 shows the suspension and resumption processes.

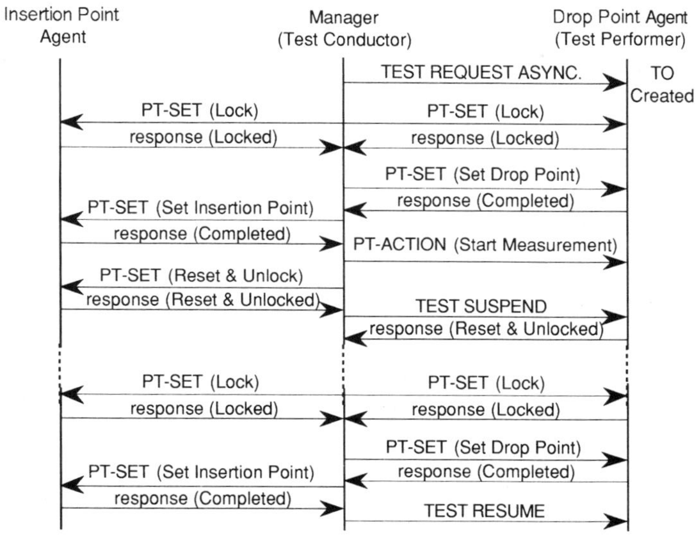

Fig. 10 An example of an application scenario (2).

6. CONCLUSION

Providing interoperable interfaces in network management systems is of great benefit to network providers. The results of this paper are therefore useful because virtual path testing, one of the most basic ATM network operations, was shown to be executable according to OSI systems management standards. First, technical requirements for achieving remote control of virtual path testing were clarified. Among these requirements, concurrency control was identified as an important issue, and several alternative concurrency control methods were identified and compared. It was concluded that the scheme of Ref. [14], which employs the wound-wait algorithm, is more advantageous in terms of control stability and the availability of priority control. Next, several decisions concerning test category definitions were made based on the clarified requirements. Next, a method of defining the TO and AOs was developed to effectively implement concurrency control and to permit measurement information to be saved and retrieved easily. Finally, application scenarios were proposed considering the appropriate order of setting resources and considering standardized communication services. Although further studies are needed on various management issues, this paper is expected to become one of the first steps in achieving a B-ISDN testing environment managed by a standardized network management scheme.

ACKNOWLEDGMENT

The authors gratefully acknowledge Kazumitsu Maki of NTT Transmission Systems Laboratories for his valuable guidance.

REFERENCES

[1] S. E. Minzer, "Broadband ISDN and Asynchronous Transfer Mode (ATM)," *IEEE Commun. Mag.*, **27**, 9, pp. 17-24, Sept. 1989.

[2] J. Burgin and D. Dorman, "Broadband ISDN resource management: The role of virtual paths," *IEEE Commun. Mag.*, **29**, 9, pp. 44-48, Sept. 1991.

[3] I. Tokizawa, T. Kanada, and K. Sato, "A new transport network architecture based on asynchronous transfer mode techniques," *in Proc. ISSLS '88*, Boston, MA, Sept. 1988.

[4] S. Ohta, K. Sato, and I. Tokizawa, "A dynamically controllable ATM transport network based on the Virtual Path concept," *in Proc. GLOBECOM '88*, Hollywood, FL, Nov. 1988.

[5] K. Sato, S. Ohta, and I. Tokizawa, "Broad-band ATM network architecture based on virtual paths," *IEEE Trans. Commun.*, **COM-38**, 8, pp. 1212-1222, Aug. 1990.

[6] R. G. Addie, J. L. Burgin, and S. L. Sutherland, "Information transfer protocols for the broadband ISDN," *in Proc. GLOBECOM '88*, Hollywood, FL, Nov. 1988.

[7] Y. Kanayama, Y. Maeda, and H. Ueda, "Virtual path management functions for broadband ATM networks," *in Proc. GLOBECOM '91*, Phoenix, Az, Dec. 1991.

[8] Bellcore, "Broadband ISDN operations: Framework generic criteria," *Framework Technical Advisory*, **FA-NWT-001248**, 1, Dec. 1991.

[9] J. Anderson and M. D. Nguyen, "ATM-layer OAM implementation issues," *IEEE Commun. Mag.*, **29**, 9, pp. 79-81, Sept. 1991.

[10] M. D. Prycker et al., "An ATM switching architecture with intrinsic multicast capabilities for the Belgian broadband experiment," *in Proc. ISS '90*, Stockholm, Sweden, May 1990.

[11] CCITT Rec. M.3010, "Principles for a telecommunications management network," 1991.

[12] ISO/IEC 10040, "Systems management overview," 1991.

[13] ISO/IEC DIS10164-12, "Test management function," 1991.

[14] I. Yoda, K. Sakae, and N. Fujii, "Configuration of local fiber optical network management system based on multiple manager systems environment," *in Proc. NOMS '92*, Memphis, Tenn., Apr. 1992.

[15] D. J. Rosenkrantz, R. E. Stearns, and P. M. Lewis II, "System level concurrency control for distributed database systems," *ACM Trans. on Database Syst.*, **3**, 2, pp. 178-198, June 1978.

[16] P. Franaszek and J. T. Robinson, "Limitations of concurrency in transaction processing," *ACM Trans. on Database Syst.*, **10**, 1, pp. 1-28, Mar. 1985.

[17] P. A. Bernstein and N. Goodman, "Concurrency control in distributed database systems," *Computing Surveys*, **13**, 2, pp. pp. 185-221, June 1981.

[18] R. Agrawal, M. J. Carey and L. W. McVoy, "The performance of alternative strategies for dealing with deadlocks in database management systems," *IEEE Trans. on Software Eng.*, **SE-13**, pp. 1348-1363, Dec. 1987.

[19] ISO/IEC 10164-1, "Object management function," 1991.

Integrated Network Management, III (C-12)
H.-G. Hegering and Y. Yemini (Editors)
Elsevier Science Publishers B.V. (North-Holland)

Damage Assessment - A Network Management Application Based On An Integrated Information Model Of The Network

Jelena Vucetic and Alexander Gersht

GTE Laboratories Incorporated, 40 Sylvan Road, Waltham, MA 02254, U.S.A

Abstract

An original integrated information model of a network is proposed to support different network management applications. The information model includes three mutually interconnected hierarchical trees, which represent network functionality, logic and equipment, respectively. The model consists of three types of managed objects (functional, logical, and physical), the relations between the objects, and the attributes which are associated to the objects. The model is a basis for a knowledge database, which stores the information necessary for the network management. The interface between the network management applications and the knowledge database is established through different types of database queries. In this paper, the focus is primarily on damage assessment. An original methodology for damage assessment is proposed.

Keyword Codes: C.2.3; D.1.5; E.2

Keywords: Network Operations; Object-Oriented Programming; Data Storage Representation

1. INTRODUCTION

An important requirement in design of contemporary networks is a high quality of service. It includes a high call-handling capacity, reliability, and maintainability of the network. These goals should be supported by efficient and reliable network management system.

The need for more complex and frequent communication services influences a rapid evolution of networks. Therefore, the evolution of network management strategies becomes unavoidable. In [Sti 92], three generations of network management (NM) systems are distinguished. The first generation represents real-time NM systems which provide collection, filtration, storage and display of raw network alarms and reports with no integration of the information. The second-generation NM systems represent a transition to the third-generation *intelligent* NM systems, which are characterized by the functional integration of different NM planes, providing a comprehensive insight into the network behavior. The intelligence of the third-generation NM systems is performed through the advanced NM applications which enable monitoring, prediction and control of the network performance in global, as well as in some particular aspect.

The problem of managing large networks is considered in [Wol91]. The authors use statistical tests, alerts, and correlation among alerts to define a model of network management functions as data-manipulation operations. They propose a language that is declarative and set-oriented.

Holter [Hol92] considers new features in future SONET network management systems. He proposes new methods of managing SONET equipment, which require new features in operating systems and new methods of collecting information from network elements. Acording to [Hol92], the new operating system features will have to support fault locating with continuous in-service performance monitoring in order to reduce reactive dispatches for network repairs.

There is an extensive research in the area of network modeling which is suitable for the third-generation NM systems. Lazar [Laz90] proposes a four-plane Integrated Reference Model (IRM) of a network to organize information on various network entities and to enable efficient utilization of these information in network management. The IRM supports three fundamental principles: the Separation Principle between communications and control in networks, the Principle of Communication Transparency, and the Principle of Asynchronous Resource Management.

A six-plane network model, which integrates different aspects of network functionality is proposed in [Khe91]. According to the network functionalities, the model consists of the following planes: user, access, signaling, switching, interoffice transmission, and infrastructure. The user plane enables the management of the user terminals. The access plane controls the network-user interface. The signaling plane controls signaling within the network. The switching plane controls switching and call processing in the network. The interoffice transmission plane controls the interoffice transmission within the network. The infrastructure plane enables management of the power and housing in the network. This six-plane network model organizes network information and enables both a local network management within a plane, and an overall network management across the planes. The model is suitable for the third-generation network management systems. In this study, we use this model as a basis for an integrated information network model, and for various network management applications. We focus on the switching plane of the model, although the proposed analysis hold for all other planes.

The implementation of the integrated network model into the real NM system represents an information model of the network. The information model consists of all information pertinent to the management of the given network. These information are organized according to their nature and behavior in the network. There are three general types of the information model constituents [TONICS]: managed objects, relations between the managed objects, and attributes associated with the managed objects. The information model is stored into the *knowledge database* [Laz90, Maz90, Maz91]. Relating to the knowledge database, different questions may arise, such as: Which types of information are sufficient to support an efficient and reliable network management [Maz90]? How should these information be organized [TONICS]? How frequently should these information be updated? Where should the corresponding sensors be set up [Maz91]? In this paper, we are interested in the possibilities of utilization of the knowledge database in a NM application called damage assessment.

The integrated information model of the network is described in Section 2 of this study. Managed objects, their relations and attributes are defined. Then, the information model, consisting of three hierarchically organized trees, is proposed. After that, it is shown how the information model can be implemented into the knowledge database of the network. In Section 3, a NM application called damage assessment, based on the integrated information model, is described. Section 4 represents the conclusion of this study.

2. INTEGRATED INFORMATION MODEL OF THE NETWORK

To describe different properties of a network, we introduce an integrated information model of the network. It consists of [TONICS]: managed objects, relations between the managed objects, and attributes associated with the managed objects.

This model provides information for fast and precise network management activities, such as: monitoring, control, diagnosis, maintenance, repair, and evaluation of the network performance and robustness. The model is primarily fault-management oriented, although it provides a considerable amount of information necessary for other aspects of network management, such as: configuration control, scheduling and buffer management, routing, flow control and admission control [Laz90]. The model is similar to the D-plane of the Integrated Reference Model (IRM), which is proposed in [Laz90].

In this section, the constituents of the model will be described. Then, their role and importance within the model will be shown.

2.1 Managed Objects

Managed objects describe network elements, such as switching and link components. In our information model, we distinguish three types of managed objects: physical, logical, and functional.

Physical objects represent real-world network elements, e.g. cabinets (frames) with electronic boards (cards), or integrated components on the boards within a network switch. Physical objects are used to physically locate network equipment which should be controlled or which may be faulty.

Logical objects represent logical properties of the network elements. Logical objects are used to provide the appropriate user interface with the operators who monitor and control the network.

Functional objects represent functionalities of the network elements. Functional objects are used to monitor the behavior of the network in real-time conditions, to detect faults in the network or to estimate the seriosity of degradation of the network performance due to these faults.

2.2 Relations between the Managed Objects

Relations between objects support monitoring and control of the network activities. In our model, we distinguish the following types of relations: physical containment, logical containment, functional containment, physical implementation, and logical implementation.

Physical containment is a hierarchical relation between physical objects in which one of the objects contains the others. For example, a switch physically contains frames (cabinets), a frame physically contains files (shelves), and a file physically contains card slots. Physical objects and associated physical containment relations form a hierarchical tree called *Physical Management Information Tree (PMIT)* [TONICS]. In our model, card slots represent the lowest level of the PMIT because it is assumed that possible faults in the network should be located on the card level.

Logical containment is a hierarchical relation between logical objects in which one of the objects contains the others. Logical objects and associated logical containment relations form a hierarchical tree called *Logical Management Information Tree (LMIT)* [TONICS]. For example, a switch logically contains a base unit, a network unit and an interface unit. These units represent logical objects which also have their own subordinate logical objects, i.e. logical constituents. The lowest level of the LMIT are logical objects which represent cards because it is assumed that network management and control, as well as fault management are based on the card level.

Functional containment is a hierarchical relation between functional objects in which one object consists of the others. Functional objects and associated functional containment relations form a hierarchical tree called *Functional Management Information Tree (FMIT)*. For example, the overall switch functionality is a functional object which represent all functions executed in the switch. The overall switch functionality functionally contains call processing functions, maintenance functions and operating system functions as subordinate functional objects. Each of these subordinate functional objects has its own subordinates. The lowest level of the FMIT are functional objects which represent functions executed in particular cards. The development of the FMIT to the card functionality level is useful particularly because it enables an efficient fault assessment and management when the fault is located on the card level.

Physical implementation is a relation between physical and logical objects. It represents a mapping between the logical properties incorporated within a logical object into the corresponding physical network equipment. The complete set of physical implementation relations represents the connection between the LMIT and the PMIT. Physical implementation is used to locate managed logical objects within the network equipment, particularly in maintenance and repair activities. For example, a telephony processor (within a switch [Jin80, And81]) is a logical object which is physically implemented in a Telephony Processor Complex Frame (TPCF).

Logical implementation is a relation between functional and logical objects. It represents a mapping between the functionality of a functional object into the corresponding manageable logical object. The complete set of logical implementation relations represents the connection between the FMIT and the LMIT. Logical implementation is particularly useful for the analysis of the network alarms. When an alarm message denotes the faulty logical object, the logical object can easily be mapped into the corresponding functional object via logical implementation relation. Thus obtained functional object shows which network functionality is deteriorated and how seriously the overall network performance is influenced. For example, call processing is a high-level functional object which is logically implemented in (i.e. executed by) the telephony processor.

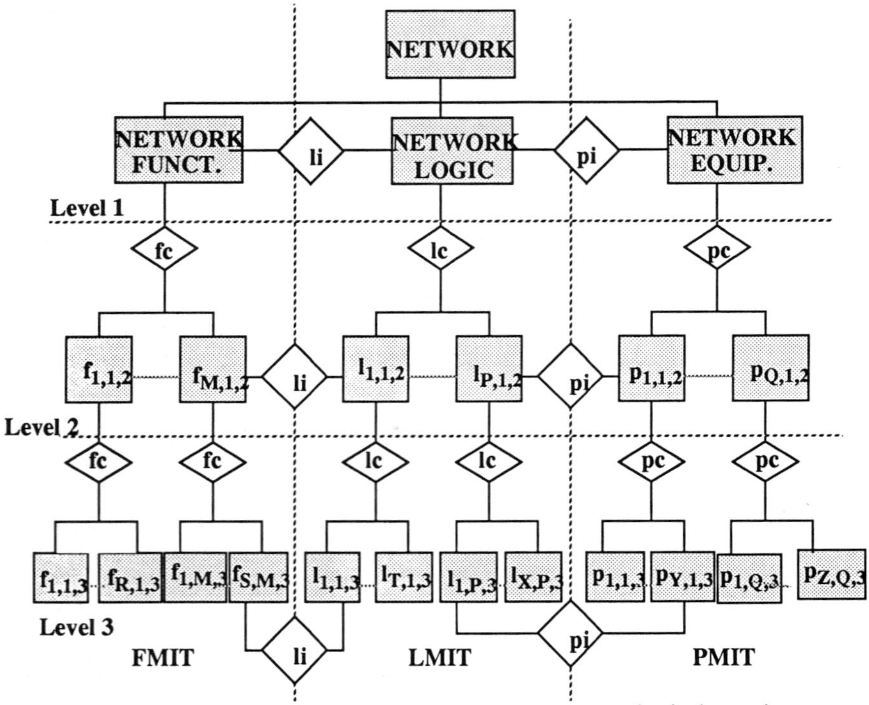

pc - physical containment
lc - logical containment
fc - functional containment
pi - physical implementation
li - logical implementation

Figure 1 - An Integrated Information Model of a Network

2.3 Integrated Information Model of a Network

Figure 1 represents a general integrated information model of a network. The model is a complex structure consisting of the FMIT, the LMIT, and the PMIT. The highest level objects of the FMIT, the LMIT, and the PMIT are the network functionality, the network logic, and the network equipment, respectively. The model presented in Figure 1 contains three levels of hierarchy, although this number generally depends on the type and complexity of the network and the strictness of the network management requirements. The more strict these requirements are, the more precise and detailed the model should be. Then, the number of hierarchical levels should be increased.

The shaded blocks denote managed objects, while the unshaded rhombi denote the relations between the objects.

Each managed object may contain several lower-level objects of the same type. In fault management applications, such hierarchical organization of information enables efficient focusing on the lowest level of the network constituents which are in trouble starting from fault detection in high-level managed objects.

Each lowest-level functional object should be logically implemented in one and only one logical object. If a lack of some functionality is detected, the appropriate alarm is generated in the network. Then, the corresponding functional object should be easily mapped into its lowest-level functional objects using the functional containment relations. After that, the lowest-level functional objects should be mapped, using the logical implementation relations, into the appropriate logical objects which may be faulty according to the initially generated alarm.

Each lowest-level logical object should be physically implemented in one and only one lowest-level physical object. This requirement enables fault location on the card level, which means that a fault can be overcome by the replacement of the card which is faulty.

2.4 Attributes of the Managed Objects

Attributes represent additional information on the static and dynamic properties of the managed objects. Static properties are characteristics of the objects which are not changed in time, or which are not changed in sufficiently long time period. For example, the identity and the location of objects are static properties. Dynamic properties are characteristics of the objects which are subject to relatively frequent changes in time. For example, the activity and the status of the objects are dynamic properties because they are changed during real-time operation of the network.

Each managed object has its unique set of attributes. Some of the attributes are:

- *Identity of the managed object* is a unique notation of the object which consists of the generic object name extended by the number assigned to the object in the populated model of the network. This attribute is static since it does not change as long as the object exists within the actual network configuration.
- *Location of the managed object* is an information on the position of the object within the network equipment, expressed in terms of the office, the aisle, the frame, the file, and the card slot in which the object resides. This attribute may be treated as semi-static since its value may vary if the configuration of the network is changed. In real-time conditions, it is assumed that this attribute does not change.
- *Type of the managed object* denotes if the object is functional, logical or physical. According to the type of the object, different attributes may be assigned. This attribute is static.
- *Status of the managed object* denotes if the object is currently healthy, vulnerable, faulty, or out of service. An object is healthy if it executes all its functions properly. An object is faulty if it does not execute at least one its function properly, but it is still not put out of service. An object is put out of service if the fault related to the object has been determined and located, while the actions for the network recovery and overcoming the problem have been initiated. An object is vulnerable if it is duplicated (primary and redundant copy), and the primary copy is faulty. Therefore, the redundant copy of the object takes over the tasks which were active in the primary copy when the primary copy became faulty. Since only the redundant copy of the object is healthy, the object is more sensitive to future possible fault than when both primary and redundant copies are healthy. This attribute is dynamic, and event-driven. Thus, it needs sufficiently fast and accurate update.
- *Activity of the managed object* denotes if the object is currently active or inactive. The value of this attribute is event-driven. Therefore, it is important to provide its efficient and accurate update in real-time conditions.
- *Average utilization of the managed object* represents an average percentage of the utilization of the object in some defined time period. This information is useful for prediction and prevention of possible "bottlenecks" in the network. Also, this information can be used in damage assessment: the higher the average utilization of a managed object is, the more sig-

nificant decrease of performance is caused by a failure of this object. The value of the average utilization attribute is obtained as a result of monitoring the event traces [Maz91] in a sufficiently long time period. The length of the time period depends on the statistics of traffic requirements within the network.

- *Statistics* is used in addition to the Average utilization attribute to describe how the managed object is utilized. While the Average utilization gives the mean percentage value of the object's utilization, the Statistics attribute shows whether the object is utilized uniformly in time or only in some "burst" time intervals. If the utilization is not uniform, this attribute shows the "peak" utilization of the object, which is helpful to predict and prevent some "bottlenecks" in the network during busy hours.
- *Reference to the responsible person or to the relevant documentation* enables an efficient addressing to the person responsible for the object in alarm situations, or the documentation relevant to the object if some information on the object are needed. This attribute may be treated as semi-static, since it is usually rarely changed.
- *Relations with the other managed objects* describes the relations between the object in question and the other objects in the network. These relations depend on the type of the object, as well as of its role within the network. This attribute is semi-static because it may be changed if the network configuration is modified.
- *List of possible alarms related to the managed object* shows all possible alarms which may be generated if the object is faulty. This attribute is important for fault-detection and decision making applications. It is a basis of the fault management learning system, which should be systematically improved to enable an accurate decision making even in various unusual and complex failure scenarios.
- *List of the alarms related to the managed object which are generated in a predefined time interval* represents all the alarms related to the object which are generated during some period of time (e.g. during busy hours), so that the responsible person may initiate proper actions afterwards, when the network is not so busy.
- *List of suggested fault recovery procedures* represents all the recovery procedures which relate to the described object, and which should be activated (automatically, or manually) in response to the relevant alarm messages.

2.5 A Knowledge Database

Managed objects and their associated attributes are implemented as *records* of a *knowledge database* which is a basis for network management applications. The database should contain accurate values of the attributes. The database capacity should correspond to the real configuration. It means that the values of the attributes and the population of the managed objects should be properly set during the network initialization according to the defined network configuration demands. During the real-time operation of the network, the values of the attributes and the population of the managed objects should be updated dynamically. Mazumdar and Lazar [Maz91] propose an objective-driven measurement strategy for establishing the dynamic database of the network. Object-driven monitoring allows the activation of sensors for data collection and abstraction based on a set of objectives. These objectives represent different requirements which are derived from the desired quality of service within the network.

The choice of the attributes associated to the managed objects should be based on the types of the operator's queries of the database. Each managed object should have a set of attributes which is flexible for eventual future extension or modification. This is necessary because the properties and population of the network elements are continuously modified and increased as the traffic demands grow.

A knowledge database record is defined as a high-level object class called a "managed_object". It includes all types of attributes which may be assigned to a managed object. From the "managed_object" object class, we derive three subclasses which describe three types of managed objects: functional, logical and physical.

The "functional_object" subclass includes the attributes which show the functional containment relations (upwards and downwards in the FMIT) and the logical implementation of the object in question. The "logical_object" subclass contains the attributes which show the logical containment relations (upwards and downwards in the LMIT), the physical implementation of, and the functionality which is logically implemented into the described logical object. The "physical_object" subclass includes the attributes which show the physical containment relations (upwards and downwards in the PMIT), and the logical object which is physically implemented in the described physical object.

Some types of the queries which are executed on the described database will be described in the next section within the description of the damage assessment application, which is based on the proposed integrated information model of a network.

3. DAMAGE ASSESSMENT

Damage assessment is a NM application which determines the deterioration level of the network functionality when a logical object of the network is faulty. The major goals of damage assessment are to provide continuity of service when faults occur, and to support prompt and accurate repair of the faulty network constituents [Hau85].

In Figure 2, the FMIT and the LMIT of the integrated network model is presented. The shaded block l_f denotes a faulty logical object. We want to estimate how severe this fault deteriorates the overall network functionality.

First, through the reverse logical implementation relation, it is possible to determine which low-level functional objects may be deteriorated due to the detected fault in the logical object. Then, through the reverse functional containment relation, we can determine which higher level functional objects may be damaged due to the improper or impossible execution of their low level functions. In Figure 2, the shaded blocks in the FMIT represent the functional objects which are in trouble due to the fault in the logical object l_f.

Let us assume that the FMIT has N levels of hierarchy, and that Level 1 represents the top of the FMIT, i.e. the highest level of the hierarchy. We also denote functional objects of the FMIT as $f_{i_k, i_{k-1}, k}$, which means that this object belongs to the *k*-th hierarchical level within the FMIT, $1 \le k \le N$, and that it is the i_k-th object functionally contained in the higher level object i_{k-1}. Therefore, this notation is relative and based on the "parent-child" functional containment relationship between the objects within the FMIT. The only exception is the top object of the FMIT. Since it does not have any superior object, it is denoted as f_1.

Note that an object may have more than one notation $f_{i_k, i_{k-1}, k}$ at the same time if it is functionally contained in more than one higher level object.

Each functional object in the FMIT has two parameters assigned to *each* of its notations for utilization in damage assessment:

1. Parameter $r_{i_k, i_{k-1}, k}$ represents the health of functional object $f_{i_k, i_{k-1}, k}$. Similarly, parameter r_1 denote the health of object f_1. If $r_{i_k, i_{k-1}, k} = 1$, functional object $f_{i_k, i_{k-1}, k}$ is damaged. Otherwise, it is healthy. If object $f_{i_k, i_{k-1}, k}$ functionally contains object $f_{i_{k+1}, i_k, k+1}$, the following statements hold:
   - If $r_{i_{k+1}, i_k, k+1} = 1$, then $r_{i_k, i_{k-1}, k} = 1$. It means that the damage in a lower level object propagates to its "parent" object.
   - If $r_{i_{k+1}, i_k, k+1} = 0$, then $r_{i_k, i_{k-1}, k} = 0$ or $r_{i_k, i_{k-1}, k} = 1$, depending on the health of other objects which are functionally contained in object $f_{i_k, i_{k-1}, k}$. If all of them are currently healthy, $r_{i_k, i_{k-1}, k} = 0$. Otherwise, $r_{i_k, i_{k-1}, k} = 1$.
   - If $r_{i_k, i_{k-1}, k} = 0$, then $r_{i_{k+1}, i_k, k+1} = 0$. It means that a healthy object functionally contains all healthy objects.
   - If $r_{i_k, i_{k-1}, k} = 1$, then $r_{i_{k+1}, i_k, k+1} = 0$ or $r_{i_{k+1}, i_k, k+1} = 1$. It means that a damaged

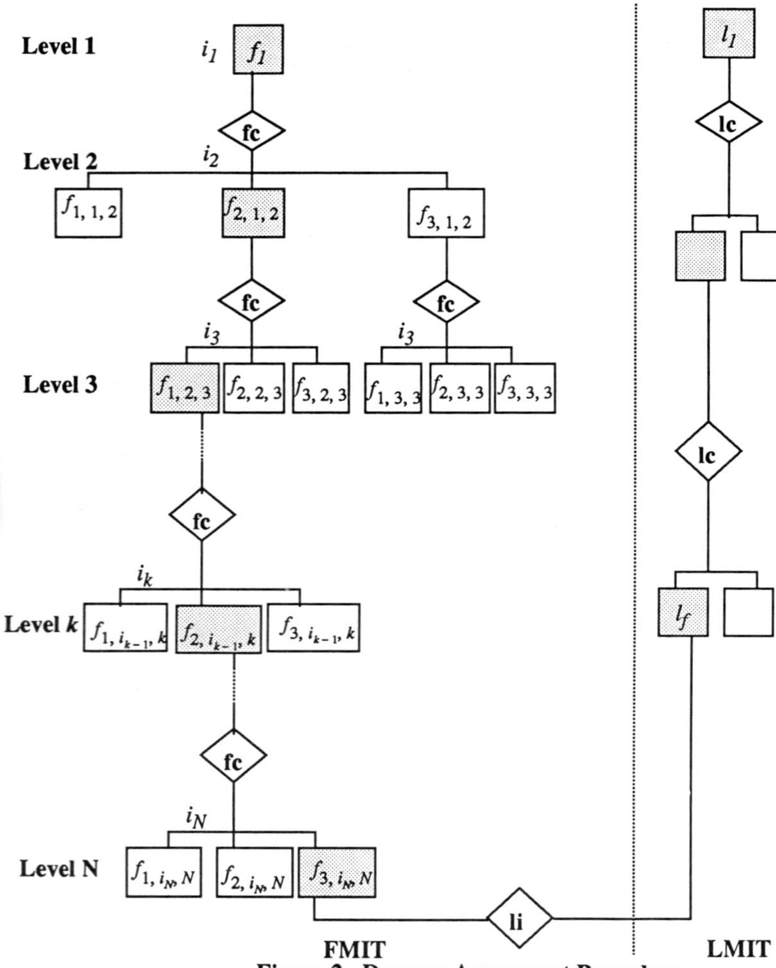

Figure 2 - Damage Assessment Procedure

object functionally contains at least one damaged lower level objects. For the simplicity of this analysis, it is assumed that the propagation of a damage is only from lower to higher levels of the FMIT. However, in practise it is possible that the damage of object $f_{i_k, i_{k-1}, k}$, caused by the damage of its functionally contained object $f_{i_{k+1}, i_k, k+1}$, may also cause the damage of some other object(s) $f_{i'_{k+1}, i_k, k+1}$, $(i'_{k+1} \neq i_{k+1})$, which are functionally contained in $f_{i_k, i_{k-1}, k}$, and which are not directly logically implemented in any faulty logical object.

2. Parameter $c_{i_k, i_{k-1}, k}$ represents the non-negative number of objects which are functionally contained in functional object $f_{i_k, i_{k-1}, k}$. Similarly, c_1 denotes the number of objects which are functionally contained in the top object f_1. Therefore, $1 \le i_k \le c_{i_{k-1}, i_{k-2}, k-1}$.

Let us introduce a notation D_t for the *Total Damage,* which represents the result of damage assessment. First, we assume that each functional object $f_{i_k, i_{k-1}, k}$ contains $c_{i_k, i_{k-1}, k}$ subordinates which have mutually equal importance for the execution of the object $f_{i_k, i_{k-1}, k}$. Therefore, the influence of each of these subordinates on D_t can be presented by the coeficient $\frac{1}{c_{i_k, i_{k-1}, k}}$. Using the parameters defined above, the following relation holds:

$$D_t = r_1 \sum_{i_2 = 1}^{c_1} \frac{r_{i_2, 1, 2}}{c_1} \times \sum_{i_3 = 1}^{c_{i_2, 1, 2}} \frac{r_{i_3, i_2, 3}}{c_{i_2, 1, 2}} \times \ldots \times \sum_{i_N = 1}^{c_{i_{N-1}, i_{N-2}, N-1}} \frac{r_{i_N, i_{N-1}, N}}{c_{i_{N-1}, i_{N-2}, N-1}}$$

where $c_{i_k, i_{k-1}, k} \neq 0$ and $1 \leq k \leq N$.

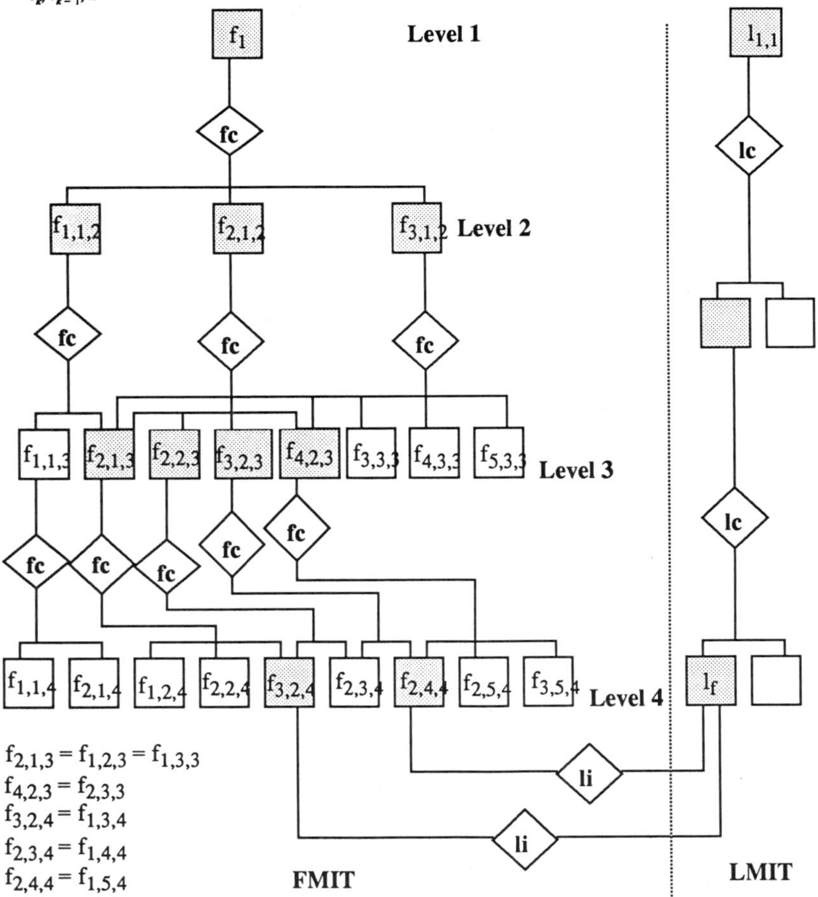

Figure 3 - An Example of the Damage Assessment

As an example, let us determine D_t for the model presented in Figure 3. Assume that logical object l_f is faulty, and that l_f represents a logical implementation of functional objects $f_{3, 2, 4}$ (or $f_{1, 3, 4}$) and $f_{2, 4, 4}$ (or $f_{1, 5, 4}$). These functional objects belong to the lowest level (in this example, level 4) of the FMIT.

The shaded blocks of the FMIT represent the functional objects which are damaged or deteriorated due to the fault in l_f. As it can be seen in Figure 3, these objects are determined through the reverse functional containment relationship from objects $f_{3,2,4}$, $f_{1,3,4}$, $f_{2,4,4}$, and $f_{1,5,4}$. Now, D_t can be determined from the following relation:

$$D_t = r_1 \sum_{i_2=1}^{c_1} \frac{r_{i_2,1,2}}{c_1} \times \sum_{i_3=1}^{c_{i_2,1,2}} \frac{r_{i_3,i_2,3}}{c_{i_2,1,2}} \times \sum_{i_4=1}^{c_{i_3,i_2,3}} \frac{r_{i_4,i_3,4}}{c_{i_3,i_2,3}}$$

According to Figure 3, parameters $r_{i_k,i_{k-1},k}$ and $c_{i_k,i_{k-1},k}$ have the following values:

$f_{i_k,i_{k-1},k}$	$c_{i_k,i_{k-1},k}$	$r_{i_k,i_{k-1},k}$
f_1	3	1
$f_{1,1,2}$	2	1
$f_{2,1,2}$	4	1
$f_{3,1,2}$	5	1
$f_{1,1,3}$	2	0
$f_{2,1,3}$	3	1
$f_{1,2,3}$	3	1
$f_{1,3,3}$	3	1
$f_{2,2,3}$	2	1
$f_{3,2,3}$	2	1
$f_{4,2,3}$	3	1
$f_{2,3,3}$	3	1
$f_{3,3,3}$	0	0
$f_{4,3,3}$	0	0
$f_{5,3,3}$	0	0
$f_{1,1,4}$	0	0
$f_{2,1,4}$	0	0
$f_{1,2,4}$	0	0
$f_{2,2,4}$	0	0
$f_{3,2,4}$	0	1
$f_{1,3,4}$	0	1
$f_{2,3,4}$	0	0
$f_{1,4,4}$	0	0
$f_{2,4,4}$	0	1
$f_{1,5,4}$	0	1
$f_{2,5,4}$	0	0
$f_{3,5,4}$	0	0

Substituting these values into the relation above, the total damage can be calculated in the following way:

$$D_t = 1 \times \left(\frac{1}{3} \times \left(\frac{1}{2} \times \frac{1}{3}\right) + \frac{1}{3} \times \left(\frac{1}{4} \times \frac{1}{3} + \frac{1}{4} \times \frac{1}{2} + \frac{1}{4} \times \frac{1}{2} + \frac{1}{4} \times \frac{1}{3}\right) + \frac{1}{3} \times \left(\frac{1}{5} \times \frac{1}{3} + \frac{1}{5} \times \frac{1}{3}\right)\right)$$

From this calculation, we obtain that $D_t = 0.2388$, which means that 23.88% of the overall network functionality is damaged when logical object l_f is faulty.

Besides total damage, we can also determine the damage of particular functionalities of the network. Let us denote the damage of functional object $f_{i_k,i_{k-1},k}$ as a *Partial Damage* (d_{i_k}). It may be expressed in the form of the following relation:

$$d_{i_k} = r_{i_k,i_{k-1},k} \sum_{i_{k+1}=1}^{c_{i_k,i_{k-1},k}} \frac{r_{i_{k+1},i_k,k+1}}{c_{i_k,i_{k-1},k}} \times \sum_{i_{k+2}=1}^{c_{i_{k+1},i_k,k+1}} \frac{r_{i_{k+2},i_{k+1},k+2}}{c_{i_{k+1},i_k,k+1}} \times \ldots \times \sum_{i_N=1}^{c_{i_{N-1},i_{N-2},N-1}} \frac{r_{i_N,i_{N-1},N}}{c_{i_{N-1},i_{N-2},N-1}}$$

In example presented in Figure 3, the damage of functional object $f_{3,1,2}$ is equal to 0.13333 because two of five objects which are functionally contained in $f_{3,1,2}$ are damaged, and each of these two objects has one damaged object out of three functionally contained objects.

The analysis given above is derived assuming that all functional objects which are contained in a higher level functional object have mutually equal *importance for the execution* of the higher level object. However, it is not always the truth. For example, the functional object called a "Functionality of a processing complex within a switch" may functionally contain lower level functions "Date administration" and "Interface with local memory". If the "Interface with local memory" is deteriorated, the damage is higher than if the "Date administration" is not executed properly. Therefore, these functional objects should have different "weights" assigned to them. The weights should be proportional to the importance/criticality of the related functional objects.

Further, these weights should also be modified according to the *utilization* of the related functional object in real-time conditions. Assume that an object is highly utilized in the real-time operation of the network. The deterioration of this object would cause more damage than if the object was rarely utilized. Therefore, the weights assigned with the objects to assess the damage, should be proportional with the utilization of the objects.

If the faulty logical object has a *stand-by copy* as a reserve, these weights are lower than if the logical object has just one copy. When a fault occurs, the stand-by copy of the faulty logical object takes over the current activities of the primary copy.

In a real network configuration, it is possible that several logical objects belong to the same generic type. For example, in a switch, there may be several hundred interfaces between the network switch and telephone lines. All of them have the same architecture and functionality and they are modeled by the same generic logical object. If one of the interfaces is faulty, any other interface which is currently available can take over the activities of the faulty one. The higher the number of these interfaces is within the switch, the higher is the probability that some of them is available to substitute the faulty one. Therefore, the total damage is lower. Thus, the weights should be reversely proportional to the *number of real objects* which are modeled by the same generic logical object.

Finally, we derive a general relation for total damage:

$$D_t = r_1 \sum_{i_2=1}^{c_1} (r_{i_2,1,2} \times a_{i_2,1,2}) \times \sum_{i_3=1}^{c_{i_2,1,2}} (r_{i_3,i_2,3} \times a_{i_3,i_2,3}) \times \ldots \times \sum_{i_N=1}^{c_{i_{N-1},i_{N-2},N-1}} (r_{i_N,i_{N-1},N} \times a_{i_N,i_{N-1},N})$$

where $a_{i_k,i_{k-1},k}$ denote the weight of functional object $f_{i_k,i_{k-1},k}$. This weight is derived from the functional importance and the utilization of the object. The following relations should hold: $0 \le a_{i_k,i_{k-1},k} \le 1$ and

$$\sum_{i_k=1}^{c_{i_{k-1},i_{k-2},k-1}} a_{i_k,i_{k-1},k} = 1$$

The functional importance of the objects may be derived from the description of the operating system and the functionality of the network constituents. The utilization of the objects may be obtained from the relevant knowledge database attributes, which are described in the previous section. The existence of the stand-by copy of a logical object depends on the generic type of the object and on the current status of the object. The number of real logical objects which are modeled by the same generic logical object depends on the real configuration of the network.

4. CONCLUSION

An original integrated information model of a network is proposed in this paper. This model consists of three hierarchical, mutually interconnected trees, which represent network functionality, logic, and equipment, respectively. The model consists of managed objects, their relations, and attributes. The managed objects are hierarchically organized, which enables either global, selective or prioritized network management on different levels. The information model is implemented into a knowledge database. The elements (entries) of the database are records which represent managed objects of the network. Each entry consists of several fields which contain information on the attributes and relations of the corresponding managed object. In this paper, a format of the entries is proposed.

The information model is suitable as a support for various kinds of network management applications. We are primarily interested in the fault management applications. In this paper, a methodology for a damage assessment based on the information model is proposed.

Besides its efficiency of searching for particular data, the information model is flexible for different types of network constituents. Its modular structure, based on defined generic objects, enables its easy and fast modifications when there are some changes in the network configuration. The information model is also extendable in terms of new network management applications.

5. REFERENCES

[And81] S. W. Anderes, S. E. Puccini, L. A. Tarr: "GTD-5 EAX - Digital Remote Units," GTE Automatic Electric Journal, January - February 1981

[Del85] J. P. Delatore, R. J. Frank, H. Oehring, L. C. Stecher: "The 5ESS Switching System: Operational Software," AT&&T Technical Journal, Vol. 64, No. 6, pp. 1357-1384, July-August 1985

[Hau85] G. Haugk, F. M. Lax, R. D. Royer, J. R. Williams: "The 5ESS Switching System: Maintenance Capabilities," AT&&T Technical Journal, Vol. 64, No. 6, pp. 1385-1416, July-August 1985

[Hol92] R. Holter: "Managing SONET Equipment," IEEE Network Magazine, January 1992

[Jin80] W. S. Jinbo, S. Magnusson: "GTD-5 EAX Hardware Description," GTE Automatic Electronic Journal, July 1980

[Joh89] E. F. Johnson, K. Reichard: "W-Window Applications Programming," Advanced Computer Books, Management Informations Source, Inc., 1989

[Khe91] S. Kheradpir, W. Stinson, G. Sundstrom: "A Network Management Architecture for SONET-Based Multi-Service Networks," Integrated Network Management II, I. Krishnan & W. Zimmer (Editors), Elsevier Science Publishers B. V., IFIP, 1991

[Laz90] A. A. Lazar: "Object-Oriented Modeling of the Architecture of Integrated Networks," CTR Technical Report #182-90-12, Columbia University, 1990

[Laz91] A. A. Lazar: "An Architecture for Real-Time Control of Broadband Networks," Proc. of the GLOBECOM'91, Phoenix, AZ, December 1991

[Maz90] S. Mazumdar, A. A. Lazar: "Monitoring Integrated Networks for Performance Management," Proc. of the IEEE International Conference on Communications/SUPERCOM '90, Atlanta, GA, April 1990

[Maz91] S. Mazumdar, A. A. Lazar: "Objective-Driven Monitoring," Proc. of the Second International Symposium on Integrated Network Management, Washington, D.C., April 1991

[Tsu91] M. Tsychida, A. A. Lazar, N. G. Aneroussis: "The integration of Real-Time Control with Management in Broadband Networks," CTR Technical Report #250-91-31, Columbia University, December 1991

[TONICS] R. Beal, G. Bossert, A. Friedman, S. Kheradpir, T. Mertzig, A. Salvador, W. Stinson, G. Sundstrom: "TONICS: TelOps Network Integrated Control System," Technical Requirements & Development Plan, Network Management Project (465), GTE Laboratories Incorporated, August 1991

[Wol91] O. Wolfson, S. Sengupta, Y. Yemini: "Managing Communication Networks by Monitoring Databases," IEEE Transactions on Software Engineering, VOL. 17, NO. 9, September 1991

VI.

MANAGEMENT APPLICATIONS

A.

PANEL

How Secure Should Network Management Be?

Panel Chair: Keith McCloghrie, Hughes LAN Systems, USA

Panelists: Jeff Case, SNMP Research, Inc., USA
John Hooder, US Navy, USA
Phil Janson, IBM Zurich, SWITZERLAND

As the importance of network management increases, so does the need to prevent its misuse. In particular, most of the reasons for network management's importance are also reasons for preventing all but authorized personnel from using its capabilities. Such prevention is of necessity based on the network security procedures. Unfortunately, security always has attendant costs, e.g., implementation complexity, or product export licensing, or additional operational procedures.

This panel will debate the tradeoffs between increased security measures and increased difficulty of use. A range of viewpoints will be represented by having a security expert, a network management expert/implementor, and a network user/planner as the three panelists.

VI.

MANAGEMENT APPLICATIONS

B.

FAULT MANAGEMENT APPLICATIONS

Integrated Network Management, III (C-12)
H.-G. Hegering and Y. Yemini (Editors)
Elsevier Science Publishers B.V. (North-Holland)

A Network Diagnostic Expert System for Acculink™ Multiplexers Based on a General Network Diagnostic Scheme

Kar-Wing Edward Lor

AT&T Bell Labs, 200 Laurel Ave., Middletown, NJ 07748, U.S.A.

Abstract

The objective of this research is to investigate the common link in various network diagnostic systems, and whether there exist any underlying common expertise and strategies that are applicable to different circuit-switch networks. If so, how is this general expertise related to the network-specific expertise? Can the general expertise provide a foundation on which most of the network diagnostic expert systems may be built?

Based on this method, we have developed an expert system to diagnose multiplexer networks. It interprets the alarms delivered to the Network Management System (NMS), determines the network's faulty behavior, and offers recovery suggestions. We classify the diagnostic expertise into general expertise and customized expertise. The former consists of certain general strategies and common sense in network diagnostics, while the latter consists of specific technology know-how of a network of multiplexers. In this work, the expert system serves as an aid and resides on top of the NMS.

1. INTRODUCTION

Expert systems for diagnostic purposes have been developed at AT&T since the early 80's. The first one was the ACE system [15] for diagnosing cables. The most representative one on network diagnostics was the Starkeeper® NMS Network Troubleshooter [9] that diagnosed individual components of the Datakit© VCS network. An experimental troubleshooting expert system for the APL modem network has also been built based on the Datakit Troubleshooter development platform. In the industry, there have been several network diagnostic expert systems including British Telecom's Advanced Maintenance Facility (AMF) [14], GTE's Central Office Maintenance Printout Analysis and Suggestion System (COMPASS) [10], BellCore's Switching Maintenance Analysis and Repair Tool (SMART) [13], DEC's Network Troubleshooting Consultant (NTC) [7], etc. The irony of these projects is that they have very little in common. Each system was tailored towards the specific application, especially in the diagnostic of specific equipment and facilities. Hence these systems offer little insight in modeling the diagnostic expertise across different applications.

Diagnosing problems requires substantial human judgement, experience and in-depth knowledge. In a large-scale network with hundreds of nodes, alarms are sent to a Network Management System (NMS) whenever an abnormal event or a fault occurs. Precise diagnostic knowledge is required to interpret one or more alarms to locate the fault. A majority of the knowledge follows established procedures or guidelines indicating that such tasks may be automated.

The separation of the core knowledge and the customized knowledge stems from two aspects of network troubleshooting. The core knowledge is for locating the general vicinity of faults and possibly a set of suspects within a large network. The customized knowledge is geared towards the precise diagnostic and recovery procedures among the selected network entities. While the former may be reusable in various circuit-switch networks, the latter is application-specific. Figure 1 shows the data interactions between the general and specific diagnostic expertise in our work.

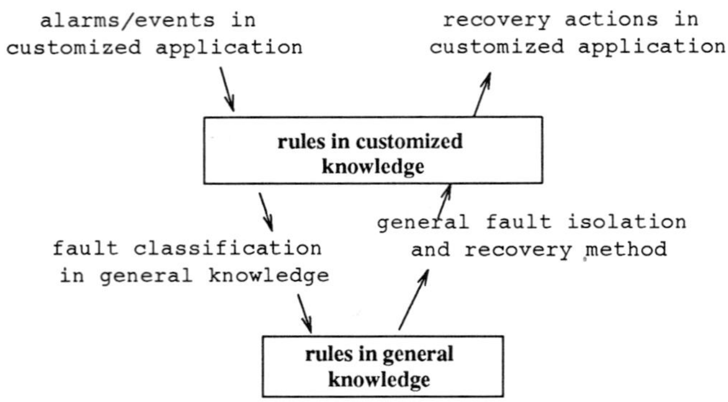

Figure 1. Interactions between the General and Customized Expertise

This paper first discusses the diagnostic expertise that makes up the core of the knowledge base, as well as certain customized knowledge for the Acculink Network Management System (ANMS) which governs a network of 740/745 Acculink™ Multiplexers. Then it presents the system configuration, particularly the interfaces among the network, the expert system, and the end-user.

2. THE GENERAL DIAGNOSTIC PRINCIPLES

The general diagnostic knowledge is mostly common sense in debugging techniques and topology problems in a connection-based network. The purpose is to isolate the faults to a certain subset of network entities. The general diagnostic guidelines can be divided into three phases. It looks for self-reporting culprits in the first phase. If it does not resolve the problem, it tries to locate a set of suspects in the second phase before pinning down the culprit from the suspects in the final phase.

As in the case of criminal investigation, the ideal scenario is that the culprit turns itself in whenever a fault in the network occurs. This is the most straightforward and easiest diagnostic. A faulty component in the network, let it be a node or a link, or any of its sub-components, reports its problem to the network management system. All attentions should be directed to this particular component.

However, if nobody turns itself in, an investigation is required to capture the culprit. In principle, all elements in the network could be examined to locate the malfunctioning one, but this approach is impractical. Thus the investigation needs some other intelligent means to isolate the culprit, or at least narrow it down to a list of suspects. That is where the fault isolation expertise helps. The knowledge in this phase forms the core expertise.

There are a wide range of strategies used to narrow down a list of suspects. One sample strategy is *Elimination of the innocents*. A faulty element in the network, let it be a node or a link, usually causes the breakdown of one or more logical circuits. On the other hand, there may also exist one or more routes bypassing the malfunctioning element, or alternate routes may be successfully established. Based on the *up* or *down* status of various routes, it is able to shrink the number of suspects, rather than dealing with all elements in the *down* routes in a large network. In the algorithm below, S represents the set of suspicious candidates, and I represents the set of innocents.

Let $RD_{i,\ 1 \le i \le N}$ be the down routes, E_{RD_i} be the set of elements in each down route, $RU_{j,\ 1 \le j \le M}$ be the reroutes or up routes, E_{RU_j} be the elements in each reroute or up route

$I \leftarrow E_{RU_1} \cup E_{RU_2} \cup \ldots \cup E_{RU_M}$;

$S \leftarrow E_{RD_1} \cap E_{RD_2} \cap \ldots \cap E_{RD_N} - I$,

elements appearing in all down routes, do not appear in any up route or reroute;

for each element e in S

count the number of down routes associated with e;

start diagnostics with the element of the highest down-route-count until the fault is located.

At this level of diagnostics, the rules only deal with the logical entities and physical entities of the network. A route or circuit is a logical entity, while the nodes and the links are physical entities. The expertise at this level is indifferent about the facts of the applications, e.g. a particular node is an Acculink 745 multiplexer, a particular route is a permanent channel group, etc.

If the culprit is not yet captured but a set of suspects are located, these suspects should be the focus of further investigations in phase three. By probability, normal network operation may be resumed at the earliest if the strategically most significant or most suspicious elements are attended to first. In other words, a ranking is needed at this stage to locate it as soon as possible. This ranking is based on both the static weights, assigned at network configuration time, and dynamic weights, calculated *on-the-fly*, of the elements. The static weights, as well as the balance between the static and dynamic weights, are determined by the human experts who are responsible for network configuration and diagnostics.

The rankings also play an important role when there are multiple faults in the network. If the faults are unrelated, i.e. the set of suspects are mutually exclusive, diagnostics will begin at the heaviest weighted suspect. Suppose the faults are related, and there are common suspects among them. The expert system will still start the diagnostics with the heaviest weighted suspect if it is not aware of multiple faults. After the full recovery of one fault, the system then frees all the suspects associated with that fault case, and turns to the suspects of any remaining faults. Unless the expert system knows explicitly the existence of mulitiple faults, as in the case of multiple *down* routes in the *Elimination of the Innocents*, that it will first examine the common suspects.

3. KNOWLEDGE CUSTOMIZED TO ACCULINK™ MULTIPLEXER NETWORK

With the core knowledge as the foundation, specific knowledge customized towards a particular application is built on top of it to make the system operational. This section presents certain customized knowledge of diagnosing a network built of AT&T's 740/745 Acculink™ Multiplexers.

Briefly, an Acculink™ Multiplexer performs time division multiplexing. Operating at T1 facilities, a 740 multiplexer usually connects one single T1 link and other data communication equipment such as a PBX or modems, while a 745 multiplexer can connect up to sixteen T1 links. It supports synchronous and asynchronous data channels, voice channels and DS1 channels. Multiple channels of identical source and destination may be grouped together to form a channel group, the equivalent of a logical route with respect to the core knowledge. Further details on the multiplexers are available in [1, 2].

Whenever a fault occurs, one or more alarms will be sent to the Acculink Network Management System (ANMS) [3], which may or may not reveal the source and/or the nature of the fault.

As illustrated in Fig. 1, the diagnostic process is bi-directional between the core expertise and the customized expertise. A specific failure invokes a rule in the customized knowledge, which in turn invokes a rule in the generic knowledge. The generic rules offer a general diagnosis and recovery plan for such a failure. The customized expertise is then invoked again to implement this plan. In an ordinary diagnostic session with more than one network faults, the diagnostic system goes back and forth between customized and general expertise many times, at least once for each failure encountered.

The diagnostic expertise is encoded as rules. In rule-based programming, the program consists of a collection of rules, the invocation, or *firing*, of which is governed by an inference engine. A rule consists of an antecedent set and a consequence set. Instances of the rules' premises and conclusions are represented by elements in a working memory, called the Working Memory Elements (WME). When there exist WMEs that match all the antecedents of a rule, indicating that certain network state and/or diagnostic status are satisfied, the corresponding consequences are carried out. In other words, this particular rule is fired. Each consequence either creates some new WMEs that eventually fire another rule, or constitutes some primitive diagnostic operations. A rule has the form of

$$\begin{array}{l}(\texttt{antecedent}_1\ \texttt{antecedent}_2\ \ldots\ \texttt{antecedent}_n \\ \ \texttt{-->} \\ \ \texttt{consequence}_1\ \texttt{consequence}_2\ \ldots\ \texttt{consequence}_m)\end{array}$$

In the following sub-sections, the antecendents and consequences are presented in natural language for comprehension purpose.

3.1 Customization of a Self-reporting Culprit

In the case of self-reporting culprit, the diagnostic is straightforward since it bypasses the general fault localization expertise. After the faulty entity signals a problem, the expert system invokes its recovery action upon retrieving the alarm in the NMS. To customize self-reporting faults in the Acculink application, a recovery procedure is encoded for every specific alarm. For example, in the power modules of the Acculink multiplexer, an alarm that reports error code of 70 is self-explanatory. The malfunction is caused by excess operating temperature. The recovery recommendation is a two-step procedure indicated by the following rule.

```
(catch power module of node N as the culprit, with error
  code = 70
-->
 recommendation #1 is to check the fan operations;
 recommendation #2 is to lower environment temperature.
)                                          .... (CK_power_rec)
```

In a case of more than one recovery recommendations, a human operator tries them one by one until the problem is resolved.

In handling self-reporting faults, the focus is on the recovery actions. Since precise, stepwise diagnostic and recovery procedures for most multiplexer internal alarms are documented in the manuals [1, 2], these procedures are encoded in the knowledge base to guide the human user to perform them step by step.

3.2 Customization of *Eliminate_the_Innocents*

Given a situation that requires the domain-specific expertise, certain customized knowledge is added on top of the core knowledge. Both sets of knowledge in this scenario are represented as rules. The core rules for this expertise, derived from the algorithm *Elimination of the Innocents* in Section 2, are given as follows:

```
(route R is faulty &
 an alternate route R1 is up
-->
 identify down-route suspects)      .... (GK_ei.1)

(identify down-route suspects &
 entity e was associated with faulty route R &
 entity e is not associated with up route R1
-->
 e is a suspect)               .... (GK_ei.2)

(e is a suspect
-->
 accumulate the number of down routes associated with e
 )                                           .... (GK_ei.3)

(e has non-zero N down routes associated with it
-->
 make a diagnostic case for element e with weight N
 )                                           .... (GK_ei.4)
```

In the multiplexer application, ANMS commands like *Multiplexer Alarms (MA)* and *Multiplexer Event Log (MEL)* review the faults, either implicitly or explicitly, in the network. In a case when two nodes are not communicating with each other but no explicit multiplexer alarms are seen, the expert system enters the *MEL* command to examine the connection and disconnection events of the channel groups. Upon detection of a channel group disconnection, it creates a WME

```
(alarm: channel group CG disconnected abnormally).
```

Coupled with the mapping of channel group *CG* to a route *R*, they represent the customized instance of the generic knowledge

```
(logical route R is faulty).
```

And the fact that *CG* is disconnected, followed immediately by another event of

```
(event: channel group CG connected)
```

represents the customized instance of

```
(an alternate route for R, namely R1, is up).
```

Therefore, on top of the core rules, there are two customized rules:

```
(alarm: channel group CG disconnected abnormally &
 channel group CG is mapped to route R
-->
 route R is faulty)                         .... (CK ei.1)
```

and

```
(alarm: channel group CG disconnected abnormally &
 event: channel group CG connected &
 connection happens shortly after disconnection &
 channel group CG is currently mapped to route R1
-->
 an alternate route R1 is up)          .... (CK ei.2)
```

The consequence of rule $GK_{ei.4}$ is only a high-level fault isolation scheme. After identifying the suspects with weights assigned, the expert system starts examining the ones weighted the heaviest. It invokes the customized rules to issue diagnostic commands to check out the multiplexers, or to notify the carrier or the switching offices to check out the facilities. If an individual component within a multiplexer is the cause of the failure, the human operator applies the customized recovery procedures as described in the previous section.

3.3 A Diagnostic Example

This section presents a simple example to show how the *Eliminate_the_Innocents* scheme works. The example lists the sequence of rules applied in this scenario. A sample network is shown in Figure 2.

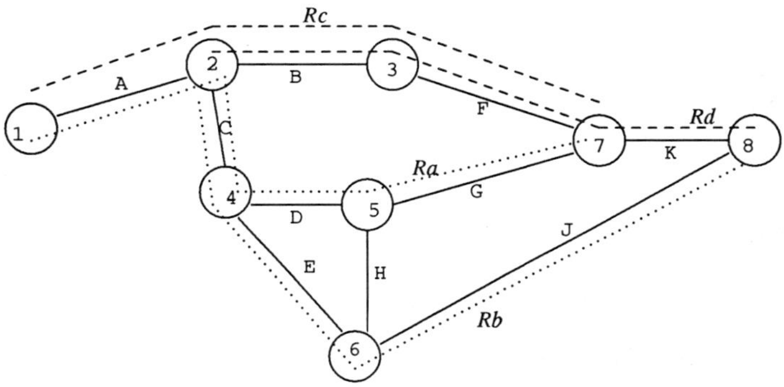

--------- newly established routes

.... routes disconnected

Figure 2. Sample network to illustrate the *Eliminate_the_Innocents*

In this network, there are two channel groups — `X`, connecting nodes `1` and `7`, and `Y`, connecting nodes `2` and `8`. Initially, channel group `X` is routed according to `Ra` and `Y` according to `Rb`. There are also mappings for all physical entities along the routes, e.g.

(`entity` *A* `is currently associated with route` *Ra*`)`,.....

Abruptly, both channel groups are disconnected, and the network is able to re-route the two channel groups automatically via routes `Rc` and `Rd`. These activities create four new WMEs:

(`alarm: channel group` *X* `disconnected abnormally at time` $t_{X.dc}$)
(`alarm: channel group` *Y* `disconnected abnormally at time` $t_{Y.dc}$)
(`event: channel group` *X* `connected at time` $t_{X.rr}$)
(`event: channel group` *X* `connected at time` $t_{Y.rr}$)

The disconnections/re-reroutes also update the WMEs concerning the old routes and creates WMEs for the new routes. These WMEs indicate the associations between the physical entities and the up/down routes. As a result, the current WMEs invoke the following sequence of rules:

rule	variable bindings
$\mathtt{CK}_{\mathtt{ei.1}}$	X/*CG*, Ra/*R*
$\mathtt{CK}_{\mathtt{ei.1}}$	Y/*CG*, Rb/*R*
$\mathtt{CK}_{\mathtt{ei.2}}$	X/*CG*, Rc/*R1*
$\mathtt{CK}_{\mathtt{ei.2}}$	Y/*CG*, Rd/*R1*
$\mathtt{GK}_{\mathtt{ei.1}}$	Ra/*R*, Rc/*R1*
$\mathtt{GK}_{\mathtt{ei.2}}$	Ra/*R*, C/*e*
$\mathtt{GK}_{\mathtt{ei.3}}$	C/*e*
.....	

After applying the rules to the sample network, the following eight elements make up the set of suspects S

$$S = \{4,\ 5,\ 6,\ C,\ D,\ E,\ G,\ J\}.$$

The rule $\mathtt{GK}_{\mathtt{ei.3}}$, being applied to the suspects repeatedly, accumulates the *down-route-count* for these elements, as given in the following table:

entities	4	5	6	C	D	E	G	J
count	2	1	1	2	1	1	1	1

Eventually, rule $\mathtt{GK}_{\mathtt{ei.4}}$ makes up a diagnostic case for each suspect with the *down-route-count* as the primary weight. The most suspicious elements in this scenario are node 4 and link C, since both of them are involved in the most down routes (2). They will be the first elements to be investigated using the application specific diagnostic procedures.

4. THE SYSTEM CONFIGURATION

To implement the diagnostic method presented in this paper, a Network Diagnostic Expert System (NDES) prototype is built on top of the Acculink Network Management System. The configuration of the system is illustrated in Fig. 3. NDES consists of four major modules, as described below.

- The System Driver

  This module is the main program of the entire system. It monitors the other modules and orders the sequence of their operations. It also carries out various miscellaneous tasks global to the system, like initializing the system, establishing dial-up connection to the network's ANMS, logging the session, doing bookkeeping upon termination, etc. The driver also serves as an interface between the entire system and the end-user. The system supports both X-window and dumb-terminal interfaces.

- The Rule-based System

  The rule-based system consists of a working memory, a knowledge base, and an inference engine [4]. The working memory is a database containing information about the

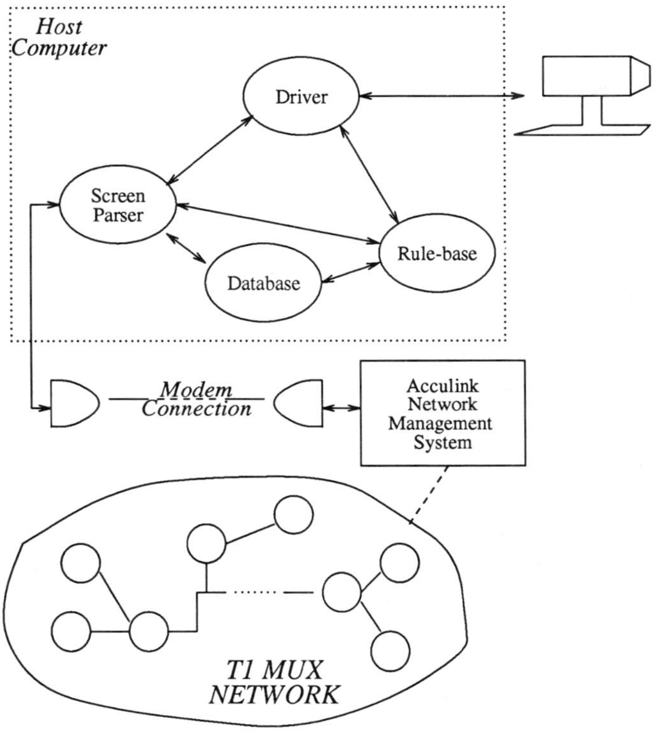

Figure 3. Configuration for the Network Diagnostic Expert System

diagnostic cases and the network. Such information consists of facts and assertions. Facts are deposited into the working memory before the diagnostic process, while assertions are created and removed during the process. The knowledge base consists of a set of production rules, representing the general and customized expertise in network diagnostics. The inference engine is the software equivalent of the CPU for an expert system. One at a time, it executes, or fires, a production rule the antecedents of which are true. It is the job of the inference engine to check out which rules, among numerous rules, have their antecedents satisfied by the current state of the working memory. In case of more than one satisfied, the engine uses some pre-determined schemes to break the tie. This tie-breaking procedure is called *conflict resolution*.

- The Screen Parser

The Network Management System collects the operating and fault statuses of the network and displays them in its own format or language. The Network Diagnostic Expert System interacts with the NMS through an interface called the Screen Parser. It translates the network information from the NMS into representations suitable for the expert system.

To obtain network information, the Screen Parser sends a command and its parameters to ANMS. Upon successful execution of a command, ANMS outputs a character stream regarding network information. The destination of this stream of control and ascii characters is supposed to be an ANMS display terminal. The purpose of the screen parser is to intercept this stream, re-construct the screen in its program's internal data structure, parse the screen, and create working memory elements out of it.

- The Database

  The Network Diagnostic Expert System keeps a database of its own for instant information retrieval. Certain network information, both static and dynamic, is needed during the diagnostic process. It is readily available in the Network Management System, but retrieving it from the NMS induces a huge overhead. The use of a local database eliminates the overhead of data transmission and parsing.

  Information stored in the database include the relationships between the logical entities (channel groups) and the physical entities (nodes or links) like routing information, attributes of the physical entities, and node/link incident relationship, etc. The physical network information is entered into the database when the network is configured, while the dynamic information, fairly unstable, should be updated periodically to match the ANMS database.

The system was developed on AT&T's 6386 computer with System V Unix. The rule-base was coded in AT&T's C5 language [11], a dialect of OPS5 [4], and all other modules were coded in C. The user-interface was based on X-window. Currently there are about 70 rules in the core expertise, 200 rules in the customized expertise, and 30 miscellaneous rules.

5. CONCLUSIONS

The Network Diagnostic Expert System formalizes the network diagnostic process, and identifies certain expertise that may be applicable to different types of networks. The knowledge in the prototype is always subject to expansion, but it provides a foundation for a marketable product, or even to be incorporated into the NMS itself.

A prototype of NDES has been installed at AT&T's NetCare Center, which monitors and maintains the networks for Acculink customers. At the current stage, NDES operates reactively upon the human operator's request. When there is a network failure, an operator brings up NDES, which retrieves the alarms from the network, interprets the alarms, and guides the operator in the troubleshooting process. However, it is not meant to replace the operator since the final decisions are still left to the human.

6. REFERENCES

[1] *740/741 Acculink™ Multiplexer User's Manual*, Document # 999-100-289IS, AT&T, 1989.

[2] *745 Acculink™ Multiplexer User's Manual*, Document # 999-100-227, AT&T, 1989.

[3] *AT&T Paradyne® 6800 Series Network Management System User Guide*, AT&T 1989.

[4] Brownston, L., R. Farrell, E. Kant, and N. Martin, *Programming Expert Systems in OPS5: An Introduction to Rule-Based Programming*, Addison-Wesley, Reading, MA, 1986.

[5] de Kleer, J., and B.C. Wiiliams, "Diagnosing Multiple Faults", *Artificial Intelligence*, 32(1), Apr. 1987, pp. 97-130.

[6] Frontini, M., J. Griffin and S. Towers, "A Knowledge-Based System for Fault Localisation in Wide Area Networks", *Proceedings of 2nd International Symposium on Integrated Network Management*, Washington, D.C., pp. 519-530, Apr. 1991

[7] Hannan, J., "Network Solutions Employing Expert Systems", *Phoenix Computers and Communications Conference*, pp. 543-547, IEEE, 1987.

[8] Mathonet, R., H. Van Cotthem, and L. Vanryckeghem, "DANTES: An Expert System for Real-Time Network Troubleshooting", *Proceedings of 10th International Joint Conference on Artificial Intelligence*, Milan, Italy, Aug. 1987, pp 527-530.

[9] Marques, T.E. A Symptom-Driven Expert System for Isolating and Correcting Network Faults", *IEEE Communications Magazine*, Vol. 26, No.3, March 1988, pp 6-14.

[10] Prerau, D., A. S. Gunderson, R.E. Reinke, and S.K. Goyal, "The COMPASS Expert System: Verification, Technology Transfer, and Expansion", *Second Conference on Artificial Intelligence Applications*, pp. 597-602, IEEE, 1985.

[11] Rowland, J.R., and G. T. Vesonder, "C5 User Manual, Release 1.0", AT&T Technical Memorandum 59346-870715-39TM, July 1987.

[12] St. Jacques, M., D. Stevens, J. Sipos, and L. Lau, "RAC & ROLE, A Knowledge-Based System for Network Trouble Administration Defined by Rapid-Prototyping", *Proceedings of 2nd International Symposium on Integrated Network Management*, Washington, D.C., pp. 531-540, Apr. 1991

[13] Sutter, M., "The Smart Project: An approach to Expert System Integration and Evaluation in the BOCs", *ICC '86*, pp 1230-1232, June 1986.

[14] Thandasseri, M., "Expert Systems Application for TXE4A Exchanges," *Electrical Communications*, Vol. 60, No. 2, 1986

[15] Wright, J.R., J.E. Zielinski and E.M. Horton "Expert System Development: The ACE System", in *Expert System Applications to Telecommunications*, John Wiley & Sons, Inc., New York, 1988.

Integrated Network Management, III (C-12)
H.-G. Hegering and Y. Yemini (Editors)
Elsevier Science Publishers B.V. (North-Holland)

A case-based reasoning approach to the resolution of faults in communications networks

Lundy Lewis

Cabletron Systems R&D Center, 9 Executive Park Drive, Merrimack, NH 03060 USA

Abstract

Most fault resolution systems for communications networks represent fault resolution expertise using a rule-based reasoning (RBR) paradigm. Although the RBR paradigm is appropriate for problem-solving tasks that are confined and well-understood, its limitations are (i) an inability to learn from experience, (ii) an inability to deal with novel problems, and (iii) the difficulty of updating the systems to keep up with rapidly changing domains such as expanding heterogeneous networks. In the paper we describe a problem-solving paradigm, case-based reasoning (CBR), that is less constrained by these limitations. Our approach is to enhance a standard fault management system with a CBR component. We review CBR techniques for retrieving, adapting, and embedding knowledge in a case library and describe CRITTER, a CBR trouble ticketing system for managing and resolving network faults.

Keyword Codes: C.2.3; I.2.4
Keywords: Network Operations, Computer-Communication Networks; Knowledge Representation Formalisms and Methods, Artificial Intelligence

1. INTRODUCTION

The management of faults in a small homogeneous network is straightforward. However, as a network becomes increasingly large and heterogeneous, fault management becomes a problem. Part of the solution is to fashion a structured fault management system with a trouble ticketing package. Using a trouble ticketing package, the network administrator can design (i) a trouble ticket that reflects the interests of the end-users and support staff and (ii) a ticket transition graph that enforces the proper escalation and routing of a ticket to closure [Olsen and Blackwell 90]. In addition to streamlining the management of network faults, this method issues a store of problem-solving expertise that is recorded when trouble tickets are resolved and logged into the trouble ticket database. The support staff can appeal to this expertise when similar problems recur.

A structured trouble ticketing system does not provide the whole solution to the fault management problem. For time-critical network services, the downtime that elapses from the observation of a network fault, the submission of a trouble ticket, to the closure of the trouble ticket can be expensive. The downtime can be reduced by providing a communication path between a network fault detection system and a trouble ticketing system. For example, the SPECTRUM network management system developed by Cabletron Systems detects and locates network faults. The Action Request (AR) System developed by the Remedy Corporation controls the management of network faults. Cabletron has developed an application, the AR System Gateway, that collects SPECTRUM fault information and transmits the information to the AR System in the form of a trouble ticket, whereupon the AR System manages the communication and workflow among the network administrator, support staff, and end users.

The synergy between a fault detection system and a trouble ticketing system comprises a considerable portion of the solution to the fault management problem, but not all of it. The time required to research and resolve a fault can remain expensive. The solution to this part of the problem is difficult because it is knowledge-intensive. In most commercial and research applications that tackle this part of the problem, fault resolution expertise is represented in a rule-based reasoning (RBR) system, and the system is embedded in existing operations software ([Lewis 92a], [Peacock 88], [Pagurek et al 88], [Callahan 88], [Tjaden et al 91], [Marques 88], [Mathonet et al 87]).

A good review of network management applications of RBR (a.k.a. production systems) is [Cronk et al 88]. A RBR system consists of a working memory (WM), a rule-base, and a control procedure. For a network application, the WM typically contains a representation of characteristics of the network, including topological and state information. The rule-base represents knowledge about what operations to perform when the network enters an undesirable state. If the network enters an undesirable state, the control procedure selects those rules that are applicable to the current situation. Of the rules that are applicable, a pre-determined control strategy selects a rule that is actually executed. A rule can perform tests on a network, query a database, provide directives to a network configuration manager, or invoke another expert system. With these results, the system updates the WM by asserting, modifying, or retracting WM elements. This cycle continues until a desirable state in WM is achieved.

There exist several variations on the basic RBR paradigm. The control procedure can be enhanced with a belief revision capability. The control procedure keeps a list of rules that were selected on each cycle, and may backtrack to a previous cycle to select an alternative rule if progress is not being made towards a desirable state (assuming that no operation has been performed that cannot be undone). In addition, the rule-base can be functionally distributed, and a meta-control strategy can be provided that selects the component RBR system that should be executed for specific kinds of task (e.g. [Sutter and Zeldin 88], [Frontini et al 91]). Alternatively, the rule-base can be distributed component-wise such that there exist specialized rule-bases for each component in the network (e.g. [Schroder and Schodl 91]).

2. THE PROBLEM WITH RBR APPROACHES TO FAULT RESOLUTION

The procedure for constructing a RBR fault resolution system is (i) to define a description language that represents the problem domain, (ii) to extract expertise from multiple domain experts and/or trouble-shooting documents, and (iii) to represent the expertise in the RBR format. The procedure can require several iterations of an interview/implement/test cycle in order to achieve a correct system. If the domain remains relatively constant, a correct system needs little maintenance (e.g. see the discussions in [Vesonder et al 83], [Williams et al 83], and [Wright et al 88]). However, if the system is used to resolve faults in unpredictable or rapidly changing domains, two problems inevitably occur: (i) the system suffers the problem of brittleness and (ii) the development process suffers the problem of knowledge acquisition bottleneck. Brittleness means that the system will fail when it is presented with a novel problem. The counterpart of the brittleness problem is the system's lack of ability to adapt existing knowledge to a novel situation or to learn from experience. A knowledge acquisition bottleneck can result when a knowledge engineer tries to devise special rules and control procedures that will cover unforeseen situations. When this happens, the system typically becomes unwieldy, unpredictable, and unmaintainable. With rapidly changing domains, the system can become obsolete quickly. The alternatives at this stage are to limit the coverage of the RBR system or to search for other approaches.

3. A CASE-BASED REASONING APPROACH

Case-based reasoning (CBR) is an alternative approach to problem-solving that offers potential solutions to the problems of brittleness and knowledge acquisition bottleneck ([Riesbeck and Schank 89], [Slade 91], [Kolodner 91], [Barletta 91], [Lewis et al 92]). The goals of CBR systems are (i) to learn from experience, (ii) to offer solutions to novel problems based on past experience, and (iii) to avoid extensive maintenance. *The basic idea of CBR is to recall, adapt, and execute episodes of former problem-solving in an attempt to deal with a current problem.* Former episodes of problem-solving are represented as cases in a case library. When confronted with a new problem, a CBR system retrieves a similar case and tries to adapt the case in an attempt to solve the outstanding problem. The experience with the proposed solution is embedded in the library for future reference. The general CBR architecture is shown in Figure 1.

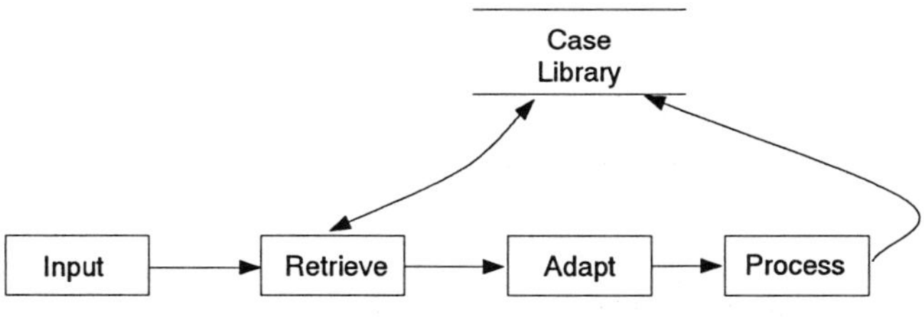

Figure 1. The general CBR architecture

Our approach to the fault resolution problem is to add a CBR problem-solving component to a standard trouble ticketing system. In CBR terminology, a resolved trouble ticket is essentially a case that records a former episode of problem-solving, and a trouble ticket database is essentially a case library. CBR enhances a trouble ticketing system by providing mechanisms to (i) retrieve a useful ticket when an outstanding trouble is submitted to the system, (ii) if necessary, adapt the ticket to recommend a solution to an outstanding trouble, and (iii) embed the experience in the trouble ticket database. Below we describe the principles that underlie these mechanisms and then describe the design for CRITTER, a CBR trouble ticketing system that integrates the management and resolution of network faults.

4. THE STRUCTURE OF A TROUBLE TICKET

CBR is well-suited for application to problem domains for which a body of knowledge with a case-like structure already exists or is easily obtainable. The structure of typical trouble ticket is much like the structure of a case, and thus trouble ticketing systems are natural candidates for enhancement with CBR techniques. An example ticket structure is shown in Figure 2. Some of the ticket fields are used for management purposes only, e.g. "Notify-method", "Assigned-priority", "Assigned-to", and "Ticket Status". Other fields describe the nature of the trouble, e.g. "Alarm Date/Time", "Device Name", "Device Type", "IP Address", and "Trouble". The field "Additional Data" contains a list of data that an expert may be interested in knowing before making an attempt to diagnose and resolve the trouble. When the trouble is resolved, the expert enters the resolution strategy in the "resolution" field, plus any additional data that turned out to be useful in the "additional data" field. In total, the resolved ticket is a good example of an episode of problem-solving that can be exploited in the future.

Entry-ID	00000000116	Submitter	SPECTRUM
Alarm ID	57	Create-date	07/24/92 08:51:08
Alarm Date/Time	07/24/92 08:51:07	Notify-method	○ None ○ Notifier ○ E-mail
Condition	○ Red ○ Orange ○ Yellow	Assigned-Priority	○ Low ○ Medium ○ High
Device Name	Ethernet-Randy	Assigned-to	
Device Type	Ethernet	Last-modified-by	SPECTRUM
IP Address		Modified-date	07/24/92 08:51:08
Trouble	file_transfer_throughput=slow		
Additional Data	network_ load=20, collision_rate=15, deferment_rate=20, users=31		
History of Trouble			
Probable Cause			
Resolution			
Resolution Status	○ Good ○ No Good ○ In Progress		
Ticket Status	○ New ○ Assigned ○ Rejected ○ Closed		

Figure 2. A trouble ticket

Note that it is advisable to develop a structured descriptive language for the information that enters fields like "Trouble", "Additional Data", and "Resolution". This is easy to do with most current trouble ticket packages. Although this approach avoids some of the difficult issues in natural language processing, it imposes bounds on the amount of learning the system can achieve. For example, a CBR system would not be good at creating brand new solutions, but it would be good at tweaking an old solution or creating a new solution from two existing solutions.

5. RETRIEVING A USEFUL TROUBLE TICKET

The usefulness of a retrieved ticket depends upon the degree to which it is similar to the open ticket. Generally, increasing similarity between a resolved ticket and an open ticket justifies more assurance that the resolution strategy of the one will work for the other. Thus, a first-order definition of similarity is prescribed by the number of matches between the attributes of two tickets, where a larger number of matches indicates increasing similarity. However, if two tickets are determined highly similar based on irrelevant attributes, the fact that the one ticket is resolved by method *M* does not warrant the inference that the other one can be resolved by *M*. For example, if two tickets match on all fields filled in for management purposes but not on fields that describe the trouble, it is not likely that the retrieved ticket will be of much use. If it is known, however, that some attributes are particularly relevant to a specific kind of trouble, and just these attributes are used to retrieve a similar ticket, then the inference of *M* as a possible resolution strategy is more reasonable. Similarity is thus defined as the larger number of matches between attributes that are relevant to a particular kind of trouble.

This idea is worked out conceptually in the theory of determinations [Davies and Russell 87]. The theory suggests that a library of tickets be augmented with a set of determinators that record relevance information among classes of network troubles and sets of ticket attributes. An example of a determinator is the following:

The solution to the trouble "file_transfer_throughput is slow" is determined by looking at bandwidth, network load, packet collision rate, and packet deferment rate

The determinator points to the relevant attributes for the trouble file_transfer_throughput. By deferring to these attributes when selecting possibly applicable tickets, the system will avoid the selection of those tickets that are similar for the wrong reasons.

Note that determinators do not serve the same function as rules in a RBR system. A RBR rule performs actions that update a working memory, and the rule set typically runs through several cycles before a solution is found. If the problem domain changes, the rule set must be modified and re-tested. In contrast, determinators allow a system to focus on trouble tickets that are likely to contain strategies that will lead to a resolution of the current trouble.

A subsidiary problem is, of course, how determinators come to exist. One approach is as follows. We have noted that network support staff employ conceptual determinators while sifting through a trouble ticket database searching for useful tickets. Typically, a set of tickets is retrieved that match on the "trouble" field of the current problem, and this set is pruned by matching on other relevant fields. With trouble ticketing packages, it is easy to record the actions of the support staff during the selection process and define the process as a macro. Essentially, these macros are determinators and may be used again for similar troubles. Alternatively, domain experts can articulate a small, albeit imperfect set of determinators at the outset. These rules could be refined automatically as the system learns the domain, and they can change as the network changes. Therefore, the knowledge acquisition bottleneck that usually occurs in rapidly changing domains becomes less problematic. A third approach is to apply an information theoretic algorithm such as ID3 to a trouble ticket database. The output of the algorithm is essentially a list of determinators that match trouble ticket attributes to fault resolution procedures [Goodman et al 91]. In our work so far, we have used the second approach. However, some synergy between the approaches could be possible.

6. ADAPTING A TICKET AND EMBEDDING THE EXPERIENCE IN THE LIBRARY

In a standard trouble ticket library, tickets are stored sequentially in flat files or as a linked list. The CBR retrieval algorithm imposes a structure on the library as follows. Determinators enable the system to retrieve a list of tickets from the library that are likely to hold strategies for resolving an outstanding trouble. These tickets are indexed with respect to increasing similarity to the current trouble. The first ticket is chosen and and its resolution strategy is considered and/or executed. If the strategy does not work, the next ticket is chosen, etc. Importantly, the experience with retrieved tickets results in new tickets that are placed in the library for future reference. This includes both fruitful and unfruitful experiences, indicated by "good" and "no good" in the "resolution status" field.

In fortunate situations, a resolution will be directly applicable to the outstanding trouble. Where this is not possible, however, a retrieved ticket can be adapted to resolve the trouble. Here, we describe simple illustrations of three ways to adapt a source ticket to an outstanding target ticket [Riesbeck and Schank 89].

6.1 Parameterized adaptation

Consider an outstanding trouble with file_transfer_throughput, and imagine that only one source ticket is retrieved from the trouble ticket database, as shown below. In the example, the resolution is adjust_network_load=*A*, where *A* is a solution variable that holds the amount by which network load is to be adjusted.

trouble: file_transfer_throughput=F
additional data: none
resolution: A=f(F), adjust_network_load=A
resolution status: good

Parameterized adaptation is a method by which the solution variable of an outstanding ticket is adjusted relative to its problem variables, based on the relation between the solution and problem variables in a source ticket ([Bain 86], [Rissland and Ashley 86]). Everything else being equal, an outstanding trouble file_transfer_throughput=*F'* should propose the the resolution adjust_network_load=*A'*, where *F'* and *A'* stand in the same relation as *F* and *A* in the source ticket. The proposed resolution to the new ticket, therefore, would look like the following:

trouble: file_transfer_throughput=F'
additional data: none
resolution: A'=f(F'), adjust_network_load=A'
resolution status: good

In this example, there are a number of ways to represent the *f* function. The simplest way is a look-up table, where values of *F* that are not in the table are calculated by interpolation. In related work, we are experimenting with representing the *f* function with fuzzy inference [Lewis 92b]. For other kinds of troubles, the resolution might be a sequence of steps or a decision tree.

6.2 Abstraction/respecialization adaptation

The idea here is that the system considers viable alternative resolutions to a trouble. If there is a prohibitive constraint on a proposed resolution, the system abstracts over the ticket that contains the proposed resolution and respecializes to a ticket that contains an alternative solution (if one is available). For example, consider again the file_transfer_throughput problem, and suppose that two source tickets are retrieved that are equally similar to the open ticket, as shown below:

trouble: file_transfer_throughput=F
additional data: none
resolution: A=f(F), adjust_network_load=A
resolution status: good

trouble: file_transfer_throughput=F
additional data: none
resolution: B=g(F), increase_bandwidth=B
resolution status: good

If a constraint on a possible resolution of an outstanding ticket is "not adjust_network_load", or if the execution of adjust_network_load does not resolve the trouble, the system would propose increase_bandwidth. This results in a new ticket added to the library, shown below.

trouble: file_transfer_throughput=F
additional data: adjust_network_load=no
resolution: A=f(F), adjust_network_load=A
resolution status: no good

A more complex version of this method involves the integration of pieces of several resolution strategies into an overall solution [Alterman 86].

6.3 Critic-based adaptation

Critic-based adaptation occurs when a critic repairs a retrieved solution to fit the target case [Hammond 86]. Examples of this technique are to add, remove, reorder, or replace steps in the retrieved solution. Given the example source ticket in Section 6.1 above, an expert could include the attribute network_load and refine the solution by providing a two-place function *f* that calculates the amount of adjustment based on the values of file_transfer_throughput and network_load. The proposed resolution to the new trouble, therefore, would look like the following:

trouble: file_transfer_throughput=F
additional data: network_load=N
resolution: A=f(F,N), adjust_network_load=A
resolution status: good

Importantly, this ticket provides new knowledge that is logged to the trouble ticket library. If a problem with file_transfer_throughput recurs and the value of network_load is available, the system would propose a value for *A* using the two-place function *f*. If the value of network_load is unavailable, the value of *A* would be determined using the one-place function *f*. Note that parameterized adaptation would be used in either case.

7. THE DESIGN OF CRITTER

The architecture in which CRITTER is embedded is shown in Figure 3. The primary components of the architecture are a network configuration manager, a fault detection system, a fault manager, and a fault resolution system. SPECTRUM provides the functions of configuration management and fault detection. The AR System provides the function of fault management. The addition of the CBR fault resolution component to the fault management system makes up CRITTER. Note that the user can intervene at three junctures in the architecture: (i) to filter trouble tickets that are submitted to CRITTER, (ii) to adapt proposed solutions, and (iii) to regulate instructions that are sent to SPECTRUM.

The CRITTER architecture consists of five modules:

1. The Input module is used to acquire information about current troubles in the form of a trouble ticket. The information can be entered manually by a user or automatically by the application that couples SPECTRUM and the AR System.
2. The Retrieve module is transparent to the user. The module uses determinators to retrieve a group of tickets from the library that are similar to an outstanding ticket. The initial set of determination rules can be provided by domain experts or extracted from a troubleshooting document. The initial rules do not have to be perfect. They become more accurate with the use of the system.
3. The Adapt module is also transparent. The module looks at the ticket that is maximally similar to the outstanding ticket. If the retrieved ticket is a perfect match in all relevant fields, the resolution is not altered. If not, the module employs either parameterized adaptation or the abstraction/respecialization method.
4. The Propose module displays potential solutions found by the reasoning module and allows the user to inspect and manually adapt these solutions.
5. The Process module updates the ticket library with the new experience. In addition, the process module may send instructions to the SPECTRUM configuration manager. Where this is possible, the fault support loop is closed and the management of faults is fully automated.

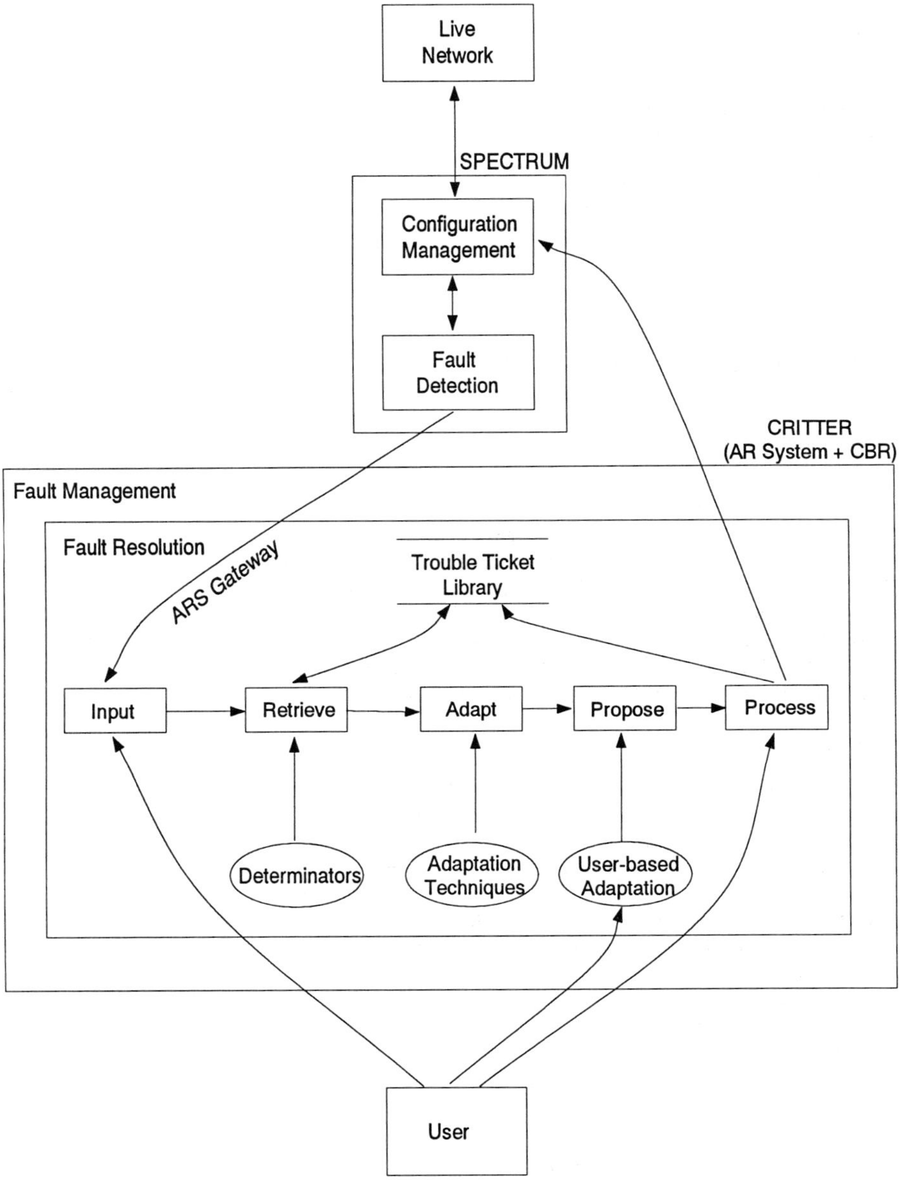

Figure 3. The structure of CRITTER

8. CURRENT STATUS

A large part of the overall system is in place, including the configuration manager and the fault detection system (SPECTRUM), the fault management system (AR System), and the gateway that connects these components (AR System Gateway). The first two of these products are at the general release stage, while the AR System Gateway is in beta release. The implementations of the algorithms to retrieve and adapt trouble tickets are in pre-beta release. As of this writing, the overall system is being used internally to manage our own enterprise networks. We hope to get the system out to other sites and to publish performance data by mid-1993.

9. CONCLUSION

The problem of fault management in expanding heterogeneous networks is expensive and hard to solve. The solution to the problem includes (i) a fault detection system, (ii) a trouble ticketing management system, and (iii) a representation of fault resolution expertise. Fault detection systems and trouble ticketing systems are relatively mature. The representation of fault resolution expertise is less mature. The current approach is to represent this expertise with the RBR paradigm. Unfortunately, RBR systems can be bottlenecks. They have inherent limitations that impede their growth in proportion to rapidly changing networks, viz. they cannot offer solutions to novel faults and they cannot learn from experience. We have described an alternative representation of fault resolution expertise, CBR, that shows promise to avoid these limitations. Fault resolution expertise is represented as cases rather than rules. We described mechanisms to retrieve similar cases, to adapt a case in order to propose a solution to a novel fault, and to embed the experience with the proposed solution in a case library. These mechanisms enhance a standard trouble ticketing system.

10. REFERENCES

[Alterman 86] R. Alterman. An Adaptive Planner. *Proceedings of the 1986 National Conference on Artificial Intelligence*. AAAI Press. pp. 65-69.

[Bain 86] W. Bain. A Case-Based Reasoning System for Subjective Assessment. *Proceedings of the 1986 National Conference on Artificial Intelligence*. AAAI Press. pp. 523-527.

[Barletta 91] R. Barletta. An Introduction to Case-Based Reasoning. *AI Expert*. August 1991. pp. 43-49.

[Callahan 88] P. Callahan. Expert Systems for AT&T Switched Network Maintenance. *AT&T Technical Journal*, Vol. 67, No.1, January-February 1988. pp. 93-103. Reprinted in [Ericson et al 89].

[Cronk et al 88] R. Cronk, P. Callahan, and L. Berstein. Rule-Based Expert Systems for Network Management and Operations: An Introduction. *IEEE Network Magazine*. Vol. 5, No. 4, September 1988. pp. 7-21. Reprinted in [Ericson et al 89].

[Davies and Russell 87] T. Davies and S. Russell. A Logical Approach to Reasoning by Analogy. *Proceedings of the 1987 International Joint Conference on Artificial Intelligence*. Morgan Kaufmann Publishers, Inc. pp. 264 - 270.

[Ericson et al 89] E. Ericson, L. Ericson, and D. Minoli (editors). *Expert Systems Applications in Integrated Network Management.* Norwood, MA: Archtech House, Inc. 1989.

[Frontini et al 91] M. Frontini, J. Griffin, and S. Towers. A Knowledge-Based System for Fault Localization in Wide Area Networks. *Integrated Network Management, II* (editors: I. Krishnan and W. Zimmer). Elsevier Science Publishers (North Holland). 1991. pp. 519-530.

[Goodman et al 91] R. M. Goodman. Automated Knowledge Acquisition from Network Management Databases. *Integrated Network Management, II* (editors: I. Krishnan and W. Zimmer). Elsevier Science Publishers (North Holland). 1991.

[Hammond 86] K. Hammond. CHEF: A Model of Case-based Planning. *Proceedings of the 1986 National Conference on Artificial Intelligence.* AAAI Press. pp. 267-271.
[Kolodner 91] J. Kolodner. Improving Human Decision-Making through Case-Based Decision Aiding. *AI Magazine*. Volume 12, Number 2. AAAI Press. Summer 91. Pages 52-68.

[Lewis 92a] L. Lewis. A Review of Rule-Based Approaches to Network Management. Technical Note ctron-lml-92-02. Cabletron Systems Research and Development Center. 1992.

[Lewis 92b] L. Lewis. Case-Based Reasoning with Fuzzy Inference: A Backend to Network Analyzers. First International Conference on Fuzzy Theory and Technology. Duke University. October 1992.

[Lewis et al 92] L. Lewis, D. Minior, and S. Brown. A Case-Based Reasoning Solution to the Problem of Redundant Engineering in Large Scale Manufacturing. *International Journal of Expert Systems: Research and Applications. Special Issue on Case-Based Reasoning.* March 1992.

[Marques 88] T. Marques. A Symptom-Driven Expert System for Isolating and Correcting Network Faults. *IEEE Communications Magazine*, Vol. 26, No. 3. March 1988. pp. 6-14. Reprinted in [Ericson et al 89].

[Mathonet et al 87] R. Mathonet, H. Van Cotthem, and L. Vanryckeghem. DANTES: An Expert System for Real-Time Network Troubleshooting. *Proceedings of the Tenth International Joint Conference on Artificial Intelligence.* Milan, Italy. August 1987. pp. 527-530. Reprinted in [Ericson et al 89].

[Olson and Blackwell] L. Olson and A. Blackwell. Understanding Network Management with OOA. *IEEE Network Magazine*. July 1990.

[Pagurek et al 88] B. Pagurek, A. Kaye, and D. Helmy. Knowledge Based Fault Location in a Data Communication Network. IEEE Document CH2538-7/88/0000-1729. pp. 1729-1733. Reprinted in [Ericson et al 89].

[Peacock 88] D. Peacock. On-Line Expertise for Telecommunication. In *Expert System Applications to Telecommunications* (Jay Liebowitz, editor). New York: John Wiley and Sons. 1988. pp. 143-159.

[Riesbeck and Schank 89] C. Riesbeck and R. Schank. *Inside Case-Based Reasoning.* Lawrence Erlbaum Associates. Hillsdale, N.J. 1989.

[Rissland and Ashley 86] E. Rissland and K. Ashley. Hypotheticals as Heuristic Device. *Proceedings of the 1986 National Conference on Artificial Intelligence.* AAAI Press. Pages 289-297.

[Schroder and Schodl 91] J. Schroder and W. Schodl. A Modular Knowledge Base for Local Area Network Diagnosis. *Integrated Network Management, II* (editors: I. Krishnan and W. Zimmer). Elsevier Science Publishers (North Holland). 1991. pp. 493-503.

[Slade 91] S. Slade. Case-Based Reasoning: A Research Paradigm. *AI Magazine.* Vol. 12, No. 1, pages 42-55. Spring 1991.

[Sutter and Zeldin 88] M. Sutter and P. Zeldin. Designing Expert Systems for Real-Time Diagnosis of Self-Correcting Networks. *IEEE Network Magazine.* Vol. 5, No. 4, September 1988. pp. 43-51. Reprinted in [Ericson et al 89].

[Tjaden et al 91] G. Tjaden, M. Wall, J. Goldman, and C. Jeromnimon. Integrated Network Management for Real-Time Operations. *IEEE Network Magazine.* March 1991. pp. 10-15.

[Vesonder et al 83] G. Vesonder, S. Stolfo, J. Zielinski, F. Miller, and D. Copp. ACE: An Expert System for Telephone Cable Maintenance. *Proceedings of the Eight International Joint Conference on Artificial Intelligence.* Karlsruhe, West Germany. August 8-12, 1983. pp. 116-121.

[Williams et al 83] T. Williams, P. Orgren, and C. Smith. Diagnosis of Multiple Faults in a Nationwide Communications Network. *Proceedings of the Eight International Joint Conference on Artificial Intelligence.* Karlsruhe, West Germany. August 8-12, 1983. pp. 179-181.

[Wright et al 88] J. Wright, J. Zielinski, and E. Horton. Expert Systems Development: The ACE System. In *Expert System Applications to Telecommunications* (editor: J. Liebowitz). John Wiley and Sons, New York. 1988. pp. 45-72.

Integrated Network Management, III (C-12)
H.-G. Hegering and Y. Yemini (Editors)
Elsevier Science Publishers B.V. (North-Holland)

Event Correlation in Heterogeneous Networks Using the OSI Management Framework

J.F. Jordaan[1] and M.E. Paterok

IBM European Networking Center, Tiergartenstraße 8, 6900 Heidelberg, Germany
email: djordaan@cs.up.ac.za, paterok@dhdibm1.bitnet

Abstract

The paper describes a prototype event correlation application developed at the IBM European Networking Center. In heterogeneous networks with integrated network management systems the symptoms of the failure of a single network resource could be detected and reported independently by many different system components. As a result, a single network failure triggers numerous event reports, with no indication which one of them (if any) reports an actual failure.

The application described here assists in finding the actual failure by analyzing and structuring such event reports. The analysis consists of correlation of reports resulting from the same failure and ordering correlated reports to indicate the resource where the failure has occurred with a high probability, thereby decreasing the complexity of the operator task. It relies on a homogeneous model of interconnected, heterogeneous networks and explores relationships among physical and logical network resources to perform its task.

The application uses the OSI standardized management framework and management communication protocol, and OSI-based managed objects.

Keyword Codes: C.2.3; D.4.4; K.6.4
Keywords: Network Operations; Communications Management; System Management

1. INTRODUCTION

Integrated management of heterogeneous networks is finally becoming a reality. Integrating architectures and standardized communication mechanisms are starting to deliver on the promises which they've been making since a few years ago. The impact of these technologies is visible on the network management console of today. Information describing, or originating from, multiple vendors' equipment is presented to the human operator in a unified fashion.

The increased availability of information has thus far not necessarily resulted in a similar decrease of the complexity of the management task. Consider the following simple scenario: A physical link failure occurs in one of several interconnected networks. The failure is detected and reported by the management component monitoring the physical resource in question. The failure also has side effects on other resources in the network, e.g., connections on the various layers which use the link will experience timeouts. (In the remainder

---

[1] Work done during a tenure at IBM as visiting scientist from: Unit for Software Engineering, Department of Computer Science, University of Pretoria, Hillcrest, 0083, South Africa

of this paper the terms *primary* and *secondary failure* will be used to refer to the actual failure and side effects, respectively.) The management components of resources such as, e.g., protocol stacks, will therefore also report secondary failures. The result is that the operator console is literally flooded with reports indicating the existence of some abnormal network condition, making it extremely difficult to determine the real cause of the problem.

This problem has been recognized by the Systems and Network Management Department (SNM) of the IBM European Networking Center (ENC)[2] and has resulted in the development of a prototype event correlation application. The development took place within the context of an ongoing larger project addressing network management requirements in heterogeneous networks. This project concentrates on management infrastructure and management information modelling aspects. The management system (and therefore the application) is based on the OSI management framework [8]. The application correlates (groups) event reports which have the same underlying cause, thereby significantly decreasing the complexity of the fault identification task.

An important requirement for the development was that the resulting application should be based on an *open* approach, to be able to function effectively in networks consisting of many different types of equipment, architectures and communication protocols. The approach chosen was that prescribed by ISO in their work on OSI Systems Management, i.e. the protocol used for exchange of management information is the Common Management Information Protocol (CMIP) [11]. The event correlation application is implemented as an OSI application process performing a *manager* role. Its sources of information are application processes performing the *agent* role, each responsible for providing information about a set of resources in the managed network. The resources in the interconnected network are modelled as *managed objects* (MOs). The application relies on generic properties of the information model, which describes heterogeneous resources in a homogeneous fashion. Agent processes relay *notifications* emitted by managed objects to the event correlation application through the use of the OSI Common Management Information Service (CMIS) M-EVENT-REPORT service [10]. In the remainder of this paper the terms *event* and *event report* are used interchangeably to refer to an M-EVENT-REPORT indication received by the application. The MO which emitted the notification is referred to as the *source instance* or *source object* of the event. Where a single failure results in multiple event reports, the one(s) closest to the primary failing resource will be referred to as *primary event(s)* and all others as *secondary events*.

The paper proceeds in the following fashion: Firstly, the key features of the algorithm used for event correlation are outlined. Subsequently the concept of a generic management information model, which forms the basis for the application, is introduced before the key algorithm features are explained in more detail. A description of the prototype environment and architecture follows. The next section contains a brief overview of related research and the last provides conclusions and guidelines for future research.

2. OUTLINE OF THE ALGORITHM

The following list highlights the key features of the algorithm:

- Event Driven

  The application is passive until an event arrives. (An event is any event report indicating an abnormal condition in the network.) Then, the correlation processing of the event starts immediately. As a result, the algorithm is *driven by incoming events* and works on a dynamic information base, where events are added during the process of correlation. The application never finishes its processing. As long as events arrive, it will attempt to correlate them to other events in the database.

---

[2] The contents of this paper reflect the result of experimental work and do not suggest any commitment by IBM to incorporate the ideas expressed here into any product development.

- Relationship Exploration

  The application correlates events by detecting special relationships between the source objects of those events. Only a small part of the information which is required to establish these relationships is included in the event reports themselves. Therefore, for each incoming event, the application sends requests to the agents in order to collect information about the source instance and about other objects with which it is in relationship. The correlation then proceeds based on this information, which is maintained in an MO cache. However, in the presence of failures, parts of the network might be unreachable, causing requests to time out. In order to avoid these excessive delays, the exploration of the various MOs is performed in *parallel.* This is realized by a thread-like mechanism using so-called *search activities.* Search activities are independent activities which make it possible to invoke information retrieval services (CMIS M-GETs) to the various agent systems in parallel. The relationship exploration relies heavily on the generic properties of the object model which is being used in the prototype system.

- Correlation and Causality

  The fault identification is performed in two steps:

  - *event correlation:* decides whether two events result from the same primary network failure.
  - *causality decision:* decides whether one of the failures indicated by two correlated events caused the other, i.e., whether one of the two lies closer to the primary failure.

  Note that the correlation step does not make any statement about causality. It only determines whether the two events belong together with respect to their cause.

The last two points of the preceding list will be explained in more detail in later sections.

3. MANAGEMENT INFORMATION MODEL

Before looking in more detail at the correlation approach, a brief overview of the concept of a generic object model is given here. The use of a generic model of a communications network for the purpose of network management and the resulting benefits in the areas of application development and interoperability have been described by a number of authors, e.g. Dupuy et. al. [4], Liao and Seret [15]. Details of the management information model which is employed in the prototype system are described in [16], which also contains an example of a specific architecture being mapped to the generic, OSI-based model. It also gives more references to work on generic models.

To be able to correlate events in heterogeneous networks, it is necessary for the application to access management information about the network. In reality, this information is contained in highly varying formats, and distributed among heterogeneous network resources. In the extreme approach, similar information would be modelled differently for each different resource type. In this case the application requires knowledge about where to find the information for every resource type separately. This clearly is a rather expensive approach in terms of the effort required to develop the application (and to extend it in the case where new types of equipment should be added). Instead, by developing and applying a generic managed object model, the application can be freed from dependencies on diverse network resources. In a generic model similar attributes are used by the application in a homogeneous fashion for managed objects representing the heterogeneous network.

A generic model of a communications network makes provision for physical and logical views of the network. It contains sufficient information to describe the network in terms of *systems*, *communication entities*, *physical links* and *logical connections.* In order to make provision for different mechanisms of interconnecting subnetworks, the model also includes generic *switch* and *gateway* object classes.

Rather than burdening application designers with the task of making provision for the detailed characteristics of different resource types, the event correlation application was designed to make use of a *generic model* of a communication network.

4. RELATIONSHIP EXPLORATION

The question about *what* has to be correlated, has to be addressed before the mechanism of correlation is described, The application under discussion here has been aimed at the processing of event reports indicating a failure or other abnormal condition, such as throughput degradation, in a network resource. The goal of *correlation* is to identify event reports which occur as symptoms of the same fault.

Event reports are assumed to be correlated if the managed objects from which they originate are directly or indirectly related. In practice almost all object instances modelling the network are related in some fashion. It would, however, be pointless to attempt to consider all such relationships in an event correlation application, because this might lead to the application correlating each event to each other event. Instead, a few relationships have been identified, which proved to be important for describing fault propagation and, therefore, for event correlation. The discussion concentrates on relationships between managed objects modelling links and connections.

The underlying assumption is, that faults often occur in lower layers of the communication architecture (e.g. the hardware level), thereby causing side effects in higher layers (e.g. connection time-outs). Thus, one direction to search for the cause of an event is 'downward'. Another important search direction is 'horizontally', i.e., between connections of the same layer. Two networks may be coupled, e.g., at the router or gateway level. When a link in one of the networks fails, connections in both networks will experience outages or throughput degradations. Downward search will of course not yield any result in the fault-free network, since the primary failure is located in the other network. The secondary events in the two networks could only be correlated at the layer of the communication architecture at which the two networks are interconnected, hence determination of this correlation could only be achieved via a horizontal search.

Using these guidelines, a number of important relationships have been identified for the exploration. These will now be described with reference to Figure 1, which shows a graphical representation of an n-layer connection between objects *a* and *b*, and corresponding underlying connections and/or links used to realize this connection.

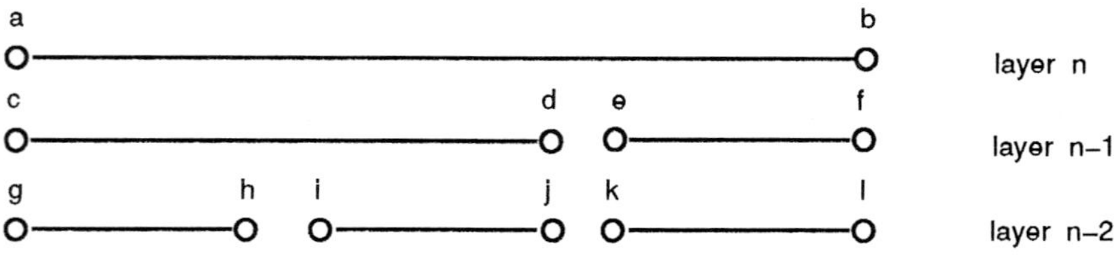

Figure 1. Layered connections

- *uses* - This relationship makes the layered organization of protocol architectures explicit by relating resources to the hard- or software resources on which they are based. For an n-layer connection, the application of this relationship will yield the local view of the (n-1)-layer connection (or link) which it uses for transmission. In Figure 1, for example, object *a* uses *c*, *c* uses *g*, etc.

- *partner connection endpoint* - A connection (or link) is modelled by two separate managed objects, representing the local view from each endpoint of the connection (or link), to the example set by OSI in their modelling work [13]. The two managed objects representing the endpoints are directly related to each other via this attribute. In Figure 1, for example, object *a* is the partner of *b*, *c* of *d*, etc.

- *matched connection* - This relationship ties together connections (links) to their peer connections (links) to which they are linked via gateways or switches. In Figure 1, for example, assuming the existence of an (n-1)-layer gateway or router, object *d* could be related to *e* via this relationship. (Note that this doesn't necessarily mean that *j* is also the matched connection of *k*, because the relay of the connection is accomplished at the higher level, between *d* and *e*.)

The application gathers information to be used in the correlation task by exploring the above mentioned relationships for the source object, and for the managed objects related to that object. Exploration for received event reports takes place in parallel by using a thread-like mechanism to investigate each report, i.e. for each new incoming event a different *search activity* is initiated to find the managed objects related to the source object instance. New search activities are also started for each new object discovered in this way. This process continues until no more unexplored objects can be discovered. For example, assume that an event report originating from object *d* in Figure 1 has been received. A search activity will be generated to first explore object *d*. Once it has been found that the objects *c*, *e* and *j* are related to object *d*, new search activities (one per object) will be generated to explore these in parallel, and so on.

The relationship exploration function relies on generic properties of the management information model used in the prototype system, which was mentioned in the previous section. The model makes it possible for the application to retrieve all the above mentioned relationships via attributes of generic managed objects.[3] The attributes are not necessarily included directly in the object from which an event notification has originally been emitted. In the case of the *matched connection* relationship, for example, the exploration function of the application has to determine whether the connection endpoint object was contained in a system which might be a switch or a gateway. If so, the switch table or gateway index table has to be retrieved to look up whether it contains an entry for the connection. Only then is it possible to determine whether the object is in fact related to another object via this relationship. It is worth noting at this point that although the application has been implemented using one specific generic object model, its concepts are applicable to any generic model from which the required relationships could be retrieved, or even to a non-generic model, at the cost of increased complexity to determine relationships between objects.

The management traffic generated as a result of the dynamic exploration of relationships through the retrieval of managed object attribute values places an additional load on the network. Its effect may be reduced by exploring relationships using static configuration information where this is available, thus only accessing the management network when the information is not readily available in local storage. The additional load can, however, not be avoided completely when highly dynamic objects, such as network connections, are considered in the correlation process (as is done in the prototype application).

5. CORRELATION

The relationship exploration discussed in the previous section provides the necessary information to the application to enable it to decide whether two events are correlated. If, during a search activity resulting (directly or indirectly) from a given event report, a relationship is discovered to an object which is already known to the application as a result of a search ac-

---

[3] Ongoing work in the area of relationships between MOs, e.g. the provision of Query Relationship actions [9], would significantly decrease the complexity of the application's exploration function.

tivity investigating *a different* event report, the two event reports are regarded as *correlated.* In practice, for each event the algorithm constructs a graph of objects related to the source object of that event. (The nodes in the graph represent managed objects and the arcs the relationships existing between them.) When two such graphs touch each other, i.e., contain at least one common object, the events which initiated their construction are regarded to be correlated. The conclusion is that these events are possibly the result of the same fault.

The actions to be taken once it has been decided that two events are correlated are heavily influenced by assumptions about the nature of the correlation. As stated earlier, two events are correlated if they result from the same failure. As an algebraic relation defined on the events, it is assumed that correlation is an equivalency relation. Therefore, where A, B, and C are events, a correlation (denoted by $\sim$) defined on these has the following properties:

reflexivity: $A \sim A$
symmetry: $A \sim B \rightarrow B \sim A$
transitivity: $A \sim B \wedge B \sim C \rightarrow A \sim C$

One has to keep in mind that correlation does not include any information about causality. Thus, reflexivity and symmetry almost trivially hold. From an intuitive viewpoint, transitivity holds too.[4]

Taking correlation as an equivalency relation, the event management is significantly eased. The set of all events can be partitioned into sets containing correlated events (further called *event sets*). Any one event belongs to exactly one event set, which may as an extreme case be a set containing only this one event. When it has been decided as a result of exploration that two events from different event sets are correlated, then all events included in the two sets are correlated. This is reflected in a merge of the two event sets.

With this structure, the correlation process becomes straightforward:

1. For each incoming event, an event set containing only this event is created.

2. The exploration process is started for the event and a search graph is built using the object instance relationships mentioned earlier. When the search graphs of two events touch each other, i.e., at least one common object instance is discovered, the two events are correlated.

3. When two events from different event sets are correlated, the two event sets are merged.

In order to avoid unlimited growth of the correlated sets, events are timestamped on receipt by the application. An additional constraint on the correlation decision (the second point above), specifies that two events are only correlated if they've occurred within a configurable time period of each other. This may cause problems with late-arriving event reports being correlated together with events which have occurred much later. The approach of using timestamping of events at the origin has also been considered, but has been rejected due to the subsequent need for synchronized clocks among all managed systems.

6. CAUSALITY

Although the grouping of related event reports (correlation) is already helpful to the operator, it still does not provide more information about *where* the primary failure has occurred. Hence the application continues to establish (where possible) an *ordering* among the events within the same event set, where the ordering indicates which event(s) is (are) most closely

---

[4] In the presence of multiple, simultaneous faults, the assumption of transitivity may lead to too many events being correlated, hence additional intelligent correlation rules are required to limit this effect. This is the topic of further study.

related to the primary failure and which ones resulted from secondary failures, i.e. from side effects of the primary failure. This ordering represents a directed relationship among correlated events, *causality*. It constitutes a first step at fault diagnosis, i.e., it tries to identify the cause of the fault, or at least to steer further diagnostic and test actions in the right direction. (The testing and repair steps which normally follow fault diagnosis in the fault management process fall outside of the scope of this discussion.) The causality relationships between correlated events could be expressed as a directed acyclic graph, a *causality graph*, where each event is represented by a node in the graph and each causality relationship is expressed as a directed arc between a pair of nodes.

The authors have identified a number of heuristics which could be used to determine the causality relationship between two related events:

1. The *level* at which the event was detected in a layered architecture.

   By level is meant the position of a layer of hard- or software in a layered network architecture. A layer does not necessarily reflect an OSI protocol layer, but any organization in which objects are related via the 'uses' relationship. For example, in the Asynchronous Transfer Mode (ATM) case, Virtual Channel Connections (VCCs) and Virtual Path Connections (VPCs) are also regarded as belonging to different layers, because a VCC might (indirectly) use a VPC to transfer data.

   Consider two related events, one for which the source object is in layer n and the other where the source object is a managed object in layer m. If $n < m$, the heuristic states that there is a high probability that the primary failure has occurred in layer n, or below. For the purpose of this heuristic, no distinction is made between objects representing physical, and those representing logical resources.

   Given that in most network architectures, errors are not reported to lower layers, this is a reasonable assumption to make. (In most implementations of (layered) architectures a lower level protocol entity or physical link would be unaware of a problem experienced by a higher level connection which 'uses' it and would thus not emit an event notification which is related to the higher level problem).

   When two events are correlated as the result of the existence of one or more 'uses' relationship between directly or indirectly related managed objects (see Section 4), this heuristic could be used to determine the causality relationship between them. For example, in Figure 1, if events from objects b, e and k are found to be correlated, this rule selects object k as being the closest to the primary failure.

2. Event *types*.

   The knowledge that some event types are more likely to indicate existence of a fault than others could be made explicit and used to determine the causality relationship between pairs of events. For example, a transient event of type "Experiencing throughput degradation" is more likely to be the (indirect) effect of an actual fault than an event of type "Operational State changed to disabled".

3. *Severity* of an event.

   The assumption here is that a high severity event might result in several lower severity events reported by other object instances. Severity is a rather subjective measure as perceived by different equipment manufacturers. The assumption of a uniform (generic) object model should, however, negate this fact and result in some common scale for event severity.

4. The *origin* of an event.

   Event reports sometimes contain information about the way in which the notification was emitted. Notifications could for example be emitted as a result of the propagation of state change among related objects. If the origin field of an event states this, it is known to be related to a secondary failure, instead of an indication that the source object might be the one where the primary fault has occurred.

The algorithm for determining causality relationships between events which are correlated, could apply the above mentioned heuristics in sequence or in combination to each pair of events. In the initial implementation of the prototype only the first heuristic is used.

If all heuristics have been applied without being able to order the events, additional information is required before the application could resolve the question. It is, however, not required that each pair of events be pairwise ordered. The ordering merely provides additional information which could be useful to the operator.

7. PROTOTYPE IMPLEMENTATION

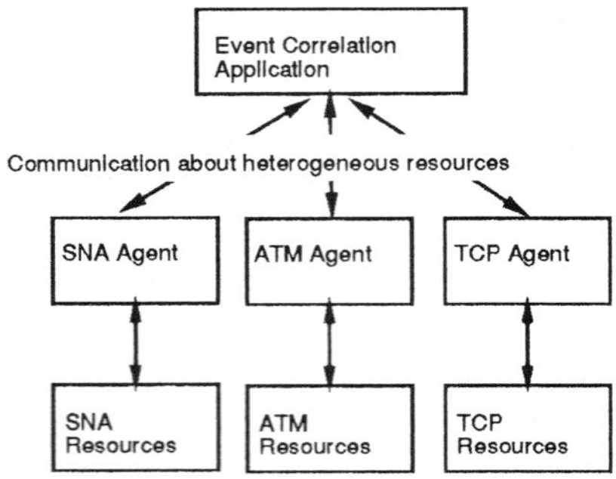

Figure 2. The Target Scenario

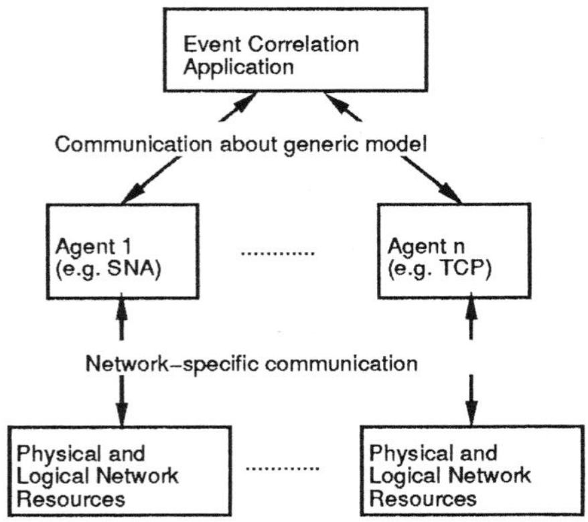

Figure 3. Generic View of a Heterogeneous Network

As mentioned previously, the event correlation application was one element of a larger network management prototype. This section provides more information about the design and implementation of the application. It also provides references to literature about other components of the prototype, such as the management platform, graphical user interface and interaction between managed objects and real network resources.

The target scenario for the prototype effort is illustrated in Figure 2. The application forms part of a centralized network management solution which integrates the information from three different management systems, each managing an architecture-specific portion of the network. The three subnetworks are not independent, but are interconnected in various ways. The scenario could now be the following: An ATM resource, say a Virtual Path Link, fails. The symptoms of this failure are reported independently by the three management systems in the form of multiple event reports to the centralized management system: The ATM agent indicates the abnormal condition as detected by various sublayers of the ATM network, the Systems Network Architecture (SNA) and Transmission Control Protocol (TCP) agents indicate problems experienced by Sessions or connections routed across that part of the ATM network. To correlate these reports the application needs to have knowledge of SNA, ATM and TCP resources, and the relationships between them.

As discussed earlier, the heterogeneity of the target scenario is overcome by the use of a generic model. The suppliers of management information to the application (its agents) are expected to supply information in a format which conforms to this model. Figure 3 illustrates how the application's view of the target scenario (Figure 2) changes as a result of this approach.

Figure 4 illustrates the architecture of the prototype application. It consists of the following major components:

Correlation This is the main control of the application. It implements the correlation algorithm outlined in Section 5 and uses the other components to support it in the correlation task.

Relationship Exploration This component explores the relationships which are being used for the correlation of events. It gets the required attribute values of managed objects from the agents via the CMIS M-GET service and updates these in a cache of managed object information.

Managed Object Cache The cache serves as a repository for information about managed objects, both for objects known as a result of incoming events and the objects related to these. It includes information about the relationships between the objects.

Event Management This provides the support for manipulating sets of correlated events. It includes the creation of a causality graph from an event set by application of the causality heuristics.

Search Control The thread-like search activities, which are started by the correlation application for each new managed object discovered, are supported by Search Control.

User Interface Support Provides the required functionality for interaction with the graphical user interface, which will be discussed later in this section.

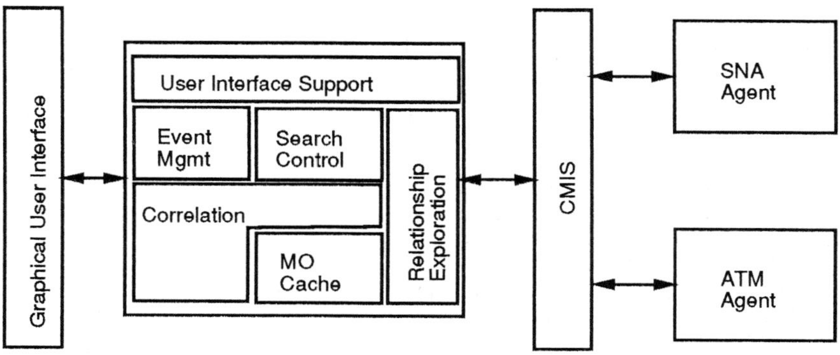

Figure 4. Event Correlation Application Architecture

In the initial prototype configuration the application accepts (via CMIS) event reports from two agent implementations, presenting views of SNA and ATM subnetworks, respectively. Because of the fact that the application only communicates with its agents about a generic object model it could also support other agents when conformant implementations of such agents become available.

The managed objects representing SNA resources were mapped to real physical and logical resources, e.g. Host, Physical and Logical Units (PUs and LUs) and LU-LU Sessions. In [5] the interactions between the SNA OSI-agent and NetView to acquire the necessary information about the SNA network resources, and to map these to the generic object model, are described. The managed objects used in the ATM agent implementation did not have any real resources attached to them, but instead interacted with a test application which was used to simulate failures in the modelled ATM resources.

The initial version of the application was set up to process event reports resulting from the use of the State Change Reporting service as defined by ISO [12], specifically reports of Operational State changes where the state has changed to Disabled. This enables the application to recognize these event reports as indications of network faults. (It could easily be extended to identify *any* type of event as a fault indication.) The first such event report to arrive triggers the application to start its execution. Although the single event report is trivially correlated to itself, the application starts its exploration algorithm to find related managed objects in preparation of further event reports which are (with a high probability) likely to result from the same network fault. For each new incoming event report the exploration algorithm is invoked again, maintaining an object graph for each and correlating events as described in Section 5.

The application supports interaction with a graphical user interface (GUI) [18]. The role of the interface is to assist the human operator by presenting a structured view of event reports requiring attention. It is used to display information about events, information about managed objects which emitted notifications and most importantly, the causality graphs of each set of correlated events.

Figure 5 contains a snapshot of a display generated by the GUI during a test run of the application, illustrating the logging of events, graphical display of uncorrelated events and the causality graph of a set of events found to be related. The icons identifying events are labelled with integer ids which are assigned as soon as the event is logged by the application. It is possible to request textual displays providing more information about an event or the managed object where it has originated. (For the implementation of the GUI a tool developed by the University of Kaiserslautern, the Object Oriented Design Editor (ODE), was used [1].)

The prototype application has been implemented on an IBM RS/6000 under the AIX operating system using the C++ language. The communication of management information is supported by an adaptation of the OSIMIS package from the University College of London [14]. (OSIMIS includes an implementation of the Common Management Information Protocol (CMIP) as defined by ISO.) [3] contains more detail about the support infrastructure and describes how its design ensures the required degree of connectivity in the prototype environment.

8. RELATED RESEARCH

There are a number of systems dealing with network fault identification and diagnosis. In this section selected examples are mentioned. Most of these fall in the category of knowledge-based systems, making use of expert systems techniques (e.g. [6] [7][17]). They rely on knowledge or rule bases which need to be updated by some knowledge acquisition process, which is not a simple task [7]. The information to be supplied generally describes the network configuration, the event reports which indicate faults, exhaustive enumeration of possible network faults and the relationships between faults, events and network resources. [7] also makes provision for uncertainties in their system.

In [2], correlation of alarms is performed by considering, for each incoming alarm, the set of possible failures which might have caused the alarm. It is not quite clear how these sets are obtained. For correlation and the causality decision, intersections of sets of possible failures are used. Conflicts are resolved by heuristics making use of pre-assigned fault probabilities.

In [19] the main focus is on modelling network management operations as data manipulation operations. One of the examples mentioned is that of event correlation. It defines a correlated event in terms of other events occurring within a minimum time interval of each other and shows how data manipulation operations could be used to detect the occurrence of a correlated event.

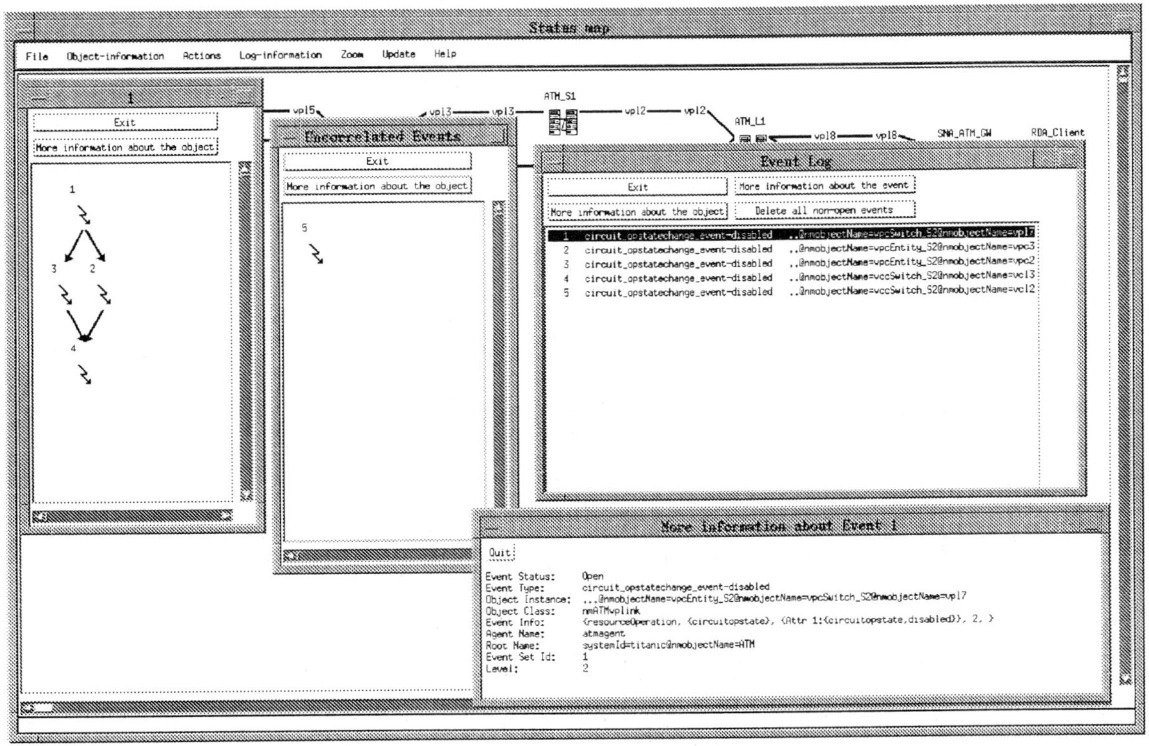

Figure 5. Event Correlation Application - Example Session

The application described here differs in the fact that it performs correlation with very little or no preconfigured knowledge by dynamically exploring relationships between managed objects. The possibility of integrating some of the techniques described in the above mentioned papers in order to achieve a stronger correlation decision is currently being investigated.

9. CONCLUSIONS AND OUTLOOK

This paper has described an event correlation application which makes use of the OSI management framework. The application has been developed as part of a prototype network management system. During testing the application has successfully correlated events originating from agents representing SNA and ATM networks, respectively. The correlation approach is based on important relationships between OSI-based managed objects. Heuristics for establishing causality among correlated events have been defined. The use of a generic model of heterogeneous networks resulted in significant simplification of the application, enabling the prototype implementation to be completed in a reasonably short time. The information model is sufficient to allow the application to homogeneously explore resources and their relationships in heterogeneous networks.

Sources of more information to optimize the search strategy and to improve the correlation and causality decisions are currently being investigated. The most important areas receiving attention are: Context information included in event report information (there is none included in event reports resulting from state change notifications in the current agent implementations), event type, severity and origin (to determine causality more accurately). The integration of an expert system shell to process this information is also receiving attention. The application needs to be tested in more problem scenarios than have been done so far.

It is not required to use the application in combination with a user interface as has been done in the prototype implementation. It is easy to suggest alternative ways of using it: For example, in a manager of managers configuration it could be used to intelligently filter event reports forwarded to higher level fault management applications, only supplying summary information about primary failures.

Acknowledgements

We would like to thank all our colleagues in the Systems and Network Management department participating in the prototype development which this application formed part of. In addition, we would like to thank Brigitte Bär, John Dinger and Klaus-Peter Keilmann for proofreading the paper and making helpful suggestions for improvement.

10. REFERENCES

[1] **Arbeitsgemeinschaft Telematik der Universität Kaiserslautern** *ODE Programmer's Manual* Kaiserlautern, 1991.

[2] **A. Bouloutas, S. Calo, A. Finkel** *Alarm Correlation and Fault Identification in Communication Networks* IBM Research Report No. 17967, May 1992.

[3] **J. Dinger, D. Jordaan** *Experiences with the Use of OSI Management Infrastructures* To appear in: Proceedings of IEEE Globecom, Orlando, Florida, U.S.A., Dec. 1992.

[4] **A. Dupuy, S. Sengupta, O. Wolfson, Y. Yemini** *NETMATE: A Network Management Environment* IEEE Network, Vol. 5, No. 2, March 1991, p. 35-43.

[5] **W. Filip, G. Kar** *NetView in a Heterogeneous Environment: Implementation of an Experimental Agent System* Appearing elsewhere in this volume.

[6] **M. Frontini, J. Griffin, S. Towers** *A Knowledge-Based System for Fault Localisation in Wide Area Networks* Integrated Network Management, II, Proceedings of the IFIP TC6/WG6.6 Second International Symposium on Integrated Network Management, Washington D.C., U.S.A., 1-5 April 1991, p. 519-530.

[7] **P. Hong, S. Sen** *Incorporated Non-deterministic Reasoning in Managing Heterogeneous Network Faults* Integrated Network Management, II, Proceedings of the IFIP TC6/WG6.6 Second International Symposium on Integrated Network Management, Washington D.C., U.S.A., 1-5 April 1991, p. 481-492.

[8] **International Standards Organization:** *OSI System Management Overview* International Standard ISO/IEC 10040, August 1991.

[9] **International Standards Organization:** *Information Retrieval, Transfer & Management for OSI - General Relationship Model - Third Working Draft* ISO/IEC JTC 1/SC 21 N 7126, 1992-07-21

[10] **International Standards Organization:** *Information Technology - Open Systems Interconnection - Common management information service specification.* International Standard ISO/IEC 9595, First Edition 1990-05-15.

[11] **International Standards Organization:** *Information Technology - Open Systems Interconnection - Common management information protocol specification.* International Standard ISO/IEC 9596, First Edition 1990-05-15.

[12] **International Standards Organization:** *Information Technology - Open Systems Interconnection - Systems Management - Part 2: State management function.* International Standard ISO/IEC 10164-2, 1991-11-04.

[13] **International Standards Organization:** *Information Technology - Open Systems Interconnection - Systems Management - Structure of Management Information - Part 5: Generic Management Information.* Draft Recommendation ISO/IEC CD 10165-5, 1991-10-01.

[14] **Knight G., Pavlou G., Walton S.** *Experience of Implementing OSI Management Facilities* Integrated Network Management, II, Proceedings of the IFIP TC6/WG6.6 Second International Symposium on Integrated Network Management, Washington D.C., U.S.A., 1-5 April 1991, p. 259-270.

[15] **Liao T., Seret D.** *Network Management: interoperability and information model* Computer Communications, Vol. 14, No. 10, December 1991, p. 588-597.

[16] **B. Schott, A. Clemm, U. Hollberg** *An ISO/OSI based Approach for Modeling Heterogeneous Networks* Proceedings of the Fourth International Conference on Information Networks and Data Communication, INDC-92, Espoo, Finland, 16-19 March 1992.

[17] **J.M. Schröder, W. Schödl** *A Modular Knowledge Base for Local Area Network Diagnosis* Integrated Network Management, II, Proceedings of the IFIP TC6/WG6.6 Second International Symposium on Integrated Network Management, Washington D.C., U.S.A., 1-5 April 1991, p. 493-503.

[18] **A. Syri** *Graphische Benutzeroberfläche zur Darstellung von Netzabläufen* Universität Kaiserslautern, Fachbereich Informatik, 25 May 1992 (German)

[19] **O. Wolfson et al** *Managing Communication Networks by Monitoring Databases* IEEE Trans. Software Engineering, Vol. 17, No. 9, September 1991.

Integrated Network Management, III (C-12)
H.-G. Hegering and Y. Yemini (Editors)
Elsevier Science Publishers B.V. (North-Holland)

A Probabilistic Approach to Fault Diagnosis in Linear Lightwave Networks

Robert H. Deng[a], Aurel A. Lazar[b] and Weiguo Wang[c]

[a]Department of Electrical Engineering, National University of Singapore, Singapore 0511, Republic of Singapore

[b]Department of Electrical Engineering, Columbia University, New York, NY 10027-6699, USA and

[c]Institute of Systems Science, National University of Singapore, Singapore 0511, Republic of Singapore

Abstract

The application of probabilistic reasoning to fault diagnosis in Linear Lightwave Networks (LLNs) is investigated. The LLN inference model is represented by a Bayesian network (or causal network). An inference algorithm is proposed that is capable of conducting fault diagnosis with incomplete evidence and in an iterative fashion. A belief updating algorithm for a multiply connected LLN inference model is proposed which is used by the inference algorithm for computing the belief function. We also introduce a *t*-fault diagnosis system and an adaptive diagnosis system to further reduce the computational complexity of the fault diagnosis process.

Keyword Codes: I.2.1; B.4.5
Keywords: Artificial Intelligence, Applications and Expert Systems; Reliability, Testing, and Fault-Tolerance

1. INTRODUCTION

The complexity of communication networks and the volume of information provided by these networks have caused an increase in demand for network management systems and personnel. In particular, the area of network fault management requires a great deal of network expertise (design, operation, management, etc.) which has proved to be difficult to acquire and maintain. The application of expert, or knowledge-based, systems to attack the inherent complexity of network fault management is a growing effort (see e.g., NDS [21], YES/MVS [6], ACE [12], Troubleshooter [10], and ISM [5]). However, most of the network fault management systems were built on an ad hoc and unstructured basis. The research on integrated network fault management is still in its infancy. There is a pressing need, therefore, for establishing a theoretical foundation of network fault management and for bridging the gap between research and working systems.

Due to the distributed nature of communication networks the relationships between fault patterns and network events are often inexact and the conclusions are uncertain [20]. Moreover, unavailability, loss or delay of network messages require the fault diagnosis system to have the ability to conduct inference with incomplete information. For reasoning with such imprecise information and relationships, it is highly desirable for the inference mechanism to have the capability to conduct inference with uncertainty.

Over the last few years, reasoning using probability has become very popular within the AI community. For introductions to this area as well as reviews on recent developments, the reader is directed to [1] and [8]. Most recently, applications of probabilistic reasoning in network fault management have begun to appear in the literature. Network fault recognition using probabilistic data and machine learning was studied by Maxion [11]. Incorporating non-deterministic reasoning in managing heterogeneous network faults has been investigated by Hong and Sen [9].

The Bayesian network (or causal network) model [16] is a popular probabilistic reasoning model developed recently. It provides certain "common sense" reasoning, e.g., *explain-away*[16], that cannot be supported in classical rule-based systems. It is iterative in nature in that after receiving new evidence, the belief about the causes of the total observed evidence can be re-computed (updated). This feature seems to suite well the network fault diagnosis process in general. Thus, we begin our investigation with applying this model to specific network fault diagnosis problems.

Fault diagnosis of the Linear Lightwave Network (LLN) [19] has been studied by Schroff and Schwartz using deterministic approaches [17]. Modeling the same problem using Bayesian networks, a special class of Bayesian networks is obtained on which the belief updating algorithm can be greatly simplified. The resulting fault diagnosis algorithm improves that of Schroff and Schwartz in that it allows multiple simultaneous faults in the network. The work cited allows only for single faults.

The organization of this paper is as follows. In Section 2, we illustrate how to derive the inference model (i.e., the Bayesian network) for fault diagnosis from a given LLN. In Section 3, we give an inference algorithm that is performed by the inference engine during the fault diagnosis process. Belief updating algorithms are discussed and presented in Section 4. To further simplify the belief function computation, we propose in Section 5 the concepts of *t*-fault diagnosis system and adaptive fault diagnosis system. Finally, Section 6 contains our concluding remarks.

2. THE INFERENCE MODEL

The inference model is a Bayesian network, i.e., a *directed acyclic graph* (DAG) where the nodes are random variables and certain independence assumptions hold [16, 13, 1]. In this section, we illustrate, by using a simple example, how a Bayesian network is constructed from a given LLN.

The Linear Lightwave Network constitutes the physical layer of ACORN [19, 17], a gigabit research network testbed. It is a new type of architecture for lightwave networks based on establishing controllable transparent optical paths among network users. The LLN is based on a single key component: a controllable *Linear Divider/ Combiner* (LDC). By appropriately controlling the settings of LDCs placed at each network node, internal connections between input and output ports of an LDC can be set up, and as a result, optical paths (also called routes) on which many signals are multiplexed can be created on demand. For a detailed introduction to LLNs, the reader is directed to [19].

A simple LLN with five LDCs is shown in Fig. 1(a). Each LDC has a number of input and output ports and LDCs are interconnected by optical links. In Fig. 1(a), the symbols i_{jk} and o_{jl} denote, respectively, the k-th input port and the l-th output port of LDC j; a capital letter with subscript p associated with each port denotes the optical input (output) power value to (from) that port (e.g., B_p denotes the output power value from port o_{21} as well as the input power value to port i_{41}); and finally, I_{ip} and O_{kp} denote, the input and output power values to/from the LLN periphery, respectively. The following assumptions are made in the model:

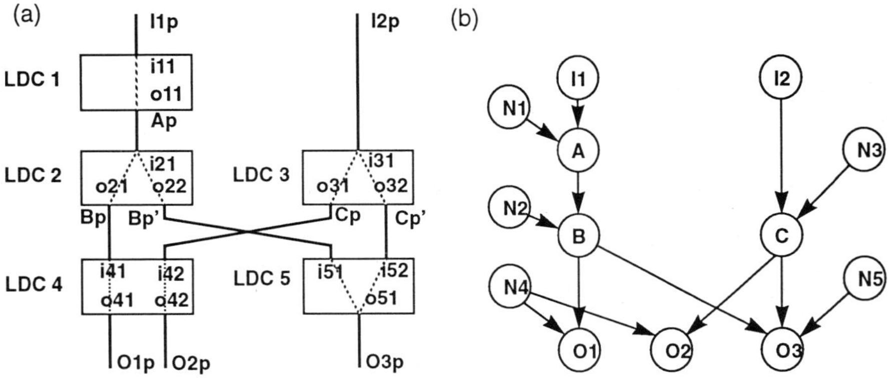

Fig. 1 An Example of an LLN (a) and its corresponding Bayesian Network (b)

1. The *observable evidence* is the LDC input and output power values. Faults are detected when the observed values deviate from the expected range of power values.
2. The *Basic Network Components* (BNCs) are the LDCs. That is, LDCs are treated as black boxes and are the elementary units for the network fault manager. If an LDC has failed or some part of it has failed, all its output power values will be incorrect (outside the expected range) but its input power values remains unaffected. We assume that the optical links between LDCs are not subject to failure. This is not really a restriction because we can model a link as a one-input/one-output "LDC".
3. The network routes are considered to be quasi-static. That is, the connections that are established remain up for a long time when compared with the time required to carry out the fault diagnosis.

Table 1: LLN Routing table

Routes	List of LDC ports in the route					
Route 1	i11	o11	i21	o21	i41	o41
Route 2	i11	o11	i21	o22	i51	o51
Route 3	i31	o31	i42	o42		
Route 4	i31	o32	i52	o51		

The LLN routes/connections determine the internal connections between the input and the output ports of every LDC. This routing information is required by the fault diagnosis system for constructing the inference model. Assuming the routing table as given in Table 1, the internal connections of each LDC are shown in Fig. 1(a) by dotted lines. Hence, Fig. 1(a) summarizes the network configuration, which together with the above three assumptions constitutes our knowledge about the LLN. The inference model is constructed based on this knowledge.

Since a power value, say *Ap*, is assumed to be either within the expected range or outside it, we define a binary-valued random variable (RV) *A* for power value *Ap*, such that $A = 1$ if and only if the value of *Ap* is within expected range. For each LDC we define a binary-valued RV, $N_j = 0$ if and only if LDC *j* is faulty. Furthermore, we assume that the N_j's are mutually independent with probability distribution

$$P(N_j = 0) = p_j,\ P(N_j = 1) = 1 - p_j = q_j. \tag{1}$$

With the above defined RVs, based on Fig. 1 (a) and the assumptions 1 and 2, the following dependence relations among the RVs are obvious:

$$\begin{aligned} &A = I_1 \wedge N_1, \quad C = C' = I_2 \wedge N_3, \quad O_1 = B \wedge N_4, \\ &B = B' = A \wedge N_2, \quad O_2 = C \wedge N_4, \quad O_3 = B \wedge C \wedge N_5, \end{aligned} \tag{2}$$

where ^ denotes the logical "AND" operation. The dependence relations in (2) can be graphically represented as a DAG in which each node represents a RV (Fig. 1(b)).

For a non-root RV *V* in the DAG of Fig. 1(b), *V* and its parents $U_1, \ldots, U_n$ are related by

$$V = U_1 \wedge U_2 \wedge \ldots \wedge U_n. \tag{3}$$

The conditional probability $P(v \mid u_1, \ldots, u_n)$ (for a RV *V*, we denote its value by the corresponding lower case letter *v*) is then given by

$$P(v \mid u_1, \ldots, u_n) = 1 \text{ if and only if } v = u_1 \wedge \ldots \wedge u_n, \tag{4}$$

which is the probabilistic representation of the dependence relation given in (3).

The DAG in Fig. 1 (b) and the probability distributions of (1) and (4) forms a Bayesian network. They constitute the inference model for the LLN of Fig. 1(a). The inference model captures the knowledge required for LLN fault diagnosis.

To recap, the procedure described above applies to the construction of an inference model for any LLN. Given an LLN, the internal connections of every LDC may be determined from the LLN's routing table. Consider an LDC, say LDC *k*, with *n* input ports and *m* output ports. Denote its input and output power values by $U_{p1}, \ldots, U_{pn}$, and $X_{p1}, \ldots, X_{pm}$, respectively. We form an *m*x*n* matrix T_k, called the internal connection matrix of LDC *k*, as follows: $t_{ij} = 1$ if there exists a connection between the *j*-th input port and the *i*-th output port and $t_{ij} = 0$ otherwise. Then the dependence relation between the corresponding binary variables X_i, U_j's and N_k is given by

$$X_i = N_k \wedge (t_{i1} U_1 \wedge t_{i2} U_2 \wedge \ldots \wedge t_{in} U_n). \tag{5}$$

Based on this dependence relationship, the corresponding inference model can be obtained as in the above example. This process can be easily automated for arbitrary network configurations.

3. THE INFERENCE ALGORITHM

The RVs in the inference model of an LLN (see Fig. 1(b)) can be classified into two types: *observable*, e.g., A, I_1, O_1 and *unobservable*, e.g., N_j (because we have treated LDCs as black boxes). The task of the LLN fault diagnosis system is to infer, based on the values of the observable RVs, the values of the unobservable RVs.

If the fault diagnosis system were supplied with power measurements at all the LDCs, fault diagnosis would be trivial. However, power measurement is expensive in that it consumes optical power and disturbs normal operations. Under normal network operations, the power measurements are performed only at the input and output ports of the LLN. (Here the LLN input ports are defined as all the LDC's input ports where network routes originate and the LLN output ports are defined as all the LDC's output ports where the network routes terminate). Further measurements are conducted only if additional evidence is needed in order to diagnose faults. Note that, by monitoring the input and output power values of an LDC, we can determine whether the LDC is a faulty. The above analysis suggest that the inference algorithm must have the following two features:

1. It should be able to conduct inference with incomplete evidence and be able to identify the faulty LDC or a plausible set.
2. It should perform inference on an iterative fashion. Initially it should identify a plausible subset based on the currently available evidence. Then with this subset as a guide, it should decide what additional evidence needs to be acquired. With the new evidence, it should narrow down the possible set of faulty LDCs and finally pinpoint the faulty ones.

With respect to the LLN inference model, the evidence is a set of instantiated RVs, $W = \{W_1 = w_1, ..., W_L = w_L\}$ which are derived from power measurements. For example, if measurement indicates that the power value W_{lp} is incorrect, then $W_l = 0$. The belief function of an LDC related RV N_j, is defined as

$$b(N_j = n_j) = P(N_j = n_j | W). \qquad (6)$$

The belief function tells us the likelihood that the RV N_j takes value 0 or 1. Therefore, $b(N_j=0)$ is the likelihood that LDC j has failed.

For the LLN in Fig. 1(a), assume that the LLN input power I_{1p} and I_{2p} are both correct, and that LDC 1 and LDC 5 have failed simultaneously. Due to these failures, the observed power values O_{1p} and O_{3p} will be incorrect but O_{2p} will be correct. It follows that $O_2 = 1$ and all the ancestors of O_2, i.e., N_4, C, N_3, and I_2, must also have value 1. Therefore, O_2 and all its ancestors can be eliminated from the inference model in Fig. 1(b). Similarly, I_1 can be deleted. In general for given the evidence, $W = \{W_1 = w_1, ..., W_L = w_L\}$, if $w_i = 1$, then the RV W_i and all its ancestors can be deleted from the inference model. We call this process the *trimming* of the inference model. We are now ready to present the inference algorithm for LLN fault diagnosis.

Algorithm 1. (LLN fault diagnosis algorithm.)

Step 1. Read the input and output power values at the periphery of the LLN:

- If all input power values are correct but some output power values are incorrect, then go to step 2, otherwise exit.

Step 2. Let W denote the evidence obtained from step 1. Trim the inference model if possible. Compute $b(N_j = 0)$, for all j, based on some belief updating algorithm (to be described in more detail later). Find the N_k with the largest belief. Iт $b(N_k = 0)$

=1, then LDC k has failed. Go to step 5; otherwise, go to step 3.

Step 3. Measure the input and output power values of LDC k:

- If some or all of the output power values are correct, LDC k is not faulty; clear "uncertain" marks if any; go to step 4.
- If all the input power values are correct and some output power values are incorrect, then LDC k has failed. Go to step 5.
- If some input and all output power values are incorrect, mark node N_k "uncertain"; go to step 4.

Step 4. Let W^* denote the updated evidence obtained from the power measurement of the selected LDC k plus W. Trim the inference model if possible. Compute $b(N_j=0)$ for all j. Find the RV, say N_i, with the largest belief and *not* marked "uncertain". let $k = i$; go to step 3;

Step 5. Repair LDC k. Clear all "uncertain" marks. Goto step 1. □

Algorithm 1 does not include the belief updating part. This is discussed in the next section. To see that the algorithm always converges, note that the number of "uncertain" marks is strictly increasing until a faulty LDC is repaired or a good LDC is identified. According to step 3, an LDC is marked uncertain if and only if it has an faulty ancestor. Since the network is finite and acyclic, there is always an LDC that does not have any faulty ancestor. Therefore, a faulty LDC without faulty ancestor will guarantee to be found as N_i in step 4 and repaired. Also follows this argument is the bound on the number of power tests needed for the algorithm to locate and repair all faults. Assuming there are L LDCs, the number of "uncertain" marks can increase to at most L, and there are at most L resets to the counting. So, the total number of power tests is bounded by $O(L^2)$.

4. BELIEF UPDATING ALGORITHMS

In this section general belief updating algorithms for Bayesian networks are discussed. In subsection 4.1, Pearl's belief updating algorithm for singly connected inference models and the ways for dealing with multiply connected ones are (only) briefly mentioned (due to space limit). A new belief updating algorithm for multiply connected LLN inference models is presented in subsection 4.2.

4. 1 Belief Updating for Singly and Multiply Connected Inference Models

Since our inference model for fault diagnosis of LLNs is a Bayesian network, the core operation in the diagnosis algorithm presented in the last section is the belief updating. Belief updating for general Bayesian networks has been proven to be NP-hard [3]. For singly connected Bayesian networks (namely, networks with no loops, here the loops are defined as undirected cycles in the graph.), the belief updating can be done in polynomial time [16, 13]. For the class of Bayesian networks involved in the LLN inference models, the belief updating can be further simplified because the conditional probabilities between nodes are in general in the form of (4). Due to space limitation, we skip the discussion on this. The reader is referred to [16, 13] for the details of the algorithm from which the simplification for LLN is rather straightforward.

Pearl's belief updating algorithm cannot be applied to multiply connected Bayesian networks, i.e., networks with loops in it. The inference model in Fig. 1(b) is multiply connected. There is one loop in this model, formed by the nodes B, O_1, N_4, O_2, C and O_3. There are three general approaches of belief updating for multiply connected networks: *clustering*, *simulation* and *conditioning*. The reader is referred to [16], [13], [7], [15], [2], [4] and [18] for the details of these techniques. These approaches all suffer exponential computational complexity one way or another.

4. 2 A Belief Updating Algorithm for LLNs

In what follows, we present a belief updating algorithm for multiply connected LLN inference models. The computational complexity of the algorithm is $O(LK^L)$ in the worst case, where K is the number of N_j's in the trimmed inference model and L is the number of nodes instantiated to value 0.

As illustrated in the last section, given the evidence $W = \{W_1 = w_1, ..., W_L = w_L\}$, if $W_l = 1$, we can always trim the inference model by detecting the node W_l and all its ancestors from the original model. Therefore, without loss of generality, we assume that $W = \{W_1 = 0, ..., W_L = 0\}$. For LDC j, we denote the event N_j=0 by n_j and N_j=1 by $\neg n_j$. From (1), we have that $P(n_j) = p_j$, $P(\neg n_j) = 1 - p_j$ and the n_j's are mutually independent events.

For each $W_l = 0$ in the evidence $W = \{W_1 = 0, ..., W_L = 0\}$, define the set

$$\bar{N}_l = \{n_j | N_j \text{ has a path to } W_l\}. \tag{7}$$

Note that any single event in $\bar{N}_l$ alone can determine the instantiation of W_l to 0. Set $\bar{N}_l$ must not be empty since there must be a cause for the faulty power output at W_l. The right-hand side of (7) can be interpreted as a disjunction of events n_j. Therefore, conjunction of $\bar{N}_l$ for l = 1, 2, ..., L gives the set of all possible joint events that each causes the instantiation of W. Let

$$V = \bar{N}_1 \wedge \bar{N}_2 \wedge \ldots \wedge \bar{N}_L. \tag{8}$$

Using laws of Boolean algebra, we can rewrite (8) into the irreducible disjunctive normal form (DNF), i.e., into a set of disjunctions of conjunctions of basic events in which each disjunct contains no identical basic events and no disjunct is a subset of the other.

The DNF form of V represents precisely the set of all joint events each of which *alone* can cause the observed evidence W. Thus, probabilistically, event W is equivalent to V. The belief function of $N_j = 0$ given W can, therefore, be expressed as:

$$b(n_j) = P(V \wedge n_j)/P(V), \tag{9}$$

since the sets V and W are probabilistically equivalent.

It is clear that the main task in the belief computation is to convert (8) into DNF form. Although there are standard ways for DNF conversion given in text-books, the problem that we have here is simpler in the sense that the formulas do not content negations. Also, for the sake of developing the *t*-fault diagnosis algorithm, we present the special DNF conversion step by step as follows.

The belief updating algorithm to be presented below consists of L+1 steps. Step 0 finds the sets $\bar{N}_l$, $l = 1, ..., L$,; steps 1 to L-1 find the irreducible set M_L recursively; and the last step computes the belief functions. The proof of the correctness of steps 1 to L-1 is straightforward but tedious; therefore, it is omitted. In what follows, we will use $A \cap B$, $A \cup B$, $A - B$ to denote set operations, and $A \wedge B$, $A \vee B$ to denote Boolean operations.

***Algorithm* 2.** (Belief Updating Algorithm for LLNs.)

Step 0. Given the inference model and the evidence $W = \{W_1 = 0, ..., W_L = 0\}$, compute $\overline{N}_l$, $l = 1, ..., L$ by finding all ancestors of W_j that are LDC RVs.

Step 1. Find the intersection of $\overline{N}_1$ and $\overline{N}_2$, $S_1 = \overline{N}_1 \cap \overline{N}_2$, let

$$V_2 = S_1 \cup [(\overline{N}_1 - S_1) \wedge (\overline{N}_2 - S_1)]. \tag{10}$$

Step l (for $1 < l < L$-1).

(a) find the *generalized intersection* of V_l and $\overline{N}_{l+1}$,

$$S_l = \overline{N}_{l+1} \otimes V_l = \{m \in V_l : \exists n \in \overline{N}_{l+1},\ m \text{ contains } n \text{ as a basic event}\}. \tag{11}$$

In other words, S_l contains those joint events in V_l that contain some events in $\overline{N}_{l+1}$.

(b) Let

$$V_{l+1} = S_l \cup [(\overline{N}_{l+1} - S_l) \wedge (V_l - S_l)]. \tag{12}$$

(c) As we shall see later, the set V_{l+1} thus obtained may not be in the irreducible DNF form because some of its joint events might contain some other joint events. The former should be eliminated. This can be done by arranging the joint events in V_{l+1} in increasing cardinality order and then scan from the smallest upwards checking for proper containment. The resulting set is V_{l+1} in the irreducible DNF form.

Step L. We have by now $V = V_L$. Compute $P(V)$, $P(V \wedge n_j)$ and $b(n_j)$ for $j = 1, ..., K$ by using (9). □

Note that Step l (c) is necessary because Step l (b) does not necessarily return an irreducible DNF form. Here is an example: Let $V_2 = \{n_1 \wedge n_2, n_2 \wedge n_3, n_1 \wedge n_3\}$, $\overline{N}_3 = \{n_1\}$. (It is easy to see that this can be the case for some LLN). Now, by (11), $S_2 = \{n_1 \wedge n_2, n_1 \wedge n_3\}$, and by (12), $V_3 = \{n_1 \wedge n_2, n_1 \wedge n_3, n_1 \wedge n_2 \wedge n_3\}$ and the joint event $n_1 \wedge n_2 \wedge n_3$ is redundant.

If an element in V_L is a joint event of k basic events, we say that the element has multiplicity k. In particular, if $k = 1$, it is called a singleton. Note that, if all the elements in V_l are all singletons, then the generalized intersection defined in (11) reduces to the ordinary intersection.

Let us estimate the worst case time complexity of this algorithm. Recall that we have K LDC nodes and L output ports whose power measurements are incorrect. In Step 0, to compute each $\overline{N}_j$, it takes at most $O(K^2)$ amount of time to traverse the DAG. Thus, this step takes at most $O(LK^2)$ amount of time. In Step l all three subsets (a), (b) and (c) take time proportional to the size V_l and $\overline{N}_{l+1}$. The former is bounded by $O(K^l)$ and the latter by $O(K)$. Thus, Steps 1 through L-1 take at most $O(LK^L)$ amount of time. Step L takes time proportional to the size of V_L which is bounded by $O(LK^L)$. Therefore, in the worst case the entire algorithm takes $O(LK^L)$ amount of time. The worst case bound is not very good, but as we will see in the next section, this method will allow us to come up with a better average case algorithm.

5. THE t-FAULT AND ADAPTIVE FAULT DIAGNOSIS SYSTEMS

A fault diagnosis system employing Algorithm 1 in conjunction with Algorithm 2 is capable of handling any number of simultaneous faults with repair. In a practical LLN, simultaneous failure of a large number of LDCs is very unlikely. Therefore, for practical purposes, it is worthwhile to consider fault diagnosis systems with diagnosis capabilities limited to certain number of faults. A fault diagnosis system is said to have *t*-fault diagnosis capability if it is designed to identify all fault patterns with *t* or fewer simultaneous faulty LDCs. We call such systems *t*-fault diagnosis systems. We observe that if all the fault patterns have *t* or fewer faults, then the corresponding element in the irreducible set *V* will be of multiplicity at most *t*. The *t*-fault diagnosis system conducts its inference using Algorithm 1 and a modified version of Algorithm 2.

The basic idea of the modified Algorithm 2 is as follows. Step 0 remains the same as in Algorithm 2. In Step *l* (b), when forming set V_{l+1} we discard all those joint events that have *more than t* basic events. We denote the resulting set of step *l* (c) by V^t_{l+1}. As the result, the final V^t as well as all intermediate V^t_l's will contain only joint events of multiplicity less or equal to *t*. Step *L* also remains the same as in Algorithm 2:

$$b(t, n_j) = P(V^t \wedge n_j) / P(V^t). \tag{13}$$

Following the same complexity analysis of Algorithm 2, we can see that the worst case time complexity of the *t*-fault diagnosis algorithm is at most $O(LK^t)$. Note that this is a significant reduction in time complexity since it is now a polynomial in the size of the LLN. On the other hand, during the steps *l*=1, 2, ..., *L*-1, set V^t_l might become empty. This means that no *t* or less faults can cause the given evidence. Thus, such fault combinations can not be identified by the *t*-fault diagnosis algorithm. Therefore, there is a trade-off between the fault-diagnosis capability and the computational complexity. As a special case, when *t*=1, finding V^1 is particularly simple: $V^1 = \bar{N}_1 \cap \bar{N}_2 \cap \ldots \cap \bar{N}_L$, where operator $\cap$ denotes the set intersection operation instead of the Boolean "and". It can be shown that if the apriori probabilities p_i are all identical, the 1-fault diagnosis system becomes the deterministic fault diagnosis system presented in [17].

To take the advantage of low computational complexity for small *t*'s, we propose the adaptive inference algorithm that runs the modified version of Algorithm 2 with increasing *t* values. More precisely, it starts with *t*=1. If the set V^t is not empty, it computes $b(t, n_j)$ and conducts the inference process as described in Algorithm 1. Only when the set V^t is found to be empty, it increases *t* by one and restarts the modified Algorithm 2 again. It repeats this until it finds a non-empty V^t.

To analyze the time complexity of this adaptive inference algorithm, we assume that in some instance of execution there are actually *T* faults. In the worst case, the algorithm will find all V^t, $t = 1, \ldots, T-1$, to be empty. But for $t = T$, V^t will be non-empty. As shown above, for each $t = 1, \ldots, T-1$, the algorithm takes $O(LK^t)$ amount of time and the last step takes $O(LK^T)$. Therefore, it takes at most $O(\sum_{t=1}^{T} LK^t) = O(LK^T)$ amount of time when there are exactly *T* faults. Normally, each LDC has limited fan-out. Therefore, the number of output ports *L* should be bounded by a constant times

K, the number of LDCs. Thus, the time complexity of the algorithm is $O(K^{T+1})$ for T faults. Assuming that the apriori failure probabilities of the LDCs are all equal to p, the average time complexity of the algorithm is bounded by

$$O\left(\sum_{T=1}^{K} K^{T+1} \frac{\binom{K}{T} p^T (1-p)^{K-T}}{1-(1-p)^K}\right) \le O\left(\sum_{T=1}^{K} (K^2 p)^T\right). \tag{14}$$

When $p < 0.5K^{-2}$, the above quantity is bounded by a constant. In other words, whenever some fault occurs, the adaptive inference algorithm would spend, on average, a constant amount of time to compute the belief function, independent of the size of the network as long as $p < 0.5K^{-2}$. As an example, when p can be guaranteed to be smaller than 5×10^{-5}, a reasonable assumption given today's optical technology, the network can be made as large as having 100 nodes.

6. SIMULATION RESULTS

We have conducted some preliminary simulation experiments to assess the performance of Algorithm 1 with our belief updating algorithm (Algorithm 2) proposed in subsection 4.2. We are mainly concerned with two performance measures, the number of power tests required to locate all faulty LDCs and the computational complexity. The former can be directly counted as the number of iterations of step 3 of Algorithm 1. We have seen in the complexity analysis earlier in subsection 4.2 that the size of the sets V_i (see Algorithm 2) is proportional to the time complexity of the algorithm. Therefore, we counted the number of terms in the sets V_i for the estimation of the computational complexity.

We have tested LLNs with various sizes from 40 nodes to 100 nodes. The LLNs and the routes were generated at random. Each LLN has four stages of LDCs with no connection within any stage.There is a fixed number of output ports. Each output port has at least two routes from some input ports.

Table 2: Average number of power tests per fault

Faults Nodes	1	2	3	4	5	6
40	2.03	2.15	2.26	2.21	2.05	1.99
50	2.02	2.10	2.19	2.10	2.11	2.05
60	2.06	2.12	2.16	2.17	2.14	2.15
70	1.91	2.05	2.09	2.13	2.07	2.02
80	1.98	2.09	2.09	2.17	2.07	2.14
90	2.01	2.15	2.12	2.15	2.16	2.08
100	1.90	2.03	1.99	1.96	2.10	2.15

The apriori failure probability is assumed to be the same for all LDCs and for the simulation it was arbitrarily chosen to be 0.001. For each of the sizes, we generated 10 sample networks with routes. Then, for each sample LLN with routes, we generate fault patterns containing one up to six faulty LDCs. For each of the fixed number of faults, we generated ten fault patterns at random. For each execution of the algorithm, the total number of power tests and the maximum number of terms in the sets V_i were recorded. The results presented in Table 2 and Table 3 are the average over the samples for each size and the number of faults.

Table 3: Average time complexity

Faults / Nodes	1	2	3	4	5	6
40	8.03	36.9	54.8	85.6	91.2	128.3
50	7.88	45.9	142.8	168.5	240.8	291.3
60	6.52	41.2	124.8	397.7	703.2	1013.8
70	7.84	32.5	151.9	427.8	1051.2	1178.3
80	8.01	41.1	165.3	471.7	1569.2	3070.2
90	6.39	42.4	184.7	640.0	2108.5	3230.0
100	7.59	40.0	204.4	782.8	2006.7	4243.8

From Table 2, we find that the average number of power tests per fault does not change as the size of the network and the total number of faults increase. The reason for the values being around two is due to the structure of the networks we chose to simulate; all the networks consist of four stages and all the LDCs have the same apriori failure probability. In the single fault case, it is easy to see that the average number of tests per fault is the half of the average length of the paths. But, it is remarkable that in multiple fault cases, our algorithm performs equally well in terms of number of extra power tests per fault.

As for the computational complexity, from Table 3, we can see that it increases slowly as the network size increases but fast when the number of faults grows. The figures seem to suggest an exponential increase in the number of faults. Fortunately, multiple simultaneous faults occur very rarely as long as the single failure probability of the LDCs is small.

7. Conclusions

In this paper we proposed a probabilistic approach to fault diagnosis in LLNs. In particular, the application of the Bayesian network inference model to fault diagnosis in LLN was investigated. A general inference algorithm and a belief update algorithm for multiply connect LLN inference models was proposed. We also proposed the *t*-fault and adaptive diagnosis concept for the trade-off between computational cost and fault-diagnosis capability.

Some preliminary simulation of our algorithm was reported. The results suggest that the algorithm performs well with single as well as multiple faults. The average number of power tests per fault does not increase as the network size and number of faults increase. It seems to depend only on the average lengths of the routes from the input to output going ports. As a future study, more pervasive experiments will be done to test the algorithm on LLNs with different topological structure and with non-uniform failure probabilities.

Acknowledgment. The authors would like to thank the referees for valuable comments.

8. References

1 E. Charniak. "Bayesian networks without tears," *AAAI AI Magazine*, pp. 50-63, Winter 1991.

2 H. L. Chin, and G. F. Cooper. "Stochastic simulation of Bayesian belief networks", in L. N. Kanal, et al. (Eds.), *Uncertainty in AI 3*, pp. 129-148, North-Holland, Amsterdam, 1989.

3 G. F. Cooper. "Probabilistic inference using belief networks is NP-hard", *Tech Report KSL-87-27*, Stanford University, Stanford, CA., 1988.

4 R. Fung and K. C. Chang. "Weighing and integrating evidence for stochastic simulation in Bayesian networks", in M. Henrion, et al. (Eds.), *Uncertainty in AI 5*, pp. 209-219, North-Holland, Amsterdam, 1990.

5 R. M. Goodman, J. Miller, and P. Smyth. "Real time autonomous expert systems in network management," *Integrated Network Management, I*, B. Meandzija and J. Westcoot (Eds.), pp. 599-624, Elsevier Science Publishers B. V., 1989.

6 J. H. Griesmer. "YES/MVS: a continuous real time expert system," *Proc. of AAAI-84*, Austin, Texas, 1984.

7 M. Henrion. "Propagating uncertainty by logic sampling in Bayes' network", in L. N. Kanal, J. F. Lemmer (eds.): *Uncertainty In AI 2*, pp. 149-163, North-Holland, Amsterdam, 1988.

8 M. Henrion, J. S. Breese, and E. J. Horvitz. "Decision analysis and expert systems," *AAAI AI Magazine*, pp. 64-91, Winter 1991.

9 P. Hong and P. Sen. "Incorporating non-deterministic reasoning in managing heterogeneous network faults," *Integrated Network Management II*, I. Krishnan and W. Zimmer (Eds.), pp. 481-492, Elsevier Science Publishers B. V.(North-Holland), 1991.

10 T. E. Marques. "A symptom driven expert system for isolating and correcting network faults," *IEEE Communications*, vol. 26, Mar. 1988.

11 R. A. Maxion. "Anomaly detection for diagnosis", *20th IEEE Intnl. Conf. on Fault Tolerant Computing*, pp. 20-27, 1990.

12 F. M. Miller, G. V. E. Otto, E. M. Siegfried, and P. E. Zeldin. "ACE: a knowledge based maintenance analyzer," *Proc. of the IEEE Automated Testing Conf.*, 1985.

13 R. E. Neapolitan. *Probabilistic Reasoning in Expert Systems: Theory and Algorithms*, John Wiley & Sons, Inc., NY, 1990.

14 A. Papoulis. *Probability, Random Variables, and Stochastic Processes*, 3rd Ed., McGraw-Hill, New York, 1991.

15 J. Pearl. "Evidential reasoning using stochastic simulation of causal models", *Artificial Intelligence 32*, pp.245-257, 1987.

16 J. Pearl. *Probabilistic Reasoning in Intelligent Systems*, 2nd edition, Morgan Kaufmann Publisher, Inc., San Mateo, CA., 1992.

17 N. Schroff and M. Schwartz. "Fault detection/identification in the linear lightwave networks", *Tech Report: CU/CTR/TR 243-91-24*, Columbia University, Dept. of Electrical Engineering, New York, 1991

18 R. D. Shachter and M. A. Peot. "Simulation approaches to general probabilistic inference on belief networks", in M. Henrion, et al. (Eds.), *Uncertainty in AI 5*, pp. 221-231, North-Holland, Amsterdam, 1990.

19 T. E. Stern. "Linear lightwave networks: how far can they go?", *Proc. of IEEE Globcomm*, pp. 1866-1872, San Diego, Dec. 2-5, 1990.

20 Z. Wang. "Model of network faults," *Integrated Network Management I*, B. Meandzija and J. Westcott (Eds.), pp. 345-352, Elsevier Science Pub. B. V.(North-Holland), 1989.

21 T. L. Williams, P. J. Orgren, and C. L. Smith. "Diagnosis of multiple faults in a nationwide communications network," *Proc. of 8th IJICAI*, Karsruhe, West Germany, Aug. 1983.

VI.

MANAGEMENT APPLICATIONS

C.

MULTIVENDOR APPLICATIONS - THE WAVE OF THE FUTURE

Integrated Network Management, III (C-12)
H.-G. Hegering and Y. Yemini (Editors)
Elsevier Science Publishers B.V. (North-Holland)

Multivendor Network Management - The Realities

Dave Mahler

Vice President of Marketing

Remedy Corporation 1965 Landings Drive, Mountain View, CA. 94043 USA

Abstract

For years, the industry has talked about the "need for" and the "theory of" multivendor network management. With the acceptance of the platform/application model in conjunction with the use of a common management protocol (SNMP), there is now a growing body of experience among vendors and users on the "realities of" building and using products that cross vendor boundaries. This paper will focus on the experience gained to date by platform and application vendors in the use of various API's and other methods of software integration.

1. What is Nirvana?

When IBM raised the issues of network management to the boardroom level in 1986 with the introduction of NetView, they did the industry at large a service. They posited that the issues associated with managing the information environment were complex and yet critically important - a dangerous combination. They presented the notion of a common approach to what was until then a collection of loosely related issues. They defined a "cockpit of the future" for design, planning, and control of a corporations information assets. They got customers dreaming.

When customers talk of the Holy Grail of network management, they often use terms like "self healing networks" and "proactive network management" meaning knowing about the problem and often automatically doing something about it before the user is impacted. They imagine the ability to step back from the network and see the "big picture" of how it is performing overall. Yet they want the ability to drill down very deep into any device remotely and diagnose, re-configure and update it on the fly. Customers want a common approach to the process that spans network, system and application management, crosses vendor boundaries transparently, is hardware and operating system independent and doesn't cost a lot. And they want it yesterday.

It has been over half a decade since IBM got customers dreaming and they have yet to find that Holy Grail. The issues are largely unchanged, though technology has evolved underneath the dream and the list of vendors on the forefront of addressing the issues has gone through a transformation. This paper will talk about where we are on the road to that nirvana. It is largely based on the experiences of vendors and customers trying to grapple with the issues in managing distributed PC and workstation networks. In that sense, it does not span the issues faced in all areas of network management , including such areas as large telecommunications network or centralized mainframe shops. It is not a detailed treatise on some single technical issue facing the product development community today. It adopts the perspective of the customer who pays for all this stuff at the end of the day. It also does represent where the largest and fastest growing segments of the network management business are now and where they are going in the coming years.

2. Starting From The Bottom

One of the interesting ironies of the evolution of network management is the dichotomy of perspective of the customers vs. the vendors. Customers, meaning the struggling network mangers, most often describe their vision from a top-down perspective. They are trying to deliver a service to a set of very demanding end users who care little about networking technology, standards (or their absence), cost or the differing histories of network vs. system management. All the end user cares about is that the business application they need is available when they need it.

The vendors that have driven the evolution of network management, on the other hand, make their living selling boxes. The box may be a hub, router, server or PC, but it has generally been some type of network or system device. To the vendors, network management has been, and remains a hindrance to selling more boxes. So the vendor community has defined network management in terms of device management. That is a decidedly bottom - up perspective in direct contrast to the users.

This helps to explain all of the focus on protocols, MIB's, and proprietary network element management systems and the absence of work on high level management applications and tools to assess and aid in the process of keeping the end users business application available to them.

Only in the last two years has the focus begun to shift to recognize and adopt the network mangers perspective and it still has a long way to go in appreciating their real dilemmas. The shift in focus is a result to two forces.

The first has been continuous market pressure from network managers that they would rather buy nothing than spend a lot of money on useless tools. The customer has spoken clearly enough here to drive a number of network management companies out of business altogether and to force the network equipment manufacturers to reassess their approach to the problem. Nearly every company that approached the network management issue has underestimated its size and complexity. The initial approach was to believe that a single vendor could solve the whole management problem even if the network contained multiple vendors equipment. Experience has shown that the complexity of the problem demands a layered approach with different companies providing expertise in the various components of a total management solution.

The second force that has helped shift the industry focus has been the growing experience of the vendor community as it has invested heavily in grappling with the network management issues. The current approach just couldn't work. Their experiences were consistent with the message from the market that a new approach was necessary.

3. Element Managers, Platforms and MOM

The initial approach to network management (Figure 1) consisted mostly of a drive upward from every device vendor to provide some management capability with their products and a drive downward from a few major industry players to control the whole problem from a central Manager of Managers (MOM) system.

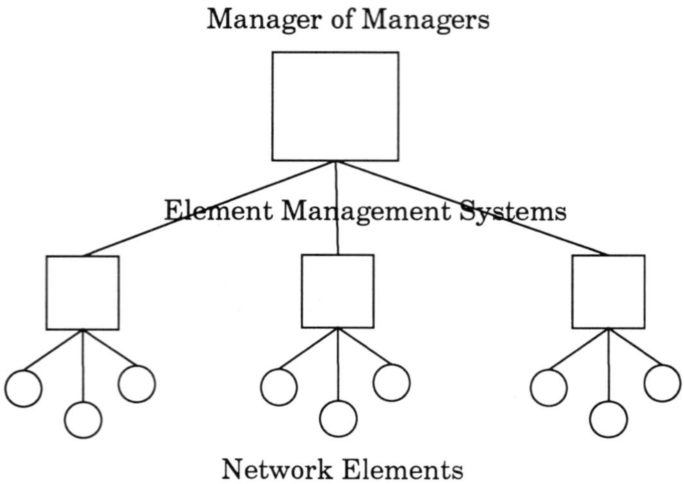

Figure 1

Both strategies failed. The element management system approach foundered on the issues of cost as each vendor required a separate system to manage their piece of equipment and complexity as a network manager that somehow acquired all these systems could never learn how to use them together to solve a problem. The MOM approach failed since there was nothing common among all of the element management systems. There was simply no way that adding yet another system , the MOM, could glue together the various pieces and present a functional, unified and by the way, distributed, interface.

The concept of a common, open management platform that could be used by the network equipment vendors as a basis for their element management systems arose in 1988 from a number of workstation and minicomputer vendors, but wasn't really delivered until 1990. The initial benefit seemed to be that the device vendor could reduce their engineering costs by buying a "standard" platform, and indeed that is the case. The second benefit is that such a standard platform could now allow companies who are not device manufacturers to contribute to the process by taking a high level approach to application development, an approach consistent with the often sited interest of the customer. A third benefit to the approach was that it allowed several of these platforms to be interconnected to form a distributed management "fabric" that could enable local data storage and administrative control while also allowing distributed access to address the class of problems that span the entire network.

The catalyst to enabling this shift of attention has been the broad acceptance of a very simple protocol.

4. The Hidden Benefit of SNMP

One of the characteristics of the network management market in the late 80's was the intense focus on the "protocol wars". It seemed as if nothing else mattered except the academic and political debate on how a management system should communicate with a device. Those debates are not over, and to some extent never should be, but the emergence of a dominant broadly accepted solution, SNMP, with all it's inherent shortcomings has allowed the industry to move forward. The question shifts from "Which protocol?" to "So what can you do if everybody speaks the same protocol?".

By mid-1991, the industry had largely adopted the platform model (Figure 2) as the new paradigm. Independent proprietary element management systems are on their way out. The concept of a MOM still exists, but has a much more limited role in the plans of most network managers, largely focusing on the mainframes and sometimes the network backbone. Several companies have begun using this kind of architecture as a new path to the Holy Grail. The progress to date is impressive, given the short history.

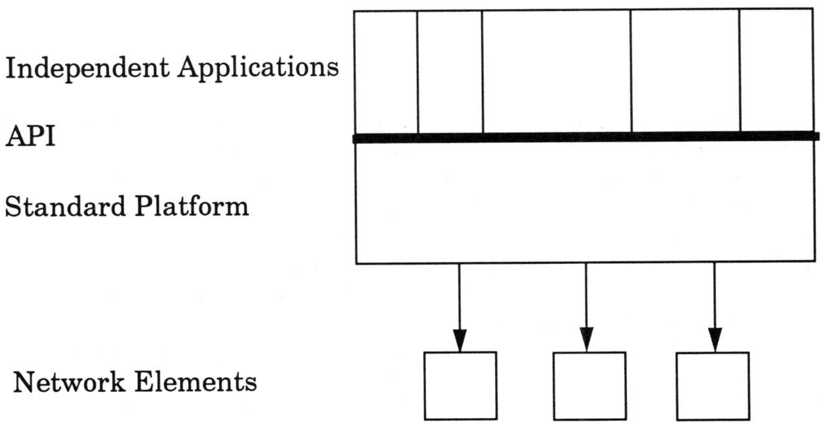

Figure 2

5. Developing the Application Model

When the platform concept was first introduced, everyone anticipated applications being written that would integrate with it. What was not clearly seen was the relative size of the two pieces. Applications were thought to be small "features" on top of the platform. The reality is quite the opposite. Much as in the case of the PC market where the user acquires an operating system and windowing system and then shifts all their attention, and money, to the applications, so to it is with network management platforms. They are quickly evolving into the role of necessary but frequently invisible. They are the means, not the end. And while they provide a common basis for application integration, groups of those same applications are forging new ties independent of the platform underneath them.

The adolescent application model consists of two types of programs; foundation applications and vendor or device unique applications.

Foundation applications are those that meet the common needs of the network manager independent of the device being managed and independent of the device vendor. Many of the management functions fall in this category. Large areas of fault management, accounting and configuration management can be best addressed from an independent perspective. Trouble ticketing is also a good example. The network manager cannot effectively use multiple trouble ticket systems. It needs to be a common element that binds together all of the components of the network, the network management staff and sometimes the users themselves.

There are those portions of configuration, fault and performance management that, at least today, are better handled on a vendor or device specific basis. Some of this may change over time if vendors agree to certain standards and conventions on how to reconfigure devices or download software updates. But device vendors will always want to use private applications to highlight those aspects of their products that differentiate them from their competitors.

6. Today's Integration Examples

It is possible today to buy a variety of applications from both of the categories described above that integrate with one of the popular management platforms and which increasingly integrate with each other too. Several examples are described below, each adding to the capability of the previous to build a set of integrated functionality that appears to the network manager as a comprehensive solution even though it was developed by many different manufacturers.. The names of the vendors supplying the products are not mentioned as they are not pertinent to the concept of what types of integration are possible.

Network management platforms generally provide a logical view of the environment. They show the network linkage between various elements independent of the physical realities. The network manager needs the physical perspective however in order to trace certain types of problems, dispatch repair personnel, and implement the moves, adds and changes that consume a large portion of their daily activity.

These physical and logical perspectives can be integrated with various products today such that the network manager can move directly from the logical representation seen on most management systems to the floorplans and wiring plans for a large building that were created during building construction. Moving "directly" means without independently launching the physical management application and without manual data exchange between the systems to identify which potions of the network are being viewed. The building plans can be scanned into a physical management system and then linked to the logical view. There is even the ability to map the color coding of device status that most management systems offer (green for device up, red for device down etc.) onto the floorplans displayed by the physical management application.

Now add the ability to have the management system, which can monitor device status (via things like pinging and SNMP traps), automatically enter trouble tickets and initiate a workflow process that may include remote notification via radio pagers, time-based escalation, signal re-routing and launching of protocol analysis and other diagnostic routines. The network manager can start with either the logical or the physical view and click on a device representation to get the number and status of "open" trouble tickets on that device without any further manual data exchange between applications.

Next add the ability for the end user to initiate the process by themselves in those cases where the management system may not detect a "fault". The user may decide that the network or a server is slow and can submit a trouble ticket directly from their desktop. Integrating a distributed protocol analysis tool into the above solution can enable the following. When the user submits a trouble ticket concerning a slow network segment, the workflow process can automatically launch the distributed protocol analyzer and configure it on the fly to collect data about the offending network segment for a period of time and then go back to sleep. There is a link formed with the trouble ticket that the user created however such that at a later point in time, when the network manager chooses to explore the problem, from wherever they may be in the network (ie. halfway across the world) , they can move directly from the trouble ticket to the protocol analyzer re-presenting the data captured.

The information necessary to configure the protocol analyzer can determined by a combination of the information the user enters on their trouble ticket and the automatic invocation of another vendors path tracing software. If the user complains about a certain server, the path tracing application can determine the all of the devices (hubs, routers, etc.) in the path from the user's desktop to the server.

This ability to capture information about a transient event without involving the network management staff is critical in providing the kind of responsiveness that users are demanding. There is no need to dispatch repair personnel with portable protocol analyzers and then wait for another slow network event. The number of people who can interpret such information is limited and leveraging their time and knowledge across widely dispersed networks is critical to letting businesses build the ever larger networks that they are becoming dependent upon.

Now move to the feature of some management systems where they can discover new devices in the network. Initially these were only devices of one protocol class, like TCP/IP. The management system could discover those devices as they were added and became active on the network and would add them to the connectivity table that drives the network topology displays.

Since these platforms are open, as opposed to the earlier generation of element management systems and MOM's, other vendors that are skilled in complementary network protocols such as IPX, SNA and DECNET are able to write independent applications that determine the connectivity of the protocol they are familiar with and add it to the platforms connectivity table dramatically increasing the percentage of the network whose configuration can be dynamically monitored.

Two different methods have been shown to invoke the above processes. First, the fatal device failure, like a down hub, that would invoke and SNMP trap and notify the platform and second, the user determined problem-the slow server. Finally, consider the class of problems that are not fatal enough to get caught by the platform monitoring and not yet visible to the end user. These are the performance degradation over time where disc or memory are being slowly consumed or arbitrary processes being independently spawned. Another vendors application that collects such long term trend data without undue network overhead by polling infrequently and keeping history records can invoke the trouble ticket/workflow process. It can even invoke it predictively such that preventive action can be taken well in advance of user disruption.

7. What Have We Learned?

The above scenarios have certain common themes. Some of the themes were anticipated, others have been learned in the process of developing the variety of applications that not only directly manage devices but that independently help the people running networks do the device independent tasks.

Perhaps the most important theme is application cooperation. This does not require a whole new architecture, a whole new body of standards or enormous amounts of engineering effort between each of the combinations of vendors possible. Quite the opposite. It can happen easily, quickly, and in many cases without the vendors ever needing to talk with each other. What must be done is to develop the applications, and the platform correctly, at the outset. Data exchange, such as was required to configure the distributed protocol analyzer on the fly, can use standard operating system methods for exchanging data, ranging from shared files to clipboard like functions. Applications should be able to be launched from a command line into any possible state. Most applications cannot do this today, it must be designed in. And the above traits of the applications and the platform must be easy to use and well documented.

The next theme is independence. One of the inhibitors to multivendor integration to date has been the reality that different applications used different databases and GUI's, ran on different hardware and integrated with different platforms. It was almost impossible to find two products that could coexist. But the ability is there, and being exercised by the leading application and platform providers to support a wide variety of configurations. Database, GUI, hardware and even management platform independence are a must in order to create the combinations that users are demanding.

8. Summary & Looking Forward

Much progress has been made in vendors understanding which products are appropriate for them to build as their part of the network management puzzle. Likewise, experience is growing in leaps and bounds on how to design applications that can be integrated quickly, easily and independently. The answers are refreshing because they suggest a level of independence is possible, even appropriate among vendors while still delivering a total customer solution that is more tightly integrated than has ever been possible before for standard off-the-shelf products.

There are however, a number of areas that will demand attention in the next few years.

Probably the most significant is the true blending of systems and network management. It is a network computing environment that must be managed. The users demand for application availability is independent of the network itself. Problems span both domains and new application tools are needed that can interpret both worlds in the diagnostic process. The network manager really cares about the health of the user's business application which requires a combination of network and system metrics be measured and interpreted. Today those worlds are addressed by different tools with little thought to interaction.

Correlating multiple events in the network computing environment to diagnose or prevent problems is an area ripe for research and development. Networks are now being heavily instrumented with agents and analyzers and the stage is set to be able to consider multiple elements before sounding an alarm condition. This is an area of great complexity but significant importance as network sizes increase and the technology complexity continues to grow.

Network Management Applications

Panel Chair: Dave Mahler, Remedy Corporation, USA

Panelists: Matt Russell, NetMetrix, USA
Bill Tracy, Cabletron, USA
Bob Emerson, Hewlett-Packard, USA

Multitudes of Network Management Applications exist and are being built for very large set of heterogeneous world of networks and network devices. In spite of this rich set of available technologies, a significant effort to raise the semantics of application has been in place for quite sometime. Several Network Management technologies exist today that operate across platforms and environments. This application independence and interoperability is the focus of this panel.

During this panel discussion, various vendors who are actively involved in pushing the leading edge of integration between platforms and applications will discuss what is possible today, and where they see themselves going in the next year to eighteen months. The panel will then open to questions and advice from the audience on what additional forms of integration users want.

VI.

MANAGEMENT APPLICATIONS

D.

MANAGEMENT SUPPORT TOOLS

Integrated Network Management, III (C-12)
H.-G. Hegering and Y. Yemini (Editors)
Elsevier Science Publishers B.V. (North-Holland)

Supporting Network Management through Declaratively Specified Data Visualizations

Mariano P. Consens
consens@db.toronto.edu

Masum Z. Hasan
zmhasan@db.toronto.edu

Computer Systems Research Institute
University of Toronto
Toronto, Canada M5S 1A1

Abstract

The complexity of managing and controlling large heterogeneous networks requires the availability of management stations equipped with sophisticated tools. A fundamental feature of an advanced network management station is the capability to present to the human manager a complete picture of the relevant scenarios. The overwhelming volume and complexity of the information involved in network management scenarios poses a major challenge.

In this paper we show how the **Hy**$^+$ *visual database system* fulfills the information presentation requirements of advanced network management stations. A visual database system is capable of manipulating data visualizations through visually expressed queries. The **Hy**$^+$ system provides a uniform framework for *hygraph* based data visualizations, queries, and their results. Visual queries, expressed in the *GraphLog* language, are interpreted as patterns that match existing visualizations and create new ones. A prime example of the functionality supported by the system consists of describing a query that *filters* a large visualization to retain the portion that is of interest for the network manager in the context of a particular task.

Examples of data visualizations that are relevant for network management are the network topology at different levels of abstraction, the presentation of network configuration information, and the display of management information bases and their history traces. Employing a fault management scenario as motivation, we demonstrate how the **Hy**$^+$ system deals with all of the above visualizations and several of their combinations, as well as entirely new ones generated through ad-hoc visual queries.

Keywords: Network Management, Management Information Databases, Data Visualization, Visual Database Systems, GraphLog Visual Query Language, **Hy**$^+$ System.

1 Introduction

Managing a large heterogeneous network is a considerably complex task. The proposed network management standards for TCP/IP based Internets [RM, CFSD, MR] and the ISO-OSI network management standard [ISOa, ISOc, ISOb] are attempts by the standard organizations to reduce the complexity involved. The manager-agent paradigm, where the manager keeps the global picture of the network, has been proposed in both standards. The management station is the control center of the network. One of the vital responsibilities of the manager is to maintain a global database containing all the management related data. The data itself is distributed over the network and accessible by the manager through the agents. Conceptually, we can think of this distributed database as one single database residing at the management station.

The complexity of managing, operating and controlling networks is reduced by systematically imposing different levels of abstraction onto the network. By doing so, the functional management can be performed in a meaningful and structured way. In [Val91], the following views at different levels of abstraction are listed: geographical view (site of components, structured by countries, areas, cities, buildings, floors, rooms), physical view (components and media with their connectivity and physical characteristics), protocol view (protocol-specific attributes for LAN-segments and subnets, etc.), application view (client-server relationships), and administrative view (structured according to administrative responsibilities within the network, e.g., domains). In addition, views may incorporate a *history view* consisting of information about the evolution of the network and its status over time [WSY91].

It is advantageous to visualize the physical view together with the protocol view. We can distinguish between two different views of the topology map: the logical map which corresponds to the network layer, and the physical map which corresponds to the datalink and physical layers. The network addressing mechanism (e.g., IP) defines a logical or network layer map of the network. Devices like bridges, repeaters and cables, together with the datalink and physical layer protocol objects define a physical layer map. Logical and physical layer maps can be synthesized to produce maps corresponding to geographical and administrative views. The clustering of network elements according to the views of the network at different levels of abstraction is emphasized in [KMG88].

The limitation in the size of the presentation media makes it difficult to visualize the status of the entire network. It might be impossible to squeeze a large and complex network with hundreds and thousands of devices and complex interconnections into one single screen, even if we use very sophisticated layout algorithms. When a problem occurs in the network, it would be helpful to see only the relevant areas of the network. It would be nice to be able to specify what we want in a simple and declarative way, filter out the problem area and visualize it in terms of the topology of the network and the views at different levels of abstraction.

In this paper we describe a database system for network management stations that manipulates not just data, but the associated visualizations as well. The **Hy**$^{+}$ system supports queries involving recursion and incorporates deductive capabilities. The user interface of the system enables network managers and operators to browse and edit the visualizations of databases, queries and query results in a uniform way.

The rest of the paper is organized as follows. An overview of the **Hy**$^{+}$ visual database system is given in Section 2. In Section 3 we describe the structure of an example network management database, and how we query and filter the initial database to produce different visualizations of the network information. We present an application of the proposed technology in the context of a realistic fault management scenario in Section 4. We conclude in Section 5.

2 Overview of the Hy^{+} System

In this section, we present an overview of the **Hy**$^{+}$ system developed at the University of Toronto in the context of its application within the CORDS project [SBF+92].

The **Hy**$^{+}$ system provides extensive support for query visualization, visualizing the input instance, and visualizing the output instance in several different modes. The visualizations manipulated by the system are labeled graphs and *hygraphs*. Hygraphs, defined in [Con92], are

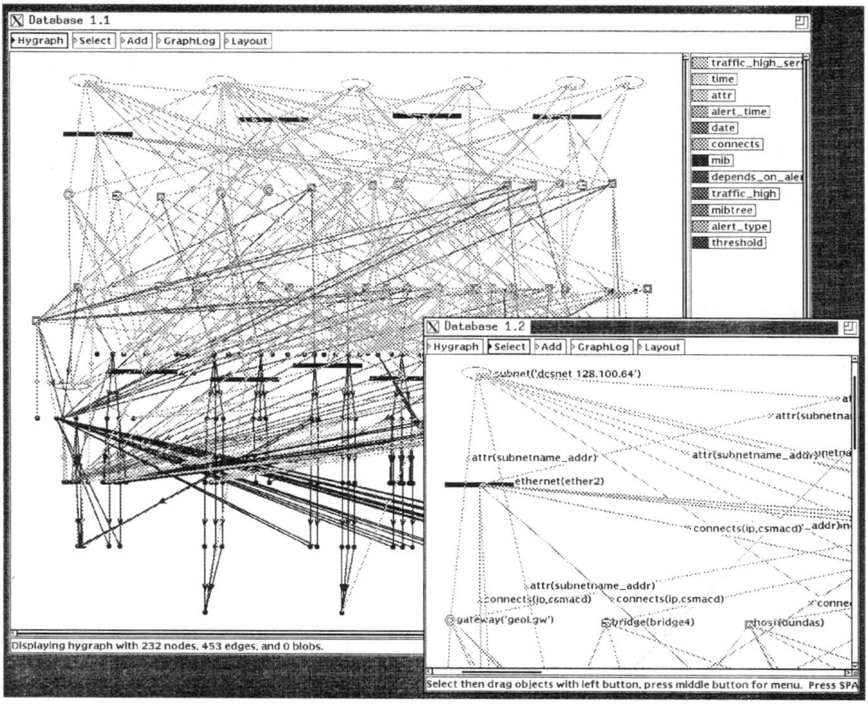

Figure 1: Browsing the example database.

a hybrid between Harel's higraphs [Har88] and directed hypergraphs. A hygraph extends the notion of a graph by incorporating *blobs* in addition to edges. A blob relates a containing node with a set of contained nodes, and is diagrammatically represented as a region associated with the container node that encloses the contained nodes.

Many databases can be naturally viewed as graphs. Object-oriented databases are especially suited to this representation, and so are hypertext databases [CM89]. A relational database can be represented by a directed multigraph having an edge labeled $r(c_1, \ldots, c_k)$ from a node labeled $(a_1, \ldots, a_i)$ to a node labeled $(b_1, \ldots, b_j)$ corresponding to each tuple $(a_1, \ldots, a_i, b_1, \ldots, b_j, c_1, \ldots, c_k)$ of each relation r in the database. Extending the representation to hygraphs allows varying levels of abstraction in the display of hierarchical data (e.g., by interactively hiding and showing blob contents) and provides a flexible mechanism for clustering information.

The $\mathbf{Hy}^+$ system has browsers with extensive facilities for editing hygraphs (e.g., copy, cut and paste; panning and zooming; executing several layout algorithms; moving hygraphs to and from files; editing of node and edge labels; etc.). These hygraph browsers can present the user with options to start browsing the structure being visualized in hypermedia-like fashion.

Figure 1 contains two $\mathbf{Hy}^+$ views of the same visualization of an example database used in the remaining of the paper. Window Database 1.1 presents an overall view (with no labels displayed) of a graph containing several hundred objects and relationships. The schema of the database is displayed in a pane of Database 1.1 that lists the colors assigned to the edges corresponding to each of the relations in the adjacent visualization. Color-coding relations constitutes a very primitive, but visually appealing, example of tailoring the visualizations to the semantics of the applications. Another instance of the customization of the visualizations supported by the system is the selection of different icons for nodes based on the properties of the database objects represented by the nodes. Window Database 1.2 contains a zoomed-in view of the top-left portion of the graph (with labels displayed). The information contained in the example database has been partly acquired using the programs described in [HHB92].

The visual queries supported by the **Hy**$^+$ system are expressions of the GraphLog query language [CM90a]. GraphLog queries are graph patterns with nodes labeled by sequences of variables and constants, and whose edges are labeled by *path regular expressions* on relations. The query evaluation process consists of finding in the database all instances of the given pattern and for each such instance performing some action, such as defining new arcs in the database graph, or extracting from the database the instances of the pattern. GraphLog has higher expressive power than SQL; in particular, it can express, with no need for recursion, queries that involve computing transitive closures or similar graph traversal operations. The language is also capable of expressing first order aggregate queries (an example involving maximizing is given later in the paper) as well as aggregation along path traversals (e.g., shortest path queries)[CM90b]. Precise theoretical characterizations of the expressive power of GraphLog and of its computational complexity can be found in the references cited above.

The **Hy**$^+$ system interprets GraphLog patterns in two fundamentally different ways:

Define is the classical database interpretation of the pattern as a query that defines relations described by the distinguished edges or blobs.

Show (formally described in [Con92]) is at the core of **Hy**$^+$ capabilities for manipulating database visualizations, as opposed to just databases. Informally, when executing a query in show mode the pattern does not define any new relation, but it filters out information from the input visualization to produce an output visualization that contains the union of all the sub-hygraphs in the input that match the query pattern.

The **Hy**$^+$ visual query system is implemented as a front-end that can communicate with multiple back-ends for the actual evaluation of the queries. The front-end, written in Smalltalk, includes the user interface of the system. Currently, there are two back-end query processors; one uses the LDL deductive database system developed at MCC [NS89], while the other is implemented in Prolog [Fuk91].

The system has the ability to invoke external programs that, for instance, browse an object being represented by a node in one of the graphs displayed by the system. This feature, which is part of the overall **Hy**$^+$ goal of providing an open architecture, can be used to support network management stations that require the co-existence of several browsers provided by different tools. The user can navigate, in hypermedia style, through the information contained in management information bases. The **Hy**$^+$ visualizations are used as overviews to locate information and then invoke third-party browsers to display the contents associated with the relevant objects. An important advantage of this approach over a purely navigational one, is the ability to use the convenience and expressive power of GraphLog patterns to retrieve the objects of interest, instead of attempting an often impractical brute force search.

The **Hy**$^+$ system has similarities to the heterogeneous network management systems proposed in [WS89, DSWY91] and to the active database approach in [WSY91], where network management functions are specified as data manipulation statements. These systems differ from

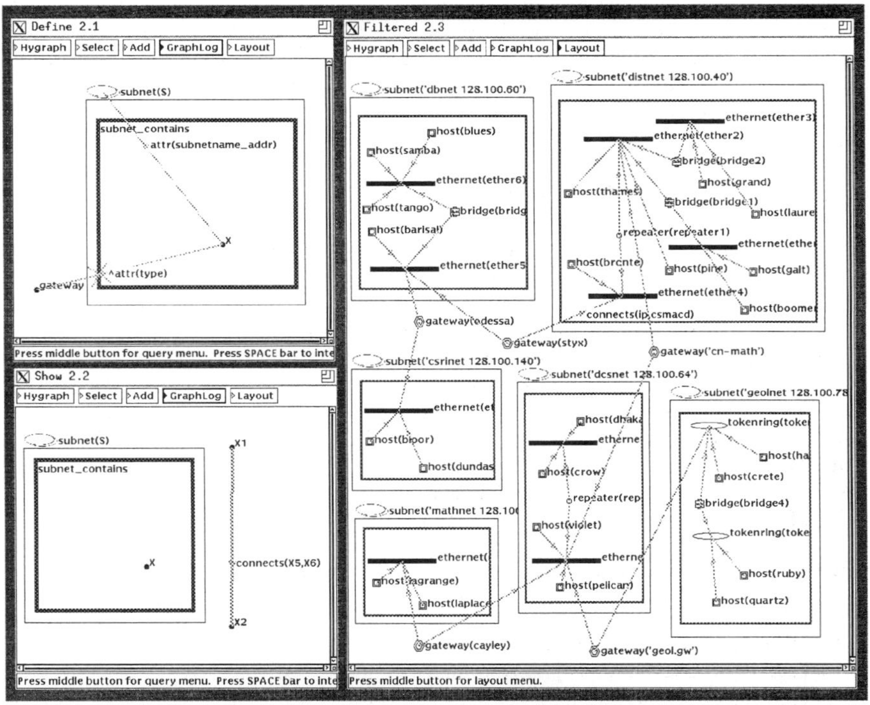

Figure 2: Defining subnets over the physical network topology.

Hy$^+$ in the emphasis given by the latter to the manipulation of database visualizations as an integral part of a toolset for network management stations. Another important difference is the expressive power of the GraphLog query language. SQL-based systems are not capable of expressing transitive closure queries, while these are directly supported in the GraphLog query language in a very intuitive way. Transitive closure queries are necessary when we want to retrieve network management information in terms of the connectivity and topology of the network.

3 Visualizing the Network Database

In this section we describe what constitutes a network database and how we obtain different views of the network using the querying capabilities provided by the **Hy**$^+$ system.

A central database that stores all the information associated with the different views of the network and the management subsystems, is a vital part of the network management station. This conceptual global database contains detailed information about the topology of the network (configuration data) and protocol related MIB objects [MR] distributed over the network. The schema for each different MIB (private, enterprise or experimental), as well as historical MIB objects, or traces, are also stored in the central database.

Let us consider an example (using the sample network database shown in Figure 1) that

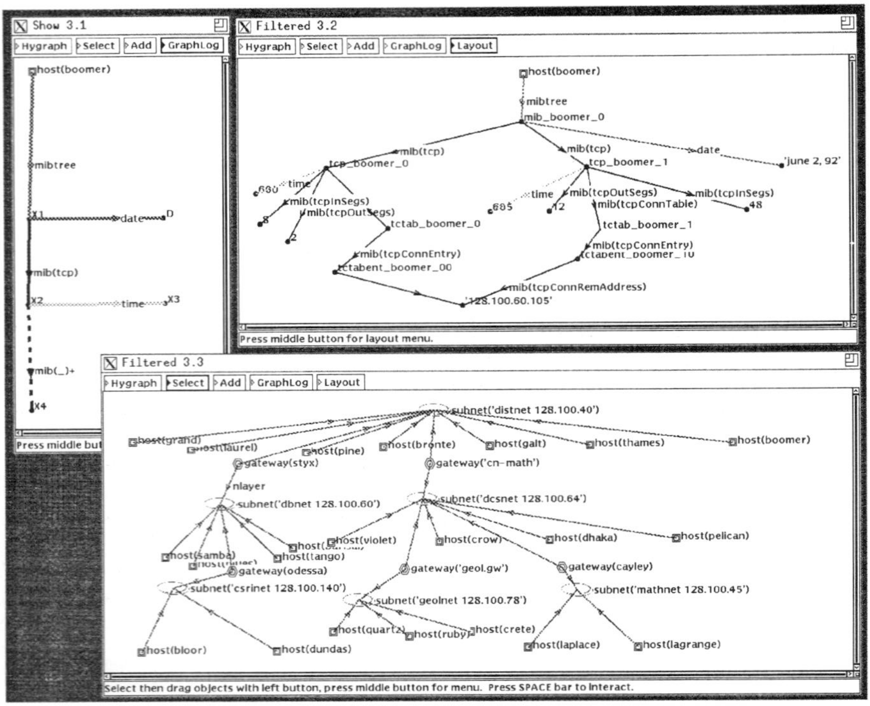

Figure 3: History trace of MIB objects for boomer.

constructs the IP subnets and displays them superimposed on top of the physical network topology. Subnets are logical objects that encapsulate the details of the datalink and physical layer. We leave gateways outside subnets, since they belong to more than one subnet. Another motivation for this query is to cluster (making use of blobs) the elements contained in a subnet.

Query Define 2.1 in Figure 2 creates a new relationship subnet_contains represented as a blob (the distinguished blob of the query) relating the container node labeled by the subnet name and address with the contained nodes that have the same value for the attribute subnetname_addr but are not nodes of type gateway. Distinguished edges or blobs are drawn thicker. The crossed out edge labeled ^attr(type) (representing negation) in query Define 2.1 excludes the gateways from being part of the subnets. Query Show 2.2 matches all subnet_contains blobs and connects edges and displays them in window Filtered 2.3.

We can browse the current values and the history traces of MIB objects. For example, query Show 3.1 in Figure 3 displays (in window Filtered 3.2) the history trace of MIB objects for the host named boomer. This involves filtering the values as well as the timestamps of the traces. The edge labeled with the regular expression mib(_)+ (shown dashed to emphasize that the edge in the query graph matches a path in the input graph) is used to return the entire MIB subtree for boomer.

The configuration database is transformed using queries (not shown for brevity) to obtain yet another different view of the network: the logical network layer map (shown in window Filtered 3.3 in Figure 3).

Administrative and geographical views of the network can be created and displayed in a similar way.

4 Case Study: Fault Management

In this section we describe how the **Hy**$^+$ system can be used in the context of a fault management scenario. The following detailed example is meant to emphasize, in the context of one typical network management activity, how the use of powerful and flexible visualization tools can contribute to the understanding of the complex information required for successfully completing the task at hand.

Fault management is a complex task. Due to the layering structure of protocol software, a fault caused at one level may show up at other levels. Additionally, the combination of protocol software incorporating mechanisms to recover from problems together with the hiding of lower level problems to the higher levels, contributes to make fault analysis and diagnosis difficult. For example, a connection failure or a long response time can be caused by congestion or it can be the result of a hardware problem. This is not to say that protocol software should not try to recover, but that fault analysis and diagnosis requires sophisticated tools. It is quite helpful to visualize the problem area and to be able to deduce possible causes by exploiting the capabilities of a powerful database querying mechanism combined with an inference engine.

To find out the exact cause of a problem we need to proceed in a structured manner. The fault diagnosis process incorporates the following steps: observing the symptoms, developing a hypothesis on what could be the cause of a symptom, testing the hypothesis, and if the hypothesis fails, iterating the above steps until a conclusion is reached.

4.1 Defining and Observing Problem Symptoms

Before we can perform fault analysis and diagnosis, we have to find out what kind of network behavior we can consider as problem symptoms. Defining potential problem symptoms is very difficult without a priori knowledge of the behavior of the network. Symptoms can be defined in terms of statistics or the expected behavior of the network. In [GL91], it is observed that a temporal cycle exists in the traffic distribution of a network. Therefore, a suitable approach consists of defining problem symptoms in terms of traffic patterns. For example, during high traffic periods, it is natural that network accesses are slowed down. But if the delays reach a certain threshold (the threshold being a function of period and traffic pattern) or the utilization of the network drops during a high traffic period, then this may indicate a possible problem symptom. Henceforth, we require access to statistics about the behavior of the network (e.g., utilization, traffic patterns, etc.).

The information about high volumes of traffic can be obtained by monitoring interesting points of the network. Traffic patterns among individual hosts can be established by monitoring the tcpInSegs and tcpOutSegs TCP MIB variables, assuming that the only activity that the clients perform is accessing the server by establishing TCP links. Current SNMP TCP MIB variables do not provide us with traffic information on a per TCP link basis. However, given SNMP's widespread usage and simplicity, we will use SNMP MIB objects in our example. Traffic patterns can also be established using the procedure mentioned in [GL91].

The traffic information between two subnets (dbnet and distnet) taken from the example network in Figure 1) is shown against the topology map in Figure 4. The server samba located

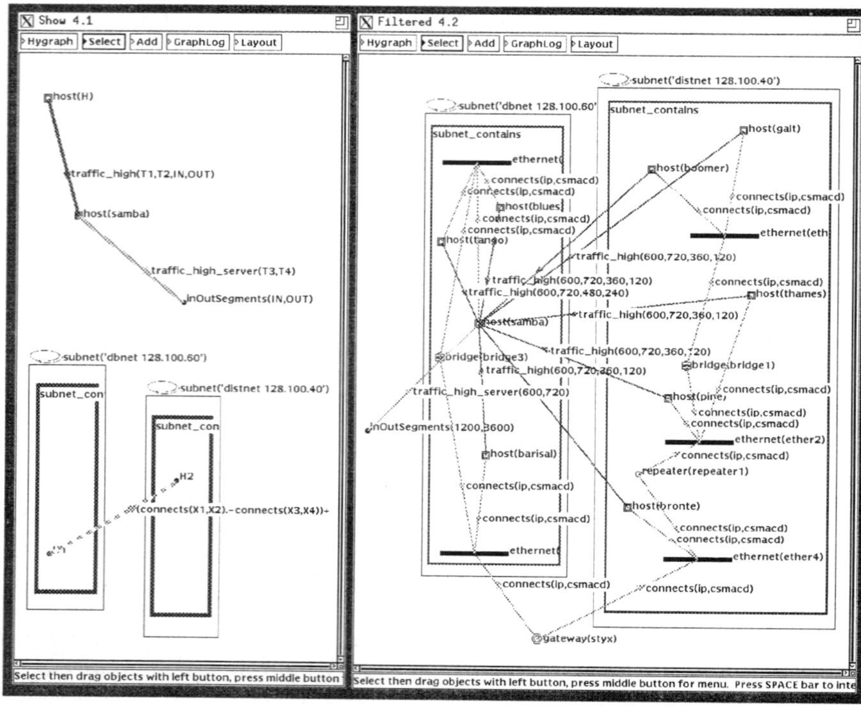

Figure 4: Traffic information displayed against the topology map.

in the subnet dbnet is accessed by clients, in the same as well as in other subnets. The edge labeled traffic_high_server(600,720) from the node labeled samba to the node labeled inOutSegments(1200,3600) tells the expected number of incoming and outgoing TCP segments from the server samba between 10 a.m (600 minutes from midnight) and 12 p.m. (720 minutes from midnight). Other edges labeled traffic_high(Time1, Time2, In, Out) between individual hosts and samba indicate the (In, Out) expected values of high volumes of traffic between the server and the hosts during the periods from Time1 to Time2.

The query Show 4.1 is used to filter the physical map to display the contents of the two subnets of interest, 'dbnet 128.100.60' and 'distnet 128.100.40', together with the traffic_high_server and traffic_high edges. In this way, high traffic spots are visualized directly against the topology, instead of presenting them as tables or rules. If we were dealing with a large network, filtering out all the subnets we are interested in, produces a more understandable and less cluttered map. The resulting display (window Filtered 4.2) helps the human network manager to understand the traffic pattern directly in terms of the topology of the network.

Now, assume that we monitor the utilization of the server samba. For this purpose we poll the tcpInSegs and tcpOutSegs TCP objects of the server. The values returned are kept in the database as history traces.

An alert can be generated if the server utilization falls below a certain threshold in a partic-

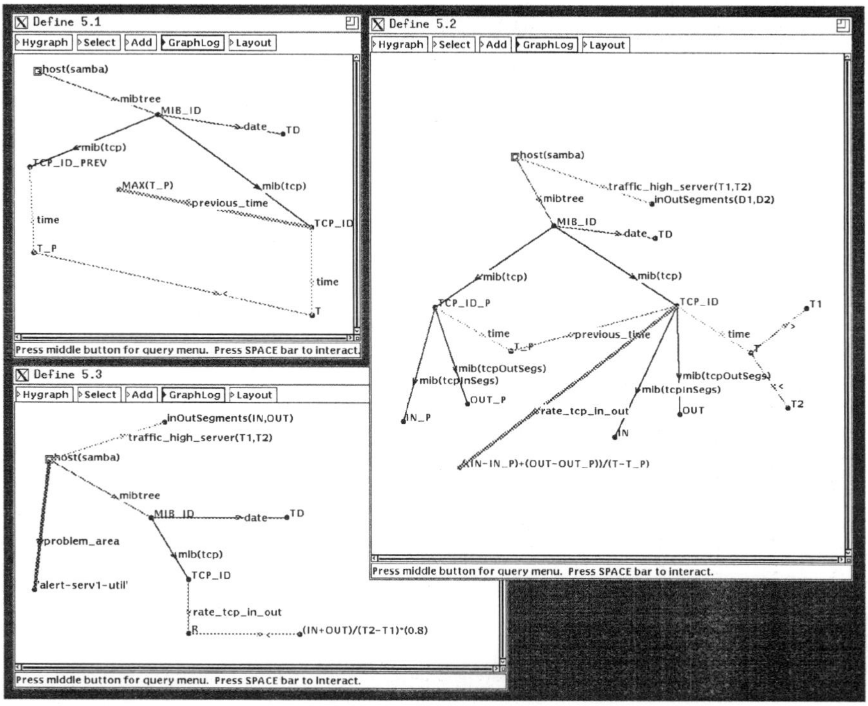

Figure 5: Defining an alert for possible problem symptoms.

ular period of time, which in this case coincides with an expected high traffic period. If the alert goes off, it indicates a possible problem symptom. The queries in Figure 5 define the alert alert-serv1-util. Query Define 5.1 finds out the time (pointed by the distinguished edge previous_time) when the previous poll of samba was taken. Query Define 5.2 computes the rate of tcpInSegs and tcpOutSegs traffic in the latest poll interval during high traffic time. Query Define 5.3 checks whether the rate computed in the previous query falls 80% below the expected traffic kept on the traffic_high_server edge of samba. If the condition is satisfied, an edge called problem-area is created from the node labeled samba to the node labeled with the name of the alert.

The queries Define 5.2 and Define 5.3 should be interpreted as two (active database) rules (see [Day88, BM91, Cer92]) that are triggered on non-empty retrieval events and perform the action of adding the distinguished edges. We are considering the addition of features to the query language that address temporal monitoring of database changes similar to the ones proposed in [WSY91].

4.2 Diagnosing a Fault

Let us consider that the alert named alert-serv1-util (the one we defined in the previous subsection) has gone off and the problem area is the server samba.

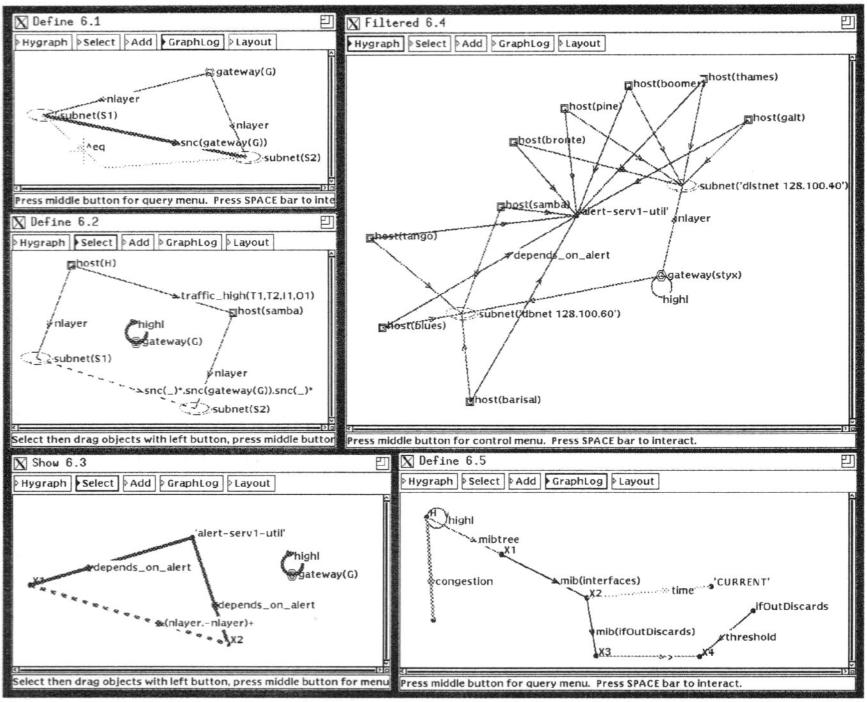

Figure 6: Highlighting congested gateways in the logical map.

Our first hypothesis is that the cause is *congestion* at nearby gateways (routers). It is possible that other sources of traffic are placing an unexpected demand on the gateways through which traffic to the server passes. As a consequence the packets destined for the server are being excessively discarded.

To prove this hypothesis we check the gateways through which packets addressed to the server are likely to flow. While investigating for congestion we are interested in focusing on the portion of the logical map of the network that is related to the alert alert-serv1-util. For each host that uses the server samba, and is therefore affected by the alert alert-serv1-util, an edge called depends-on-alert is created from the host to the node containing the value alert-serv1-util.

When we are presented with the logical map, we can easily identify the gateways that should be checked for congestion. In addition, the gateways can be identified automatically by querying and highlighting them. The queries and the resulting network layer map are shown in Figure 6, with the potentially congested gateway styx highlighted.

Query Define 6.1 creates a direct edge called snc(gateway(G)) between two subnets S1 and S2 with the gateway information carried along the edge. Query Define 6.2 finds the *simple paths* [MW89] between high traffic nodes and highlights the gateways traversed along the paths.

As a result of the filter query Show 6.3, window Filtered 6.4 shows the subset of the logical map of the network that depends on the alert alert-serv1-util, with a loop labeled highl around the

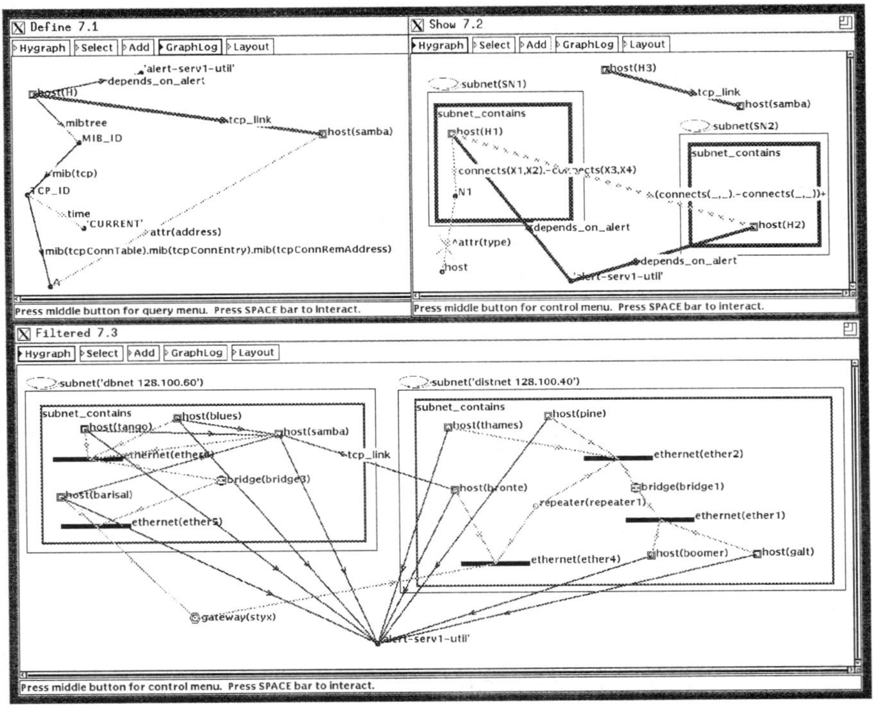

Figure 7: TCP links superimposed on the physical topology map.

gateways we are interested in highlighting. Filtering out the problem area through a powerful querying mechanism as shown above, definitely helps.

After identifying the gateways we check for congestion. The query Define 6.5 checks the current value of the ifOutDiscards variable of the interfaces MIB object for the highlighted gateways. If the value exceeds the specified threshold, then the distinguished edge called congestion is created.

If no congestion is detected at the gateway styx, we have to come up with a second hypothesis. For now, assume that there is indeed congestion at the gateway styx. The source of the congestion could be the clients of the server samba, or it could be originated somewhere else. To find out the source of the congestion we check whether the hosts using the server are placing the expected traffic to the server samba. As discussed earlier, the expected traffic patterns between the server and its clients have been established before. We can use a query to check whether the sum of the current values of tcpOutSegs for all the clients is 20% less or more than the sum of the expected values that are kept on the traffic_high edges.

If the answer to the above query is no, we may hypothesize that other sources are causing the congestion, since the clients of samba are generating traffic as expected. We could continue tracing the source of the congestion by looking at the other hosts in the network. This can be done by querying the current tcpConnTable for those hosts and creating visualizations to help locate the TCP links that have routes through the congested gateway styx.

Assume that no congestion was detected, but it was found that the clients of samba are not generating the expected traffic to the server. Our second hypothesis is that the cause of the alert is due to a *hardware problem.* In this case the map is brought up showing physical details like repeaters, bridges, datalink layer protocols (CSMA/CD, tokenring), cable types (coax, fiber), and so on. Before looking for the physical causes of the problem, we superimpose the currently active TCP links on the portion of the physical map that depends on the alert.

Query Define 7.1 in Figure 7 can be interpreted as follows: create an edge called tcp_link between the hosts H that depend on the alert alert-serv1-util and the server samba, if the IP address found in the tcpConnRemAddress of H's TCP connection table matches the address of samba (which is stored in the attribute attr(address) of the configuration database). Query Show 7.2 produces, as a result, the visualization in window Filtered 7.3 showing the portion of the physical topology map that depends on the alert, together with the previously defined tcp_link edges.

The visualization of TCP links allows us to pin-point the problem area at the portion of the network beyond ether4. The problem could be at the bridge bridge1, at the repeater repeater1, or at the ethernet segments ether1 or ether2. This follows from noticing that no TCP connection originates at ether1 or ether2. The SNMP manager could not access the current TCP objects from that part of the network.

The example shows how, when isolating faults in a network, it is advantageous to view network maps at different abstract levels, while proceeding in a structured manner to pin-point the problem area. If we were limited to work only with the entire physical map of a large network, the complexity of the fault analysis and diagnosis procedure would increase significantly. This is just one particular instance that illustrates how the capabilities of the **Hy**$^+$ system can be exploited by human network managers.

5 Conclusions

The **Hy**$^+$ visual database system described in the paper constitutes a powerful and flexible tool for visualizing the information required to manage large heterogeneous networks. Below we summarize the main characteristics of the system.

- A consistent hygraph-based framework for dealing with a wide variety of network related data visualizations is provided.
- Data visualizations are manipulated by means of visually expressed queries. Visual queries, expressed in the GraphLog query language, are interpreted as patterns matched against existing visualizations to either create new visualizations or to filter existing visualizations to retain only their relevant portions.
- The expressive power of the GraphLog query language is adequate for network management applications. Several examples of transitive closure queries were presented and extensively motivated in the previous sections (and these queries cannot be expressed in SQL).

An extensive fault management example has been used to illustrate how the capabilities of the system can be exploited by human network managers.

Future research is focused on two areas. At the back-end level we are concerned with the incorporation of heterogeneous and active database features to support the integration of network management services. The **Hy**$^+$ approach of translating GraphLog patterns into an underlying target database language constitutes a flexible architecture for the incorporation of these additions (like the contemplated addition of a translation incorporating SNMP queries). At the front-end, we are extending the presentation capabilities of the **Hy**$^+$ system to incorporate several domain specific visualizations and corresponding layout algorithms that profit from knowledge of the application domain.

Acknowledgments

We would like to thank Alberto Mendelzon for his fruitful suggestions and support. We are also indebted to the other members of the **Hy**$^+$ development group: Frank Eigler, Yanni Jew, Emanuel Noik, Dimitra Vista, and Annie Yeung. The CORDS project at the IBM Canada Centre for Advanced Studies, headed by Jacob Slonim, provided a stimulating research environment.

The first author was supported by an IBM Canada Research Fellowship and by PEDECIBA (United Nations Program for the Development of Basic Sciences in Uruguay).

References

[BM91] Catriel Beeri and Tova Milo. A model for active object oriented database. In *Proceedings of the 17th International Conference on Very Large Data Bases*, pages 337–349, September 1991.

[Cer92] Stefano Ceri. A declarative approach to active databases. In *Proceedings of the Data Engineering*, pages 452–456, 1992.

[CFSD] J. D. Case, M. S. Fedor, M. L. Schoffstal, and J. R. Davin. A simple network management protocol (SNMP). RFC 1157, SNMP Research, Performance Systems International, MIT Laboratory for Computer Science, May 1990.

[CM89] Mariano Consens and Alberto Mendelzon. Expressing structural hypertext queries in GraphLog. In *Proceedings of the Second ACM Hypertext Conference*, pages 269–292, 1989.

[CM90a] Mariano Consens and Alberto Mendelzon. GraphLog: a visual formalism for real life recursion. In *Proceedings of the Ninth ACM SIGACT-SIGMOD Symposium on Principles of Database Systems*, pages 404–416, 1990.

[CM90b] Mariano Consens and Alberto Mendelzon. Low complexity aggregation in GraphLog and Datalog. In *Proceedings of the Third International Conference on Database Theory, Lecture Notes in Computer Science Nr. 470*, pages 379–394. Springer-Verlag, 1990. A revised version has been accepted for publication in TCS.

[Con92] Mariano P. Consens. Visual manipulations of database visualizations. PhD Thesis Proposal, 1992.

[Day88] U. Dayal. Active database management systems. In *Proceedings of the third International Conference on Management of Data*, pages 225–234, June 1988.

[DSWY91] A. Dupuy, S. Sengupta, O. Wolfson, and Y. Yemini. NETMATE: A network management environment. *IEEE Network Magazine*, pages 35–43, October 1991.

[Fuk91] Milan Fukar. Translating GraphLog into Prolog. Technical report, Center for Advanced Studies IBM Canada Limited, October 1991.

[GL91] Mario Gerla and Ying-Dar Lin. Network management using database discovery tools. *IEEE proceedings on Distributed Computing Systems 1991*, pages 378–385, 1991.

[Har88] David Harel. On visual formalisms. *Communications of the ACM*, 31(5):514–530, 1988.

[HHB92] Masum Z. Hasan, James W. Hong, and James P. Black. Topology and device discovery in an Internet environment. Submitted for Publication, 1992.

[ISOa] ISO 7498, information processing systems, open systems interconnection, basic reference model part 4, OSI management framework. International Organization for Standardization, October 1986.

[ISOb] ISO 9596, information processing systems, open systems interconnection, management information protocol specification, common management information protocol. International Organization for Standardization, November 1990.

[ISOc] ISO/IEC DIS 10165-1, 10165-2 and 10165-4, information processing systems, open systems interconnection, structure of management information, parts 1, 2 and 4. International Organization for Standardization, 1990.

[KMG88] G. Kar, B. Madden, and R. S. Gilbert. Heuristic layout algorithms for network management presentation services. *IEEE Network Magazine*, pages 29–36, November 1988.

[MR] K. McCloghrie and M. T. Rose. Management information base for network management of TCP/IP based internets - MIB-II. RFC 1213, Hughes LAN Systems, Performance Systems International, March, 1991.

[MW89] A.O. Mendelzon and P.T. Wood. Finding regular simple paths in graph databases. In *Proc. 15th International Conference on Very Large Data Bases*, pages 185–194, 1989.

[NS89] Shamim Naqvi and Tsur Shalom. *A logical language for data and knowledge bases*. Computer Science Press, New York, 1989.

[RM] M. Rose and K. McCloghrie. Structure and indentification of management information for TCP/IP-based internets. RFC 1155, Performance Systems International, Hughes LAN Systems, May, 1990.

[SBF+92] J. Slonim, M. Bauer, P. Finnigan, P. Larson, A. Mendelzon, R. McBride, T. Teorey, Y. Yemini, and S. Yemini. Distributed programming environments: Challenges. In *Proceedings of the IFIP TC6/WG6.4 International Workshop on Open Distributed Processing*. North-Holland, October 1992.

[Val91] R. F. Valta. Design concepts for a global network management database. *Integrated Network Management, II*, pages 777–788, 1991.

[WS89] U. S. Warrier and Carl A. Sunshine. A platform for heterogeneous interconnection network management. In B. Meandzija and J. Westcott, editors, *Integrated Network Management, I, IFIP*, chapter 1, pages 13–24. North-Holland, 1989.

[WSY91] O. Wolfson, S. Sengupta, and Y. Yemini. Managing communication networks by monitoring databases. *IEEE Transactions on Software Engineering*, 17(9):944–953, September 1991.

Integrated Network Management, III (C-12)
H.-G. Hegering and Y. Yemini (Editors)
Elsevier Science Publishers B.V. (North-Holland)

A Tool supporting the Management Information Modeling process

Marcus Wittig and Martin Pfeiler

GMD FOKUS, Hardenbergplatz 2, 1000 Berlin 12, Federal Republic of Germany
E-mail: wittig@fokus.berlin.gmd.dbp.de

Abstract

Managed Objects represent an abstract view on logical and real resources of OSI systems. The structure and behaviour of Managed Objects have to be described unambiguously to enable the interchange of management information between open systems. The definition of Managed Objects is a rather time consuming and error prone task because numerous conditions and rules defines in the *Guidelines for the Definition of Management Information* have to be taken into consideration. This paper introduces a tool called DAMOCLES which facilitates the development of management applications by accelerating the definition and administration of Managed Object Classes.

1. Introduction

BERKOM (**Ber**liner **Kom**munikationssystem, [1]) is a joint project of the Deutsche Bundespost (the German PTT) and the State of Berlin which was started in 1986 in order to provide experience and support in using the high speed B-ISDN network. Application projects have been sponsored in a variety of application areas in order to investigate the feasibility of using B-ISDN as a communication medium for the exchange of, e.g. multimedia information.

All these projects need to be concerned with the management of their particular application. In the first version of the BERKOM reference model [2] it was acknowledged that a unified management concept should be developed for the BERKOM integrated services. In order to prevent each project from developing its own, incompatible management solutions, the BERKOM project "*Management of Distributed Applications in B-ISDN*" (BERMAN) was launched in the autumn of 1988 by GMD-FOKUS' Open Management Architecture Group.

The goal of the BERMAN project is to develop a model for managing distributed applications in an open broadband ISDN environment and to provide the *BERKOM Administration Infrastructure* (BAI). The BAI consists of the *BERKOM Directory Service* [3] and the *BERKOM Management Platform* [4]. The core of this platform is the *Basic Management Support System* (BMSS) whose basic task is to enable the exchange of management information and operations between open systems.

During the development of the BMSS it was recognised that the definition of management information is a rather tedious and error prone task because numerous conditions and rules defined in the *Guidelines for the Definition of Managed Objects* (GDMO [5]) have to be taken into consideration.

A tool which provides comfortable editing functions and which partially or completely automatizes the checking of schema definitions is therefore highly desirable. Also, the retrieval of existing definitions using a GDMO editing and browsing tool is much faster than the manual lookup of the respective definitions in the standard documents. In the BERKOM Management Platform editing and browsing support functionality is provided by a tool called DAMOCLES. It supports developers in the administration and definition of Managed Object Classes. Moreover, DAMOCLES provides an automatic check for all consistency rules which have been defined in the GDMO.

The development of DAMOCLES is still continuing. Among the future development goals is the automatic generation of code for the implementation of *Managed Objects*. Due to its characteristics, DAMOCLES can be seen as a major contribution to the development of an environment which supports the implementation of OSI-based management systems. Even in its current development stage, the use of DAMOCLES facilitates the development of management applications by accelerating the definition and administration of MO Classes significantly.

Section 2 of this paper outlines the main steps of system development from the information viewpoint. Section 3 presents the features of the DAMOCLES tool according to the requirements outlined in Section 2. Finally, Section 4 summarises how DAMOCLES is to be further developed within the context of the overall development of the BERKOM Management Platform.

2. The Information Modeling process

In general the development of an application can be divided into three basic steps that should always be present independently from the used modeling technique. The first step should be a requirement analysis which describes the given problem and identifies functional requirements which the application must meet. Subsequently a specification has to be made in which the structure and behaviour of the application is described using formal description methods. In the third step the application is implemented.

Seen from the management information viewpoint, the development process can be described as follows (see Figure 1):

- In the first step the management aspects of an application have to be identified. In particular the developer should analyse which management information is required for the identified purposes, which operations have to be performed by management, and which events are relevant from the management perspective. As management information modeling is based on object-oriented design principles, an important issue within this step is to check whether any existing definitions can be reused, either directly or by means of inheritance and specialisation. Further it is recommend to reuse definitions from the OSI standards in order to gain interoperability with other management systems. The resulting requirements catalogue should contain an informal description of the management information. For this step it is useful to have a tool which provides comfortable query and browsing functions to the management schema provided by a Management Schema Library. The library should already contain MO Class definitions prescribed in the OSI standards, e.g. the *Definition of Management Information* (DMI, [6]), and all those definitions which have already been specified in prior developments.

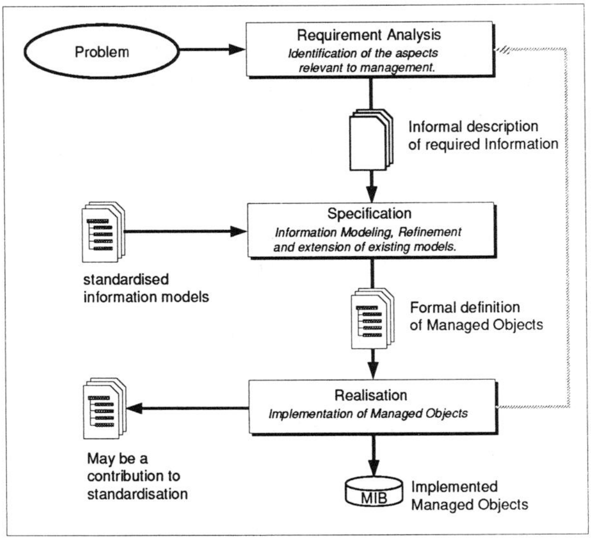

Figure 1. The Information Modeling process

- Subsequently a specification can be made based on the identified requirements. In particular the developer has to transform the informal description of the required management information into a formal definition of MO Classes. Here, it is recommended to use the structure and notation prescribed in the Guidelines for the Definition of Management Objects (GDMO, [5]). This step can be supported by a tool which simplifies the editing of definitions and provides check functions for the syntax and consistency rules which are prescribed in the GDMO.
- In the final step, the Managed Objects have to be implemented for a particular managed system and integrated into the managed application of. This could be supported by a tool which automatically produces code from the formal definitions for the implementation of Managed Objects on a particular managed system.

3. DAMOCLES

A first prototype of DAMOCLES was developed in mid 1990 in cooperation with the Technical University in Berlin, Fachgebiet für offene Kommunikationssysteme (OKS). During the last two years, the tool was further developed by BERMAN. A complete redesign was undertaken and the functionality was considerably improved [7]. The currently implemented version offers a simple line-oriented interface which is suitable for any ordinary terminal and a graphical user interface on the basis of X11R5 and OSF/Motif 1.2, which simplifies working with the tool [8]. The interface includes a MO Schema Browser, a comfortable GDMO Template Editor, and various retrieval and browsing functions.

DAMOCLES is implemented in ANSI C and runs on SUN sparc and NeXT workstations. In order to evaluate the functionality and usability of DAMOCLES, a beta test is performed.

3.1 MO Schema Browser

In principal, a new MO Class can be defined through the specialization and refinement of already existing MO Classes. The use of inheritance relations is one of the basic methods defined in the OSI Management Information Model [9]. Therefore, the MO Schema Browser was designed which gives a structured overview of all existing MO Classes and their inheritance relations.

For the presentation of the MO Class tree a simple textual list of labels is used (see Figure 2). In future versions this may be replaced by a graphic display. During the conceptual work on the browser, the problem of multiple inheritance was also taken into consideration. Since the concept of multiple inheritance allows a MO Class to derive characteristics from more than one class, the textual display of the tree may be rather complex. To increase the readability, an individual subtree is displayed for each superior class.

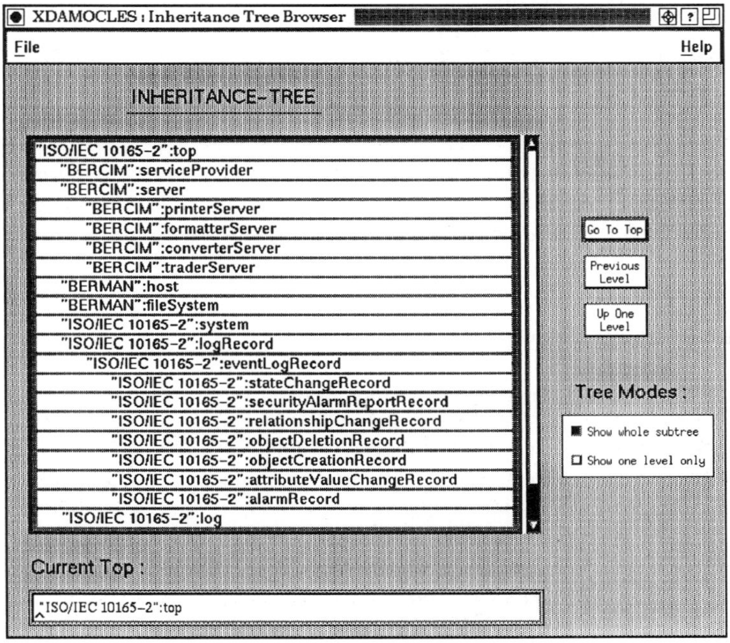

Figure 2. The MO Schema Browser

Browsing techniques for the inheritance tree

In reality, the inheritance tree may contain a large number of MO Classes. Due to the fact that screen space is limited, only a small part of the tree can be displayed on the screen at a time. Therefore it is reasonable to provide browsing techniques which allow to focus on the relevant information.

Several functions are provided to traverse through the inheritance tree. The following moves are possible:

- Return to top (*Go To Top*)
  The browser returns to the root of the inheritance tree.
- Restore the previous level (*Previous Level*)
  The browser restores the level which was previously displayed.

- Move to the superior node (*Up One Level*)
  The superior node and its subordinates are displayed.
- Move to the subordinate node
  Moving to a subordinate node is done by double-clicking on an item with the left mouse button.

Additionally two browsing modes are provided. The mode *Show whole subtree* will display all subordinates of the current top level class. Alternatively, the mode *Show one level only* will only display the direct subordinates of the current top level class.

Each class of the inheritance tree can be displayed or edited by simply double-clicking the right mouse button on the corresponding list item of the Schema Browser. The MO Class will be displayed in the Template Editor which is described in the following section.

3.2 GDMO Template Editor

The GDMO Template Editor can either be used to edit already existing definitions or to create new definitions. The structured editor offers an entry form for each definition template defined in the GDMO. The forms already contain all top level keywords (e.g. DERIVED FROM in the MO Class Template) thereby eliminating the possibility that the corresponding definition elements are forgotten or typed incorrectly (see Figure 3). Next to the keywords are the input fields for the corresponding elements of the definition. There are two different kinds of input fields which are used according to the expected amount of text for the respective definition element. With a definition element such as DERIVED FROM a list of references each separated by a comma is inserted. Simple, one-line fields are used for such parts.

In other cases, such as the CONDITIONAL PACKAGES part of an MO Class definition, a longer text spanning several lines can be expected as input. The second kind of text field is provided for such a case. Although it initially appears as one line, it can be enlarged by clicking a button next to the template (*More* button). Horizontal scrolling is allowed for both types of input fields. Vertical scrolling, however, is only possible with the second field type.

The input fields can be cleared in each template (*Clear Template*). Also, explanatory texts that provide information on the syntax of the corresponding input fields (*Reset Template*) can be added. Apart from the structure defined in the GDMO the Template Editor allows the definition of additional library information (*Show Lib-Infos*) such as the definition's author, creation date, and further comments.

The Template Editor provides a set of editing and utility functions. Already existing definitions may be deleted, modified or used as a template for creating new definitions. It is also possible to print out definitions to ASCII files and to

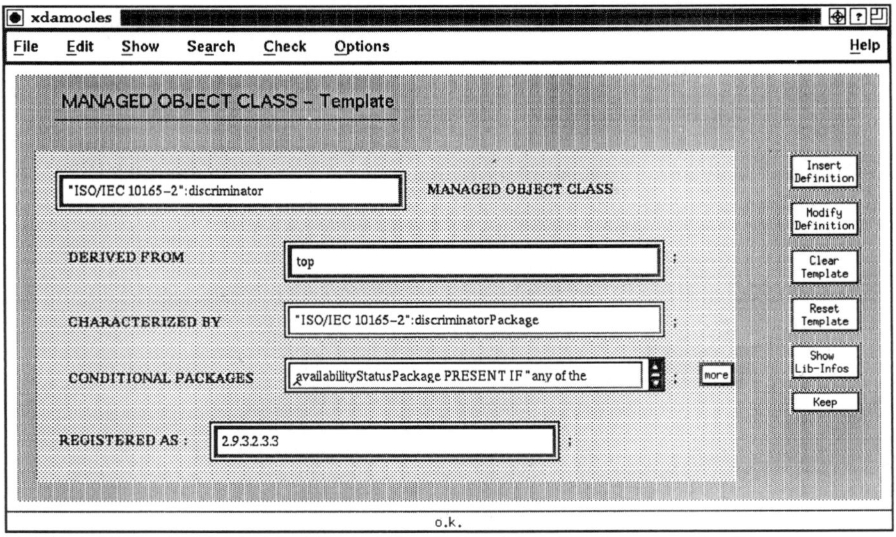

Figure 3. GDMO Template Editor showing a MO Class template

read in sets of definitions, which are stored in files. The complete notation defined in the latest version of the GDMO including the inline notation for definitions is supported by DAMOCLES.

According to the structure defined in the GDMO, an MO class definition usually consists of several partial definitions which are related by references that form a kind of usage hierarchy. If the consistency and especially the completeness of the global management schema need to be guaranteed permanently, the multiple MO Class definitions have to be inserted in a sequential order which is determined by the usage hierarchy. In order to simplify the definition in this respect, it is sensible to divide the process of defining new MO Classes into two phases (see Figure 4):

- In the first phase new definitions are first added to a *local* management schema which may be incomplete. New definitions can be entered interactively with the Template Editor, or read in from an ASCII file. In both cases, a syntax check is performed to find out whether the structure of the definitions corresponds to the notation defined in the GDMO. During the subsequent semantic analysis, the uniqueness of the specified labels and object identifiers is ensured. Additionally, a type check of the references is made. For example it is proved that a label which appears within the ATTRIBUTE clause of an MO class definition really references an attribute definition. Since the local management schema may be incomplete, i.e. the contained definitions may reference other definitions which do not exist yet, this can only be done if the referenced definition already exists.

- In the second phase schema definitions from the local schema are added to the *global* management schema represented by the Management Schema Library. However, this is only successful if all referenced definitions are now available either in the local or in the global management schema and if the reference check does not indicate any type errors. In order to ensure the consistency of the global management schema at all times, the definitions from the local management schema are inserted into the global management schema in a specific order as already mentioned. First of all, the behaviour definitions and all definitions which do not reference any other definitions are stored. Then, definitions which exclusively reference definitions already available within the global management schema are stored.

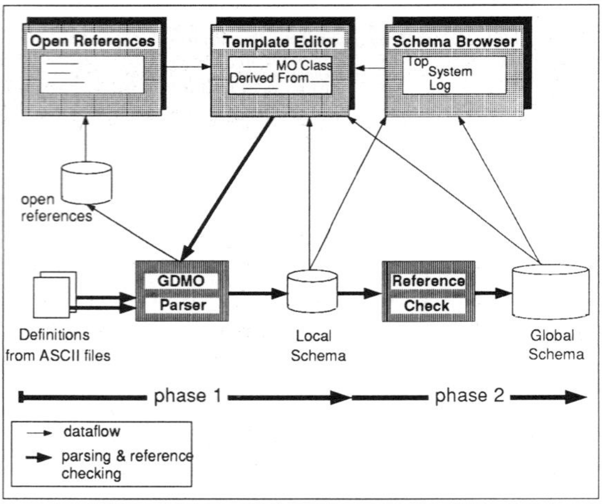

Figure 4. Logical view on the architecture

3.3 Further Retrieval and Browsing Functions

Since definitions may be incomplete in the first phase there may be references to definitions which do not already exist. These references are called *open references* and can be displayed in a separate browser window (see Figure 5). Each time a new definition is added to the local schema the list of open references will be updated. At the same time, the definition that has just been completed is removed from the list. This procedure assures that the user can easily monitor the completion of the local schema.

The list of open references may also be used to complete the local schema more quickly. By simply double-clicking an item with the left mouse button, the Template Editor is invoked with the appropriate template according to the reference's type. Also the label of the open reference is inserted into the template-label field. If the list of open references is empty, all local definitions are in a consistent state and may be added to the global schema.

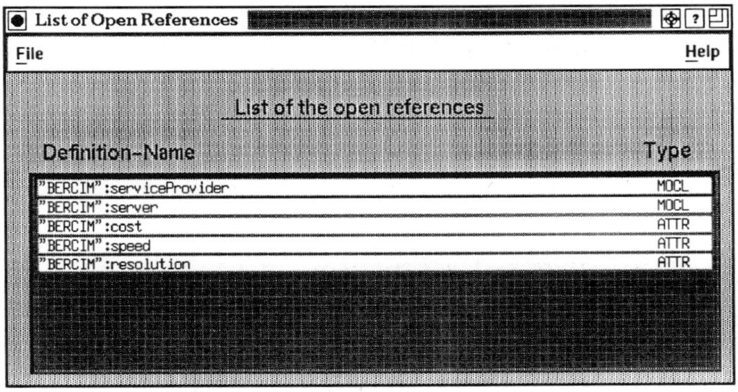

Figure 5: List of Open References

Several modes for the retrieval of existing definitions in the local or global schema are provided. A user can search for a definition by simply double-clicking on a label or OID contained in any text field. Another possibility is to invoke a browser with a list of all local or global definitions. Each definition can be easily displayed in the Template Editor by double clicking on the corresponding item in the list.

Further search dialogs are provided which allow queries for definitions by entering their labels or OID and by selecting a search scope. The search scope specifies wether the query should be performed on local (global) definitions only, or on all available definitions. Another retrieval function searches for all the superior classes (i.e. the inheritance relations to classes) of a specified MO Class.

4. Future Developments

In the context of the BERMAN project, DAMOCLES is to evolve into a user-friendly development environment for the BERKOM Management Platform. The goal of these developments is to reduce the costs of implementing management systems and the provision of an extensive MO Class Library for standardised (e.g. ISO, CCITT) and private MO Class definitions.

Besides interactive access to the management schema, a programming interface (API) will be provided. This API is to allow components of the Management Platform to query information from the schema. Such a behaviour may be sensible, for example, in order to check the validity of management operation calls. The following functions have already been identified.

- Component Identification
  This function will check whether the specified component, i.e. attribute, action or notification, is defined for the specified MO Class. Thereby the corresponding MO Class, as well as all superior MO Classes are retrieved.
- MO Operation Availability Check
  This function will examine whether the specified operation, i.e. CREATE, DELETE, can be performed on an MO of a particular MO Class.
- Attribute Operation Validity Check
  This function will examine whether the specified operation, i.e. ADD, REPLACE, can be performed on a particular attribute.
- Name Binding Retrieval
  By applying this function, the name for the instance of an MO Class (defined by a Name Binding rule) will be retrieved.

In a further step it is planned to develop a *GDMO compiler* which will automatically produce code from the management schema for C++ based Management Information Bases (MIB). This code may be used as a skeleton for the implementation of specific MOs. It will also include data structures and the associated encoding and decoding functions for parameters and attributes. In order to gain the amount and efficiency of code which can be produced by such a compiler it is studied how the specification of MOs can be extended by constructs describing implementation aspects of storage and distribution which are not considered in the Management Information Model. For example constructs to describe the binding of MOs with the real resources and the distribution of MOs will be provided which are based on the concepts of object fragments and mapping types [10, 11]. Another issue which is currently studied deals with question how parts of the object's behaviour can be described in a formal manner. However, the extent to which automatic code generation is possible for specific MOs must be still studied in more detail.

Currently, DAMOCLES is a single-user tool. An important work item is the development of a "*Distributed DAMOCLES*" which allows multiple users to work simultaneously with DAMOCLES. This will provide access to standardised and other public schema knowledge for users at different locations all over the world. Beyond that an issue studied is whether MO schema definitions could be published globally or in a closed user group using the standardised *Directory System* [12]. The basic idea is to exploit the global uniqueness of Directory names and store schema information in a globally accessible subtree of the *Directory Information Tree* (DIT). Instead of accessing uncoordinated, geographically distributed databases, both, schema definition tools and administration tools can use the information stored in the global accessible Directory. To what extent the Directory can provide the means to coordinate such global definitions will be subject of a planned feasibility study.

In addition to the work items mentioned above the functionality of the user interfaces is to be considerably improved in order to further simplify the use of DAMOCLES. The work in this area will be based upon the results of the ongoing beta test. For example, it is planned to extend the Template Editor so that the production of more complex parts of a definition (as, for example, the "Conditional Packages" clause of the Managed Object class) is supported by individual forms.

Acknowledgement

The authors would like to thank S. Krause and all members of the Open Management Architecture Group at GMD FOKUS for valuable discussions and improvements.

References

[1] H. Ricke and J. Kanzow, editors. *BERKOM. Broadband communication within the optical fibre network.* R.v.Decker's Verlag, Heidelberg, 1992.

[2] R. Popescu-Zeletin, B. Butscher, P. Egloff, and J. Kanzow. "A Global Architecture for Broadband Communication Systems: The BERKOM Approach". *in Proceedings of the Workshop on Future Trends of Distributed Computing in the 1990's*, pages 366–373. Hong Kong, IEEE Computer Society Press., September 1988.

[3] BERMAN. "Guide to the BERKOM Directory". *Part 1 of the fourth BERMAN deliverable*, December 1991. Version 2.0.

[4] BERMAN. "Guide to the BERKOM Management Platform". *Part 2 of the fourth BERMAN deliverable*, December 1991. Version 2.0.

[5] ISO/IS 10165-4. "Information processing systems — Open Systems Interconnection (OSI) — Structure of Management Information". *Part 4: Guidelines for the Definition of Managed Objects*, Mai 1992.

[6] ISO/IS 10165-2. "Information processing systems — Open Systems Interconnection (OSI) — Structure of Management Information". *Part 2: Definition of Management Information*, August 1991.

[7] M. Wittig. *Ein Werkzeug zur konsistenten Definition und Verwaltung von Management-Schemainformationen.* Technische Universität Berlin, 1991. Studienarbeit.

[8] Martin Pfeiler. *Gestaltung und Erstellung einer graphischen Benutzeroberfläche zur Definition und Darstellung von Managed Object Klassen.* Technische Universität Berlin, 1991. Studienarbeit.

[9] ISO/IS 10165-1. “Information processing systems — Open Systems Interconnection (OSI) — Structure of Management Information”. *Part 1: Management Information Model*, November 1991.

[10] S. Mark Klerer and Roberta S. Cohen. “Distribution of Managed Object Fragments and Managed Object Replication: The Data Distribution View of Management Information”. *Proceedings of IFIP 6.6 - 2nd International Symposium on Integrated Network Management*, pages 763–774. North-Holland, ISBN 0 444 89028 9, April 1991.

[11] H.-G. Hegering, S. Abeck, and Th. Boehnke. “Converting MIB Descriptions into MIB Implementations”. *Proceedings of IFIP 6.6 Workshop on Distributed Distributed Systems: Operations and Management, Santa Barbara*, October 1991.

[12] ISO 9594. “Information processing systems – Open Systems Interconnection (OSI) – The Directory”. 1989.

Integrated Network Management, III (C-12)
H.-G. Hegering and Y. Yemini (Editors)
Elsevier Science Publishers B.V. (North-Holland)

Deploying Group Communication Techniques in Network Management

Robert Valta
Leibniz-Rechenzentrum, Barer Strasse 21, 8000 Muenchen 2, Germany, Tel. +49 89 2105-7484.
Email: valta@lrz.lrz-muenchen.dbp.de

Riaan de Jager[1]
Department Computer Science, University of Pretoria, P.O. Box 0083 Hillcrest, South Africa.
Email: jager@lrz.lrz-muenchen.dbp.de

Abstract

In this paper we propose to employ Computer Supported Cooperative Work (CSCW) techniques in network management. A study conducted at the Leibniz Supercomputer Center revealed that many of our current network management problems are caused by communication or coordination problems within the provider staff. A trouble ticket system is used as an example for describing the use of CSCW. We also discuss an approach to integrating CSCW-techniques in a network management platform and outline the current status of our work.

1 Introduction

Network management relies not only on tasks performed by individual persons, but also on how such tasks influence each other. Groups of persons usually work together to perform a task, such as recognizing problems and repairing faults. Based on a study done within the Leibniz Supercomputer Center (LRZ) we determine the requirements for tools which could support group work within a network management environment. In Section 2 we give a short introduction to groupware. Section 3 demonstrates how CSCW techniques could be employed in network management with a trouble ticket system as an example. In Section 4 we propose a method of how CSCW techniques could be integrated into a network management environment. Finally, in Section 5 we describe the status of our work and outline future directions.

2 Aspects of Groupwork in Network Management

2.1 Network Management problems in the LRZ

The Leibniz Supercomputing Center provides computing and networking services for several universities and research institutes in the south-eastern part of Germany. Currently central computing power includes a CRAY YMP, parallel computers, CDC mainframes, workstation

[1] This work was done as a guest researcher at the Leibniz-Rechenzentrum

farms, PC networks and a variety of I/O-devices. Furthermore it runs a large heterogenous communication network mainly based on Ethernet with a variety of protocol stacks above it and provides access to networks and services outside the universities.

Management of the network is impeded by various factors:

- User organizations (departments, labs, etc.) are autonomous. Users decide which computer systems they purchase and connect to the network. This leads to a wide variety of different computer equipment throughout the network;
- The network is expanding rapidly in size and complexity;
- The number of people which work in the networking department of the LRZ have not increased sufficiently;
- The way the network is used changed. While in former times it was mainly used to access central computers it now forms the basis for a large distributed client-server-environment.

To cope with such factors it is a necessity for the provider organization to adapt and improve its organizational structuring and processes, which lead to an analysis of infomation flow between staff members being conducted in the LRZ. This was performed by use of Veriks, a PC-based information flow analysis tool. Staff members were requested to specify the information necessary for their work, where they obtain such information and to whom they supply information. This analysis revealed that for only 10 % of all communication computer-based tools are used, such as electronic mail and Unix-talk. Most inter-personal communication is done using traditional non-computer-based techniques like group meetings, notes, personal interactions etc. This causes several drawbacks:

- People spend a lot of time in meetings. Meetings are often used to exchange experience or to plan and coordinate future activities;
- People spend a lot of time in searching partners they want to communicate with;
- People are often disturbed in their work by others who want to communicate with them, especially experts feel interrupted a lot of times by persons bothering them with bagatelles;
- Information exchange is not documented. A lot of information gets lost or is not accessible for someone who did not take part in a meeting or conversation.

Another result from the analysis shows many divergences between the real and the planned flow of information within the provider organization. For example, users tend to contact experts directly when they have networking problems instead of informing the network operator. The consequence is that experts are interrupted in their work by users with minor problems which could be solved by non-experts. Furthermore, information flow bypasses non-experts and prevents them from learning about user problems and how to solve them. The analysis therefore pointed out the importance of providing coordination and support for communication between staff members.

2.2 Requirements

An increasing number of management activities require several persons from different groups to work together, for instance interoperability problems between components of different vendors, like when an IP-Router is connected to a X.25-switch. Another example is the introduction of new network technology, such as a high-speed network in the computing center, where networking experts and host experts must work together.

The use of a computer-based tool which provides for the specification of tasks and the coordination of tasks influencing each other could assist staff members in performing their work. Such a tool should also support a description of the structure of the organization along with the responsibilities of persons. A further requirement would be the definition of groups, for instance static groups which are established for a longer period of time or dynamic groups created by a staff member to obtain help or advice on an urgent problem. Different communication patterns must also be supported, such as discussion on how to diagnose complex errors or to improve network performance. Coordination of management activities must be supported in order to prevent duplication of work.

When people communicate or work together, many exceptions can occur which must be handled in an adequate way by a groupware system. People can be busy or absent, or they can fail to obey predefined rules.

An important requirement is the integration of this tool into the network management environment.

Although work is made easier for single experts by advanced tools such as expert systems, cable management systems and trouble ticket systems, experience within the LRZ shows that introducing such tools requires a lot of organizational work. There will still be a need for coordination of expert work to manage the whole complex network environment.

3 General Groupware Concepts

Technological growth in the area of electronic communication offers possibilities for people to interact in a variety of different and new ways. This gives rise to the emergence of systems which incorporate information processing as well as mechanisms to support activities of communication between people. This field of study is referred to as Computer Supported Cooperative Work (CSCW). This discipline tries to find and define ways in which groups of people communicate within organizations. It addresses not only the issue of creating new technology to accomplish this, but also social aspects that need to be considered when involving groups of people in a specific task ([1]). We are currently exploring this field to find concepts and ideas for group communication in network management, and to integrate them into the LRZ environment.

3.1 Example Systems

An easy way to explain the issues involving Groupware is probably to describe some types of applications which are considered to be Groupware or CSCW applications. These include systems such as the well known electronic mail to more advanced ones which attempt to assist office teams in coordinating their activities ([1]).

- Message Systems - These are systems which provide users with the capability to exchange electronic mail messages. The method of communication used is asynchronous. Systems have become available which supports the user in automatically forwarding messages, filing messages and deleting so-called "junkmail" by acting upon the contents of the messages.
- Multiuser Editors - These applications allow users in a group to simultaneously create and edit a document. In some cases, the document is divided into parts which only one user can edit at a time. The other users are allowed to view what is being done - in real-time.

- Group Hypertext Systems - One example is MCC's rIBIS [8]. This is a Hypertext system which provides users with the possibility to work on the same software system design. rIBIS supports both loosely and tightly-coupled work modes. In the loosely-coupled mode, all of the actions that one user takes are not immediately visible to other users. In the tightly-coupled mode, all of the users joining in a session see the actions that others are taking in real-time. One person is in control of the session, and the others follow what he/she is doing.

Other examples include group decision support systems, computer conferencing systems and intelligent systems which assist users in taking steps and performing group decisions. Most systems incorporate one or more of the following concepts: Coordination, Communication or Collaboration. These three concepts are thus considered to be the key areas on which groupware systems are based. Our main focus is on concepts used in coordination systems, as we will describe later.

3.2 Coordination Systems

Coordinating work people are doing to accomplish a common goal means bringing together the work of each individual and making it available to the rest of the group. These systems typically allow one user to perform actions which are then made known to the rest of the group, as well as viewing the actions taken by other group members. These systems might also inform group members of actions which are required of them and of certain conditions occuring in the work process. Different coordination approaches exist:

- Form-oriented - An example of this type of model is the "Electronic Circulation Folders (ECF)" used by the ProMInanD system [6]. In this system, folders which contain documents, notes and task descriptions migrate automatically through an organization. Migration specifications describe the steps that have to be taken on a circulation folder, as well as the people who are allowed to take these steps or influence the flow of the ECF. This migration specification makes provision for exceptions, for example when a person is not available to perform a step, some alternatives should be available for the ECF to continue it's migration and for the task to be completed;
- Conversation-oriented - This type of model is based on the fact that people inherently coordinate their activities by communication. The Coordinator ([2]) is one system which is based on this theory. It takes the focus that people use language to perform actions. This suggests, for example, that a person makes a request for a certain action. The next legal move is that another person makes a promise to perform this action, or rejects the request. Different actions can be taken by different people at different stages, resulting in the initial request being performed or rejected at some point;
- Communication structure-oriented - This model describes the activities within an organization by describing the roles that people play. Diplans [4] is a graphical language for describing and expressing plans. A plan, in the sense of coordinating activities, is seen as an "organization machine" which describes the activities being undertaken and how people take part in them.

Groupware systems in general are not necessarily based on one of these models alone. The ProMInanD system for example incorporates a communication-structure oriented approach to describe the migration of electronic circulation folders.

4 Employing Groupware concepts in Fault Management

Our opinion drawn from our work so far is that the area of fault management exhibits some properties very closely related to the problems found in environments where groupware concepts are employed. Our research is therefore concentrated on this area. In the event of an error ocurring on the network or an end system, certain steps must usually be taken by different people to handle it. Coordination of such activities and providing a means of communication is of utmost importance for effective fault-handling procedures. Our current work includes the development of a trouble ticket system incorporating CSCW-concepts. The development of the system is also aimed at creating building blocks that can be used in further systems aimed at assisting communication and coordination in a management environment.

4.1 Trouble Ticket Example

Trouble ticket systems have become an important tool to assist network operations centres (NOC) in the handling of problems. They do not only provide a way of documenting errors but also provide a limited basis for communication between the provider staff. A trouble ticket acts as "short-term memory" on a problem which means that it is used to make information on a problem available in the NOC ([5]). Our current development is based on the form-oriented approach of coordinating office tasks used in the ProMInanD system ([6]). The concepts used in this approach are in our opinion best suited for the trouble ticket area.

A trouble ticket represents the nature of, and work done to repair, an error. Different people perform steps or sub-tasks in the process of handling a problem. In ProMinanD, different people are required to perform steps on an electronic circulation folder (ECF) containing documents related to the office task . These steps are part of a migration path of a formalized task.

As in ProMInanD, a step can be part of this predefined migration path, or it can be created in real-time. The ProMInanD system is however targeted towards the office environment. Migration specifications for office tasks are generally quite fixed, but can be extended with newly required steps if necessary. In the diagnosis and repair of network and end system problems, many actions are taken which can not be predicted. Enhancement of the migration specification by inserting new steps as needed is thus a very important requirement.

The ability to cope with exceptions, such as people not being available to perform certain steps, is very important in the fault-handling environment. ProMInanD provides for such unforseen circumstances by extensive exception handling mechanisms. Substitutes can be defined for people playing certain roles, the predefined circulation of work can be altered and steps can be cancelled as a result of errors made by people.

4.2 Description of Organizational Procedure

A migration path is predefined by a migration specification. It includes the sequence of steps that should be taken to perform a task. Figure 1 shows the migration specification of a trouble ticket.

Every trouble ticket that is created inherits this migration specification. It specifies the possible route that a trouble ticket takes in terms of steps or sub-tasks. A step defines an action that is

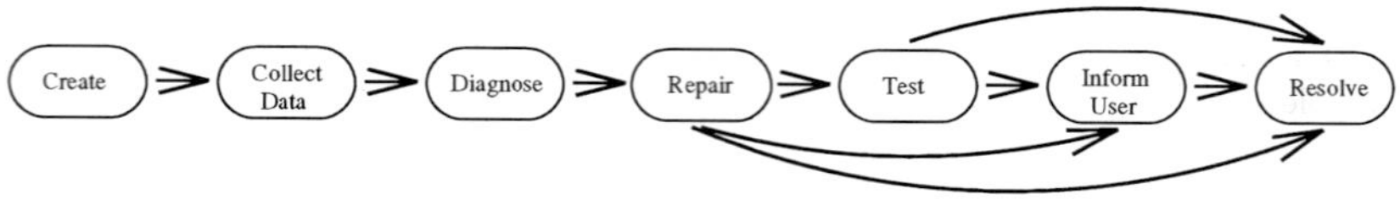

Figure 1: Trouble Ticket migration

required to be performed by a person playing a certain role in the organization. A sub-task is a complex step that is internally broken down into several other steps. The `Create` step causes a new trouble ticket to be created as result of a reported problem. This can involve a new record being created in a database, or a file which can be used to store the trouble ticket data. The means of storing this data is not considered important here. Information related to service user details is gathered in the `Collect-Data` step.

The staff member performing the first two steps usually does some initial diagnosis of the problem by initiating the `Diagnose` sub-task. A relevant expert person must now take responsibility for the ticket, depending on the nature of the problem (e.g. hardware, X.25, or IP routing), and commence work on diagnosis. After the problem has been diagnosed, the `Repair` step is requested from another staff member, containing a reference to the actual fault causing the problem. The outcome of the `Repair` step can be tested through the `Test` step, and the user can be informed through initiating `Inform-User`. The `Resolve` step involves confirmation by the person who accepted initial responsibility for the problem for verification of repair.

More than one staff member is usually involved in diagnosis. The `Diagnosis` sub-task can be broken down into an arbitrary number of `Action-Request` and `Dispatch` steps. These steps are defined dynamically by staff members as required, and are not visible to staff members not involved with diagnosis. Possibilities are:

- The initial responsible person wants to defer responsibility to someone else by requesting a `Dispatch` step. This involves inserting such a step in the migration specification;
- The responsible person needs to have a diagnosis action performed by some other staff member. An `Action` step can be inserted for this. In this case, the ticket will automatically be returned to the person requesting the action by a `Return` step being inserted automatically. This will be the next step following `Action`. Multiple actions can be requested similtaneously. In ProMInanD, if more than one step is to be taken in parallel, a so-called *dependent ECF* is started. In this light, a dependent trouble ticket is started, containing the relevant information required in the step. A mechanism for defining the relationship between such dependent trouble tickets and their "superior" is responsible for exercising control in such a case.

The `Test` and `Inform-User` steps need only be performed if indicated so in the `Repair` step. These steps can be skipped by taking the alternative routes as indicated in Figure 1.

4.3 Description of Organizational Structure

In ProMInanD, office workers are addressed by the function they perform in the organization, or their role. People in the organizational hierarchy are assigned such roles.

As mentioned earlier, steps on a trouble ticket are performed by people playing certain roles in the provider organization. A role in the context of fault management defines an expert area. Each role has an expert who is the person who is responsible for this area under normal circumstances. In [9], a proposal is made of the roles that people play in network management. Substitutes for an expert can be defined additionally to provide for exceptional cases where the expert is not available. Various degrees of substitutes can be defined for a role to make such a mechanism more effective. An organizational description of roles in terms of people needs to be available for resolving a person necessary to perform a step.

4.4 Contents and State of a Trouble Ticket

In office tasks, the contents of an ECF has no influence on migration. Migration specifications are altered by people. In the trouble ticket environment, the contents and state of a ticket affects migration. Figure 2 shows the state machine of a trouble ticket. At any given time, the state of a trouble ticket indicates the current status of work done on the problem. A ticket for instance being in the `Reported` state indicates that no action has been taken to diagnose a problem, however static information on the problem has been collected. A ticket in the `External-Wait` state indicates that the problem can not be repaired because of an external action required, e.g. a new component is required from a vendor.

The dashed arrows in the Figure indicate where steps forming part of the migration specification affect the state machine. Some of these steps affect the state of a ticket. The `Diagnose` step will for instance cause the state of the ticket to change to `Working`.

Initiation of the `Repair` step causes the ticket to change to the `Diagnosed` state. This step will only be allowed if information entered during the `Diagnose` step was sufficient to consider the diagnosis completed. This is controlled by checking whether the fault causing a problem has been identified, which is part of the information contained in the ticket.

The state machine also determines which steps can be taken, or which types of steps can be inserted into the migration specification at certain points during migration. Therefore, it is not possible for instance to perform the `Repair` step if `Diagnose` has not been done. It is also not possible to insert a `Request-Action` or `Dispatch` step into the migration specification of a trouble ticket which is not yet in the `Working` state (i.e. the `Diagnose` step has not yet been initiated). Similar control is exercised by the state machine on tickets in the `External-Wait` state. All external actions, as described earlier, have to be indicated as completed before a next step is possible (`Test` or `Resolve`).

4.5 Exceptions

Exceptions are operations done on, or conditions occuring in a trouble ticket which deviate from the normal predefined and inserted steps. They affect either the migration or the state of a trouble ticket.

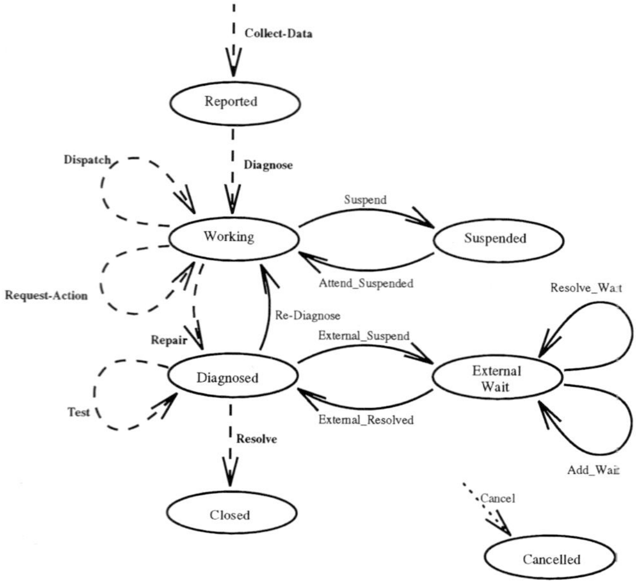

Figure 2: Trouble ticket state machine

Provider Staff induced Exceptions

Some exceptions can occur as a result of operations performed on a trouble ticket. These are operations done by staff members which do not form part of the migration specification.

A staff member can perform a **Cancel-Step** operation to cancel all work done in the current step. All work done in this step then has to be deleted.

It may happen that a staff member does not want to perform the current requested step. The **Not-me** operation can be performed to indicate this reaction. A valid reason has to be supplied, whereafter the trouble ticket is taken back into the previous step to inform the person who requested the step.

The **Refer-back** operation can be performed to indicate that more information is required before the step can be commenced. The ticket is taken back into the previous step with an indication of this requirement. It is important to note how these actions can affect the state machine of the ticket. If the step requested was `Repair`, the ticket would have moved into the `Diagnosed` state. It is now necessary to move it back into the `Working` state.

A staff member can initiate an **Inform** operation to obtain information on the current whereabouts and status of work on a ticket. A requested step can be cancelled by the **Fetch-back** operation. The current step must automatically be cancelled which means that all the work done in this step must also be cancelled.

A ticket can be cancelled completely by issuing the **Cancel** operation. If the problem is in diagnosis and dependent trouble tickets have been created, they must also be cancelled automatically by the system to suspend all work being done on the problem. There are no constraints imposed by the state machine on this operation, meaning that it can be taken at any time during migration, as shown in Figure 2.

In the event of a problem being considered irrepairable, the **Terminate** operation can be applied. In this case, all work must be terminated and the ticket must be archived.

Automatic Exception-Handling

An expert might not be available, or take too long to commence a requested step, resulting in an exception. In this case, a substitute should be evaluated to whom the step can be assigned. If no such substitute is found, the person requesting the step should be informed.

Other exceptions may occur as a result of conditions monitored by the trouble ticket state machine. These arise mainly as a result of time constraints. In Figure 2, the Suspend and External_Suspend operations are also considered exceptions on the normal migration of a trouble ticket. A time limit should be specified by the person responsible for performing this operation to indicate how long the ticket may be in either the `Suspend` or `External-Wait` states. If such a time limit is exceeded, the staff member responsible for initiating the operation should be notified, or a substitute should be searched for in case this staff member is unavailable.

5 Integration with Management Platforms

We now turn to an explanation of how group communication techniques can be integrated into a network management platform, namely Hewlett Packard's Openview product which is in use at the LRZ. Management platforms provide a set of basic management functions which includes a uniform user interface, event management services, data management services, communication services, applications for managing components and an Application Programming Interface (API) for adding new functions to the platform. In the following we discuss two methods of how a groupware system could be integrated in such a platform.

5.1 User Interface Integration

The easiest way to integrate a groupware system into a management platform is to make it accessible through the user interface of the platform (Figure 3), for example by adding correspondent entries to the menu structure.

Such an integration makes sense if a groupware system already exists and is in use within the organization, e.g., within the office environment. However, there are some drawbacks with such an approach:

- The groupware system itself is an independent subsystem with its own internal user interface which does not match design principles and guidelines of the management platforms user interface.
- The groupware system maintains its own database. However, some of its information is also

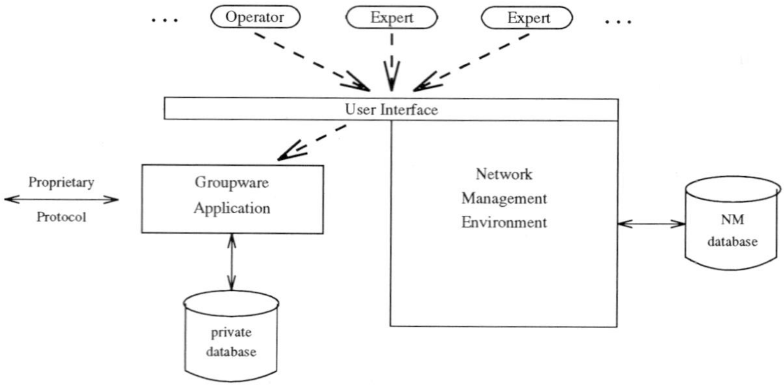

Figure 3: User Interface Integration

part of the platform's database, like personal or organizational information. As a result, two overlapping databases must be maintained with possible inconsistencies between them.

- The groupware system uses its own communication mechanisms, for example an enhanced electronic mail service.
- Management activities specified within the groupware system can not be used as a direct input to the management system. Thus, later automization of activities and steps are made more difficult.

To overcome these drawbacks it is necessary to implement the functions of the groupware system by using the services of the management platform.

5.2 Full Integration

To fully integrate a groupware application into the network management environment, three basic steps have to be done. First of all, object classes have to be defined which represent groupware-related information within the network management environment. For our prototype implementation we defined the following managed objects classes (**MOC**):

- `Step` managed object class which models the details of a step in terms of the required actions, who is responsible for the step (in terms of roles described earlier) and what criteria should be satisfied to accept the step as successfully completed. `Step` managed objects will in most cases be created dynamically at request of an application, although predefined steps are also possible;
- `Person` managed object class which describes a person involved in the organizational environment by identifying such a person and describing the possible roles that such a person can fulfil and to what a degree;
- `Role` managed object class which models roles as described earlier and encompasses the categorization of specialist areas with the people able to perform them;

As a second step, an object manager has to be implemented which is responsible for managing the object instances and which implements coordination and communication procedures. The object manager maintains the mapping of roles to persons, the evaluation of steps, the notification of staff members etc. For storing data it uses the data management services of the network management environment. As a third step, an application has to be implemented which allows users to participate in the groupware system. Every user who wants to do this has to start such an application. All communication between application and object manager is realized using the CMIS-like API of the network management environment. Within this environment, a so-called postmaster is responsible for routing all CMIS-primitives to their destination ([3]).

The following example shows how applications and the object manager work together. We discuss the simple case where an expert `A` requests an action, e.g. diagnosing a particular network component, to be performed by someone else. (The interactions taking place are indicated by the numbers in the figure.)

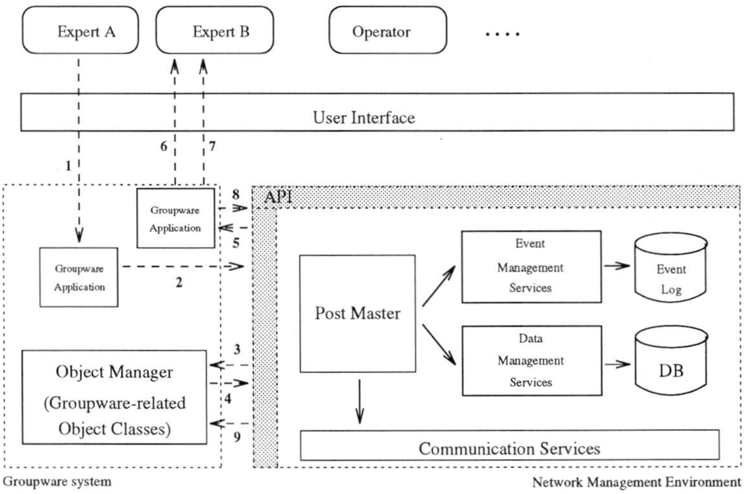

Figure 4: Integrated solution

- An expert specifies that a step of type `action` is required from some other staff member in role `role`. He specifies the role which has to perform this step, the action which is required to be performed by this role and some criteria which defines when the step can be considered successful from the his point of view (**1**).
- The group application processes this request by issueing a M_CREATE operation to create a new `step` object (**2**). This M_CREATE is forwarded to the object manager (**3**).
- The object manager creates the object and maps the step's role to a person (Expert `B` in our example.
- After that, the object manager generates an event - which is part of the definition of the MOC `step` - calling the confirmed M_EVENT service. The event data includes the name

of the person from whom the action is requested and information describing the action (4).
- The event is then delivered to each application which is registered to it by creating an appropriate event discrimator (or event sieve object as it is called in HP OpenView terms). To make sure that this event arrives at the correct person, each groupware application process must register for step events which contain the caller's userid in the event data.
- The application receives the event (5) and indicates the receipt of a step-request event to the user (6). The user (Expert B) decides to accept the step-request (7) and starts working on it. The application confirms the event (8).
- The object manager receives the confirmation (9) and assumes that Expert B is commencing the step.

The above example shows an ideal flow of information. However, many exceptions can occur which must be handled by the object manager. The expert might for instance not be available or refuse to do the step-request. The object manager will notice that as no event confirmation comes along and will start exception handling (e.g., to find a substitute who fits the role).

Our first attempt at implementing a groupware system based on the HP OpenView network management platform is currently under way. As a first step, we concentrate on implementing a basic communication mechanism as described in the above example. When this is finished we will enhance the object manager by adding new functions like managing sequences of steps or recognizing termination of steps. Another important function will be to allow management of groups of roles. For this purpose, domain concepts have to be developed similar to those which are used in [7] to describe delegation of authority within network management. Furthermore a mail-based user interface will be added to allow users to participate in the groupware system without having to use the interface of the management platform.

6 Conclusion

We have shown that many network management problems are caused by communication and coordination problems within the network provider organization. We suggest to solve such problems by employing CSCW techniques to the network management field. As an example we selected a trouble ticket system which is currently under development in the LRZ and demonstrated how such a system could be enhanced using CSCW-techniques. However, our approach is not restricted to the area of fault management but can be easily applied to other functional areas like configuration management or network planning which also require cooperation between people.

Aspects we discussed in this paper seem very important to us but are not sufficiently dealt with in management architectures, products, or research work. We intend to make the integration of CSCW-techniques one of our future research topics in collaboration with a research group at the Technical University of Munich which works on Distributed Applications and CSCW. Our future work will include implementing a prototype as described in this paper which will serve as a basis to study acceptance of groupware concepts by network management staff. Application of the CSCW concepts in other functional areas, especially configuration and change management, is planned. We are currently analyzing other techniques from CSCW to see whether they can be employed in network management, such as the use of hypertext systems to store management knowledge or to support discussion groups.

7 Acknowledgement

We gratefully thank the staff of the LRZ for supporting our efforts to determine how they communicate and cooperate. We also wish to thank the members of the Munich Network Management (MNM) Team for helpful discussions and valuable comments on previous versions of the paper. The MNM Team is a group of researchers from the Munich Universities and the Bavarian Academy of Science. It is directed by Prof. Hegering. We thank Prof. Schlichter from the Department of Computer Science at the Technical University of Munich for an introduction into CSCW and valuable hints on how to employ concepts in the field of network management.

References

[1] C.A. Ellis, S.J. Gibbs, and G.L. Rein. Groupware - some issues and experiences. *Communications of the ACM*, 34(1):39 – 58, January 1988.

[2] Fernando Flores, Michael Graves, Brad Hartfield, and Terry Winograd. Computer systems and the design of organizational interaction. *ACM Transactions on Office Information Systems*, 6(2):153 – 172, April 1988.

[3] Hewlett Packard. *HP Openview DM Platform Programmer's Guide*, 1992.

[4] Anatol W. Holt. Diplans: A new language for the study and implementation of coordination. *ACM Transactions on Office Information Systems*, 6(2):109 – 125, April 1988.

[5] Internet Activities Board. *NOC Internal Trouble Ticket System Functional Specification Wishlist ("NOC TT REQUIREMENTS")*, 1992.

[6] Bernard H. Karbe and Norbert G. Ramsperger. Influence of exception handling on the support of cooperative office work. In A.A. Verrijn-Stuart S. Gibbs, editor, *Proceedings of the Conference on Multi-User Interfaces and Applications*, pages 355 – 370, Crete,Greece, September 1990. Elsevier Science Publishers B.V. (North Holland).

[7] Jonathan D. Moffet and Morris S. Sloman. Delegation of authority. In W. Zimmer I. Krishnan, editor, *Proceedings of the IFIP TC6/WG6.6 Second International Symposium on Integrated Network Management*, pages 595 – 606, Wasington D.C., U.S.A., April 1991. North-Holland.

[8] Gail L. Rein and Clarence A. Ellis. ribis: a real-time group hypertext system. *International Journal on Man-Machine studies*, 34:349 – 367, 1991.

[9] Kornel Terplan. *Communication Networks Management*. Prentice-Hall, 2nd edition, 1992.

AUTHOR INDEX

IFIP

The INTERNATIONAL FEDERATION FOR INFORMATION PROCESSING is a multinational federation of professional and technical organisations (or national groupings of such organisations) concerned with information processing. From any one country, only one such organisation – which must be representative of the national activities in the field of information processing – can be admitted as a Full Member. In addition a regional group of developing countries can be admitted as a Full Member. On 1 October 1992, 45 organisations were Full Members of the Federation, representing 62 countries.

The aims of IFIP are to promote information science and technology by:

- fostering international co-operation in the field of information processing;
- stimulating research, development and the application of information processing in science and human activity;
- furthering the dissemination and exchange of information about the subject;
- encouraging education in information processing.

IFIP is dedicated to improving worldwide communication and increased understanding among practitioners of all nations about the role information processing can play in all walks of life.

Information technology is a potent instrument in today's world, affecting people in everything from their education and work to their leisure and in their homes. It is a powerful tool in science and engineering, in commerce and industry, in education and adminstration. It is truly international in its scope and offers a significant opportunity for developing countries. IFIP helps to bring together workers at the leading edge of the technology to share their knowledge and experience, and acts as a catalyst to advance the state of the art.

IFIP came into official existence in January, 1960. It was established to meet a need identified at the first International Conference on Information Processing which was held in Paris in June, 1959, under the sponsorship of UNESCO.

Organisational Structure

The Federation is governed by a GENERAL ASSEMBLY, which meets once every year and consists of one representative from each Member organisation. The General Assembly decides on all important matters, such as general policy, the programme of activities, admissions, elections and budget.

IFIP ORGANISATION CHART

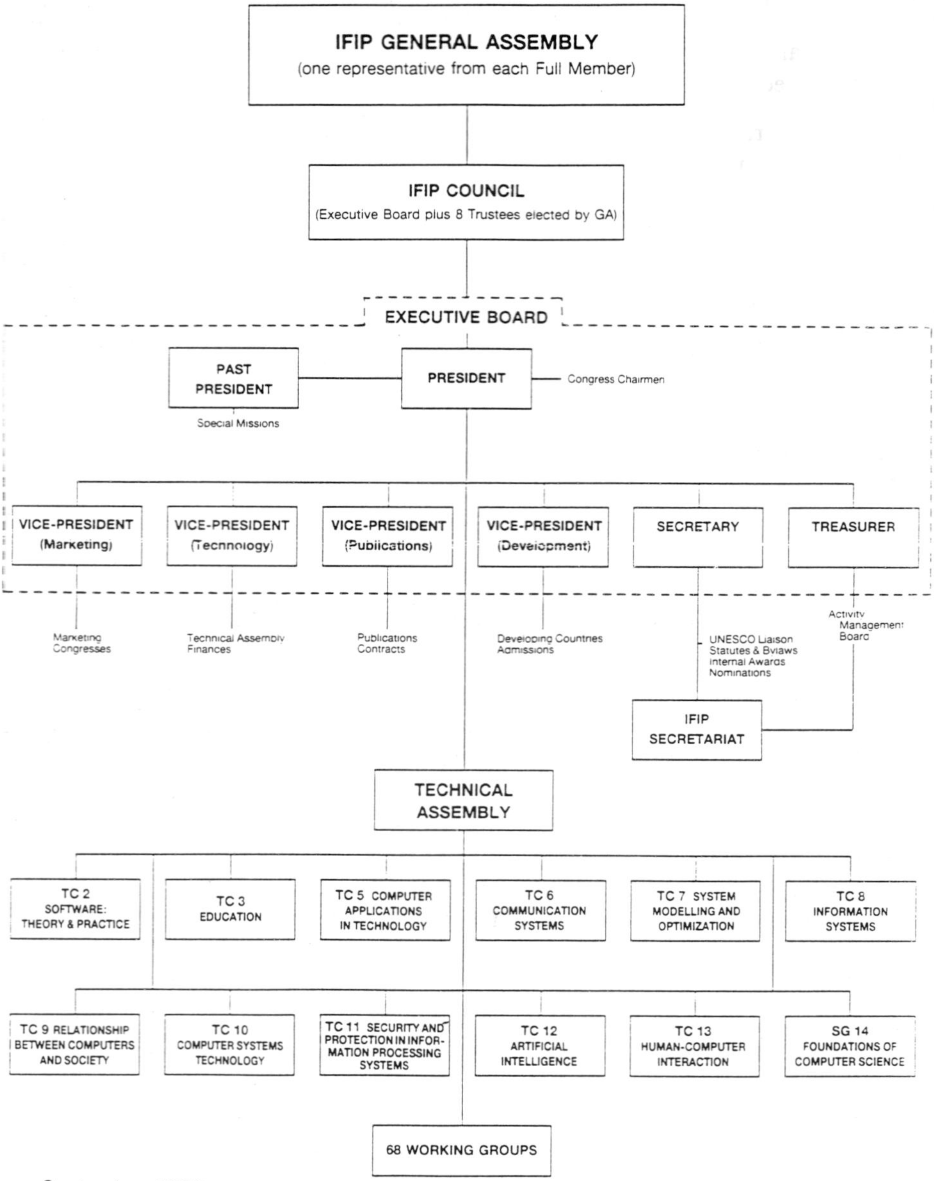

September 1992

The day-to-day work of IFIP is directed by its Officers: the President, Vice-Presidents, Secretary and Treasurer, who are elected by the General Assembly and together constitute the EXECUTIVE BOARD.

The COUNCIL, consisting of the Officers and up to eight Trustees elected from the General Assembly, meets twice a year and takes decisions which become necessary between General Assembly meetings.

The headquarters of the Federation are in Geneva, Switzerland where the IFIP Secretariat administers its affairs.

For further information please contact:

IFIP Secretariat
attn. Mme. GWYNETH ROBERTS
16 Place Longemalle
CH-1204 Geneva, Switzerland
telephone: 41 (22) 28 26 49
facsimile: 41 (22) 781 23 22
Bitnet: ifip@cgeuge51

IFIP's MISSION STATEMENT

IFIP's mission is to be the leading, truly international, apolitical organisation which encourages and assists in the development, exploitation and application of Information Technology for the benefit of all people.

Principal Elements

1. To stimulate, encourage and participate in research, development and application of Information Technology (IT) and to foster international co-operation in these activities.
2. To provide a meeting place where national IT Societies can discuss and plan courses of action on issues in our field which are of international significance and thereby to forge increasingly strong links between them and with IFIP.
3. To promote international co-operation directly and through national IT Societies in a free environment between individuals, national and international governmental bodies and kindred scientific and professional organisations.
4. To pay special attention to the needs of developing countries and to assist them in appropriate ways to secure the optimum benefit from the application of IT.
5. To promote professionalism, incorporating high standards of ethics and conduct, among all IT practitioners.
6. To provide a forum for assessing the social consequences of IT applications; to campaign for the safe and beneficial development and use of IT and the protection of people from abuse through its improper application.

7. To foster and facilitate co-operation between academics, the IT industry and governmental bodies and to seek to represent the interest of users.
8. To provide a vehicle for work on the international aspects of IT development and application including the necessary preparatory work for the generation of international standards.
9. To contribute to the formulation of the education and training needed by IT practitioners, users and the public at large.

Note to Conference Organizers

Organizers of upcoming IFIP Working Conferences are urged to contact the Publisher. Please send full details of the Conference to:

Mrs. STEPHANIE SMIT
Administrative Editor – IFIP Publications
ELSEVIER SCIENCE PUBLISHERS
P.O. Box 103, 1000 AC Amsterdam
The Netherlands
telephone: 31 (20) 5862481
facsimile: 31 (20) 5862616
email: s.smit@elsevier.nl

IFIP TRANSACTIONS

IFIP TRANSACTIONS is a serial consisting of 15,000 pages of valuable scientific information from leading researchers, published in 35 volumes per year. The serial includes contributed volumes, proceedings of the IFIP World Conferences, and conferences at Technical Committee and Working Group level. Mainstream areas in the IFIP TRANSACTIONS can be found in Computer Science and Technology, Computer Applications in Technology, and Communication Systems.

From 1993 onwards the IFIP TRANSACTIONS are only available as a full set.

IFIP TRANSACTIONS A:
Computer Science and Technology
1992: Volumes A1–A19
1993: Volumes A20–A40
ISSN 0926-5473

IFIP Technical Committees that are involved in IFIP TRANSACTIONS A
Software: Theory and Practice (TC2)
Education (TC3)
System Modelling and Optimization (TC7)
Information Systems (TC8)
Relationship Between Computers and Society (TC9)

Computer Systems Technology (TC10)
Security and Protection in Information Processing Systems (TC11)
Artificial Intelligence (TC12)
Human-Computer Interaction (TC13)
Foundations of Computer Science (SG14)

IFIP TRANSACTIONS B:
Applications in Technology
1992: Volumes B1–B8
1993: Volumes B9–B14
ISSN 0926-5481

IFIP Technical Committee that is involved in IFIP TRANSACTIONS B
Computer Applications in Technology (TC5)

IFIP TRANSACTIONS C:
Communication Systems
1992: Volumes C1–C8
1993: Volumes C9–C16
ISSN 0926-549X

IFIP Technical Committee that is involved in IFIP TRANSACTIONS C
Communication Systems (TC6)

IFIP TRANSACTIONS FULL SET: A, B & C
1992: 35 Volumes. US $ 1892.00/Dfl. 3500.00
1993: 35 Volumes. US $ 2340.00/Dfl. 3885.00
The Dutch Guilder prices (Dfl.) are definitive. The US $ prices mentioned above are for your guidance only and are subject to exchange rate fluctuations. Prices include postage and handling charges.

The volumes are also available separately in book form.

Please address all orders and correspondence to:

ELSEVIER SCIENCE PUBLISHERS
attn. PETRA VAN DER MEER
P.O. Box 103, 1000 AC Amsterdam
The Netherlands
telephone: 31 (20) 5862602
facsimile: 31 (20) 5862616
email: m.haccou@elsevier.nl